Nanotribology and Nanomechanics

Bharat Bhushan
Editor

# Nanotribology and Nanomechanics

## Measurement Techniques and Nanomechanics

Volume 1

Springer

*Editor*
Bharat Bhushan
Ohio State University
Nanoprobe Laboratory for Bio- &
Nanotechnology &
Biomimetics (NLB$^2$)
201 West 19th Avenue
Columbus Ohio 43210-1142
USA
bhushan.2@osu.edu

ISBN 978-3-642-42799-2 ISBN 978-3-642-15283-2 (eBook)
DOI 10.1007/978-3-642-15283-2
Springer Heidelberg Dordrecht London New York



*Cover design:* WMXDesign GmbH, Heidelberg, Germany

Printed on acid-free paper

Springer is part of Springer Science+Business Media (www.springer.com)

# Foreword

The invention of the scanning tunneling microscope in 1981 has led to an explosion of a family of instruments called scanning probe microscopes (SPMs). One of the most popular instruments in this family is the atomic force microscope (AFM), which was introduced to the scientific community in 1986. The application of SPMs has penetrated numerous science and engineering fields. Proliferation of SPMs in science and technology labs is similar to optical microscopes 50 years ago. SPMs have even made it into some high school science labs. Evolution of nanotechnology has accelerated the use of SPMs and vice versa. The scientific and industrial applications include quality control in the semiconductor industry and related research, molecular biology and chemistry, medical studies, materials science, and the field of information storage systems.

AFMs were developed initially for imaging with atomic or near atomic resolution. After their invention, they were modified for tribological studies. AFMs are now intensively used in this field and have lead to the development of the field of nanotribology. Researchers can image single lubricant molecules and their agglomeration and measure surface topography, adhesion, friction, wear, lubricant film thickness, and mechanical properties all on a micrometer to nanometer scale. SPMs are also used for nanofabrication and nanomachining. Beyond as an analytical instrument, SPMs are being developed as industrial tools such as for data storage.

With the advent of more powerful computers, atomic-scale simulations have been conducted of tribological phenomena. Simulations have been able to predict the observed phenomena. Development of the field of nanotribology and nanomechanics has attracted numerous physicists and chemists. I am very excited that SPMs have had such an immense impact on the field of tribology.

I congratulate Professor Bharat Bhushan in helping to develop this field of nanotribology and nanomechanics. Professor Bhushan has harnessed his own knowledge and experience, gained in several industries and universities, and has assembled a large number of internationally recognized authors. The authors come from both academia and industry.

Professor Bharat Bhushan's comprehensive book is intended to serve both as a textbook for university courses as well as a reference for researchers. It is a timely addition to the literature on nanotribology and nanomechanics, which I anticipate will stimulate further interest in this important new field. I expect that it will be well received by the international scientific community.

IBM Research Division
Rueschlikon, Switzerland
Nobel Laureate Physics, 1986

Prof. Dr. Gerd Binnig

# Preface

Tribology is the science and technology of interacting surfaces in relative motion and of related subjects and practices. The nature and consequences of the interactions that take place at the moving interface control its friction, wear and lubrication behavior. Understanding the nature of these interactions and solving the technological problems associated with the interfacial phenomena constitute the essence of tribology. The importance of friction and wear control cannot be overemphasized for economic reasons and long-term reliability.

The recent emergence and proliferation of proximal probes, in particular tip-based microscopies and the surface force apparatus and of computational techniques for simulating tip-surface interactions and interfacial properties, has allowed systematic investigations of interfacial problems with high resolution as well as ways and means for modifying and manipulating nanostructures. These advances provide the impetus for research aimed at developing a fundamental understanding of the nature and consequences of the interactions between materials on the atomic scale, and they guide the rational design of material for technological applications. In short, they have led to the appearance of the new field of nanotribology and nanomechanics.

The field of tribology is truly interdisciplinary. Until 1980s, it had been dominated by mechanical and chemical engineers who conduct macro tests to predict friction and wear lives in machine components and devise new lubricants to minimize friction and wear. Development of the field of nanotribology has attracted many more physicists, chemists, and material scientists who have significantly contributed to the fundamental understanding of friction and wear processes and lubrication on an atomic scale. Thus, tribology and mechanics are now studied by both engineers and scientists. The nanotribology and nanomechanics fields are growing rapidly and it has become fashionable to call oneself a "tribologist." The tip-based microscopies have also been used for materials characterization as well as for measurement of mechanical and electrical properties all on nanoscale. Since 1991, international conferences and courses have been organized on this new field of nanotribology, nanomechanics, and nanomaterials characterization.

There are also new applications which require detailed understanding of the tribological and mechanics processes on macro- to nanoscales. Since early 1980s, tribology of magnetic storage systems (rigid disk drives, flexible disk drives, and tape drives) has become one of the important parts of tribology. Microelectromechanical Systems (MEMS)/ Nanoelectromechanical Systems (NEMS) and biodevices, all part of nanotechnology, have appeared in the marketplace in the 1990s which present new tribological challenges. Another emerging area of importance is biomimetics. It involves taking ideas from nature and implementing them in an application. Examples include Lotus effect and gecko adhesion. Tribology of processing systems such as copiers, printers, scanners, and cameras is important although it has not received much attention. Along with the new industrial applications, there has been development of new materials, coatings and treatments such as synthetic diamond, diamondlike carbon films, self assembled monolayers, and chemically grafted films, to name a few with nanoscale thicknesses.

It is clear that the general field of tribology has grown rapidly in the last 30 years. Conventional tribology is well established but nanotribology and nanomechanics are evolving rapidly and have taken the center stage. New materials are finding use. Furthermore, new industrial applications continue to evolve with their unique challenges.

Very few tribology handbooks exist and these are dated. They have focused on conventional tribology, traditional materials and matured industrial applications. No mechanics handbook exists. Nanotribology, nanomechanics, and nanomaterial characterization are becoming important in many nanotechnology applications. A primer to nanotribology, nanomechanics, and nanomaterial characterization is needed. The purpose of this book is to present the principles of nanotribology and nanomechanics and applications to various applications. The appeal of the subject book is expected to be broad. The first edition was published in 2005 and the second in 2008. This third edition is an update based on recent developments.

The chapters in the book have been written by internationally recognized experts in the field, from academia, national research labs and industry, and from all over the world. The book integrates the knowledge of the field from mechanics and materials science points of view. In each chapter, we start with macroconcepts leading to microconcepts. We assume that the reader is not expert in the field of nanotribology and nanomechanics, but has some knowledge of macrotribology/ mechanics. It covers various measurement techniques and their applications, and theoretical modeling of interfaces. Organization of the book is straightforward. The first part of the book covers fundamental experimental and theoretical studies. The latter part covers applications.

The book is intended for three types of readers: graduate students of nanotribology/ nanomechanics/nanotechnology, research workers who are active or intend to become active in this field, and practicing engineers who have encountered a tribology and mechanics problem and hope to solve it as expeditiously as possible. The book should serve as an excellent text for one or two semester graduate courses in scanning probe microscopy/applied scanning probe methods, nanotribology/

nanomechanics/nanotechnology in mechanical engineering, materials science, or applied physics.

I would like to thank the authors for their excellent contributions in a timely manner. And I wish to thank my wife, Sudha, my son, Ankur, and my daughter, Noopur, who have been very forbearing during the preparation of this book.

Powell, OH
May 17, 2010

Bharat Bhushan

# Contents

# Contributors

**Boris Anczykowski** NanoAnalytics GmbH, Gievenbecker Weg 11, 48149 Münster, Germany, anczykowski@nanoanalytics.com

**Roland Bennewitz** Leibniz Institute for New Materials, Campus D2 2, 66123 Saarbrücken, Germany, roland.bennewitz@inm-gmbh.de

**Bharat Bhushan** Nanoprobe Laboratory for Bio- & Nanotechnology and Biomimetics, W390 Scott Laboratory, The Ohio State University, 201 W. 19th Avenue, Columbus, OH 43210–1142, USA, bhushan.2@osu.edu

**Donald W. Brenner** Department of Materials Science and Engineering, North Carolina State University, Raleigh, NC 27695-7909, USA, brenner@ncsu.edu

**Edin (I-Chen) Chen** Institute of Materials Science and Engineering, National Central University, 300 Jung-Da Rd, Chung-Li, Taoyuan 320, Taiwan, ichen@ncu.edu.tw

**Andreas Ebner** Institute for Biophysics, Altenbergerstr. 69, 4040 Linz, Austria, andreas.ebner@jku.at

**Harald Fuchs** Physikalisches Institut, Whilhem-Klemm-Str.10, 48149 Münster, Germany, fuchsh@uni-muenster.de

**Franz J. Giessibl** Institute of Experimental and Applied Physics, University of Regensburg, Physics Building, Room PHY 1.1.22, Universitätsstrasse 31, 93053 Regensburg, Germany, franz.giessibl@physik.uni-regensburg.de

**Enrico Gnecco** Institute of Physics, University of Basel, Klingelbergstr. 82, 4056 Basel, Switzerland, Enrico.Gnecco@unibas.ch

**Steve Granick** Department of Materials Science and Engineering, University of Illinois at Urbana-Champaign, 104 S. Goodwin Ave, Urbana, IL 61801, USA, sgranick@uiuc.edu

**Hermann Gruber** Institute of Biophysics, University of Linz, Altenberger Str. 69, 4040 Linz, Austria, hermann.gruber@jku.at

**Jason H. Hafner** Physics and Astronomy – MS61, Rice University, PO Box 1892, Houston, TX 77251–1892, USA, hafner@rice.edu

**Judith A. Harrison** Chemistry Department, MS 9B, U.S. Naval Academy, 572 Holloway Road, Annapolis, MD 21402, USA, jah@usna.edu

**Seong-Jun Heo** Department of Materials Science and Engineering, Univeristy of Florida, 100 Rhines Hall, P.O. Box 116400, Gainesville, FL 32611-6400, USA, heogyver@ufl.edu

**Peter Hinterdorfer** Institute for Biophysics, University of Linz, Altenberger Str. 69, 4040 Linz, Austria, peter.hinterdorfer@jku.at

**Hendrick Hölscher** Institute for Microscructure Technology, Forschungszentrum Karlsruhe, PO Box 36 40, 76021 Karlsruhe, Germany, Hendrik.Hoelscher@imt.fzk.de

**Hirotaka Hosoi** Creative Research Initiative Sousei, Hokkaido University, Kita 21, Nishi 10, Kita-ku, Sapporo, Japan, hosoi@cris.hokudai.ac.jp

**Douglas L. Irving** Department of Materials Science and Engineering, North Carolina State University, Raleigh, NC 27695-7909, USA, dlirving@ncsu.edu

**Jacob N. Israelachvili** Department of Chemical Engineering and Materials Department, University of California, Santa Barbara, CA 93106, USA, Jacob@engineering.ucsb.edu

**Sungho Jin** Department of Mechanical and Aerospace Engineering, University of California, 9500 Gilman Drive, La Jolla, San Diego, CA 92093, USA, jin@ucsd.edu

**Yong Chae Jung** Samsung Electronics C., Ltd, San #16 Banwol-Dong, Hwasung-City, Gyeonggi-Do 445–701, Korea, yc423.jung@samsung.com

**Harold Kahn** Department of Materials Science and Engineering, Case Western Reserve University, 10900 Euclid Ave, Cleveland, OH 44106–7204, USA, hxk29@po.cwru.edu

**Ruti Kapon** Department of Biological Chemistry, Weizmann Institute of Science, Rehovot 76100, Israel, ruti.kapon@weizmann.ac.il

**Ratnesh Lal** University of California, San Diego PFBH Room 219 9500 Gilman Drive, MC 0412, La Jolla, San Diego, CA 92093–0412, USA, rlal@ucsd.edu

**Carmen LaTorre** Owens Corning, Insulating Systems Business, 2790 Columbus Road, Route 16 (Bldg 20–1), Granville, OH 43023, USA, carmen.latorre@owenscorning.com

**Adrian B. Mann** Department of Ceramic and Materials Engineering and Biomedical Engineering, Rutgers University, 607 Taylor Road, Piscataway, NJ 08854, USA, abmann@rci.rutgers.edu

**Othmar Marti** Abteilung Experimentelle Physik, Universitaet Ulm, Albert-Einstein-Allee 11, 89069 Ulm, Germany, othmar.marti@uni-ulm.de

**Ernst Meyer** Institute of Physics, University of Basel, Klingelbergstr. 82, 4056 Basel, Switzerland, Ernst.Meyer@unibas.ch

**Markus Morgenstern** II. Institute of Physics B, RWTH Aachen University, D-52056 Aachen, Germany, mmorgens@physik.rwth-aachen.de

**Seizo Morita** Department of Electronic Engineering, Graduate School of Engineering, Osaka University, Yamada-Oka 2–1, Suita 565–0871, Japan, smorita@ele.eng.osaka-u.ac.jp

**Koichi Mukasa** Nanoelectronics Laboratory, Graduate School of Engineering, Hokkaido University, Nishi-8, Kita-13, Kita-ku, Sapporo 060–8628, Japan, mukasa@nano.eng.hokudai.ac.jp

**Ashis Mukhopadhyay** Department of Physics, Wayne State University, Detroit, MI, USA, ashis@physics.wayne.edu

**Michael Nosonovsky** Department of Mechanical Engineering, University of Wisconsin, EMS Building, Room E371G, Milwaukee, WI 53201–0413, USA, nosonovs@uwm.edu

**Hiroshi Onishi** Department of Chemistry, Kobe University, Rokko-dai, Nada-ku, Kobe 657–8501, Japan, oni@kobe-u.ac.jp

**Manuel L.B. Palacio** Nanoprobe Laboratory for Bio- & Nanotechnology and Biomimetics, The Ohio State University, 201 W. 19th Avenue, Columbus, OH 43210–1142, USA, palacio.1@osu.edu

**Oliver Pfeiffer** Institute of Physics, University of Basel, Klingelbergstr. 82, 4056 Basel, Switzerland, Oliver.Pfeiffer@unibas.ch

**Ziv Reich** Department of Biological Chemistry, Weizmann Institute of Science, Rehovot 76100, Israel, ziv.reich@weizmann.ac.il

**Marina Ruths** Department of Chemistry, University of Massachusetts Lowell, 1 University Avenue, Lowell, MA 01854, USA, marina_ruths@uml.edu

**Akira Sasahara** School of Materials Science, Japan Advanced Institute of Science and Technology, 1–1 Asahidai, Nomi 923–1292, Japan, sasahara@jaist.ac.jp

**André Schirmeisen** Physikalisches Institut, University of Muenster, Wilhelm-Klemm-Str.10, 48149 Muenster, Germany, schira@uni-muenster.de

**Alexander Schwarz** Institute of Applied Physics, University of Hamburg, Jungiusstr. 11, 20355 Hamburg, Germany, aschwarz@physnet.uni-hamburg.de

**Udo D. Schwarz** Department of Mechanical Engineering, Yale University, P O Box 208284, 15 Prospect Street, Rm. 213, New Haven, CT 06520–8284, USA, udo.schwarz@yale.edu

**Susan B. Sinnott** Dept. of Materials Science and Engineering, University of Florida, 154 Rhines Hall, P.O. Box 116400, Gainesville, FL 32611–6400, USA, sinnott@mse.ufl.edu

**Anisoara Socoliuc** SPECS Zurich GmbH, Technoparkstrasse 1, 8005 Zurich, Switzerland, socoliuc@nanonis.com

**Yasuhiro Sugawara** Department of Applied Physics, Graduate School of Engineering, Osaka University, Yamada-Oka 2–1, Suita 565–0871, Japan, sugawara@ap.eng.osaka-u.ac.jp

**Y. Elaine Zhu** Department of Chemical and Biomolecular Engineering, University of Notre Dame, 182 Fitzpatrick Hall, Notre Dame, IN 46556, USA, yzhu3@nd.edu

# Chapter 1
# Introduction – Measurement Techniques and Applications

**Bharat Bhushan**

In this introductory chapter, the definition and history of tribology and their industrial significance and origins and significance of an emerging field of micro/nanotribology are described. Next, various measurement techniques used in micro/nanotribological and micro/nanomechanical studies are described. The interest in micro/nanotribology field grew from magnetic storage devices and latter the applicability to emerging field micro/nanoelectromechanical systems (MEMS/NEMS) became clear. A few examples of magnetic storage devices and MEMS/NEMS are presented where micro/nanotribological and micro/nanomechanical tools and techniques are essential for interfacial studies. Finally, reasons why micro/nanotribological and micro/nanomechanical studies are important in magnetic storage devices and MEMS/NEMS are presented. In the last section, organization of the book is presented.

## 1.1 Definition and History of Tribology

The word tribology was first reported in a landmark report by Jost [1]. The word is derived from the Greek word tribos meaning rubbing, so the literal translation would be "the science of rubbing". Its popular English language equivalent is friction and wear or lubrication science, alternatively used. The latter term is hardly all-inclusive. Dictionaries define tribology as the science and technology of interacting surfaces in relative motion and of related subjects and practices. Tribology is the art of applying operational analysis to problems of great economic significance, namely, reliability, maintenance, and wear of technical equipment, ranging from spacecraft to household appliances. Surface interactions in a tribological interface are highly complex, and their understanding requires knowledge of various disciplines including physics, chemistry, applied mathematics, solid mechanics, fluid mechanics, thermodynamics, heat transfer, materials science, rheology, lubrication, machine design, performance and reliability.

It is only the name tribology that is relatively new, because interest in the constituent parts of tribology is older than recorded history [2]. It is known that

B. Bhushan (ed.), *Nanotribology and Nanomechanics*,
DOI 10.1007/978-3-642-15283-2_1, © Springer-Verlag Berlin Heidelberg 2011

drills made during the Paleolithic period for drilling holes or producing fire were fitted with bearings made from antlers or bones, and potters' wheels or stones for grinding cereals, etc., clearly had a requirement for some form of bearings [3]. A ball thrust bearing dated about AD 40 was found in Lake Nimi near Rome.

Records show the use of wheels from 3500 BC, which illustrates our ancestors' concern with reducing friction in translationary motion. The transportation of large stone building blocks and monuments required the know-how of frictional devices and lubricants, such as water-lubricated sleds. Figure 1.1 illustrates the use of a sledge to transport a heavy statue by egyptians circa 1880 BC [4]. In this transportation, 172 slaves are being used to drag a large statue weighing about 600 kN along a wooden track. One man, standing on the sledge supporting the statue, is seen pouring a liquid (most likely water) into the path of motion; perhaps he was one of the earliest lubrication engineers. (Dowson [2] has estimated that each man exerted a pull of about 800 N. On this basis, the total effort, which must at least equal the friction force, becomes 172 × 800 N. Thus, the coefficient of friction is about 0.23.) A tomb in Egypt that was dated several thousand years BC provides the evidence of use of lubricants. A chariot in this tomb still contained some of the original animal-fat lubricant in its wheel bearings.

During and after the glory of the Roman empire, military engineers rose to prominence by devising both war machinery and methods of fortification, using tribological principles. It was the renaissance engineer-artist Leonardo da Vinci (1452–1519), celebrated in his days for his genius in military construction as well as for his painting and sculpture, who first postulated a scientific approach to friction. Da Vinci deduced the rules governing the motion of a rectangular block sliding over a flat surface. He introduced for the first time, the concept of coefficient of friction as the ratio of the friction force to normal load. His work had no historical influence, however, because his notebooks remained unpublished for hundreds of years. In 1699, the French physicist Guillaume Amontons rediscovered the rules of friction after he studied dry sliding between two flat surfaces [5]. First, the friction force that resists sliding at an interface is directly proportional to the normal load.

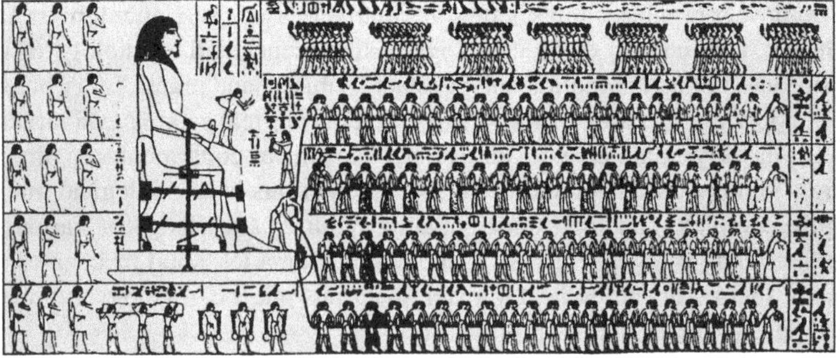

**Fig. 1.1** Egyptians using lubricant to aid movement of colossus, El-Bersheh, circa 1800 BC

Second, the amount of friction force does not depend on the apparent area of contact. These observations were verified by French physicist Charles-Augustin Coulomb (better known for his work on electrostatics [6]). He added a third rule that the friction force is independent of velocity once motion starts. He also made a clear distinction between static friction and kinetic friction.

Many other developments occurred during the 1500s, particularly in the use of improved bearing materials. In 1684, Robert Hooke suggested the combination of steel shafts and bell-metal bushes as preferable to wood shod with iron for wheel bearings. Further developments were associated with the growth of industrialization in the latter part of the eighteenth century. Early developments in the petroleum industry started in Scotland, Canada, and the United States in the 1850s [2–7].

Though essential laws of viscous flow were postulated by Sir Isaac Newton in 1668; scientific understanding of lubricated bearing operations did not occur until the end of the nineteenth century. Indeed, the beginning of our understanding of the principle of hydrodynamic lubrication was made possible by the experimental studies of Tower [8] and the theoretical interpretations of Reynolds [9] and related work by Petroff [10]. Since then developments in hydrodynamic bearing theory and practice were extremely rapid in meeting the demand for reliable bearings in new machinery.

Wear is a much younger subject than friction and bearing development, and it was initiated on a largely empirical basis. Scientific studies of wear developed little until the mid-twentieth century. Holm made one of the earliest substantial contributions to the study of wear [11].

The industrial revolution (1750–1850 A.D.) is recognized as a period of rapid and impressive development of the machinery of production. The use of steam power and the subsequent development of the railways in the 1830s led to promotion of manufacturing skills. Since the beginning of the twentieth century, from enormous industrial growth leading to demand for better tribology, knowledge in all areas of tribology has expanded tremendously [11–17].

## 1.2 Industrial Significance of Tribology

Tribology is crucial to modern machinery which uses sliding and rolling surfaces. Examples of productive friction are brakes, clutches, driving wheels on trains and automobiles, bolts, and nuts. Examples of productive wear are writing with a pencil, machining, polishing, and shaving. Examples of unproductive friction and wear are internal combustion and aircraft engines, gears, cams, bearings, and seals.

According to some estimates, losses resulting from ignorance of tribology amount in the United States to about 4% of its gross national product (or about $ 200 billion dollars per year in 1966), and approximately one-third of the world's energy resources in present use appear as friction in one form or another. Thus, the importance of friction reduction and wear control cannot be overemphasized for economic reasons and long-term reliability. According to Jost [1, 18], savings of about 1% of gross national product of an industrialized nation can be realized by research and better tribological practices. According to recent studies, expected savings are expected to be on the order of 50 times the research costs. The savings are both substantial and significant, and these savings can be obtained without the deployment of large capital investment.

The purpose of research in tribology is understandably the minimization and elimination of losses resulting from friction and wear at all levels of technology where the rubbing of surfaces is involved. Research in tribology leads to greater plant efficiency, better performance, fewer breakdowns, and significant savings.

Tribology is not only important to industry, it also affects day-to-day life. For example, writing is a tribological process. Writing is accomplished by a controlled transfer of lead (pencil) or ink (pen) to the paper. During writing with a pencil there should be good adhesion between the lead and paper so that a small quantity of lead transfers to the paper and the lead should have adequate toughness/hardness so that it does not fracture/break. Objective during shaving is to remove hair from the body as efficiently as possible with minimum discomfort to the skin. Shaving cream is used as a lubricant to minimize friction between a razor and the skin. Friction is helpful during walking and driving. Without adequate friction, we would slip and a car would skid! Tribology is also important in sports. For example, a low friction between the skis and the ice is desirable during skiing.

## 1.3 Origins and Significance of Micro/Nanotribology

At most interfaces of technological relevance, contact occurs at numerous asperities. Consequently, the importance of investigating a single asperity contact in studies of the fundamental tribological and mechanical properties of surfaces has been long recognized. The recent emergence and proliferation of proximal probes, in particular tip-based microscopies (e.g., the scanning tunneling microscope and the atomic force microscope) and of computational techniques for simulating tip-surface interactions and interfacial properties, has allowed systematic investigations of interfacial problems with high resolution as well as ways and means for modifying and manipulating nanoscale structures. These advances have led to the development of the new field of microtribology, nanotribology, molecular tribology, or atomic-scale tribology [15, 16, 19–22]. This field is concerned with experimental and theoretical investigations of processes ranging from atomic and molecular scales to microscales, occurring during adhesion, friction, wear, and thin-film lubrication at sliding surfaces.

The differences between the conventional or macrotribology and micro/nanotribology are contrasted in Fig. 1.2. In macrotribology, tests are conducted on components with relatively large mass under heavily loaded conditions. In these tests, wear is inevitable and the bulk properties of mating components dominate the tribological performance. In micro/nanotribology, measurements are made on components, at least one of the mating components, with relatively small mass

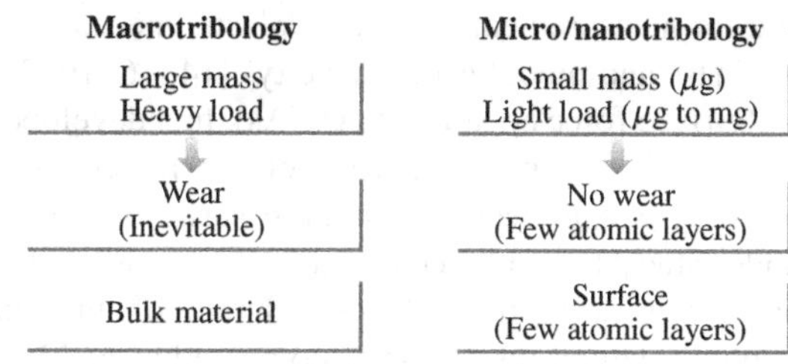


**Fig. 1.2** Comparisons between macrotribology and micro/nanotribology

under lightly loaded conditions. In this situation, negligible wear occurs and the surface properties dominate the tribological performance.

The micro/nanotribological studies are needed to develop fundamental understanding of interfacial phenomena on a small scale and to study interfacial phenomena involving ultrathin films (as low as 1–2 nm) and in micro/nanostructures, both used in magnetic storage systems, micro/nanoelectromechanical systems (MEMS/NEMS) and other industrial applications. The components used in micro- and nanostructures are very light (on the order of few micrograms) and operate under very light loads (smaller than 1 μg to a few milligrams). As a result, friction and wear (on a nanoscale) of lightly loaded micro/nanocomponents are highly dependent on the surface interactions (few atomic layers). These structures are generally lubricated with molecularly thin films. Micro/nanotribological techniques are ideal to study the friction and wear processes of ultrathin films and micro/nanostructures. Although micro/nanotribological studies are critical to study ultrathin films and micro/nanostructures, these studies are also valuable in fundamental understanding of interfacial phenomena in macrostructures to provide a bridge between science and engineering.

The probe-based microscopes (scanning tunneling microscope, the atomic force and friction force microscopes) and the surface force apparatus are widely used for micro/nanotribological studies [16, 19–23]. To give a historical perspective of the field, the scanning tunneling microscope (STM) developed by Binnig and Rohrer and their colleagues in 1981 at the IBM Zurich Research Laboratory, Forschungslabor, is the first instrument capable of directly obtaining three-dimensional (3-D) images of solid surfaces with atomic resolution [24]. STMs can only be used to study surfaces which are electrically conductive to some degree. Based on their design of STM, in 1985, Binnig et al. [25, 26] developed an atomic force microscope (AFM) to measure ultrasmall forces (less than 1 μN) present between the AFM tip surface and the sample surface. AFMs can be used for measurement of all engineering surfaces which may be either electrically conducting or insulating. AFM has become a popular surface profiler for topographic measurements on micro – to nanoscale. AFMs modified to measure both normal and friction forces, generally called friction force microscopes (FFMs) or lateral force microscopes (LFMs), are used to measure friction on micro- and nanoscales. AFMs are also used for studies of adhesion, scratching, wear, lubrication, surface temperatures, and for measurements of elastic/plastic mechanical properties (such as indentation hardness and modulus of elasticity) [14, 16, 19, 21].

Surface force apparatuses (SFAs), first developed in 1969, are used to study both static and dynamic properties of the molecularly thin liquid films sandwiched between two molecularly smooth surfaces [16, 20–22, 27]. However, the liquid under study has to be confined between molecularly-smooth optically-transparent or sometimes opaque surfaces with radii of curvature on the order of 1 mm (leading to poorer lateral resolution as compared to AFMs). Only AFMs/FFMs can be used to study engineering surfaces in the dry and wet conditions with atomic resolution.

Meanwhile, significant progress in understanding the fundamental nature of bonding and interactions in materials, combined with advances in computer-based modeling and simulation methods, have allowed theoretical studies of complex interfacial phenomena with high resolution in space and time [16, 20–22]. Such simulations provide insights into atomic-scale energetics, structure, dynamics, thermodynamics, transport and rheological aspects of tribological processes. Furthermore, these theoretical approaches guide the interpretation of experimental data and the design of new experiments, and enable the prediction of new phenomena based on atomistic principles.

## 1.4 Measurement Techniques

### *1.4.1 Scanning Probe Microscopy*

Family of instruments based on STMs and AFMs, called Scanning Probe Microscopes (SPMs), have been developed for various applications of scientific and industrial interest. These include – STM, AFM, FFM (or LFM), scanning electrostatic force microscopy (SEFM), scanning force acoustic microscopy (SFAM) (or atomic force acoustic microscopy, AFAM), scanning magnetic microscopy (SMM) (or magnetic force microscopy, MFM), scanning near field optical microscopy (SNOM), scanning thermal microscopy (SThM) scanning electrochemical microscopy (SEcM), scanning Kelvin Probe microscopy (SKPM), scanning chemical potential microscopy (SCPM), scanning ion conductance microscopy (SICM), and scanning capacitance microscopy (SCM). Family of instruments which measure forces (e.g. AFM, FFM, SEFM, SFAM, and SSM) are also referred to as scanning force microscopies (SFM). Although these instruments offer atomic resolution and are ideal for basic research, yet these are used for cutting edge industrial applications which do not require atomic resolution.

STMs, AFMs and their modifications can be used at extreme magnifications ranging from $10^3\times$ to $10^9\times$ in $x$-, $y$-, and $z$-directions for imaging macro to atomic dimensions with high-resolution information and for spectroscopy. These instruments can be used in any environment such as ambient air, various gases, liquid, vacuum, low temperatures, and high temperatures. Imaging in liquid allows the study of live biological samples and it also eliminates water capillary forces present in ambient air present at the tip–sample interface. Low temperature imaging is useful for the study of biological and organic materials and the study of low-temperature phenomena such as superconductivity or charge-density waves. Low-temperature operation is also advantageous for high-sensitivity force mapping due to the reduction in thermal vibration. These instruments also have been used to image liquids such as liquid crystals and lubricant molecules on graphite surfaces. While the pure imaging capabilities of SPM techniques dominated the application of these methods at their early development stages, the physics and chemistry of probe–sample interactions and the quantitative analyses of tribological, electronic, magnetic, biological, and chemical surfaces are commonly carried out. Nanoscale science and technology are strongly driven by SPMs which allow investigation and manipulation of surfaces down to the atomic scale. With growing understanding of the underlying interaction mechanisms, SPMs have found applications in many fields outside basic research fields. In addition, various derivatives of all these methods have been developed for special applications, some of them targeting far beyond microscopy.

A detailed overview of scanning probe microscopy – principle of operation, instrumentation, and probes is presented in a later chapter (also see [16, 20–23]). Here, a brief description of commercial STMs and AFMs follows.

## Commercial STMs

There are a number of commercial STMs available on the market. Digital Instruments, Inc. located in Santa Barbara, CA introduced the first commercial STM, the Nanoscope I, in 1987. In a recent Nanoscope IV STM for operation in ambient air, the sample is held in position while a piezoelectric crystal in the form of a cylindrical tube (referred to as PZT tube scanner) scans the sharp metallic probe over the surface in a raster pattern while sensing and outputting the tunneling current to the control station, Fig. 1.3. The digital signal processor

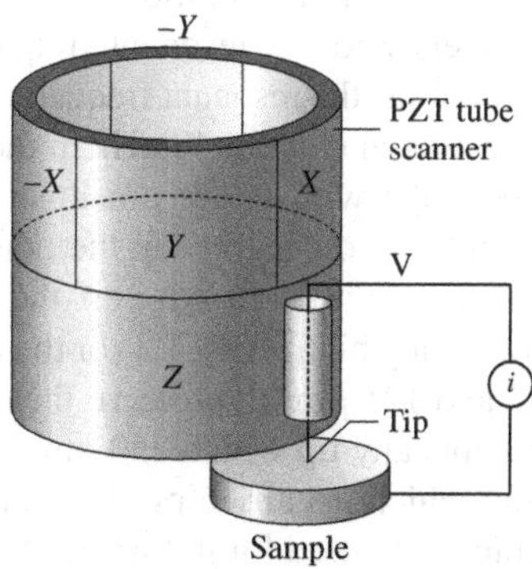


**Fig. 1.3** Principle of operation of a commercial STM, a sharp tip attached to a piezoelectric tube scanner is scanned on a sample (From [16])

(DSP) calculates the desired separation of the tip from the sample by sensing the tunneling current flowing between the sample and the tip. The bias voltage applied between the sample and the tip encourages the tunneling current to flow. The DSP completes the digital feedback loop by outputting the desired voltage to the piezoelectric tube. The STM operates in both the "constant height" and "constant current" modes depending on a parameter selection in the control panel. In the constant current mode, the feedback gains are set high, the tunneling tip closely tracks the sample surface, and the variation in the tip height required to maintain constant tunneling current is measured by the change in the voltage applied to the piezo tube. In the constant height mode, the feedback gains are set low, the tip remains at a nearly constant height as it sweeps over the sample surface, and the tunneling current is imaged.

Physically, the Nanoscope STM consists of three main parts: the head which houses the piezoelectric tube scanner for three dimensional motion of the tip and the preamplifier circuit (FET input amplifier) mounted on top of the head for the tunneling current, the base on which the sample is mounted, and the base support, which supports the base and head [16, 21]. The base accommodates samples up to 10 mm by 20 mm and 10 mm in thickness. Scan sizes available for the STM are 0.7 μm × 0.7 μm (for atomic resolution), 12 μm × 12 μm, 75 μm × 75 μm and 125 μm × 125 μm.

The scanning head controls the three dimensional motion of tip. The removable head consists of a piezo tube scanner, about 12.7 mm in diameter, mounted into an invar shell used to minimize vertical thermal drifts because of good thermal match between the piezo tube and the Invar. The piezo tube has separate electrodes for $X$, $Y$ and $Z$ which are driven by separate drive circuits. The electrode configuration (Fig. 1.3) provides $x$ and $y$ motions which are perpendicular to each other, minimizes horizontal and vertical coupling, and provides good sensitivity. The vertical motion of the tube is controlled by the $Z$ electrode which is driven by the feedback loop. The $x$ and $y$ scanning motions are each controlled by two electrodes which are driven by voltages of same magnitudes, but opposite signs. These electrodes are called $-Y$, $-X$, $+Y$, and $+X$. Applying complimentary voltages allows a short, stiff tube to provide a good scan range without large voltages. The motion of the tip due to external vibrations is proportional to the square of the ratio of vibration frequency to the resonant frequency of the tube. Therefore, to minimize the tip vibrations, the resonant frequencies of the tube are high about 60 kHz in the vertical direction and about 40 kHz in the horizontal direction. The tip holder is a stainless steel tube with a 300 μm inner diameter for 250 μm diameter tips, mounted in ceramic in order to keep the mass on the end of the tube low. The tip is mounted either on the front edge of the tube (to keep mounting mass low and resonant frequency high) (Fig. 1.3) or the center of the tube for large range scanners, namely 75 and 125 μm (to preserve the symmetry of the scanning.) This commercial STM accepts any tip with a 250 μm diameter shaft. The piezotube requires $X$–$Y$ calibration which is carried out by imaging an appropriate calibration standard. Cleaved graphite is used for the small-scan length head while two dimensional grids (a gold plated ruling) can be used for longer range heads.

The Invar base holds the sample in position, supports the head, and provides coarse *x–y* motion for the sample. A spring-steel sample clip with two thumb screws holds the sample in place. An *x–y* translation stage built into the base allows the sample to be repositioned under the tip. Three precision screws arranged in a triangular pattern support the head and provide coarse and fine adjustment of the tip height. The base support consists of the base support ring and the motor housing. The stepper motor enclosed in the motor housing allows the tip to be engaged and withdrawn from the surface automatically.

Samples to be imaged with STM must be conductive enough to allow a few nanoamperes of current to flow from the bias voltage source to the area to be scanned. In many cases, nonconductive samples can be coated with a thin layer of a conductive material to facilitate imaging. The bias voltage and the tunneling current depend on the sample. Usually they are set at a standard value for engagement and fine tuned to enhance the quality of the image. The scan size depends on the sample and the features of interest. Maximum scan rate of 122 Hz can be used. The maximum scan rate is usually related to the scan size. Scan rate above 10 Hz is used for small scans (typically 60 Hz for atomic-scale imaging with a 0.7 μm scanner). The scan rate should be lowered for large scans, especially if the sample surfaces are rough or contain large steps. Moving the tip quickly along the sample surface at high scan rates with large scan sizes will usually lead to a tip crash. Essentially, the scan rate should be inversely proportional to the scan size (typically 2–4 Hz for 1 μm, 0.5–1 Hz for 12 μm, and 0.2 Hz for 125 μm scan sizes). Scan rate in length/time, is equal to scan length divided by the scan rate in Hz. For example, for 10 μm × 10 μm scan size scanned at 0.5 Hz, the scan rate is 10 μm/s. The 256 × 256 data formats are most commonly used. The lateral resolution at larger scans is approximately equal to scan length divided by 256.

### Commercial AFM

A review of early designs of AFMs is presented by Bhushan [21]. There are a number of commercial AFMs available on the market. Major manufacturers of AFMs for use in ambient environment are: Digital Instruments Inc., a subsidiary of Veeco Instruments, Inc., Santa Barbara, California; Topometrix Corp., a subsidiary of Veeco Instruments, Inc., Santa Clara, California; and other subsidiaries of Veeco Instruments Inc., Woodbury, New York; Molecular Imaging Corp., Phoenix, Arizona; Quesant Instrument Corp., Agoura Hills, California; Nanoscience Instruments Inc., Phoenix, Arizona; Seiko Instruments, Japan; and Olympus, Japan. AFM/STMs for use in UHV environment are primarily manufactured by Omicron Vakuumphysik GMBH, Taunusstein, Germany.

We describe here two commercial AFMs – small sample and large sample AFMs – for operation in the contact mode, produced by Digital Instruments, Inc., Santa Barbara, CA, with scanning lengths ranging from about 0.7 μm (for atomic resolution) to about 125 μm [28–31]. The original design of these AFMs comes from Meyer and Amer [32]. Basically the AFM scans the sample in a raster pattern while

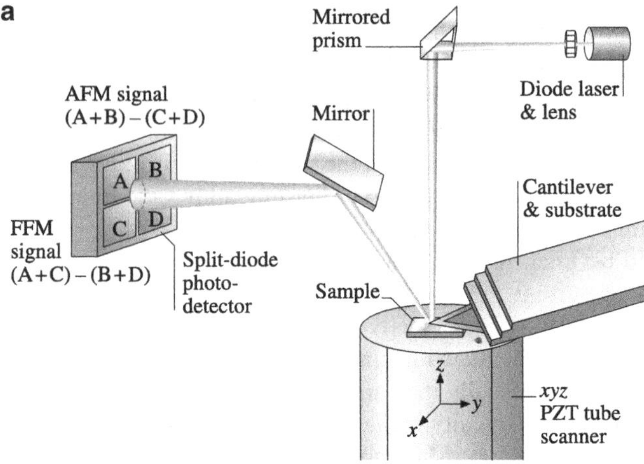


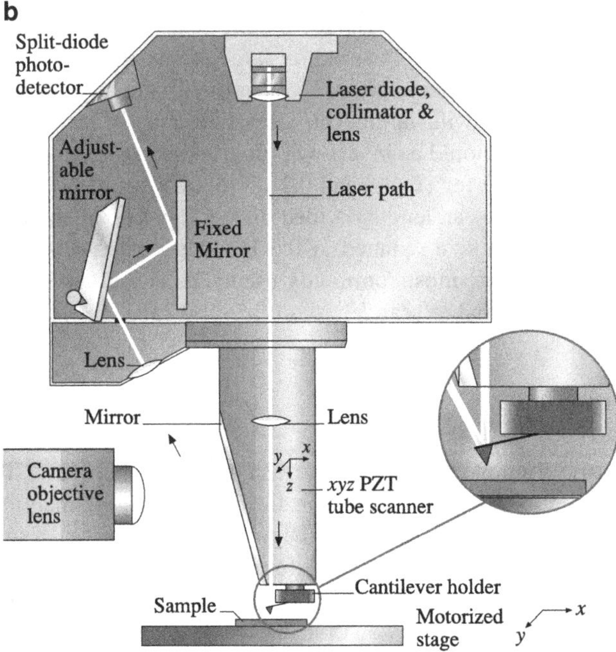


**Fig. 1.4** Principles of operation of (**a**) a commercial small sample AFM/FFM, and (**b**) a large sample AFM/FFM (From [16])

outputting the cantilever deflection error signal to the control station. The cantilever deflection (or the force) is measured using laser deflection technique, Fig. 1.4. The DSP in the workstation controls the $z$-position of the piezo based on the cantilever deflection error signal. The AFM operates in both the "constant height" and "constant force" modes. The DSP always adjusts the height of the sample under the tip based on the cantilever deflection error signal, but if the feedback gains are low the piezo remains at a nearly "constant height" and the cantilever deflection data is collected. With the high gains, the piezo height changes to keep the cantilever deflection nearly constant (therefore the force is constant) and the change in piezo height is collected by the system.

To further describe the principle of operation of the commercial small sample AFM shown in Fig. 1.4a, the sample, generally no larger than 10 mm $\times$ 10 mm, is mounted on a PZT tube scanner which consists of separate electrodes to scan precisely the sample in the *x*-*y* plane in a raster pattern and to move the sample in the vertical (*z*) direction. A sharp tip at the free end of a flexible cantilever is brought in contact with the sample. Features on the sample surface cause the cantilever to deflect in the vertical and lateral directions as the sample moves under the tip. A laser beam from a diode laser (5 mW max peak output at 670 nm) is directed by a prism onto the back of a cantilever near its free end, tilted downward at about 10° with respect to the horizontal plane. The reflected beam from the vertex of the cantilever is directed through a mirror onto a quad photodetector (split photodetector with four quadrants, commonly called position-sensitive detector or PSD, produced by Silicon Detector Corp., Camarillo, California). The differential signal from the top and bottom photodiodes provides the AFM signal which is a sensitive measure of the cantilever vertical deflection. Topographic features of the sample cause the tip to deflect in the vertical direction as the sample is scanned under the tip. This tip deflection will change the direction of the reflected laser beam, changing the intensity difference between the top and bottom sets of photodetectors (AFM signal). In the AFM operating mode called the height mode, for topographic imaging or for any other operation in which the applied normal force is to be kept a constant, a feedback circuit is used to modulate the voltage applied to the PZT scanner to adjust the height of the PZT, so that the cantilever vertical deflection (given by the intensity difference between the top and bottom detector) will remain constant during scanning. The PZT height variation is thus a direct measure of the surface roughness of the sample.

In a large sample AFM, both force sensors using optical deflection method and scanning unit are mounted on the microscope head, Fig. 1.4b. Because of vibrations added by cantilever movement, lateral resolution of this design is somewhat poorer than the design in Fig. 1.4a in which the sample is scanned instead of cantilever beam. The advantage of the large sample AFM is that large samples can be measured readily.

Most AFMs can be used for topography measurements in the so-called tapping mode (intermittent contact mode), also referred to as dynamic force microscopy. In the tapping mode, during scanning over the surface, the cantilever/tip assembly is sinusoidally vibrated by a piezo mounted above it, and the oscillating tip slightly taps the surface at the resonant frequency of the cantilever (70–400 Hz) with a constant (20–100 nm) oscillating amplitude introduced in the vertical direction with a feedback loop keeping the average normal force constant, Fig. 1.5. The oscillating amplitude is kept large enough so that the tip does not get stuck to the

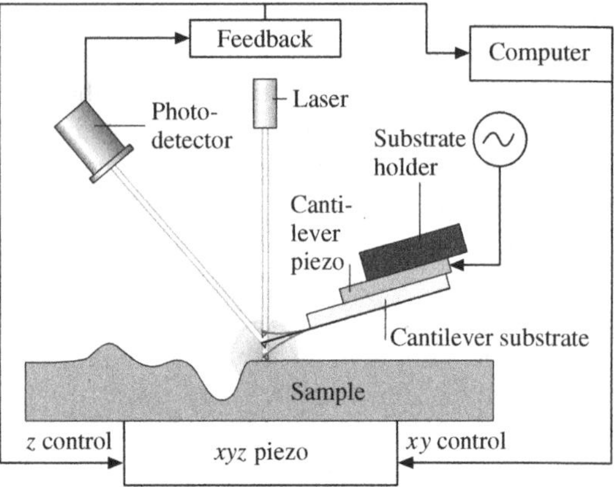


**Fig. 1.5** Schematic of tapping mode used for surface roughness measurement (From [16])

sample because of adhesive attractions. The tapping mode is used in topography measurements to minimize effects of friction and other lateral forces and/or to measure topography of soft surfaces.

Topographic measurements are made at any scanning angle. At a first instance, scanning angle may not appear to be an important parameter. However, the friction force between the tip and the sample will affect the topographic measurements in a parallel scan (scanning along the long axis of the cantilever). Therefore a perpendicular scan may be more desirable. Generally, one picks a scanning angle which gives the same topographic data in both directions; this angle may be slightly different than that for the perpendicular scan.

For measurement of friction force being applied at the tip surface during sliding, left hand and right hand sets of quadrants of the photodetector are used. In the so-called friction mode, the sample is scanned back and forth in a direction orthogonal to the long axis of the cantilever beam. A friction force between the sample and the tip will produce a twisting of the cantilever. As a result, the laser beam will be reflected out of the plane defined by the incident beam and the beam reflected vertically from an untwisted cantilever. This produces an intensity difference of the laser beam received in the left hand and right hand sets of quadrants of the photodetector. The intensity difference between the two sets of detectors (FFM signal) is directly related to the degree of twisting and hence to the magnitude of the friction force. This method provides three-dimensional maps of friction force. One problem associated with this method is that any misalignment between the laser beam and the photodetector axis would introduce error in the measurement. However, by following the procedures developed by Ruan and Bhushan [30], in which the average FFM signal for the sample scanned in two opposite directions is subtracted from the friction profiles of each of the two scans, the misalignment effect is eliminated. By following the friction force calibration procedures developed by Ruan and Bhushan [30], voltages corresponding to friction forces can be converted to force unites. The coefficient of friction is obtained from the slope of friction force data measured as a function of normal loads typically ranging from

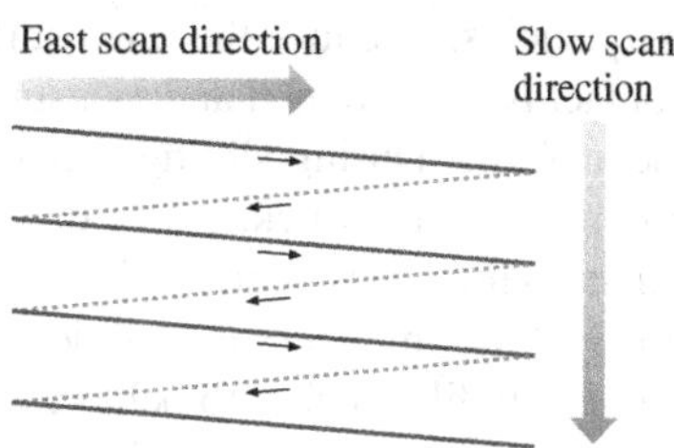


**Fig. 1.6** Schematic of triangular pattern trajectory of the AFM tip as the sample is scanned in two dimensions. During imaging, data are recorded only during scans along the solid scan lines (From [16])

10 to 150 nN. This approach eliminates any contributions due to the adhesive forces [33]. For calculation of the coefficient of friction based on a single point measurement, friction force should be divided by the sum of applied normal load and intrinsic adhesive force. Furthermore, it should be pointed out that for a single asperity contact, the coefficient of friction is not independent of load.

The tip is scanned in such a way that its trajectory on the sample forms a triangular pattern, Fig. 1.6. Scanning speeds in the fast and slow scan directions depend on the scan area and scan frequency. Scan sizes ranging from less than 1 nm × 1 nm to 125 μm × 125 μm and scan rates from less than 0.5–122 Hz typically can be used. Higher scan rates are used for smaller scan lengths. For example, scan rates in the fast and slow scan directions for an area of 10 μm × 10 μm scanned at 0.5 Hz are 10 μm/s and 20 nm/s, respectively.

## *1.4.2 Surface Force Apparatus (SFA)*

Surface Force Apparatuses (SFAs) are used to study both static and dynamic properties of the molecularly-thin liquid films sandwiched between two molecularly smooth surfaces. The SFAs were originally developed by Tabor and Winterton [27] and later by Israelachvili and Tabor [34] to measure van der Waals forces between two mica surfaces as a function of separation in air or vacuum. Israelachvili and Adams [35] developed a more advanced apparatus to measure normal forces between two surfaces immersed in a liquid so thin that their thickness approaches the dimensions of the liquid molecules themselves. A similar apparatus was also developed by Klein [36]. The SFAs, originally used in studies of adhesive and static interfacial forces were first modified by Chan and Horn [37] and later by Israelachvili et al. [38] and Klein et al. [39] to measure the dynamic shear (sliding) response of liquids confined between molecularly smooth optically-transparent mica surfaces. Optically transparent surfaces are required because the surface separation is measured using an optical interference technique. Van Alsten and Granick [40] and Peachey et al. [41] developed a new friction attachment which allow for the two surfaces to be sheared past each other at varying sliding speeds or oscillating frequencies while simultaneously measuring both the friction force and normal force between them. Israelachvili [42] and Luengo et al. [43] also presented modified SFA designs for dynamic measurements including friction at oscillating frequencies. Because the mica surfaces are molecularly smooth, the actual area of contact is well defined and measurable, and asperity deformation do not complicate the analysis. During sliding experiments, the area of parallel surfaces is very large compared to the thickness of the sheared film and this provides an ideal condition for studying shear behavior because it permits one to study molecularly-thin liquid films whose thickness is well defined to the resolution of an angstrom. Molecularly thin liquid films cease to behave as a structural continuum with properties different from that of the bulk material [40, 44–47].

Tonck et al. [48] and Georges et al. [49] developed a SFA used to measure the static and dynamic forces (in the normal direction) between a smooth fused borosilicate glass against a smooth and flat silicon wafer. They used capacitance technique to measure surface separation; therefore, use of optically-transparent surfaces was not required. Among others, metallic surfaces can be used at the interface. Georges et al. [50] modified the original SFA so that a sphere can be moved towards and away from a plane and can be sheared at constant separation from the plane, for interfacial friction studies.

For a detailed review of various types of SFAs, see Israelachvili [42, 51], Horn [52], and Homola [53]. SFAs based on their design are commercially available from SurForce Corporation, Santa Barbara, California.

### Israelachvili's and Granick's Design

Following review is primarily based on the papers by Israelachvili [42] and Homola [53]. Israelachvili et al.'s design later followed by Granick et al. for oscillating shear studies, is most commonly used by researchers around the world.

*Classical SFA* The classical apparatus developed for measuring equilibrium or static intersurface forces in liquids and vapors by Israelachvili and Adams [35], consists of a small, air-tight stainless steel chamber in which two molecularly smooth curved mica surfaces can be translated towards or away from each other, see Fig. 1.7. The distance between the two surfaces can also be independently controlled to within ±0.1 nm and the force sensitivity is about 10 nN. The technique utilizes two molecularly smooth mica sheets, each about 2 μm thick, coated with a semi reflecting 50–60 nm layer of pure silver, glued to rigid cylindrical silica disks of radius about 10 mm (silvered side down) mounted facing each other with their axes mutually at right angles (crossed cylinder position), which is geometrically equivalent to a sphere contacting a flat surface. The adhesive glue which is used to affix the mica to the support is sufficiently compliant, so the mica will flatten under the action of adhesive forces or applied load to produce a contact zone in which the surfaces are locally parallel and planar. Outside of this contact zone the separation between surfaces increases and the liquid, which is effectively in a bulk state, makes a negligible contribution to the overall response. The lower surface is supported on a cantilever spring which is used to push the two surfaces together with a known load. When the surfaces are forced into contact, they flatten elastically so that the contact zone is circular for duration of the static or sliding

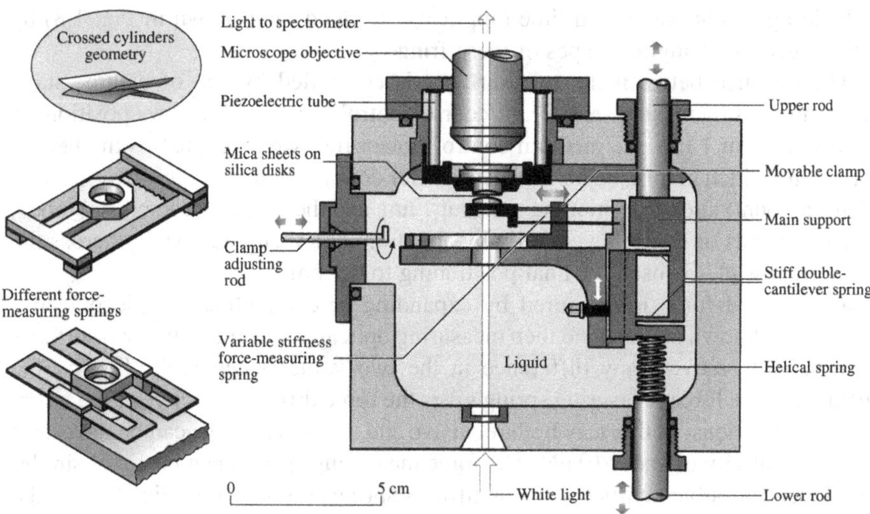


**Fig. 1.7** Schematic of the surface force apparatus that employs the cross cylinder geometry [35, 42]

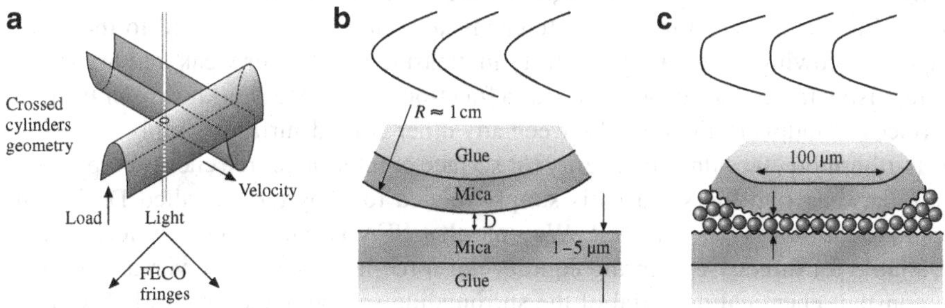


**Fig. 1.8** (**a**) Cross cylinder configuration of mica sheet, showing formation of contact area. Schematic of the fringes of equal chromatic order (FECO) observed when two mica surfaces are (**b**) separated by distance $D$ and (**c**) are flattened with a monolayer of liquid between them [54]

interactions. The surface separation is measured using optical interference fringes of equal chromatic order (FECO) which enables the area of molecular contact and the surface separation to be measured to within 0.1 nm. For measurements, white light is passed vertically up through the two mica surfaces and the emerging beam is then focused onto the slit of a grating spectrometer. From the positions and shapes of the colored FECO fringes in the spectrogram, the distance between the two surfaces and the exact shape of the two surfaces can be measured (as illustrated in Fig. 1.8), as can the refractive index of the liquid (or material) between them. In particular, this allows for reasonably accurate determinations of the quantity of material deposited or adsorbed on the surfaces and the area of contact between two molecularly smooth surfaces. Any changes may be readily observed in both static and sliding conditions in real time (applicable to the design shown in Fig. 1.8) by monitoring the changing shapes of these fringes.

The distance between the two surfaces is controlled by use of a three-stage mechanism of increasing sensitivity: coarse control (upper rod) allows positioning of within about 1 μm, the medium control (lower rod, which depresses the helical spring and which in turn, bends the much stiffer double-cantilever spring by 1/1,000 of this amount) allows positioning to about 1 nm, and the piezoelectric crystal tube – which expands or controls vertically by about 0.6 nm/V applied axially across the cylindrical wall – is used for final positioning to 0.1 nm.

The normal force is measured by expanding or contracting the piezoelectric crystal by a known amount and then measuring optically how much the two surfaces have actually moved; any difference in the two values when multiplied by the stiffness of the force measuring spring gives the force difference between the initial and final positions. In this way both repulsive and attractive forces can be measured with a sensitivity of about 10 nN. The force measuring springs can be either single-cantilever or double-cantilever fixed-stiffness springs (as shown in Fig. 1.7), or the spring stiffness can be varied during an experiment (by up to a factor of 1,000) by shifting the position of the dovetailed clamp using the adjusting rod. Other spring attachments, two of which are shown at the top of the figure, can replace the variable stiffness spring attachment (top right: nontilting nonshearing spring of fixed stiffness). Each of these springs are interchangeable and can be attached to the main support, allowing for greater versatility in measuring strong or weak and attractive or repulsive forces. Once the force $F$ as a function of distance $D$ is known for the two surfaces of radius $R$, the force between any other curved surfaces simply scales by $R$. Furthermore, the adhesion energy (or surface or interfacial free energy) $E$ per unit area between two flat surfaces is simply related to $F$ by the so-called Derjaguin approximation [51] $E = F/2\pi R$. We note that SFA is one of the few techniques available for directly measuring equilibrium force-laws (i.e., force versus distance at constant chemical potential of the surrounding solvent medium) [42]. The SFA allows for both weak or strong and attractive or repulsive forces.

Mostly the molecularly smooth surface of mica is used in these measurements [55], however, silica [56] and sapphire [57] have also been used. It is also possible to deposit or coat each mica surface with metal films [58, 59], carbon and metal oxides [60], adsorbed polymer layers [61], surfactant monolayers and bilayers [51, 58, 62, 63]. The range of liquids and vapors that can be used is almost endless.

*Sliding Attachments for Tribological Studies* So far we have described a measurement technique which allows measurements of the normal forces between surfaces, that is, those occurring when two surfaces approach or separate from each other. However, in tribological situations, it is the transverse or shear forces that are of primary interest when two surfaces slide past each other. There are essentially two approaches used in studying the shear response of confined liquid films. In the first approach (constant velocity friction or steady-shear attachment), the friction is measured when one of the surfaces is traversed at a constant speed over a distance of several hundreds of microns [38, 39, 45, 54, 60, 64, 65]. The second approach (oscillatory shear attachment) relies on the measurement of viscous dissipation and elasticity of confined liquids by using periodic sinusoidal oscillations over a range of amplitudes and frequencies [40, 41, 44, 66, 67].

For the constant velocity friction (steady-shear) experiments, the surface force apparatus was outfitted with a lateral sliding mechanism [38, 42, 46, 54, 64, 65] allowing measurements of both normal and shearing forces (Fig. 1.9). The piezo-electric crystal tube mount supporting the upper silica disk of the basic apparatus shown in Fig. 1.7, is replaced. Lateral motion is initiated by a variable speed motor-driven micrometer screw that presses against the translation stage, which is connected via two horizontal double-cantilever strip springs to the rigid mounting plate. The translation stage also supports two vertical double-cantilever springs (Fig. 1.10) that at their lower end are connected to a steel plate supporting the upper silica disk. One of the vertical springs acts as a frictional force detector by having four resistance strain gages attached to it, forming the four arms of a Wheatstone bridge and electrically connected to a chart recorder. Thus, by rotating the micrometer,

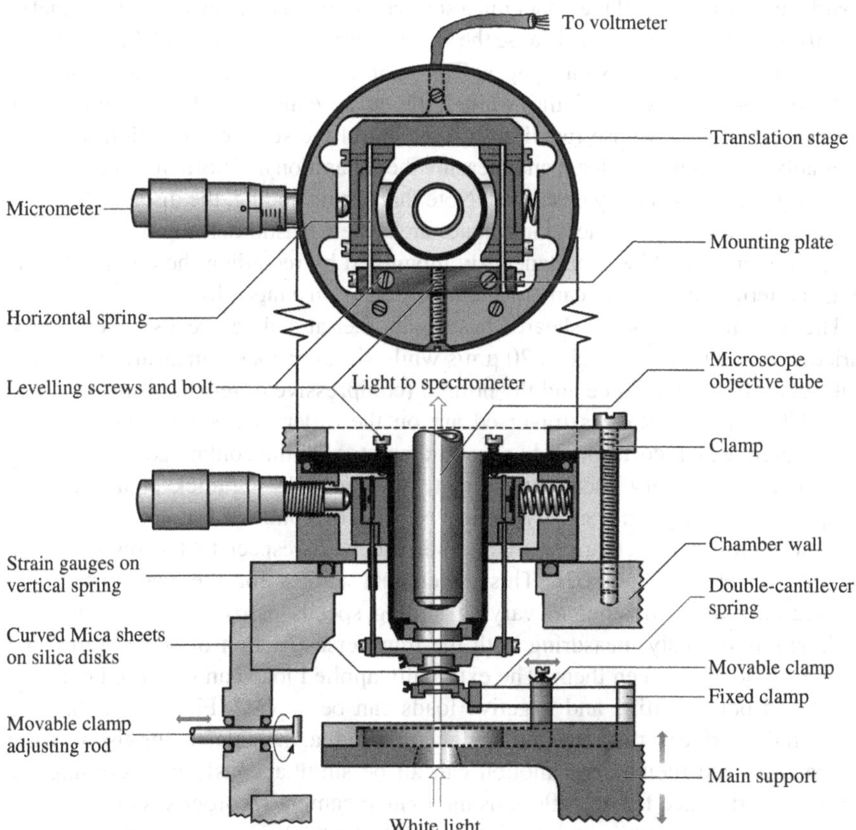


**Fig. 1.9** Schematic of shear force apparatus. Lateral motion is initiated by a variable speed motor-driven micrometer screw that presses against the translation stage which is connected through two horizontal double-cantilever strip springs to the rigid mounting plate [38, 42]

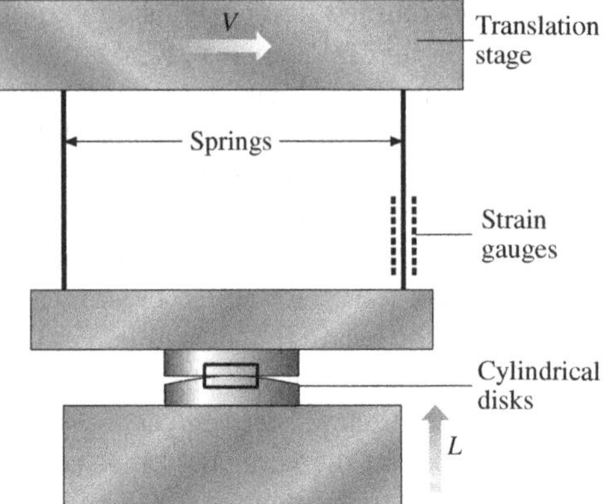


**Fig. 1.10** Schematic of the sliding attachment. The translation stage also supports two vertical double-cantilever springs, which at their lower end are connected to a steel plate supporting the upper silica disk [46]

the translation stage deflects, causing the upper surface to move horizontally and linearly at a steady rate. If the upper mica surface experiences a transverse frictional or viscous shearing force, this will cause the vertical springs to deflect, and this deflection can be measured by the strain gages. The main support, force-measuring double-cantilever spring, movable clamp, white light, etc., are all parts of the original basic apparatus (Fig. 1.7), whose functions are to control the surface separation, vary the externally applied normal load, and measure the separation and normal force between the two surfaces, as already described. Note that during sliding, the distance between the surfaces, their true molecular contact area, their elastic deformation, and their lateral motion can all be simultaneously monitored by recording the moving FECO fringe pattern using a video camera and recording it on a tape [46].

The two surfaces can be sheared past each other at sliding speeds which can be varied continuously from 0.1 to 20 μm/s while simultaneously measuring both the transverse (frictional) force and the normal (compressive or tensile) force between them. The lateral distances traversed are on the order of a several hundreds of micrometers which correspond to several diameters of the contact zone.

With an oscillatory shear attachment, developed by Granick et al., viscous dissipation and elasticity and dynamic viscosity of confined liquids by applying periodic sinusoidal oscillations of one surface with respect to the other can be studied [40, 41, 44, 66, 67]. This attachment allows for the two surfaces to be sheared past each other at varying sliding speeds or oscillating frequencies while simultaneously measuring both the transverse (friction or shear) force and the normal load between them. The externally applied load can be varied continuously, and both positive and negative loads can be applied. Finally the distance between the surfaces, their true molecular contact area, their elastic (or viscoelastic) deformation and their lateral motion can all be simultaneously by recording the moving interference fringe pattern using a video camera-recorder system.

To produce shear while maintaining constant film thickness or constant separation of the surfaces, the top mica surface is suspended from the upper portion of the apparatus by two piezoelectric bimorphs. A schematic description of the surface force apparatus with the installed shearing device is shown in Fig. 1.11 [40, 41, 44, 66, 67].

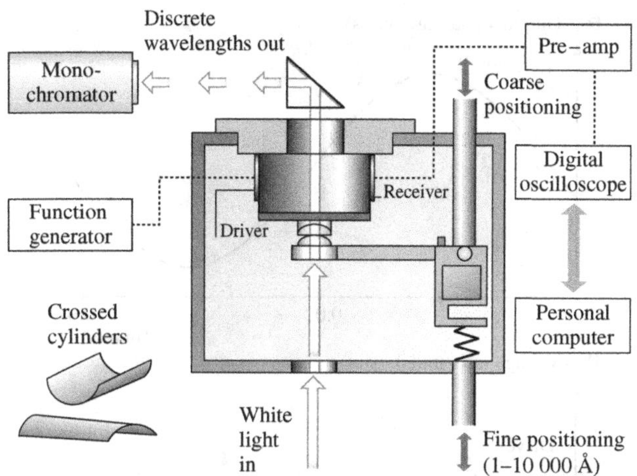


**Fig. 1.11** Schematic of the oscillatory shearing apparatus [40]

Israelachvili [42] and Luengo et al. [43] have also presented similar designs. The lower mica surface, as in the steady-shear sliding attachment, is stationary and sits at the tip of a double cantilever spring attached at the other end to a stiff support. The externally applied load can be varied continuously by displacing the lower surface vertically. An AC voltage difference applied by a signal generator (driver) across one of the bimorphs tends to bend it in oscillatory fashion while the frictional force resists that motion. Any resistance to sliding induces an output voltage across the other bimorphs (receiver) which can be easily measured by a digital oscilloscope. The sensitivity in measuring force is on the order of a few μN and the amplitudes of measured lateral displacement can range from a few nm to 10 μm. The design is flexible and allows to induce time-varying stresses with different characteristic wave shapes simply by changing the wave form of the input electrical signal. For example, when measuring the apparent viscosity, a sine wave input is convenient to apply. Figure 1.12a shows an example of the raw data, obtained with a hexadecane film at a moderate pressure, when a sine wave was applied to one of the bimorphs [66]. By comparing the calibration curve with the response curve, which was attenuated in amplitude and lagged in phase, an apparent dynamic viscosity can be inferred. On the other hand, a triangular waveform is more suitable when studying the yield stress behavior of solid-like films as of Fig. 1.12b. The triangular waveform, showing a linear increase and decrease of the applied force with time, is proportional to the driving force acting on the upper surface. The response waveform, which represents a resistance of the interface to shear, remains very small indicating that the surfaces are in a stationary contact with respect to each other until the applied stress reaches a yield point. At the yield point the slope of the response curve increases dramatically, indicating the onset of sliding.

Homola [53] compared the two approaches – steady shear attachment and oscillating shear attachment. In experiments conducted by Israelachvili and his

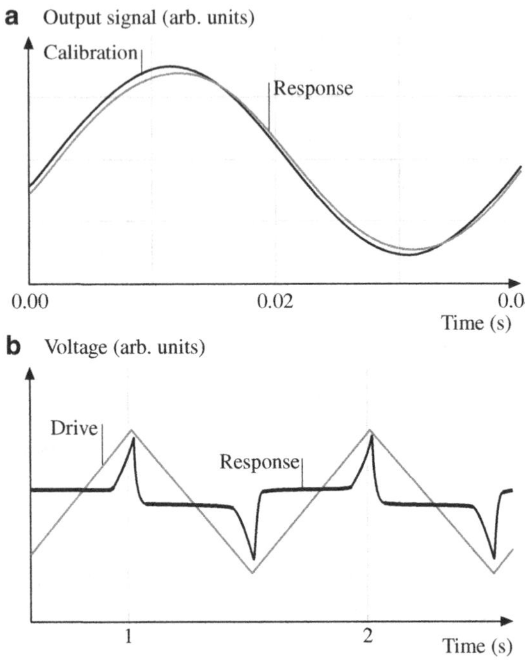


**Fig. 1.12** (**a**) Two output signals induced by an applied sine wave (not shown) are displaced. The "calibration" waveform is obtained with the mica sheets completely separated. The response waveform is obtained with a thin liquid film between the sheets, which causes it to lag the calibration waveform, (**b**) the oscilloscope trace of the drive and response voltages used to determine critical shear stress. The drive waveform shows voltage proportional to induced stress on the sheared film and the response waveform shows voltage proportional to resulting velocity. Spikes in the response curve correspond to the stick–slip event [66]

co-workers, the steady-shear attachment was employed to focus on the dynamic frictional behavior of the film after a sufficiently high shear stress was applied to exceed the yield stress and to produce sliding at a constant velocity. In these measurements, the film was subjected to a constant shearing force for a time sufficiently long to allow them to reach a dynamic equilibrium, i.e., the molecules, within the film, had enough time to order and align with respect to the surface, both normally and tangentially. Under these conditions, dynamic friction was observed to be "quantized" according to the number of liquid layers between the solid surfaces and independent of the shear rate [38]. Clearly, in this approach, the molecular ordering is optimized by a steady shear which imposes a preferred orientation on the molecules in the direction of shear.

The above mode of sliding is particularly important when the sheared film is made of a long chain lubricant molecules requiring a significantly long sliding time to order and align and even a longer time to relax (disorder) when sliding stops. This suggests that a steady-state friction is realized only when the duration of sliding exceeds the time required for an ensemble of the molecules to fully order in a specific shear field. It also suggests, that static friction should depend critically on the sliding time and the extend of the shear induced ordering [53].

In contrast, the oscillatory shear method, which utilizes periodic sinusoidal oscillations over a range of amplitudes and frequencies, addresses a response of the system to rapidly varying strain rates and directions of sliding. Under these conditions, the molecules, especially those exhibiting a solid-like behavior, cannot respond sufficiently fast to stress and are unable to order fully during duration of a single pass, i.e., their dynamic and static behavior reflects and oscillatory shear induced ordering which might or might not represent an equilibrium dynamic state. Thus, the response of the sheared film will depend critically on the conditions of shearing, i.e., the strain, the pressure, and the sliding conditions (amplitude and frequency of oscillations) which in turn will determine a degree of molecular ordering. This may explain the fact that the layer structure and "quantization" of the dynamic and static friction was not observed in these experiments in contrast to results obtained when velocity was kept constant. Intuitively, this behavior is expected considering that the shear-ordering tendency of the system is frequently disturbed by a shearing force of varying magnitude and direction. Nonetheless, the technique is capable of providing an invaluable insight into the shear behavior of molecularly thin films subjected to non-linear stresses as it is frequently encountered in practical applications. This is especially true under conditions of boundary lubrication where interacting surface asperities will be subjected to periodic stresses of varying magnitudes and frequencies [53].

#### Georges et al.'s Design

The SFA developed by Tonck et al. [48] and Georges et al. [49] to measure static and dynamic forces in the normal direction, between surfaces in close proximity, is shown in Fig. 1.13. In their apparatus, a drop of liquid is introduced between a macroscopic spherical body and a plane. The sphere is moved towards and away from a plane using the expansion and the vibration of a piezoelectric crystal. Piezoelectric crystal is vibrated at low amplitude around an average separation for dynamic measurements to provide dynamic function of the interface. The plane specimen is supported by a double-cantilever spring. Capacitance sensor $C_1$ measures the elastic deformation of the cantilever and thus the force transmitted through the liquid to the plane. Second capacitance sensor $C_2$ is designed to measure the relative displacement between the supports of the two solids. The reference displacement signal is the sum of two signals: first, a ramp provides a constant normal speed from 50 to 0.01 nm/s, and, second, the piezoelectric crystal is designed to provide a small sinusoidal motion, in order to determine the dynamic behavior of sphere-plane interactions. A third capacitance sensor C measures the electrical capacitance between the sphere and the plane. In all cases, the capacitance is determined by incorporating the signal of an oscillator in the inductive–capacitance (*L*–*C*) resonant input stage of an oscillator to give a signal-dependent frequency in the range of 5–12 MHz. The resulting fluctuations in oscillation frequency are detected using a low noise frequency discriminator.

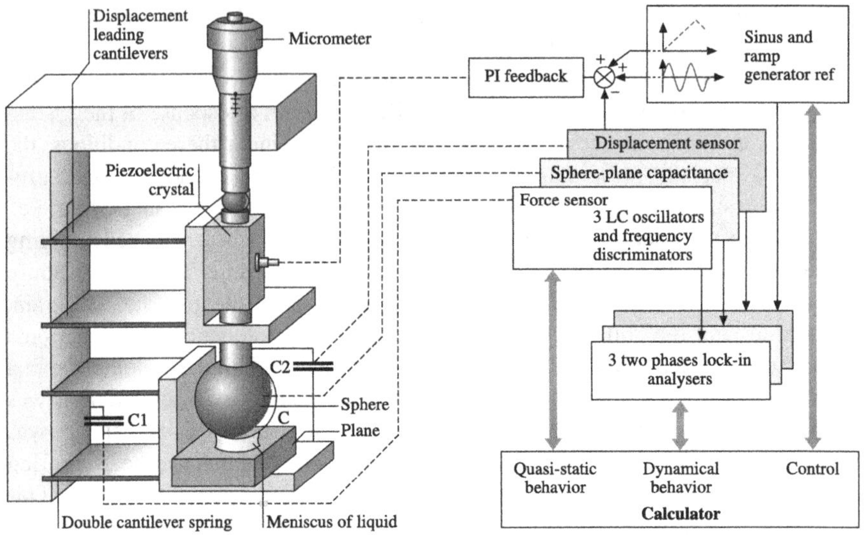


**Fig. 1.13** Schematic of the surface force apparatus that employs a sphere–plane arrangement [49]

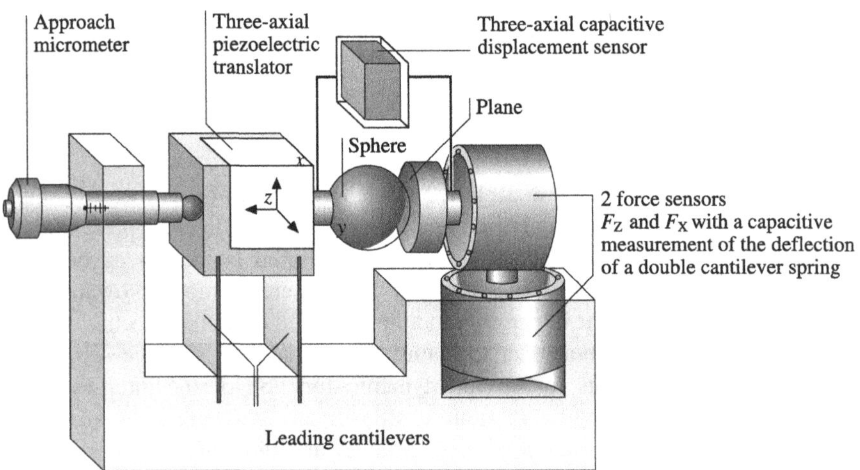


**Fig. 1.14** Schematic of shear force apparatus [50]

Simultaneous measurements of sphere-plane displacement, surface force, and the damping of the interface allows an analysis of all regimes of the interface [49]. Loubet et al. [68] used SFA in the crossed-cylinder geometry using two freshly-cleaved mica sheets similar to the manner used by Israelachvili and coworkers.

Georges et al. [50] modified their original SFA to measure friction forces. In this apparatus, in addition to having sphere move normal to the plane, sphere can be sheared at constant separation from the plane. The shear force apparatus is shown in Fig. 1.14. Three piezoelectric elements controlled by three capacitance sensors permit accurate motion control and force measurement along three orthogonal axes with displacement sensitivity of $10^{-3}$ nm and force sensitivity of $10^{-8}$ N. Adhesion and normal deformation experiments are conducted in the normal approach ($z$-axis). Friction experiments are conducted by introducing displacement in the $X$-direction at a constant normal force. In one of the experiment, Georges et al. [50] used 2.95 mm diameter sphere made of cobalt-coated fused borosilicate glass and a silicon wafer for the plane.

### 1.4.3 Vibration Isolation

STM, AFM and SFA should be isolated from sources of vibration in the acoustic and sub-acoustic frequencies especially for atomic-scale measurements. Vibration isolation is generally provided by placing the instrument on a vibration isolation air table. For further isolation, the instrument should be placed on a pad of soft silicone rubber. A cheaper alternative consists of a large mass of 100 N or more, suspended from elastic "bungee" cords. The mass should stretch the cords at least 0.3 m, but not so much that the cords reach their elastic limit. The instrument should be placed on the large mass. The system, including the microscope, should have a natural frequency of about 1 Hz or less both vertically and horizontally. Test this by gently pushing on the mass and measure the rate at which its swings or bounces.

## 1.5 Magnetic Storage Devices and MEMS/NEMS

### 1.5.1 Magnetic Storage Devices

Magnetic storage devices used for storage and retrieval are tape, flexible (floppy) disk and rigid disk drives. These devices are used for audio, video and data storage applications. Magnetic storage industry is some $ 60 billion a year industry with $ 20 billion for audio and video recording (almost all tape drives/media) and $ 40 billion for data storage. In the data storage industry, magnetic rigid disk drives/media, tape drives/media, flexible disk drives/media, and optical disk drive/media account for about $ 25 B, $ 6 B, $ 3 B, and $ 6 B, respectively. Magnetic recording and playback involves the relative motion between a magnetic medium (tape or disk) against a read-write magnetic head. Heads are designed so that they develop a (load-carrying) hydrodynamic air film under steady operating conditions to minimize head–medium contact. However, physical contact between the medium and head occurs during starts and stops, referred to as contact-start-stops (CSS) technology [13, 14, 69]. In the modern magnetic storage devices, the flying heights (head-to-medium separation) are on the order of 5–20 nm and roughnesses of head and medium surfaces are on the order of 1–2 nm RMS. The need for ever-increasing recording densities requires that surfaces be as smooth as possible and the flying heights be as low as possible. High stiction (static friction) and wear are the limiting technology to future of this industry. Head load/unload (L/UL) technology has recently been used as an alternative to CSS technology in rigid disk drives that eliminates stiction and wear failure mode associated with CSS. Several contact or near contact recording devices are at various stages of development. High stiction and wear are the major impediments to the commercialization of the contact recording.

Magnetic media fall into two categories: particulate media, where magnetic particles ($\gamma$-$Fe_2O_3$, Co-$\gamma Fe_2O_3$, $CrO_2$, Fe or metal (MP), or barium ferrite) are dispersed in a polymeric matrix and coated onto a polymeric substrate for flexible media (tape and flexible disks); thin-film media, where continuous films of magnetic materials are deposited by vacuum deposition techniques onto a polymer substrate for flexible media or onto a rigid substrate (typically aluminium and more recently glass or glass ceramic) for rigid disks. The most commonly used thin magnetic films for tapes are evaporated Co–Ni (82–18 at.%) or Co–O dual layer. Typical magnetic films for rigid disks are metal films of cobalt-based alloys (such as sputtered Co–Pt–Ni, Co–Ni, Co–Pt–Cr, Co–Cr and Co–NiCr). For high recording densities, trends have been to use thin-film media. Magnetic heads used to date are either conventional thin-film inductive, magnetoresistive (MR) and giant MR (GMR) heads. The air-bearing surfaces (ABS) of tape heads are generally cylindrical in shape. For dual-sided flexible-disk heads, two heads are either spherically contoured and slightly offset (to reduce normal pressure) or are flat and loaded against each other. The rigid-disk heads are supported by a leaf spring (flexure) suspension. The ABS of heads are almost made of Mn–Zn ferrite, Ni–Zn ferrite, $Al_2O_3$–TiC and calcium titanate. The ABS of some conventional heads are made of plasma sprayed coatings of hard materials such as $Al_2O_3$–$TiO_2$ and $ZrO_2$ [13, 14, 69].

Figure 1.15 shows the schematic illustrating the tape path with details of tape guides in a data-processing linear tape drive (IBM LTO Gen1) which uses a rectangular tape cartridge. Figure 1.16a shows the sectional views of particulate and thin-film magnetic tapes. Almost exclusively, the base film is made of

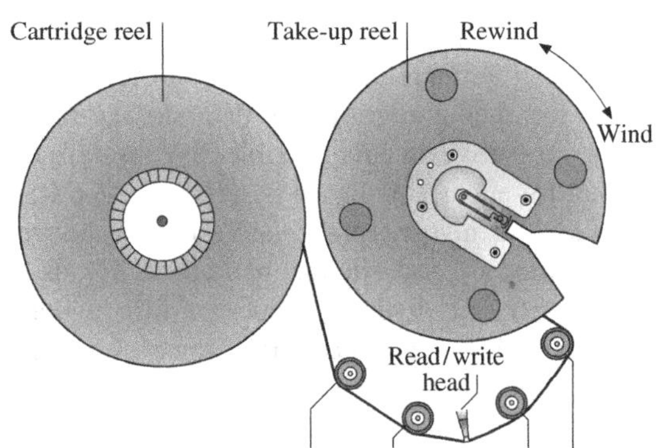


**Fig. 1.15** Schematic of tape path in an IBM Linear Tape Open (*LTO*) tape drive

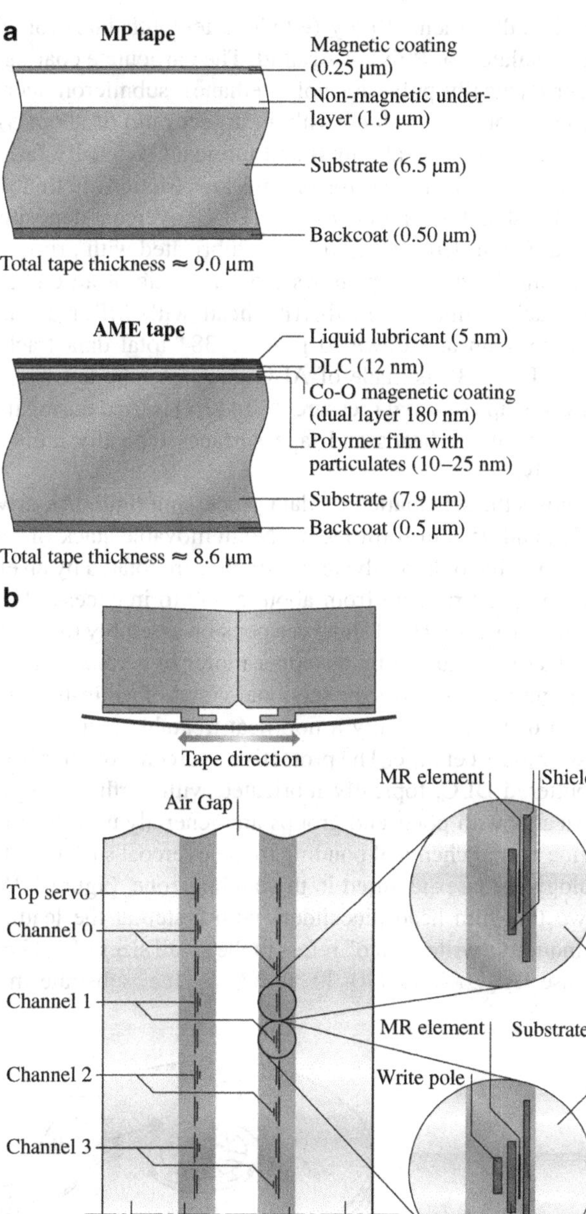


**Fig. 1.16** (**a**) Sectional views of particulate and thin-film magnetic tapes, and (**b**) schematic of a magnetic thin-film read/write head for an IBM LTO Gen 1 tape drive

semicrystalline biaxially-oriented poly (ethylene terephthalate) (or PET) or poly (ethylene 2,6 naphthalate) (or PEN) or Aramid. The particulate coating formulation consists of binder (typically polyester polyurethane), submicron accicular shaped magnetic particles (about 50 nm long with an aspect ratio of about 5), submicron head cleaning agents (typically alumina) and lubricant (typically fatty acid ester). For protection against wear and corrosion and low friction/stiction, the thin-film tape is first coated with a diamondlike carbon (DLC) overcoat deposited by plasma enhanced chemical vapor deposition, topically lubricated with primarily a perfluoropolyether lubricant. Figure 1.16b shows the schematic of an eight-track (along with two servo tracks) thin-film read-write head with MR read and inductive write. The head steps up and down to provide 384 total data tracks across the width of the tape. The ABS is made of $Al_2O_3$–TiC. A tape tension of about 1 N over a 12.7 mm wide tape (normal pressure $\approx$ 14 kPa) is used during use. The RMS roughnesses of ABS of the heads and tape surfaces typically are 1–1.5 nm and 5–8 nm, respectively.

Figure 1.17 shows the schematic of a data processing rigid disk drive with 21.6, 27.4, 48, 63.5, 75, and 95 mm form factor. Nonremovable stack of multiple disks mounted on a ball bearing or hydrodynamic spindle, are rotated by an electric motor at constant angular speed ranging from about 5,000 to in excess of 15,000 RPM, dependent upon the disk size. Head slider-suspension assembly (allowing one slider for each disk surface) is actuated by a stepper motor or a voice coil motor using a rotary actuator. Figure 1.18a shows the sectional views of a thin-film rigid disk. The substrate for rigid disks is generally a non heat-treatable aluminium–magnesium alloy 5086, glass or glass ceramic. The protective overcoat commonly used for thin-film disks is sputtered DLC, topically lubricated with perfluoropolyether type of lubricants. Lubricants with polar-end groups are generally used for thin-film disks in order to provide partial chemical bonding to the overcoat surface. The disks used for CSS technology are laser textured in the landing zone. Figure 1.18b shows the schematic of two thin-film head picosliders with a step at the leading edge, and GMR read and inductive write. "Pico" refers to the small sizes of 1.25 mm × 1 mm. These sliders use $Al_2O_3$–TiC (70–30 wt%) as the substrate material with

**Fig. 1.17** Schematic of a data-processing magnetic rigid disk drive

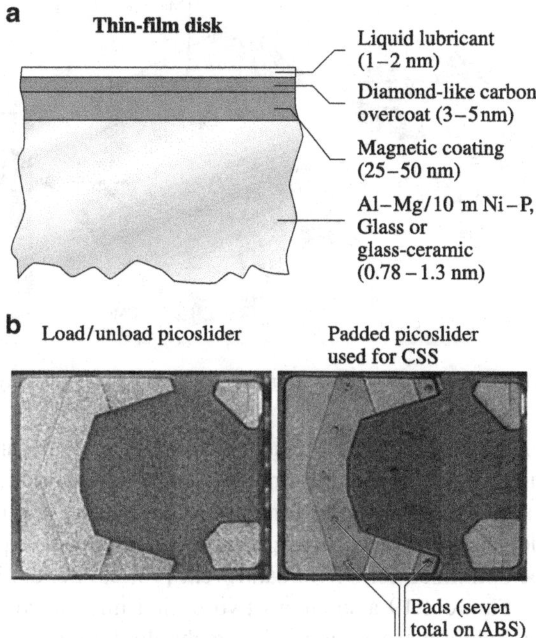


**Fig. 1.18** (**a**) Sectional views of a thin-film magnetic rigid disk, and (**b**) schematic of two picosliders – load/unload picoslider and padded picoslider used for CSS

multilayered thin-film head structure coated and with about 3.5 nm thick DLC coating to prevent the thin film structure from electrostatic discharge. The seven pads on the padded slider are made of DLC and are about 40 μm in diameter and 50 nm in height. A normal load of about 3 g is applied during use.

## *1.5.2 MEMS/NEMS*

The advances in silicon photolithographic process technology led to the development of MEMS in the mid-1980s [16]. More recently, lithographic and nonlithographic processes have been developed to process nonsilicon (plastics or ceramics) materials. MEMS for mechanical applications include acceleration, pressure, flow, and gas sensors, linear and rotary actuators, and other microstructures of microcomponents such as electric motors, gear trains, gas turbine engines, nozzles, fluid pumps, fluid valves, switches, grippers, and tweezers. MEMS for chemical applications include chemical sensors and various analytical instruments. Microoptoelectromechanical systems (or MOEMS) include micromirror arrays and fiber optic connectors. Radio frequency MEMS or RF-MEMS include inductors, capacitors, and antennas. High-aspect ratio MEMS (HARMEMS) have also been introduced. BioMEMS include biofluidic chips (microfluidic chips or bioflips or simple biochips) for chemical and biochemical analyses (biosensors in medical diagnostics,

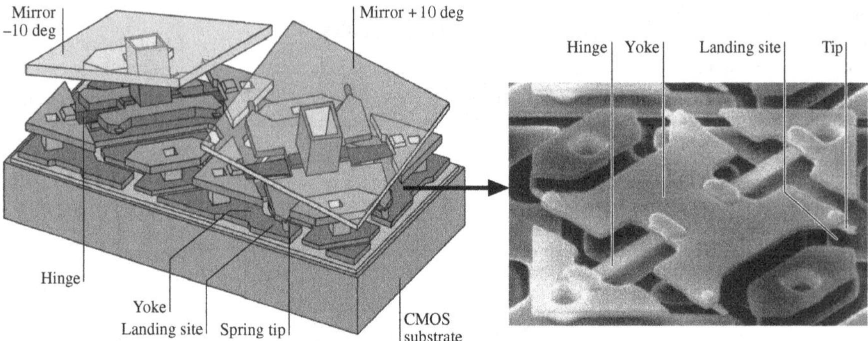


**Fig. 1.19** Digital micromirror devices for projection displays (From [16])

e.g., DNA, RNA, proteins, cells, blood pressure and assays, and toxin identification), and implantable drug delivery. Killer applications include capacitive-type silicon accelerometers for automotive sensory applications and digital micromirror devices for projection displays. Any component requiring relative motions needs to be optimized for stiction and wear [16, 20, 22, 69, 70].

Figure 1.19 also shows two digital micromirror device (DMD) pixels used in digital light processing (DLP) technology for digital projection displays in portable and home theater projectors as well as table top and projection TVs [16, 71, 72]. The entire array (chip set) consists of a large number of rotatable aluminium micromirrors (digital light switches) which are fabricated on top of a CMOS static random access memory integrated circuit. The surface micromachined array consists of half of a million to more than two million of these independently controlled reflective, micromirrors (mirror size on the order of 14 μm × 14 μm and 15 μm pitch) which flip backward and forward at a frequency of on the order of 5,000 times a second. For the binary operation, micromirror/yoke structure mounted on torsional hinges is rotated ±10° (with respect to the horizontal plane) as a result of electrostatic attraction between the micromirror structure and the underlying memory cell, and is limited by a mechanical stop. Contact between cantilevered spring tips at the end of the yoke (four present on each yoke) with the underlying stationary landing sites is required for true digital (binary) operation. Stiction and wear during a contact between aluminium alloy spring tips and landing sites, hinge memory (metal creep at high operating temperatures), hinge fatigue, shock and vibration failure, and sensitivity to particles in the chip package and operating environment are some of the important issues affecting the reliable operation of a micromirror device. Perfluorodecanoic acid (PFDA) self-assembled monolayers are used on the tip and landing sites to reduce stiction and wear. The spring tip is used in order to use the spring stored energy to pop up the tip during pull-off. A lifetime estimate of over 100,000 operating hours with no degradation in image quality is the norm.

NEMS are produced by nanomachining in a typical top-down approach (from large to small) and bottom-up approach (from small to large) largely relying on nanochemistry [16]. The top-down approach relies on fabrication methods including advanced integrated-circuit (IC) lithographic methods – electron-beam lithography, and STM writing by removing material atom by atom. The bottom-up approach includes chemical synthesis, the spontaneous "self-assembly" of molecular clusters (molecular self-assembly) from simple reagents in solution, or biological molecules (e.g., DNA) as building blocks to produce three dimensional nanostructures, quantum dots (nanocrystals) of arbitrary diameter (about 10–$10^5$ atoms), molecular beam epitaxy (MBE) and organometallic vapor phase epitaxy (OMVPE) to create specialized crystals one atomic or molecular layer at a time, and manipulation of individual atoms by an atomic force microscope or atom optics. The self-assembly must be encoded, that is, one must be able to precisely assemble one object next to another to form a designed pattern. A variety of nonequilibrium plasma chemistry techniques are also used to produce layered nanocomposites, nanotubes, and nanoparticles. NEMS field, in addition to fabrication of nanosystems, has provided impetus to development of experimental and computation tools.

Examples of NEMS include nanocomponents, nanodevices, nanosystems, and nanomaterials such as microcantilever with integrated sharp nanotips for STM and AFM, AFM array (Millipede) for data storage, AFM tips for nanolithography, dip-pen nanolithography for printing molecules, biological (DNA) motors, molecular gears, molecularly-thick films (e.g., in giant magnetioresistive or GMR heads and magnetic media), nanoparticles (e.g., nanomagnetic particles in magnetic media), nanowires, carbon nanotubes, quantum wires (QWRs), quantum boxes (QBs), and quantum transistors [16]. BIONEMS include nanobiosensors – microarray of silicon nanowires, roughly few nm in size, to selectively bind and detect even a single biological molecule such as DNA or protein by using nanoelectronics to detect the slight electrical charge caused by such binding, or a microarray of carbon nanotubes to electrically detect glucose, implantable drug-delivery devices – e.g., micro/nanoparticles with drug molecules encapsulated in functionized shells for a site-specific targeting applications, and a silicon capsule with a nanoporous membrane filled with drugs for long term delivery, nanodevices for sequencing single molecules of DNA in the Human Genome Project, cellular growth using carbon nanotubes for spinal cord repair, nanotubes for nanostructured materials for various applications such as spinal fusion devices, organ growth, and growth of artificial tissues using nanofibers.

Figure 1.20 shows AFM based nanoscale data storage system for ultrahigh density magnetic recording which experiences tribological problems [73]. The system uses arrays of several thousand silicon microcantilevers ("Millipede") for thermomechanical recording and playback on an about 40-nm thick polymer (PMMA) medium with a harder Si substrate. The cantilevers are integrated with integrated tip heaters with tips of nanoscale dimensions. Thermomechanical recording is a combination of applying a local force to the polymer layer and softening it by local heating. The tip heated to about 400°C is brought in contact with the polymer for recording. Imaging and reading are done using the heater cantilever, originally used for recording, as a thermal readback sensor by exploiting its temperature-dependent resistance. The principle of thermal sensing is based on the fact that the thermal conductivity between the heater and the storage substrate changes according to the spacing between them. When the spacing between the

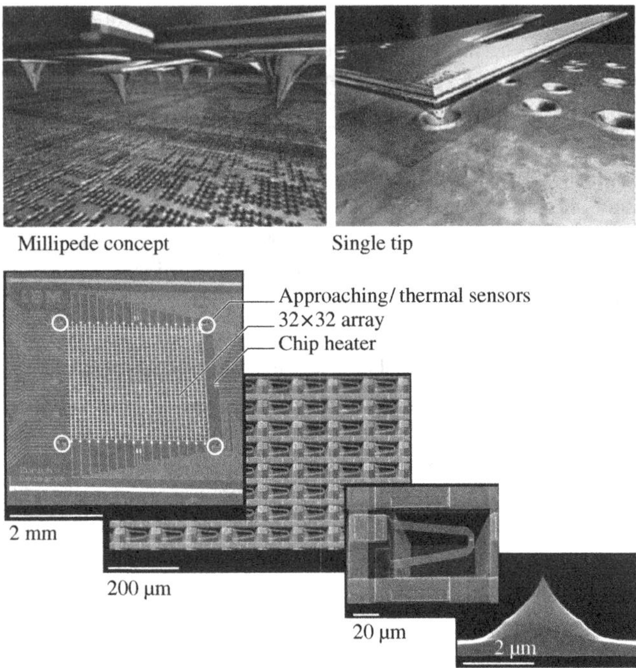


**Fig. 1.20** AFM based nanoscale data storage system with 32 × 32 tip array – that experiences a tribological problem (From [16])

heater and sample is reduced as the tip moves into a bit, the heater's temperature and hence its resistance will decrease. Thus, changes in temperature of the continuously heated resistor are monitored while the cantilever is scanned over data bits, providing a means of detecting the bits. Erasing for subsequent rewriting is carried out by thermal reflow of the storage field by heating the medium to 150°C for a few seconds. The smoothness of the reflown medium allows multiple rewriting of the same storage field. Bit sizes ranging between 10 and 50 nm have been achieved by using a 32 × 32 (1,024) array write/read chip (3 mm × 3 mm). It has been reported that tip wear occurs by the contact between tip and Si substrate during writing. Tip wear is considered a major concern for the device reliability.

## 1.6 Role of Micro/Nanotribology and Micro/Nanomechanics in Magnetic Storage Devices and MEMS/NEMS

The magnetic storage devices and MEMS/NEMS are the two examples where micro/nanotribological and micro/nanomechanical tools and techniques are essential for studies of micro/nano scale phenomena. Magnetic storage components continue to shrink in physical dimensions. Thicknesses of hard solid coating and liquid lubricant coatings on the magnetic disk surface continue to decrease. Number of contact recording devices are at various stages of development. Surface roughnesses of the storage components continue to decrease and are expected to approach to about 0.5 nm RMS or lower. Interface studies of components with ultra-thin coatings can be ideally performed using micro/nanotribological and micro/nanomechanical tools and techniques.

In the case of MEMS/NEMS, the friction and wear problems of ultrasmall moving components generally made of single-crystal silicon, polysilicon films or polymers need to be addressed for high performance, long life, and reliability. Molecularly-thin films of solid and/or liquids are used for low friction and wear in many applications. Again, interfacial phenomena in MEMS/NEMS can be ideally studied using micro/nanotribological and micro/nanomechanical tools and techniques.

## 1.7 Organization of the Book

The introductory book integrates knowledge of nanotribology and nanomechanics. The book starts with the definition of tribology, history of tribology and micro/nanotribology, its industrial significance, various measurement techniques employed, followed by various industrial applications. The remaining book is divided into four parts. The first part introduces scanning probe microscopy. The second part provides an overview of nanotechnology and nanomechanics. The third part provides an overview of molecularly-thick films for lubrication. And the last part focuses on nanotribology and nanomechanics studies conducted for various industrial applications.

## References

1. P. Jost, *Lubrication (Tribology) – A Report on the Present Position and Industry's Needs.* (Department. of Education and Science, H.M. Stationary Office, 1966)
2. D. Dowson, *History of Tribology*, 2nd edn. (Institute of Mechanical Engineers, London, 1998)
3. C.S.C. Davidson, Bearings since the Stone Age. Engineering **183**, 2–5 (1957)
4. A.G. Layard, *Discoveries in the Ruins of Nineveh and Babylon, volume I, II* (Murray, London, 1853)
5. G. Amontons, De la resistance causee dans les machines. Mem. Acad. R. A. **1706**, 257–282 (1699)
6. C.A. Coulomb, Theorie des machines simples, en ayant regard an frottement de leurs parties et a la roideur des cordages. Mem. Math. Phys. X, Paris **10**, 161–342 (1785)
7. W.F. Parish, Three thousand years of progress in the development of machinery and lubricants for the hand crafts. Mill Factory **16**, 17 (1935)
8. B. Tower, Report on friction experiments. Proc. Inst. Mech. Eng. **632**, 29–35 (1884)
9. O.O. Reynolds, On the theory of lubrication and its applications to Mr. Beauchamp tower's experiments. Philos. Trans. R. Soc. Lond. **117**, 157–234 (1886)
10. N.P. Petroff, Friction in machines and the effects of the lubricant. Eng. J. 71–140, 228–279, 377–436, 535–564 (1883)

11. R. Holm, *Electrical Contacts* (Springer, Berlin, 1946)
12. F.P. Bowden, D. Tabor, *The Friction and Lubrication of Solids*, vols 1, 2. (Clarendon, Oxford, 1950, 1964)
13. B. Bhushan, *Tribology and Mechanics of Magnetic Storage Devices*, 2nd edn. (Springer, New York, 1996)
14. B. Bhushan, *Mechanics and Reliability of Flexible Magnetic Media*, 2nd edn. (Springer, New York, 2000)
15. B. Bhushan, *Introduction to Tribology* (Wiley, New York, 2002)
16. B. Bhushan, *Springer Handbook of Nanotechnology*, 1st edn. (Springer, Berlin, 2004)
17. B. Bhushan, B.K. Gupta, *Handbook of Tribology: Materials, Coatings, and Surface Treatments* (McGraw-Hill, New York, 1991)
18. P. Jost, *Economic Impact of Tribology*, vol 423 of NBS Spec. Pub. Proc. Mechanical Failures Prevention Group, 1976
19. B. Bhushan, J.N. Israelachvili, U. Landman, Nanotribology: Friction, wear and lubrication at the atomic scale. Nature **374**, 607–616 (1995)
20. B. Bhushan, *Micro/Nanotribology and its Applications*. NATO ASI Series E: Applied Sciences, vol. 330 (Kluwer, Dordrecht, 1997)
21. B. Bhushan, *Handbook of Micro/Nanotribology*, 2nd edn. (CRC, Boca Raton, 1999)
22. B. Bhushan, *Fundamentals of Tribology and Bridging the Gap Between the Macro- and Micro/Nanoscales*. NATO Science Series II: Mathematics, Physics, and Chemistry, vol. 10 (Kluwer, Dordrecht, 2001)
23. B. Bhushan, H. Fuchs, S. Hosaka, *Applied Scanning Probe Methods* (Springer, Berlin, 2004)
24. G. Binnig, H. Rohrer, Ch Gerber, E. Weibel, Surface studies by scanning tunnelling microscopy. Phys. Rev. Lett. **49**, 57–61 (1982)
25. G. Binnig, C.F. Quate, Ch Gerber, Atomic force microscope. Phys. Rev. Lett. **56**, 930–933 (1986)
26. G. Binnig, Ch Gerber, E. Stoll, T.R. Albrecht, C.F. Quate, Atomic resolution with atomic force microscope. Europhys. Lett. **3**, 1281–1286 (1987)
27. D. Tabor, R.H.S. Winterton, The direct measurement of normal and retarded van der Waals forces. Proc. R. Soc. Lond. A **312**, 435–450 (1969)
28. S. Alexander, L. Hellemans, O. Marti, J. Schneir, V. Elings, P.K. Hansma, An atomic-resolution atomic-force microscope implemented using an optical lever. J. Appl. Phys. **65**, 164–167 (1989)
29. B. Bhushan, J. Ruan, Atomic-scale friction measurements using friction force microscopy: Part II – application to magnetic media. ASME J. Tribol. **116**, 389–396 (1994)
30. J. Ruan, B. Bhushan, Atomic-scale friction measurements using friction force microscopy: Part I – general principles and new measurement techniques. ASME J. Tribol **116**, 378–388 (1994)
31. J. Ruan, B. Bhushan, Atomic-scale and microscale friction of graphite and diamond using friction force microscopy. J. Appl. Phys. **76**, 5022–5035 (1994)
32. G. Meyer, N.M. Amer, Novel optical approach to atomic force microscopy. Appl. Phys. Lett. **53**, 1045–1047 (1988)
33. B. Bhushan, V.N. Koinkar, J. Ruan, Microtribology of magnetic media. Proc. IME J J. Eng. Tribol **208**, 17–29 (1994)
34. J.N. Israelachvili, D. Tabor, The measurement of van der Waals dispersion forces in the range of 1.5 to 130 nm. Proc. Roy. Soc. Lond. A **331**, 19–38 (1972)
35. J.N. Israelachvili, G.E. Adams, Measurement of friction between two mica surfaces in aqueous electrolyte solutions in the range 0100 nm. J. Chem. Soc. Faraday Trans. I **74**, 975–1001 (1978)
36. J. Klein, Forces between mica surfaces bearing layers of adsorbed polystyrene in cyclohexane. Nature **288**, 248–250 (1980)
37. D.Y.C. Chan, R.G. Horn, The drainage of thin liquid films between solid surfaces. J. Chem. Phys. **83**, 5311–5324 (1985)

38. J.N. Israelachvili, P.M. McGuiggan, A.M. Homola, Dynamic properties of molecularly thin liquid films. Science **240**, 189–190 (1988)
39. J. Klein, D. Perahia, S. Warburg, Forces between polymer-bearing surfaces undergoing shear. Nature **352**, 143–145 (1991)
40. J. van Alsten, S. Granick, Molecular tribology of ultrathin liquid films. Phys. Rev. Lett. **61**, 2570–2573 (1988)
41. J. Peachey, J. van Alsten, S. Granick, Design of an apparatus to measure the shear response of ultrathin liquid films. Rev. Sci. Instrum. **62**, 463–473 (1991)
42. J.N. Israelachvili, Techniques for direct measurements of forces between surfaces in liquids at the atomic scale. Chemtracts. Anal. Phys. Chem **1**, 1–12 (1989)
43. G. Luengo, F.J. Schmitt, R. Hill, J.N. Israelachvili, Thin film bulk rheology and tribology of confined polymer melts: contrasts with build properties. Macromolecules **30**, 2482–2494 (1997)
44. J. van Alsten, S. Granick, Shear rheology in a confined geometry – polysiloxane melts. Macromolecules **23**, 4856–4862 (1990)
45. A.M. Homola, J.N. Israelachvili, M.L. Gee, P.M. McGuiggan, Measurement of and relation between the adhesion and friction of two surfaces separated by thin liquid and polymer films. ASME J. Tribol **111**, 675–682 (1989)
46. M.L. Gee, P.M. McGuiggan, J.N. Israelachvili, A.M. Homola, Liquid to solid-like transitions of molecularly thin films under shear. J. Chem. Phys. **93**, 1895–1906 (1990)
47. S. Granick, Motions and relaxations of confined liquids. Science **253**, 1374–1379 (1991)
48. A. Tonck, J.M. Georges, J.L. Loubet, Measurements of intermolecular forces and the rheology of dodecane between alumina surfaces. J. Colloid Interface Sci. **126**, 1540–1563 (1988)
49. J.M. Georges, S. Millot, J.L. Loubet, A. Tonck, Drainage of thin liquid films between relatively smooth surfaces. J. Chem. Phys. **98**, 7345–7360 (1993)
50. J.M. Georges, A. Tonck, D. Mazuyer, Interfacial friction of wetted monolayers. Wear **175**, 59–62 (1994)
51. J.N. Israelachvili, *Intermolecular and Surface Forces*, 2nd edn. (Academic, London, 1992)
52. R.G. Horn, Surface forces and their action in ceramic materials. J. Am. Ceram. Soc. **73**, 1117–1135 (1990)
53. A.M. Homola, *Interfacial Friction of Molecularly Thin Liquid Films* (World Scientific, Singapore, 1993), pp. 271–298
54. A.M. Homola, J.N. Israelachvili, P.M. McGuiggan, M.L. Gee, Fundamental experimental studies in tribology: the transition from interfacial friction of undamaged molecularly smooth surfaces. Wear **136**, 65–83 (1990)
55. R.M. Pashley, Hydration forces between solid surfaces in aqueous electrolyte solutions. J. Colloid Interface Sci. **80**, 153–162 (1981)
56. R.G. Horn, D.T. Smith, W. Haller, Surface forces and viscosity of water measured between silica sheets. Chem. Phys. Lett. **162**, 404–408 (1989)
57. R.G. Horn, J.N. Israelachvili, Molecular organization and viscosity of a thin film of molten polymer between two surfaces as probed by force measurements. Macromolecules **21**, 2836–2841 (1988)
58. H.K. Christenson, Adhesion between surfaces in unsaturated vapors – a reexamination of the influence of meniscus curvature and surface forces. J. Colloid Interface Sci. **121**, 170–178 (1988)
59. C.P. Smith, M. Maeda, L. Atanasoska, H.S. White, Ultrathin platinum films on mica and measurement of forces at the platinum/water interface. J. Phys. Chem. **95**, 199–205 (1988)
60. S.J. Hirz, A.M. Homola, G. Hadzioannou, S.W. Frank, Effect of substrate on shearing properties of ultrathin polymer films. Langmuir **8**, 328–333 (1992)
61. S.S. Patel, M. Tirrell, Measurement of forces between surfaces in polymer fluids. Annu. Rev. Phys. Chem. **40**, 597–635 (1989)
62. J.N. Israelachvili, Solvation forces and liquid structure – as probed by direct force measurements. Acc. Chem. Res. **20**, 415–421 (1987)

63. J.N. Israelachvili, P.M. McGuiggan, Forces between surface in liquids. Science **241**, 795–800 (1988)
64. A.M. Homola, Measurement of and relation between the adhesion and friction of two surfaces separated by thin liquid and polymer films. ASME J. Tribol. **111**, 675–682 (1989)
65. A.M. Homola, H.V. Nguyen, G. Hadzioannou, Influence of monomer architecture on the shear properties of molecularly thin polymer melts. J. Chem. Phys. **94**, 2346–2351 (1991)
66. J. van Alsten, S. Granick, Tribology studied using atomically smooth surfaces. Tribol. Trans. **33**, 436–446 (1990)
67. W.W. Hu, G.A. Carson, S. Granick, Relaxation time of confined liquids under shear. Phys. Rev. Lett. **66**, 2758–2761 (1991)
68. J.L. Loubet, M. Bauer, A. Tonck, S. Bec, B. Gauthier-Manuel, *Nanoindentation with a Surface Force Apparatus* (Kluwer, Dordrecht, 1993), pp. 429–447
69. B. Bhushan, *Macro- and microtribology of magnetic storage devices*, vol. 2 (CRC, Boca Raton, 2001), pp. 1413–1513
70. B. Bhushan, *Tribology Issues and Opportunities in MEMS* (Kluwer, Dordrecht, 1998)
71. L.J. Hornbeck, W.E. Nelson, *Bistable Deformable Mirror Devices, OSA Technical Digest Series*, vol. 8. (OSA, Washington, DC, 1988), pp. 107–110
72. L.J. Hornbeck, A digital light processing™ update – status and future applications. Proc. SPIE **3634**, 158–170 (1999)
73. P. Vettinger, J. Brugger, M. Despont, U. Dreschier, U. Duerig, W. Haeberie, Ultrahigh density, high data-rate NEMS based AFM data storage systems. Microelectron. Eng. **46**, 11–27 (1999)

# Part I
# Scanning Probe Microscopy

# Chapter 2
# Scanning Probe Microscopy – Principle of Operation, Instrumentation, and Probes

**Bharat Bhushan and Othmar Marti**

**Abstract** Since the introduction of the STM in 1981 and the AFM in 1985, many variations of probe-based microscopies, referred to as SPMs, have been developed. While the pure imaging capabilities of SPM techniques initially dominated applications of these methods, the physics of probe–sample interactions and quantitative analyses of tribological, electronic, magnetic, biological, and chemical surfaces using SPMs have become of increasing interest in recent years. SPMs are often associated with nanoscale science and technology, since they allow investigation and manipulation of surfaces down to the atomic scale. As our understanding of the underlying interaction mechanisms has grown, SPMs have increasingly found application in many fields beyond basic research fields. In addition, various derivatives of all these methods have been developed for special applications, some of them intended for areas other than microscopy.

This chapter presents an overview of STM and AFM and various probes (tips) used in these instruments, followed by details on AFM instrumentation and analyses.

The scanning tunneling microscope (STM), developed by Binnig and his colleagues in 1981 at the IBM Zurich Research Laboratory in Rüschlikon (Switzerland), was the first instrument capable of directly obtaining three-dimensional (3-D) images of solid surfaces with atomic resolution [1]. Binnig and Rohrer received a Nobel Prize in Physics in 1986 for their discovery. STMs can only be used to study surfaces which are electrically conductive to some degree. Based on their design of the STM, in 1985, Binnig et al. developed an atomic force microscope (AFM) to measure ultrasmall forces (less than 1 μN) between the AFM tip surface and the sample surface [2] (also see [3]). AFMs can be used to measure any engineering surface, whether it is electrically conductive or insulating. The AFM has become a popular surface profiler for topographic and normal force measurements on the micro- to nanoscale [4]. AFMs modified in order to measure both normal and lateral forces are called lateral force microscopes (LFMs) or friction force microscopes (FFMs) [5, 6, 7, 8, 9, 10, 11]. FFMs have been further modified to measure lateral forces in two orthogonal directions [12, 13, 14, 15, 16]. A number of researchers have modified and improved the original AFM and FFM designs, and have used

B. Bhushan (ed.), *Nanotribology and Nanomechanics*,
DOI 10.1007/978-3-642-15283-2_2, © Springer-Verlag Berlin Heidelberg 2011 

these improved systems to measure the adhesion and friction of solid and liquid surfaces on micro- and nanoscales [4, 17, 18, 19, 20, 21, 22, 23, 24, 25, 26, 27, 28, 29, 30]. AFMs have been used to study scratching and wear, and to measure elastic/plastic mechanical properties (such as indentation hardness and the modulus of elasticity) [4, 10, 11, 21, 23, 26, 27, 28, 29, 31, 32, 33, 34, 35, 36]. AFMs have been used to manipulate individual atoms of xenon [37], molecules [38], silicon surfaces [39] and polymer surfaces [40]. STMs have been used to create nanofeatures via localized heating or by inducing chemical reactions under the STM tip [41, 42, 43] and through nanomachining [44]. AFMs have also been used for nanofabrication [4, 10, 45, 46, 47] and nanomachining [48].

STMs and AFMs are used at extreme magnifications ranging from $10^3$ to $10^9$ in the $x$-, $y$- and $z$-directions in order to image macro to atomic dimensions with high resolution and for spectroscopy. These instruments can be used in any environment, such as ambient air [2, 49], various gases [17], liquids [50, 51, 52], vacuum [1, 53], at low temperatures (lower than about 100 K) [54, 55, 56, 57, 58] and at high temperatures [59, 60]. Imaging in liquid allows the study of live biological samples and it also eliminates the capillary forces that are present at the tip–sample interface when imaging aqueous samples in ambient air. Low-temperature (liquid helium temperatures) imaging is useful when studying biological and organic materials and low-temperature phenomena such as superconductivity or charge-density waves. Low-temperature operation is also advantageous for high-sensitivity force mapping due to the reduced thermal vibration. They also have been used to image liquids such as liquid crystals and lubricant molecules on graphite surfaces [61, 62, 63, 64]. While applications of SPM techniques initially focused on their pure imaging capabilities, research into the physics and chemistry of probe–sample interactions and SPM-based quantitative analyses of tribological, electronic, magnetic, biological, and chemical surfaces have become increasingly popular in recent years. Nanoscale science and technology is often tied to the use of SPMs since they allow investigation and manipulation of surfaces down to the atomic scale. As our understanding of the underlying interaction mechanisms has grown, SPMs and their derivatives have found applications in many fields beyond basic research fields and microscopy.

Families of instruments based on STMs and AFMs, called scanning probe microscopes (SPMs), have been developed for various applications of scientific and industrial interest. These include STM, AFM, FFM (or LFM), scanning electrostatic force microscopy (SEFM) [65, 66], scanning force acoustic microscopy (SFAM) (or atomic force acoustic microscopy (AFAM)) [21, 22, 36, 67, 68, 69], scanning magnetic microscopy (SMM) (or magnetic force microscopy (MFM)) [70, 71, 72, 73], scanning near-field optical microscopy (SNOM) [74, 75, 76, 77], scanning thermal microscopy (SThM) [78, 79, 80], scanning electrochemical microscopy (SEcM) [81], scanning Kelvin probe microscopy (SKPM) [82, 83, 84, 85, 86], scanning chemical potential microscopy (SCPM) [79], scanning ion conductance microscopy (SICM) [87, 88] and scanning capacitance microscopy (SCM) [82, 89, 90, 91]. When the technique is used to measure forces (as in AFM, FFM, SEFM, SFAM and SMM) it is also referred to as scanning force microscopy

**Table 2.1** Comparison of various conventional microscopes with SPMs

| | Optical | SEM/TEM | Confocal | SPM |
|---|---|---|---|---|
| Magnification | $10^3$ | $10^7$ | $10^4$ | $10^9$ |
| Instrument price (US$) | $10k | $250k | $30k | $100k |
| Technology age | 200 y | 40 y | 20 y | 20 y |
| Applications | Ubiquitous | Science and technology | New and unfolding | Cutting edge |
| Market 1993 | $800 M | $400 M | $80 M | $100 M |
| Growth rate | 10% | 10% | 30% | 70% |

(SFM). Although these instruments offer atomic resolution and are ideal for basic research, they are also used for cutting-edge industrial applications which do not require atomic resolution. The commercial production of SPMs started with the STM in 1987 and the AFM in 1989 by Digital Instruments, Inc. (Santa Barbara, USA). For comparisons of SPMs with other microscopes, see Table 2.1 (Veeco Instruments, Inc., Santa Barbara, USA). Numbers of these instruments are equally divided between the US, Japan and Europe, with the following split between industry/university and government laboratories: 50/50, 70/30, and 30/70, respectively. It is clear that research and industrial applications of SPMs are expanding rapidly.

## 2.1 Scanning Tunneling Microscope

The principle of electron tunneling was first proposed by Giaever [93]. He envisioned that if a potential difference is applied to two metals separated by a thin insulating film, a current will flow because of the ability of electrons to penetrate a potential barrier. To be able to measure a tunneling current, the two metals must be spaced no more than 10 nm apart. Binnig et al. [1] introduced vacuum tunneling combined with lateral scanning. The vacuum provides the ideal barrier for tunneling. The lateral scanning allows one to image surfaces with exquisite resolution – laterally to less than 1 nm and vertically to less than 0.1 nm – sufficient to define the position of single atoms. The very high vertical resolution of the STM is obtained because the tunnel current varies exponentially with the distance between the two electrodes; that is, the metal tip and the scanned surface. Typically, the tunneling current decreases by a factor of 2 as the separation is increased by 0.2 nm. Very high lateral resolution depends upon sharp tips. Binnig et al. overcame two key obstacles by damping external vibrations and moving the tunneling probe in close proximity to the sample. Their instrument is called the scanning tunneling microscope (STM). Today's STMs can be used in ambient environments for atomic-scale imaging of surfaces. Excellent reviews on this subject have been presented by Hansma and Tersoff [92], Sarid and Elings [94], Durig et al. [95]; Frommer [96], Güntherodt and Wiesendanger [97], Wiesendanger and Güntherodt [98], Bonnell [99], Marti and Amrein [100], Stroscio and Kaiser [101], and Güntherodt et al. [102].

The principle of the STM is straightforward. A sharp metal tip (one electrode of the tunnel junction) is brought close enough (0.3–1 nm) to the surface to be investigated (the second electrode) to make the tunneling current measurable at a convenient operating voltage (10 mV–1 V). The tunneling current in this case varies from 0.2 to 10 nA. The tip is scanned over the surface at a distance of 0.3–1 nm, while the tunneling current between it and the surface is measured. The STM can be operated in either the constant current mode or the constant height mode (Fig. 2.1). The left-hand column of Fig. 2.1 shows the basic constant current mode of operation. A feedback network changes the height of the tip $z$ to keep the current constant. The displacement of the tip, given by the voltage applied to the piezoelectric drive, then yields a topographic map of the surface. Alternatively, in the constant height mode, a metal tip can be scanned across a surface at nearly constant height and constant voltage while the current is monitored, as shown in the right-hand column of Fig. 2.1. In this case, the feedback network responds just rapidly enough to keep the average current constant. The current mode is generally used for atomic-scale images; this mode is not practical for rough surfaces. A three-dimensional picture [$z(x, y)$] of a surface consists of multiple scans [$z(x)$] displayed laterally to each other in the $y$-direction. It should be noted that if different atomic species are present in a sample, the different atomic species within a sample may produce different tunneling currents for a given bias voltage. Thus the height data may not be a direct representation of the topography of the surface of the sample.

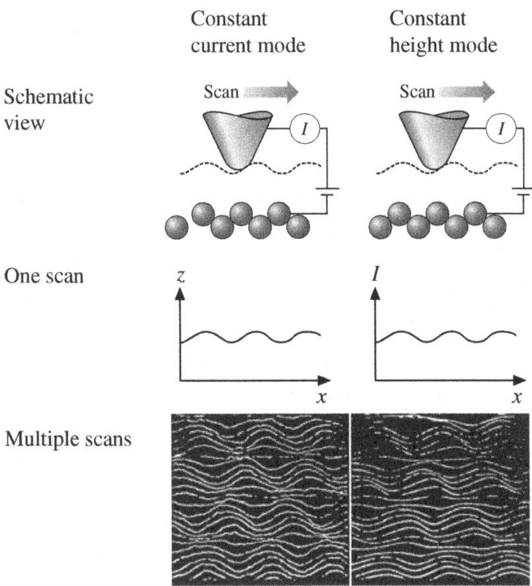


**Fig. 2.1** An STM can be operated in either the constant-current or the constant-height mode. The images are of graphite in air (After [92])

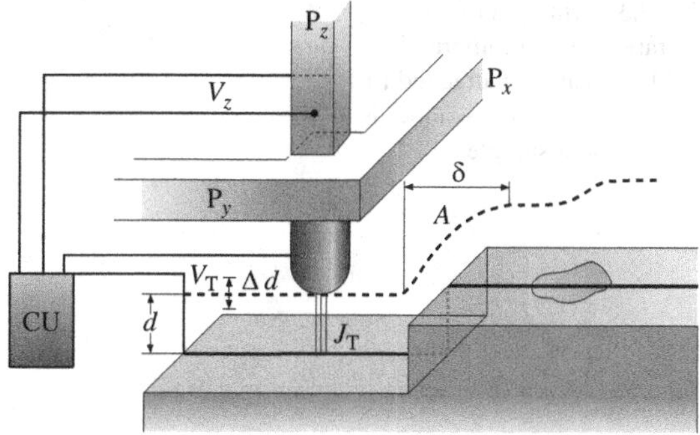


**Fig. 2.2** Principle of operation of the STM, from Binnig and Rohrer [103]

## 2.1.1 The STM Design of Binnig et al.

Figure 2.2 shows a schematic of an AFM designed by Binnig and Rohrer and intended for operation in ultrahigh vacuum [1, 103]. The metal tip was fixed to rectangular piezodrives $P_x$, $P_y$, and $P_z$ made out of commercial piezoceramic material for scanning. The sample is mounted via either superconducting magnetic levitation or a two-stage spring system to achieve a stable gap width of about 0.02 nm. The tunnel current $J_T$ is a sensitive function of the gap width $d$ where $J_T \propto V_T \exp(-A\phi^{1/2}d)$. Here $V_T$ is the bias voltage, $\phi$ is the average barrier height (work function) and the constant $A = 1.025\ \text{eV}^{-1/2}\ \text{Å}^{-1}$. With a work function of a few eV, $J_T$ changes by an order of magnitude for an angstrom change in $d$. If the current is kept constant to within, for example, 2%, then the gap $d$ remains constant to within 1 pm. For operation in the constant current mode, the control unit CU applies a voltage $V_z$ to the piezo $P_z$ such that $J_T$ remains constant when scanning the tip with $P_y$ and $P_x$ over the surface. At a constant work function $\phi$, $V_z(V_x, V_y)$ yields the roughness of the surface $z(x, y)$ directly, as illustrated by a surface step at $A$. Smearing the step, $\delta$ (lateral resolution) is on the order of $(R)^{1/2}$, where $R$ is the radius of the curvature of the tip. Thus, a lateral resolution of about 2 nm requires tip radii on the order of 10 nm. A 1 mm diameter solid rod ground at one end at roughly 90° yields overall tip radii of only a few hundred nanometers, the presence of rather sharp microtips on the relatively dull end yields a lateral resolution of about 2 nm. In situ sharpening of the tips, achieved by gently touching the surface, brings the resolution down to the 1 nm range; by applying high fields (on the order of $10^8$ V/cm) for, say, half an hour, resolutions considerably below 1 nm can be reached. Most experiments have been performed with tungsten wires either ground or etched to a typical radius of 0.1–10 μm. In some cases, in situ processing of the tips has been performed to further reduce tip radii.

## 2.1.2 Commercial STMs

There are a number of commercial STMs available on the market. Digital Instruments, Inc., introduced the first commercial STM, the Nanoscope I, in 1987. In the

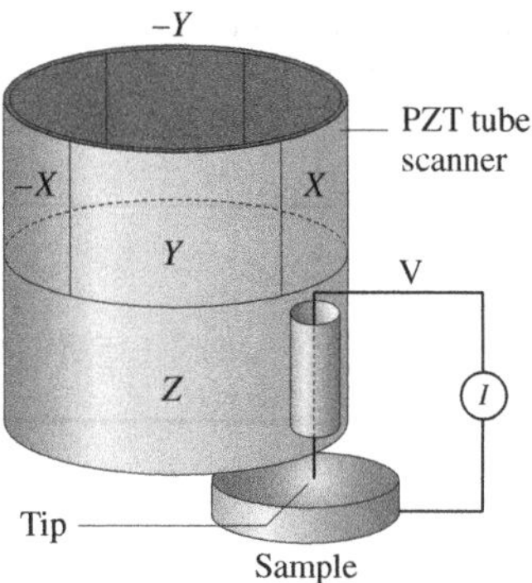


**Fig. 2.3** Principle of operation of a commercial STM. A sharp tip attached to a piezoelectric tube scanner is scanned on a sample

recent Nanoscope IV STM, intended for operation in ambient air, the sample is held in position while a piezoelectric crystal in the form of a cylindrical tube (referred to as a PZT tube scanner) scans the sharp metallic probe over the surface in a raster pattern while sensing and relaying the tunneling current to the control station (Fig. 2.3). The digital signal processor (DSP) calculates the tip–sample separation required by sensing the tunneling current flowing between the sample and the tip. The bias voltage applied between the sample and the tip encourages the tunneling current to flow. The DSP completes the digital feedback loop by relaying the desired voltage to the piezoelectric tube. The STM can operate in either the *constant height* or the *constant current* mode, and this can be selected using the control panel. In the constant current mode, the feedback gains are set high, the tunneling tip closely tracks the sample surface, and the variation in the tip height required to maintain constant tunneling current is measured by the change in the voltage applied to the piezo tube. In the constant height mode, the feedback gains are set low, the tip remains at a nearly constant height as it sweeps over the sample surface, and the tunneling current is imaged.

Physically, the Nanoscope STM consists of three main parts: the head, which houses the piezoelectric tube scanner which provides three-dimensional tip motion and the preamplifier circuit for the tunneling current (FET input amplifier) mounted on the top of the head; the base on which the sample is mounted; and the base support, which supports the base and head [4]. The base accommodates samples which are up to 10 mm by 20 mm and 10 mm thick. Scan sizes available for the STM are 0.7 μm (for atomic resolution), 12 μm, 75 μm and 125 μm square.

The scanning head controls the three-dimensional motion of the tip. The removable head consists of a piezo tube scanner, about 12.7 mm in diameter, mounted into an Invar shell, which minimizes vertical thermal drift because of the good thermal match between the piezo tube and the Invar. The piezo tube has separate electrodes for $x$-, $y$- and $z$-motion, which are driven by separate drive circuits. The electrode configuration (Fig. 2.3) provides $x$- and $y$-motions which are perpendicular to each other, it minimizes horizontal and vertical coupling, and it provides good sensitivity. The vertical motion of the tube is controlled by the

Z-electrode, which is driven by the feedback loop. The $x$- and $y$-scanning motions are each controlled by two electrodes which are driven by voltages of the same magnitude but opposite signs. These electrodes are called $-y$, $-x$, $+y$, and $+x$. Applying complimentary voltages allows a short, stiff tube to provide a good scan range without the need for a large voltage. The motion of the tip that arises due to external vibrations is proportional to the square of the ratio of vibration frequency to the resonant frequency of the tube. Therefore, to minimize the tip vibrations, the resonant frequencies of the tube are high: about 60 kHz in the vertical direction and about 40 kHz in the horizontal direction. The tip holder is a stainless steel tube with an inner diameter of 300 μm when 250 μm diameter tips are used, which is mounted in ceramic in order to minimize the mass at the end of the tube. The tip is mounted either on the front edge of the tube (to keep the mounting mass low and the resonant frequency high) (Fig. 2.3) or the center of the tube for large-range scanners, namely 75 and 125 μm (to preserve the symmetry of the scanning). This commercial STM accepts any tip with a 250 μm diameter shaft. The piezotube requires $x$–$y$-calibration, which is carried out by imaging an appropriate calibration standard. Cleaved graphite is used for heads with small scan lengths while two-dimensional grids (a gold-plated rule) can be used for long-range heads.

The Invar base holds the sample in position, supports the head, and provides coarse $x$–$y$-motion for the sample. A sprung-steel sample clip with two thumb screws holds the sample in place. An $x$–$y$-translation stage built into the base allows the sample to be repositioned under the tip. Three precision screws arranged in a triangular pattern support the head and provide coarse and fine adjustment of the tip height. The base support consists of the base support ring and the motor housing. The stepper motor enclosed in the motor housing allows the tip to be engaged and withdrawn from the surface automatically.

Samples to be imaged with the STM must be conductive enough to allow a few nanoamperes of current to flow from the bias voltage source to the area to be scanned. In many cases, nonconductive samples can be coated with a thin layer of a conductive material to facilitate imaging. The bias voltage and the tunneling current depend on the sample. Usually they are set to a standard value for engagement and fine tuned to enhance the quality of the image. The scan size depends on the sample and the features of interest. A maximum scan rate of 122 Hz can be used. The maximum scan rate is usually related to the scan size. Scan rates above 10 Hz are used for small scans (typically 60 Hz for atomic-scale imaging with a 0.7 μm scanner). The scan rate should be lowered for large scans, especially if the sample surfaces are rough or contain large steps. Moving the tip quickly along the sample surface at high scan rates with large scan sizes will usually lead to a tip crash. Essentially, the scan rate should be inversely proportional to the scan size (typically 2–4 Hz for a scan size of 1 μm, 0.5–1 Hz for 12 μm, and 0.2 Hz for 125 μm). The scan rate (in length/time) is equal to the scan length divided by the scan rate in Hz. For example, for a scan size of 10 μm $\times$ 10 μm scanned at 0.5 Hz, the scan rate is 10 μm/s. 256 $\times$ 256 data formats are the most common. The lateral resolution at larger scans is approximately equal to scan length divided by 256.

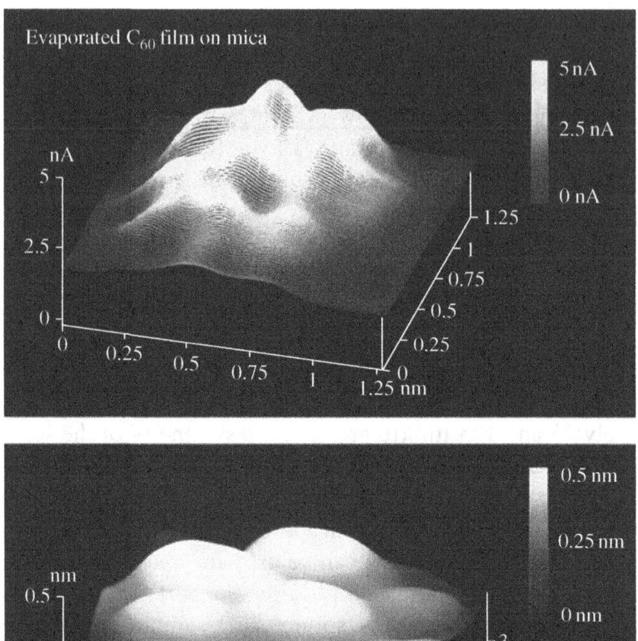

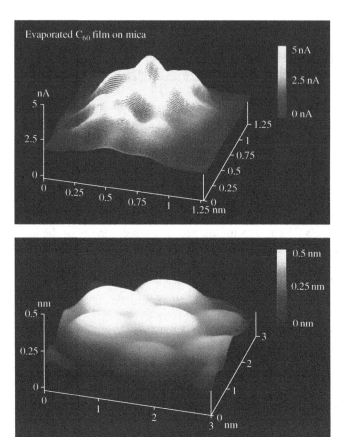

**Fig. 2.4** STM images of evaporated $C_{60}$ film on gold-coated freshly cleaved mica obtained using a mechanically sheared Pt-Ir (80/20) tip in constant height mode (After [104])

Figure 2.4 shows sample STM images of an evaporated $C_{60}$ film on gold-coated freshly-cleaved mica taken at room temperature and ambient pressure [104]. Images were obtained with atomic resolution at two scan sizes. Next we describe some STM designs which are available for special applications.

## Electrochemical STM

The electrochemical STM is used to perform and monitor the electrochemical reactions inside the STM. It includes a microscope base with an integral potentiostat, a short head with a 0.7 μm scan range and a differential preamp as well as the software required to operate the potentiostat and display the result of the electrochemical reaction.

**Standalone STM**

Standalone STMs are available to scan large samples. In this case, the STM rests directly on the sample. It is available from Digital Instruments in scan ranges of 12 and 75 μm. It is similar to the standard STM design except the sample base has been eliminated.

### *2.1.3 STM Probe Construction*

The STM probe has a cantilever integrated with a sharp metal tip with a low aspect ratio (tip length/tip shank) to minimize flexural vibrations. Ideally, the tip should be atomically sharp, but in practice most tip preparation methods produce a tip with a rather ragged profile that consists of several asperities where the one closest to the surface is responsible for tunneling. STM cantilevers with sharp tips are typically fabricated from metal wires (the metal can be tungsten (W), platinum-iridium (Pt-Ir), or gold (Au)) and are sharpened by grinding, cutting with a wire cutter or razor blade, field emission/evaporation, ion milling, fracture, or electrochemical polishing/etching [105, 106]. The two most commonly used tips are made from either Pt-Ir (80/20) alloy or tungsten wire. Iridium is used to provide stiffness. The Pt-Ir tips are generally formed mechanically and are readily available. The tungsten tips are etched from tungsten wire by an electrochemical process, for example by using 1 M KOH solution with a platinum electrode in a electrochemical cell at about 30 V. In general, Pt-Ir tips provide better atomic resolution than tungsten tips, probably due to the lower reactivity of Pt. However, tungsten tips are more uniformly shaped and may perform better on samples with steeply sloped features. The tungsten wire diameter used for the cantilever is typically 250 μm, with the radius of curvature ranging from 20 to 100 nm and a cone angle ranging from 10° to 60° (Fig. 2.5). The wire can be bent in an L shape, if so required, for use in the instrument. For calculations of the normal spring constant and the natural frequency of round cantilevers, see Sarid and Elings [94].

High aspect ratio, controlled geometry (CG) Pt-Ir probes are commercially available to image deep trenches (Fig. 2.6). These probes are electrochemically etched from Pt-Ir (80/20) wire and are polished to a specific shape which is consistent from tip to tip. The probes have a full cone angle of ≈15°, and a tip radius of less than 50 nm. To image very deep trenches (>0.25 μm) and nanofeatures, focused ion beam (FIB)-milled CG probes with extremely sharp

**Fig. 2.5** Schematic of a typical tungsten cantilever with a sharp tip produced by electrochemical etching

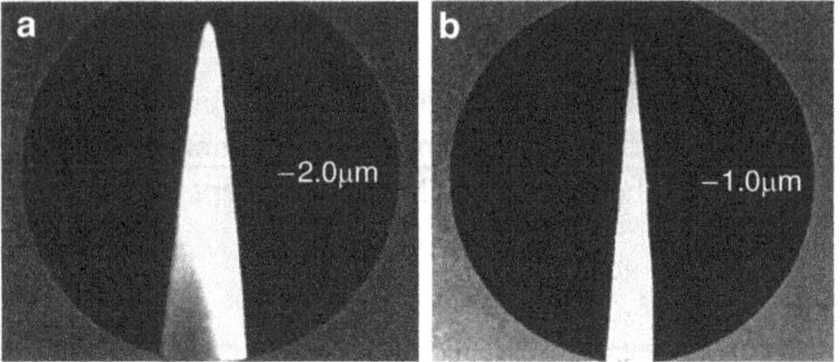


**Fig. 2.6** Schematics of **(a)** CG Pt-Ir probe, and **(b)** CG Pt-Ir FIB milled probe

tips (radii <5 nm) are used. The Pt-Ir probes are coated with a nonconducting film (not shown in the figure) for electrochemistry. These probes are available from Materials Analytical Services (Raleigh, USA).

Pt alloy and W tips are very sharp and give high resolution, but are fragile and sometimes break when contacting a surface. Diamond tips have been used by Kaneko and Oguchi [107]. Diamond tips made conductive by boron ion implantation were found to be chip-resistant.

## 2.2 Atomic Force Microscope

Like the STM, the AFM relies on a scanning technique to produce very high resolution 3-D images of sample surfaces. The AFM measures ultrasmall forces (less than 1 nN) present between the AFM tip surface and a sample surface. These small forces are measured by measuring the motion of a very flexible cantilever beam with an ultrasmall mass. While STMs require the surface being measured be electrically conductive, AFMs are capable of investigating the surfaces of both conductors and insulators on an atomic scale if suitable techniques for measuring the cantilever motion are used. During the operation of a high-resolution AFM, the sample is generally scanned instead of the tip (unlike for STM) because the AFM measures the relative displacement between the cantilever surface and the reference surface and any cantilever movement from scanning would add unwanted vibrations. However, for measurements of large samples, AFMs are available where the tip is scanned and the sample is stationary. As long as the AFM is operated in the so-called contact mode, little if any vibration is introduced.

The AFM combines the principles of the STM and the stylus profiler (Fig. 2.7). In an AFM, the force between the sample and tip is used (rather than the tunneling current) to sense the proximity of the tip to the sample. The AFM can be used either in the static or the dynamic mode. In the static mode, also referred to as the repulsive or contact mode [2], a sharp tip at the end of the cantilever is brought into contact with the surface of the sample. During initial contact, the atoms at the

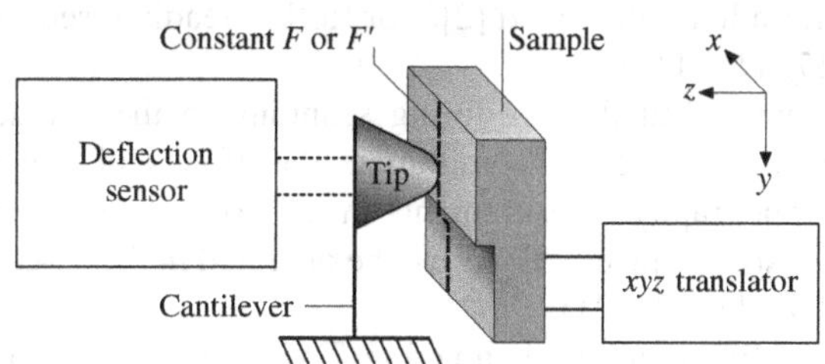


**Fig. 2.7** Principle of operation of the AFM. Sample mounted on a piezoelectric scanner is scanned against a short tip and the cantilever deflection is usually measured using a laser deflection technique. The force (in contact mode) or the force gradient (in noncontact mode) is measured during scanning

end of the tip experience a very weak repulsive force due to electronic orbital overlap with the atoms in the surface of the sample. The force acting on the tip causes the cantilever to deflect, which is measured by tunneling, capacitive, or optical detectors. The deflection can be measured to within 0.02 nm, so a force as low as 0.2 nN (corresponding to a normal pressure of $\approx$200 MPa for a $Si_3N_4$ tip with a radius of about 50 nm against single-crystal silicon) can be detected for typical cantilever spring constant of 10 N/m. (To put these number in perspective, individual atoms and human hair are typically a fraction of a nanometer and about 75 μm in diameter, respectively, and a drop of water and an eyelash have masses of about 10 μN and 100 nN, respectively.) In the dynamic mode of operation, also referred to as attractive force imaging or noncontact imaging mode, the tip is brought into close proximity to (within a few nanometers of), but not in contact with, the sample. The cantilever is deliberately vibrated in either amplitude modulation (AM) mode [65] or frequency modulation (FM) mode [65, 94, 108, 109]. Very weak van der Waals attractive forces are present at the tip–sample interface. Although the normal pressure exerted at the interface is zero in this technique (in order to avoid any surface deformation), it is slow and difficult to use, and is rarely used outside of research environments. The surface topography is measured by laterally scanning the sample under the tip while simultaneously measuring the separation-dependent force or force gradient (derivative) between the tip and the surface (Fig. 2.7). In the contact (static) mode, the interaction force between tip and sample is measured by monitoring the cantilever deflection. In the noncontact (or dynamic) mode, the force gradient is obtained by vibrating the cantilever and measuring the shift in the resonant frequency of the cantilever. To obtain topographic information, the interaction force is either recorded directly, or used as a control parameter for a feedback circuit that maintains the force or force derivative at a constant value. Using an AFM operated in the contact mode, topographic images with a vertical resolution of less than 0.1 nm (as low as 0.01 nm) and a lateral resolution of about 0.2 nm have been obtained [3, 50, 110, 111, 112, 113, 114]. Forces of 10 nN to 1 pN are measurable with a displacement sensitivity of 0.01 nm. These forces are comparable to the forces associated with chemical bonding, for example 0.1 μN for an ionic bond and 10 pN for a hydrogen bond [2]. For further reading, see [94, 95, 96, 100, 102, 115, 116, 117, 118, 119].

Lateral forces applied at the tip during scanning in the contact mode affect roughness measurements [120]. To minimize the effects of friction and other lateral forces on topography measurements in the contact mode, and to measure the topographies of soft surfaces, AFMs can be operated in the so-called tapping or force modulation mode [32, 121].

The STM is ideal for atomic-scale imaging. To obtain atomic resolution with the AFM, the spring constant of the cantilever should be weaker than the equivalent spring between atoms. For example, the vibration frequencies $\omega$ of atoms bound in a molecule or in a crystalline solid are typically $10^{13}$ Hz or higher. Combining this with an atomic mass $m$ of $\approx 10^{25}$ kg gives an interatomic spring constant $k$, given by $\omega^2 m$, of around 10 N/m [115]. (For comparison, the spring constant of a piece of household aluminium foil that is 4 mm long and 1 mm wide is about 1 N/m.) Therefore, a cantilever beam with a spring constant of about 1 N/m or lower is desirable. Tips must be as sharp as possible, and tip radii of 5–50 nm are commonly available.

Atomic resolution cannot be achieved with these tips at normal loads in the nN range. Atomic structures at these loads have been obtained from lattice imaging or by imaging the crystal's periodicity. Reported data show either perfectly ordered periodic atomic structures or defects on a larger lateral scale, but no well-defined, laterally resolved atomic-scale defects like those seen in images routinely obtained with a STM. Interatomic forces with one or several atoms in contact are 20–40 or 50–100 pN, respectively. Thus, atomic resolution with an AFM is only possible with a sharp tip on a flexible cantilever at a net repulsive force of 100 pN or lower [122]. Upon increasing the force from 10 pN, Ohnesorge and Binnig [122] observed that monoatomic steplines were slowly wiped away and a perfectly ordered structure was left. This observation explains why mostly defect-free atomic resolution has been observed with AFM. Note that for atomic-resolution measurements, the cantilever should not be so soft as to avoid jumps. Further note that performing measurements in the noncontact imaging mode may be desirable for imaging with atomic resolution.

The key component in an AFM is the sensor used to measure the force on the tip due to its interaction with the sample. A cantilever (with a sharp tip) with an extremely low spring constant is required for high vertical and lateral resolutions at small forces (0.1 nN or lower), but a high resonant frequency is desirable (about 10–100 kHz) at the same time in order to minimize the sensitivity to building vibrations, which occur at around 100 Hz. This requires a spring with an extremely low vertical spring constant (typically 0.05–1 N/m) as well as a low mass (on the order of 1 ng). Today, the most advanced AFM cantilevers are microfabricated from silicon or silicon nitride using photolithographic techniques. Typical lateral dimensions are on the order of 100 μm, with thicknesses on the order of 1 μm. The force on the tip due to its interaction with the sample is sensed by detecting the deflection of the compliant lever with a known spring constant. This cantilever deflection (displacement smaller than 0.1 nm) has been measured by detecting a tunneling current similar to that used in the STM in the pioneering work of Binnig et al. [2] and later used by Giessibl et al. [56], by capacitance detection [123, 124], piezoresistive detection [125, 126], and by four optical techniques, namely (1) optical interferometry [5, 6, 127, 128] using optical fibers [57, 129] (2) optical polarization detection [72, 130], (3) laser diode feedback [131] and (4) optical (laser) beam deflection [7, 8, 53, 111, 112]. Schematics of the four more commonly used detection systems are shown in Fig. 2.8. The tunneling method originally used by Binnig et al. [2] in the first version of the AFM uses a second tip to monitor the deflection of the cantilever with its force sensing tip. Tunneling is rather sensitive to contaminants and the interaction between the tunneling tip and the rear side of the cantilever can become comparable to the interaction between the tip and sample. Tunneling is rarely used and is mentioned mainly for historical reasons. Giessibl et al. [56] have used it for a low-temperature AFM/STM design. In contrast to tunneling, other deflection sensors are placed far from the cantilever, at distances of micrometers to tens of millimeters. The optical techniques are believed to be more sensitive, reliable and easily implemented detection methods than the others [94, 118]. The optical beam deflection method has the largest working distance, is insensitive to distance changes and is capable of measuring angular changes (friction forces); therefore, it is the most commonly used in commercial SPMs.

Almost all SPMs use piezo translators to scan the sample, or alternatively to scan the tip. An electric field applied across a piezoelectric material causes a change in the crystal structure, with expansion in some directions and contraction in others. A net change in volume also occurs [132]. The first STM used a piezo tripod for scanning [1]. The piezo tripod is one way to generate three-dimensional movement of a tip attached at its center. However, the tripod needs to be fairly large ($\approx$50 mm) to get a suitable range. Its size and asymmetric shape makes it susceptible to thermal drift. Tube scanners are widely used in AFMs [133]. These provide ample scanning range with a small size. Electronic control systems for AFMs are

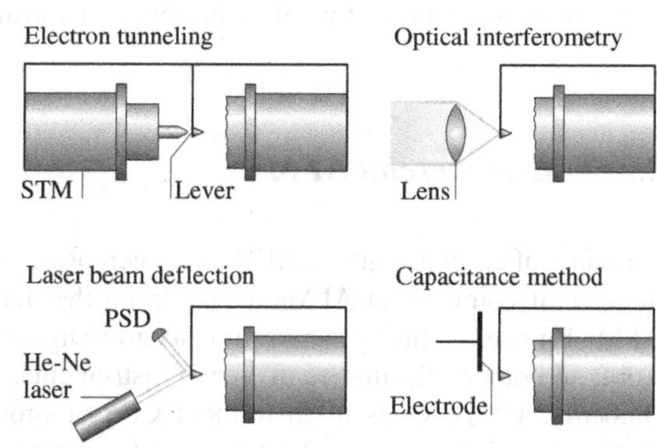


**Fig. 2.8** Schematics of the four detection systems to measure cantilever deflection. In each set-up, the sample mounted on piezoelectric body is shown *on the right*, the cantilever *in the middle*, and the corresponding deflection sensor *on the left* (After [118])

based on either analog or digital feedback. Digital feedback circuits are better suited for ultralow noise operation.

Images from the AFMs need to be processed. An ideal AFM is a noise-free device that images a sample with perfect tips of known shape and has a perfectly linear scanning piezo. In reality, scanning devices are affected by distortions and these distortions must be corrected for. The distortions can be linear and nonlinear. Linear distortions mainly result from imperfections in the machining of the piezo translators, causing cross-talk between the $Z$-piezo to the $x$- and $y$-piezos, and vice versa. Nonlinear distortions mainly result from the presence of a hysteresis loop in piezoelectric ceramics. They may also occur if the scan frequency approaches the upper frequency limit of the $x$- and $y$-drive amplifiers or the upper frequency limit of the feedback loop ($z$-component). In addition, electronic noise may be present in the system. The noise is removed by digital filtering in real space [134] or in the spatial frequency domain (Fourier space) [135].

Processed data consists of many tens of thousand of points per plane (or data set). The outputs from the first STM and AFM images were recorded on an $x$–$y$-chart recorder, with the $z$-value plotted against the tip position in the fast scan direction. Chart recorders have slow responses, so computers are used to display the data these days. The data are displayed as wire mesh displays or grayscale displays (with at least 64 shades of gray).

### 2.2.1 The AFM Design of Binnig et al.

In the first AFM design developed by Binnig et al. [2], AFM images were obtained by measuring the force exerted on a sharp tip created by its proximity to the surface of a sample mounted on a 3-D piezoelectric scanner. The tunneling current between the STM tip and the backside of the cantilever beam to which the tip was attached was measured to obtain the normal force. This force was kept at a constant level with a feedback mechanism. The STM tip was also mounted on a piezoelectric element to maintain the tunneling current at a constant level.

### 2.2.2 Commercial AFMs

A review of early designs of AFMs has been presented by Bhushan [4]. There are a number of commercial AFMs available on the market. Major manufacturers of AFMs for use in ambient environments are: Digital Instruments, Inc., Topometrix Corp. and other subsidiaries of Veeco Instruments, Inc., Molecular Imaging Corp. (Phoenix, USA), Quesant Instrument Corp. (Agoura Hills, USA), Nanoscience Instruments, Inc. (Phoenix, USA), Seiko Instruments (Chiba, Japan); and Olympus

(Tokyo, Japan). AFM/STMs for use in UHV environments are manufactured by Omicron Vakuumphysik GmbH (Taunusstein, Germany).

We describe here two commercial AFMs – small-sample and large-sample AFMs – for operation in the contact mode, produced by Digital Instruments, Inc., with scanning lengths ranging from about 0.7 μm (for atomic resolution) to about 125 μm [9, 111, 114, 136]. The original design of these AFMs comes from Meyer and Amer [53]. Basically, the AFM scans the sample in a raster pattern while outputting the cantilever deflection error signal to the control station. The cantilever deflection (or the force) is measured using a laser deflection technique (Fig. 2.9). The DSP in the workstation controls the $z$-position of the piezo based on the cantilever deflection error signal. The AFM operates in both *constant height* and *constant force* modes. The DSP always adjusts the distance between the sample and the tip according to the cantilever deflection error signal, but if the feedback gains are low the piezo remains at an almost *constant height* and the cantilever deflection data is collected. With high gains, the piezo height changes to keep the cantilever deflection nearly constant (so the force is constant), and the change in piezo height is collected by the system.

In the operation of a commercial small-sample AFM (as shown in Fig. 2.9a), the sample (which is generally no larger than 10 mm × 10 mm) is mounted on a PZT tube scanner, which consists of separate electrodes used to precisely scan the sample in the $x$–$y$-plane in a raster pattern and to move the sample in the vertical ($z$-) direction. A sharp tip at the free end of a flexible cantilever is brought into contact with the sample. Features on the sample surface cause the cantilever to deflect in the vertical and lateral directions as the sample moves under the tip. A laser beam from a diode laser (5 mW max. peak output at 670 nm) is directed by a prism onto the back of a cantilever near its free end, tilted downward at about 10° with respect to the horizontal plane. The reflected beam from the vertex of the cantilever is directed through a mirror onto a quad photodetector (split photodetector with four quadrants) (commonly called a position-sensitive detector or PSD, produced by Silicon Detector Corp., Camarillo, USA). The difference in signal between the top and bottom photodiodes provides the AFM signal, which is a sensitive measure of the cantilever vertical deflection. The topographic features of the sample cause the tip to deflect in the vertical direction as the sample is scanned under the tip. This tip deflection will change the direction of the reflected laser beam, changing the intensity difference between the top and bottom sets of photodetectors (AFM signal). In a mode of operation called the height mode, used for topographic imaging or for any other operation in which the normal forceapplied is to be kept constant, a feedback circuit is used to modulate the voltage applied to the PZT scanner in order to adjust the height of the PZT, so that the cantilever vertical deflection (given by the intensity difference between the top and bottom detector) will remain constant during scanning. The PZT height variation is thus a direct measure of the surface roughness of the sample.

In a large-sample AFM, force sensors based on optical deflection methods or scanning units are mounted on the microscope head (Fig. 2.9b). Because of the

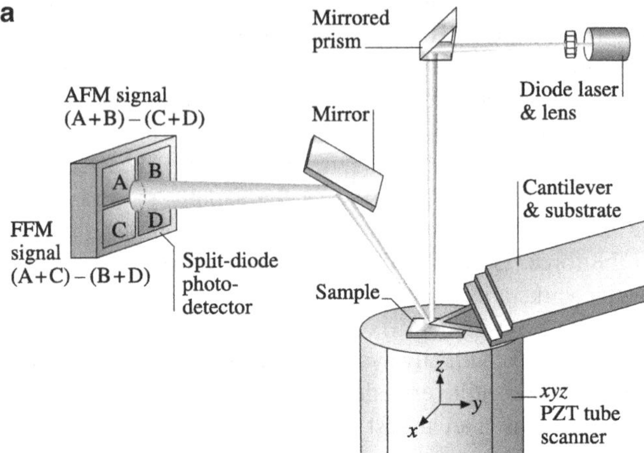


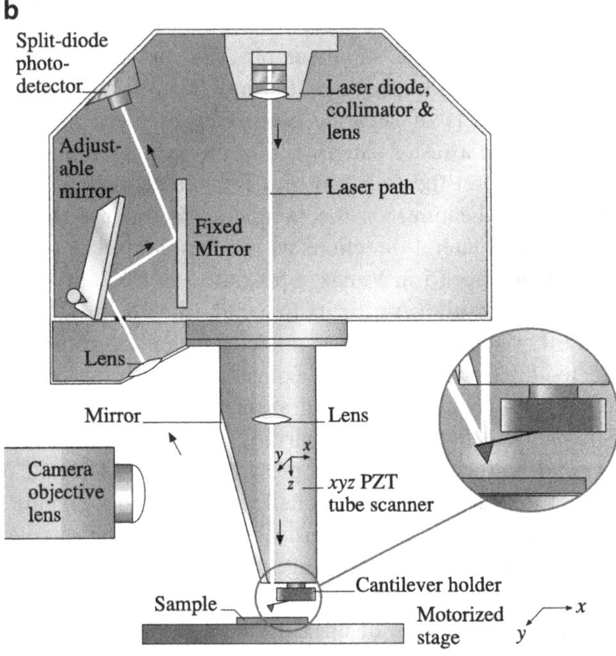


**Fig. 2.9** Principles of operation of (**a**) a commercial small-sample AFM/FFM, and (**b**) a large-sample AFM/FFM

unwanted vibrations caused by cantilever movement, the lateral resolution of this design is somewhat poorer than the design in Fig. 2.9a in which the sample is scanned instead of the cantilever beam. The advantage of the large-sample AFM is that large samples can be easily measured.

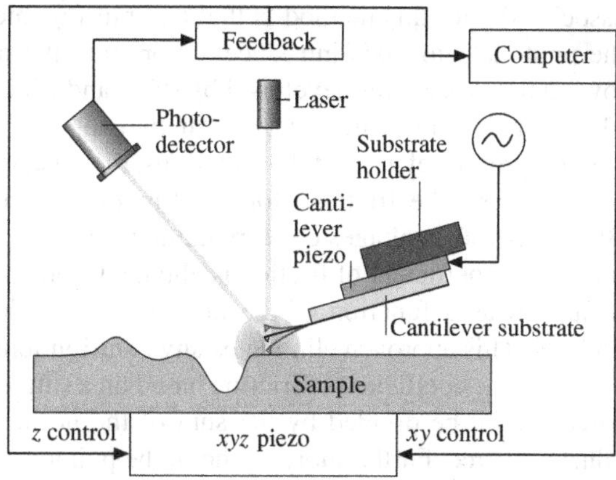


**Fig. 2.10** Schematic of tapping mode used for surface roughness measurements

Most AFMs can be used for topography measurements in the so-called tapping mode (intermittent contact mode), in what is also referred to as dynamic force microscopy. In the tapping mode, during the surface scan, the cantilever/tip assembly is sinusoidally vibrated by a piezo mounted above it, and the oscillating tip slightly taps the surface at the resonant frequency of the cantilever (70–400 kHz) with a constant (20–100 nm) amplitude of vertical oscillation, and a feedback loop keeps the average normal force constant (Fig. 2.10). The oscillating amplitude is kept large enough that the tip does not get stuck to the sample due to adhesive attraction. The tapping mode is used in topography measurements to minimize the effects of friction and other lateral forces to measure the topography of soft surfaces.

Topographic measurements can be made at any scanning angle. At first glance, the scanning angle may not appear to be an important parameter. However, the friction force between the tip and the sample will affect the topographic measurements in a parallel scan (scanning along the long axis of the cantilever). This means that a perpendicular scan may be more desirable. Generally, one picks a scanning angle which gives the same topographic data in both directions; this angle may be slightly different to that for the perpendicular scan.

The left-hand and right-hand quadrants of the photodetector are used to measure the friction force applied at the tip surface during sliding. In the so-called friction mode, the sample is scanned back and forth in a direction orthogonal to the long axis of the cantilever beam. Friction force between the sample and the tip will twist the cantilever. As a result, the laser beam will be deflected out of the plane defined by the incident beam and the beam is reflected vertically from an untwisted cantilever. This produces a difference in laser beam intensity between the beams received by the left-hand and right-hand sets of quadrants of the photodetector. The intensity difference between the two sets of detectors (FFM signal) is directly related to the degree of twisting and hence to the magnitude of the friction force. This method provides three-dimensional maps of the friction force. One problem associated with this method is that any misalignment between the laser beam and the photodetector axis introduces errors into the measurement. However, by following the procedures developed by Ruan and Bhushan [136], in which the average FFM signal for the sample scanned in two opposite directions is subtracted from the friction profiles of each of the two scans, the misalignment effect can be eliminated. By following the friction force calibration procedures developed by Ruan and Bhushan [136], voltages corresponding to friction forces can be converted to force units. The coefficient of friction is obtained from the slope of the friction force data measured as a function of the normal load, which typically ranges from 10 to 150 nN. This approach eliminates any contributions from adhesive forces [10]. To calculate the coefficient of friction based on a single point measurement, the friction force should be divided by the sum of the normal load applied and the intrinsic adhesive force. Furthermore, it should be pointed out that the coefficient of friction is not independent of load for single-asperity contact. This is discussed in more detail later.

The tip is scanned in such a way that its trajectory on the sample forms a triangular pattern (Fig. 2.11). Scanning speeds in the fast and slow scan directions depend on the scan area and scan frequency. Scan sizes ranging from less than 1 nm × 1 nm to 125 μm × 125 μm and scan rates of less than 0.5–122 Hz are typically used. Higher scan rates are used for smaller scan lengths. For example, the scan rates in the fast and slow scan directions for an area of 10 μm × 10 μm scanned at 0.5 Hz are 10 μm/s and 20 nm/s, respectively.

We now describe the construction of a small-sample AFM in more detail. It consists of three main parts: the optical head which senses the cantilever deflection; a PZT tube scanner which controls the scanning motion of the sample mounted on one of its ends; and the base, which supports the scanner and head and includes circuits for the deflection signal (Fig. 2.12a). The AFM connects directly to a control system. The optical head consists of a laser diode stage, a photodiode stage preamp board, the cantilever mount and its holding arm, and the deflected beam reflecting mirror, which reflects the deflected beam toward the photodiode (Fig. 2.12b). The laser diode stage is a tilt stage used to adjust the position of the laser beam relative to the cantilever. It consists of the laser diode, collimator, focusing lens, baseplate, and the $x$- and $y$-laser diode positioners. The positioners are used to place the laser spot on the end of the cantilever. The photodiode stage is an adjustable stage used to position the photodiode elements relative to the reflected

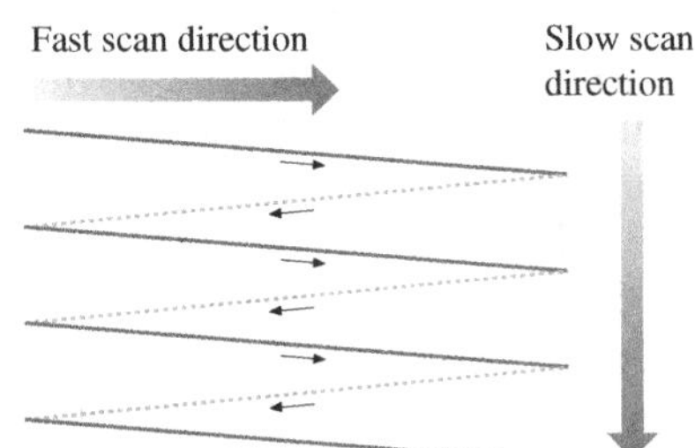


**Fig. 2.11** Schematic of triangular pattern trajectory of the AFM tip as the sample is scanned in two dimensions. During imaging, data are only recorded during scans along the *solid scan lines*

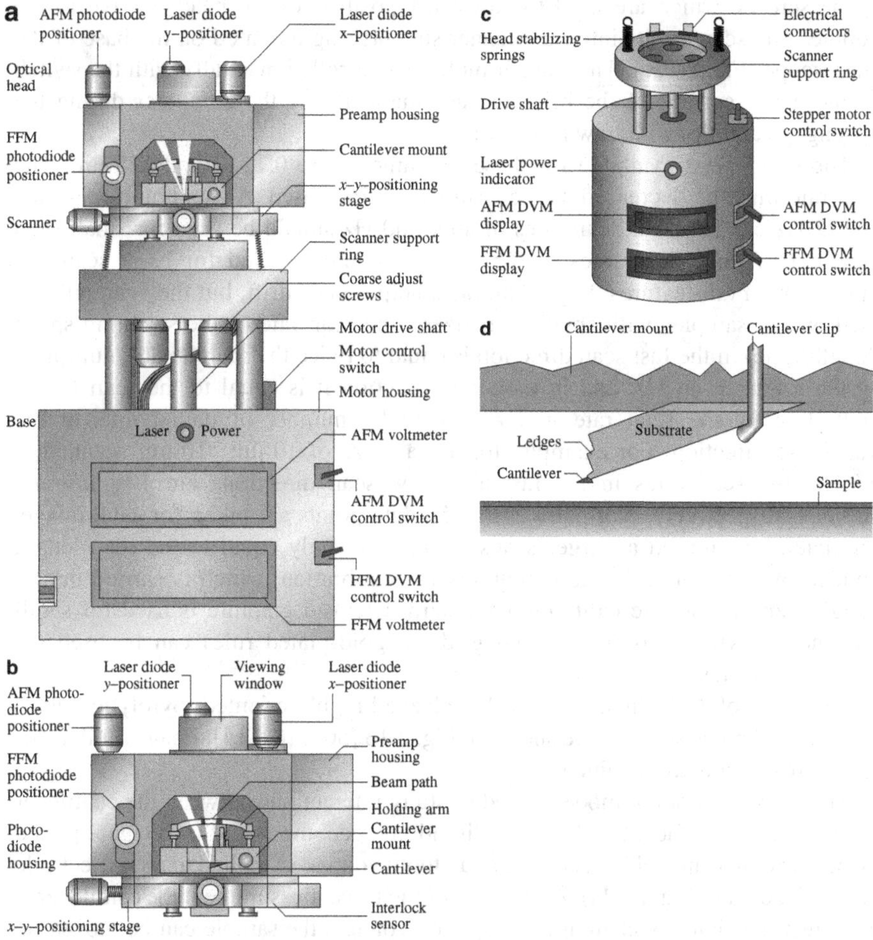


**Fig. 2.12** Schematics of a commercial AFM/FFM made by Digital Instruments, Inc. (**a**) Front view, (**b**) optical head, (**c**) base, and (**d**) cantilever substrate mounted on cantilever mount (not to scale)

laser beam. It consists of the split photodiode, the base plate, and the photodiode positioners. The deflected beam reflecting mirror is mounted on the upper left in the interior of the head. The cantilever mount is a metal (for operation in air) or glass (for operation in water) block which holds the cantilever firmly at the proper angle (Fig. 2.12d). Next, the tube scanner consists of an Invar cylinder holding a single tube made of piezoelectric crystal which imparts the necessary three-dimensional motion to the sample. Mounted on top of the tube is a magnetic cap on which the steel sample puck is placed. The tube is rigidly held at one end with the sample mounted on the other end of the tube. The scanner also contains three fine-pitched screws which form the mount for the optical head. The optical head rests on the tips of the screws, which are used to adjust the position of the head relative to the sample. The scanner fits into the scanner support ring mounted on the base of the microscope (Fig. 2.12c). The stepper motor is controlled manually with the switch on the upper surface of the base and automatically by the computer during the tip–engage and tip–withdraw processes.

The scan sizes available for these instruments are 0.7, 12 and 125 μm. The scan rate must be decreased as the scan size is increased. A maximum scan rate of 122 Hz can be used. Scan rates of about 60 Hz should be used for small scan lengths (0.7 μm). Scan rates of 0.5–2.5 Hz should be used for large scans on samples with tall features. High scan rates help reduce drift, but they can only be used on flat samples with small scan sizes. The scan rate or the scanning speed (length/time) in the fast scan direction is equal to twice the scan length multiplied by the scan rate in Hz, and in the slow direction it is equal to the scan length multiplied by the scan rate in Hz divided by number of data points in the transverse direction. For example, for a scan size of 10 μm × 10 μm scanned at 0.5 Hz, the scan rates in the fast and slow scan directions are 10 μm/s and 20 nm/s, respectively. Normally 256 × 256 data points are taken for each image. The lateral resolution at larger scans is approximately equal to the scan length divided by 256. The piezo tube requires $x$–$y$-calibration, which is carried out by imaging an appropriate calibration standard. Cleaved graphite is used for small scan heads, while two-dimensional grids (a gold-plated rule) can be used for long-range heads.

Examples of AFM images of freshly cleaved highly oriented pyrolytic (HOP) graphite and mica surfaces are shown in Fig. 2.13 [50, 110, 114]. Images with near-atomic resolution are obtained.

The force calibration mode is used to study interactions between the cantilever and the sample surface. In the force calibration mode, the $x$- and $y$-voltages applied to the piezo tube are held at zero and a sawtooth voltage is applied to the $z$-electrode of the piezo tube (Fig. 2.14a). At the start of the force measurement the cantilever is in its rest position. By changing the applied voltage, the sample can be moved up and down relative to the stationary cantilever tip. As the piezo moves the sample up and down, the cantilever deflection signal from the photodiode is monitored. The force–distance curve, a plot of the cantilever tip deflection signal as a function of the voltage applied to the piezo tube, is obtained. Figure 2.14b shows the typical features of a force–distance curve. The arrowheads indicate the direction of piezo travel. As the piezo extends, it approaches the tip, which is in mid-air at this point and hence shows no deflection. This is indicated by the flat portion of the curve. As the tip approaches the sample to within a few nanometers (point A), an attractive force kicks in between the atoms of the tip surface and the atoms of the surface of the sample. The tip is pulled towards the sample and contact occurs at point B on the graph. From this point on, the tip is in contact with the surface, and as the piezo extends further, the tip gets deflected further. This is represented by the sloped portion of the curve. As the piezo retracts, the tip moves beyond the zero deflection (flat) line due to attractive forces (van der Waals forces and long-range meniscus forces), into the adhesive regime. At point C in the graph, the tip snaps free of the

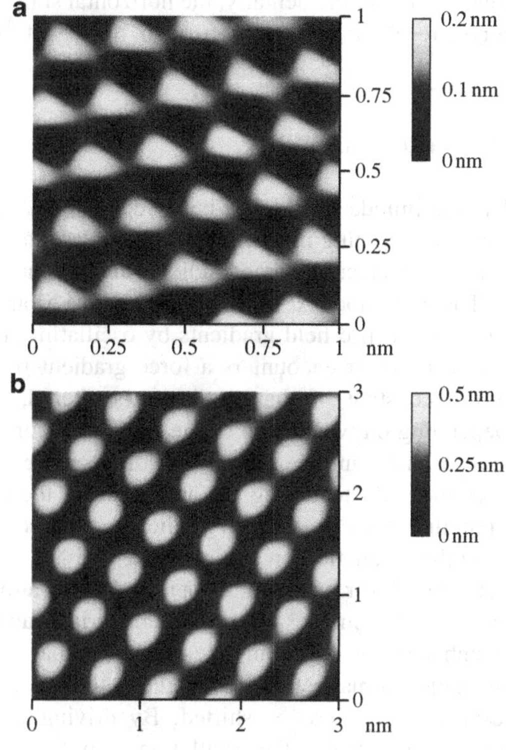


**Fig. 2.13** Typical AFM images of freshly-cleaved (**a**) highly oriented pyrolytic graphite and (**b**) mica surfaces taken using a square pyramidal $Si_3N_4$ tip

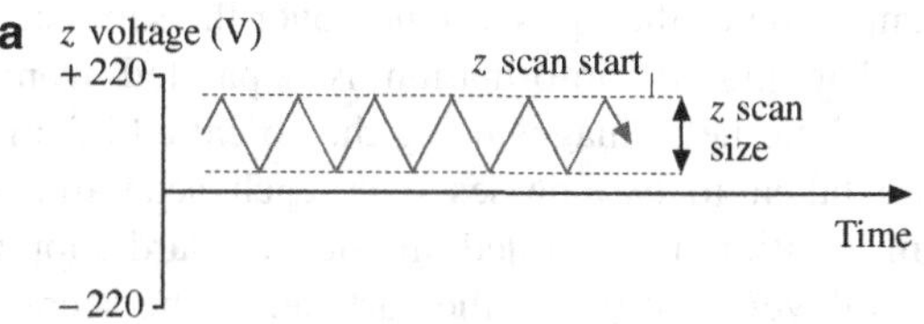


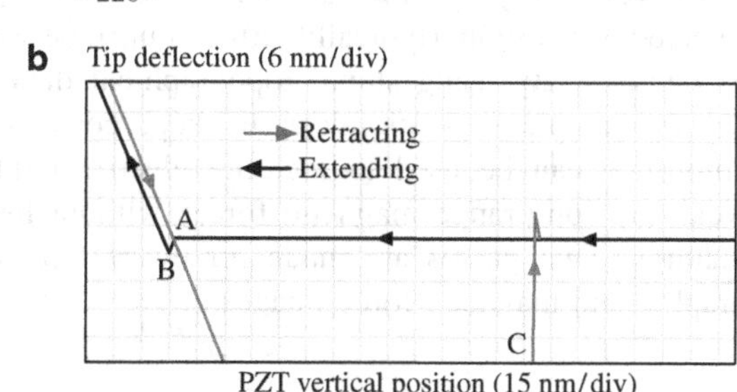


**Fig. 2.14** (**a**) Force calibration Z waveform, and (**b**) a typical force–distance curve for a tip in contact with a sample. Contact occurs at point B; tip breaks free of adhesive forces at point C as the sample moves away from the tip

adhesive forces, and is again in free air. The horizontal distance between points B and C along the retrace line gives the distance moved by the tip in the adhesive regime. Multiplying this distance by the stiffness of the cantilever gives the adhesive force. Incidentally, the horizontal shift between the loading and unloading curves results from the hysteresis in the PZT tube [4].

### Multimode Capabilities

The multimode AFM can be used for topography measurements in the contact mode and tapping mode, described earlier, and for measurements of lateral (friction) force, electric force gradients and magnetic force gradients.

The multimode AFM, when used with a grounded conducting tip, can be used to measure electric field gradients by oscillating the tip near its resonant frequency. When the lever encounters a force gradient from the electric field, the effective spring constant of the cantilever is altered, changing its resonant frequency. Depending on which side of the resonance curve is chosen, the oscillation amplitude of the cantilever increases or decreases due to the shift in the resonant frequency. By recording the amplitude of the cantilever, an image revealing the strength of the electric field gradient is obtained.

In the magnetic force microscope (MFM), used with a magnetically coated tip, static cantilever deflection is detected when a magnetic field exerts a force on the tip, and MFM images of magnetic materials can be obtained. MFM sensitivity can be enhanced by oscillating the cantilever near its resonant frequency. When the tip encounters a magnetic force gradient, the effective spring constant (and hence the resonant frequency) is shifted. By driving the cantilever above or below the resonant frequency, the oscillation amplitude varies as the resonance shifts. An image of the magnetic field gradient is obtained by recording the oscillation amplitude as the tip is scanned over the sample.

Topographic information is separated from the electric field gradient and magnetic field images using the so-called lift mode. In lift mode, measurements are taken in two passes over each scan line. In the first pass, topographical information is recorded in the standard tapping mode, where the oscillating cantilever lightly taps the surface. In the second pass, the tip is lifted to a user-selected separation (typically 20–200 nm) between the tip and local surface topography. By using stored topographical data instead of standard feedback, the tip–sample separation can be kept constant. In this way, the cantilever amplitude can be used to measure electric field force gradients or relatively weak but long-range magnetic forces without being influenced by topographic features. Two passes are made for every scan line, producing separate topographic and magnetic force images.

### Electrochemical AFM

This option allows one to perform electrochemical reactions on the AFM. The technique involves a potentiostat, a fluid cell with a transparent cantilever holder and electrodes, and the software required to operate the potentiostat and display the results of the electrochemical reaction.

### 2.2.3 AFM Probe Construction

Various probes (cantilevers and tips) are used for AFM studies. The cantilever stylus used in the AFM should meet the following criteria: (1) low normal spring constant (stiffness); (2) high resonant frequency; (3) high cantilever quality factor $Q$; (4) high lateral spring constant (stiffness); (5) short cantilever length; (6) incorporation of components (such as mirror) for deflection sensing; and (7) a sharp protruding tip [137]. In order to register a measurable deflection with small forces, the cantilever must flex with a relatively low force (on the order of few nN), requiring vertical spring constants of $10^2$–$10^2$ N/m for atomic resolution in the contact profiling mode. The data rate or imaging rate in the AFM is limited by the mechanical resonant frequency of the cantilever. To achieve a large imaging bandwidth, the AFM cantilever should have a resonant frequency of more than about 10 kHz (30–100 kHz is preferable), which makes the cantilever the least sensitive part of the system. Fast imaging rates are not just a matter of convenience, since the effects of thermal drifts are more pronounced with slow scanning speeds. The combined requirements of a low spring constant and a high resonant frequency are met by reducing the mass of the cantilever. The quality factor $Q$ (= $\omega_R/(c/m)$, where $\omega_R$ is the resonant frequency of the damped oscillator, $c$ is the damping constant and $m$ is the mass of the oscillator) should have a high value for some applications. For example, resonance curve detection is a sensitive modulation technique for measuring small force gradients in noncontact imaging. Increasing the $Q$ increases the sensitivity of the measurements. Mechanical $Q$ values of 100–1,000 are typical. In contact modes, the $Q$ value is of less importance. A high lateral cantilever spring constant is desirable in order to reduce the effect of lateral forces in the AFM, as frictional forces can cause appreciable lateral bending of the cantilever. Lateral bending results in erroneous topography measurements. For friction measurements, cantilevers with reduced lateral rigidity are preferred. A sharp protruding tip must be present at the end of the cantilever to provide a well-defined interaction with the sample over a small area. The tip radius should be much smaller than the radii of the corrugations in the sample in order for these to be measured accurately. The lateral spring constant depends critically on the tip length. Additionally, the tip should be centered at the free end.

In the past, cantilevers have been cut by hand from thin metal foils or formed from fine wires. Tips for these cantilevers were prepared by attaching diamond fragments to the ends of the cantilevers by hand, or in the case of wire cantilevers, electrochemically etching the wire to a sharp point. Several cantilever geometries for wire cantilevers have been used. The simplest geometry is the L-shaped cantilever, which is usually made by bending a wire at a 90° angle. Other geometries include single-V and double-V geometries, with a sharp tip attached at the apex of the V, and double-X configuration with a sharp tip attached at the intersection [31, 138]. These cantilevers can be constructed with high vertical spring constants. For example, a double-cross cantilever with an effective spring constant of 250 N/m was used by Burnham and Colton [31]. The small size and low mass needed in the AFM make hand fabrication of the cantilever a difficult process with poor reproducibility. Conventional microfabrication techniques are ideal for constructing planar thin-film structures which have submicron lateral dimensions. The triangular (V-shaped) cantilevers have improved (higher) lateral spring constants in comparison to rectangular cantilevers. In terms of spring constants, the triangular cantilevers are approximately equivalent to two rectangular cantilevers placed in parallel [137]. Although the macroscopic radius of a photolithographically patterned corner is seldom much less than about 50 nm, microscopic asperities on the etched surface provide tips with near-atomic dimensions.

Cantilevers have been used from a whole range of materials. Cantilevers made of $Si_3N_4$, Si, and diamond are the most common. The Young's modulus and the density are the material parameters that determine the resonant frequency, aside from the geometry. Table 2.2 shows the relevant properties and the speed of sound, indicative of the resonant frequency for a given shape. Hardness is an important indicator of the durability of the cantilever, and is also listed in the table. Materials used for STM cantilevers are also included.

Silicon nitride cantilevers are less expensive than those made of other materials. They are very rugged and well suited to imaging in almost all environments. They are especially compatible with organic and biological materials. Microfabricated triangular silicon nitride beams with integrated square pyramidal tips made using plasma-enhanced chemical vapor deposition (PECVD) are the most common [137]. Four cantilevers, marketed by Digital Instruments, with different sizes and spring constants located on cantilever substrate made of boron silicate glass (Pyrex), are shown in Figs. 2.15a and 2.16. The two pairs of cantilevers on each substrate measure about 115 and 193 μm from the substrate to the apex of the triangular cantilever, with base widths of 122 and 205 μm, respectively. The cantilever legs, which are of the same thickness (0.6 μm) in all the cantilevers, are available in wide and narrow forms. Only one cantilever is selected and used from each substrate. The calculated spring constants and measured natural frequencies for each of the configurations are listed in Table 2.3. The most commonly used cantilever beam is the 115 μm long, wide-legged cantilever (vertical spring

**Table 2.2** Relevant properties of materials used for cantilevers

| Property | Young's modulus ($E$) (GPa) | Density ($\rho g$) (kg/m$^3$) | Microhardness (GPa) | Speed of sound ($\sqrt{E/\rho}$) (m/s) |
|---|---|---|---|---|
| Diamond | 900–1,050 | 3,515 | 78.4–102 | 17,000 |
| $Si_3N_4$ | 310 | 3,180 | 19.6 | 9,900 |
| Si | 130–188 | 2,330 | 9–10 | 8,200 |
| W | 350 | 19,310 | 3.2 | 4,250 |
| Ir | 530 | – | ≈3 | 5,300 |

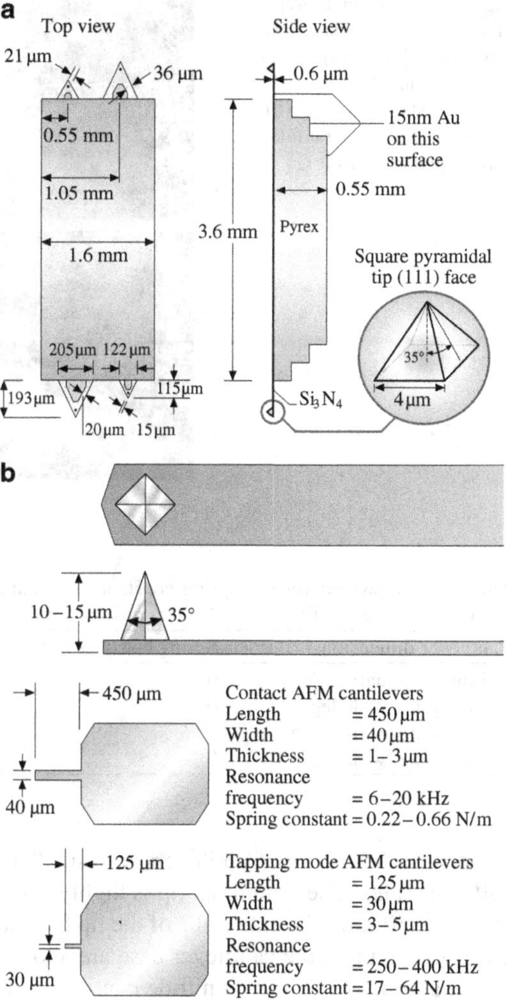


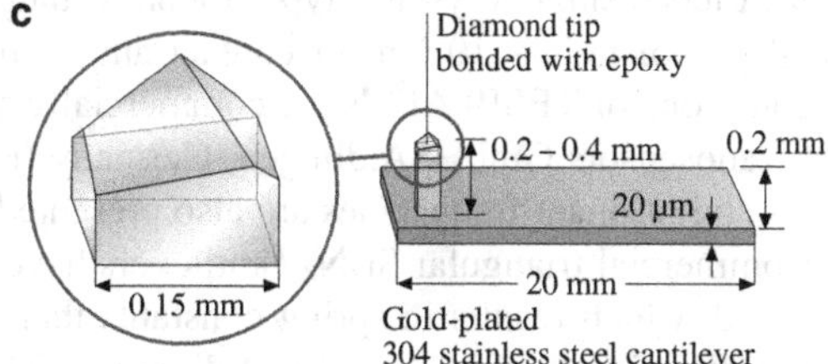


**Fig. 2.15** Schematics of (**a**) triangular cantilever beam with square-pyramidal tips made of PECVD $Si_3N_4$, (**b**) rectangular cantilever beams with square-pyramidal tips made of etched single-crystal silicon, and (**c**) rectangular cantilever stainless steel beam with three-sided pyramidal natural diamond tip

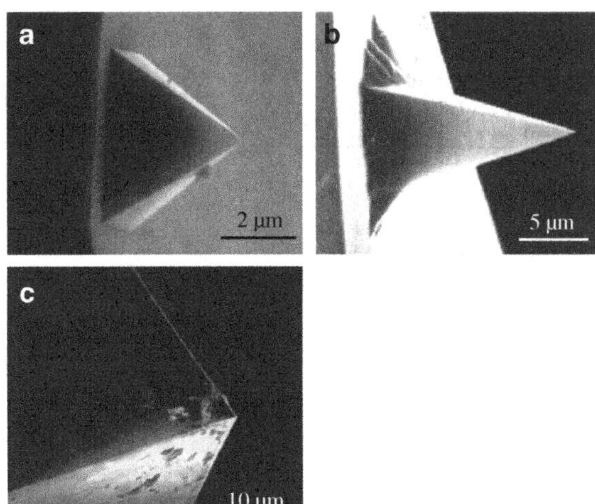


**Fig. 2.16** SEM micrographs of a square-pyramidal PECVD $Si_3N_4$ tip (**a**), a square-pyramidal etched single-crystal silicon tip (**b**), and a three-sided pyramidal natural diamond tip (**c**)

**Table 2.3** Measured vertical spring constants and natural frequencies of triangular (V-shaped) cantilevers made of PECVD $Si_3N_4$ (data provided by Digital Instruments, Inc.)

| Cantilever dimension | Spring constant ($k_z$) (N/m) | Natural frequency ($\omega_0$) (kHz) |
|---|---|---|
| 115 μm long, narrow leg | 0.38 | 40 |
| 115 μm long, wide leg | 0.58 | 40 |
| 193 μm long, narrow leg | 0.06 | 13–22 |
| 193 μm long, wide leg | 0.12 | 13–22 |

constant = 0.58 N/m). Cantilevers with smaller spring constants should be used on softer samples. The pyramidal tip is highly symmetric, and the end has a radius of about 20–50 nm. The side walls of the tip have a slope of 35° and the lengths of the edges of the tip at the cantilever base are about 4 μm.

An alternative to silicon nitride cantilevers with integrated tips are microfabricated single-crystal silicon cantilevers with integrated tips. Si tips are sharper than $Si_3N_4$ tips because they are formed directly by anisotropic etching of single-crystal Si, rather than through the use of an etch pit as a mask for the deposited material [139]. Etched single-crystal n-type silicon rectangular cantilevers with square pyramidal tips of radii <10 nm for contact and tapping mode (tapping-mode etched silicon probe or TESP) AFMs are commercially available from Digital Instruments and Nanosensors GmbH, Aidlingen, Germany (Figs. 2.15b and 2.16). Spring constants and resonant frequencies are also presented in the Fig. 2.15b.

Commercial triangular $Si_3N_4$ cantilevers have a typical width:thickness ratio of 10 to 30, which results in spring constants that are 100–1000 times stiffer in the lateral direction than in the normal direction. Therefore, these cantilevers are not well suited for torsion. For friction measurements, the torsional spring constant should be minimized in order to be sensitive to the lateral force. Rather long

**Table 2.4** Vertical ($k_z$), lateral ($k_y$), and torsional ($k_{yT}$) spring constants of rectangular cantilevers made of Si (IBM) and PECVD $Si_3N_4$ (source: Veeco Instruments, Inc.)

| Dimensions/stiffness | Si cantilever | $Si_3N_4$ cantilever |
|---|---|---|
| Length $L$ (μm) | 100 | 100 |
| Width $b$ (μm) | 10 | 20 |
| Thickness $h$ (μm) | 1 | 0.6 |
| Tip length (μm) | 5 | 3 |
| $k_z$ (N/m) | 0.4 | 0.15 |
| $k_y$ (N/m) | 40 | 175 |
| $k_{yT}$ (N/m) | 120 | 116 |
| $\omega_0$ (kHz) | ≈90 | ≈ 65 |

Note: $k_z = Ebh^3/(4L^3)$, $k_y = Eb^3h/(4\ell^3)$, $k_{yT} = Gbh^3/(3L\ell^2)$, and $\omega_0 = [k_z/(m_c + 0.24bhL\rho)]^{1/2}$, where $E$ is Young's modulus, $G$ is the modulus of rigidity [$= E/2(1 + \nu)$, $\nu$ is Poisson's ratio], $\rho$ is the mass density of the cantilever, and $m_c$ is the concentrated mass of the tip (≈4 ng) [94]. For Si, $E = 130$ GPa, $\rho g = 2{,}300$ kg/m$^3$, and $\nu = 0.3$. For $Si_3N_4$, $E = 150$ GPa, $\rho g = 3{,}100$ kg/m$^3$, and $\nu = 0.3$

cantilevers with small thicknesses and large tip lengths are most suitable. Rectangular beams have smaller torsional spring constants than the triangular (V-shaped) cantilevers. Table 2.4 lists the spring constants (with the full length of the beam used) in three directions for typical rectangular beams. We note that the lateral and torsional spring constants are about two orders of magnitude larger than the normal spring constants. A cantilever beam required for the tapping mode is quite stiff and may not be sensitive enough for friction measurements. Meyer et al. [140] used a specially designed rectangular silicon cantilever with length = 200 μm, width = 21 μm, thickness = 0.4 μm, tip length = 12.5 μm and shear modulus = 50 GPa, giving a normal spring constant of 0.007 N/m and a torsional spring constant of 0.72 N/m, which gives a lateral force sensitivity of 10 pN and an angle of resolution of $10^{-7}$ rad. Using this particular geometry, the sensitivity to lateral forces can be improved by about a factor of 100 compared with commercial V-shaped $Si_3N_4$ or the rectangular Si or $Si_3N_4$ cantilevers used by Meyer and Amer [8], with torsional spring constants of ≈100 N/m. Ruan and Bhushan [136] and Bhushan and Ruan [9] used 115 μm long, wide-legged V-shaped cantilevers made of $Si_3N_4$ for friction measurements.

For scratching, wear and indentation studies, single-crystal natural diamond tips ground to the shape of a three-sided pyramid with an apex angle of either 60° or 80° and a point sharpened to a radius of about 100 nm are commonly used [4, 10] (Figs. 2.15c and 2.16). The tips are bonded with conductive epoxy to a gold-plated 304 stainless steel spring sheet (length = 20 mm, width = 0.2 mm, thickness = 20–60 μm) which acts as a cantilever. The free length of the spring is varied in order to change the beam stiffness. The normal spring constant of the beam ranges from about 5 to 600 N/m for a 20 μm thick beam. The tips are produced by R-DEC Co., Tsukuba, Japan.

High aspect ratio tips are used to image within trenches. Examples of two probes used are shown in Fig. 2.17. These high aspect ratio tip (HART) probes are

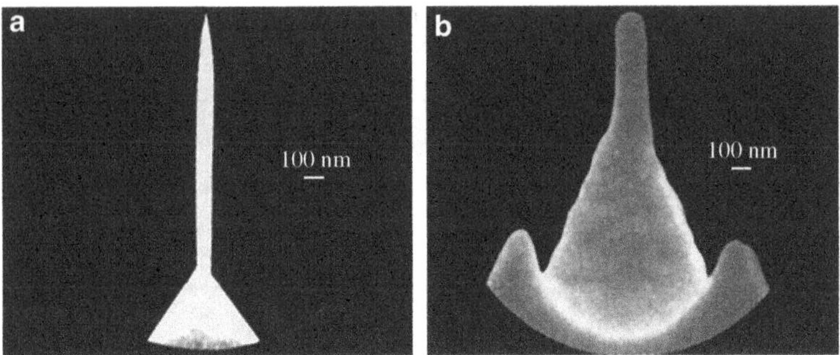


**Fig. 2.17** Schematics of (**a**) HART $Si_3N_4$ probe, and (**b**) an FIB-milled $Si_3N_4$ probe

produced from conventional $Si_3N_4$ pyramidal probes. Through a combination of focused ion beam (FIB) and high-resolution scanning electron microscopy (SEM) techniques, a thin filament is grown at the apex of the pyramid. The probe filament is $\approx 1$ μm long and 0.1 μm in diameter. It tapers to an extremely sharp point (with a radius that is better than the resolutions of most SEMs). The long thin shape and sharp radius make it ideal for imaging within *vias* of microstructures and trenches ($>0.25$ μm). This is, however, unsuitable for imaging structures at the atomic level, since probe flexing can create image artefacts. A FIB-milled probe is used for atomic-scale imaging, which is relatively stiff yet allows for closely spaced topography. These probes start out as conventional $Si_3N_4$ pyramidal probes, but the pyramid is FIB-milled until a small cone shape is formed which has a high aspect ratio and is 0.2–0.3 μm in length. The milled probes permit nanostructure resolution without sacrificing rigidity. These types of probes are manufactured by various manufacturers including Materials Analytical Services.

Carbon nanotube tips with small diameters and high aspect ratios are used for high-resolution imaging of surfaces and of deep trenches, in the tapping mode or the noncontact mode. Single-wall carbon nanotubes (SWNTs) are microscopic graphitic cylinders that are 0.7–3 nm in diameter and up to many microns in length. Larger structures called multiwall carbon nanotubes (MWNTs) consist of nested, concentrically arranged SWNTs and have diameters of 3–50 nm. MWNT carbon nanotube AFM tips are produced by manual assembly [141], chemical vapor deposition (CVD) synthesis, and a hybrid fabrication process [142]. Figure 2.18 shows a TEM micrograph of a carbon nanotube tip, ProbeMax, commercially produced by mechanical assembly by Piezomax Technologies, Inc. (Middleton, USA). To fabricate these tips, MWNTs are produced using a carbon arc and they are physically attached to the single-crystal silicon, square-pyramidal tips in the SEM, using a manipulator and the SEM stage to independently control the nanotubes and the tip. When the nanotube is first attached to the tip, it is usually too long to image with. It is shortened by placing it in an AFM and applying voltage between the tip and the sample. Nanotube tips are also commercially produced by CVD synthesis by NanoDevices (Santa Barbara, USA).

**Fig. 2.18** SEM micrograph of a multiwall carbon nanotube (MWNT) tip physically attached to a single-crystal silicon, square-pyramidal tip (Courtesy of Piezomax Technologies, Inc.)

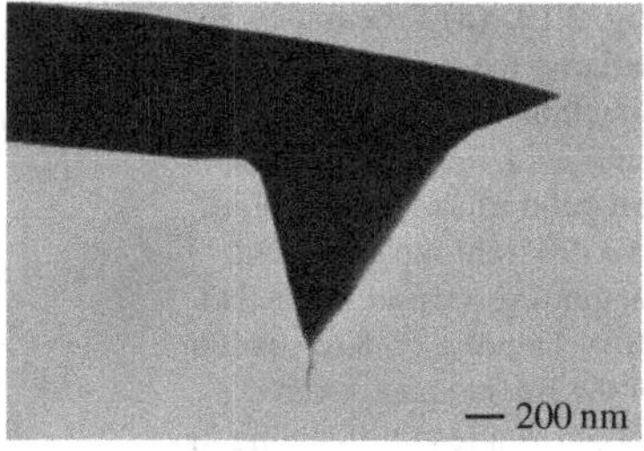


## 2.2.4 Friction Measurement Methods

The two methods for performing friction measurements that are based on the work by Ruan and Bhushan [136] are now described in more detail (also see [8]). The scanning angle is defined as the angle relative to the $y$-axis in Fig. 2.19a. This is also the long axis of the cantilever. The zero-degree scanning angle corresponds to the sample scan in the $y$-direction, and the 90° scanning angle corresponds to the sample scan perpendicular to this axis in the $x$–$y$-plane (along $x$-axis). If both the $y$- and $-y$-directions are scanned, we call this a *parallel scan*. Similarly, a *perpendicular scan* means that both the $x$- and $-x$-directions are scanned. The direction of sample travel for each of these two methods is illustrated in Fig. 2.19b.

Using method 1 (*height* mode with parallel scans) in addition to topographic imaging, it is also possible to measure friction force when the sample scanning direction is parallel to the $y$-direction (parallel scan). If there was no friction force between the tip and the moving sample, the topographic feature would be the only factor that would cause the cantilever to be deflected vertically. However, friction force does exist on all surfaces that are in contact where one of the surfaces is moving relative to the other. The friction force between the sample and the tip will also cause the cantilever to be deflected. We assume that the normal force between the sample and the tip is $W_0$ when the sample is stationary ($W_0$ is typically 10–200 nN), and the friction force between the sample and the tip is $W_f$ as the sample is scanned by the tip. The direction of the friction force ($W_f$) is reversed as the scanning direction of the sample is reversed from the positive ($y$) to the negative ($-y$) direction ($W_{f(y)} = -W_{f(-y)}$).

When the vertical cantilever deflection is set at a constant level, it is the total force (normal force and friction force) applied to the cantilever that keeps the cantilever deflection at this level. Since the friction force is directed in the opposite direction to the direction of travel of the sample, the normal force will have to be adjusted accordingly when the sample reverses its traveling direction, so that the total deflection of the cantilever will remain the same. We can calculate the difference in the normal force between the two directions of travel for a given

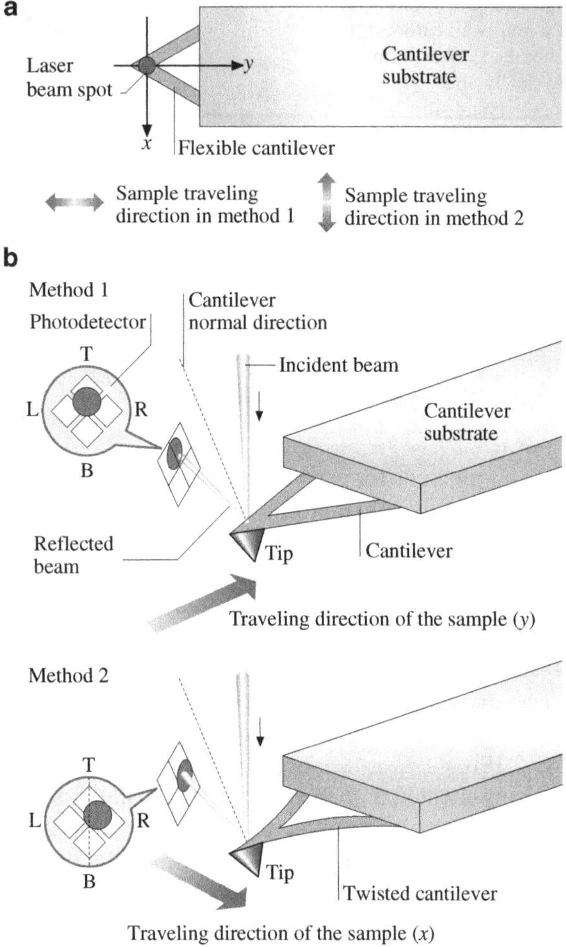


**Fig. 2.19** (**a**) Schematic defining the $x$- and $y$-directions relative to the cantilever, and showing the direction of sample travel in two different measurement methods discussed in the text. (**b**) Schematic of deformation of the tip and cantilever shown as a result of sliding in the $x$- and $y$-directions. A twist is introduced to the cantilever if the scanning is performed in the $x$-direction ((**b**), *lower part*) (After [136])

friction force $W_f$. First, since the deflection is constant, the total moment applied to the cantilever is constant. If we take the reference point to be the point where the cantilever joins the cantilever holder (substrate), point P in Fig. 2.20, we have the following relationship

$$(W_0 - \Delta W_1)L + W_f\ell = (W_0 + \Delta W_2)L - W_f\ell \tag{2.1}$$

or

$$(\Delta W_1 + \Delta W_2)L = 2W_f\ell. \tag{2.2}$$

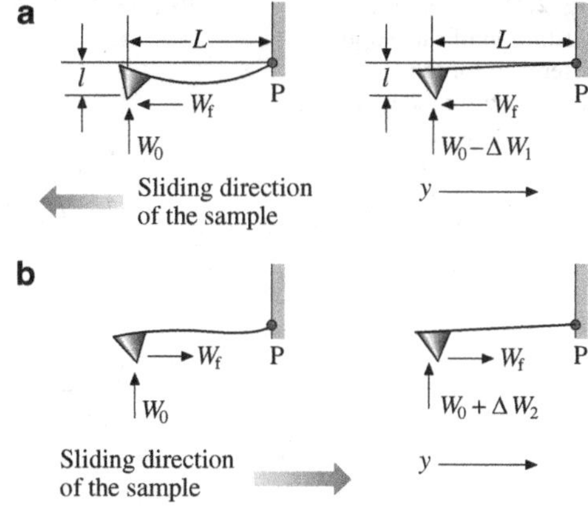


**Fig. 2.20** (**a**) Schematic showing an additional bending of the cantilever due to friction force when the sample is scanned in the $y$- or $-y$-directions (*left*). (**b**) This effect can be canceled out by adjusting the piezo height using a feedback circuit (*right*) (After [136])

Thus

$$W_f = (\Delta W_1 + \Delta W_2)L/(2\ell), \tag{2.3}$$

where $\Delta W_1$ and $\Delta W_2$ are the absolute values of the changes in normal force when the sample is traveling in the $-y$- and $y$-directions, respectively, as shown in Fig. 2.20; $L$ is the length of the cantilever; $\ell$ is the vertical distance between the end of the tip and point P. The coefficient of friction ($\mu$) between the tip and the sample is then given as

$$\mu = \frac{W_f}{W_0} = \left(\frac{(\Delta W_1 + \Delta W_2)}{W_0}\right)\left(\frac{L}{2\ell}\right). \tag{2.4}$$

There are adhesive and interatomic attractive forces between the cantilever tip and the sample at all times. The adhesive force can be due to water from the capillary condensation and other contaminants present at the surface, which form meniscus bridges [4, 143, 144] and the interatomic attractive force includes van der Waals attractions [18]. If these forces (and the effect of indentation too, which is usually small for rigid samples) can be neglected, the normal force $W_0$ is then equal to the initial cantilever deflection $H_0$ multiplied by the spring constant of the cantilever. ($\Delta W_1 + \Delta\, W_2$) can be derived by multiplying the same spring constant by the change in height of the piezo tube between the two traveling directions ($y$- and $-y$-directions) of the sample. This height difference is denoted as ($\Delta H_1 + \Delta H_2$), shown schematically in Fig. 2.21. Thus, (2.4) can be rewritten as

$$\mu = \frac{W_f}{W_0} = \left(\frac{(\Delta H_1 + \Delta H_2)}{H_0}\right)\left(\frac{L}{2\ell}\right). \tag{2.5}$$

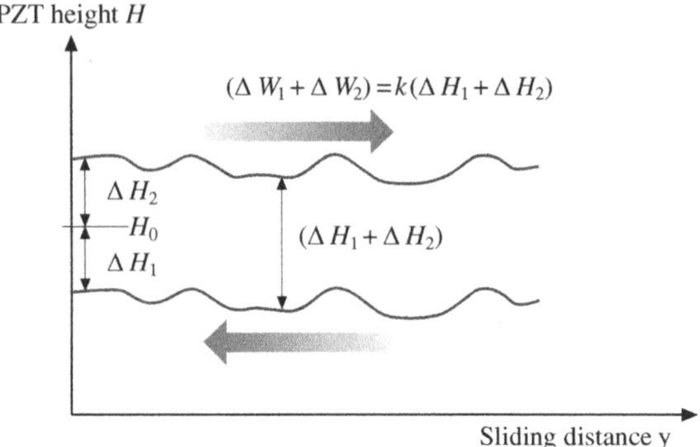


**Fig. 2.21** Schematic illustration of the height difference for the piezoelectric tube scanner as the sample is scanned in the $y$- and $-y$-directions

Since the vertical position of the piezo tube is affected by the topographic profile of the sample surface in addition to the friction force being applied at the tip, this difference must be found point-by-point at the same location on the sample surface, as shown in Fig. 2.21. Subtraction of point-by-point measurements may introduce errors, particularly for rough samples. We will come back to this point later. In addition, precise measurements of $L$ and $\ell$ (which should include the cantilever angle) are also required.

If the adhesive force between the tip and the sample is large enough that it cannot be neglected, it should be included in the calculation. However, determinations of this force can involve large uncertainties, which is introduced into (2.5). An alternative approach is to make the measurements at different normal loads and to use $\Delta(H_0)$ and $\Delta(\Delta H_1 + \Delta H_2)$ in (2.5). Another comment on (2.5) is that, since only the ratio between $(\Delta H_1 + \Delta H_2)$ and $H_0$ enters this equation, the vertical position of the piezo tube $H_0$ and the difference in position $(\Delta H_1 + \Delta H_2)$ can be in volts as long as the vertical travel of the piezo tube and the voltage applied to have a linear relationship. However, if there is a large nonlinearity between the piezo tube traveling distance and the applied voltage, this nonlinearity must be included in the calculation.

It should also be pointed out that (2.4) and (2.5) are derived under the assumption that the friction force $W_f$ is the same for the two scanning directions of the sample. This is an approximation, since the normal force is slightly different for the two scans and the friction may be direction-dependent. However, this difference is much smaller than $W_0$ itself. We can ignore the second-order correction.

Method 2 (*aux* mode with perpendicular scan) of measuring friction was suggested by Meyer and Amer [8]. The sample is scanned perpendicular to the long axis of the cantilever beam (along the $x$- or $-x$-direction in Fig. 2.19a) and the outputs from the two horizontal quadrants of the photodiode detector are measured. In this arrangement, as the sample moves under the tip, the friction force will cause the cantilever to twist. Therefore, the light intensity between the left and right (L and R in Fig. 2.19b, right) detectors will be different. The differential signal between the left and right detectors is denoted the FFM signal [(L − R)/(L + R)]. This signal can be related to the degree of twisting, and hence to the magnitude of friction force. Again, because possible errors in measurements of the normal force due to the presence of adhesive force at the tip–sample interface, the slope of the friction data (FFM signal versus normal load) needs to be measured for an accurate value of the coefficient of friction.

While friction force contributes to the FFM signal, friction force may not be the only contributing factor in commercial FFM instruments (for example, NanoScope IV). One can see this if we simply engange the cantilever tip with the sample. The left and right detectors can be balanced beforehand by adjusting the positions of the detectors so that the intensity difference between these two detectors is zero (FFM signal is zero). Once the tip is engaged with the sample, this signal is no longer zero, even if the sample is not moving in the $x$–$y$-plane with no friction force applied. This would be a detrimental effect. It has to be understood and eliminated from the data acquisition before any quantitative measurement of friction force is made.

One of the reasons for this observation is as follows. The detectors may not have been properly aligned with respect to the laser beam. To be precise, the vertical axis of the detector assembly (the line joining T–B in Fig. 2.22) is not in the plane defined by the incident laser beam and the beam reflected from the untwisted cantilever (we call this plane the *beam plane*). When the cantilever vertical deflection changes due to a change in the normal force applied (without the sample being scanned in the $x$–$y$-plane), the laser beam will be reflected up and down and form a projected trajectory on the detector. (Note that this trajectory is

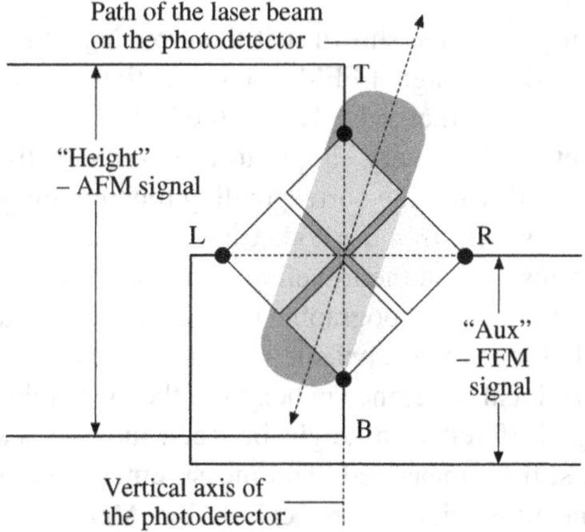


**Fig. 2.22** The trajectory of the laser beam on the photodetectors as the cantilever is vertically deflected (with no torsional motion) with respect to the laser beam for a misaligned photodetector. For a change of normal force (vertical deflection of the cantilever), the laser beam is projected to a different position on the detector. Due to a misalignment, the projected trajectory of the laser beam on the detector is not parallel with the detector vertical axis (the line T–B) (After [136])

in the defined beam plane.) If this trajectory is not coincident with the vertical axis of the detector, the laser beam will not evenly bisect the left and right quadrants of the detectors, even under the condition of no torsional motion of the cantilever (Fig. 2.22). Thus, when the laser beam is reflected up and down due a change in the normal force, the intensity difference between the left and right detectors will also change. In other words, the FFM signal will change as the normal force applied to the tip is changed, even if the tip is not experiencing any friction force. This (FFM) signal is unrelated to friction force or to the actual twisting of the cantilever. We will call this part of the FFM signal $\mathrm{FFM_F}$, and the part which is truly related to friction force $\mathrm{FFM_T}$.

The $\mathrm{FFM_F}$ signal can be eliminated. One way of doing this is as follows. First the sample is scanned in both the $x$- and the $-x$-directions and the FFM signals for scans in each direction are recorded. Since the friction force reverses its direction of action when the scanning direction is reversed from the $x$- to the $-x$-direction, the $\mathrm{FFM_T}$ signal will change signs as the scanning direction of the sample is reversed ($\mathrm{FFM_T}(x) = -\mathrm{FFM_T}(-x)$). Hence the $\mathrm{FFM_T}$ signal will be canceled out if we take the sum of the FFM signals for the two scans. The average value of the two scans will be related to $\mathrm{FFM_F}$ due to the misalignment,

$$\mathrm{FFM}(x) + \mathrm{FFM}(-x) = 2\mathrm{FFM_F}. \tag{2.6}$$

This value can therefore be subtracted from the original FFM signals of each of these two scans to obtain the true FFM signal ($\mathrm{FFM_T}$). Or, alternately, by taking the difference of the two FFM signals, one gets the $\mathrm{FFM_T}$ value directly

$$\mathrm{FFM}(x) - \mathrm{FFM}(-x) = \mathrm{FFM_T}(x) - \mathrm{FFM_T}(-x) = 2\mathrm{FFM_T}(x). \tag{2.7}$$

Ruan and Bhushan [136] have shown that the error signal ($\mathrm{FFM_F}$) can be very large compared to the friction signal $\mathrm{FFM_T}$, so correction is required.

Now we compare the two methods. The method of using the *height* mode and parallel scanning (method 1) is very simple to use. Technically, this method can provide 3-D friction profiles and the corresponding topographic profiles. However, there are some problems with this method. Under most circumstances, the piezo scanner displays hysteresis when the traveling direction of the sample is reversed. Therefore, the measured surface topographic profiles will be shifted relative to each other along the $y$-axis for the two opposite ($y$ and $-y$) scans. This would make it difficult to measure the local difference in height of the piezo tube for the two scans. However, the average difference in height between the two scans and hence the average friction can still be measured. The measurement of average friction can serve as an internal means of friction force calibration. Method 2 is a more desirable approach. The subtraction of the $\mathrm{FFM_F}$ signal from FFM for the two scans does not introduce any error into local friction force data. An ideal approach when using this method would be to add the average values of the two profiles in order to get the error component ($\mathrm{FFM_F}$) and then subtract this component from either profile to get true friction profiles in either directions. By performing measurements at various loads, we can get the average value of the coefficient of friction which then can be used to convert the friction profile to the coefficient of friction profile. Thus, any directionality and local variations in friction can be easily measured. In this method, since topography data are not affected by friction, accurate topography data can be measured simultaneously with friction data and a better localized relationship between the two can be established.

### 2.2.5 *Normal Force and Friction Force Calibrations of Cantilever Beams*

Based on Ruan and Bhushan [136], we now discuss normal force and friction force calibrations. In order to calculate the absolute values of normal and friction forces in Newtons using the measured AFM and $FFM_T$ voltage signals, it is necessary to first have an accurate value of the spring constant of the cantilever ($k_c$). The spring constant can be calculated using the geometry and the physical properties of the cantilever material [8, 94, 137]. However, the properties of the PECVD $Si_3N_4$ (used to fabricate cantilevers) can be different from those of the bulk material. For example, using ultrasonics, we found the Young's modulus of the cantilever beam to be about 238 ± 18 GPa, which is less than that of bulk $Si_3N_4$ (310 GPa). Furthermore, the thickness of the beam is nonuniform and difficult to measure precisely. Since the stiffness of a beam goes as the cube of thickness, minor errors in precise measurements of thickness can introduce substantial stiffness errors. Thus one should measure the spring constant of the cantilever experimentally. Cleveland et al. [145] measured normal spring constants by measuring resonant frequencies of beams.

For normal spring constant measurement, Ruan and Bhushan [136] used a stainless steel spring sheet of known stiffness (width = 1.35 mm, thickness = 15 μm, free hanging length = 5.2 mm). One end of the spring was attached to the sample holder and the other end was made to contact with the cantilever tip during the measurement (Fig. 2.23). They measured the piezo travel for a given cantilever deflection. For a rigid sample (such as diamond), the piezo travel $Z_t$ (measured from the point where the tip touches the sample) should equal the cantilever deflection. To maintain the cantilever deflection at the same level using a flexible spring sheet, the new piezo travel $Z_{t'}$ would need to be different from $Z_t$. The difference between $Z_{t'}$ and $Z_t$ corresponds to the deflection of the spring sheet. If the spring constant of the spring sheet is $k_s$, the spring constant of the cantilever $k_c$ can be calculated by

$$(Z_{t'} - Z_t)k_s = Z_t k_c$$

or

$$k_c = k_s(Z_{t'} - Z_t)/Z_t. \tag{2.8}$$

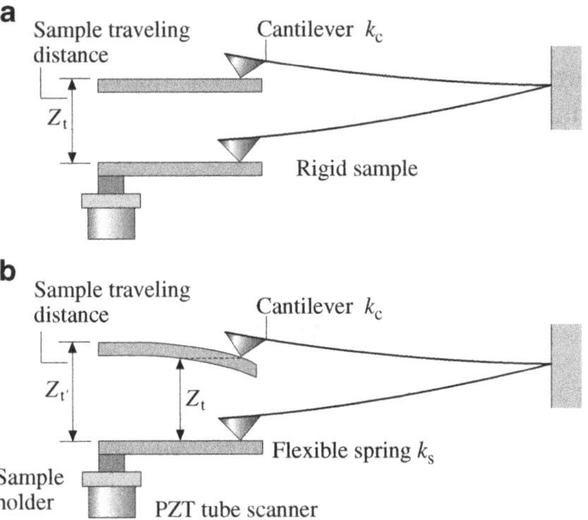


**Fig. 2.23** Illustration showing the deflection of the cantilever as it is pushed by **(a)** a rigid sample, **(b)** a flexible spring sheet (After [136])

The spring constant of the spring sheet ($k_s$) used in this study is calculated to be 1.54 N/m. For the wide-legged cantilever used in our study (length = 115 μm, base width = 122 μm, leg width = 21 μm and thickness = 0.6 μm), $k_c$ was measured to be 0.40 N/m instead of the 0.58 N/m reported by its manufacturer – Digital Instruments, Inc. To relate the photodiode detector output to the cantilever deflection in nanometers, they used the same rigid sample to push against the AFM tip. Since the cantilever vertical deflection equals the sample traveling distance measured from the point where the tip touches the sample for a rigid sample, the photodiode output observed as the tip is pushed by the sample can be converted directly to the cantilever deflection. For these measurements, they found the conversion factor to be 20 nm/V.

The normal force applied to the tip can be calculated by multiplying the cantilever vertical deflection by the cantilever spring constant for samples that have very small adhesion with the tip. If the adhesive force between the sample and the tip is large, it should be included in the normal force calculation. This is particularly important in atomic-scale force measurements, because the typical normal force that is measured in this region is in the range of a few hundreds of nN to a few mN. The adhesive force could be comparable to the applied force.

The conversion of friction signal (from $FFM_T$) to friction force is not as straightforward. For example, one can calculate the degree of twisting for a given friction force using the geometry and the physical properties of the cantilever [53, 144]. One would need information about the detector such as its quantum efficiency, laser power, gain and so on in order to be able convert the signal into the degree of twisting. Generally speaking, this procedure can not be accomplished without having some detailed information about the instrument. This information is not usually provided by the manufacturer. Even if this information is readily available, errors may still occur when using this approach because there will always be variations as a result of the instrumental set-up. For example, it has been noticed that the measured $FFM_T$ signal varies for the same sample when different AFM microscopes from the same manufacturer are used. This means that one can not calibrate the instrument experimentally using this calculation. O'Shea et al. [144] did perform a calibration procedure in which the torsional signal was measured as the sample was displaced a known distance laterally while ensuring that the tip did not slide over the surface. However, it is difficult to verify that tip sliding does not occur.

A new method of calibration is therefore required. There is a simpler, more direct way of doing this. The first method described above (method 1) of measuring friction can provide an absolute value of the coefficient of friction directly. It can therefore be used as an internal calibration technique for data obtained using method 2. Or, for a polished sample, which introduces the least error into friction measurements taken using method 1, method 1 can be used to calibrate the friction force for method 2. Then this calibration can be used for measurements taken using method 2. In method 1, the length of the cantilever required can be measured using an optical microscope; the length of the tip can be measured using a scanning electron microscope. The relative angle between the cantilever and the horizontal sample surface can be measured directly. This enables the coefficient of friction to be measured with few unknown parameters. The friction force can then be calculated by multiplying the coefficient of friction by the normal load. The $\mathrm{FFM_T}$ signal obtained using method 2 is then converted into the friction force. For their instrument, they found the conversion to be 8.6 nN/V.

## 2.3 AFM Instrumentation and Analyses

The performance of AFMs and the quality of AFM images greatly depend on the instrument available and the probes (cantilever and tips) in use. This section describes the mechanics of cantilevers, instrumentation and analysis of force detection systems for cantilever deflections, and scanning and control systems.

### *2.3.1 The Mechanics of Cantilevers*

#### Stiffness and Resonances of Lumped Mass Systems

All of the building blocks of an AFM, including the body of the microscope itself and the force-measuring cantilevers, are mechanical resonators. These resonances can be excited either by the surroundings or by the rapid movement of the tip or the sample. To avoid problems due to building- or air-induced oscillations, it is of paramount importance to optimize the design of the AFM for high resonant frequencies. This usually means decreasing the size of the microscope [146]. By using cube-like or sphere-like structures for the microscope, one can considerably increase the lowest eigenfrequency. The fundamental natural frequency $\omega_0$ of any spring is given by

$$\omega_0 = \frac{1}{2\pi}\sqrt{\frac{k}{m_{\mathrm{eff}}}}, \tag{2.9}$$

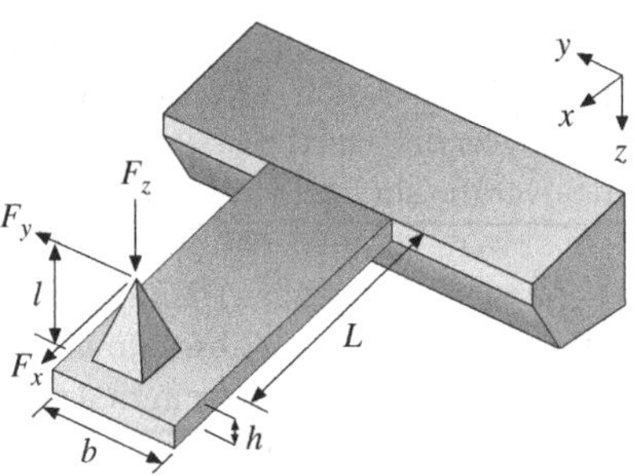


**Fig. 2.24** A typical AFM cantilever with length $L$, width $b$, and height $h$. The height of the tip is $\ell$. The material is characterized by the Young's modulus $E$, the shear modulus $G$ and the mass density $\rho$. Normal ($F_z$), axial ($F_x$) and lateral ($F_y$) forces exist at the end of the tip

where $k$ is the spring constant (stiffness) in the normal direction and $m_{\text{eff}}$ is the effective mass. The spring constant $k$ of a cantilever beam with uniform cross section (Fig. 2.24) is given by [147]

$$k = \frac{3EI}{L^3}, \tag{2.10}$$

where $E$ is the Young's modulus of the material, $L$ is the length of the beam and $I$ is the moment of inertia of the cross section. For a rectangular cross section with a width $b$ (perpendicular to the deflection) and a height $h$ one obtains the following expression for $I$

$$I = \frac{bh^3}{12}. \tag{2.11}$$

Combining (2.9)–(2.11), we get an expression for $\omega_0$

$$\omega_0 = \sqrt{\frac{Ebh^3}{4L^3 m_{\text{eff}}}}. \tag{2.12}$$

The effective mass can be calculated using Raleigh's method. The general formula using Raleigh's method for the kinetic energy $T$ of a bar is

$$T = \frac{1}{2}\int_0^L \frac{m}{L}\left(\frac{\partial z(x)}{\partial t}\right)^2 \mathrm{d}x. \tag{2.13}$$

For the case of a uniform beam with a constant cross section and length $L$, one obtains for the deflection $z(x) = z_{\max}[1 - (3x/2L) + (x^3/2L^3)]$ Inserting $z_{\max}$ into (2.13) and solving the integral gives

$$T = \frac{1}{2}\int_0^L \frac{m}{L}\left[\frac{\partial z_{\max}(x)}{\partial t}\left(1 - \frac{3x}{2L}\right) + \left(\frac{x^3}{L^3}\right)\right]^2 \mathrm{d}x = \frac{1}{2} m_{\text{eff}}(z_{\max} t)^2,$$

which gives

$$m_{\text{eff}} = \frac{9}{20} m. \tag{2.14}$$

Substituting (2.14) into (2.12) and noting that $m = \rho L b h$, where $\rho$ is the mass density, one obtains the following expression

$$\omega_0 = \left( \frac{\sqrt{5}}{3} \sqrt{\frac{\mathrm{E}}{\rho}} \right) \frac{h}{L^2}. \tag{2.15}$$

It is evident from (2.15) that one way to increase the natural frequency is to choose a material with a high ratio $E/\rho$; see Table 2.2 for typical values of $\sqrt{E/\rho}$ for various commonly used materials. Another way to increase the lowest eigenfrequency is also evident in (2.15). By optimizing the ratio $h/L^2$, one can increase the resonant frequency. However, it does not help to make the length of the structure smaller than the width or height. Their roles will just be interchanged. Hence the optimum structure is a cube. This leads to the design rule that long, thin structures like sheet metal should be avoided. For a given resonant frequency, the quality factor $Q$ should be as low as possible. This means that an inelastic medium such as rubber should be in contact with the structure in order to convert kinetic energy into heat.

### Stiffness and Resonances of Cantilevers

Cantilevers are mechanical devices specially shaped to measure tiny forces. The analysis given in the previous section is applicable. However, to better understand the intricacies of force detection systems, we will discuss the example of a cantilever beam with uniform cross section (Fig. 2.24). The bending of a beam due to a normal load on the beam is governed by the Euler equation [147]

$$M = EI(x) \frac{\mathrm{d}^2 z}{\mathrm{d}x^2}, \tag{2.16}$$

where $M$ is the bending moment acting on the beam cross section. $I(x)$ is the moment of inertia of the cross section with respect to the neutral axis, defined by

$$I(x) = \int_z \int_y z^2 \, \mathrm{d}y \mathrm{d}z. \tag{2.17}$$

For a normal force $F_z$ acting at the tip,

$$M(x) = (L - x) F_z \tag{2.18}$$

since the moment must vanish at the endpoint of the cantilever. Integrating (2.16) for a normal force $F_z$ acting at the tip and observing that $EI$ is a constant for beams with a uniform cross section, one gets

$$z(x) = \frac{L^3}{6EI}\left(\frac{x}{L}\right)^2\left(3 - \frac{x}{L}\right)F_z. \tag{2.19}$$

The slope of the beam is

$$z'(x) = \frac{Lx}{2EI}\left(2 - \frac{x}{L}\right)F_z. \tag{2.20}$$

From (2.19) and (2.20), at the end of the cantilever (for $x = L$), for a rectangular beam, and by using an expression for $I$ in (2.11), one gets

$$z(L) = \frac{4}{Eb}\left(\frac{L}{h}\right)^3 F_z, \tag{2.21}$$

$$z'(L) = \frac{3}{2}\left(\frac{z}{L}\right). \tag{2.22}$$

Now, the stiffness in the normal ($z$) direction $k_z$ is

$$k_z = \frac{F_z}{z(L)} = \frac{Eb}{4}\left(\frac{h}{L}\right)^3. \tag{2.23}$$

and the change in angular orientation of the end of cantilever beam is

$$\Delta\alpha = \frac{3}{2}\frac{z}{L} = \frac{6}{Ebh}\left(\frac{L}{h}\right)^2 F_z. \tag{2.24}$$

Now we ask what will, to a first-order approximation, happen if we apply a lateral force $F_y$ to the end of the tip (Fig. 2.24). The cantilever will bend sideways and it will twist. The stiffness in the lateral ($y$) direction $k_y$ can be calculated with (2.23) by exchanging $b$ and $h$

$$k_y = \frac{Eh}{4}\left(\frac{b}{L}\right)^3. \tag{2.25}$$

Therefore, the bending stiffness in the lateral direction is larger than the stiffness for bending in the normal direction by $(b/h)^2$. The twisting or torsion on the other hand is more complicated to handle. For a wide, thin cantilever ($b \gg h$) we obtain torsional stiffness along $y$-axis $k_{y\mathrm{T}}$

$$k_{y\mathrm{T}} = \frac{Gbh^3}{3L\ell^2}, \tag{2.26}$$

where $G$ is the modulus of rigidity ($= E/2(1 + \nu)$; $\nu$ is Poisson's ratio). The ratio of the torsional stiffness to the lateral bending stiffness is

$$\frac{k_{y\mathrm{T}}}{k_y} = \frac{1}{2}\left(\frac{\ell b}{hL}\right)^2, \tag{2.27}$$

where we assume $\nu = 0.333$. We see that thin, wide cantilevers with long tips favor torsion while cantilevers with square cross sections and short tips favor bending. Finally, we calculate the ratio between the torsional stiffness and the normal bending stiffness,

$$\frac{k_{y\mathrm{T}}}{k_z} = 2\left(\frac{L}{\ell}\right)^2. \tag{2.28}$$

Equations (2.26)–(2.28) hold in the case where the cantilever tip is exactly in the middle axis of the cantilever. Triangular cantilevers and cantilevers with tips which are not on the middle axis can be dealt with by finite element methods.

The third possible deflection mode is the one from the force on the end of the tip along the cantilever axis, $F_x$ (Fig. 2.24). The bending moment at the free end of the cantilever is equal to $F_x\ell$. This leads to the following modification of (2.18) for forces $F_z$ and $F_x$

$$M(x) = (L - x)F_z + F_x\ell. \tag{2.29}$$

Integration of (2.16) now leads to

$$z(x) = \frac{1}{2EI}\left[Lx^2\left(1 - \frac{x}{3L}\right)F_z + \ell x^2 F_x\right] \tag{2.30}$$

and

$$z'(x) = \frac{1}{EI}\left[\frac{Lx}{2}\left(2 - \frac{x}{L}\right)F_z + \ell x F_x\right]. \tag{2.31}$$

Evaluating (2.30) and (2.31) at the end of the cantilever, we get the deflection and the tilt

$$\begin{aligned} z(L) &= \frac{L^2}{EI}\left(\frac{L}{3}F_z - \frac{\ell}{2}F_x\right),\\ z'(L) &= \frac{L}{EI}\left(\frac{L}{2}F_z + \ell F_x\right). \end{aligned} \tag{2.32}$$

From these equations, one gets

$$\begin{aligned} F_z &= \frac{12EI}{L^3}\left[z(L) - \frac{Lz'(L)}{2}\right],\\ F_x &= \frac{2EI}{\ell L^2}\left[2Lz'(L) - 3z(L)\right]. \end{aligned} \tag{2.33}$$

A second class of interesting properties of cantilevers is their resonance behavior. For cantilever beams, one can calculate the resonant frequencies [147, 148]

$$\omega_n^{\text{free}} = \frac{\lambda_n^2}{2\sqrt{3}} \frac{h}{L^2} \sqrt{\frac{E}{\rho}} \tag{2.34}$$

with $\lambda_0 = (0.596864\ldots)\pi$, $\lambda_1 = (1.494175\ldots)\pi$, $\lambda_n \to (n + 1/2)\pi$. The subscript $n$ represents the order of the frequency, such as the fundamental, the second mode, and the $n$th mode.

A similar equation to (2.34) holds for cantilevers in rigid contact with the surface. Since there is an additional restriction on the movement of the cantilever, namely the location of its endpoint, the resonant frequency increases. Only the terms of $\lambda_n$ change to [148]

$$\lambda_0' = (1.2498763\ldots)\pi, \quad \lambda_1' = (2.2499997\ldots)\pi, \quad \lambda_n' \to (n + 1/4)\pi. \tag{2.35}$$

The ratio of the fundamental resonant frequency during contact to the fundamental resonant frequency when not in contact is 4.3851.

For the torsional mode we can calculate the resonant frequencies as

$$\omega_0^{\text{tors}} = 2\pi \frac{h}{Lb} \sqrt{\frac{G}{\rho}}. \tag{2.36}$$

For cantilevers in rigid contact with the surface, we obtain the following expression for the fundamental resonant frequency [148]

$$\omega_0^{\text{tors,contact}} = \frac{\omega_0^{\text{tors}}}{\sqrt{1 + 3(2L/b)^2}}. \tag{2.37}$$

The amplitude of the thermally induced vibration can be calculated from the resonant frequency using

$$\Delta z_{\text{therm}} = \sqrt{\frac{k_{\text{B}} T}{k}}, \tag{2.38}$$

where $k_{\text{B}}$ is Boltzmann's constant and $T$ is the absolute temperature. Since AFM cantilevers are resonant structures, sometimes with rather high $Q$ values, the thermal noise is not as evenly distributed as (2.38) suggests. The spectral noise density below the peak of the response curve is [148]

$$z_0 = \sqrt{\frac{4 k_{\text{B}} T}{k \omega_0 Q}} \quad (\text{in m}/\sqrt{\text{Hz}}), \tag{2.39}$$

where $Q$ is the quality factor of the cantilever, described earlier.

### *2.3.2 Instrumentation and Analyses of Detection Systems for Cantilever Deflections*

A summary of selected detection systems was provided in Fig. 2.8. Here we discuss the pros and cons of various systems in detail.

#### Optical Interferometer Detection Systems

Soon after the first papers on the AFM [2] appeared, which used a tunneling sensor, an instrument based on an interferometer was published [149]. The sensitivity of the interferometer depends on the wavelength of the light employed in the apparatus. Figure 2.25 shows the principle of such an interferometeric design. The light incident from the left is focused by a lens onto the cantilever. The reflected light is collimated by the same lens and interferes with the light reflected at the flat. To separate the reflected light from the incident light, a $\lambda/4$ plate converts the linearly polarized incident light into circularly polarized light. The reflected light is made linearly polarized again by the $\lambda/4$-plate, but with a polarization orthogonal to that of the incident light. The polarizing beam splitter then deflects the reflected light to the photodiode.

Homodyne Interferometer

To improve the signal-to-noise ratio of the interferometer, the cantilever is driven by a piezo near its resonant frequency. The amplitude $\Delta z$ of the cantilever as a function of driving frequency $\Omega$ is

$$\Delta z(\Omega) = \Delta z_0 \frac{\Omega_0^2}{\sqrt{\left(\Omega^2 - \Omega_0^2\right)^2 + \frac{\Omega^2 \Omega_0^2}{Q^2}}}, \qquad (2.40)$$

where $\Delta z_0$ is the constant drive amplitude and $\Omega_0$ the resonant frequency of the cantilever. The resonant frequency of the cantilever is given by the effective potential

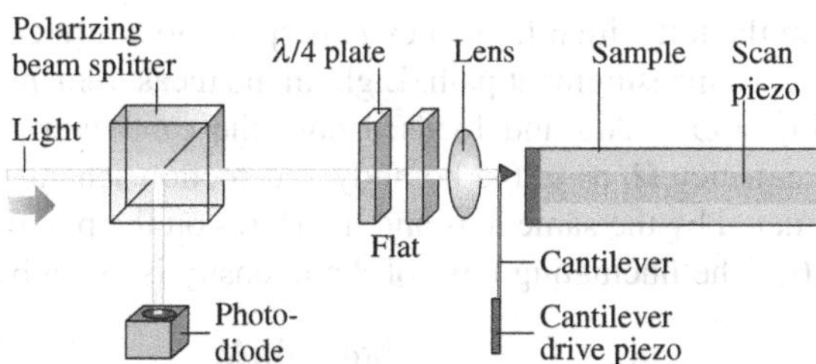


**Fig. 2.25** Principle of an interferometric AFM. The light from the laser light source is polarized by the polarizing beam splitter and focused onto the back of the cantilever. The light passes twice through a quarter-wave plate and is hence orthogonally polarized to the incident light. The second arm of the interferometer is formed by the flat. The interference pattern is modulated by the oscillating cantilever

$$\Omega_0 = \sqrt{\left(k + \frac{\partial^2 U}{\partial z^2}\right)\frac{1}{m_{\text{eff}}}}, \tag{2.41}$$

where $U$ is the interaction potential between the tip and the sample. Equation (2.41) shows that an attractive potential decreases $\Omega_0$. The change in $\Omega_0$ in turn results in a change in $\Delta z$ (2.40). The movement of the cantilever changes the path difference in the interferometer. The light reflected from the cantilever with amplitude $A_{\ell,0}$ and the reference light with amplitude $A_{\text{r},0}$ interfere on the detector. The detected intensity $I(t) = [A_\ell(t) + A_\text{r}(t)]^2$ consists of two constant terms and a fluctuating term

$$2A_\ell(t)A_\text{r}(t) = A_{\ell,0}A_{\text{r},0}\sin\left[\omega t + \frac{4\pi\delta}{\lambda} + \frac{4\pi\Delta z}{\lambda}\sin(\Omega t)\right]\sin(\omega t). \tag{2.42}$$

Here $\omega$ is the frequency of the light, $\lambda$ is the wavelength of the light, $\delta$ is the path difference in the interferometer, and $\Delta z$ is the instantaneous amplitude of the cantilever, given according to (2.40) and (2.41) as a function of $\Omega$, $k$, and $U$. The time average of (2.42) then becomes

$$\begin{aligned}\langle 2A_\ell(t)A_\text{r}(t)\rangle_T &\propto \cos\left[\frac{4\pi\delta}{\lambda} + \frac{4\pi\Delta z}{\lambda}\sin(\Omega t)\right] \\ &\approx \cos\left(\frac{4\pi\delta}{\lambda}\right) - \sin\left[\frac{4\pi\Delta z}{\lambda}\sin(\Omega t)\right] \approx \cos\left(\frac{4\pi\delta}{\lambda}\right) - \frac{4\pi\Delta z}{\lambda}\sin(\Omega t).\end{aligned} \tag{2.43}$$

Here all small quantities have been omitted and functions with small arguments have been linearized. The amplitude of $\Delta z$ can be recovered with a lock-in technique. However, (2.43) shows that the measured amplitude is also a function of the path difference $\delta$ in the interferometer. Hence, this path difference $\delta$ must be very stable. The best sensitivity is obtained when $\sin\,(4\delta/\lambda) \approx 0$.

#### Heterodyne Interferometer

This influence is not present in the heterodyne detection scheme shown in Fig. 2.26. Light incident from the left with a frequency $\omega$ is split into a reference path (upper path in Fig. 2.26) and a measurement path. Light in the measurement path is shifted in frequency to $\omega_1 = \omega + \Delta\omega$ and focused onto the cantilever. The cantilever oscillates at the frequency $\Omega$, as in the homodyne detection scheme. The reflected light $A_\ell(t)$ is collimated by the same lens and interferes on the photodiode with the reference light $A_\text{r}(t)$. The fluctuating term of the intensity is given by

$$2A_\ell(t)A_\text{r}(t) = A_{\ell,0}A_{\text{r},0}\sin\left[(\omega + \Delta\omega)t + \frac{4\pi\delta}{\lambda} + \frac{4\pi\Delta z}{\lambda}\sin(\Omega t)\right]\sin(\omega t), \tag{2.44}$$

where the variables are defined as in (2.42). Setting the path difference $\sin\,(4\pi\delta/\lambda) \approx 0$ and taking the time average, omitting small quantities and linearizing functions with small arguments, we get

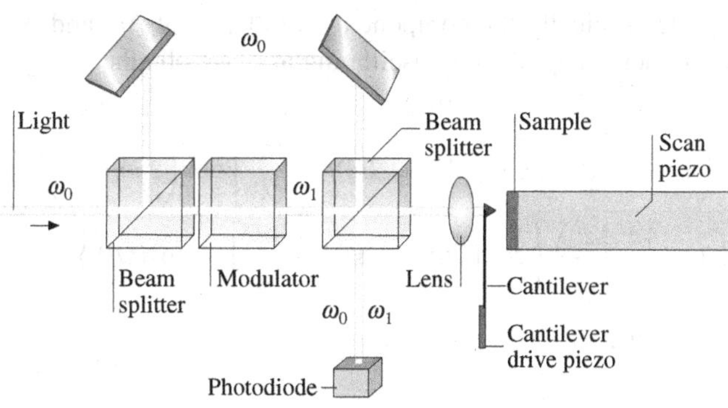


**Fig. 2.26** Principle of a heterodyne interferometric AFM. Light with frequency $\omega_0$ is split into a reference path (upper path) and a measurement path. The light in the measurement path is frequency shifted to $\omega_1$ by an acousto-optical modulator (or an electro-optical modulator). The light reflected from the oscillating cantilever interferes with the reference beam on the detector

$$
\begin{aligned}
&\langle 2A_\ell(t)A_\mathrm{r}(t)\rangle_\mathrm{T} \propto \cos\left[\Delta\omega t + \frac{4\pi\delta}{\lambda} + \frac{4\pi\Delta z}{\lambda}\sin(\Omega t)\right] \\
&= \cos\left(\Delta\omega t + \frac{4\pi\delta}{\lambda}\right)\cos\left[\frac{4\pi\Delta z}{\lambda}\sin(\Omega t)\right] - \sin\left(\Delta\omega t + \frac{4\pi\delta}{\lambda}\right)\sin\left[\frac{4\pi\Delta z}{\lambda}\sin(\Omega t)\right] \\
&\approx \cos\left(\frac{4\pi\delta}{\lambda}\right) - \sin\left[\frac{4\pi\Delta z}{\lambda}\sin(\Omega t)\right] \approx \cos\left(\Delta\omega t + \frac{4\pi\delta}{\lambda}\right)\left[1 - \frac{8\pi^2\Delta z^2}{\lambda^2}\sin(\Omega t)\right] \\
&- \frac{4\pi\Delta z}{\lambda}\sin\left(\Delta\omega t + \frac{4\pi\delta}{\lambda}\right)\sin(\Omega t) \\
&= \cos\left(\Delta\omega t + \frac{4\pi\delta}{\lambda}\right) - \frac{8\pi^2\Delta z^2}{\lambda^2}\cos\left(\Delta\omega t + \frac{4\pi\delta}{\lambda}\right) \\
&\times \sin(\Omega t) - \frac{4\pi\Delta z}{\lambda}\sin\left(\Delta\omega t + \frac{4\pi\delta}{\lambda}\right)\sin(\Omega t) \\
&= \cos\left(\Delta\omega t + \frac{4\pi\delta}{\lambda}\right) - \frac{4\pi^2\Delta z^2}{\lambda^2}\cos\left(\Delta\omega t + \frac{4\pi\delta}{\lambda}\right) + \frac{4\pi^2\Delta z^2}{\lambda^2}\cos\left(\Delta\omega t + \frac{4\pi\delta}{\lambda}\right) \\
&\cos(2\Omega t) - \frac{4\pi\Delta z}{\lambda}\sin\left(\Delta\omega t + \frac{4\pi\delta}{\lambda}\right)\sin(\Omega t) \\
&= \cos\left(\Delta\omega t + \frac{4\pi\delta}{\lambda}\right)\left(1 - \frac{4\pi^2\Delta z^2}{\lambda^2}\right) + \frac{2\pi^2\Delta z^2}{\lambda^2}\left\{\cos\left[(\Delta\omega + 2\Omega)t + \frac{4\pi\delta}{\lambda}\right]\right. \\
&\left. + \cos\left[(\Delta\omega - 2\Omega)t + \frac{4\pi\delta}{\lambda}\right]\right\} + \frac{2\pi\Delta z}{\lambda}\left\{\cos\left[(\Delta\omega + \Omega)t + \frac{4\pi\delta}{\lambda}\right]\right. \\
&\left. + \cos\left[(\Delta\omega - \Omega)t + \frac{4\pi\delta}{\lambda}\right]\right\}.
\end{aligned}
\tag{2.45}
$$

Multiplying electronically the components oscillating at $\Delta\omega$ and $\Delta\omega + \Omega$ and rejecting any product except the one oscillating at $\Omega$ we obtain

$$\begin{aligned} A &= \frac{2\Delta z}{\lambda}\left(1 - \frac{4\pi^2\Delta z^2}{\lambda^2}\right)\cos\left[(\Delta\omega + 2\Omega)t + \frac{4\pi\delta}{\lambda}\right]\cos\left(\Delta\omega t + \frac{4\pi\delta}{\lambda}\right) \\ &= \frac{\Delta z}{\lambda}\left(1 - \frac{4\pi^2\Delta z^2}{\lambda^2}\right)\left\{\cos\left[(2\Delta\omega + \Omega)t + \frac{8\pi\delta}{\lambda}\right] + \cos(\Omega t)\right\} \\ &\approx \frac{\pi\Delta z}{\lambda}\cos(\Omega t). \end{aligned} \tag{2.46}$$

Unlike in the homodyne detection scheme, the recovered signal is independent from the path difference $\delta$ of the interferometer. Furthermore, a lock-in amplifier with the reference set sin $(\Delta\omega t)$ can measure the path difference $\delta$ independent of the cantilever oscillation. If necessary, a feedback circuit can keep $\delta = 0$.

### Fiber-Optical Interferometer

The fiber-optical interferometer [129] is one of the simplest interferometers to build and use. Its principle is sketched in Fig. 2.27. The light of a laser is fed into an optical fiber. Laser diodes with integrated fiber pigtails are convenient light sources. The light is split in a fiber-optic beam splitter into two fibers. One fiber is terminated by index-matching oil to avoid any reflections back into the fiber. The end of the other fiber is brought close to the cantilever in the AFM. The emerging light is partially reflected back into the fiber by the cantilever. Most of the light, however, is lost. This is not a big problem since only 4% of the light is reflected at the end of the fiber, at the glass–air interface. The two reflected light waves interfere with each other. The product is guided back into the fiber coupler and again split into two parts. One half is analyzed by the photodiode. The other half is fed back into the laser. Communications grade laser diodes are sufficiently resistant to feedback to be operated in this environment. They have, however, a bad coherence length, which in this case does not matter, since the optical path difference is in any case no larger than 5 μm. Again the end of the fiber has to be positioned on a piezo drive to set the distance between the fiber and the cantilever to $\lambda(n + 1/4)$.

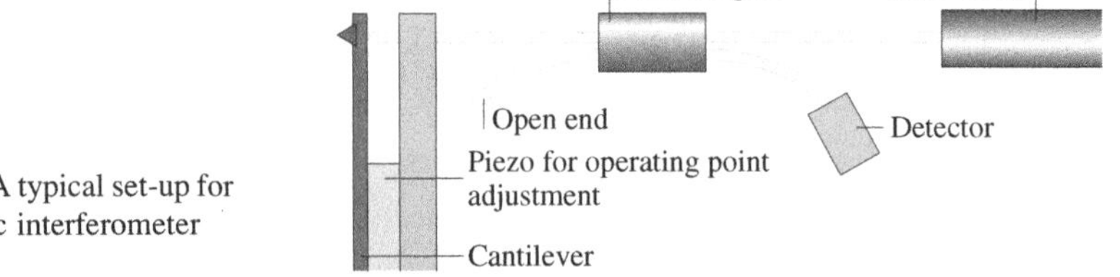


**Fig. 2.27** A typical set-up for a fiber-optic interferometer readout

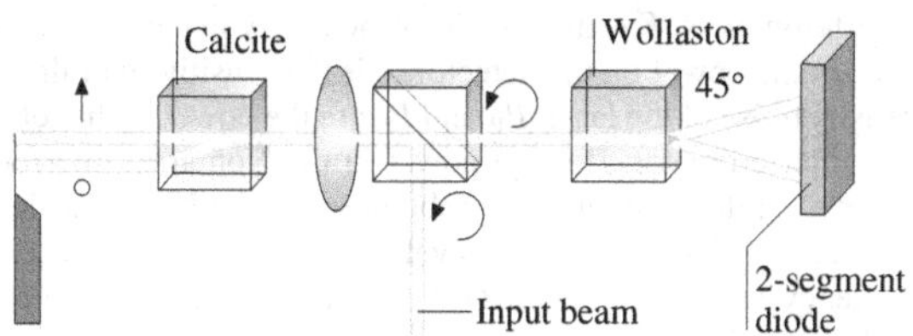


**Fig. 2.28** Principle of Nomarski AFM. The circularly polarized input beam is deflected to the left by a nonpolarizing beam splitter. The light is focused onto a cantilever. The calcite crystal between the lens and the cantilever splits the circular polarized light into two spatially separated beams with orthogonal polarizations. The two light beams reflected from the lever are superimposed by the calcite crystal and collected by the lens. The resulting beam is again circularly polarized. A Wollaston prism produces two interfering beams with a $\pi/2$ phase shift between them. The minimal path difference accounts for the excellent stability of this microscope

### Nomarski-Interferometer

Another way to minimize the optical path difference is to use the Nomarski interferometer [130]. Figure 2.28 shows a schematic of the microscope. The light from a laser is focused on the cantilever by lens. A birefringent crystal (for instance calcite) between the cantilever and the lens, which has its optical axis 45° off the polarization direction of the light, splits the light beam into two paths, offset by a distance given by the length of the crystal. Birefringent crystals have varying indices of refraction. In calcite, one crystal axis has a lower index than the other two. This means that certain light rays will propagate at different speeds through the crystal than others. By choosing the correct polarization, one can select the ordinary ray or the extraordinary ray or one can get any mixture of the two rays. A detailed description of birefringence can be found in textbooks (e.g., [150]). A calcite crystal deflects the extraordinary ray at an angle of 6° within the crystal. Any separation can be set by choosing a suitable length for the calcite crystal.

The focus of one light ray is positioned near the free end of the cantilever while the other is placed close to the clamped end. Both arms of the interferometer pass through the same space, except for the distance between the calcite crystal and the lever. The closer the calcite crystal is placed to the lever, the less influence disturbances like air currents have.

Sarid [116] has given values for the sensitivities of different interferometeric detection systems. Table 2.5 presents a summary of his results.

### Optical Lever

The most common cantilever deflection detection system is the optical lever [53, 111]. This method, depicted in Fig. 2.29, employs the same technique as light beam deflection galvanometers. A fairly well collimated light beam is reflected off a

**Table 2.5** Noise in interferometers. $F$ is the finesse of the cavity in the homodyne interferometer, $P_i$ the incident power, $P_d$ is the power on the detector, $\eta$ is the sensitivity of the photodetector and RIN is the relative intensity noise of the laser. $P_R$ and $P_S$ are the power in the reference and sample beam in the heterodyne interferometer. $P$ is the power in the Nomarski interferometer, $\delta\theta$ is the phase difference between the reference and the probe beam in the Nomarski interferometer. $B$ is the bandwidth, $e$ is the electron charge, $\lambda$ is the wavelength of the laser, $k$ the cantilever stiffness, $\omega_0$ is the resonant frequency of the cantilever, $Q$ is the quality factor of the cantilever, $T$ is the temperature, and $\delta i$ is the variation in current $i$

| | Homodyne interferometer, fiber-optic interferometer | Heterodyne interferometer | Nomarski interferometer |
|---|---|---|---|
| Laser noise $\delta i^2_L$ | $\frac{1}{4}\eta^2 F^2 P_i^2$ RIN | $\eta^2 (P_R^2 + P_S^2)$ RIN | $\frac{1}{16}\eta^2 P^2 \delta\theta$ |
| Thermal noise $\delta i^2_T$ | $\frac{16\pi^2}{\lambda^2}\eta^2 F^2 P_i^2 \frac{4k_B TBQ}{\omega_0 k}$ | $\frac{4\pi^2}{\lambda^2}\eta^2 P_d^2 \frac{4k_B TBQ}{\omega_0 k}$ | $\frac{\pi^2}{\lambda^2}\eta^2 P^2 \frac{4k_B TBQ}{\omega_0 k}$ |
| Shot noise $\delta i^2_S$ | $4e\eta P_d B$ | $2e\eta(P_R + P_S)B$ | $\frac{1}{2}e\eta PB$ |

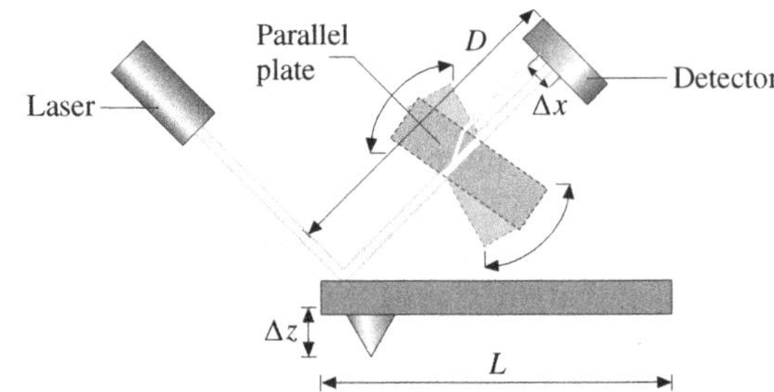


**Fig. 2.29** Set-up for an optical lever detection microscope

mirror and projected to a receiving target. Any change in the angular position of the mirror will change the position where the light ray hits the target. Galvanometers use optical path lengths of several meters and scales projected onto the target wall are also used to monitor changes in position.

In an AFM using the optical lever method, a photodiode segmented into two (or four) closely spaced devices detects the orientation of the end of the cantilever. Initially, the light ray is set to hit the photodiodes in the middle of the two subdiodes. Any deflection of the cantilever will cause an imbalance of the number of photons reaching the two halves. Hence the electrical currents in the photodiodes will be unbalanced too. The difference signal is further amplified and is the input signal to the feedback loop. Unlike the interferometeric AFMs, where a modulation technique is often necessary to get a sufficient signal-to-noise ratio, most AFMs employing the optical lever method are operated in a static mode. AFMs based on the optical lever method are universally used. It is the simplest method for constructing an optical readout and it can be confined in volumes that are smaller than 5 cm in side length.

The optical lever detection system is a simple yet elegant way to detect normal and lateral force signals simultaneously [7, 8, 53, 111]. It has the additional advantage that it is a remote detection system.

Implementations

Light from a laser diode or from a super luminescent diode is focused on the end of the cantilever. The reflected light is directed onto a quadrant diode that measures the direction of the light beam. A Gaussian light beam far from its waist is characterized by an opening angle $\beta$. The deflection of the light beam by the cantilever surface tilted by an angle $\alpha$ is $2\alpha$. The intensity on the detector then shifts to the side by the product of $2\alpha$ and the separation between the detector and the cantilever. The readout electronics calculates the difference in the photocurrents. The photocurrents, in turn, are proportional to the intensity incident on the diode.

The output signal is hence proportional to the change in intensity on the segments

$$I_{\text{sig}} \propto 4\frac{\alpha}{\beta} I_{\text{tot}}. \tag{2.47}$$

For the sake of simplicity, we assume that the light beam is of uniform intensity with its cross section increasing in proportion to the distance between the cantilever and the quadrant detector. The movement of the center of the light beam is then given by

$$\Delta x_{\text{Det}} = \Delta z \frac{D}{L}. \tag{2.48}$$

The photocurrent generated in a photodiode is proportional to the number of incoming photons hitting it. If the light beam contains a total number of $N_0$ photons, then the change in difference current becomes

$$\Delta(I_{\text{R}} - I_{\text{L}}) = \Delta I = \text{const}\ \Delta z\ D\ N_0. \tag{2.49}$$

Combining (2.48) and (2.49), one obtains that the difference current $\Delta I$ is independent of the separation of the quadrant detector and the cantilever. This relation is true if the light spot is smaller than the quadrant detector. If it is greater, the difference current $\Delta I$ becomes smaller with increasing distance. In reality, the light beam has a Gaussian intensity profile. For small movements $\Delta x$ (compared to the diameter of the light spot at the quadrant detector), (2.49) still holds. Larger movements $\Delta x$, however, will introduce a nonlinear response. If the AFM is operated in a constant force mode, only small movements $\Delta x$ of the light spot will occur. The feedback loop will cancel out all other movements.

The scanning of a sample with an AFM can twist the microfabricated cantilevers because of lateral forces [5, 7, 8] and affect the images [120]. When the tip is subjected to lateral forces, it will twist the cantilever and the light beam reflected from the end of the cantilever will be deflected perpendicular to the ordinary deflection direction. For many investigations this influence of lateral forces is unwanted. The design of the triangular cantilevers stems from the desire to minimize the torsion effects. However, lateral forces open up a new dimension in force measurements. They allow, for instance, two materials to be distinguished because of their different friction coefficients, or adhesion energies to be determined. To measure lateral forces, the original optical lever AFM must be modified. The only modification compared with Fig. 2.29 is the use of a quadrant detector photodiode instead of a two-segment photodiode and the necessary readout electronics (Fig. 2.9a). The electronics calculates the following signals

$$\begin{aligned} U_{\text{normal force}} &= \alpha[(I_{\text{upper left}} + I_{\text{upper right}}) - (I_{\text{lower left}} + I_{\text{lower right}})], \\ U_{\text{lateral force}} &= \beta[(I_{\text{upper left}} + I_{\text{lower left}}) - (I_{\text{upper right}} + I_{\text{lower right}})]. \end{aligned} \quad (2.50)$$

The calculation of the lateral force as a function of the deflection angle does not have a simple solution for cross sections other than circles. An approximate formula for the angle of twist for rectangular beams is [151]

$$\theta = \frac{M_t L}{\beta G b^3 h}, \quad (2.51)$$

where $M_t = F_y \ell$ is the external twisting moment due to lateral force $F_y$ and $\beta$ a constant determined by the value of $h/b$. For the equation to hold, $h$ has to be larger than $b$.

Inserting the values for a typical microfabricated cantilever with integrated tips

$$\begin{aligned} b &= 6 \times 10^{-7}\,\text{m}, \\ h &= 10^{-5}\,\text{m}, \\ L &= 10^{-4}\,\text{m}, \\ \ell &= 3.3 \times 10^{-6}\,\text{m}, \\ G &= 5 \times 10^{10}\,\text{Pa}, \\ \beta &= 0.333 \end{aligned} \quad (2.52)$$

into (2.51) we obtain the relation

$$F_y = 1.1 \times 10^{-4}\,\text{N} \times \theta. \quad (2.53)$$

Typical lateral forces are of the order of $10^{-10}$ N.

### Sensitivity

The sensitivity of this set-up has been calculated in various papers [116, 148, 152]. Assuming a Gaussian beam, the resulting output signal as a function of the deflection angle is dispersion-like. Equation (2.47) shows that the sensitivity can be increased by increasing the intensity of the light beam $I_{\text{tot}}$ or by decreasing the divergence of the laser beam. The upper bound of the intensity of the light $I_{\text{tot}}$ is given by saturation effects on the photodiode. If we decrease the divergence of a laser beam we automatically increase the beam waist. If the beam waist becomes larger than the width of the cantilever we start to get diffraction. Diffraction sets a lower bound on the divergence angle. Hence one can calculate the optimal beam waist $w_{\text{opt}}$ and the optimal divergence angle $\beta$ [148, 152]

$$\begin{aligned} w_{\text{opt}} &\approx 0.36b, \\ \theta_{\text{opt}} &\approx 0.89\frac{\lambda}{b}. \end{aligned} \tag{2.54}$$

The optimal sensitivity of the optical lever then becomes

$$\varepsilon[\text{mW/rad}] = 1.8\frac{b}{\lambda}I_{\text{tot}}[\text{mW}]. \tag{2.55}$$

The angular sensitivity of the optical lever can be measured by introducing a parallel plate into the beam. Tilting the parallel plate results in a displacement of the beam, mimicking an angular deflection.

Additional noise sources can be considered. Of little importance is the quantum mechanical uncertainty of the position [148, 152], which is, for typical cantilevers at room temperature

$$\Delta z = \sqrt{\frac{\hbar}{2m\omega_0}} = 0.05\ \text{fm}, \tag{2.56}$$

where $\hbar$ is the Planck constant ($=6.626\times10^{34}$ J s). At very low temperatures and for high-frequency cantilevers this could become the dominant noise source. A second noise source is the shot noise of the light. The shot noise is related to the particle number. We can calculate the number of photons incident on the detector using

$$n = \frac{I\tau}{\hbar\omega} = \frac{I\lambda}{2\pi B\hbar c} = 1.8\times10^{9}\frac{I[\text{W}]}{B[\text{Hz}]}, \tag{2.57}$$

where $I$ is the intensity of the light, $\tau$ the measurement time, $B=1/\tau$ the bandwidth, and $c$ the speed of light. The shot noise is proportional to the square root of the number of particles. Equating the shot noise signal with the signal resulting from the deflection of the cantilever one obtains

$$\Delta z_{\text{shot}} = 68\frac{L}{\omega}\sqrt{\frac{B[\text{kHz}]}{I[\text{mW}]}}[\text{fm}]. \tag{2.58}$$

where $w$ is the diameter of the focal spot. Typical AFM set-ups have a shot noise of 2 pm. The thermal noise can be calculated from the equipartition principle. The amplitude at the resonant frequency is

$$\Delta z_{\text{therm}} = 129\sqrt{\frac{B}{k[\text{N/m}]\omega_0 Q}}\,[\text{pm}]. \tag{2.59}$$

A typical value is 16 pm. Upon touching the surface, the cantilever increases its resonant frequency by a factor of 4.39. This results in a new thermal noise amplitude of 3.2 pm for the cantilever in contact with the sample.

### Piezoresistive Detection

#### Implementation

A piezoresistive cantilever is an alternative detection system which is not as widely used as the optical detection schemes [125, 126, 132]. This cantilever is based on the fact that the resistivities of certain materials, in particular Si, change with the applied stress. Figure 2.30 shows a typical implementation of a piezo-resistive cantilever. Four resistances are integrated on the chip, forming a Wheatstone bridge. Two of the resistors are in unstrained parts of the cantilever, and the other two measure the bending at the point of the maximal deflection. For instance, when an AC voltage is applied between terminals a and c, one can measure the detuning of the bridge between terminals b and d. With such a connection the output signal only varies due to bending, not due to changes in the ambient temperature and thus the coefficient of the piezoresistance.

#### Sensitivity

The resistance change is [126]

$$\frac{\Delta R}{R_0} = \Pi\delta. \tag{2.60}$$

where $\Pi$ is the tensor element of the piezo-resistive coefficients, $\delta$ the mechanical stress tensor element and $R_0$ the equilibrium resistance. For a single resistor, they

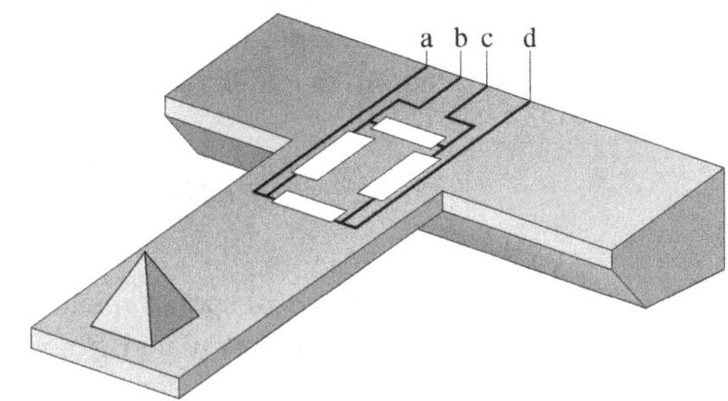


**Fig. 2.30** A typical set-up for a piezoresistive readout

separate the mechanical stress and the tensor element into longitudinal and transverse components

$$\frac{\Delta R}{R_0} = \Pi_t \delta_t + \Pi_l \delta_l. \tag{2.61}$$

The maximum values of the stress components are $\Pi_t = 64.0 \times 10^{11}$ m$^2$/N and $\Pi_l = 71.4 \times 10^{11}$ m$^2$/N for a resistor oriented along the (110) direction in silicon [126]. In the resistor arrangement of Fig. 2.30, two of the resistors are subject to the longitudinal piezo-resistive effect and two of them are subject to the transversal piezo-resistive effect. The sensitivity of that set-up is about four times that of a single resistor, with the advantage that temperature effects cancel to first order. The resistance change is then calculated as

$$\frac{\Delta R}{R_0} = \Pi \frac{3Eh}{2L^2} \Delta z = \Pi \frac{6L}{bh^2} F_z, \tag{2.62}$$

where $\Pi = 67.7 \times 10^{11}$ m$^2$/N is the averaged piezo-resistive coefficient. Plugging in typical values for the dimensions (Fig. 2.24) ($L = 100$ μm, $b = 10$ μm, $h = 1$ μm), one obtains

$$\frac{\Delta R}{R_0} = \frac{4 \times 10^{-5}}{\text{nN}} F_z. \tag{2.63}$$

The sensitivity can be tailored by optimizing the dimensions of the cantilever.

### Capacitance Detection

The capacitance of an arrangement of conductors depends on the geometry. Generally speaking, the capacitance increases for decreasing separations. Two parallel plates form a simple capacitor (Fig. 2.31, upper left), with capacitance

$$C = \frac{\varepsilon \varepsilon_0 A}{x}, \tag{2.64}$$

where $A$ is the area of the plates, assumed equal, and $x$ is the separation. Alternatively one can consider a sphere versus an infinite plane (Fig. 2.31, lower left). Here the capacitance is [116]

$$C = 4\pi\varepsilon_0 R \sum_{n=2}^{\infty} \frac{\sinh(\alpha)}{\sinh(n\alpha)} \tag{2.65}$$

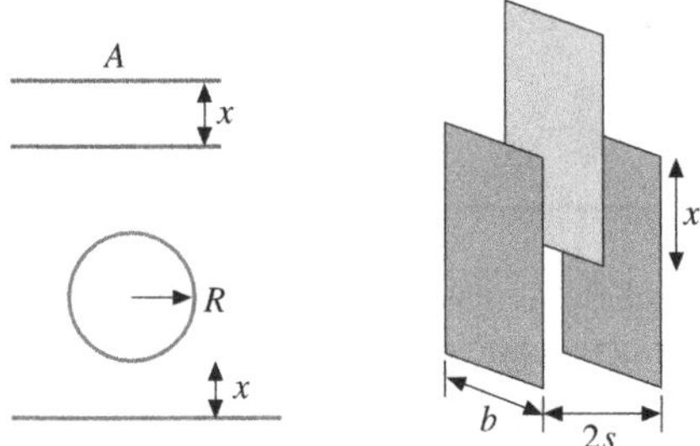


**Fig. 2.31** Three possible arrangements of a capacitive readout. The *upper left* diagram shows a cross section through a parallel plate capacitor. The *lower left* diagram shows the geometry of a sphere versus a plane. The *right-hand* diagram shows the linear (but more complicated) capacitive readout

**Fig. 2.32** Measuring the capacitance. (**a**) Low pass filter, (**b**) capacitive divider. $C$ (*left*) and $C_2$ (*right*) are the capacitances under test

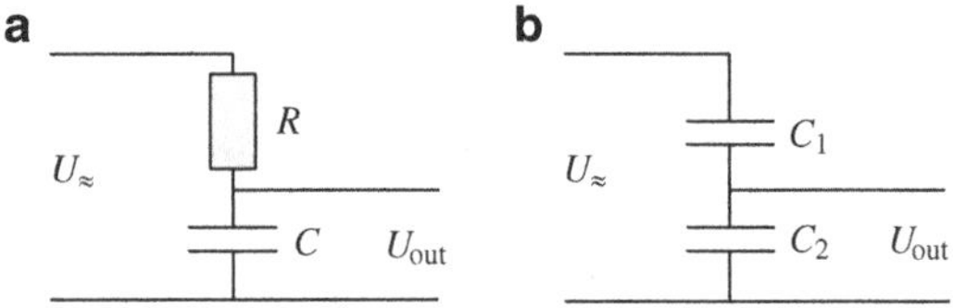


where $R$ is the radius of the sphere, and $\alpha$ is defined by

$$\alpha = \ln\left(1 + \frac{z}{R} + \sqrt{\frac{z^2}{R^2} + 2\frac{z}{R}}\right). \tag{2.66}$$

One has to bear in mind that the capacitance of a parallel plate capacitor is a nonlinear function of the separation. One can circumvent this problem using a voltage divider. Figure 2.32a shows a low-pass filter. The output voltage is given by

$$\begin{aligned} U_{\text{out}} &= U \approx \frac{\frac{1}{\mathrm{j}\omega C}}{R + \frac{1}{\mathrm{j}\omega C}} = U_{\approx} \frac{1}{\mathrm{j}\omega CR + 1} \\ &\cong \frac{U_{\approx}}{\mathrm{j}\omega CR}. \end{aligned} \tag{2.67}$$

Here $C$ is given by (2.64), $\omega$ is the excitation frequency and j is the imaginary unit. The approximate relation at the end is true when $\omega CR \gg 1$. This is equivalent to the statement that $C$ is fed by a current source, since $R$ must be large in this set-up. Plugging (2.64) into (2.67) and neglecting the phase information, one obtains

$$U_{\text{out}} = \frac{U_{\approx} x}{\omega R \varepsilon \varepsilon_0 A}, \tag{2.68}$$

which is linear in the displacement $x$.

Figure 2.32b shows a capacitive divider. Again the output voltage $U_{\text{out}}$ is given by

$$U_{\text{out}} = U_{\approx}\frac{C_1}{C_2 + C_1} = U_{\approx}\frac{C_1}{\frac{\varepsilon\varepsilon_0 A}{x} + C_1}. \tag{2.69}$$

If there is a stray capacitance $C_s$ then (2.69) is modified as

$$U_{\text{out}} = U_{\approx}\frac{C_1}{\frac{\varepsilon\varepsilon_0 A}{x} + C_s + C_1}. \tag{2.70}$$

Provided $C_s + C_1 \ll C_2$, one has a system which is linear in $x$. The driving voltage $U_{\approx}$ must be large (more than 100 V) to gave an output voltage in the range of 1 V. The linearity of the readout depends on the capacitance $C_1$ (Fig. 2.33).

Another idea is to keep the distance constant and to change the relative overlap of the plates (Fig. 2.31, right side). The capacitance of the moving center plate versus the stationary outer plates becomes

$$C = C_s + 2\frac{\varepsilon\varepsilon_0 bx}{s}, \tag{2.71}$$

where the variables are defined in Fig. 2.31. The stray capacitance comprises all effects, including the capacitance of the fringe fields. When the length $x$ is comparable to the width $b$ of the plates, one can safely assume that the stray capacitance is constant and independent of $x$. The main disadvantage of this set-up is that it is not as easily incorporated into a microfabricated device as the others.

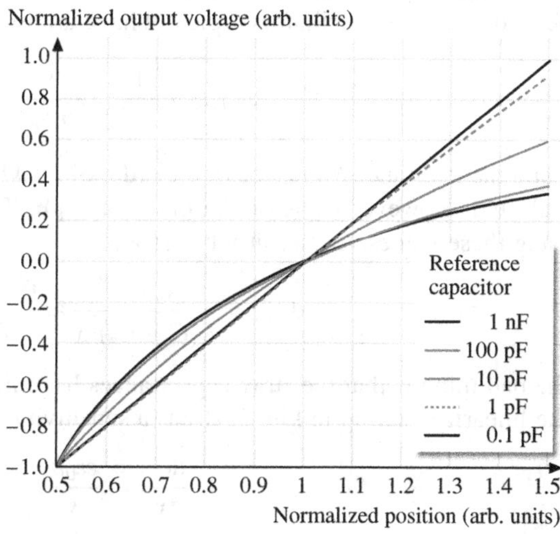


**Fig. 2.33** Linearity of the capacitance readout as a function of the reference capacitor

Sensitivity

The capacitance itself is not a measure of the sensitivity, but its derivative is indicative of the signals one can expect. Using the situation described in Fig. 2.31 (upper left) and in (2.64), one obtains for the parallel plate capacitor

$$\frac{\mathrm{d}C}{\mathrm{d}x} = -\frac{\varepsilon\varepsilon_0 A}{x^2}. \tag{2.72}$$

Assuming a plate area $A$ of 20 μm by 40 μm and a separation of 1 μm, one obtains a capacitance of 31 fF (neglecting stray capacitance and the capacitance of the connection leads) and a $\mathrm{d}C/\mathrm{d}x$ of $3.1\times10^{-8}$ F/m =31 fF/μm. Hence it is of paramount importance to maximize the area between the two contacts and to minimize the distance $x$. The latter however is far from being trivial. One has to go to the limits of microfabrication to achieve a decent sensitivity.

If the capacitance is measured by the circuit shown in Fig. 2.32, one obtains for the sensitivity

$$\frac{\mathrm{d}U_{\mathrm{out}}}{U_{\approx}} = \frac{\mathrm{d}x}{\omega R\varepsilon\varepsilon_0 A}. \tag{2.73}$$

Using the same value for $A$ as above, setting the reference frequency to 100 kHz, and selecting $R$ =1 GΩ, we get the relative change in the output voltage $U_{\mathrm{out}}$ as

$$\frac{\mathrm{d}U_{\mathrm{out}}}{U_{\approx}} = \frac{22.5\times10^{-6}}{\text{Å}} \times \mathrm{d}x. \tag{2.74}$$

A driving voltage of 45 V then translates to a sensitivity of 1 mV/Å. A problem in this set-up is the stray capacitances. They are in parallel to the original capacitance and decrease the sensitivity considerably.

Alternatively, one could build an oscillator with this capacitance and measure the frequency. $RC$-oscillators typically have an oscillation frequency of

$$f_{\mathrm{res}} \propto \frac{1}{RC} = \frac{x}{R\varepsilon\varepsilon_0 A}. \tag{2.75}$$

Again the resistance $R$ must be of the order of 1 GΩ when stray capacitances $C_{\mathrm{s}}$ are neglected. However $C_{\mathrm{s}}$ is of the order of 1 pF. Therefore one gets $R$ =10 MΩ. Using these values, the sensitivity becomes

$$\mathrm{d}f_{\mathrm{res}} = \frac{C\,\mathrm{d}x}{R(C+C_{\mathrm{s}})^2 x} \approx \frac{0.1\,\mathrm{Hz}}{\text{Å}}\mathrm{d}x. \tag{2.76}$$

The bad thing is that the stray capacitances have made the signal nonlinear again. The linearized set-up in Fig. 2.31 has a sensitivity of

$$\frac{\mathrm{d}C}{\mathrm{d}x} = 2\frac{\varepsilon\varepsilon_0 b}{s}. \tag{2.77}$$

Substituting typical values ($b$ =10 μm, $s$ =1 μm), one gets $\mathrm{d}C/\mathrm{d}x = 1.8\times10^{10}$ F/m. It is noteworthy that the sensitivity remains constant for scaled devices.

Implementations

Capacitance readout can be achieved in different ways [123, 124]. All include an alternating current or voltage with frequencies in the 100 kHz to 100 MHz range. One possibility is to build a tuned circuit with the capacitance of the cantilever determining the frequency. The resonance frequency of a high-quality $Q$ tuned circuit is

$$\omega_0 = (LC)^{-1/2}. \tag{2.78}$$

where $L$ is the inductance of the circuit. The capacitance $C$ includes not only the sensor capacitance but also the capacitance of the leads. The precision of a frequency measurement is mainly determined by the ratio of $L$ and $C$

$$Q = \left(\frac{L}{C}\right)^{1/2} \frac{1}{R}. \tag{2.79}$$

Here $R$ symbolizes the losses in the circuit. The higher the quality, the more precise the frequency measurement. For instance, a frequency of 100 MHz and a capacitance of 1 pF gives an inductance of 250 μH. The quality then becomes $2.5\times10^8$. This value is an upper limit, since losses are usually too high.

Using a value of $\mathrm{d}C/\mathrm{d}x$ =31 fF/μm, one gets $\Delta C$/Å =3.1 aF/Å. With a capacitance of 1 pF, one gets

$$\begin{aligned} \frac{\Delta\omega}{\omega} &= \frac{1}{2}\frac{\Delta C}{C}, \\ \Delta\omega &= 100\,\mathrm{MHz} \times \frac{1}{2}\frac{3.1\mathrm{aF}}{1\,\mathrm{pF}} = 155\,\mathrm{Hz}. \end{aligned} \tag{2.80}$$

This is the frequency shift for a deflection of 1 Å. The calculation shows that this is a measurable quantity. The quality also indicates that there is no physical reason why this scheme should not work.

### 2.3.3 Combinations for 3-D Force Measurements

Three-dimensional force measurements are essential if one wants to know all of the details of the interaction between the tip and the cantilever. The straightforward attempt to measure three forces is complicated, since force sensors such as interferometers or capacitive sensors need a minimal detection volume, which is often too large. The second problem is that the force-sensing tip has to be held in some way. This implies that one of the three Cartesian axes is stiffer than the others.

However, by combining different sensors it is possible to achieve this goal. Straight cantilevers are employed for these measurements, because they can be handled analytically. The key observation is that the optical lever method does not determine the position of the end of the cantilever. It measures the orientation. In the previous sections, one has always made use of the fact that, for a force along one of the orthogonal symmetry directions at the end of the cantilever (normal force, lateral force, force along the cantilever beam axis), there is a one-to-one correspondence of the tilt angle and the deflection. The problem is that the force along the cantilever beam axis and the normal force create a deflection in the same direction. Hence, what is called the normal force component is actually a mixture of two forces. The deflection of the cantilever is the third quantity, which is not considered in most of the AFMs. A fiber-optic interferometer in parallel with the optical lever measures the deflection. Three measured quantities then allow the separation of the three orthonormal force directions, as is evident from (2.27) and (2.33) [12, 13, 14, 15, 16].

Alternatively, one can put the fast scanning direction along the axis of the cantilever. Forward and backward scans then exert opposite forces $F_x$. If the piezo movement is linearized, both force components in AFM based on optical lever detection can be determined. In this case, the normal force is simply the average of the forces in the forward and backward direction. The force $F_x$ is the difference in the forces measured in the forward and backward directions.

### 2.3.4 Scanning and Control Systems

Almost all SPMs use piezo translators to scan the tip or the sample. Even the first STM [1, 103] and some of its predecessors [153, 154] used them. Other materials or set-ups for nanopositioning have been proposed, but they have not been successful [155, 156].

#### Piezo Tubes

A popular solution is tube scanners (Fig. 2.34). They are now widely used in SPMs due to their simplicity and their small size [133, 157]. The outer electrode is segmented into four equal sectors of 90°. Opposite sectors are driven by signals of the same magnitude, but opposite sign. This gives, through bending, two-dimensional movement on (approximately) a sphere. The inner electrode is normally driven by the $z$-signal. It is possible, however, to use only the outer electrodes for scanning and for the $z$-movement. The main drawback of applying the $z$-signal to the outer electrodes is that the applied voltage is the sum of both the $x$- or $y$-movements and the $z$-movement. Hence a larger scan size effectively reduces the available range for the $z$-control.

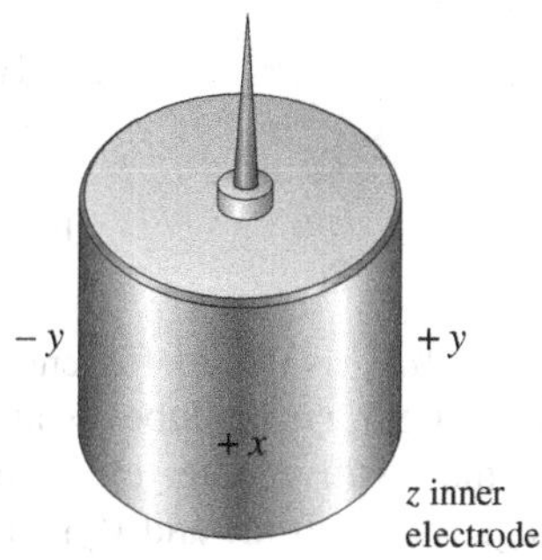


**Fig. 2.34** Schematic drawing of a piezoelectric tube scanner. The piezo ceramic is molded into a tube form. The outer electrode is separated into four segments and connected to the scanning voltage. The $z$-voltage is applied to the inner electrode

## Piezo Effect

An electric field applied across a piezoelectric material causes a change in the crystal structure, with expansion in some directions and contraction in others. Also, a net volume change occurs [132]. Many SPMs use the transverse piezo electric effect, where the applied electric field $\boldsymbol{E}$ is perpendicular to the expansion/contraction direction.

$$\Delta L = L(\boldsymbol{E} \cdot \boldsymbol{n})d_{31} = L\frac{V}{t}d_{31}, \tag{2.81}$$

where $d_{31}$ is the transverse piezoelectric constant, $V$ is the applied voltage, $t$ is the thickness of the piezo slab or the distance between the electrodes where the voltage is applied, $L$ is the free length of the piezo slab, and $\boldsymbol{n}$ is the direction of polarization. Piezo translators based on the transverse piezoelectric effect have a wide range of sensitivities, limited mainly by mechanical stability and breakdown voltage.

## Scan Range

The scanning range of a piezotube is difficult to calculate [157, 158, 159]. The bending of the tube depends on the electric fields and the nonuniform strain induced. A finite element calculation where the piezo tube was divided into 218 identical elements was used [158] to calculate the deflection. On each node, the mechanical stress, the stiffness, the strain and the piezoelectric stress were calculated when a voltage was applied on one electrode. The results were found to be linear on the first iteration and higher order corrections were very small even for large electrode voltages. It was found that, to first order, the $x$- and $z$-movement of the tube could be reasonably well approximated by assuming that the piezo tube is a segment of a torus. Using this model, one obtains

$$\mathrm{d}x = (V_+ - V_-)|d_{31}|\frac{L^2}{2td}, \tag{2.82}$$

$$\mathrm{d}z = (V_+ + V_- - 2V_z)|d_{31}|\frac{L}{2t}, \tag{2.83}$$

where $|d_{31}|$ is the coefficient of the transversal piezoelectric effect, $L$ is the tube's free length, $t$ is the tube's wall thickness, $d$ is the tube's diameter, $V_+$ is the voltage on the positive outer electrode, while $V$ is the voltage of the opposite quadrant negative electrode and $V_z$ is the voltage of the inner electrode.

The cantilever or sample mounted on the piezotube has an additional lateral movement because the point of measurement is not in the endplane of the piezotube. The additional lateral displacement of the end of the tip is $\ell \sin\varphi \approx \ell\varphi$, where  is the tip length and $\varphi$ is the deflection angle of the end surface. Assuming that the sample or cantilever is always perpendicular to the end of the walls of the tube, and calculating with the torus  model, one gets for the angle

$$\varphi = \frac{L}{R} = \frac{2dx}{L}, \tag{2.84}$$

where $R$ is the radius of curvature of the piezo tube. Using the result of (2.84), one obtains for the additional $x$-movement

$$\mathrm{d}x_{\mathrm{add}} = \ell\varphi = \frac{2\mathrm{d}x\ell}{L} = (V_+ - V_-)|d_{31}|\frac{\ell L}{td} \tag{2.85}$$

and for the additional $z$-movement due to the $x$-movement

$$\mathrm{d}z_{\mathrm{add}} = \ell - \ell\cos\varphi = \frac{\ell\varphi^2}{2} = \frac{2\ell(\mathrm{d}x)^2}{L^2} = (V_+ - V_-)^2|d_{31}|^2\frac{\ell L^2}{2t^2d^2}. \tag{2.86}$$

*Carr* [158] assumed for his finite element calculations that the top of the tube was completely free to move and, as a consequence, the top surface was distorted, leading to a deflection angle that was about half that of the geometrical model. Depending on the attachment of the sample or the cantilever, this distortion may be smaller, leading to a deflection angle in-between that of the geometrical model and the one from the finite element calculation.

### Nonlinearities and Creep

Piezo materials with a high conversion ratio (a large $d_{31}$ or small electrode separations with large scanning ranges) are hampered by substantial hysteresis  resulting in a deviation from linearity by more than 10%. The sensitivity of the piezo ceramic material (mechanical displacement divided by driving voltage) decreases with reduced scanning range, whereas the hysteresis is reduced. Careful selection of the material used for the piezo scanners, the design of the scanners, and of the operating conditions is necessary to obtain optimum performance.

### Passive Linearization: Calculation

The analysis of images affected by piezo nonlinearities [160, 161, 162, 163] shows that the dominant term is

$$x = AV + BV^2, \tag{2.87}$$

where $x$ is the excursion of the piezo, $V$ is the applied voltage and $A$ and $B$ are two coefficients describing the sensitivity of the material. Equation (2.87) holds for scanning from $V = 0$ to large $V$. For the reverse direction, the equation becomes

$$x = \tilde{A}V - \tilde{B}(V - V_{\max})^2, \tag{2.88}$$

where $\tilde{A}$ and $\tilde{B}$ are the coefficients for the back scan and $V_{\max}$ is the applied voltage at the turning point. Both equations demonstrate that the true $x$-travel is small at the beginning of the scan and becomes larger towards the end. Therefore, images are stretched at the beginning and compressed at the end.

Similar equations hold for the slow scan direction. The coefficients, however, are different. The combined action causes a greatly distorted image. This distortion can be calculated. The data acquisition systems record the signal as a function of $V$. However the data is measured as a function of $x$. Therefore we have to distribute the $x$-values evenly across the image. This can be done by inverting an approximation of (2.87). First we write

$$x = AV\left(1 - \frac{B}{A}V\right). \tag{2.89}$$

For $B \ll A$ we can approximate

$$V = \frac{x}{A}. \tag{2.90}$$

We now substitute (2.90) into the nonlinear term of (2.89). This gives

$$\begin{aligned} x &= AV\left(1 + \frac{Bx}{A^2}\right), \\ V &= \frac{x}{A}\frac{1}{(1 + Bx/A^2)} \approx \frac{x}{A}\left(1 - \frac{Bx}{A^2}\right). \end{aligned} \tag{2.91}$$

Hence an equation of the type

$$\begin{aligned} x_{\text{true}} &= x(\alpha - \beta x / x_{\max}) \\ \text{with } 1 &= \alpha - \beta \end{aligned} \tag{2.92}$$

takes out the distortion of an image. $\alpha$ and $\beta$ are dependent on the scan range, the scan speed and on the scan history, and have to be determined with exactly the same settings as for the measurement. $x_{\max}$ is the maximal scanning range. The condition for $\alpha$ and $\beta$ guarantees that the image is transformed onto itself.

Similar equations to the empirical one shown above (2.92) can be derived by analyzing the movements of domain walls in piezo ceramics.

### Passive Linearization: Measuring the Position

An alternative strategy is to measure the positions of the piezo translators. Several possibilities exist.

1. The interferometers described above can be used to measure the elongation of the piezo elongation. The fiber-optic interferometer is especially easy to implement. The coherence length of the laser only limits the measurement range. However, the signal is of a periodic nature. Hence direct use of the signal in a feedback circuit for the position is not possible. However, as a measurement tool and, especially, as a calibration tool, the interferometer is without competition. The wavelength of the light, for instance that in a He-Ne laser, is so well defined that the precision of the other components determines the error of the calibration or measurement.
2. The movement of the light spot on the quadrant detector can be used to measure the position of a piezo [164]. The output current changes by 0.5 A/cm $\times P$(W)/$R$ (cm). Typical values ($P$ =1 mW, $R$ =0.001 cm) give 0.5 A/cm. The noise limit is typically 0.15 nm $\times \sqrt{\Delta f(\text{Hz})/H(\text{W/cm}^2)}$. Again this means that the laser beam above would have a 0.1 nm noise limitation for a bandwidth of 21 Hz. The advantage of this method is that, in principle, one can linearize two axes with only one detector.
3. A knife-edge blocking part of a light beam incident on a photodiode can be used to measure the position of the piezo. This technique, commonly used in optical shear force detection [75, 165], has a sensitivity of better than 0.1 nm.
4. The capacitive detection [166, 167] of the cantilever deflection can be applied to the measurement of the piezo elongation. Equations (2.64)–(2.79) apply to the problem. This technique is used in some commercial instruments. The difficulties lie in the avoidance of fringe effects at the borders of the two plates. While conceptually simple, one needs the latest technology in surface preparation to get a decent linearity. The electronic circuits used for the readout are often proprietary.
5. Linear variable differential transformers (LVDT) are a convenient way to measure positions down to 1 nm. They can be used together with a solid state joint set-up, as often used for large scan range stages. Unlike capacitive

detection, there are few difficulties in implementation. The sensors and the detection circuits LVDTs are available commercially.

6. A popular measurement technique is the use of strain gauges. They are especially sensitive when mounted on a solid state joint where the curvature is maximal. The resolution depends mainly on the induced curvature. A precision of 1 nm is attainable. The signals are low – a Wheatstone bridge is needed for the readout.

#### Active Linearization

Active linearization is done with feedback systems. Sensors need to be monotonic. Hence all of the systems described above, with the exception of the interferometers, are suitable. The most common solutions include the strain gauge approach, capacitance measurement or the LVDT, which are all electronic solutions. Optical detection systems have the disadvantage that the intensity enters into the calibration.

### Alternative Scanning Systems

The first STMs were based on piezo tripods [1]. The piezo tripod (Fig. 2.35) is an intuitive way to generate the three-dimensional movement of a tip attached to its center. However, to get a suitable stability and scanning range, the tripod needs to be fairly large (about 50 mm). Some instruments use piezo stacks instead of monolithic piezoactuators. They are arranged in a tripod. Piezo stacks are thin layers of piezoactive materials glued together to form a device with up to 200 μm of actuation range. Preloading with a suitable metal casing reduces the nonlinearity.

If one tries to construct a homebuilt scanning system, the use of linearized scanning tables is recommended. They are built around solid state joints and actuated by piezo stacks. The joints guarantee that the movement is parallel with little deviation from the predefined scanning plane. Due to the construction it is easy to add measurement devices such as capacitive sensors, LVDTs or strain gauges, which are essential for a closed loop linearization. Two-dimensional tables

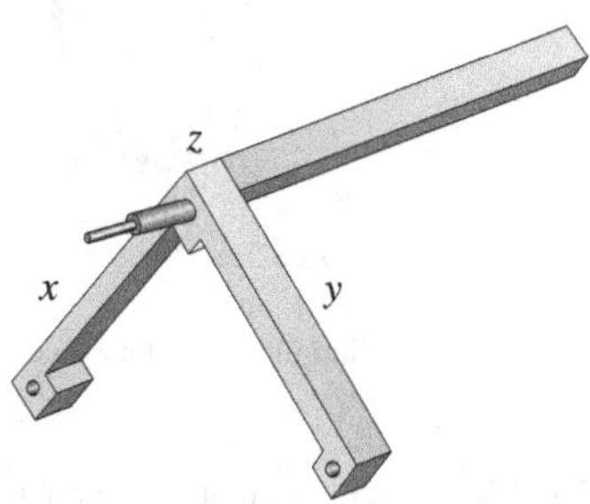


**Fig. 2.35** An alternative type of piezo scanner: the tripod

can be bought from several manufacturers. They have linearities of better than 0.1% and a noise level of $10^{-4} - 10^{-5}$ for the maximal scanning range.

### Control Systems

#### Basics

The electronics and software play an important role in the optimal performance of an SPM. Control electronics and software are supplied with commercial SPMs. Electronic control systems can use either analog or digital feedback. While digital feedback offers greater flexibility and ease of configuration, analog feedback circuits might be better suited for ultralow noise operation. We will describe here the basic set-ups for AFMs.

Figure 2.36 shows a block schematic of a typical AFM feedback loop. The signal from the force transducer is fed into the feedback loop, which consists mainly of a subtraction stage to get an error signal and an integrator. The gain of the integrator (high gain corresponds to short integration times) is set as high as possible without generating more than 1% overshoot. High gain minimizes the error margin of the current and forces the tip to follow the contours of constant density of states as well as possible. This operating mode is known as constant force mode. A high-voltage amplifier amplifies the outputs of the integrator. As AFMs using piezotubes usually require ±150 V at the output, the output of the integrator needs to be amplified by a high-voltage amplifier.

In order to scan the sample, additional voltages at high tension are required to drive the piezo. For example, with a tube scanner, four scanning voltages are required, namely $+V_x$, $-V_x$, $+V_y$ and $-V_y$. The $x$- and $y$-scanning voltages are generated in a scan generator (analog or computer-controlled). Both voltages are input to the two respective power amplifiers. Two inverting amplifiers generate the input voltages for the other two power amplifiers. The topography of the sample

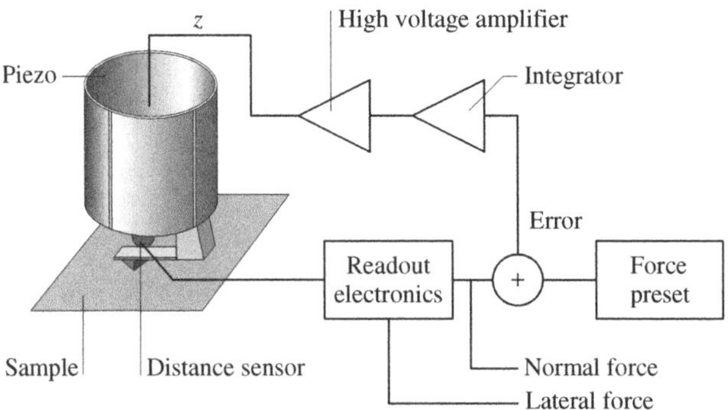


**Fig. 2.36** Block schematic of the feedback control loop of an AFM

surface is determined by recording the input voltage to the high-voltage amplifier for the $z$-channel as a function of $x$ and $y$ (constant force mode).

Another operating mode is the variable force mode. The gain in the feedback loop is lowered and the scanning speed increased such that the force on the cantilever is no longer constant. Here the force is recorded as a function of $x$ and $y$.

### Force Spectroscopy

Four modes of spectroscopic imaging are in common use with force microscopes: measuring lateral forces, $\partial F/\partial z$, $\partial F/\partial x$ spatially resolved, and measuring force versus distance curves. Lateral forces can be measured by detecting the deflection of a cantilever in a direction orthogonal to the normal direction. The optical lever deflection method does this most easily. Lateral force measurements give indications of adhesion forces between the tip and the sample.

$\partial F/\partial z$ measurements probe the local elasticity of the sample surface. In many cases the measured quantity originates from a volume of a few cubic nanometers. The $\partial F/\partial z$ or local stiffness signal is proportional to Young's modulus, as far as one can define this quantity. Local stiffness is measured by vibrating the cantilever by a small amount in the $z$-direction. The expected signal for very stiff samples is zero: for very soft samples one also gets, independent of the stiffness, a constant signal. This signal is again zero for the optical lever deflection and equal to the driving amplitude for interferometric measurements. The best sensitivity is obtained when the compliance of the cantilever matches the stiffness of the sample.

A third spectroscopic quantity is the lateral stiffness. It is measured by applying a small modulation in the $x$-direction on the cantilever. The signal is again optimal when the lateral compliance of the cantilever matches the lateral stiffness of the sample. The lateral stiffness is, in turn, related to the shear modulus of the sample.

Detailed information on the interaction of the tip and the sample can be gained by measuring force versus distance curves. The cantilevers need to have enough compliance to avoid instabilities due to the attractive forces on the sample.

### Using the Control Electronics as a Two-Dimensional Measurement Tool

Usually the control electronics of an AFM is used to control the $x$- and $y$-piezo signals while several data acquisition channels record the position-dependent signals. The control electronics can be used in another way: they can be viewed as a two-dimensional function generator. What is normally the $x$- and $y$-signal can be used to control two independent variables of an experiment. The control logic of the AFM then ensures that the available parameter space is systematically probed at equally spaced points. An example is friction force curves measured along a line across a step on graphite.

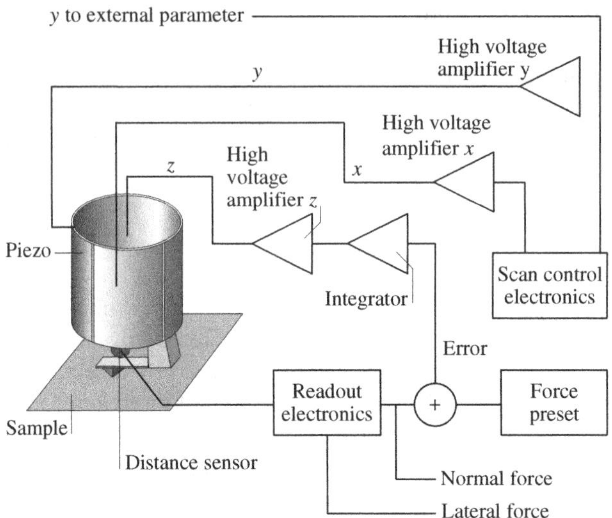


**Fig. 2.37** Wiring of an AFM to measure friction force curves along a line

Figure 2.37 shows the connections. The $z$-piezo is connected as usual, like the $x$-piezo. However, the $y$-output is used to command the desired input parameter. The offset of the $y$-channel determines the position of the tip on the sample surface, together with the $x$-channel.

## Some Imaging Processing Methods

The visualization and interpretation of images from AFMs is intimately connected to the processing of these images. An ideal AFM is a noise-free device that images a sample with perfect tips of known shape and has perfect linear scanning piezos. In reality, AFMs are not that ideal. The scanning device in an AFM is affected by distortions. The distortions are both linear and nonlinear. Linear distortions mainly result from imperfections in the machining of the piezotranslators causing crosstalk from the $z$-piezo to the $x$- and $y$-piezos, and vice versa. Among the linear distortions, there are two kinds which are very important. First, scanning piezos invariably have different sensitivities along the different scan axes due to variations in the piezo material and uneven electrode areas. Second, the same reasons might cause the scanning axes to be nonorthogonal. Furthermore, the plane in which the piezoscanner moves for constant height $z$ is hardly ever coincident with the sample plane. Hence, a linear ramp is added to the sample data. This ramp is especially bothersome when the height $z$ is displayed as an intensity map.

The nonlinear distortions are harder to deal with. They can affect AFM data for a variety of reasons. First, piezoelectric ceramics do have a hysteresis loop, much like ferromagnetic materials. The deviations of piezoceramic materials from linearity increase with increasing amplitude of the driving voltage. The mechanical position for one voltage depends on the previously applied voltages to the piezo. Hence, to get the best positional accuracy, one should always approach a point on the sample from the same direction. Another type of nonlinear distortion of images occurs when the scan frequency approaches the upper frequency limits of the $x$- and $y$-drive amplifiers or the upper frequency limit of the feedback loop ($z$-component). This distortion, due to the feedback loop, can only be minimized by reducing the scan frequency. On the other hand, there is a simple way to reduce distortions due to the $x$- and $y$-piezo drive amplifiers. To keep the system as simple as possible, one normally uses a triangular waveform to drive the scanning piezos. However, triangular waves contain frequency components as multiples of the scan frequency. If the cut-off frequencies of the $x$- and $y$-drive electronics or of the feedback loop are too close to the scanning frequency (two or three times the scanning frequency), the triangular drive voltage is rounded off at the turning points. This rounding error causes, first, a distortion of the scan linearity and, second, through phase lags, the projection of part of the backward scan onto the forward scan. This type of distortion can be minimized by carefully selecting the scanning frequency and by using driving voltages for the $x$- and $y$-piezos with waveforms like trapezoidal waves, which are closer to a sine wave. The values measured for $x$-, $y$- or $z$-piezos are affected by noise. The origin of this noise can be either electronic, disturbances, or a property of the sample surface due to adsorbates. In addition to this incoherent noise, interference with main and other equipment nearby might be present. Depending on the type of noise, one can filter it in real space or in Fourier space. The most important part of image processing is to visualize the measured data. Typical AFM data sets can consist of many thousands to over a million points per plane. There may be more than one image plane present. The AFM data represents a topography in various data spaces.

Most commercial data acquisition systems implicitly use some kind of data processing. Since the original data is commonly subject to slopes on the surface, most programs use some kind of slope correction. The least disturbing way is to subtract a plane $z(x, y) = Ax + By + C$ from the data. The coefficients are determined by fitting $z(x, y)$ to the data. Another operation is to subtract a second-order function such as $z(x, y) = Ax^2 + By^2 + Cxy + Dx + Ey + F$. Again, the parameters are determined with a fit. This function is appropriate for almost planar data, where the nonlinearity of the piezos caused the distortion.

In the image processing software from Digital Instruments, up to three operations are performed on the raw data. First, a zero-order flatten is applied. The flatten operation is used to eliminate image bow in the slow scan direction (caused by a physical bow in the instrument itself), slope in the slow scan direction, and bands in the image (caused by differences in the scan height from one scan line to the next). The flattening operation takes each scan line and subtracts the average value of the height along each scan line from each point in that scan line. This brings each scan line to the same height. Next, a first-order plane fit is applied in the fast scan direction. The plane-fit operation is used to eliminate bow and slope in the fast scan direction. The plane fit operation calculates a best fit plane for the image and subtracts it from the image. This plane has a constant nonzero slope in the fast scan direction. In some cases a higher order polynomial *plane* may be required. Depending upon the quality of the raw data, the flattening operation and/or the plane fit operation may not be required at all.

## References


1. G. Binnig, H. Rohrer, C. Gerber, E. Weibel, Surface studies by scanning tunneling microscopy, Phys. Rev. Lett. **49**, 57–61 (1982)
2. G. Binnig, C.F. Quate, C. Gerber, Atomic force microscope, Phys. Rev. Lett. **56**, 930–933 (1986)
3. G. Binnig, C. Gerber, E. Stoll, T.R. Albrecht, C.F. Quate, Atomic resolution with atomic force microscope, Europhys. Lett. **3**, 1281–1286 (1987)
4. B. Bhushan, *Handbook of Micro-/Nanotribology*, 2nd edn. (CRC, Boca Raton 1999)
5. C.M. Mate, G.M. McClelland, R. Erlandsson, S. Chiang, Atomic-scale friction of a tungsten tip on a graphite surface, Phys. Rev. Lett. **59**, 1942–1945 (1987)
6. R. Erlandsson, G.M. McClelland, C.M. Mate, S. Chiang, Atomic force microscopy using optical interferometry, J. Vac. Sci. Technol. A **6**, 266–270 (1988)
7. O. Marti, J. Colchero, J. Mlynek, Combined scanning force and friction microscopy of mica, Nanotechnology **1**, 141–144 (1990)
8. G. Meyer, N.M. Amer, Simultaneous measurement of lateral and normal forces with an optical-beam-deflection atomic force microscope, Appl. Phys. Lett. **57**, 2089–2091 (1990)
9. B. Bhushan, J. Ruan, Atomic-scale friction measurements using friction force microscopy: Part II – Application to magnetic media, ASME J. Tribol. **116**, 389–396 (1994)
10. B. Bhushan, V.N. Koinkar, J. Ruan, Microtribology of magnetic media, Proc. Inst. Mech. Eng. Part J **208**, 17–29 (1994)
11. B. Bhushan, J.N. Israelachvili, U. Landman, Nanotribology: Friction, wear, and lubrication at the atomic scale, Nature **374**, 607–616 (1995)
12. S. Fujisawa, M. Ohta, T. Konishi, Y. Sugawara, S. Morita, Difference between the forces measured by an optical lever deflection and by an optical interferometer in an atomic force microscope, Rev. Sci. Instrum. **65**, 644–647 (1994)
13. S. Fujisawa, E. Kishi, Y. Sugawara, S. Morita, Fluctuation in 2-dimensional stick-slip phenomenon observed with 2-dimensional frictional force microscope, Jpn. J. Appl. Phys. **33**, 3752–3755 (1994)
14. S. Grafstrom, J. Ackermann, T. Hagen, R. Neumann, O. Probst, Analysis of lateral force effects on the topography in scanning force microscopy, J. Vac. Sci. Technol. B **12**, 1559–1564 (1994)
15. R.M. Overney, H. Takano, M. Fujihira, W. Paulus, H. Ringsdorf, Anisotropy in friction and molecular stick-slip motion, Phys. Rev. Lett. **72**, 3546–3549 (1994)
16. R.J. Warmack, X.Y. Zheng, T. Thundat, D.P. Allison, Friction effects in the deflection of atomic force microscope cantilevers, Rev. Sci. Instrum. **65**, 394–399 (1994)
17. N.A. Burnham, D.D. Domiguez, R.L. Mowery, R.J. Colton, Probing the surface forces of monolayer films with an atomic force microscope, Phys. Rev. Lett. **64**, 1931–1934 (1990)
18. N.A. Burham, R.J. Colton, H.M. Pollock, Interpretation issues in force microscopy, J. Vac. Sci. Technol. A **9**, 2548–2556 (1991)
19. C.D. Frisbie, L.F. Rozsnyai, A. Noy, M.S. Wrighton, C.M. Lieber, Functional group imaging by chemical force microscopy, Science **265**, 2071–2074 (1994)
20. V.N. Koinkar, B. Bhushan, Microtribological studies of unlubricated and lubricated surfaces using atomic force/friction force microscopy, J. Vac. Sci. Technol. A **14**, 2378–2391 (1996)

21. V. Scherer, B. Bhushan, U. Rabe, W. Arnold, Local elasticity and lubrication measurements using atomic force and friction force microscopy at ultrasonic frequencies, IEEE Trans. Magn. **33**, 4077–4079 (1997)
22. V. Scherer, W. Arnold, B. Bhushan, Lateral force microscopy using acoustic friction force microscopy, Surf. Interface Anal. **27**, 578–587 (1999)
23. B. Bhushan, S. Sundararajan, Micro-/nanoscale friction and wear mechanisms of thin films using atomic force and friction force microscopy, Acta Mater. **46**, 3793–3804 (1998)
24. U. Krotil, T. Stifter, H. Waschipky, K. Weishaupt, S. Hild, O. Marti, Pulse force mode: A new method for the investigation of surface properties, Surf. Interface Anal. **27**, 336–340 (1999)
25. B. Bhushan, C. Dandavate, Thin-film friction and adhesion studies using atomic force microscopy, J. Appl. Phys. **87**, 1201–1210 (2000)
26. B. Bhushan, *Micro-/Nanotribology and Its Applications* (Kluwer, Dordrecht 1997)
27. B. Bhushan, *Principles and Applications of Tribology* (Wiley, New York 1999)
28. B. Bhushan, *Modern Tribology Handbook – Vol. 1: Principles of Tribology* (CRC, Boca Raton 2001)
29. B. Bhushan, *Introduction to Tribology* (Wiley, New York 2002)
30. M. Reinstädtler, U. Rabe, V. Scherer, U. Hartmann, A. Goldade, B. Bhushan, W. Arnold, On the nanoscale measurement of friction using atomic force microscope cantilever torsional resonances, Appl. Phys. Lett. **82**, 2604–2606 (2003)
31. N.A. Burnham, R.J. Colton, Measuring the nanomechanical properties and surface forces of materials using an atomic force microscope, J. Vac. Sci. Technol. A **7**, 2906–2913 (1989)
32. P. Maivald, H.J. Butt, S.A.C. Gould, C.B. Prater, B. Drake, J.A. Gurley, V.B. Elings, P.K. Hansma, Using force modulation to image surface elasticities with the atomic force microscope, Nanotechnology **2**, 103–106 (1991)
33. B. Bhushan, A.V. Kulkarni, W. Bonin, J.T. Wyrobek, Nano/picoindentation measurements using capacitive transducer in atomic force microscopy, Philos. Mag. A **74**, 1117–1128 (1996)
34. B. Bhushan, V.N. Koinkar, Nanoindentation hardness measurements using atomic force microscopy, Appl. Phys. Lett. **75**, 5741–5746 (1994)
35. D. DeVecchio, B. Bhushan, Localized surface elasticity measurements using an atomic force microscope, Rev. Sci. Instrum. **68**, 4498–4505 (1997)
36. S. Amelio, A.V. Goldade, U. Rabe, V. Scherer, B. Bhushan, W. Arnold, Measurements of mechanical properties of ultra-thin diamond-like carbon coatings using atomic force acoustic microscopy, Thin Solid Films **392**, 75–84 (2001)
37. D.M. Eigler, E.K. Schweizer, Positioning single atoms with a scanning tunnelling microscope, Nature **344**, 524–528 (1990)
38. A.L. Weisenhorn, J.E. MacDougall, J.A.C. Gould, S.D. Cox, W.S. Wise, J. Massie, P. Maivald, V.B. Elings, G.D. Stucky, P.K. Hansma, Imaging and manipulating of molecules on a zeolite surface with an atomic force microscope, Science **247**, 1330–1333 (1990)
39. I.W. Lyo, P. Avouris, Field-induced nanometer-to-atomic-scale manipulation of silicon surfaces with the STM, Science **253**, 173–176 (1991)
40. O.M. Leung, M.C. Goh, Orientation ordering of polymers by atomic force microscope tip-surface interactions, Science **225**, 64–66 (1992)
41. D.W. Abraham, H.J. Mamin, E. Ganz, J. Clark, Surface modification with the scanning tunneling microscope, IBM J. Res. Dev. **30**, 492–499 (1986)
42. R.M. Silver, E.E. Ehrichs, A.L. de Lozanne, Direct writing of submicron metallic features with a scanning tunnelling microscope, Appl. Phys. Lett. **51**, 247–249 (1987)
43. A. Kobayashi, F. Grey, R.S. Williams, M. Ano, Formation of nanometer-scale grooves in silicon with a scanning tunneling microscope, Science **259**, 1724–1726 (1993)
44. B. Parkinson, Layer-by-layer nanometer scale etching of two-dimensional substrates using the scanning tunneling microscopy, J. Am. Chem. Soc. **112**, 7498–7502 (1990)

45. A. Majumdar, P.I. Oden, J.P. Carrejo, L.A. Nagahara, J.J. Graham, J. Alexander, Nanometer-scale lithography using the atomic force microscope, Appl. Phys. Lett. **61**, 2293–2295 (1992)
46. B. Bhushan, Micro-/nanotribology and its applications to magnetic storage devices and MEMS, Tribol. Int. **28**, 85–96 (1995)
47. L. Tsau, D. Wang, K.L. Wang, Nanometer scale patterning of silicon(100) surface by an atomic force microscope operating in air, Appl. Phys. Lett. **64**, 2133–2135 (1994)
48. E. Delawski, B.A. Parkinson, Layer-by-layer etching of two-dimensional metal chalcogenides with the atomic force microscope, J. Am. Chem. Soc. **114**, 1661–1667 (1992)
49. B. Bhushan, G.S. Blackman, Atomic force microscopy of magnetic rigid disks and sliders and its applications to tribology, ASME J. Tribol. **113**, 452–458 (1991)
50. O. Marti, B. Drake, P.K. Hansma, Atomic force microscopy of liquid-covered surfaces: atomic resolution images, Appl. Phys. Lett. **51**, 484–486 (1987)
51. B. Drake, C.B. Prater, A.L. Weisenhorn, S.A.C. Gould, T.R. Albrecht, C.F. Quate, D.S. Cannell, H.G. Hansma, P.K. Hansma, Imaging crystals, polymers and processes in water with the atomic force microscope, Science **243**, 1586–1589 (1989)
52. M. Binggeli, R. Christoph, H.E. Hintermann, J. Colchero, O. Marti, Friction force measurements on potential controlled graphite in an electrolytic environment, Nanotechnology **4**, 59–63 (1993)
53. G. Meyer, N.M. Amer, Novel optical approach to atomic force microscopy, Appl. Phys. Lett. **53**, 1045–1047 (1988)
54. J.H. Coombs, J.B. Pethica, Properties of vacuum tunneling currents: Anomalous barrier heights, IBM J. Res. Dev. **30**, 455–459 (1986)
55. M.D. Kirk, T. Albrecht, C.F. Quate, Low-temperature atomic force microscopy, Rev. Sci. Instrum. **59**, 833–835 (1988)
56. F.J. Giessibl, C. Gerber, G. Binnig, A low-temperature atomic force/scanning tunneling microscope for ultrahigh vacuum, J. Vac. Sci. Technol. B **9**, 984–988 (1991)
57. T.R. Albrecht, P. Grutter, D. Rugar, D.P.E. Smith, Low temperature force microscope with all-fiber interferometer, Ultramicroscopy **42–44**, 1638–1646 (1992)
58. H.J. Hug, A. Moser, T. Jung, O. Fritz, A. Wadas, I. Parashikor, H.J. Güntherodt, Low temperature magnetic force microscopy, Rev. Sci. Instrum. **64**, 2920–2925 (1993)
59. C. Basire, D.A. Ivanov, Evolution of the lamellar structure during crystallization of a semicrystalline-amorphous polymer blend: Time-resolved hot-stage SPM study, Phys. Rev. Lett. **85**, 5587–5590 (2000)
60. H. Liu, B. Bhushan, Investigation of nanotribological properties of self-assembled monolayers with alkyl and biphenyl spacer chains, Ultramicroscopy **91**, 185–202 (2002)
61. J. Foster, J. Frommer, Imaging of liquid crystal using a tunneling microscope, Nature **333**, 542–547 (1988)
62. D. Smith, H. Horber, C. Gerber, G. Binnig, Smectic liquid crystal monolayers on graphite observed by scanning tunneling microscopy, Science **245**, 43–45 (1989)
63. D. Smith, J. Horber, G. Binnig, H. Nejoh, Structure, registry and imaging mechanism of alkylcyanobiphenyl molecules by tunnelling microscopy, Nature **344**, 641–644 (1990)
64. Y. Andoh, S. Oguchi, R. Kaneko, T. Miyamoto, Evaluation of very thin lubricant films, J. Phys. D **25**, A71–A75 (1992)
65. Y. Martin, C.C. Williams, H.K. Wickramasinghe, Atomic force microscope-force mapping and profiling on a sub 100Åscale, J. Appl. Phys. **61**, 4723–4729 (1987)
66. J.E. Stern, B.D. Terris, H.J. Mamin, D. Rugar, Deposition and imaging of localized charge on insulator surfaces using a force microscope, Appl. Phys. Lett. **53**, 2717–2719 (1988)
67. K. Yamanaka, H. Ogisco, O. Kolosov, Ultrasonic force microscopy for nanometer resolution subsurface imaging, Appl. Phys. Lett. **64**, 178–180 (1994)
68. K. Yamanaka, E. Tomita, Lateral force modulation atomic force microscope for selective imaging of friction forces, Jpn. J. Appl. Phys. **34**, 2879–2882 (1995)
69. U. Rabe, K. Janser, W. Arnold, Vibrations of free and surface-coupled atomic force microscope: Theory and experiment, Rev. Sci. Instrum. **67**, 3281–3293 (1996)

70. Y. Martin, H.K. Wickramasinghe, Magnetic imaging by force microscopy with 1000Å resolution, Appl. Phys. Lett. **50**, 1455–1457 (1987)
71. D. Rugar, H.J. Mamin, P. Güthner, S.E. Lambert, J.E. Stern, I. McFadyen, T. Yogi, Magnetic force microscopy – General principles and application to longitudinal recording media, J. Appl. Phys. **63**, 1169–1183 (1990)
72. C. Schönenberger, S.F. Alvarado, Understanding magnetic force microscopy, Z. Phys. B **80**, 373–383 (1990)
73. U. Hartmann, Magnetic force microscopy, Annu. Rev. Mater. Sci. **29**, 53–87 (1999)
74. D.W. Pohl, W. Denk, M. Lanz, Optical stethoscopy-image recording with resolution lambda/20, Appl. Phys. Lett. **44**, 651–653 (1984)
75. E. Betzig, J.K. Troutman, T.D. Harris, J.S. Weiner, R.L. Kostelak, Breaking the diffraction barrier – optical microscopy on a nanometric scale, Science **251**, 1468–1470 (1991)
76. E. Betzig, P.L. Finn, J.S. Weiner, Combined shear force and near-field scanning optical microscopy, Appl. Phys. Lett. **60**, 2484 (1992)
77. P.F. Barbara, D.M. Adams, D.B. O'Connor, Characterization of organic thin film materials with near-field scanning optical microscopy (NSOM), Annu. Rev. Mater. Sci. **29**, 433–469 (1999)
78. C.C. Williams, H.K. Wickramasinghe, Scanning thermal profiler, Appl. Phys. Lett. **49**, 1587–1589 (1986)
79. C.C. Williams, H.K. Wickramasinghe, Microscopy of chemical-potential variations on an atomic scale, Nature **344**, 317–319 (1990)
80. A. Majumdar, Scanning thermal microscopy, Annu. Rev. Mater. Sci. **29**, 505–585 (1999)
81. O.E. Husser, D.H. Craston, A.J. Bard, Scanning electrochemical microscopy – High resolution deposition and etching of materials, J. Electrochem. Soc. **136**, 3222–3229 (1989)
82. Y. Martin, D.W. Abraham, H.K. Wickramasinghe, High-resolution capacitance measurement and potentiometry by force microscopy, Appl. Phys. Lett. **52**, 1103–1105 (1988)
83. M. Nonnenmacher, M.P. O'Boyle, H.K. Wickramasinghe, Kelvin probe force microscopy, Appl. Phys. Lett. **58**, 2921–2923 (1991)
84. J.M.R. Weaver, D.W. Abraham, High resolution atomic force microscopy potentiometry, J. Vac. Sci. Technol. B **9**, 1559–1561 (1991)
85. D. DeVecchio, B. Bhushan, Use of a nanoscale Kelvin probe for detecting wear precursors, Rev. Sci. Instrum. **69**, 3618–3624 (1998)
86. B. Bhushan, A.V. Goldade: Measurements and analysis of surface potential change during wear of single-crystal silicon (100) at ultralow loads using Kelvin probe microscopy, Appl. Surf. Sci. **157**, 373–381 (2000)
87. P.K. Hansma, B. Drake, O. Marti, S.A.C. Gould, C.B. Prater, The scanning ion-conductance microscope, Science **243**, 641–643 (1989)
88. C.B. Prater, P.K. Hansma, M. Tortonese, C.F. Quate, Improved scanning ion-conductance microscope using microfabricated probes, Rev. Sci. Instrum. **62**, 2634–2638 (1991)
89. J. Matey, J. Blanc, Scanning capacitance microscopy, J. Appl. Phys. **57**, 1437–1444 (1985)
90. C.C. Williams, Two-dimensional dopant profiling by scanning capacitance microscopy, Annu. Rev. Mater. Sci. **29**, 471–504 (1999)
91. D.T. Lee, J.P. Pelz, B. Bhushan, Instrumentation for direct, low frequency scanning capacitance microscopy, and analysis of position dependent stray capacitance, Rev. Sci. Instrum. **73**, 3523–3533 (2002)
92. P.K. Hansma, J. Tersoff, Scanning tunneling microscopy, J. Appl. Phys. **61**, R1–R23 (1987)
93. I. Giaever, Energy gap in superconductors measured by electron tunneling, Phys. Rev. Lett. **5**, 147–148 (1960)
94. D. Sarid, V. Elings, Review of scanning force microscopy, J. Vac. Sci. Technol. B **9**, 431–437 (1991)
95. U. Durig, O. Zuger, A. Stalder, Interaction force detection in scanning probe microscopy: Methods and applications, J. Appl. Phys. **72**, 1778–1797 (1992)

96. J. Frommer, Scanning tunneling microscopy and atomic force microscopy in organic chemistry, Angew. Chem. Int. Ed. **31**, 1298–1328 (1992)
97. H.J. Güntherodt, R. Wiesendanger (Eds.), *Scanning Tunneling Microscopy I: General Principles and Applications to Clean and Adsorbate-Covered Surfaces* (Springer, Berlin, Heidelberg 1992)
98. R. Wiesendanger, H.J. Güntherodt (Eds.), *Scanning Tunneling Microscopy II: Further Applications and Related Scanning Techniques* (Springer, Berlin, Heidelberg 1992)
99. D.A. Bonnell (Ed.), *Scanning Tunneling Microscopy and Spectroscopy – Theory, Techniques, and Applications* (VCH, New York 1993)
100. O. Marti, M. Amrein (Eds.), *STM and SFM in Biology* (Academic, San Diego 1993)
101. J.A. Stroscio, W.J. Kaiser (Eds.), *Scanning Tunneling Microscopy* (Academic, Boston 1993)
102. H.J. Güntherodt, D. Anselmetti, E. Meyer (Eds.), *Forces in Scanning Probe Methods* (Kluwer, Dordrecht 1995)
103. G. Binnig, H. Rohrer, Scanning tunnelling microscopy, Surf. Sci. **126**, 236–244 (1983)
104. B. Bhushan, J. Ruan, B.K. Gupta, A scanning tunnelling microscopy study of fullerene films, J. Phys. D **26**, 1319–1322 (1993)
105. R.L. Nicolaides, W.E. Yong, W.F. Packard, H.A. Zhou, Scanning tunneling microscope tip structures, J. Vac. Sci. Technol. A **6**, 445–447 (1988)
106. J.P. Ibe, P.P. Bey, S.L. Brandon, R.A. Brizzolara, N.A. Burnham, D.P. DiLella, K.P. Lee, C.R.K. Marrian, R.J. Colton, On the electrochemical etching of tips for scanning tunneling microscopy, J. Vac. Sci. Technol. A **8**, 3570–3575 (1990)
107. R. Kaneko, S. Oguchi, Ion-implanted diamond tip for a scanning tunneling microscope, Jpn. J. Appl. Phys. **28**, 1854–1855 (1990)
108. F.J. Giessibl, Atomic resolution of the silicon(111)–(7×7) surface by atomic force microscopy, Science **267**, 68–71 (1995)
109. B. Anczykowski, D. Krüger, K.L. Babcock, H. Fuchs, Basic properties of dynamic force spectroscopy with the scanning force microscope in experiment and simulation, Ultramicroscopy **66**, 251–259 (1996)
110. T.R. Albrecht, C.F. Quate, Atomic resolution imaging of a nonconductor by atomic force microscopy, J. Appl. Phys. **62**, 2599–2602 (1987)
111. S. Alexander, L. Hellemans, O. Marti, J. Schneir, V. Elings, P.K. Hansma, An atomic-resolution atomic-force microscope implemented using an optical lever, J. Appl. Phys. **65**, 164–167 (1989)
112. G. Meyer, N.M. Amer, Optical-beam-deflection atomic force microscopy: The NaCl(001) surface, Appl. Phys. Lett. **56**, 2100–2101 (1990)
113. A.L. Weisenhorn, M. Egger, F. Ohnesorge, S.A.C. Gould, S.P. Heyn, H.G. Hansma, R.L. Sinsheimer, H.E. Gaub, P.K. Hansma, Molecular resolution images of Langmuir–Blodgett films and DNA by atomic force microscopy, Langmuir **7**, 8–12 (1991)
114. J. Ruan, B. Bhushan, Atomic-scale and microscale friction of graphite and diamond using friction force microscopy, J. Appl. Phys. **76**, 5022–5035 (1994)
115. D. Rugar, P.K. Hansma, Atomic force microscopy, Phys. Today **43**, 23–30 (1990)
116. D. Sarid, *Scanning Force Microscopy* (Oxford Univ. Press, Oxford 1991)
117. G. Binnig, Force microscopy, Ultramicroscopy **42–44**, 7–15 (1992)
118. E. Meyer, Atomic force microscopy, Surf. Sci. **41**, 3–49 (1992)
119. H.K. Wickramasinghe, Progress in scanning probe microscopy, Acta Mater. **48**, 347–358 (2000)
120. A.J. den Boef, The influence of lateral forces in scanning force microscopy, Rev. Sci. Instrum. **62**, 88–92 (1991)
121. M. Radmacher, R.W. Tillman, M. Fritz, H.E. Gaub, From molecules to cells: Imaging soft samples with the atomic force microscope, Science **257**, 1900–1905 (1992)
122. F. Ohnesorge, G. Binnig, True atomic resolution by atomic force microscopy through repulsive and attractive forces, Science **260**, 1451–1456 (1993)

123. G. Neubauer, S.R. Coben, G.M. McClelland, D. Horne, C.M. Mate, Force microscopy with a bidirectional capacitance sensor, Rev. Sci. Instrum. **61**, 2296–2308 (1990)
124. T. Goddenhenrich, H. Lemke, U. Hartmann, C. Heiden, Force microscope with capacitive displacement detection, J. Vac. Sci. Technol. A **8**, 383–387 (1990)
125. U. Stahl, C.W. Yuan, A.L. Delozanne, M. Tortonese, Atomic force microscope using piezoresistive cantilevers and combined with a scanning electron microscope, Appl. Phys. Lett. **65**, 2878–2880 (1994)
126. R. Kassing, E. Oesterschulze, Sensors for scanning probe microscopy. In: *Micro-/Nanotribology and Its Applications*, ed. by B. Bhushan (Kluwer, Dordrecht 1997) pp.35–54
127. C.M. Mate, Atomic-force-microscope study of polymer lubricants on silicon surfaces, Phys. Rev. Lett. **68**, 3323–3326 (1992)
128. S.P. Jarvis, A. Oral, T.P. Weihs, J.B. Pethica, A novel force microscope and point contact probe, Rev. Sci. Instrum. **64**, 3515–3520 (1993)
129. D. Rugar, H.J. Mamin, P. Güthner, Improved fiber-optical interferometer for atomic force microscopy, Appl. Phys. Lett. **55**, 2588–2590 (1989)
130. C. Schönenberger, S.F. Alvarado, A differential interferometer for force microscopy, Rev. Sci. Instrum. **60**, 3131–3135 (1989)
131. D. Sarid, D. Iams, V. Weissenberger, L.S. Bell, Compact scanning-force microscope using laser diode, Opt. Lett. **13**, 1057–1059 (1988)
132. N.W. Ashcroft, N.D. Mermin, *Solid State Physics* (Holt Reinhart and Winston, New York 1976)
133. G. Binnig, D.P.E. Smith, Single-tube three-dimensional scanner for scanning tunneling microscopy, Rev. Sci. Instrum. **57**, 1688 (1986)
134. S.I. Park, C.F. Quate, Digital filtering of STM images, J. Appl. Phys. **62**, 312 (1987)
135. J.W. Cooley, J.W. Tukey, An algorithm for machine calculation of complex Fourier series, Math. Comput. **19**, 297 (1965)
136. J. Ruan, B. Bhushan, Atomic-scale friction measurements using friction force microscopy: Part I – General principles and new measurement techniques, ASME J. Tribol. **116**, 378–388 (1994)
137. T.R. Albrecht, S. Akamine, T.E. Carver, C.F. Quate, Microfabrication of cantilever styli for the atomic force microscope, J. Vac. Sci. Technol. A **8**, 3386–3396 (1990)
138. O. Marti, S. Gould, P.K. Hansma, Control electronics for atomic force microscopy, Rev. Sci. Instrum. **59**, 836–839 (1988)
139. O. Wolter, T. Bayer, J. Greschner, Micromachined silicon sensors for scanning force microscopy, J. Vac. Sci. Technol. B **9**, 1353–1357 (1991)
140. E. Meyer, R. Overney, R. Luthi, D. Brodbeck, Friction force microscopy of mixed Langmuir–Blodgett films, Thin Solid Films **220**, 132–137 (1992)
141. H.J. Dai, J.H. Hafner, A.G. Rinzler, D.T. Colbert, R.E. Smalley, Nanotubes as nanoprobes in scanning probe microscopy, Nature **384**, 147–150 (1996)
142. J.H. Hafner, C.L. Cheung, A.T. Woolley, C.M. Lieber, Structural and functional imaging with carbon nanotube AFM probes, Prog. Biophys. Mol. Biol. **77**, 73–110 (2001)
143. G.S. Blackman, C.M. Mate, M.R. Philpott, Interaction forces of a sharp tungsten tip with molecular films on silicon surface, Phys. Rev. Lett. **65**, 2270–2273 (1990)
144. S.J. O'Shea, M.E. Welland, T. Rayment, Atomic force microscope study of boundary layer lubrication, Appl. Phys. Lett. **61**, 2240–2242 (1992)
145. J.P. Cleveland, S. Manne, D. Bocek, P.K. Hansma, A nondestructive method for determining the spring constant of cantilevers for scanning force microscopy, Rev. Sci. Instrum. **64**, 403–405 (1993)
146. D.W. Pohl, Some design criteria in STM, IBM J. Res. Dev. **30**, 417 (1986)
147. W.T. Thomson, M.D. Dahleh, *Theory of Vibration with Applications*, 5th edn. (Prentice Hall, Upper Saddle River 1998)
148. J. Colchero, Reibungskraftmikroskopie. Ph.D. Thesis (University of Konstanz, Konstanz 1993), in German

149. G.M. McClelland, R. Erlandsson, S. Chiang, Atomic force microscopy: General principles and a new implementation. In: *Review of Progress in Quantitative Nondestructive Evaluation*, Vol.6B, ed. by D.O. Thompson, D.E. Chimenti (Plenum, New York 1987) pp.1307–1314
150. Y.R. Shen, *The Principles of Nonlinear Optics* (Wiley, New York 1984)
151. T. Baumeister, S.L. Marks, *Standard Handbook for Mechanical Engineers*, 7th edn. (McGraw-Hill, New York 1967)
152. J. Colchero, O. Marti, H. Bielefeldt, J. Mlynek, Scanning force and friction microscopy, Phys. Status Solidi (a) **131**, 73–75 (1991)
153. R. Young, J. Ward, F. Scire, Observation of metal-vacuum-metal tunneling, field emission, and the transition region, Phys. Rev. Lett. **27**, 922 (1971)
154. R. Young, J. Ward, F. Scire, The topographiner: An instrument for measuring surface microtopography, Rev. Sci. Instrum. **43**, 999 (1972)
155. C. Gerber, O. Marti, Magnetostrictive positioner, IBM Tech. Discl. Bull. **27**, 6373 (1985)
156. R. Garcìa Cantù, M.A. Huerta Garnica, Long-scan imaging by STM, J. Vac. Sci. Technol. A **8**, 354 (1990)
157. C.J. Chen, In situ testing and calibration of tube piezoelectric scanners, Ultramicroscopy **42–44**, 1653–1658 (1992)
158. R.G. Carr, Finite element analysis of PZT tube scanner motion for scanning tunnelling microscopy, J. Microsc. **152**, 379–385 (1988)
159. C.J. Chen, Electromechanical deflections of piezoelectric tubes with quartered electrodes, Appl. Phys. Lett. **60**, 132 (1992)
160. N. Libioulle, A. Ronda, M. Taborelli, J.M. Gilles, Deformations and nonlinearity in scanning tunneling microscope images, J. Vac. Sci. Technol. B **9**, 655–658 (1991)
161. E.P. Stoll, Restoration of STM images distorted by time-dependent piezo driver aftereffects, Ultramicroscopy **42–44**, 1585–1589 (1991)
162. R. Durselen, U. Grunewald, W. Preuss, Calibration and applications of a high precision piezo scanner for nanometrology, Scanning **17**, 91–96 (1995)
163. J. Fu, In situ testing and calibrating of Z-piezo of an atomic force microscope, Rev. Sci. Instrum. **66**, 3785–3788 (1995)
164. R.C. Barrett, C.F. Quate, Optical scan-correction system applied to atomic force microscopy, Rev. Sci. Instrum. **62**, 1393 (1991)
165. R. Toledo-Crow, P.C. Yang, Y. Chen, M. Vaez-Iravani, Near-field differential scanning optical microscope with atomic force regulation, Appl. Phys. Lett. **60**, 2957–2959 (1992)
166. J.E. Griffith, G.L. Miller, C.A. Green, A scanning tunneling microscope with a capacitance-based position monitor, J. Vac. Sci. Technol. B **8**, 2023–2027 (1990)
167. A.E. Holman, C.D. Laman, P.M.L.O. Scholte, W.C. Heerens, F. Tuinstra, A calibrated scanning tunneling microscope equipped with capacitive sensors, Rev. Sci. Instrum. **67**, 2274–2280 (1996)

# Chapter 3
# General and Special Probes in Scanning Microscopies

**Jason Hafner, Edin (I-Chen) Chen, Ratnesh Lal, and Sungho Jin**

**Abstract** Scanning probe microscopy (SPM) provides nanometer-scale mapping of numerous sample properties in essentially any environment. This unique combination of high resolution and broad applicability has led to the application of SPM to many areas of science and technology, especially those interested in the structure and properties of materials at the nanometer scale. SPM images are generated through measurements of a tip–sample interaction. A well-characterized tip is the key element to data interpretation and is typically the limiting factor.

Commercially available atomic force microscopy (AFM) tips, integrated with force-sensing cantilevers, are microfabricated from silicon and silicon nitride by lithographic and anisotropic etching techniques. The performance of these tips can be characterized by imaging nanometer-scale standards of known dimension, and the resolution is found to roughly correspond to the tip radius of curvature, the tip aspect ratio, and the sample height. Although silicon and silicon nitride tips have a somewhat large radius of curvature, low aspect ratio, and limited lifetime due to wear, the widespread use of AFM today is due in large part to the broad availability of these tips. In some special cases, small asperities on the tip can provide resolution much higher than the tip radius of curvature for low-$Z$ samples such as crystal surfaces and ordered protein arrays.

Several strategies have been developed to improve AFM tip performance. Oxide sharpening improves tip sharpness and enhances tip asperities. For high-aspect-ratio samples such as integrated circuits, silicon AFM tips can be modified by focused ion beam (FIB) milling. FIB tips reach 3° cone angles over lengths of several microns and can be fabricated at arbitrary angles.

Other high resolution and high-aspect-ratio tips are produced by electron-beam deposition (EBD), in which a carbon spike is deposited onto the tip apex from the background gases in an electron microscope. Finally, carbon nanotubes have been employed as AFM tips. Their nanometer-scale diameter, long length, high stiffness, and elastic buckling properties make them possibly the ultimate tip material for AFM. Nanotubes can be manually attached to silicon or silicon nitride AFM tips or *grown* onto tips by chemical vapor deposition (CVD), which should soon make them widely available. In scanning tunneling microscopy (STM), the electron tunneling signal decays exponentially with tip–sample separation, so that in principle only the last few atoms contribute to the signal. STM tips are, therefore, not as sensitive to the

B. Bhushan (ed.), *Nanotribology and Nanomechanics*,
DOI 10.1007/978-3-642-15283-2_3, 

nanoscale tip geometry and can be made by simple mechanical cutting or electrochemical etching of metal wires. In choosing tip materials, one prefers hard, stiff metals that will not oxidize or corrode in the imaging environment.

## Abbreviations

| | |
|---|---|
| AC | alternating-current |
| AC | amorphous carbon |
| AFM | atomic force microscope |
| AFM | atomic force microscopy |
| CCD | charge-coupled device |
| CNF | carbon nanofiber |
| CNT | carbon nanotube |
| CVD | chemical vapor deposition |
| DC-PECVD | direct-current plasma-enhanced CVD |
| DC | direct-current |
| EBD | electron beam deposition |
| EBID | electron-beam-induced deposition |
| FIB | focused ion beam |
| LPCVD | low-pressure chemical vapor deposition |
| MEMS | microelectromechanical system |
| MWNT | multiwall nanotube |
| NSOM | near-field scanning optical microscopy |
| PECVD | plasma-enhanced chemical vapor deposition |
| PMMA | poly(methyl methacrylate) |
| SEM | scanning electron microscope |
| SEM | scanning electron microscopy |
| SPM | scanning probe microscope |
| SPM | scanning probe microscopy |
| STM | scanning tunneling microscope |
| STM | scanning tunneling microscopy |
| SWNT | single wall nanotube |
| SWNT | single-wall nanotube |
| TEM | transmission electron microscope |
| TEM | transmission electron microscopy |
| UHV | ultrahigh vacuum |

In scanning probe microscopy (SPM), an image is created by raster-scanning a sharp probe tip over a sample and measuring some highly localized tip–sample interaction as a function of position. SPMs are based on several interactions, the major types including scanning tunneling microscopy (STM), which measures an electronic tunneling current; atomic force microscopy (AFM), which measures force interactions; and near-field scanning optical microscopy (NSOM), which measures local optical properties by exploiting near-field effects (Fig. 3.1). These

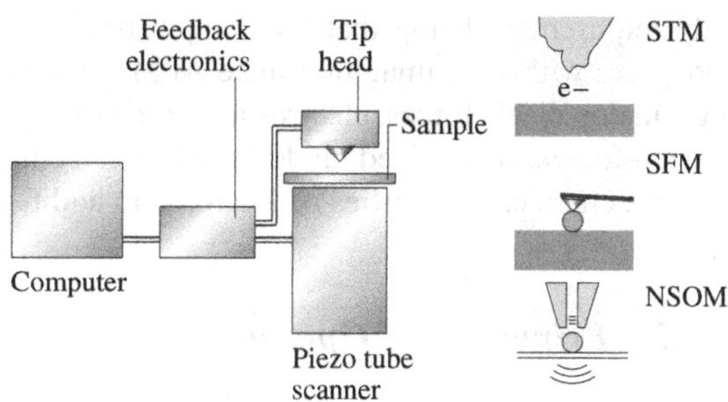


**Fig. 3.1** A schematic of the components of a scanning probe microscope and the three types of signals observed: STM senses electron tunneling currents, AFM measures forces, and NSOM measures near-field optical properties via a subwavelength aperture

methods allow the characterization of many properties (structural, mechanical, electronic, optical) on essentially any material (metals, semiconductors, insulators, biomolecules) and in essentially any environment (vacuum, liquid, ambient air conditions). The unique combination of nanoscale resolution, previously the domain of electron microscopy, *and broad applicability* has led to the proliferation of SPM into virtually all areas of nanometer-scale science and technology.

Several enabling technologies have been developed for SPM, or borrowed from other techniques. Piezoelectric tube scanners allow accurate, subangstrom positioning of the tip or sample in three dimensions. Optical deflection systems and microfabricated cantilevers can detect forces in AFM down to the piconewton range. Sensitive electronics can measure STM currents $<1$ pA. High-transmission fiber optics and sensitive photodetectors can manipulate and detect small optical signals of NSOM. Environmental control has been developed to allow SPM imaging in ultrahigh vacuum (UHV), cryogenic temperatures, at elevated temperatures, and in fluids. Vibration and drift have been controlled such that a probe tip can be held over a single molecule for hours of observation. Microfabrication techniques have been developed for the mass production of probe tips, making SPMs commercially available and allowing the development of many new SPM modes and combinations with other characterization methods. However, of all this SPM development over the past 20 years, what has received the least attention is perhaps the most important aspect: the probe tip.

Interactions measured in SPMs occur at the tip–sample interface, which can range in size from a single atom to tens of nanometers. The size, shape, surface chemistry, and electronic and mechanical properties of the tip apex will directly influence the data signal and the interpretation of the image. Clearly, the better characterized the tip, the more useful the image information. In this chapter, the fabrication and performance of AFM and STM probes will be described.

## 3.1 Atomic Force Microscopy

AFM is the most widely used form of SPM, since it requires neither an electrically conductive sample, as in STM, nor an optically transparent sample or substrate, as in most NSOMs. Basic AFM modes measure the topography of a sample, with the only requirement being that the sample be deposited on a flat surface and rigid enough to withstand imaging. Since AFM can measure a variety of forces, including van der Waals forces, electrostatic forces, magnetic forces, adhesion forces, and friction forces, specialized modes of AFM can characterize the electrical, mechanical, and chemical properties of a sample in addition to its topography.

### 3.1.1 Principles of Operation

In AFM, a probe tip is integrated with a microfabricated force-sensing cantilever. A variety of silicon and silicon nitride cantilevers are commercially available with micrometer-scale dimensions, spring constants ranging from 0.01 to 100 N/m, and resonant frequencies ranging from 5 kHz to over 300 kHz. The cantilever deflection is detected by optical beam deflection, as illustrated in Fig. 3.2. A laser beam bounces off the back of the cantilever and is centered on a split photodiode. Cantilever deflections are proportional to the difference signal $V_A - V_B$. Subangstrom deflections can be detected, and therefore forces down to tens of piconewtons can be measured. A more recently developed method of cantilever deflection measurement is through a piezoelectric layer on the cantilever that registers a voltage upon deflection [1].

A piezoelectric scanner rasters the sample under the tip while the forces are measured through deflections of the cantilever. To achieve more controlled imaging conditions, a feedback loop monitors the tip–sample force and adjusts the sample *z*-position to hold the force constant. The topographic image of the sample is then taken from the sample *z*-position data. The mode described is called the contact mode, in which the tip is deflected by the sample due to repulsive forces, or *contact*. It is generally only used for flat samples that can withstand lateral forces during scanning. To minimize lateral forces and sample damage, two alternating-current (AC) modes have been developed. In these, the cantilever is driven into AC oscillation near its resonant frequency (tens to hundreds of kHz) with desired

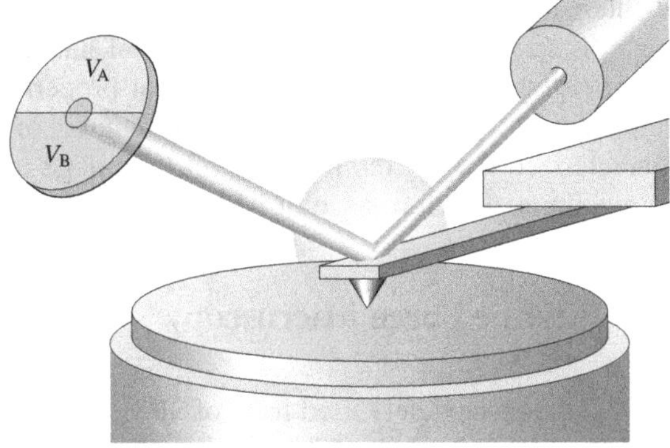


**Fig. 3.2** An illustration of the optical beam deflection system that detects cantilever motion in the AFM. The voltage signal $V_A - V_B$ is proportional to the deflection

amplitudes. When the tip approaches the sample, the oscillation is damped, and the reduced amplitude is the feedback signal, rather than the direct-current (DC) deflection. Again, topography is taken from the varying $Z$-position of the sample required to keep the tip oscillation amplitude constant. The two AC modes differ only in the nature of the interaction. In intermittent contact mode, also called tapping mode, the tip contacts the sample on each cycle, so the amplitude is reduced by ionic repulsion as in contact mode. In noncontact mode, long-range van der Waals forces reduce the amplitude by effectively shifting the spring constant experienced by the tip and changing its resonant frequency.

### *3.1.2 Standard Probe Tips*

In early AFM work, cantilevers were made by hand from thin metal foils or small metal wires. Tips were created by gluing diamond fragments to the foil cantilevers or electrochemically etching the wires to a sharp point. Since these methods were labor intensive and not highly reproducible, they were not amenable to large-scale production. To address this problem, and the need for smaller cantilevers with higher resonant frequencies, batch fabrication techniques were developed (Fig. 3.3). Building on existing methods to batch-fabricate $Si_3N_4$ cantilevers, *Albrecht* et al. [2] etched an array of small square openings in an $SiO_2$ mask layer over a (100) silicon surface. The exposed square (100) regions were etched with KOH, an anisotropic etchant that terminates at the (111) planes, thus creating pyramidal etch pits in the silicon surface. The etch pit mask was then removed and another was applied to define the cantilever shapes with the pyramidal etch pits at the end. The Si wafer was then coated with a low-stress $Si_3N_4$ layer by low-pressure chemical vapor deposition (LPCVD). The $Si_3N_4$ fills the etch pit, using it as a mold to create a pyramidal tip. The silicon was later removed by etching to free the cantilevers and tips. Further steps resulting in the attachment of the cantilever to a macroscopic piece of glass are not described here. The resulting pyramidal tips were highly symmetric and had a tip radius of <30 nm, as determined by scanning

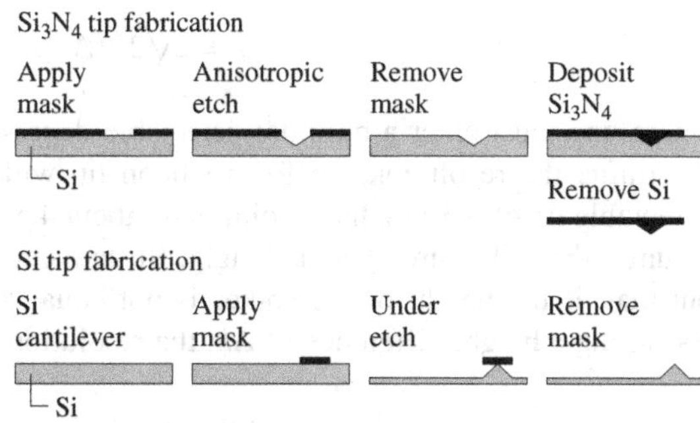


**Fig. 3.3** A schematic overview of the fabrication of Si and $Si_3N_4$ tip fabrication as described in the text

electron microscopy (SEM). This procedure has likely not changed significantly, since commercially available $Si_3N_4$ tips are still specified to have a radius of curvature of 30 nm.

Wolter et al. [3] developed methods to batch-fabricate single-crystal Si cantilevers with integrated tips. Microfabricated Si cantilevers were first prepared using previously described methods, and a small mask was formed at the end of the cantilever. The Si around the mask was etched by KOH, so that the mask was undercut. This resulted in a pyramidal silicon tip under the mask, which was then removed. Again, this partial description of the full procedure only describes tip fabrication. With some refinements the silicon tips were made in high yield with radii of curvature of less than 10 nm. Si tips are sharper than $Si_3N_4$ tips, because they are directly formed by the anisotropic etch in single-crystal Si, rather than using an etch pit as a mask for deposited material. Commercially available silicon probes are made by similar refined techniques and provide a typical radius of curvature of $<10$ nm.

### 3.1.3 Probe Tip Performance

In atomic force microscopy the question of resolution can be a rather complicated issue. As an initial approximation, resolution is often considered strictly in geometrical terms that assume rigid tip–sample contact. The topographical image of a feature is broadened or narrowed by the size of the probe tip, so the resolution is approximately the width of the tip. Therefore, the resolution of AFM with standard commercially available tips is on the order of 5–10 nm. Bustamante and Keller [4] carried the geometrical model further by drawing an analogy to resolution in optical systems. Consider two sharp spikes separated by a distance $d$ to be point objects imaged by AFM (Fig. 3.4). Assume the tip has a parabolic shape with an end radius $R$. The tip-broadened image of these spikes will appear as inverted parabolas. There will be a small depression between the images of depth $\Delta z$. The two spikes are considered *resolved* if $\Delta z$ is larger than the instrumental noise in the $z$-direction. Defined in this manner, the resolution $d$, the minimum separation at which the spikes are resolved, is

$$d = 2\sqrt{2R}(\Delta z), \qquad (3.1)$$

where one must enter a minimal detectable depression for the instrument ($\Delta z$) to determine the resolution. So for a silicon tip with radius 5 nm and a minimum detectable $\Delta z$ of 0.5 nm, the resolution is about 4.5 nm. However, the above model assumes the spikes are of equal height. Bustamante and Keller [4] went on to point out that, if the height of the spikes is not equal, the resolution will be affected. Assuming a height difference of $\Delta h$, the resolution becomes

$$d = \sqrt{2R}(\sqrt{\Delta z} + \sqrt{\Delta z} + \Delta h). \qquad (3.2)$$

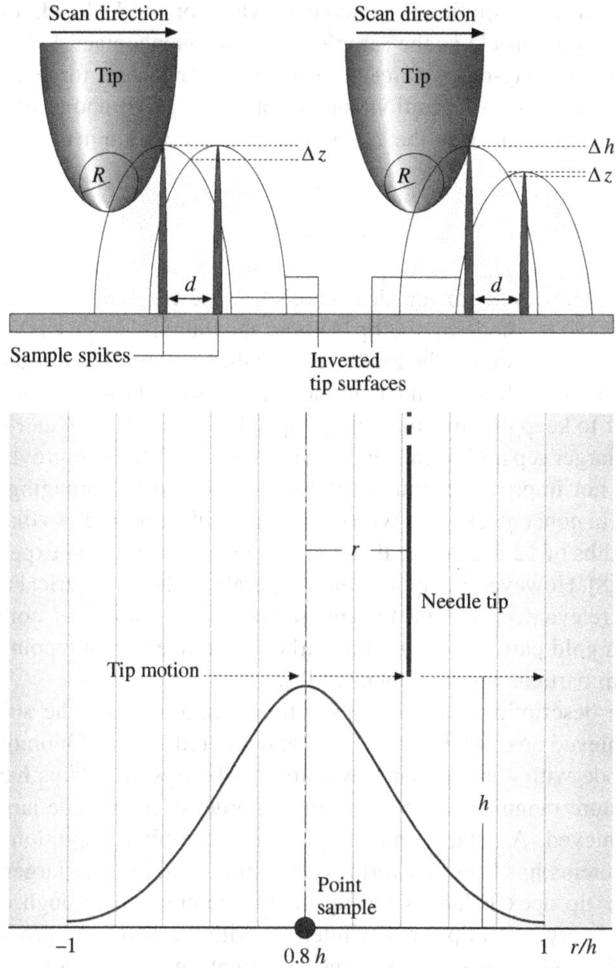


**Fig. 3.4** The factors that determine AFM imaging resolution in contact mode (*top*) and noncontact mode (*bottom*) (After [4])

For a pair of spikes with a 2 nm height difference, the resolution drops to 7.2 nm for a 5 nm tip and 0.5 nm minimum detectable $\Delta z$. While geometrical considerations are a good starting point for defining resolution, they ignore factors such as the possible compression and deformation of the tip and sample. Vesenka et al. [5] confirmed a similar geometrical resolution model by imaging monodisperse gold nanoparticles with tips characterized by transmission electron microscopy (TEM).

Noncontact AFM contrast is generated by long-range interactions such as van der Waals forces, so resolution will not simply be determined by geometry because the tip and sample are not in rigid contact. Bustamante and Keller [4] have derived an expression for the resolution in noncontact AFM for an idealized, infinitely thin *line* tip and a point particle as the sample (Fig. 3.4). Noncontact AFM is sensitive to the gradient of long-range forces, so the van der Waals force gradient was calculated as a function of position for the tip at height $h$ above the surface. If the resolution $d$ is defined as the full-width at half-maximum of this curve, the resolution is

$$d = 0.8h \tag{3.3}$$

This shows that, even for an ideal geometry, the resolution is fundamentally limited in noncontact mode by the tip–sample separation. Under UHV conditions, the tip–sample separation can be made very small, so atomic resolution is possible on flat, crystalline surfaces. Under ambient conditions, however, the separation must be larger to keep the tip from being trapped in the ambient water layer on the surface. This larger separation can lead to a point where further improvements in tip sharpness do not improve resolution. It has been found that imaging 5 nm gold nanoparticles in noncontact mode with carbon nanotube tips of 2 nm diameter leads to particle widths of 12 nm, larger than the 7 nm width one would expect assuming rigid contact [8]. However, in tapping-mode operation, the geometrical definition of resolution is relevant, since the tip and sample come into rigid contact. When imaging 5 nm gold particles with 2 nm carbon nanotube tips in tapping mode, the expected 7 nm particle width is obtained [9].

The above descriptions of AFM resolution cannot explain the subnanometer resolution achieved on crystal surfaces [10] and ordered arrays of biomolecules [11] in contact mode with commercially available probe tips. Such tips have nominal radii of curvature ranging from 5 to 30 nm, an order of magnitude larger than the resolution achieved. A detailed model to explain the high resolution on ordered membrane proteins has been put forth by [6]. In this model, the larger part of the silicon nitride tip apex balances the tip–sample interaction through electrostatic forces, while a very small tip asperity interacts with the sample to provide contrast (Fig. 3.5). This model is supported by measurements at varying salt concentrations to vary the electrostatic interaction strength and the observation of defects in the ordered samples. However, the existence of such asperities has never been confirmed by independent electron microscopy images of the tip. Another model,

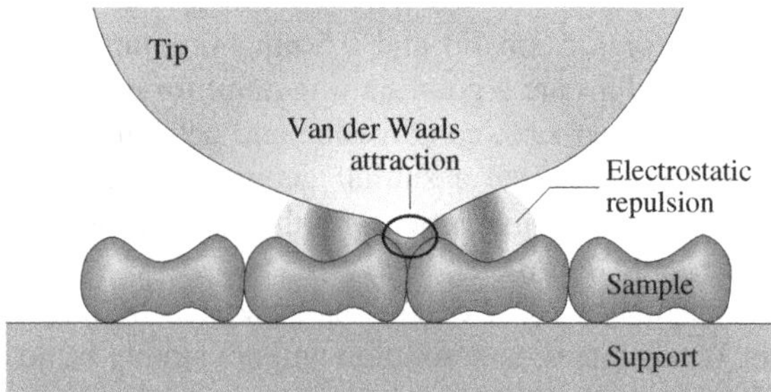


**Fig. 3.5** A tip model to explain the high resolution obtained on ordered samples in contact mode (After [6])

considered especially applicable to atomic resolution on crystal surfaces, assumes that the tip is in contact with a region of the sample much larger than the observed resolution, and that force components matching the periodicity of the sample are transmitted to the tip, resulting in an *averaged* image of the periodic lattice. Regardless of the mechanism, the structures determined are accurate and make this a highly valuable method for membrane proteins. However, this level of resolution should not be expected for most biological systems.

### 3.1.4 Oxide-Sharpened Tips

Both Si and $Si_3N_4$ tips with increased aspect ratio and reduced tip radius can be fabricated through oxide sharpening of the tip. If a pyramidal or cone-shaped silicon tip is thermally oxidized to $SiO_2$ at low temperature (<1050°C), Si–$SiO_2$ stress formation reduces the oxidation rate at regions of high curvature. The result is a sharper, higher-aspect-ratio cone of silicon at the high-curvature tip apex inside the outer pyramidal layer of $SiO_2$ (Fig. 3.6). Etching the $SiO_2$ layer with HF then leaves tips with aspect ratios up to 10:1 and radii down to 1 nm [7], although 5–10 nm is the nominal specification for most commercially available tips. This oxide-sharpening technique can also be applied to $Si_3N_4$ tips by oxidizing the silicon etch pits that are used as molds. As with tip fabrication, oxide sharpening is not quite as effective for $Si_3N_4$. $Si_3N_4$ tips were reported to have an 11 nm radius of curvature [12], while commercially available oxide-sharpened $Si_3N_4$ tips have a nominal radius of <20 nm.

### 3.1.5 Focused Ion Beam Tips

A common AFM application in integrated circuit manufacture and microelectromechanical systems (MEMS) is to image structures with very steep sidewalls such as trenches. To image these features accurately, one must consider the micrometer-scale tip structure, rather than the nanometer-scale structure of the tip

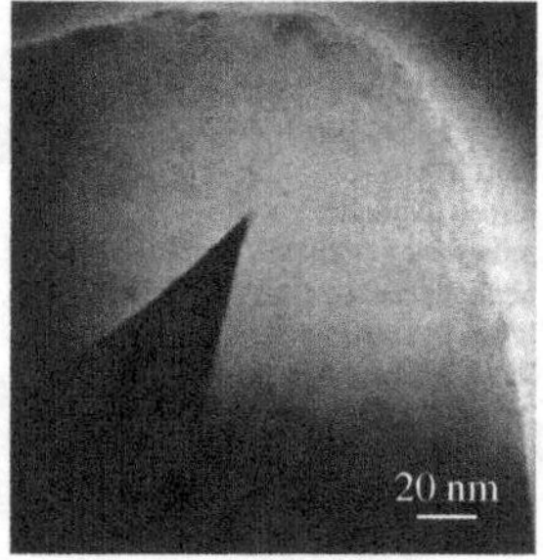


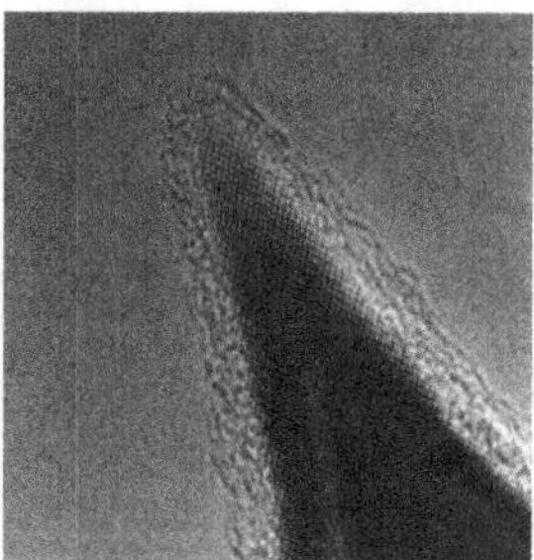

**Fig. 3.6** Oxide sharpening of silicon tips. The *left image* shows a sharpened core of silicon in an outer layer of $SiO_2$. The *right image* is a higher magnification view of such a tip after the $SiO_2$ is removed (After [7])

apex. Since tip fabrication processes rely on anisotropic etchants, the cone half-angles of pyramidal tips are approximately 20°. Images of deep trenches taken with such tips display slanted sidewalls and may not reach the bottom of the trench due to the tip broadening effects. To image such samples more faithfully, high-aspect-ratio tips are fabricated by focused ion beam (FIB) machining a Si tip to produce a sharp spike at the tip apex. Commercially available FIB tips have half-cone angles of <3° over lengths of several micrometers, yielding aspect ratios of approximately 10:1. The radius of curvature at the tip end is similar to that of the tip before the FIB machining. Another consideration for high-aspect-ratio tips is the tip tilt. To ensure that the pyramidal tip is the lowest part of the tip–cantilever assembly, most AFM designs tilt the cantilever about 15° from parallel. Therefore, even an ideal *line tip* will not give an accurate image of high steep sidewalls, but will produce an image that depends on the scan angle. Due to the versatility of FIB machining, tips are available with the spikes at an angle to compensate for this effect.

### 3.1.6 Electron-Beam Deposition Tips

Another method of producing high-aspect-ratio tips for AFM is called electron-beam deposition (EBD). First developed for STM tips [13, 14], EBD tips were introduced for AFM by focusing an SEM onto the apex of a pyramidal tip arranged so that it pointed along the electron beam axis (Fig. 3.7). Carbon material was deposited by the dissociation of background gases in the SEM vacuum chamber. Schiffmann [15] systematically studied the following parameters and how they affected EBD tip geometry:

| | |
|---|---|
| Deposition time: | 0.5–8 min |
| Beam current: | 3–300 pA |
| Beam energy: | 1–30 keV |
| Working distance: | 8–48 mm |

EBD tips were cylindrical with end radii of 20–40 nm, lengths of 1–5 μm, and diameters of 100–200 nm. Like FIB tips, EBD tips were found to achieve improved

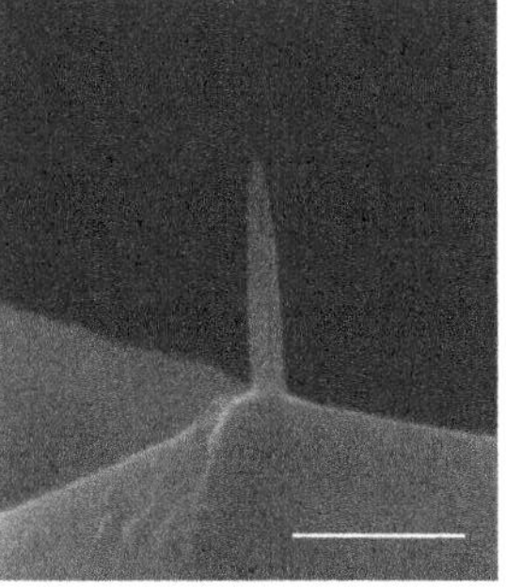

**Fig. 3.7** A pyramidal tip before (*left*, 2 μm-scale bar) and after (*right*, 1 μm-scale bar) electron beam deposition (After [13])

imaging of steep features. By controlling the position of the focused beam, the tip geometry can be further controlled. Tips were fabricated with lengths over 5 μm and aspect ratios greater than 100:1, yet these were too fragile to use as AFM tips [13].

## 3.1.7 Single- and Multiwalled Carbon Nanotube Tips

Carbon nanotubes (CNTs), which were discovered in 1991, are composed of graphene sheets that are rolled up into tubes. Due to their high-aspect-ratio geometry, small tip diameter, and excellent mechanical properties, CNTs have become a promising candidate for new AFM probes to replace standard silicon or silicon nitride probes. CNT tips could offer high-resolution images, while the length of CNT tips allows the tracing of steep and deep features.

### Structures of Carbon Nanotubes

CNTs are seamless cylinders formed by the honeycomb lattice of a single layer of crystalline graphite, called a graphene sheet. In general, CNTs are divided into two types, single-walled nanotubes (SWNTs) and multiwalled nanotubes (MWNTs). Figure 3.8 shows the structures of CNTs explored by a high-resolution TEM [16, 17]. A SWNT is composed of only one rolled-up grapheme, whereas a MWNT consists of a number of concentric tubes. Multiwalled CNTs grown by the thermal CVD process generally exhibit concentric cylinder shape (Fig. 3.9a), while those grown by direct-current plasma-enhanced CVD (DC-PECVD) often exhibit a stacked cone structure (also known as herringbone- or bamboo-like structures, a cross-section of which is illustrated in Fig. 3.9b). Herringbone-like CNTs are also called carbon nanofibers (CNFs) since they are not made of perfect graphene tube cylinders.

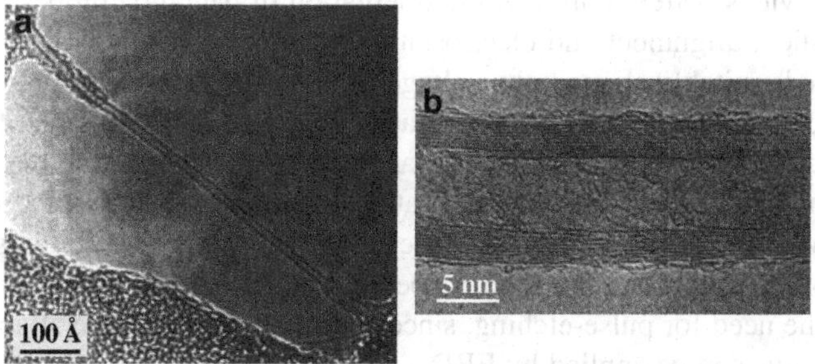


**Fig. 3.8** The structure of carbon nanotubes. **(a)** TEM image of SWNTs (after [16]). **(b)** TEM image of MWNTs (After [17])

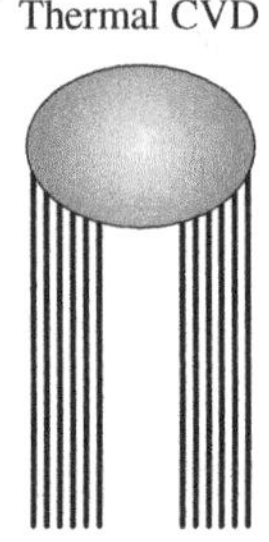


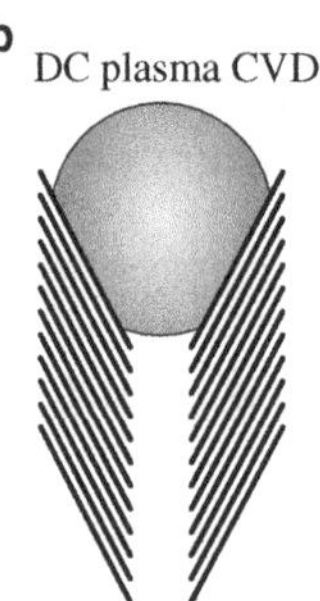


**Fig. 3.9** Schematic structures of (**a**) tubelike carbon nanotubes and (**b**) stacked-cone nanotubes

## Carbon Nanotube Probes by Attachment Approaches

CNTs have been attached onto AFM cantilever pyramid tips by various approaches. The first CNT AFM probes [18] were fabricated by techniques developed for assembling single-nanotube field-emission tips [19]. This process, illustrated in Fig. 3.10, used a purified MWNT material synthesized by the carbon arc procedure. The raw material must contain at least a few percent of long nanotubes ($>10\ \mu m$), purified by oxidation to $\approx 1\%$ of its original mass. A torn edge of the purified material was attached to a micromanipulator by carbon tape and viewed under an optical microscope. Individual nanotubes and nanotube bundles were visible as filaments under dark-field illumination. A commercially available AFM tip was attached to another micromanipulator opposing the nanotube material. Glue was applied to the tip apex from high-vacuum carbon tape supporting the nanotube material. Nanotubes were then manually attached to the tip apex by micromanipulation. As assembled, MWNT tips were often too long for imaging due to thermal vibration during their use as AFM probes. Nanotubes tips were shortened by applying 10 V pulses to the tip while near a sputtered niobium surface. This process etched 100 nm lengths of nanotubes per pulse.

Since the nanotube orientation cannot be well controlled during manual attachment processes, the transfer procedure from the nanotube probe cartridge to the Si tips was operated under an electric field [20]. When applying a low voltage, the nanotube is attracted to the cantilever tip and aligned with the apex of the tip. This approach provides better control of the orientation of nanotube probes because of the electric-field alignment and electrostatic attraction of nanotube probes. When the nanotube is suitably aligned, the voltage is increased to induce an arc discharge in which the nanotube is energetically disassociated and the formation of a carbide may occur at the contact site. Thus, the nanotube can be attached to the cantilever free from the cartridge. The mechanical attachment method has also been carried out in a SEM rather than an optical microscope [21]. This process allows selecting a single nanotube and attaching it to a specific site on the Si tip. This approach eliminates the need for pulse-etching, since short nanotubes can be attached to the tip, and the *glue* can be applied by EBD.

A method to attach CNTs onto AFM tips using magnetic-field alignment has been developed [23]. The experimental apparatus is designed to introduce a

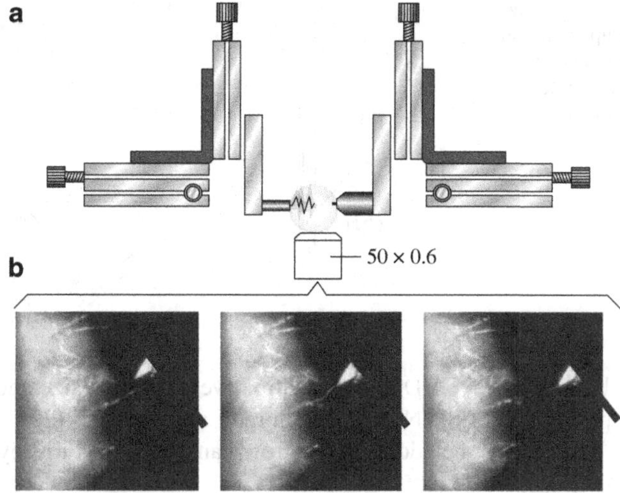


**Fig. 3.10** Schematic drawing of the setup for manual assembly of carbon nanotube tips (**a**) and (**b**) optical microscopy images of the assembly process (the cantilever was drawn in for clarity)

magnetic field onto a single AFM probe and a nanotube suspension. With this apparatus, the anisotropic properties of the CNT cause the nanotubes that come into contact with the probe tip to be preferentially oriented parallel to the tip direction and hence protrude down from the end. Another attachment method based on liquid deposition of CNTs onto AFM probes is the dielectrophoresis process [24, 25]. A Si AFM probe and a metal plate are used as electrodes to apply the AC electric field. A charge-coupled device (CCD) connected to a computer could be used to monitor the process. With in situ observation using the CCD image, the counterelectrode was slowly moved close to the AFM probe until the suspension surface touches its apex. The electrode was then gradually withdrawn until a CNT tip with the desired length was assembled. In this dielectrophoresis process, the length of the CNT probe is controlled by the distance that the counterelectrode is translated under the AC field.

### Nanotube Probe Synthesis by Thermal CVD

The mechanical attachment approaches are tedious and time consuming since nanotube tips are made individually. So, these methods cannot be applied for mass production. The problems of manual assembly of nanotube probes discussed above can largely be solved by directly growing nanotubes onto AFM tips by metal-catalyzed chemical vapor deposition (CVD). Nanometer-scale metal catalyst particles are heated in a gas mixture containing hydrocarbon or CO. The gas molecules dissociate on the metal surface, and carbon is adsorbed into the catalyst particle. When this carbon precipitates, it nucleates a nanotube of similar diameter to the

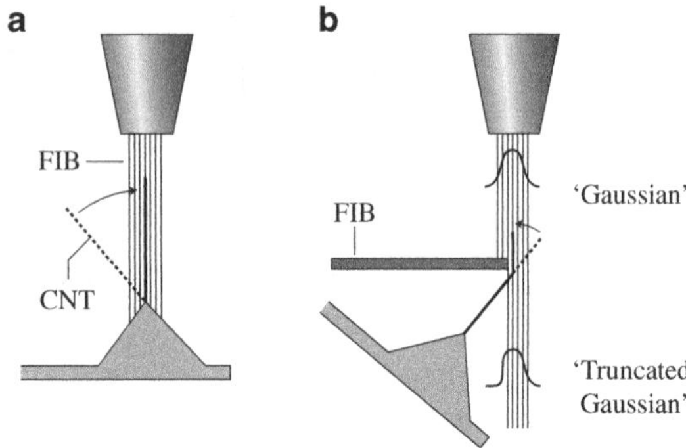


**Fig. 3.11** Schematics for two experimental setup conditions using a focused ion beam for (**a**) aligning a nanotube tip and (**b**) bending the tip (After [22])

catalyst particle. Therefore, CVD allows control over nanotube size and structure, including the production of SWNTs [26] with radii as low as 3.5 [27].

Several key issues must be addressed to grow nanotube AFM tips by CVD:

1. The alignment of the nanotubes at the tip
2. The number of nanotubes that grow at the tip
3. The length of the nanotube tip.

Li et al. [28] found that nanotubes grow perpendicular to a porous surface containing embedded catalyst. This approach was exploited to fabricate nanotube tips by CVD [29] with the proper alignment, as illustrated in Fig. 3.11. A flattened area of $\approx$1–5 $\mu m^2$ was created on Si tips by scanning in contact mode at high load (1 μN) on a hard, synthetic diamond surface. The tip was then anodized in HF to create 100 nm-diameter pores in this flat surface [30]. It is important to anodize only the last 20–40 μm of the cantilever, which includes the tip, so that the rest of the cantilever is still reflective for use in the AFM. This was achieved by anodizing the tip in a small drop of HF under the view of an optical microscope. Next, iron was electrochemically deposited into the pores to form catalyst particles [31]. Tips prepared in this way were heated in low concentrations of ethylene at 800 °C, which is known to favor the growth of thin nanotubes [26]. When imaged by SEM, nanotubes were found to grow perpendicular to the surface from the pores as desired, and TEM revealed that the nanotubes were thin, individual, multiwalled nanotubes with typical radii of 3–5 nm.

These *pore-growth* CVD nanotube tips were typically several micrometers in length – too long for imaging – and were pulse-etched to usable length of <500 nm. The tips exhibited elastic buckling behavior and were very robust in imaging. The pore-growth method demonstrated the potential of CVD to simplify the fabrication of nanotube tips, although there were still limitations. In particular, the porous layer was difficult to prepare and rather fragile.

An alternative approach to CVD fabrication of nanotube tips involves direct growth of SWNTs on the surface of a pyramidal AFM tip [32, 33]. In this *surface-growth* approach, an alumina/iron/molybdenum powdered catalyst known to produce SWNT [26] was dispersed in ethanol at 1 mg/ml. Silicon tips were dipped in this solution and allowed to dry, leaving a sparse layer of $\approx$ 100 nm catalyst clusters on the tip. When CVD conditions were applied, single-walled nanotubes grew along the silicon tip surface. At a pyramid edge, nanotubes can either bend to align with the edge or protrude from the surface. If the energy required to bend the tube and follow the edge is less than the attractive nanotube surface energy, then the nanotube will follow the pyramid edge to the apex. Therefore, nanotubes were effectively steered toward the tip apex by the pyramid edges. At the apex, the nanotube protruded from the tip, since the energetic cost of bending around the sharp silicon tip was too high. The high aspect ratio at the oxide-sharpened silicon tip apex was critical for good nanotube alignment. These *surface-growth* nanotube tips exhibit a high aspect ratio and high-resolution imaging, as well as elastic buckling. This method has been expanded to include wafer-scale production of nanotube tips with high yields [34], yet one obstacle remains to the mass production of nanotube probe tips. Nanotubes protruding from the tip are several micrometers long, and since they are so thin, they must be etched to less than 100 nm.

### Hybrid Nanotube Tip Fabrication: Pick-Up Tips

Another method of creating nanotube tips is something of a hybrid between assembly and CVD. The motivation was to create AFM probes that have an *individual* SWNT at the tip to achieve the ultimate imaging resolution. In order to synthesize isolated SWNT, isolated catalyst particles were formed by dipping a silicon wafer in an isopropyl alcohol solution of $Fe(NO_3)_3$. When heated in a CVD furnace, the iron became mobile and aggregated to form isolated iron particles. By controlling the reaction time, the SWNT lengths were kept shorter than their typical separation, so that the nanotubes never had a chance to form bundles. In the *pick-up tip* method, these isolated SWNT substrates were imaged by AFM with silicon tips in air [9]. When the tip encountered a vertical SWNT, the oscillation amplitude was damped, so the AFM pulled the sample away from the tip. This procedure pulled the SWNT into contact with the tip along its length, so that it became attached to the tip. This assembly process happened automatically when imaging in tapping mode; no special tip manipulation was required.

### PECVD-Grown Nanotube Probe

The attachment methods are time consuming and often result in nonreproducible CNT configuration and placement. While thermal CVD approaches can potentially lead to wafer-scale production of AFM tips, the number, orientation, and length of CNTs are difficult to control. At the end of the fabrication, these processes usually require a one-at-a-time manipulation approach to remove extra CNTs and/or to shorten the remaining CNTs for SPM applications.

The key process for CVD-grown CNT probe fabrication is catalyst patterning, which determines the position, number, and diameter of the probe. Electrophoretically deposited or spin-coated colloidal catalyst particles on Si pyramid tips cannot provide reliable control of the position and number of catalyst particles. MWNT probes on tipless cantilevers have been fabricated based on conventional Si fabrication process in which the catalyst pattering was proceeded by typical e-beam lithography and lift-off of spin-coated poly(methyl methacrylate) (PMMA) layer, and plasma-enhanced chemical vapor deposition (PECVD) was used for CNT growth [35, 36]. The fabrication method described in [35] allows CNT tips to be grown directly on silicon cantilevers at the wafer scale. CNT tip locations and diameters are defined by e-beam lithography. CNT length and orientation are controlled by the growth conditions of the PECVD method. Therefore, there is no need to shorten the CNT after the growth. In PECVD, an electric field is present in the plasma discharge to direct the nanotubes to grow parallel to the electric field. A tilted probe is desirable as it compensates for the operating tilt angle of the AFM cantilever so that the probe itself is close to vertical for stable imaging.

A spin-coated PMMA layer cannot be uniformly conformal on the relatively small piece of tipless cantilevers or on the Si pyramid tip. For e-beam lithography-based processes, the patterned catalyst dots either have to be formed before the fabrication of the cantilevers (although then a protection layer is needed) [35] or lithography steps have to be applied twice to remove extra catalyst on commercial tipless-cantilever chips [36]. Therefore, the electron-beam-induced deposition (EBID) technique has been developed to make catalyst patterns for CNT probe fabrication [37, 38]. EBID is a simple and fast technique to make patterns and deposit materials simultaneously without using any e-beam resist. Its resist-free nature makes EBID a good choice for the fabrication of patterns on the edge of the substrate. A schematic diagram of probe fabrication based on EBID patterning and PECVD is shown in Fig. 3.12. No special carbonaceous precursor molecules were introduced, as the residual carbon-containing molecules naturally present in the SEM chamber were sufficient for EBID processing to form amorphous carbon dots on the cantilever surface. A single carbon dot with a diameter of $\approx$400 nm was deposited near the front-end edge of the cantilever by EBID. The carbon dot serves as a convenient etch mask for chemical etching of the catalyst film. The removal of the carbon dot mask after catalyst patterning was performed with oxygen reactive-ion etch, which exposed the catalyst island. The cantilever with the catalyst island was then transferred to the DC-PECVD system for subsequent growth of the CNT. Figure 3.13 shows SEM images of a CNT probe grown on a tipless cantilever.

### 3.1.8 Bent Carbon Nanotube Tips

The orientation of CNT tips can be manipulated by FIB treatments, utilizing the interaction between the ion beam and the CNT tip [22, 39]. Figure 3.11 shows a schematic of the process of aligning and bending the CNT by using FIB. The aligning and bending phenomena were observed in both as-grown CNT and metal-coated CNT tips. The aligning process is faster with larger values of beam current and acceleration voltage. Under the same voltage, a greater current or longer process time is needed for straightening compared with bending. By using this process, CNT tips can be aligned in any specified direction with precision of less than 1°. Precise

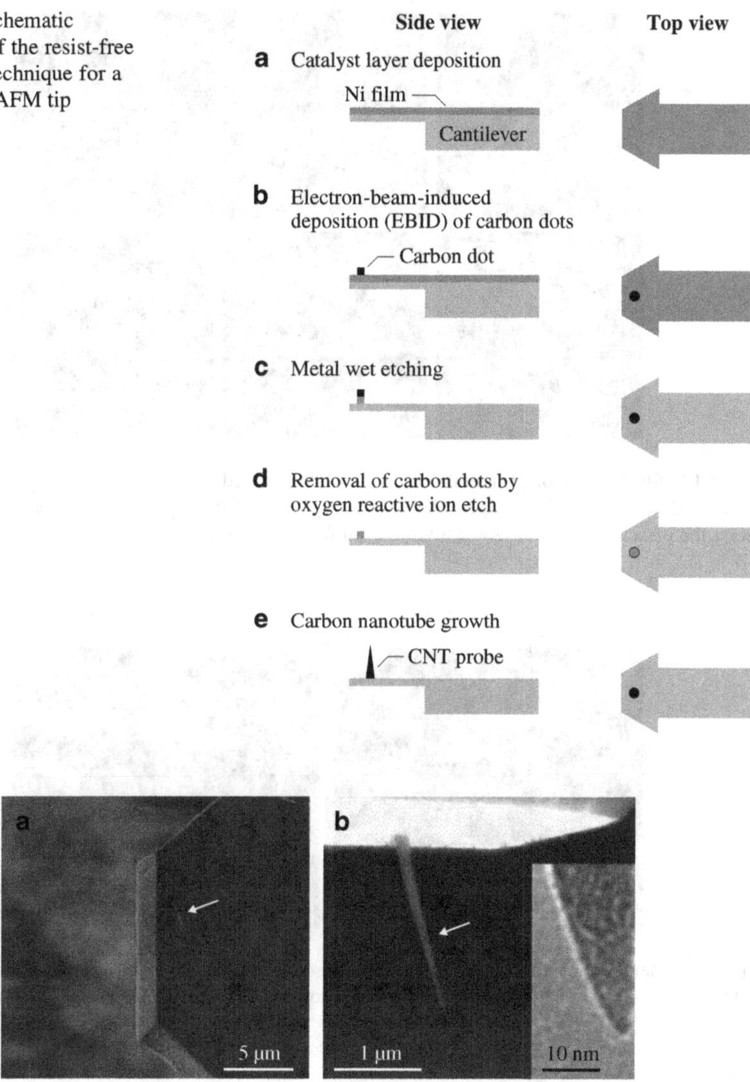


**Fig. 3.12** Schematic illustration of the resist-free fabrication technique for a single CNT AFM tip

**Fig. 3.13** (**a**) Top view SEM image of the very sharp single CNT probe. (**b**) Side view SEM image of the CNT probe (After [37]). The *arrow* indicates a very sharp, single CNT tip grown on the cantilever

control over the orientation of a metal-coated nanotube using a FIB is shown in Fig. 3.14. Figure 3.15 illustrates bending of the end of a CNT tip, which is expected to have potential applications for sidewall measurements in AFM imaging.

CNT tip bending can also be accomplished by changing the direction of the applied bias electric field during DC plasma-enhanced CVD growth [40, 41]. As depicted in Fig. 3.16, the nanotube tip can be bent either slightly, by ≈45° or by ≈90° using various electric-field angles during the growth process.

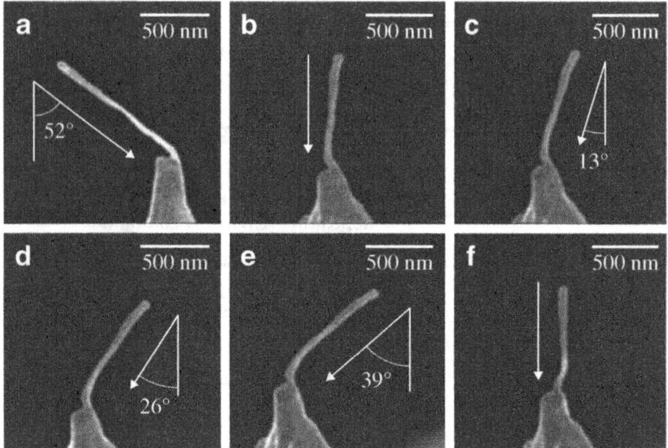


**Fig. 3.14** SEM image of (**a**) metal-coated nanotube aligned at 52° with respect to the axis of the pyramidal tip. (**b–f**) SEM images of the same tip after being exposed to the ion beam incident along the direction of the *arrow* drawn in each image (After [39])

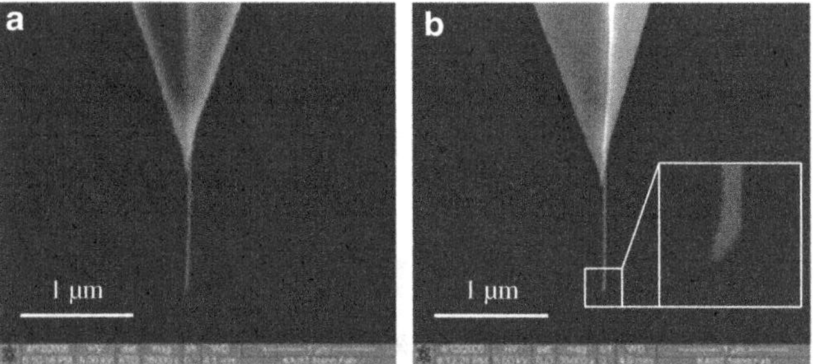


**Fig. 3.15** Bending the end of the CNT with focused ion beam. (**a**) CNT as attached to a Si probe. (**b**) CNT end slightly bent after the FIB process toward the source (After [22])

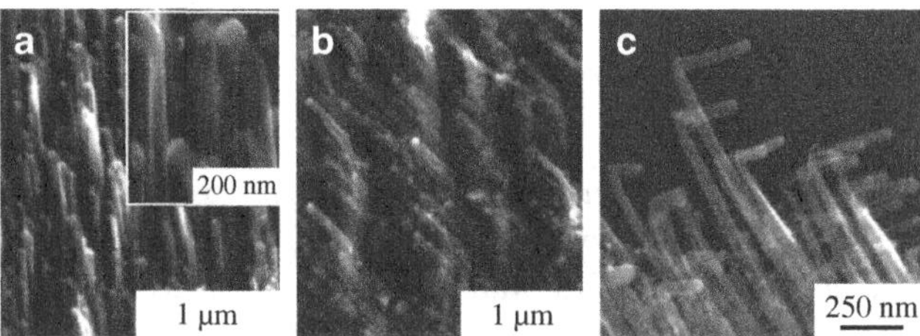


**Fig. 3.16** Carbon nanotube bending using tilted bias electric field during plasma enhanced CVD growth. The nanotube tip can be bent (**a**) either slightly, (**b**) by ≈ 45°, or (**c**) by ≈ 90° using various electric field angles during the growth process (After [40, 41])

### 3.1.9 Low-Stiffness Cantilevers with Carbon Nanotube Tips

Direct growth of a CNT probe on a low-stiffness cantilever by PECVD is desirable for AFM imaging on soft or fragile materials. As introduced in Sect. 3.1.7, by combining an electron-beam lithography approach for catalyst patterning with PECVD for CNT growth, the location, length, and diameter of CNTs can be well controlled. The plasma-induced stresses and damages introduced during PECVD growth of nanotubes, however, result in severely bent cantilevers when a thin, low-stiffness cantilever is utilized as the substrate. If the bend is sufficiently large, the AFM laser spot focused at their end will be deflected off of the position-sensitive detector, rendering the cantilevers unusable for AFM measurements.

An in situ process to control the deflection of cantilever beams during CNT growth has been demonstrated by introducing hydrogen gas into the (acetylene + ammonia) feed gas and adjusting the ammonia-to-hydrogen flow ratio [42]. The total flow rate of $NH_3$ and $H_2$ was kept constant during growth, while the gas mix ratio ($R$), defined as $NH_3/(NH_3 + H_2)$, was varied in the range $0 \leq R \leq 1$. Figure 3.17 shows comparative, cross-sectional cantilever images for three different CNT

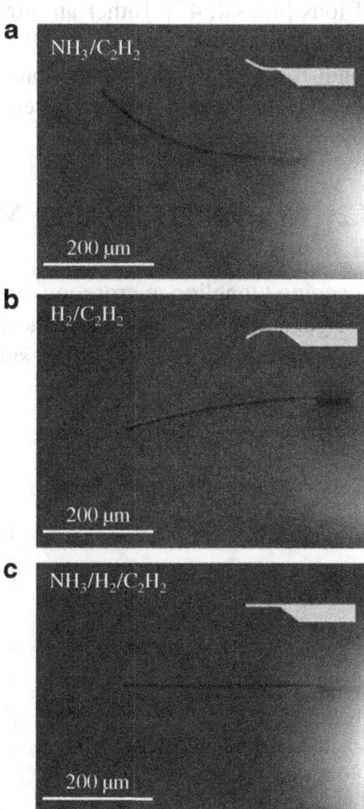


**Fig. 3.17** Optical microscope images of cantilevers bending after plasma treatments with $C_2H_2$ gas and **(a)** $NH_3$ gas ($R = 1$). **(b)** $H_2$ gas ($R = 0$). **(c)** Mixed $NH_3/H_2$ gas ($R = 0.5$) (After [42])

growth conditions using different feed gas compositions. A large upward or downward bending of the cantilever is observed for $R = 1$ and $R = 0$, respectively. By employing a particular gas ratio of $R = 0.5$, a nearly flat cantilever beam can be obtained after PECVD growth of a CNT probe.

### *3.1.10 Conductive Probe Tips*

Conductive AFM probes are useful for the study of electrical or ionic properties of nanostructures, especially for the investigation of biological nanofeatures such as ion channels and receptors, the key regulators of cellular homeostasis and sustenance. Disturbed ion-channel behavior in cell membranes such as in the transport of $Ca^{2+}$, $K^+$, $Na^+$ or $Cl^-$ ions leads to a variety of channelopathies such as Alzheimer's, Parkinson's, cystic fibrosis, cardiac arrhythmias, and other systemic diseases. Real-time structure–activity relation of these channels and their (patho)physiological controls can be studied using conductive AFM. An integrated conductive AFM will allow simultaneous acquisition of structure and activity data and to correlate three-dimensional (3-D) nanostructure of individual ion channels and real-time transport of ions [43, 44, 45]. Either an intrinsically conductive and stable probe such as a carbon nanotube tip or a metal-coated silicon nitride tip can be utilized. The conductive AFM tip serves as one of the electrodes, measuring the ionic currents between the tip and a reference electrode, as illustrated in Fig. 3.18.

## 3.2 Scanning Tunneling Microscopy

Scanning tunneling microscopy (STM) was the original scanning probe microscopy and generally produces the highest-resolution images, routinely achieving atomic resolution on flat, conductive surfaces. In STM, the probe tip consists of a

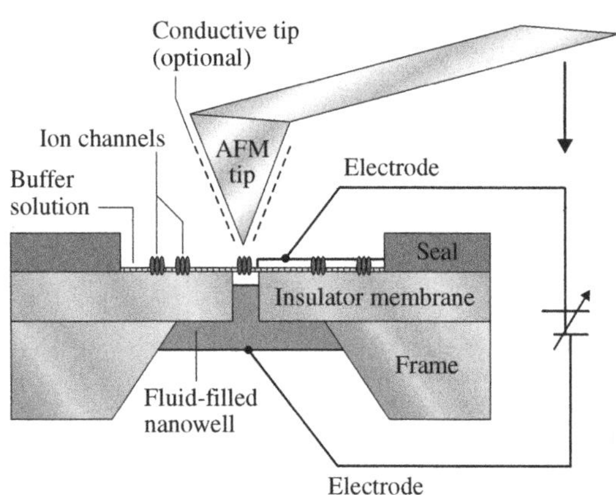


**Fig. 3.18** Schematic illustration of the use of conductive AFM probe tip for ion channel conductivity study

sharpened metal wire that is held 0.3–1 nm from the sample. A potential difference of 0.1–1 V between the tip and sample leads to tunneling currents on the order of 0.1–1 nA. As in AFM, a piezo-scanner rasters the sample under the tip, and the $z$-position is adjusted to hold the tunneling current constant. The $z$-position data represents the *topography*, or in this case the surface of constant electron density. As with other SPMs, the tip properties and performance greatly depend on the experiment being carried out. Although it is nearly impossible to prepare a tip with known atomic structure, a number of factors are known to affect tip performance, and several preparation methods that produce good tips have been developed.

The nature of the sample being investigated and the scanning environment will affect the choice of the tip material and how the tip is fabricated. Factors to consider are mechanical properties – a hard material that will resist damage during tip–sample contact is desired. Chemical properties should also be considered – formation of oxides or other insulating contaminants will affect tip performance. Tungsten is a common tip material because it is very hard and will resist damage, but its use is limited to ultrahigh-vacuum (UHV) conditions, since it readily oxidizes. For imaging under ambient conditions an inert tip material such as platinum or gold is preferred. Platinum is typically alloyed with iridium to increase its stiffness.

### 3.2.1 Mechanically Cut STM Tips

STM tips can be fabricated by simple mechanical procedures such as grinding or cutting metal wires. Such tips are not formed with highly reproducible shapes and have a large opening angle and a large radius of curvature in the range of 0.1–1 μm (Fig. 3.19a). They are not useful for imaging samples with surface roughness above a few nanometers. However, on atomically flat samples, mechanically cut tips can achieve atomic resolution due to the nature of the tunneling signal, which drops exponentially with tip–sample separation. Since mechanically cut tips contain many small asperities on the larger tip structure, atomic resolution is easily achieved as long as one atom of the tip is just a few angstroms lower than all of the others.

### 3.2.2 Electrochemically Etched STM Tips

For samples with more than a few nanometers of surface roughness, the tip structure in the nanometer size range becomes an issue. Electrochemical etching can provide tips with reproducible and desirable shapes and sizes (Fig. 3.19), although the exact atomic structure of the tip apex is still not well controlled. The parameters of electrochemical etching depend greatly on the tip material and the desired tip shape. The following is an entirely general description. A fine metal wire (0.1–1 mm diameter) of the tip material is immersed in an appropriate electrochemical etchant solution. A bias voltage of 1–10 V is applied between the tip and a

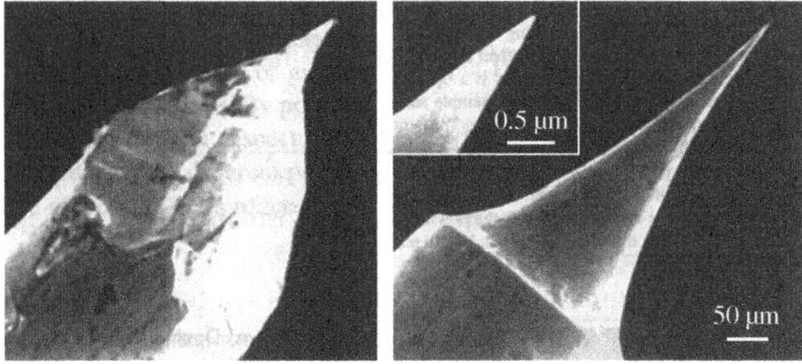


**Fig. 3.19** A mechanically cut STM tip (*left*) and an electrochemically etched STMtip (*right*) (After [46])

counterelectrode such that the tip is etched. Due to the enhanced etch rate at the electrolyte–air interface, a neck is formed in the wire. This neck is eventually etched thin enough that it cannot support the weight of the part of the wire suspended in the solution, and it breaks to form a sharp tip. The widely varying parameters and methods will be not be covered in detail here, but many recipes can be found in the literature for common tip materials [47, 48, 49, 50, 51].

## References

1. R. Linnemann, T. Gotszalk, I.W. Rangelow, P. Dumania, E. Oesterschulze, Atomic force microscopy and lateral force microscopy using piezoresistive cantilevers. J. Vac. Sci. Technol. B **14**(2), 856–860 (1996).
2. T.R. Albrecht, S. Akamine, T.E. Carver, C.F. Quate, Microfabrication of cantilever styli for the atomic force microscope. J. Vac. Sci. Technol. A **8**(4), 3386–3396 (1990).
3. O. Wolter, T. Bayer, J. Greschner, Micromachined silicon sensors for scanning force microscopy. J. Vac. Sci. Technol. B **9**(2), 1353–1357 (1991).
4. C. Bustamante, D. Keller, Scanning force microscopy in biology. Phys. Today **48**(12), 32–38 (1995).
5. J. Vesenka, S. Manne, R. Giberson, T. Marsh, E. Henderson, Colloidal gold particles as an incompressible atomic force microscope imaging standard for assessing the compressibility of biomolecules. Biophys. J. **65**, 992–997 (1993).
6. D.J. Müller, D. Fotiadis, S. Scheuring, S.A. Müller, A. Engel, Electrostatically balanced subnanometer imaging of biological specimens by atomic force microscope. Biophys. J. **76**(2), 1101–1111 (1999).
7. R.B. Marcus, T.S. Ravi, T. Gmitter, K. Chin, D.J. Liu, W. Orvis, D.R. Ciarlo, C.E. Hunt, J. Trujillo, Formation of silicon tips with $< 1$ nm radius. Appl. Phys. Lett. **56**(3), 236–238 (1990).
8. J.H. Hafner, C.L. Cheung, C.M. Lieber, unpublished results (2001).
9. J.H. Hafner, C.L. Cheung, T.H. Oosterkamp, C.M. Lieber, High-yield assembly of individual single-walled carbon nanotube tips for scanning probe microscopies. J. Phys. Chem. B **105**(4), 743–746 (2001).
10. F. Ohnesorge, G. Binnig, True atomic resolution by atomic force microscopy through repulsive and attractive forces. Science **260**, 1451–1456 (1993).

11. D.J. Müller, D. Fotiadis, A. Engel, Mapping flexible protein domains at subnanometer resolution with the atomic force microscope. FEBS Letters **430**(1/2), 105–111 (1998), Special Issue SI.
12. S. Akamine, R.C. Barrett, C.F. Quate, Improved atomic force microscope images using microcantilevers with sharp tips. Appl. Phys. Lett. **57**(3), 316–318 (1990).
13. D.J. Keller, C. Chih-Chung, Imaging steep, high structures by scanning force microscopy with electron beam deposited tips. Surf. Sci. **268**, 333–339 (1992).
14. T. Ichihashi, S. Matsui, In situ observation on electron beam induced chemical vapor deposition by transmission electron microscopy. J. Vac. Sci. Technol. B **6**(6), 1869–1872 (1988).
15. K.I. Schiffmann, Investigation of fabrication parameters for the electron-beam-induced deposition of contamination tips used in atomic force microscopy. Nanotechnology **4**, 163–169 (1993).
16. D.S. Bethune, C.H. Kiang, M.S. de Vries, G. Gorman, R. Savoy, J. Vazquez, R. Beyers, Cobalt-catalysed growth of carbon nanotubes with single-atomic-layer walls. Nature **363** (6430), 605–607 (1993).
17. E.T. Thostenson, Z. Ren, T.W. Chou, Advances in the science and technology of carbon nanotubes and their composites: A review. Compos. Sci. Technol. **61**(13), 1899–1912 (2001).
18. H.J. Dai, J.H. Hafner, A.G. Rinzler, D.T. Colbert, R.E. Smalley, Nanotubes as nanoprobes in scanning probe microscopy. Nature **384**(6605), 147–150 (1996).
19. A.G. Rinzler, Y.H. Hafner, P. Nikolaev, L. Lou, S.G. Kim, D. Tomanek, D.T. Colbert, R.E. Smalley, Unraveling nanotubes: Field emission from atomic wire. Science **269**, 1550 (1995).
20. R. Stevens, C. Nguyen, A. Cassell, L. Delzeit, M. Meyyappan, J. Han, Improved fabrication approach for carbon nanotube probe devices. Appl. Phys. Lett. **77**, 3453–3455 (2000).
21. H. Nishijima, S. Kamo, S. Akita, Y. Nakayama, K.I. Hohmura, S.H. Yoshimura, K. Takeyasu, Carbon-nanotube tips for scanning probe microscopy: Preparation by a controlled process and observation of deoxyribonucleic acid. Appl. Phys. Lett. **74**, 4061–4063 (1999).
22. B.C. Park, K.Y. Jung, W.Y. Song, O. Beom-Hoan, S.J. Ahn, Bending of a carbon nanotube in vacuum using a focused ion beam. Adv. Mater. **18**, 95–98 (2006).
23. A. Hall, W.G. Matthews, R. Superfine, M.R. Falvo, S. Washburna, Simple and efficient method for carbon nanotube attachment to scanning probes and other substrates. Appl. Phys. Lett. **82**, 2506–2508 (2003).
24. J. Tang, G. Yang, Q. Zhang, A. Parhat, B. Maynor, J. Liu, L.C. Qin, O. Zhou, Rapid and reproducible fabrication of carbon nanotube AFM probes by dielectrophoresis. Nano Lett. **5**, 11–14 (2005).
25. J.-E. Kim, J.-K. Park, C.-S. Han, Use of dielectrophoresis in the fabrication of an atomic force microscope tip with a carbon nanotube: Experimental investigation. Nanotechnology **17**, 2937–2941 (2006).
26. J.H. Hafner, M.J. Bronikowski, B.R. Azamian, P. Nikolaev, A.G. Rinzler, D.T. Colbert, K.A. Smith, R.E. Smalley, Catalytic growth of single-wall carbon nanotubes from metal particles. Chem. Phys. Lett. **296**(1/2), 195–202 (1998).
27. P. Nikolaev, M.J. Bronikowski, R.K. Bradley, F. Rohmund, D.T. Colbert, K.A. Smith, R.E. Smalley, Gas-phase catalytic growth of single-walled carbon nanotubes from carbon monoxide. Chem. Phys. Lett. **313**(1/2), 91–97 (1999).
28. W.Z. Li, S.S. Xie, L.X. Qian, B.H. Chang, B.S. Zou, W.Y. Zhou, R.A. Zhao, G. Wang, Large-scale synthesis of aligned carbon nanotubes. Science **274**(5293), 1701–1703 (1996).
29. J.H. Hafner, C.L. Cheung, C.M. Lieber, Growth of nanotubes for probe microscopy tips. Nature **398**(6730), 761–762 (1999).
30. V. Lehmann, The physics of macroporous silicon formation. Thin Solid Films **255**, 1–4 (1995).
31. F. Ronkel, J.W. Schultze, R. Arensfischer, Electrical contact to porous silicon by electrodeposition of iron. Thin Solid Films **276**(1–2), 40–43 (1996).
32. J.H. Hafner, C.L. Cheung, C.M. Lieber, Direct growth of single-walled carbon nanotube scanning probe microscopy tips. J. Am. Chem. Soc. **121**(41), 9750–9751 (1999).

33. E.B. Cooper, S.R. Manalis, H. Fang, H. Dai, K. Matsumoto, S.C. Minne, T. Hunt, C.F. Quate, Terabit-per-square-inch data storage with the atomic force microscope. Appl. Phys. Lett. **75**(22), 3566–3568 (1999).
34. E. Yenilmez, Q. Wang, R.J. Chen, D. Wang, H. Dai, Wafer scale production of carbon nanotube scanning probe tips for atomic force microscopy. Appl. Phys. Lett. **80**(12), 2225–2227 (2002).
35. Q. Ye, A.M. Cassell, H.B. Liu, K.J. Chao, J. Han, M. Meyyappan, Large-scale fabrication of carbon nanotube probe tips for atomic force microscopy critical dimension imaging applications. Nano Lett. **4**, 1301–1308 (2004).
36. H. Cui, S.V. Kalinin, X. Yang, D.H. Lowndes, Growth of carbon nanofibers on tipless cantilevers for high resolution topography and magnetic force imaging. Nano Lett. **4**, 2157–2161 (2004).
37. I.-C. Chen, L.-H. Chen, X.-R. Ye, C. Daraio, S. Jin, C.A. Orme, A. Quist, R. Lal, Extremely sharp carbon nanocone probes for atomic force microscopy imaging. Appl. Phys. Lett. **88**, 153102 (2006).
38. I.-C. Chen, L.-H. Chen, C.A. Orme, A. Quist, R. Lal, S. Jin, Fabrication of high-aspect-ratio carbon nanocone probes by electron beam induced deposition patterning. Nanotechnology **17**, 4322 (2006).
39. Z.F. Deng, E. Yenilmez, A. Reilein, J. Leu, H. Dai, K.A. Moler, Nanotube manipulation with focused ion beam. Appl. Phys. Lett. **88**, 023119 (2006).
40. J.F. AuBuchon, L.-H. Chen, S. Jin, Control of carbon capping for regrowth of aligned carbon nanotubes. J. Phys. Chem. B **109**, 6044–6048 (2005).
41. J.F. AuBuchon, L.-H. Chen, A.I. Gapin, S. Jin, electric-field-guided growth of carbon nanotubes during DC plasma-enhanced CVD. Chem. Vap. Depos. **12**(6), 370–374 (2006).
42. I.-C. Chen, L.-H. Chen, C.A. Orme, S. Jin, Control of curvature in highly compliant probe cantilevers during carbon nanotube growth. Nano Lett. **7**(10), 3035–3040 (2007).
43. A. Quist, I. Doudevski, H. Lin, R. Azimova, D. Ng, B. Frangione, B. Kagan, J. Ghiso, R. Lal, Amyloid ion channels: A common structural link for protein-misfolding disease. Proc. Natl. Acad. Sci. USA **102**, 10427 (2005).
44. A.P. Quist, A. Chand, S. Ramachandran, C. Daraio, S. Jin, R. Lal, AFM imaging and electrical recording of lipid bilayers supported over microfabricated silicon chip nanopores, A lab on-chip system for lipid membrane and ion channels. Langmuir **23**(3), 1375 (2007).
45. J. Thimm, A. Mechler, H. Lin, S.K. Rhee, R. Lal, Calcium dependent open-closed conformations and interfacial energy maps of reconstituted individual hemichannels. J. Biol. Chem. **280**, 10646 (2005).
46. A. Stemmer, A. Hefti, U. Aebi, A. Engel, Scanning tunneling and transmission electron microscopy on identical areas of biological specimens. Ultramicroscopy **30**(3), 263 (1989).
47. J.H. Hafner, C.L. Cheung, A.T. Woolley, C.M. Lieber, Structural and functional imaging with carbon nanotube AFM probes. Prog. Biophys. Mol. Biol. **77**(1), 73–110 (2001).
48. R. Nicolaides, L. Yong, W.E. Packard, W.F. Zhou, H.A. Blackstead, K.K. Chin, J.D. Dow, J.K. Furdyna, M.H. Wei, R.C.J. Jaklevic, W. Kaiser, A.R. Pelton, M.V. Zeller, J. Bellina Jr., Scanning tunneling microscope tip structures. J. Vac. Sci. Technol. A **6**(2), 445–447 (1988).
49. J.P. Ibe, P.P. Bey, S.L. Brandow, R.A. Brizzolara, N.A. Burnham, D.P. DiLella, K.P. Lee, C. R.K. Marrian, R.J. Colton, On the electrochemical etching of tips for scanning tunneling microscopy. J. Vac. Sci. Technol. A **8**, 3570–3575 (1990).
50. L. Libioulle, Y. Houbion, J.-M. Gilles, Very sharp platinum tips for scanning tunneling microscopy. Rev. Sci. Instrum. **66**(1), 97–100 (1995).
51. A.J. Nam, A. Teren, T.A. Lusby, A.J. Melmed, Benign making of sharp tips for STM and FIM: Pt, Ir, Au, Pd, and Rh, J., Vac. Sci. Technol. B **13**(4), 1556–1559 (1995).

# Chapter 4
# Calibration of Normal and Lateral Forces in Cantilevers Used in Atomic Force Microscopy

**Manuel L.B. Palacio and Bharat Bhushan**

**Abstract** Atomic force microscopy (AFM) is an indispensable technique for nanoscale topographic imaging as well as quantification of normal and lateral forces exerted on the AFM tip while interacting with the surface of materials. In order to measure these forces, an accurate determination of the normal and lateral forces exerted on the AFM cantilever is necessary. In this chapter, we present a critical review of various techniques for measuring cantilever stiffness in the normal and lateral/torsional directions in order to calibrate the normal and lateral forces exerted on AFM cantilevers. The key concepts of each technique are presented, along with a discussion of their advantages and disadvantages.

## 4.1 Introduction

The popularity and ease of force measurements and imaging at the molecular level can be traced to the development of the surface force apparatus (SFA) and scanning probe microscopy (SPM) techniques [1, 2]. Schematics showing the principle behind these experimental tools are shown in Fig. 4.1. The SFA, first introduced in 1968, enables the measurement of the normal forces between two curved molecularly smooth surfaces (such as mica and silica) immersed in liquid or vapor. Aside from normal forces, friction forces are also measurable at varying sliding speeds or oscillating frequencies with the use of attachments.

Meanwhile, scanning probe microscopy techniques, developed in the 1980s, have been primarily used to obtain high resolution three-dimensional (3-D) images. These techniques rely on the use of a probe tip, which scans a specimen in order to generate an image. The most widely-used variations of SPM are the scanning tunneling microscope (STM) and the atomic force microscope (AFM), first introduced in 1981, and 1985, respectively. In STM, a bias voltage is applied between a metallic probe and the sample, allowing tunneling current flow. This tunneling current is monitored while the tip is scanned over the sample in order to generate a topographic image. In AFM, the force between the tip and sample is used (instead of the tunneling current) to produce an image. The tip can be in contact with the sample during imaging in AFM, whereas in STM, the tip is not in contact. As

B. Bhushan (ed.), *Nanotribology and Nanomechanics*,

DOI 10.1007/978-3-642-15283-2_4, 

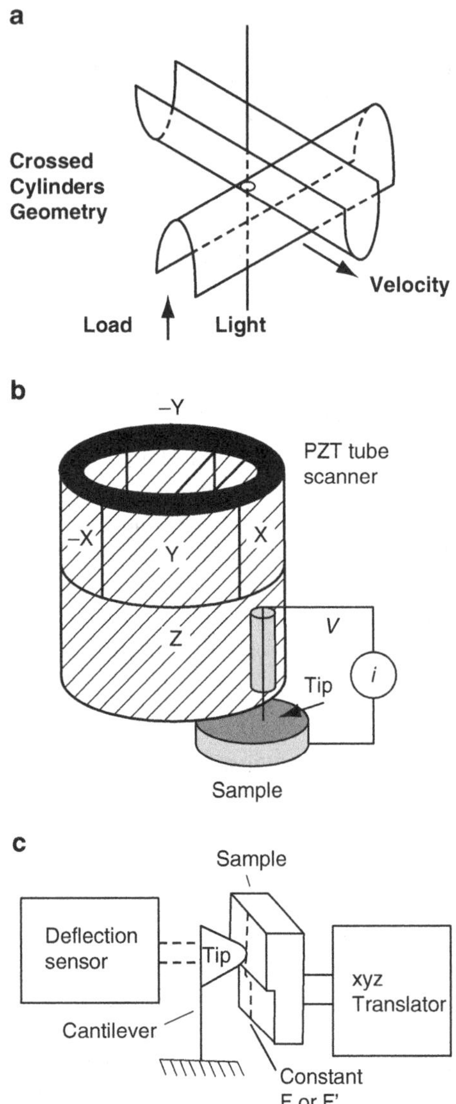


**Fig. 4.1** Schematics of the (**a**) surface force apparatus, (**b**) scanning tunneling microscope, and (**c**) atomic force microscope

a consequence, the AFM can be used for a variety of physical measurements. In addition, while STMs can only be used to investigate surfaces which are electrically conductive to some degree, AFMs do not suffer this limitation and can be used to study any material type [1, 2]. Over the years, the use of AFMs have expanded beyond surface profiling to other capabilities such as the measurement of adhesion, friction, elastic/plastic mechanical properties, electrical, magnetic and thermal properties, as well as in situ nanofeature creation and nanomanipulation. Out of the multiple capabilities of atomic force microscopy, novel technologies, such as probe-based data storage and various other microelectromechanical systems (MEMS)-based technologies, have emerged [1–3].

Multiple AFM designs exist and examples of commercial small-sample and large-sample AFMs are shown in Fig. 4.2a, b, respectively. In the small-sample AFM design, the sample, generally no larger than 10 mm $\times$ 10 mm, is mounted onto a lead zirconate titanate (PZT) tube scanner which consists of separate electrodes to precisely scan the sample in the x-y plane in a raster pattern and to move the sample in the vertical (z) direction, while in the large-sample AFM design, the sample is stationary while the tip is scanned. A sharp tip at the free end of a flexible cantilever is brought into contact with the sample. Features on the sample surface cause the cantilever to deflect vertically and laterally during scanning. A laser beam from a diode laser (5 mW max peak output at 670 nm wavelength) is directed by a prism onto the back of a cantilever near its free end, tilted downward at about 12° with respect to the horizontal plane. The reflected beam from the vertex of the cantilever is directed through a mirror onto a split photodetector with four quadrants (commonly called position-sensitive detector or PSD). The differential signal from the top and bottom photodiodes provides the AFM signal which is a sensitive measure of the cantilever vertical deflection. For surface imaging, the tip is scanned either along or transverse to the longitudinal axis of the cantilever. Topographical features of the sample cause the tip to deflect in the vertical direction as the sample is scanned. This tip deflection will change the direction of the reflected laser beam, changing the intensity difference between the top and bottom sets of photodetectors (AFM signal). In the AFM operating mode called the height mode, for topographical imaging or for any other operation in which the applied normal force is to be kept constant, a feedback circuit is used to modulate the voltage applied to the PZT scanner to adjust the height of the PZT, so that the cantilever vertical deflection (given by the intensity difference between the top and bottom detector) will remain constant during scanning. The PZT height variation is thus a direct measure of the surface roughness of the sample. The imaging is carried out either in contact mode or tapping mode. In contact mode, the tip is in contact at all times as it slides on the sample surface with the applied normal force kept constant. In tapping mode, the tip is oscillating as it slides on the sample surface, and is therefore not in contact at all times. The amplitude of oscillation is kept constant during the scan.

The friction force being applied at the tip during sliding can be measured using the quadrants on the left and right sides of the photodetector. In the so-called friction mode, the sample is scanned back and forth in a direction transverse to the long axis of the cantilever beam. Friction force between the sample and the tip will produce a twisting of the cantilever. As a result, the laser beam will be reflected out of the plane defined by the incident beam and the vertically reflected beam from an untwisted cantilever. This produces an intensity difference of the laser beam received in the left hand and right hand sets of quadrants of the photodetector. The intensity difference between the two sets of detectors (friction force

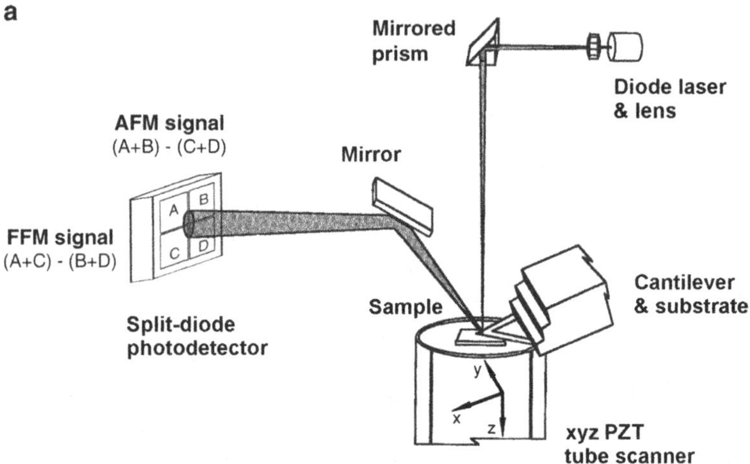


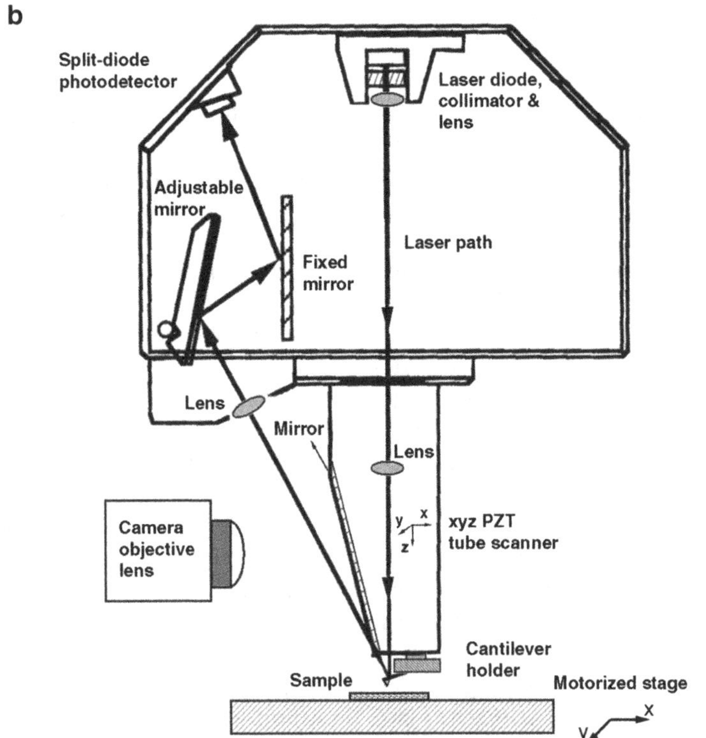


**Fig. 4.2** Principle of operation of commercial (**a**) small-sample AFM, and (**b**) large-sample AFM

microscope or FFM signal) is directly related to the degree of twisting and hence to the magnitude of the friction force. This method provides maps of the friction force [1, 2, 4 ].

Illustrations of the two common cantilever configurations, triangular (or V-shaped) and rectangular are shown in Fig. 4.3. These cantilevers are made from a wide range of materials, the most common being $Si_3N_4$ (triangular) and silicon (rectangular). Silicon nitride cantilevers are less expensive than those made of silicon. They are very rugged and well suited to imaging in various conditions [1, 2]. Microfabricated silicon nitride triangular beams with integrated square pyramidal tips made by plasma-enhanced chemical vapor deposition (PECVD) are most commonly used [5]. These cantilevers are typically coated with a thin gold film to increase the laser signal reflected to the photodetector. AFM cantilevers are commercially available from Veeco (Santa Barbara, CA), Nanosensors GmbH (Aidlingen, Germany), and NT-MDT (Moscow, Russia), among others. Typical specifications of a silicon nitride cantilever are 115–196 μm length, 17–41 μm width and 0.6 μm thickness (NP series, Veeco). The pyramidal tips are highly symmetric with the end having a radius of about 20–50 nm. The tip side walls have a slope of 35°, and the height of the tip is about 3 μm.

Microfabricated single-crystal silicon cantilevers with integrated tips are also used. Si tips are sharper than $Si_3N_4$ tips because they are directly formed by the anisotropic etch in single-crystal Si rather than using an etch pit as a mask for deposited materials [6]. Etched single-crystal n-type silicon rectangular cantilevers with square pyramidal tips typically have an end radius of less than 10 nm for contact and tapping mode. These cantilevers may be coated with a thin aluminum reflective film to increase the laser signal. Typical specifications for a rectangular cantilever

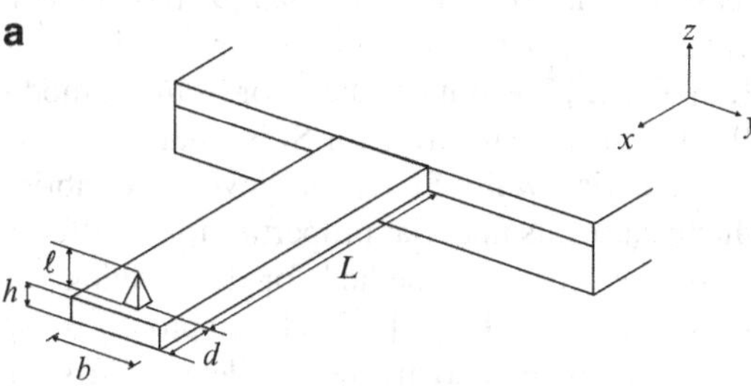


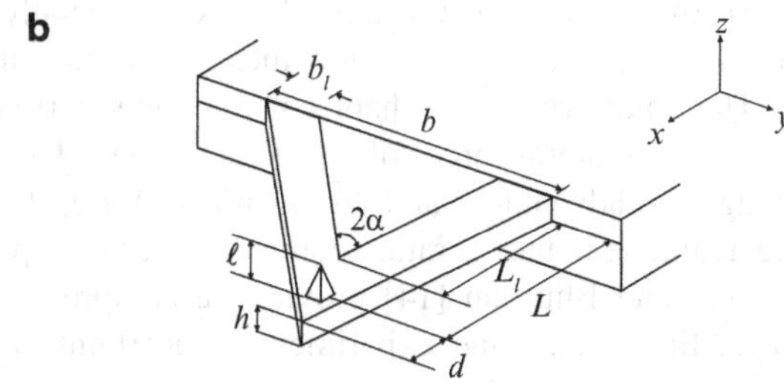


**Fig. 4.3** Typical AFM cantilevers with (**a**) rectangular, and (**b**) triangular geometries. The cantilevers have length $L$, width $b$, and height (or thickness) $h$. The height of the tip is $\ell$. The cantilever material is characterized by Young's modulus $E$, the shear modulus $G$ and a mass density $\rho$

for tapping mode applications are 125 μm length, 30 μm width, and 4 μm thickness. The tip side walls have a slope of 22.5°, and the tip height ranges from 10 to 15 μm (TESP series, Veeco). Since $Si_3N_4$ cantilevers are produced by thin film deposition, thinner cantilevers can be produced compared to those made with Si. $Si_3N_4$ is used for cantilevers with a normal stiffness of up to about 1 N/m and Si is used for higher stiffness applications.

An accurate determination of normal, bending, and torsion forces is necessary in order to measure the interaction forces between the surface and the AFM tip. This requires determining the stiffness of the cantilever used, as well as the PSD calibration in order to convert the measured signal (in Volts) into a force (in Newtons) [1, 2]. There are a number of techniques reported in the literature to obtain the PSD calibration.

With regards to the estimation of cantilever stiffness in general, it will be shown in the next section that the stiffnesses are directly proportional to the cube of the thickness of the cantilever. This presents a source of uncertainty for the stiffness determined by using fundamental beam theory. Microfabricated cantilevers often have non-uniformity in their dimensions, including their thickness. Average measurement errors for the cantilever length, width, and thickness are about 1%, 4%, and 5%, respectively [7]. The non-uniformity of the reflective coating used (Au or Al) also adds to the thickness uncertainty. Scanning electron microscopy (SEM) should be used instead of optical microscopy in order to measure these dimensions accurately. However, this could be time-consuming and difficult to implement in a routine manner [8].

There is also some variation coming from the mechanical properties of the cantilever material used in the calculations. For example, by using an ultrasonic measurement, Ruan and Bhushan [9] found the Young's modulus of the cantilever beam to be about 238 ± 18 GPa, which is less than that of bulk $Si_3N_4$, 310 GPa [1, 2]. Ohler [7] estimates the error from the modulus measurement at 5% for Si and 20% for $Si_3N_4$. The error for Si is smaller because the cantilever is made from the bulk material, while $Si_3N_4$ cantilevers are fabricated through a deposition process which causes its material properties to be different from the bulk.

Another common method for determining cantilever stiffness is through finite element analysis (FEA) [10]. This approach is especially useful for non-rectangular configurations, e.g., triangular cantilevers. FEA suffers from inaccuracy of the dimensions of the cantilever, as well as the mechanical properties needed. Since there are uncertainties in stiffness determination inherent from the beam theory-based approach and FEA methods, there is a need to experimentally determine the stiffness of AFM cantilevers in order to accurately measure normal and lateral forces.

The objective of this chapter is to review various techniques for calibrating the normal and lateral forces of AFM cantilevers. The key concepts will be discussed, along with advantages and disadvantages for each technique. A number of articles are available in the literature comparing various specific techniques (e.g., [11–14]). Palacio and Bhushan [14] provided a comprehensive review of the theory, features, limitations, and experimental uncertainties, which is needed to guide the experimentalist in the proper implementation of the various normal and lateral force calibration techniques proposed to date. In this chapter, Sect. 4.2 discusses the analytical methods used to determine the normal and lateral stiffness. Section 4.3 reviews the various experimental techniques proposed for evaluating the normal stiffness, while Sect. 4.4 reviews the different techniques for calibrating the lateral force and/or lateral stiffness of AFM cantilevers. This chapter is based in part on the review article by Palacio and Bhushan [14] on this subject.

## 4.2 Analytical Approaches for Determining the Cantilever Normal and Lateral Stiffness

The analytical methods used to determine the normal and lateral stiffness of the cantilever is briefly reviewed in this section.

The normal stiffness can be calculated using the geometrical and physical properties of the cantilever material [4, 5, 15]. By definition, the normal stiffness $k_z$ (also commonly referred to as the spring constant or spring stiffness) is given by $F_z = k_z \Delta z$, where $F_z$ is the normal force, and $\Delta z$ is the deflection of the cantilever beam. From fundamental beam theory, the expression for the stiffness of a rectangular cantilever beam with a uniform cross section, with the load applied on its end and experiencing small deflections, is given by [16, 17]

$$k_z = \frac{3EI}{L^3} \tag{4.1}$$

where $E$ is the Young's modulus of the material, $L$ is the length of the beam, and $I$ is the area moment of inertia of the cross section. Equation (4.1) assumes that the cantilever is homogeneous, isotropic, and exhibits linear elastic behavior. For a rectangular cross section with a width $b$ and a height (or thickness) $h$, one obtains an expression for $I$ [16, 17]

$$I = \frac{bh^3}{12} \tag{4.2}$$

$$k_z = \frac{Eb}{4}\left(\frac{h}{L}\right)^3 \tag{4.3}$$

Equation (4.3) is the intrinsic stiffness of the cantilever. For the purpose of calculating the stiffness of an AFM cantilever, a correction has to be made which accounts for the cantilever's tilt (on the order of 12°) relative to the horizontal axis. The "effective" stiffness is given by [7, 18]

$$k_{z,eff} = \frac{k_z}{\cos^2\gamma} \tag{4.4}$$

where, $\gamma$ is the tilt angle of the cantilever.

If a lateral force $F_y$ is applied to the end of the cantilever beam, the cantilever will bend sideways. The bending stiffness in the lateral direction, $k_{yB}$, is given by $F_y = k_{yB}\Delta y$, where $\Delta y$ is the deflection in the y-direction. The quantity $k_{yB}$ can be calculated with (4.3) by exchanging $b$ and $h$ [16, 17]

$$k_{yB} = \frac{Eh}{4}\left(\frac{b}{L}\right)^3 \tag{4.5}$$

Therefore, the bending stiffness in the lateral direction is larger than the stiffness for bending in the normal direction by $(b/h)^2$.

When the lateral force $F_y$ is applied at the end of the tip, as in the case of the AFM cantilever, lateral deflection is accompanied by twisting or torsion along the cantilever axis. Instead of $k_{yb}$, we now define $k_{yT}$ from $F_y = k_{yT}\Delta y$. For a wide, thin cantilever ($b >> h$) experiencing a rotation $\phi$, the torsional stiffness of the beam, $k_\phi$, is given as $M = k_\phi \phi$, where $M$ is the torque or torsion moment. From beam theory, the torsional stiffness is defined as follows [16, 19]

$$k_\phi = \frac{Gbh^3}{3L} \tag{4.6}$$

where $G$ is the modulus of rigidity or shear modulus [$= E/2(1 + \nu)$, where $\nu$ is the Poisson's ratio]. The relationship between $E$ and $G$ is valid for homogeneous, isotropic, and linear elastic materials. The lateral stiffness of the cantilever-tip assembly with the lateral load applied at the end of the tip (torque) is defined as [1, 2, 16]

$$k_{yT} = \frac{k_\phi}{\ell^2} \tag{4.7}$$

where $\ell$ is the length of the tip mounted at the end of the cantilever. For completeness, we note that the ratio $k_{yT}/k_z$ is independent of the cantilever thickness and width, and it is much larger than one.

The normal stiffness of triangular cantilevers has been approximated by assuming that the cantilever is composed of two rectangular cantilevers in parallel. This "parallel beam approximation" (PBA) was first proposed by Albrecht et al. [5], and modifications were suggested by Butt et al. [20] and Sader [21]. In the latter, Sader [21] argued that inappropriate width and length of the two rectangular cantilevers were used by Albrecht et al. [5] and Butt et al. [20]. Based on finite element analysis, the formula for the normal stiffness for the triangular cantilever is of the form [21]

$$k_z = \frac{Eh^3 b_l}{2(L+d)^3}\left(1 + \frac{4b_l^3}{b^3}(3\cos\alpha - 2)\right)^{-1} \tag{4.8}$$

where $\alpha$ is one-half the included angle between the legs of the cantilever.

An approach considered to be more accurate than the PBA was proposed by Neumeister and Ducker [10], who modeled the triangular cantilever by subdividing it into two parts, namely, a triangular plate corresponding to the front part and two prismatic beams corresponding to the legs of the cantilever. They derived expressions for the normal, lateral, and torsional stiffnesses. This approach was further modified by Clifford and Seah [11].

The normal stiffness of commercial triangular $Si_3N_4$ cantilevers is typically from 0.01 to 0.6 N/m [22]. These cantilevers have a typical width to thickness ratio of 10–30. The width to thickness ratio and the triangular geometry results in 100–1,000 times greater stiffness values in the lateral direction compared to the normal direction. Using the Young's modulus value of Si(111) which is 181 GPa, the rectangular Si cantilever described earlier (TESP series, Veeco) has stiffness values of typically 40 N/m, 2,500 N/m, 2,230 N/m and $0.36 \times 10^{-6}$ Nm in the normal ($k_z$), lateral bending ($k_{yB}$), lateral bending due to torque ($k_{yT}$), and torsional ($k_\phi$) directions, respectively. A cantilever beam required for tapping mode is quite stiff and may not be sensitive enough and is therefore not well suited for measuring torsion. For friction measurements, the torsional stiffness should be minimized in order to be sensitive to the lateral forces. Long cantilevers with small thickness and large tip length are most suitable. An example of the dimensions for a rectangular silicon cantilever for high lateral force sensitivity is 200 μm length, 21 μm width, 0.4 μm thickness, and 12.5 μm tip length, which gives stiffness values of 0.007 N/m, 21 N/m, 1 N/m and $1.6 \times 10^{-10}$ N m in the normal ($k_z$), lateral bending ($k_{yB}$), lateral bending due to torque ($k_{yT}$), and torsional ($k_\phi$) directions, respectively. This cantilever has a lateral force sensitivity of 10 pN, assuming an angular resolution of $10^{-7}$ rad. With this particular geometry, sensitivity to lateral forces is improved by about a factor of 100 or more compared with more commonly used triangular $Si_3N_4$ or rectangular Si or $Si_3N_4$ cantilevers [1, 2].

Triangular cantilevers have been assumed to have a high lateral bending stiffness (relative to rectangular cantilevers), which minimizes lateral deflection of the cantilever during imaging [5, 23]. This assumption has been questioned by Sader and co-workers, who conducted theoretical analysis and experiments to test this assumption [24, 25]. Their theoretical approach involved the determination of the lateral resistance $R$, defined as follows

$$R_x = \frac{k_x}{k_z} \tag{4.9}$$

$$R_y = \frac{k_y}{k_z} \tag{4.10}$$

where $k_x$ is the lateral stiffness in the direction parallel to the longitudinal axis of the cantilever. The lateral resistances of the rectangular and triangular cantilevers are comparable only when the latter has narrow legs (i.e., $b_l/b$ of the triangular cantilever shown in Fig. 4.3 is close to 0.1). However, typical triangular cantilevers have $b_l/b > 0.1$. In general, it was found that the rectangular cantilever is stiffer (less prone to bending) than the comparable triangular cantilever by as much as seven and four times in the x- and y-directions, respectively [24]. Experimentally, this was confirmed on model macroscopic cantilevers where a rod was attached on the ends. By applying a torque and measuring the rotation angle, it was found that rectangular cantilevers are less prone to rotation, as predicted by theory [25]. These results imply that the assumption stated above is incorrect as rectangular cantilevers are actually less susceptible to lateral forces [24, 25]. It appears that the use of triangular cantilevers is historical with no obvious advantage.

As indicated in the Introduction, errors in the measured cantilever dimensions and the mechanical properties of the cantilever material will lead to uncertainties in calculating the stiffness by using the analytical approaches presented in this section, such that both the normal and lateral cantilever stiffness have to be determined experimentally. Sections 4.3 and 4.4 describe the various techniques for calibrating the normal stiffness and the lateral force/stiffness, respectively.

## 4.3 Normal Stiffness Calibration Techniques

Experimental methods for measuring the normal stiffness of cantilevers are described in this section. The methods are mainly divided into static and dynamic techniques. Static methods rely on the cantilever deflection, either through a reference cantilever or reference mass. Dynamic methods depend on the oscillation of the cantilever. For the dynamic methods, a common parameter is the frequency $f$, of the cantilever, defined as [17]

$$f = \frac{\omega}{2\pi} = \frac{1}{2\pi}\sqrt{\frac{k_z}{m_{eff}}} \tag{4.11}$$

where ω is the angular frequency and $m_{eff}$ is the effective mass, which varies depending on the cantilever used. For instance, for a rectangular cantilever with $L/b > 5$, its effective mass is approximated as $m^* \approx 0.2427\ m$, where $m$ is the mass of the cantilever (with $m = \rho_c bhL$ and $\rho_c$ is the density of the cantilever). Another important parameter is the quality factor, $Q$ (also known as the Q-factor). The Q-factor is a metric which describes energy loss during an oscillation. A high Q-factor value indicates low energy dissipation per cycle, and is therefore desirable for cantilevers.

### 4.3.1 *Static Methods*

#### Reference Cantilever Method

One way of determining the cantilever stiffness is to use another larger cantilever with a known stiffness, i.e., a reference cantilever. Ruan and Bhushan [9] used a stainless steel spring sheet of known stiffness (width = 1.35 mm, thickness = 15 μm, free hanging length = 5.2 mm). One end of the spring was attached to the sample holder, and the other end was made to contact the cantilever tip during the measurement, see Fig. 4.4. They measured the piezo traveling distance for a given cantilever deflection. For a rigid sample (such as diamond, Fig. 4.4a), the piezo traveling distance $Z_t$ (measured from the point where the tip touches the sample) should equal the cantilever deflection. $Z_t$ should be determined beforehand in order to differentiate between the deflection of the cantilever and the spring sheet, as shown in Fig. 4.4b. To keep the cantilever deflection at the same level using a flexible spring sheet, the new piezo traveling distance $Z_t'$ would be different from $Z_t$. The difference between $Z_t'$ and $Z_t$ corresponds to the deflection of the spring sheet. If the stiffness of the reference is $k_s$, the stiffness of the cantilever $k_z$ can be calculated by [9]

$$(Z_t' - Z_t)k_s = Z_t k_z$$

or

$$k_z = k_s(Z_t' - Z_t)/Z_t \tag{4.12}$$

A source of uncertainty in this method is the offset from positioning the cantilever relative to the end of the reference. Since the reference gets stiffer as load is applied farther from the end, the cantilever should contact the reference as close to its end as possible. If there is an offset in the positioning (such as that shown schematically in Fig. 4.3), $k_z$ can be corrected, e.g., for a rectangular beam by using [8]

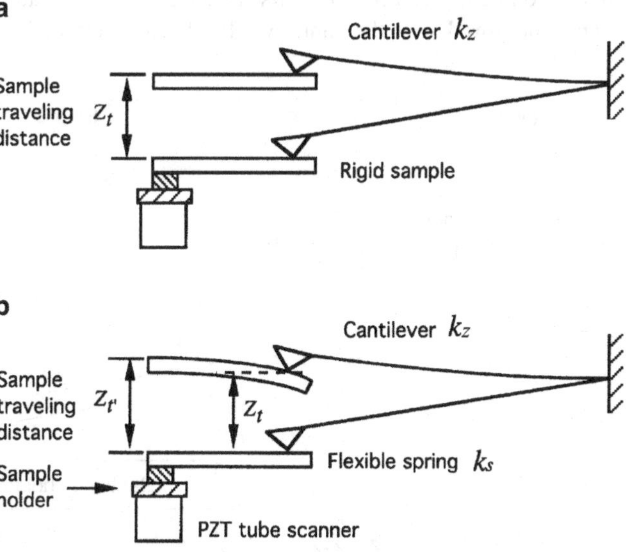


**Fig. 4.4** Illustration showing the deflection of cantilever as it is pushed by (**a**) a rigid sample, or by (**b**) a flexible spring sheet [9]

$$k_z = k_{z,off}\left(\frac{L+d}{L}\right)^3 \tag{4.13}$$

where $k_{z,off}$ is the normal stiffness measured with the offset, $L + d$ is the total length of the reference cantilever, and $d$ is the distance away from the end of the reference that the load is applied.

The advantage of the reference cantilever method is that it is experimentally simple. The deflection voltage signals can be used without the need to convert into meters as the calibration constant will cancel out in (4.12). However, this technique requires accurate positioning of the cantilever of interest relative to the reference cantilever, as discussed above. In addition, since the sharp tip is in contact with a single asperity, extreme roughness on the reference surface may cause variations on the measured piezo travel distance.

Commercial reference cantilevers referred to as a "force calibration chip" contain three rectangular cantilevers on single crystal silicon, and are available from Park Scientific Instruments [26]. The cantilevers have lengths of 97, 197, and 397 μm, and have equal thickness of 2 μm and equal width of 29 μm. For these cantilevers, the stiffness ranged from 0.16 to 10 N/m, as determined using the resonance method (to be discussed in Sect. 4.3.2.2).

### Inverted Loaded Cantilever Method

In this method, particles are attached to the end of the cantilever. As shown in Fig. 4.5, the deflection of the cantilever is first measured after the addition of the particle and then remeasured after the cantilever has been inverted. The difference

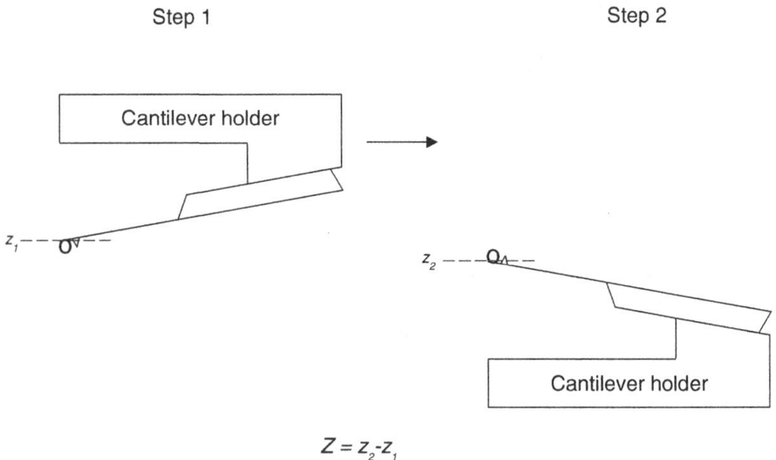


**Fig. 4.5** Schematic of the inverted cantilever method

in the deflection, $Z$ (in volts), is twice the deflection due to gravity. The stiffness is expressed as [27]

$$k_z = \frac{2m_s g}{z} = \frac{8\pi r^3 \rho_s g}{3z} \tag{4.14}$$

where the mass of the spherical particle is given in terms of $r$ and $\rho_s$, its radius and density, and $g$ is the gravitational constant. In (4.14), $z = \Omega Z$, where $z$ and $Z$ correspond to the deflection difference in meters and volts, respectively, and $\Omega$ is the deflection calibration which relates the cantilever deflection to the detector voltage reading. Senden and Ducker [27] used tungsten spheres that are 10–50 μm in diameter.

The attachment of a particle and the inversion procedure both pose a risk of damage to the cantilever being analyzed. In addition, uncertainties in the radius and density of the particle will lead to errors in the measured $k_z$ value. SEM would be a more preferable technique to optical microscopy in obtaining accurate particle dimensions. Optical microscopy is known to give an average error of about 5–10% in determining the dimensions of the particle [7].

## Pendulum Method

A schematic illustrating this technique is shown in Fig. 4.6. As the name implies, the stiffness is determined by pushing the tip against the pendulum and measuring the deflection of the pendulum and the cantilever, $z_p$ and $z_c$, respectively. The force exerted by the cantilever and the component of the gravitational force of the pendulum in the direction of motion are the same at equilibrium [20]

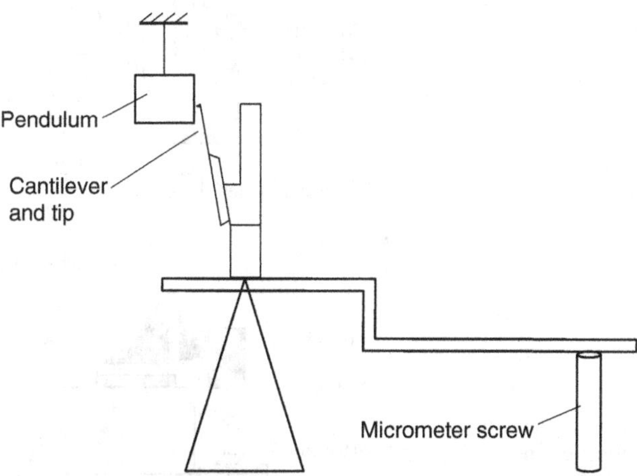


**Fig. 4.6** Schematic of the pendulum method

$$F = k_z z_c = \frac{m_p g z_p}{l_p} \tag{4.15}$$

where $m_p$ is the mass of the pendulum, and $l_p$ is the length of the pendulum. The stiffness is given by

$$k_z = \frac{m_p g z_p}{l_p z_c} \tag{4.16}$$

The main disadvantage of this technique is the need for the pendulum set-up. The experiment itself presents a risk of damaging the cantilever. Calibration of the pendulum deflection is necessary, and this could be a source of measurement errors.

### Miscellaneous Methods

The following are examples of methods where additional instrumentation needs to be implemented to determine the normal stiffness. The Electrostatic Force Balance (EFB), its schematic shown in Fig. 4.7a, was developed at the United States National Institute of Standards and Technology [28, 30]. Electrostatic force acting

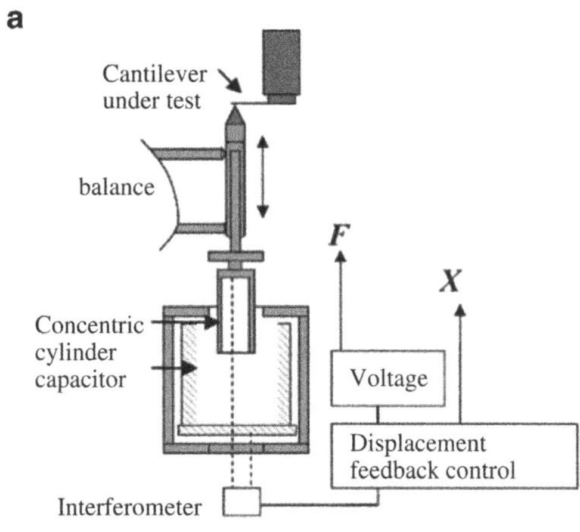


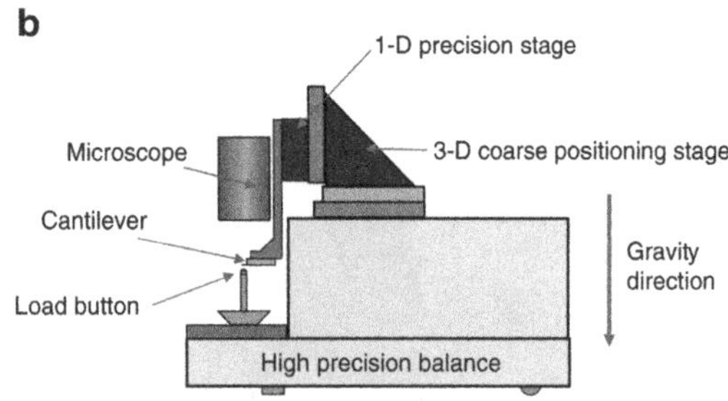


**Fig. 4.7** Schematics of the (**a**) electrostatic force balance (Adapted from [28]), and (**b**) nano force calibrator [29]

along the vertical axis is generated when voltages are applied to the coaxial cylinders. The displacement is then monitored with an interferometer. Another example is the so-called Nano Force Calibrator (NFC), its schematic shown in Fig. 4.7b, was developed at the Korea Research Institute of Standards and Science [29, 31]. In this setup, the cantilever is placed in contact with a precision balance, and controlled displacement is applied by a stage capable of moving in the nanometer range. It should be mentioned that other comparable force balances, as well as stiffness artifacts (standard cantilevers) have been developed in the National Physical Laboratory in the United Kingdom [32, 33] and in the Physikalisch-Teknische Bundesanstalt (PTB) in Germany [34, 35]. The main advantage of these techniques is that the measured force is traceable to the Systeme International d Unites (SI units). The disadvantage is that the necessity for additional instrumentation, and especially in the case of the NFC, environmental factors such as acoustic noise, thermal fluctuations, and air flow fluctuations will cause measurement inaccuracies.

## *4.3.2 Dynamic Methods*

### Added Mass Method

The added mass method was proposed by Cleveland et al. [36] and is also referred to as the "frequency scaling" technique. In this method, the resonance frequencies of cantilevers are measured before and after the addition of small masses (such as tungsten particles with diameter 7–16 μm and mass 2.8–44 ng) at the tip of the cantilever. A schematic is shown in Fig. 4.8a.

The normal stiffness of the cantilever can be obtained from its effective mass and angular resonance frequency. By using (4.3) and (4.11), and with $m_{eff} = m^*$, the angular resonance frequency can be expressed as

$$\omega_0 \approx \frac{h}{2L^2}\left(\frac{E}{0.2427\rho_c}\right)^{1/2} \approx \frac{h}{L^2}\left(\frac{E}{\rho_c}\right)^{1/2} \tag{4.17}$$

Equation (4.17) and the approximation for the effective mass given above both require the dimensions and material properties of the cantilever, which as described earlier are susceptible to measurement errors. In addition, the resonance frequency defined in (4.17) is for vacuum, not in air. (A correction that accounts for the surrounding medium is discussed in Sect. 4.3.2.2 below.) Therefore, there is a need to determine the resonance frequency and effective mass directly from the measurements in order to obtain $k_z$ more accurately.

In the added mass method, a spherical particle with mass $m_s$ is added to the cantilever, such that the effective mass is $m_{eff} = m^* + m_s$. The angular resonance frequency $\omega$ changes correspondingly as a result of particle addition ($\omega \neq \omega_0$).

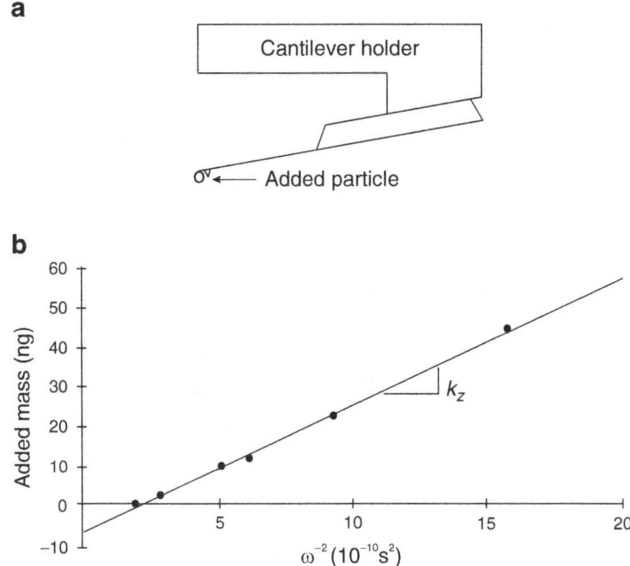


**Fig. 4.8** (**a**) Schematic of the side view of a cantilever with an added mass, and (**b**) plot of added mass vs. $\omega^{-2}$ for a single cantilever, where the slope gives the stiffness (adapted from [36])

The stiffness expression in (4.11) can be modified to account for the added mass as follows [36]

$$m_s = \frac{k_z}{\omega^2} - m^* \tag{4.18}$$

The resonance frequency is determined during the tuning procedure, where the cantilever is vibrated at a certain frequency range, usually between 10 and 400 kHz, and the frequency which produces maximum amplitude response is determined. By adding a number of different masses to the cantilever and measuring the resonance frequency at each time, a plot of $m_s$ vs. $\omega^{-2}$ yields a straight line with a slope of $k_z$ and a y-intercept of $m^*$, as shown in Fig. 4.8b.

In the method described in the previous paragraph, multiple masses are used, and $k_z$ is obtained from the line fit. One can also obtain $k_z$ using a single mass. In this case, $k_z$ can be solved for by expressing (4.11) with and without the added mass, such that

$$k_z = \frac{m_s}{\left(1/\omega^2 - 1/\omega_o^2\right)} \tag{4.19}$$

The data obtained with the added mass method was found to follow a scaling law where the stiffness is proportional to the cube of the unloaded resonance frequency

$$k_z = b\left(\omega_0 L\sqrt{\rho_c}\right)^3/4\sqrt{E} \tag{4.20}$$

where the length $L$, width $w$, elastic modulus $E$, and density $\rho_c$ are known. The stiffness can also be evaluated with (4.20); however, the use of cantilever dimensions introduces error. Therefore, the result is not as precise as that obtained by the addition of various end masses (4.18).

A correction could be made to account for the distance that the added particle is offset from the end of the cantilever by $d$ (Fig. 4.3). As an example, for a rectangular beam, $m_s$ can be corrected by [8]

$$m_s = m_{s,off}\left(\frac{L}{L+d}\right)^3 \tag{4.21}$$

where $m_{s,off}$ is the effective mass of the particle measured with the offset, which can then be used in either (4.18) or (4.19) in place of $m_s$.

This method is time-intensive and has the risk of damaging the cantilever during the addition of the particles. In practice, the difficulty of positioning the particle on the cantilever is also an issue [7, 8].The measured stiffness has an uncertainty of approximately 15–30% .The variation occurs mainly from errors in measuring the diameter of the added particle. The particles may not be a perfect sphere, so the volume should be calculated as an ellipsoid and careful measurements must be taken with the SEM. Another source of error is the mass of the cantilever, as its dimensions and density are used in the calculations.

### Resonance Method

Sader et al. [8] suggested calculating $k_z$ from the measurement of cantilever resonance frequency in vacuum (in the absence of a fluid medium) $\omega_{\mathrm{vac}}$, using (4.11) with $m_{\mathrm{eff}} = m^*$.

As stated earlier, the resonance frequency can be obtained during the cantilever tuning procedure. This approach poses some limitations. Measurement of cantilever thickness, density and mass are sources of error. Moreover, this approach assumes a cantilever frequency in vacuum and does not account for the surrounding medium (air or liquid). Sader [37] proposed an approach which accounts for the fluid medium and eliminates the need for measuring the density of the cantilever. The resonance frequency for a rectangular cantilever in vacuum is related to the resonance frequency in fluid $\omega_f$ by

$$\omega_{vac} = \omega_f\left(1+\frac{\pi\rho_f b}{4\rho_c h}\Gamma_r(\omega_f)\right)^{1/2} \tag{4.22}$$

The cantilever density is given by the following

$$\rho_c h = \frac{\pi\rho_f}{4}\left[Q\Gamma_i(\omega_f) - \Gamma_r(\omega_f)\right] \tag{4.23}$$

In (4.22 and 4.23), $\rho_f$ is the density of the fluid medium, and $\Gamma_r$ and $\Gamma_i$ are the real and imaginary components of the hydrodynamic function, $\Gamma$, which, in turn, depends on the Reynolds number $\mathrm{Re} = \rho_f\, \omega_f\, b^2/4\eta$. The hydrodynamic function is obtained using an analytical expression applicable to rectangular beams. The quantity $\eta$ is the viscosity of the fluid medium, which is independent of cantilever thickness and density. Substituting (4.22) and (4.23) into (4.11), the stiffness is given by [38]

$$k_z = 0.1906\rho_f b^2 L Q \omega_f^2 \Gamma_i(\omega_f) \tag{4.24}$$

This technique is not prone to the experimental uncertainties inherent in determining thickness and density of the cantilever, but it still requires knowledge of its width and length. In addition, the model requires knowledge of the Reynolds number of the fluid medium, which is a disadvantage as it could change a lot depending on the elevation and must therefore be determined accurately. The hydrodynamic function calculation is not straightforward and can only be solved for by approximate analytical solutions such as the one by Sader [37]. However, for relatively stiff cantilevers (where the expected $k_z > 1$ N/m), this method is highly desirable. Since measurement of the resonance frequency is highly accurate (average error ~0.1%), the uncertainty of the resulting $k_z$ is low, ~4% [7].

It must be noted that the resonance method discussed here is based on the geometry of a rectangular cantilever beam. However, the analytical solution to the hydrodynamic function needs to be extended to the geometry of triangular cantilevers. The normal stiffness obtained from the resonance method on a rectangular cantilever can be used to calibrate a triangular cantilever as long as both rectangular and triangular cantilevers are present on a single wafer. The assumption is that the rigidity $Eh^3$ of the rectangular and triangular cantilevers is the same since they are on the same wafer [8]. For the rectangular cantilever, $Eh^3$ is solved for by using (4.3), with $b$ and $L$ being known and $k_z$ obtained from the resonance method. Afterwards, this value for $Eh^3$ is used on the triangular cantilever by using the appropriate expression for the normal stiffness, provided that the length, width, and included angle of the triangular cantilever are known. Wafers containing rectangular and triangular cantilevers are not widely available. Most commercial cantilevers are fabricated such that the wafer only contains either rectangular or triangular cantilevers, not both.

More recently, Sader et al. [39] proposed a general expression that enables one to determine the stiffness for any geometry of the cantilever. The stiffness is derived based on its relationship to the energy dissipation of the cantilever during oscillation. Through dimensional analysis, the stiffness is related to the density of the fluid, the cantilever length, the Q factor, resonance frequency, and the Reynolds number of the fluid as follows

$$k_z = \rho_f L^3 \Omega(\mathrm{Re}) Q \omega_f^2 \tag{4.25}$$

The dimensionless function $\Omega(\mathrm{Re})$ is defined depending on the geometry. For a rectangular cantilever, an analytical solution is known such that $\Omega(\mathrm{Re})$ can be expressed in terms of the hydrodynamic function $\Gamma(\omega_f)$, which leads to the expression in (4.24). For a triangular cantilever, an analytical solution is not available and $\Omega(\mathrm{Re})$ has to be determined by performing experiments in a fluid of interest to measure the resonance frequency and Q factor as a function of fluid pressure. Then, an expression for the normal stiffness is obtained from a curvefit.

The resonance method can also be implemented using a Laser Doppler vibrometer (LDV) instead of an AFM [40]. This provides independent validation of the results obtained from AFM. However, the disadvantage of LDV is that this requires additional instrumentation aside from the AFM, which might not be easily accessible.

### Thermal Noise Method

A harmonic oscillator in equilibrium with its surroundings will fluctuate in response to thermal noise [41]. As shown in Fig. 4.9a, the AFM tip-sample system is modeled as a spherical tip held at a distance from the sample surface by a spring (representing the cantilever). The spring is assumed to behave like a simple harmonic oscillator.

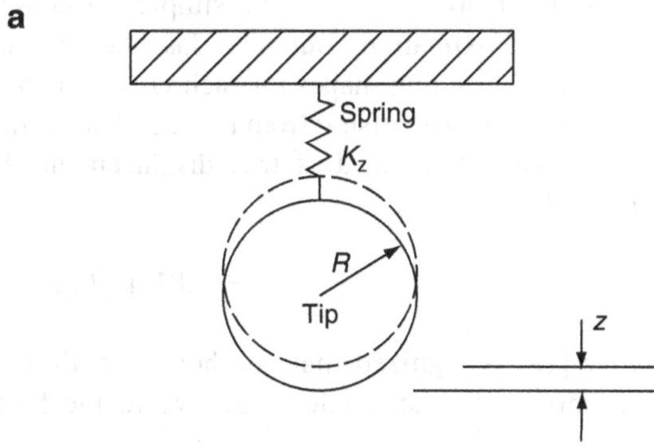


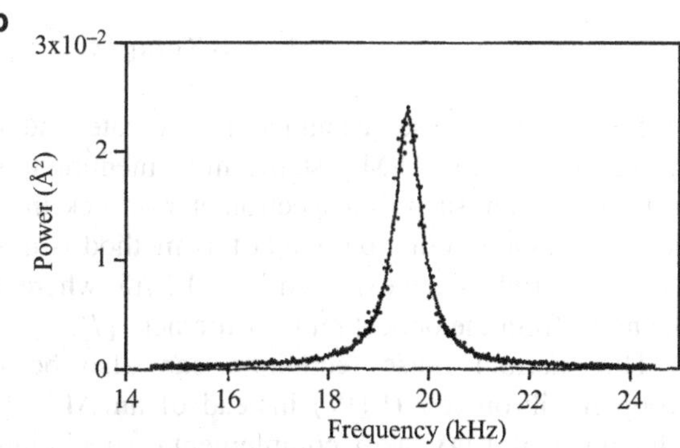


**Fig. 4.9** (**a**) Illustration of the principle behind the thermal noise method, and (**b**) power spectral density plot of the cantilever deflection fluctuations (Adapted from [41])

The normal stiffness of the AFM cantilever can be related to its thermal energy during its vibration by the equipartition theorem, leading to the following relationship

$$\frac{1}{2}k_z\langle z^2\rangle = \frac{1}{2}\mathrm{k}_B T \tag{4.26}$$

where $\mathrm{k_B}$ is the Boltzmann constant, $T$ is the temperature, and $<z^2>$ is the mean square deflection of the cantilever, which fluctuates due to thermal noise.

An example of the power spectrum of cantilever deflection fluctuations is shown in Fig. 4.9b. From experimental data, $p$, the area of the power spectrum of the thermal fluctuations is equal to $<z^2>$, so the stiffness can be expressed as [41]

$$k_z = \mathrm{k_B}T/p \tag{4.27}$$

The thermal noise method (also referred to as the thermal tune method) for determining the normal stiffness is widely regarded as being less prone to experimental uncertainties, as it eliminates the need for the dimensions and the mechanical properties of the cantilever. However, a disadvantage of this method is that cantilevers do not behave perfectly like simple harmonic oscillators, such that (4.27) is only an approximation. Butt and Jaschke [42] proposed a correction which accounts for the bending shapes for each vibration mode. Moreover, they accounted for the additional error arising from the fact that the inclination of the cantilever ($dz(L)/dx$) is measured instead of true displacement. They proposed the following equation [42]

$$k_z = 0.817\mathrm{k_B}T/p \tag{4.28}$$

Hutter [18] recognized that another correction is necessary since the cantilever is mounted at an angle $\gamma$ relative to the horizontal axis, leading to this equation

$$k_z = 0.817\mathrm{k_B}T\cos^2\gamma/p \tag{4.29}$$

The thermal noise technique is accurate and relatively simple to perform. However, for some AFM systems, implementing this technique requires additional instrumentation, such as a spectrometer or lock-in amplifier to collect the thermal noise data. One limitation is that this method is most suitable for calibrating soft cantilevers (where the expected $k_z < 1$ N/m) where the thermal noise is higher than the noise from the deflection measurement [7].

The thermal noise technique can also be implemented using a Laser Doppler vibrometer (LDV) instead of an AFM [40]. The advantage and disadvantage of LDV as a complementary technique has been discussed above (Sect. 4.3.2.2).

### 4.3.3 Discussion

The main highlights of the normal stiffness calibration methods discussed here are shown in Table 4.1. Three static measurement techniques were reviewed, namely, the reference, inverted cantilever, and pendulum methods. The reference cantilever method is simple and straightforward. Its only limitation is that it requires accurate positioning of the cantilever relative to the reference during the experiment in order to obtain accurate results. In the inverted loaded cantilever experiment, the addition of a sphere presents the risk of damaging the cantilever, rendering it unsuitable for further use. In addition, the necessity for calibrating the observed deflection signal corresponding to cantilever motion further complicates the experiment. The pendulum method is based on a concept similar to the inverted cantilever technique in the sense that it is also a gravity-based experiment. In this case, the mass of the pendulum is required, and this quantity can be accurately determined. However, calibrating the pendulum deflection is needed, which could be a source of measurement uncertainty. The Electrostatic Force Balance and the Nano Force Calibrator are examples of experimental normal stiffness calibration techniques where the force application is traceable to SI units. However, these setups need to be built and additional instrumentation is necessary.

The dynamic experiments discussed above are the added mass, resonance, and thermal noise methods. Similar to the static inverted cantilever method, the added mass method involves adding a particle to the cantilever, which is time-consuming and has the risk of damaging the cantilever if the particle is added improperly. In addition, uncertainties in the dimensions of the cantilever and errors in placement of the particle lead to significant error. The resonance method has been shown to be accurate for stiffer cantilevers, i.e. normal stiffness higher than 1 N/m. It requires measurement of the cantilever width, which gives the result some uncertainty. Also, since the model is derived for a rectangular cantilever beam, there is a need to use the extended theoretical treatment so that it can be applied to triangular cantilevers as well.

The thermal noise method does not require any parameters related to the cantilever beam dimension, which makes it accurate. The precision of reproducibility of the obtained stiffness using the thermal noise method has been reported by Hutter and Bechhoefer [41] and Matei et al. [43] to be as high as 5%. However, this method is mostly limited for calibrating soft cantilevers where the thermal noise is higher than the noise from the deflection measurement. Depending on the AFM system, the technique may require additional instrumentation such as a spectrometer or lock-in amplifier (which are commonly available instruments) in order to obtain the thermal noise spectra.

The tilt of the AFM cantilever affects the normal stiffness, and a correction could be necessary depending on the calibration method used. The added mass and resonance methods yield the intrinsic stiffness, i.e., the value that one would obtain assuming no tilt, so results from these methods should be modified by using (4.4) (introduced in Sect. 4.2). This correction is not necessary for the reference cantilever method since the "effective" stiffness is obtained in the measurement [7].

**Table 4.1** Summary of techniques for calibrating the normal stiffness

| Technique | Principle/key equation | Pros | Cons |
|---|---|---|---|
| *Static methods* | | | |
| Reference cantilever [9] 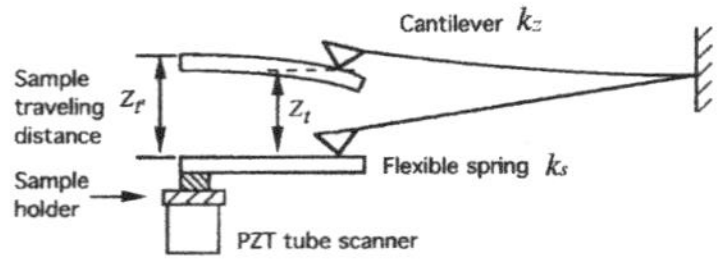<br>Sample traveling distance $Z_r$; $Z_t$; Cantilever $k_z$; Flexible spring $k_s$; Sample holder; PZT tube scanner | A calibrated reference is used to calibrate probes<br>$k_z = k_s(Z_t' - Z_t)/Z_t$ | Simple; easy to implement | Requires accurate positioning of cantilever relative to reference |
| Inverted loaded cantilever [27] 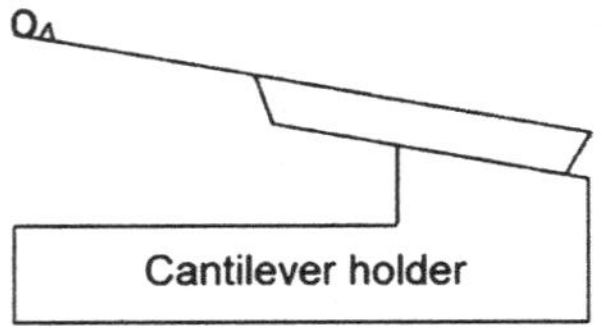<br>Cantilever holder | A particle is attached to cantilever and the deflection of the inverted cantilever is measured<br>$k_z = \dfrac{8\pi r^3 \rho_s g}{3z}$ | – | Requires attachment of a particle to cantilever; risk of cantilever damage |
| Pendulum [20] 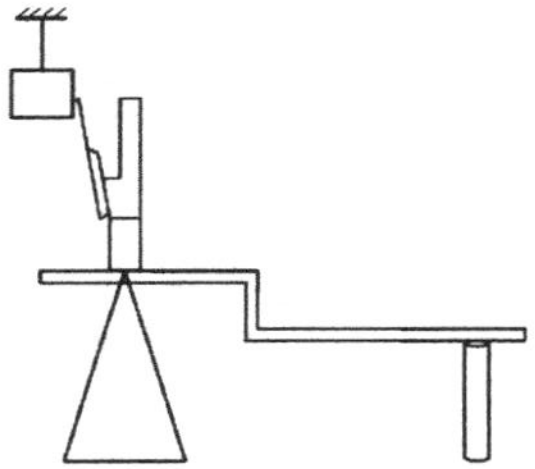 | Tip is pushed against a pendulum, displacements of tip and pendulum are measured<br>$k_z = \dfrac{m_p g z_p}{l_p z_c}$ | – | Requires pendulum set-up; risk of cantilever damage; calibration of pendulum needed |

| | | | |
|---|---|---|---|
| Electrostatic force balance [30] 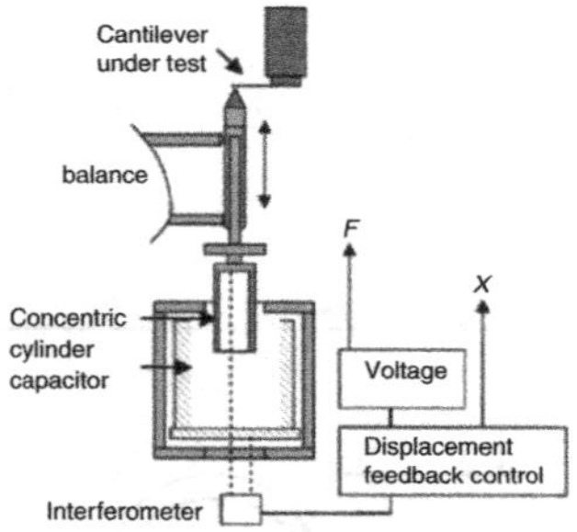 | Electrostatic force is applied to the cantilever and displacement is measured by an interferometer | – | Additional instrumentation is required |
| Nano force calibrator [29] 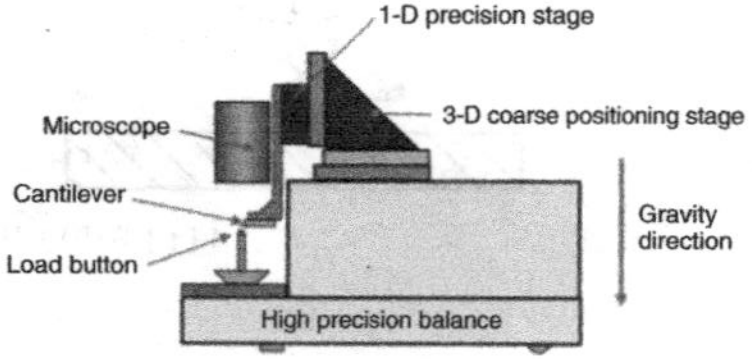 | Precision stage controls the displacement while a balance measures the force | – | Additional instrumentation is required |
| *Dynamic methods* | | | |
| Added mass [36] 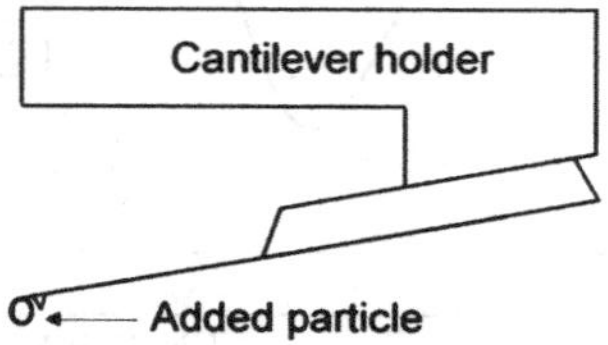 | A number of particles are individually attached to the end of the cantilever and the resonance frequency is measured $m_s = \frac{k_z}{\omega^2} - m^*$ | – | Requires attachment of particles to cantilever; risk of cantilever damage; uncertainty in dimensions and errors in placement of added particle can lead to significant error |



(*continued*)

**Table 4.1** (continued)

| Technique | Principle/key equation | Pros | Cons |
|---|---|---|---|
| Resonance [8, 38] | The resonance frequency, quality factor and dimensions are measured<br>For a rectangular beam,<br>$k_z = 0.1906\rho_f b^2 L Q \omega^2 \Gamma_i(\omega)$ | Relatively simple; desirable for stiff cantilevers | Method relies on cantilever dimensions; requires Reynolds number of the fluid medium and calculation of the hydrodynamic function |
| Thermal noise [41] | Thermal fluctuation of the cantilever is measured<br>$k_z = \mathrm{k_B} T/p$ | Accurate; relatively simple; desirable for soft cantilevers | Potential inaccuracy when applied to stiff cantilevers |

Selecting a technique for calibrating the normal stiffness depends on both the available instrumentation and the user's experience in data analysis. For high accuracy measurements, the resonance and thermal noise methods are the most preferable for stiff and soft cantilevers, respectively. A simpler technique such as the reference cantilever method is highly recommended for calibrating the normal stiffness as well.

## 4.4 Lateral Force and Stiffness Calibration Techniques

Static and dynamic methods for calibrating the lateral force and stiffness are reviewed in this section. Static methods require either bending or torsion of the cantilever, while dynamic methods involve the determination of the torsional vibration characteristics of the cantilever. We describe various methods for measuring lateral forces by using a direct method for measuring the coefficient of friction, where the lateral force is calculated from (Sects. 4.4.1.1–4.4.1.2). We also describe methods for the determination of either the torsional or torsional and lateral stiffness (Sects. 4.4.1.3, 4.4.1.4, 4.4.2.1 and 4.4.2.2). A discussion on calculating the lateral force from the torsional stiffness is given in Sect. 4.4.2.3.

### *4.4.1 Static Methods*

#### Axial Sliding Method

Based on the work by Ruan and Bhushan [9], the axial friction measurement method is described. A scanning angle is defined as the angle relative to the x-axis in Fig. 4.10a. This is also the long axis of the cantilever. A 0° scanning angle corresponds to the sample scanning in the x direction, and a 90° scanning angle corresponds to the sample scanning perpendicular to this axis in the xy plane (in y axis). If the scanning direction is in both x and -x directions, this is called a "parallel scan". Similarly, a "perpendicular scan" means scanning is done in the y and -y directions. The sample traveling direction for each of these two scanning directions is illustrated in Fig. 4.10b. Parallel scanning is discussed as "method 1," where the coefficient of friction is obtained. Perpendicular scanning is described below as "method 2," where lateral forces are determined based on the measured signal due to sliding in the lateral direction. In order to convert the measured lateral signal to a force, a conversion factor is calculated based on the coefficient of friction determined using method 1. This is then used to obtain a friction force in three dimensions.

In method 1, aside from topographic imaging, it is also possible to measure friction force. If no friction force existed between the tip and the moving sample, the topographic feature would be the only factor which causes the cantilever to be

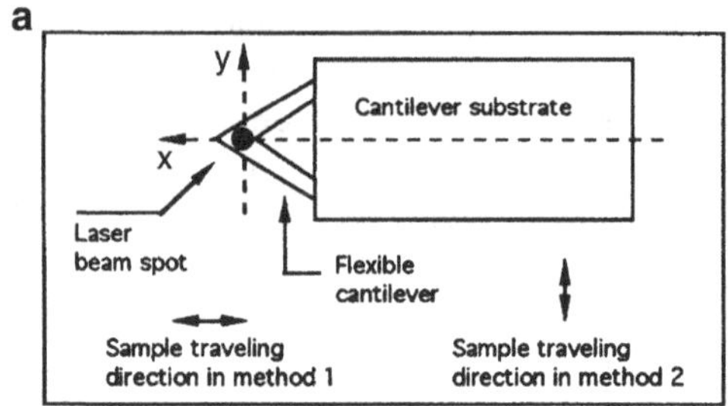


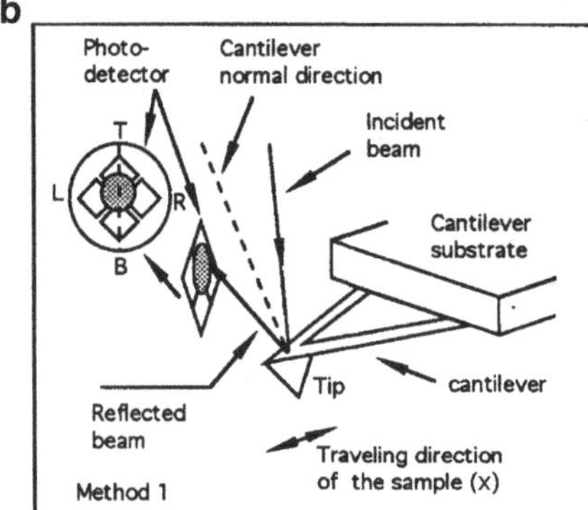


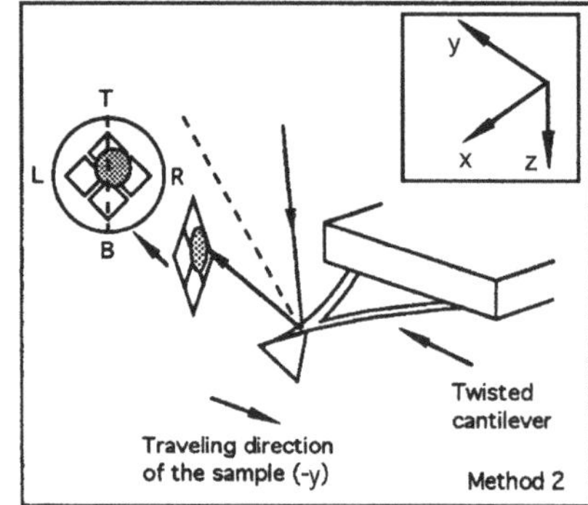


**Fig. 4.10** (**a**) Schematic defining the x- and y-directions relative to the cantilever, and showing the sample traveling direction in two different measurement methods discussed in the text, and (**b**) schematic of the deformation of the tip and cantilever shown as a result of sliding in the x- and y- directions. A twist is introduced to the cantilever if the scanning is in the y- direction [(**b**), *right*] [9]

deflected vertically. However, friction force does exist on all contact surfaces where one object is moving relative to another. The friction force between the sample and the tip will also cause cantilever deflection. We assume that the normal force between the sample and the tip is $W$ when the sample is stationary ($W$ is typically in the range of 1–200 nN), and the friction force (lateral force) between the sample and the tip is $F$ as the sample scans against the tip (Fig. 4.10). The direction of friction force is reversed as the scanning direction of the sample is reversed from positive (x) to negative (-x) directions, i.e., $\vec{F}_x = -\vec{F}_{-x}$.

When the vertical cantilever deflection is set at a constant level, it is both normal and friction forces applied to the cantilever that keeps the cantilever deflection at this level. Since the friction force is in opposite directions as the traveling direction of the sample is reversed, the normal force is adjusted accordingly when the sample reverses its traveling direction, so that the cantilever deflection remains the same. In order to maintain constant deflection, the bending of the cantilever (left side of Fig. 4.11) is canceled by adjusting the piezotube height by a feedback circuit (right side of Fig. 4.11). The observed cantilever deflection is the contribution of two bending moments. The first contribution is from the friction force, and the second is

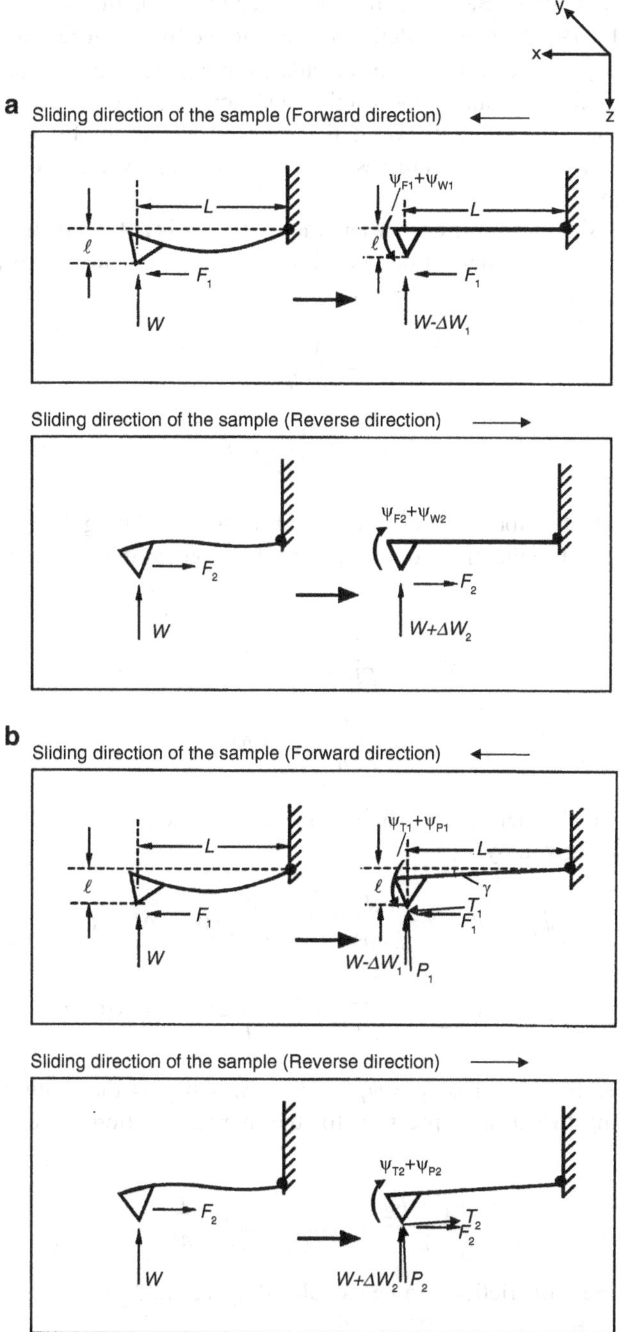


**Fig. 4.11** Schematic showing an additional bending of the cantilever due to friction force when the sample is scanned in the x or -x direction (*left*), and this effect will be canceled by adjusting the piezo height by a feedback circuit (*right*) for (**a**) horizontal tip, and (**b**) tip tilted at an angle $\gamma$

from the normal force. Based on the basic definition for the angular deflection $\psi = \int \frac{M}{EI} dx$ [16, 19], the angular deflections due to the friction force and the normal load, $\psi_F$ and $\psi_W$, respectively, will be calculated. It will be further assumed that the total angular deflection due to the friction and normal force is constant in either sliding direction when the cantilever deflection is kept constant. Friction force may have a directionality effect, and forces in the forward and reverse directions may be slightly different.

We first consider a simple case where the cantilever is not tilted, as illustrated in Fig. 4.10a. The angular deflection contributed by the moment due to the friction force ($\psi_F$) is given by

$$\psi_{F1} = \frac{12L}{Ebh^3} F_1 \ell \tag{4.30}$$

$$\psi_{F2} = -\frac{12L}{Ebh^3} F_2 \ell$$

in the forward (subscript "1") and reverse (subscript "2") sliding directions, respectively. The angular deflection contributed by the moment due to the normal force ($\psi_W$)is

$$\psi_{W1} = \frac{6L}{Ebh^3} (W - \Delta W_1) L \tag{4.31}$$

$$\psi_{W2} = \frac{6L}{Ebh^3} (W + \Delta W_2) L$$

The total angular deflection, which is the sum of the friction and normal force contributions, is given by

$$\psi_{F1} + \psi_{W1} = \frac{12L}{Ebh^3} F_1 \ell + \frac{6L}{Ebh^3} (W - \Delta W_1) L \tag{4.32}$$

$$\psi_{F2} + \psi_{W2} = -\frac{12L}{Ebh^3} F_2 \ell + \frac{6L}{Ebh^3} (W + \Delta W_2) L$$

As stated earlier, $\psi_{F1} + \psi_{W1} = \psi_{F2} + \psi_{W2}$. Simplifying the resulting equation, and rearranging to get an expression for the average friction force $(F_1 + F_2)/2$, leads to

$$\frac{F_1 + F_2}{2} = (\Delta W_1 + \Delta W_2) \frac{L}{4\ell} \tag{4.33}$$

The coefficient of friction can be calculated using (4.33) as

$$\mu = \frac{F_1 + F_2}{2W} = \left[ \frac{(\Delta W_1 + \Delta W_2)}{W} \right] \frac{L}{4\ell} \tag{4.34}$$

In all circumstances, there are adhesive and interatomic attractive forces between the cantilever tip and the sample. The adhesive force can be due to water from capillary condensation and other contaminants present at the surface which form meniscus bridges and the interatomic attractive force includes van der Waals attraction [1, 2]. There is an indentation effect as well, which is usually small for rigid samples. If these forces can be neglected, the normal force $W$ is then equal to the initial cantilever deflection $H_0$ multiplied by the stiffness of the cantilever. $(\Delta W_1 + \Delta W_2)$ can be measured by multiplying the same stiffness by the height difference of the piezo tube between the two traveling directions (forward and reverse) of the sample. This height difference is denoted as $(\Delta H_1 + \Delta H_2)$, shown schematically in Fig. 4.12a. Thus, (4.34) can be rewritten as

$$\mu = \frac{F_1 + F_2}{2W} = \left[\frac{(\Delta H_1 + \Delta H_2)}{H_0}\right]\left(\frac{L}{4\ell}\right) \tag{4.35}$$

Since the piezo tube vertical position is affected by the surface topographic profile of the sample in addition to the friction force being applied at the tip, this difference has to be taken point by point at the same location on the sample surface as shown in Fig. 4.12a. Subtraction of point by point measurements may introduce errors, particularly for rough samples. In addition, precise measurements of $L$ and $\ell$ (which should include the cantilever angle) are also required. Since only the ratio between $(\Delta H_1 + \Delta H_2)$ and $H_0$ comes into (4.35), the piezo tube vertical position $H_0$ and its position difference $(\Delta H_1 + \Delta H_2)$ can be in the units of volts as long as the vertical traveling distance of the piezo tube and the voltage applied to it has a linear relationship. However, if there is a large nonlinearity between the piezo tube traveling distance and the applied voltage, this nonlinearity must be included in the calculation [9].

If the adhesive forces between the tip and the sample are large enough that it cannot be neglected, one should include it in the calculation. However, there could be a large uncertainty in determining this force, and thus an uncertainty in using (4.35). An alternative approach is to measure the height difference of the PZT $(\Delta H_1 + \Delta H_2)$ at different normal loads $(H_0)$ and to use the slope of $(\Delta H_1 + \Delta H_2)$ vs. the slope of $(H_0)$ from the measurements in (4.35) to calculate μ. Figure 4.12b shows the data from three sets of measurements at various loads on a Pt sample, whereit is seen that a linear fit is obtained. The coefficient of friction for this sample was found to be 0.054 [9].

Now we consider the cantilever tilt angle $\gamma$ relative to the horizontal axis, as illustrated in Fig. 4.11b. In this case, the friction force $(T)$ and normal load $(P)$ components are resolved in terms of the measured horizontal force $(F)$ and normal load $(W)$ along the x and z axes [44].

The resolved horizontal and normal force components along x and z axes in the forward sliding direction are given by

$$\begin{aligned} T_1 &= F_1 \cos\gamma - (W - \Delta W_1)\sin\gamma \\ P_1 &= F_1 \sin\gamma + (W - \Delta W_1)\cos\gamma \end{aligned} \tag{4.36}$$

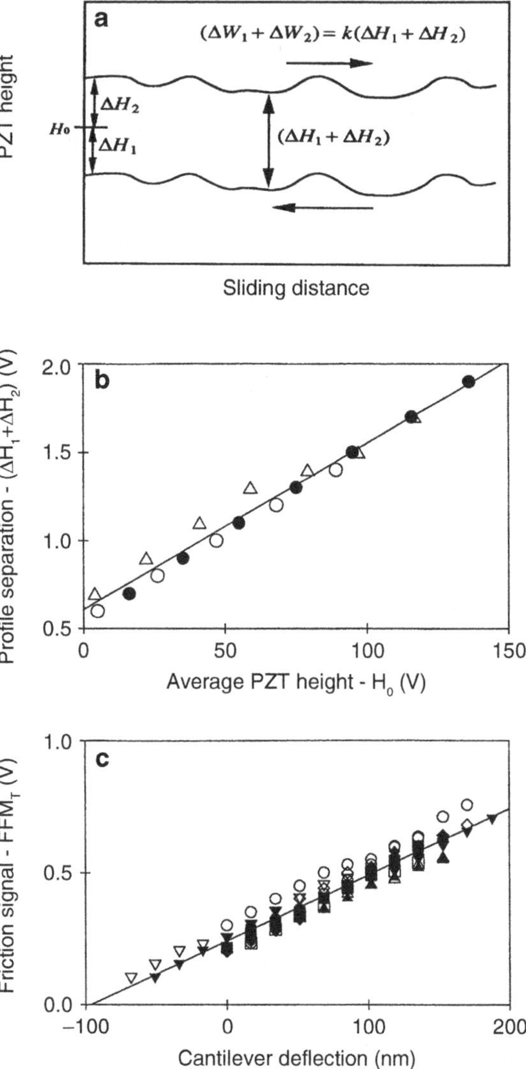


**Fig. 4.12** (**a**) Schematic of the height difference of the piezoelectric tube scanner as the sample is scanned in y and - y directions, (**b**) the vertical height difference as a function of the PZT center position between the two sliding directions on a Pt sample (method 1). The three symbols represent three sets of repeated measurements. The slope of the linear fit is proportional to the coefficient of friction between the $Si_3N_4$ tip and Pt. (**c**) Friction signal as a function of cantilever vertical deflection for Pt (method 2). Different symbols represent 11 sets of repeated measurements. The slope of the linear fit is proportional to the coefficient of friction between the $Si_3N_4$ tip and Pt [9]

and in the reverse direction by

$$\begin{aligned} T_2 &= F_2 \cos\gamma + (W + \Delta W_2)\sin\gamma \\ P_2 &= -F_2 \sin\gamma + (W + \Delta W_2)\cos\gamma \end{aligned} \tag{4.37}$$

Using (4.30), the angular deflection contributed by the force along the longitudinal axis of the cantilever is given by

$$\psi_{T1} = \frac{12L}{Ebh^3}(F_1 \cos\gamma - (W - \Delta W_1)\sin\gamma)\ell \tag{4.38}$$
$$\psi_{T2} = -\frac{12L}{Ebh^3}(F_2 \cos\gamma + (W + \Delta W_2)\sin\gamma)\ell$$

Based on (4.31), the angular deflection contributed by the normal load perpendicular to the longitudinal axis of the cantilever is given by

$$\psi_{P1} = \frac{6L}{Ebh^3}(F_1 \sin\gamma + (W - \Delta W_1)\cos\gamma)L \tag{4.39}$$
$$\psi_{P2} = \frac{6L}{Ebh^3}(-F_2 \sin\gamma + (W + \Delta W_2)\cos\gamma)L$$

The total angular deflection, obtained by adding the friction force and normal load contributions is given by

$$\begin{aligned}\psi_{T1} + \psi_{P1} &= \frac{12L\ell}{Ebh^3}(\mu(W - \Delta W_1)\cos\gamma - (W - \Delta W_1)\sin\gamma)\\ &+ \frac{6L^2}{Ebh^3}(\mu(W - \Delta W_1)\sin\gamma + (W - \Delta W_1)\cos\gamma)\end{aligned} \tag{4.40}$$

$$\begin{aligned}\psi_{T2} + \psi_{P2} &= -\frac{12L\ell}{Ebh^3}(\mu(W + \Delta W_2)\cos\gamma + (W + \Delta W_2)\sin\gamma)\\ &+ \frac{6L^2}{Ebh^3}(-\mu(W + \Delta W_2)\sin\gamma + (W + \Delta W_2)\cos\gamma)\end{aligned}$$

where μ is defined as $F_1/(W - \Delta W_1) = F_2/(W + \Delta W_2)$. Since it is assumed that $\psi_{T1} + \psi_{P1} = \psi_{T2} + \psi_{P2}$, by simplifying the resulting equation, the expression for μ is obtained as follows

$$\begin{aligned}\mu &= \frac{\Delta W_1 + \Delta W_2}{2W - \Delta W_1 + \Delta W_2}\frac{\left(-\frac{12L\ell}{Ebh^3}\sin\gamma + \frac{6L^2}{Ebh^3}\cos\gamma\right)}{\left(\frac{12L\ell}{Ebh^3}\cos\gamma + \frac{6L^2}{Ebh^3}\sin\gamma\right)}\\ &= \frac{\Delta W_1 + \Delta W_2}{2W - \Delta W_1 + \Delta W_2}\frac{(-2\ell\sin\gamma + L\cos\gamma)}{(2\ell\cos\gamma + L\sin\gamma)}\end{aligned} \tag{4.41}$$

This can be simplified by expressing $(P_1 + P_2)$ and $(P_1 - P_2)$ in terms of the height difference of the piezotube.

$$\begin{aligned}\mu &= \frac{\Delta H_1 + \Delta H_2}{2H_o - \Delta H_1 + \Delta H_2}\frac{(-2\ell\sin\gamma + L\cos\gamma)}{(2\ell\cos\gamma + L\sin\gamma)}\\ &\approx \frac{\Delta H_1 + \Delta H_2}{2H_o}\frac{(-2\ell\sin\gamma + L\cos\gamma)}{(2\ell\cos\gamma + L\sin\gamma)}\end{aligned} \tag{4.42}$$

For $\gamma = 0$, (4.41) reduces to (4.34), and (4.42) reduces to (4.35).

Method 2 ("aux" mode with perpendicular scan) is used to obtain 3D maps of friction. The sample is scanned perpendicular to the long axis of the cantilever beam (i.e., to scan along the y or –y direction in Fig. 4.10a), and the output of the two horizontal quadrants of the photodiode-detector is measured. In this arrangement, as the sample moves under the tip, the friction force will cause the cantilever to twist. Therefore, the light intensity between the left and right (L and R in Fig. 4.10b, right) detectors will be different. The differential signal between the left and right detectors is denoted as the friction force microscopy or FFM signal $[(L - R)/(L + R)]$. This signal can be related to the degree of twisting, hence to the magnitude of friction force. Again, because of a possible error in determining normal force due to the presence of an adhesive force at the tip-sample interface, the slope of the friction data (FFM signal vs. normal load) needs to be taken for an accurate value of coefficient of friction.

While friction force contributes to the FFM signal, it may not be the only contributing factor in commercial FFM instruments. One can notice this by engaging the cantilever tip with the sample. Before engaging, the left and right detectors can be balanced by adjusting the position of the detectors so that the intensity difference between these two detectors is zero (FFM signal is zero). Once the tip is engaged with the sample, this signal is no longer zero even if the sample is not moving in the xy plane with no friction force applied. This would be a detrimental effect. It has to be understood and eliminated from the data acquisition before any quantitative measurement of friction force becomes possible.

When the laser beam is reflected up and down due to a change of the normal force, the intensity difference between the left and right detectors will also change [9]. In other words, the FFM signal will change as the normal force applied to the tip is changed, even if the tip is not experiencing any friction force. This FFM signal is unrelated to friction force or to the actual twisting of the cantilever, but rather to cross-talk effects, which can either be optical or mechanical in nature. We will call this part of the FFM signal "$FFM_F$", and the part which is truly related to friction force "$FFM_T$". The $FFM_F$ signal can be eliminated. One way of doing this is as follows. First the sample is scanned in both y and –y directions, and the FFM signal for scans in each direction is recorded. Since friction force reverses its direction when the scanning direction is reversed from y to –y direction, the $FFM_T$ signal will have opposite signs as the scanning direction of the sample is reversed ($FFM_T(y) = -FFM_T(-y)$). Hence the $FFM_T$ signal will be canceled out if we take the sum of the FFM signals for the two scans. The average value of the two scans will be related to $FFM_F$ due to the misalignment,

$$FFM(y) + FFM(-y) = 2FFM_F \tag{4.43}$$

This value can therefore be subtracted from the original FFM signals of each of these two scans to obtain the true FFM signal ($FFM_T$). Alternately, by taking the difference of the two FFM signals, one directly gets the $FFM_T$ value

$$FFM(y) - FFM(-y) = FFM_T(y) - FFM_T(-y) = 2FFM_T(y) \tag{4.44}$$

Ruan and Bhushan [9] have shown that an error signal ($FFM_F$) can be very large compared to the friction signal $FFM_T$, thus correction is required. An implementation of this method is shown in Fig. 4.12c, where the true friction signal as a function of cantilever vertical deflection is shown for Pt. The coefficient of friction obtained using method 1 is then used to convert the friction force into nN.

One can measure $L$ and $\ell$ for a full implementation of the axial sliding method (4.35). Instead of doing this, we use the following procedure. It has been observed that the coefficient of friction for a $Si_3N_4$ or Si tip on single crystal silicon in ambient air is approximately 0.05. Since the coefficient of friction is an interface property, it is expected to be the same for $Si_3N_4$ or Si tips with varying dimensions [45]. Hence, for a given $Si_3N_4$ or Si tip, measurements are first made on a single crystal silicon sample in a perpendicular scan, and the conversion factor for the lateral voltage signal is obtained based on a coefficient of friction of 0.05. This conversion factor is then used to analyze friction data on the test sample. The normal load is calculated from the normal stiffness and the photodetector sensitivity (in V/nm). Since both the normal and lateral forces are known, the coefficient of friction can be calculated.

The advantages of this technique are simplicity and ease of implementation. However, the need for measuring the cantilever length and tip height provides a source of measurement errors for this method.

### Wedge Method

A schematic for the wedge method is shown in Fig. 4.13a. In the original work by Ogletree et al. [46], the tip is scanned across a calibration sample containing triangular features with two well-defined slopes. This is based on the knowledge that when the tip is scanned across the sample surface, the measured friction forces are generated by both material effects as well as topography-induced effects [4, 47–50].

The approach for obtaining the coefficient of friction is as follows [46, 51]. At any given load, the friction, $F$, and normal, $W$, forces depend on the direction of motion, either uphill (denoted by subscript "u") or downhill (denoted by subscript "d") and can be resolved in terms of the applied load, $P$, and horizontal force, $T$, components. The total normal force includes the external applied force plus the intrinsic adhesive force, $A$. For uphill motion

$$W_u = P\cos\theta + T_u\sin\theta + A_u$$

$$F_u = -P\sin\theta + T_u\cos\theta \tag{4.45}$$

For downhill motion

$$W_d = P\cos\theta - T_d\sin\theta + A_d$$

$$F_d = P\sin\theta + T_d\cos\theta \tag{4.46}$$

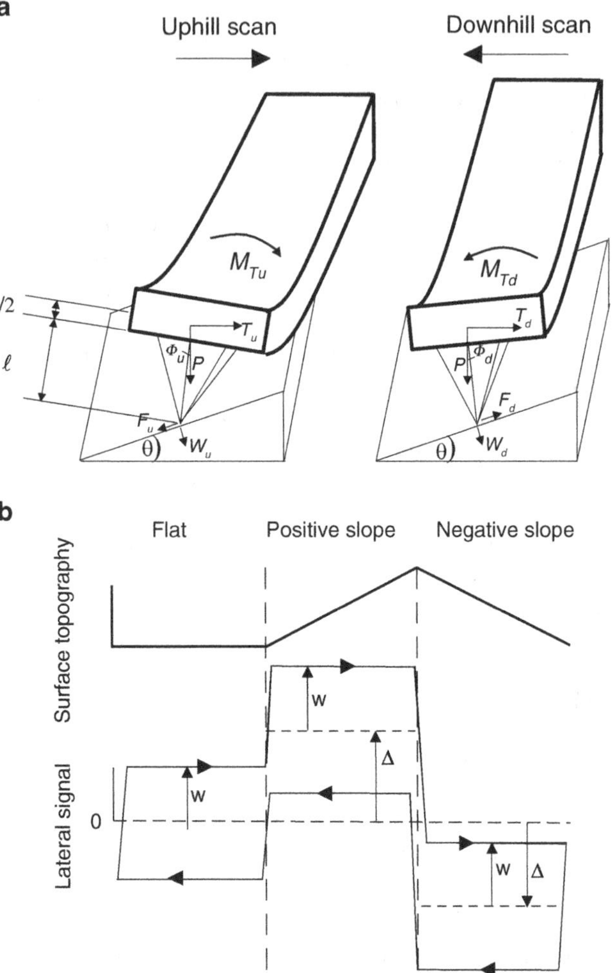


**Fig. 4.13** Schematics of (**a**) cantilever torsion while sliding up and down on an inclined surface, the basis of the wedge method, and (**b**) friction loops for flat, positively sloped and negatively sloped surfaces

It is assumed that the friction force is linearly related to the total normal force [52, 53], such that

$$\mu = \frac{F_u}{W_u} = \frac{F_d}{W_d} \tag{4.47}$$

Substituting (4.45 and 4.46) into (4.47) and assuming that $A_u = A_d = A$ leads to

$$\mu = \frac{-P\sin\theta + T_u\cos\theta}{P\cos\theta + T_u\sin\theta + A} = \frac{P\cos\theta - T_d\sin\theta}{P\cos\theta - T_d\sin\theta + A} \tag{4.48}$$

In order to solve for $\mu$, $T_u$ and $T_d$ will be related to experimentally measurable parameters.

Equations (4.45) and (4.46) can be rearranged into

$$T_u = \frac{P\sin\theta + \mu P\cos\theta + \mu A}{\cos\theta - \mu\sin\theta}$$

$$T_d = \frac{-P\sin\theta + \mu P\cos\theta + \mu A}{\cos\theta + \mu\sin\theta} \tag{4.49}$$

Furthermore, $T_u$ and $T_d$ can be related to the torsion moment $M$ by

$$M_{Tu} = T_u(\ell + h/2)$$

$$M_{Td} = T_d(\ell + h/2) \tag{4.50}$$

where $\ell$ and $h$ pertain to the tip length and cantilever thickness, respectively (Fig. 4.3). Next, the friction loops for the flat, positively sloped and negatively sloped surfaces are considered. Focusing on the two sloped regions shown in Fig. 4.13b, the half-width of the friction loop, $w$, and the friction loop offset, $\Delta$, are measured experimentally. These two quantities are related to the torsion moment as follows

$$w = \frac{M_{Tu} - M_{Td}}{2} \tag{4.51}$$

$$\Delta = \frac{M_{Tu} + M_{Td}}{2} \tag{4.52}$$

Substituting (4.49) and (4.50) into (4.51) and (4.52) yields the following

$$w = \left(\ell + \frac{h}{2}\right)\frac{\mu P + \mu A\cos\theta}{\cos^2\theta - \mu^2\sin^2\theta} \tag{4.53}$$

$$\Delta = \left(\ell + \frac{h}{2}\right)\frac{P\sin\theta\cos\theta + \mu^2 P\sin\theta\cos\theta + \mu^2 A\sin\theta}{\cos^2\theta - \mu^2\sin^2\theta} \tag{4.54}$$

The lateral forces vary with the applied load such that it is necessary to define the slopes $w' = dw/dP$ and $\Delta' = d\Delta/dP$, obtained by taking the first derivative of (4.53) and (4.54)

$$w' = \alpha_c w_o' = \frac{dw}{dP} = \left(\ell + \frac{h}{2}\right)\frac{\mu}{\cos^2\theta - \mu^2\sin^2\theta} \tag{4.55}$$

$$\Delta' = \alpha_c \Delta_o' = \frac{d\Delta}{dP} = \left(\ell + \frac{h}{2}\right) \frac{(1+\mu^2)\sin\theta\cos\theta}{\cos^2\theta - \mu^2\sin^2\theta} \tag{4.56}$$

where $w_o$ and $\Delta_o$ are the experimentally-determined half-width of the friction loop and the friction loop offset, respectively (both in volts), and $\alpha_c$ is the lateral force calibration factor (in N/V). In the limiting case of no friction, $w' \rightarrow 0$ and $\Delta' \rightarrow \tan\theta$, as expected. It should be noted that by taking the derivatives (4.55 and 4.56), the adhesive force is eliminated from the equations needed for determining $\mu$. Finally, the expression for the coefficient of friction is obtained by dividing (4.56) by (4.55).

$$\mu + \frac{1}{\mu} = \frac{2\Delta'}{w' \sin 2\theta} \tag{4.57}$$

After the coefficient of friction is determined, it can be used in either (4.50) or (4.51) to calculate $\alpha_c$, and to obtain $w$ and $\Delta$ in their proper units. Two values of $\mu$ are obtained from the quadratic equation in (4.57), and both roots are considered to be equally good solutions. However, one of the roots could give a value for $\mu_s$ that is large enough such that the denominator in either (4.55) or (4.56) becomes negative (i.e., negative $\alpha_c$) As this is artificial, then this spurious value is disregarded, leaving only one acceptable $\mu$.

An image of the $SrTiO_3$ calibration sample with an inclination of 54°44′ used by Ogletree et al. [46], along with an example of the data they obtained using the wedge method are presented in Fig. 4.14a. A limitation of the method is that it is only suitable for sharp tips because large, blunt tips (such as colloidal probes) will give unreliable data while sliding on the $SrTiO_3$ surface, which is relatively steep. In addition, crosstalk between deflection and torsion signals, signal drift, and laser or cantilever misalignment, cause an uncertainty in identifying the zero point in the torsion signal, affecting the value of the friction loop offset [51].

Varenberg et al. [51] applied the wedge method to a silicon calibration grating instead of $SrTiO_3$, shown in Fig. 4.14b. The use of the Si grating enables the measurement of friction on a flat surface, aside from the sloped surfaces (with slope of 54°44′) measured in the original method. This allows measurements on tips with large radii. Taking the measurement on a flat surface eliminates the uncertainty of determining the zero point in the torsion signal, which is a limitation of the original wedge method. In the following equations, the distinction between the sloped and flat surface is made by the subscripts $s$ and $f$. The coefficient of friction on the sloped surface $\mu_s$ is obtained by taking the ratio of (4.53) and (4.54) such that

$$\frac{\Delta_s}{w_s} = \frac{P\sin\theta\cos\theta + \mu_s^2 P\sin\theta\cos\theta + \mu_s^2 A\sin\theta}{\mu_s P + \mu_s A\cos\theta}$$

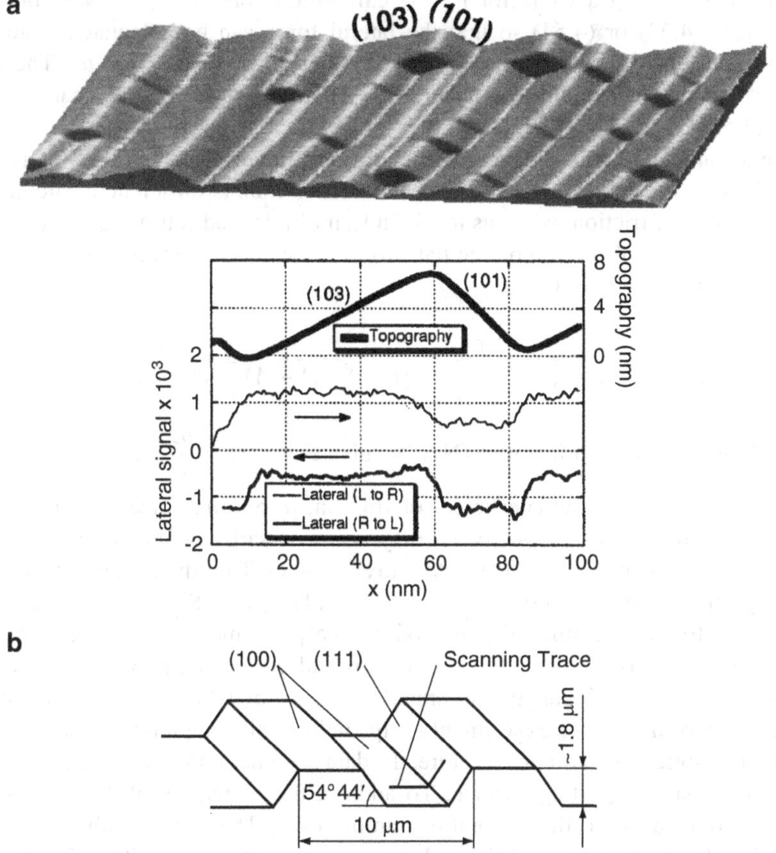


**Fig. 4.14** (**a**) AFM image of the $SrTiO_3$ calibration surface used in the original wedge method (*top*), and experimental data for the wedge method (*bottom*). Lateral deflection signal for each direction and topography measured on the (101) and (103) facets of the $SrTiO_3(305)$ calibration sample [46], and (**b**) schematic of the silicon calibration grating used by Varenberg et al. [51]

or

$$\sin\theta(P\cos\theta + A)\mu_s^2 - \frac{\Delta_s}{w_s}(P + A\cos\theta)\mu_s + P\sin\theta\cos\theta = 0 \qquad (4.58)$$

In (4.58), $\theta$, $P$ and $A$ are known. It should be recalled that the voltage outputs $\Delta_{o,s}$ and $w_{o,s}$ are being measured instead of $\Delta_s$ and $w_s$. Since the ratio $\Delta_s/w_s$ is taken in (4.58), $\alpha_c$ cancels out and the measured voltage outputs are sufficient. Also, $\Delta_{o,s} = \Delta_{o,s,measured} - \Delta_{o,f}$ because the friction loop offset voltage of the flat surface $\Delta_{o,f}$, is nonzero and must therefore be subtracted from the measured value on the sloped surface.

After determining a value for $\mu_s$, the calibration constant $\alpha_c$ is then obtained from either (4.53) or (4.54) so that the lateral force can be calculated from the voltage data. The quadratic equation (4.58) yields two values for $\mu_s$. The real solution is one that would not result in a negative value for the quantities in either (4.53) or (4.54).

The advantage of using (4.58) over the ratio of the derivatives (4.57) is that by using the silicon grating, one set of load and friction loop is sufficient in calculating the coefficient of friction, whereas for $SrTiO_3$, multiple load settings are necessary.

The coefficient of friction on the flat surface $\mu_f$, can be determined by substituting $\theta = 0°$ into (4.53) such that

$$\mu_f = \frac{1}{\left(\ell + \frac{h}{2}\right)} \frac{w_f}{(P+A)} = \frac{\beta_c}{\left(\ell + \frac{h}{2}\right)} \frac{w_o}{(P+A)} = \frac{\alpha_c w_o}{(P+A)} \tag{4.59}$$

where $\beta_c$ is a calibration factor in (N m/V) given as $\alpha_c = \dfrac{\beta_c}{\left(\ell + \frac{h}{2}\right)}$.

The two values for the coefficient of friction, $\mu_s$ and $\mu_f$, corresponding to the sloped and flat surfaces, respectively, may not necessarily be equal, but has been shown by Varenberg et al. [51] to be close to each other. The advantages of this using the Si grating over the wedge method with $SrTiO_3$ are: Si is a commercially-available calibration grating, the method can be performed at any single applied load, and that all types of cantilevers (sharp and colloidal tips with a radius of curvature up to 2 μm) can be calibrated. The use of silicon grating limits the applicability of the wedge experiment to tips with small cone angles. For tips with larger cone angles and radii of curvature, the data becomes unreliable due to the high slope of the silicon grating surface. To address this, Tocha et al. [54] proposed another modification to the calibration sample used. They used a silicon surface milled by a focused ion beam (FIB) such that notches with slopes of 20°, 25°, 30° and 35° (relative to the wafer surface) were present. The advantage of scanning on less steep slopes is that larger tips, such as colloidal probes with radius of curvature greater than 2 μm, can be calibrated, in addition to the sharp integrated tips.

The wedge method is not as convenient as the axial sliding technique (Sect. 4.4.1.1). For a flat portion of the calibration grating, it depends not only on the signal output of the AFM, but also requires the cantilever dimensions (4.59). The wedge method requires a calibration standard (either $SrTiO_3$ or Si), and in addition, the method presented above is computationally not straightforward. It has also been shown that the pull-off force value used in the calculations is a major source of measurement errors in the implementation of the wedge method [55]. Expressions for the friction loop parameters were derived to take into account adhesive forces. Other sources of error related to the detection of cantilever deflection have been identified, which affects the measured μ. This includes a non-zero lateral deflection of the cantilever in the absence of applied torque, variation in cantilever deflection (which in turn affects lateral deflection) due to feedback response limitations, and the susceptibility of the lateral signal to optical interference effects. The latter two are relevant to other techniques as well.

### Lever Method

The lever method for torsional stiffness determination is shown schematically in Fig. 4.15, where a lever assembly consisting of a glass fiber and a silica sphere is attached to a tipless cantilever [56]. It should be noted that a similar technique was proposed by Bogdanovic et al. [57], but is only suitable for rectangular cantilevers. A hammerhead cantilever configuration has been proposed by Reitsma [58] which allows in situ calibration and friction measurements. Placement of a colloidal probe off the center of the cantilever [59] is another adaptation of the lever method that has been proposed.

In the lever method, the force-distance curve is obtained using the cantilever of interest prior to lever attachment in order to get the vertical detector sensitivity of the cantilever, $c_z$ (in m/V) which is the slope of the constant compliance region [56]

$$c_z = \frac{z^0}{\Delta V_z} \tag{4.60}$$

where $z^0$ is the cantilever deflection without the lever, and $\Delta V_z$ is the change in the vertical voltage. Next, the cantilever deflection after attachment of a lever with length $l$ is considered. If a cantilever with the attached lever is pressed against the sample, the calibration factor for the vertical deflection with an attached lever, $c_{z,L}$ (in m/V) is given by

$$c_{z,L} = \frac{z}{\Delta V_z} \tag{4.61}$$

where $z$ is the resulting deflection. The calibration factor $\varepsilon$, which converts the measured lateral voltage $\Delta V_y$ into an applied torque $M_T$, is defined as

$$M_T = \varepsilon \Delta V_y \tag{4.62}$$

The torque is from the applied force $F$ acting on the lever, i.e., $M_T = Fl$. This applied force causes a change in both the vertical and lateral voltage signals. In the vertical direction, $F = k_z z^0$. Substituting this and (4.60) into the basic definition for torque yields

$$M_T = k_z \, c_z \, \Delta V_z \, l \tag{4.63}$$

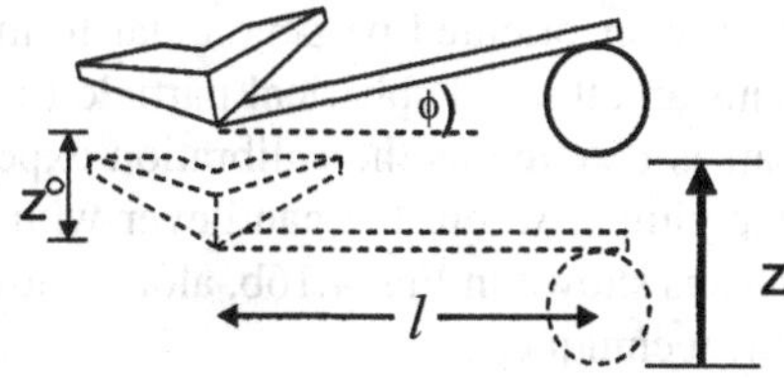


Fig. 4.15 Schematic of the lever method (adapted from [56])

By combining (4.62) and (4.63), an expression for ε is obtained

$$\varepsilon = c_z k_z l \frac{\Delta V_z}{\Delta V_y} \tag{4.64}$$

By applying a known torque to the cantilever while measuring the change in the angular deflection of the cantilever $\phi$, the torsional stiffness can be directly calculated from $M_T = k_\phi \phi$. The angular deflection is obtained from the difference between the vertical movement of the lever and that of the cantilever itself, divided by the lever length such that

$$\phi = \frac{(z - z^o)}{l} = \frac{(z - c_z \Delta V_z)}{l} = \frac{z(1 - c_z/c_{z,L})}{l} \tag{4.65}$$

By substituting this to (4.63), the expression for $k_\phi$is obtained as

$$k_\phi = \frac{k_z l^2}{(c_{z,L}/c_z - 1)} \tag{4.66}$$

In this method, the friction force acting on the tip with height $\ell$ (with the lever no longer present) can be determined independently from the torsional stiffness as long as the calibration factor ε (4.64) is known. In this case, $M_T = F\ell$. By relating this to (4.64), the following is obtained

$$F = \frac{\varepsilon \Delta V_y}{\ell} \tag{4.67}$$

The main disadvantage of this technique is that it is only practically applicable to tipless cantilevers due to the necessity of attaching a lever to the cantilever of interest in the calibration procedure, which in itself is difficult and time consuming in practice. It can be used on cantilevers with integrated tips as long as the tip is not affected when the glass lever is attached to the cantilever of interest.

### Particle Interaction Apparatus

The so-called particle interaction apparatus relies on the presence of a reference cantilever oriented perpendicular to the cantilever of interest, which in turn, contains an attached spherical particle (colloidal probe). As shown in Fig. 4.16a, this part is denoted as the calibration experiment [60]. The second part is the friction experiment where the cantilever with the colloidal probe slides on a flat surface. This is shown in Fig. 4.16b, along with the definition of the relevant parameters for this technique.

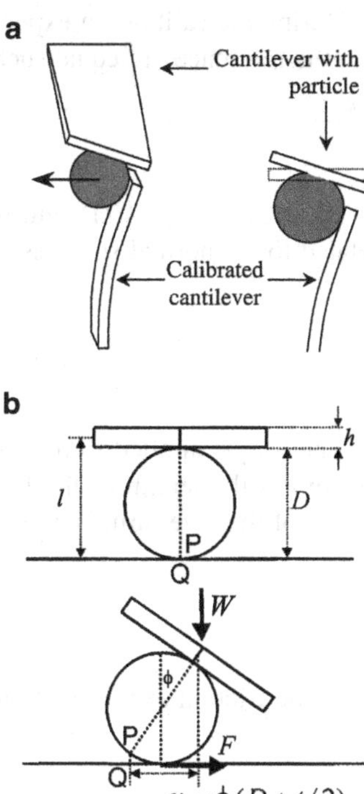


**Fig. 4.16** (**a**) Schematic of the particle and reference cantilever in the particle interaction apparatus, and (**b**) schematic of the front view of a particle attached to a cantilever in contact with the surface in the particle interaction apparatus. The equilibrium position is shown on top. At the bottom illustration, normal force has been applied, leading to rotation of the particle by angle $\theta$ (Adapted from [60])

The lateral sensitivity $c_y$ (in m/V) is defined as the ratio of the total lateral displacement on the lateral voltage $\Delta V_y$. In this method, two lateral sensitivities are determined. First, the calibration experiment is performed where the reference cantilever is pushed against the cantilever of interest, and the lateral sensitivity of the reference $c_y^{ref}$ is obtained as [60]

$$c_y^{ref} = \frac{y_o + y_{ref}}{\Delta V_y} \tag{4.68}$$

where the displacement comes from both the tip $y_o$ and the reference $y_{ref}$. Afterwards, the friction experiment is performed on a rigid flat surface instead of the reference cantilever in order to obtain the lateral sensitivity of the cantilever on the rigid surface $c_y^0$

$$c_y^0 = \frac{y_o}{\Delta V_y} \tag{4.69}$$

where the displacement is solely from the tip. By substituting (4.69) into (4.68), the sensitivity of the reference cantilever is expressed as

$$c_y^{ref} = c_y^0 + \frac{y_{ref}}{\Delta V_y} \tag{4.70}$$

During the calibration experiment the forces acting on the cantilever of interest and the reference are equal such that

$$k'_{yT}y' = k_z^{ref} y_{ref} \tag{4.71}$$

By substituting (4.71) into (4.70), the lateral stiffness of the cantilever, $k'_{yT}$, is related to the normal stiffness of the reference cantilever, $k_z^{ref}$, by

$$k'_{yT} = \frac{c_y^{ref} - c_y^0}{c_y^0} k_z^{ref} \tag{4.72}$$

The torque applied to the cantilever by a lateral force $F$ is given by $M_T = Fl$, where $l$ is the length of the lever arm. During calibration with the reference, the length of the lever arm is given by

$$l' = \frac{1}{2}(D + h) \tag{4.73}$$

In the friction experiment on the rigid flat surface, $l$ is given by

$$l = \left(D + \frac{h}{2}\right) \tag{4.74}$$

where $h/2$ takes into consideration that the twisting axis is located in the middle of the thickness of the cantilever.

Since the torque is also related to the twist of the cantilever $\phi$ (Fig. 4.16b) by $M_T = k_\phi \phi$, the lateral stiffness can be expressed as

$$k_{yT} = \frac{k_\phi}{l^2} \tag{4.75}$$

Equation (4.75) can be used to relate the lateral force constant from the calibration $k'_{yT}$ and that from the friction experiment, $k_{yT}$, as follows

$$\frac{k_{yT}}{k'_{yT}} = \left(\frac{l'}{l}\right)^2 \tag{4.76}$$

The expressions for the length of the lever arm can then be substituted into (4.76) such that

$$k_{yT} = k'_{yT}\left(\frac{D + h}{2D + h}\right)^2 \tag{4.77}$$

By substituting (4.77) into (4.72) the lateral stiffness is obtained

$$k_{yT} = \frac{c_y^{ref} - c_y^0}{c_y^0} k_z^{ref} \left( \frac{D+h}{2D+h} \right)^2 \tag{4.78}$$

The lateral force during the friction experiment is obtained once $k_{yT}$ is known by using

$$F = k_{yT} c_y \Delta V_L \tag{4.79}$$

where

$$c_y = c_y^{o} \left( \frac{D+h}{2D+h} \right) \tag{4.80}$$

As presented above, this particle interaction approach only gives a value for the lateral stiffness, and the torsional stiffness is determined indirectly by using the relationship $k_{yT} = k_{\phi}/l^2$. A disadvantage of this method is that it is limited in applicability; it is only suited for colloidal probes. In addition, the normal stiffness of the reference cantilever is required. Another limitation is that this method requires the measurement of the cantilever thickness and the diameter of the attached particle, which requires SEM for highest accuracy.

### Miscellaneous Methods

The following are examples of methods where either limitations in the technique or the need for additional instrumentation for measuring the lateral force or the torsional stiffness may not facilitate their routine implementation.

Stiernstedt et al. [61] proposed a method for measuring the coefficient of friction while sliding in the axial direction. This method is based on the observed hysteresis of the slope of the photodetector signal versus piezo distance during approach and withdrawal of the tip on the surface [61–64]. The hysteresis is caused by the tilt of the cantilever (approx. 10°–15°) relative to the horizontal axis, which causes the tip to slide on the substrate during the acquisition of the curve for the normal load ("force curve"). As shown in the left side of Fig. 4.17a, $L_0$ is the length of the flexible part of the cantilever, and $L_1$ is the length of the adhesive holding the probe. A tip (or colloidal probe) with length $L_2$ is attached at the point $L_1 + L_0$ from the base of the cantilever. $\gamma_0$ is the angle of the cantilever relative to the horizontal axis. Before contact, the vertical separation between tip and substrate is $h$. During contact, a normal force $W$ and torque $M_T$ applied to the end of the cantilever causes a deflection at the end of the beam, $y$, and an additional deflection, $\gamma$, such that the actual angle of the tip is $\gamma_0 + \gamma$ (right side of Fig. 4.17a). This compliance hysteresis analysis accounts for the presence of friction between the tip and substrate during contact,

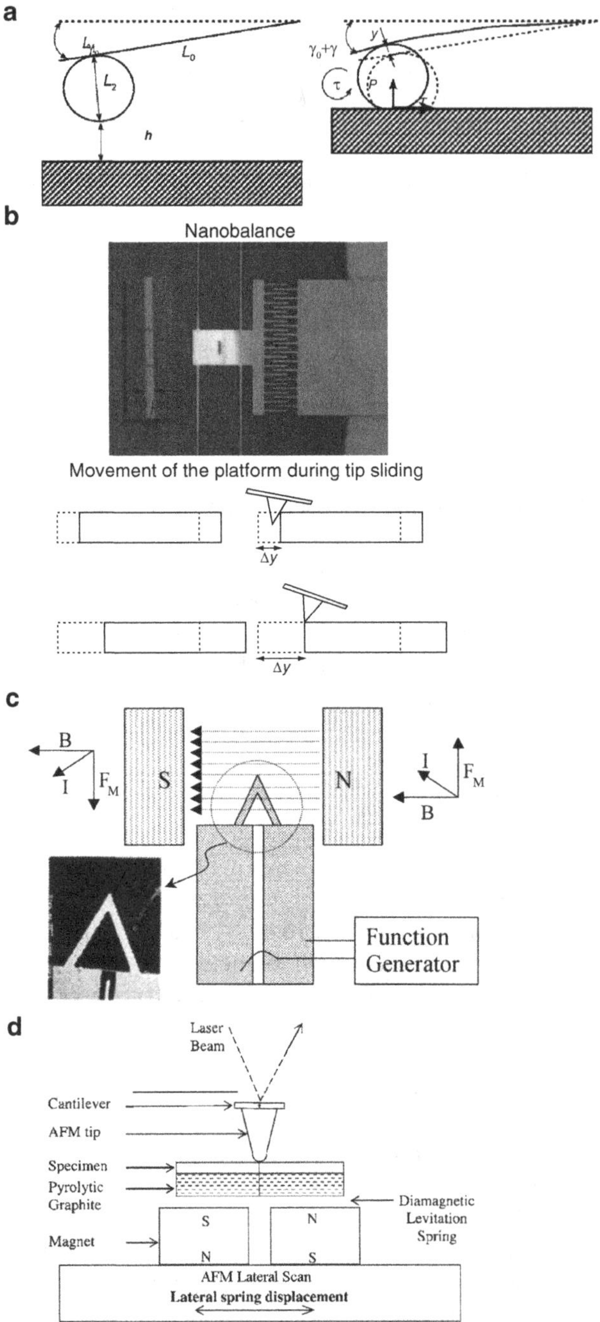


**Fig. 4.17** Schematics of (**a**) the tip-surface contact showing horizontal sliding of the tip in the compliance hysteresis method (Adapted from [61]), (**b**) lateral electrical nanobalance and movement of the platform (Δy) during the tip sliding [65], (**c**) Lorentz force induction technique [66], and (**d**) diamagnetic levitation spring system [67]

and removes the effect of friction from the measured forces. One main disadvantage of this method is that it is not a direct means for converting the observed deflection signal into the lateral force. The authors used the lever method (discussed in Sect. 4.4.1.3) to calibrate their cantilever. On the other hand, this method provides a direct way of determining the coefficient of friction, provided that the angle of tip tilt relative to the horizontal axis is accurately known. This method is also unsuitable for materials exhibiting viscoelasticity, where the observed hysteresis is from relaxation of the stresses in the material and not from friction. In addition, the method outlined above is computationally not straightforward.

The lateral electrical nanobalance (LEN), shown in Fig. 4.17b, was developed at the National Physical Laboratory in the United Kingdom. It is a comb actuator-based MEMS device fabricated by silicon-on-insulator micromachining [65]. The nanobalance consists of a gold-coated silicon platform suspended on cantilever beams. The platform contains a 3 μm wide slit, which the AFM tip will enter as it scans the surface. The displacement of the platform as the tip enters the slit is a measure of lateral displacement. Prior to a cantilever calibration, the nanobalance displacement is calibrated in order to obtain its static and dynamic displacement properties, and to separate actual mechanical displacement from parasitic capacitances. In the cantilever calibration experiment, the tip is scanned in and out of the nanobalance platform in this device. The lateral force is the product of the lateral spring constant and the lateral displacement of the platform as during the scan. The torsional stiffness is obtained by taking the ratio of the stiffness of the device and the slope of the recorded lateral force signal when scanning on the platform. The authors reported that the precision of the measured lateral force is around $\pm 7\%$. However, a limitation of this technique is that the deflection signal is susceptible to nonlinear behavior and crosstalk between lateral and normal force signals.

Jeon et al. [66] measured the torsional stiffness without the application of a normal load by inducing the Lorentz force when the cantilever of interest was positioned between two strong permanent magnets (Fig. 4.17c). The Lorentz force ($F_M$) was developed by applying an electric current to the cantilever in a magnetic field. The torsional stiffness of the cantilever of interest can be obtained by using the relationship between the magnetic torque and the torque induced by the twisting of the cantilever

$$k_\phi = \frac{1}{\phi} NISB \sin\theta \tag{4.81}$$

where $N$ is the number of current loops (1), $I$ is the current and $S$ is the area of the cantilever tip, $B$ is the magnetic field, and $\theta$ is the angle between the applied magnetic field and the normal to the cantilever plane. Unfortunately, the technique is not straightforward to implement, as it requires knowledge of the cantilever geometry, as well as the need for a magnetic field source, and the requirement that the cantilever should be metal-coated in order to form a circuit.

Another instrumentation-based lateral force calibration technique was proposed by Li et al. [67]. As shown in Fig. 4.17d, the calibration of the lateral force constant was performed by placing the cantilever on a set-up where the specimen is mounted on a pyrolytic graphite sheet and four magnets. Since the graphite sheet levitates in a magnetic field, it acts as a spring system. When the tip is in contact with the surface, the magnets and the AFM base are scanned laterally, with the normal load held constant. Then, the photodetector voltage is recorded relative to the lateral force displacement. The stiffness of the levitation system is evaluated from the natural vibration frequencies of the system. In the cantilever calibration experiment, the lateral photodetector output is monitored as a function of the lateral spring displacement, which is assumed to be predominantly from the magnetic spring (tip displacement assumed to be negligible). This allows the measurement of the lateral stiffness with a reported accuracy on the order of 0.1%. The major disadvantage of this technique is the need for additional instrumentation for the diamagnetic levitation system.

### *4.4.2 Dynamic Methods*

#### Torsional Added Mass Method

The added mass method for determining stiffness was extended to enable the calibration of the torsional stiffness [68]. They used an approach similar to that used in the determination of the normal stiffness. In this case, the radial resonance frequency of torsional vibration, $\omega_T$, changes upon addition of a mass

$$\omega_T^2 = \frac{k_\phi}{J_s + J_e} \tag{4.82}$$

where $k_\phi$ is the torsional stiffness, $J_e$ is the effective mass moment of inertia, and $J_s$ is the mass moment of inertia of the added mass $m_s$, where $J_s = \frac{7}{5} m_s r^2$. The mass of the sphere can be expressed in terms of the sphere's radius $r$ and density $\rho_s$ such that

$$J_s = \frac{7}{5} m_s r^2 = \frac{28}{15} \pi \rho_s r^5 \tag{4.83}$$

Substitution of (4.83) into (4.82) gives an expression for the torsional stiffness [68]

$$\frac{28 \pi \rho_s r^5}{15} = \frac{k_\phi}{\omega_T^2} - J_e \tag{4.84}$$

A plot of $J_s$ vs. ${\omega_T}^{-2}$ will give a straight line with a slope equal to $k_\phi$. An example of the application of the torsional added mass method is shown in Fig. 4.18.

Similar to the added mass method for normal stiffness determination, this technique can potentially damage the cantilever, and errors in determining the

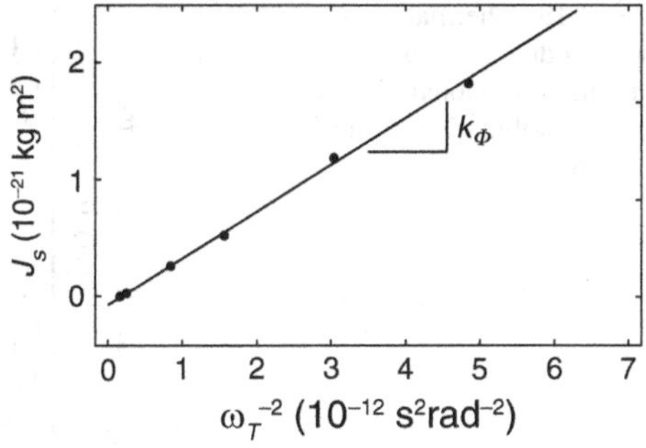


**Fig. 4.18** Mass moment of inertia as a function of the square of the angular frequency of torsional vibration, where the slope is the torsional stiffness (Adapted from [68])

radius and density of the added mass directly affect the measured torsional stiffness. In addition, placement of the added mass away from the axis or away from the free end of the cantilever can change the $k_\phi$ by as much as 25% [68].

## Torsional Resonance Method

In this technique by Green et al. [68], the resonance method for normal stiffness determination [38, 39] was extended for determining the torsional stiffness. On the assumption that $L >> b >> h$, $k_\phi$ for a rectangular beam is given by [69]

$$k_\phi = \frac{1}{3\pi}\rho_c b^3 h L \omega_{T,vac}^2 \tag{4.85}$$

where $\omega_{T,vac}$ is the resonance frequency of torsional vibration in vacuum. The quantity $\omega_{T,vac}$ can be related to its corresponding value in a fluid medium (e.g., air) $\omega_T$ by [68]

$$\omega_{T,vac} = \omega_T \left(1 + \frac{3\pi\rho_f b}{2\rho h}\Gamma_r^T(\omega_T)\right)^{1/2} \tag{4.86}$$

where $\Gamma_r^T$ is the real part of the hydrodynamic function. The density of the cantilever is given by the following

$$\rho_c h = \frac{3\pi\rho_f b}{2}\left[Q_T \Gamma_i^T(\omega_T) - \Gamma_r^T(\omega_T)\right] \tag{4.87}$$

where $Q_T$ is the quality factor of the resonance peak of torsional vibration in a fluid medium. By substituting (4.86) and (4.87) into (4.85), the torsional stiffness is given by [68]

$$k_\phi = 0.1592\rho_f b^4 L Q_T \omega_T^2 \Gamma_i^T(\omega_T) \tag{4.88}$$

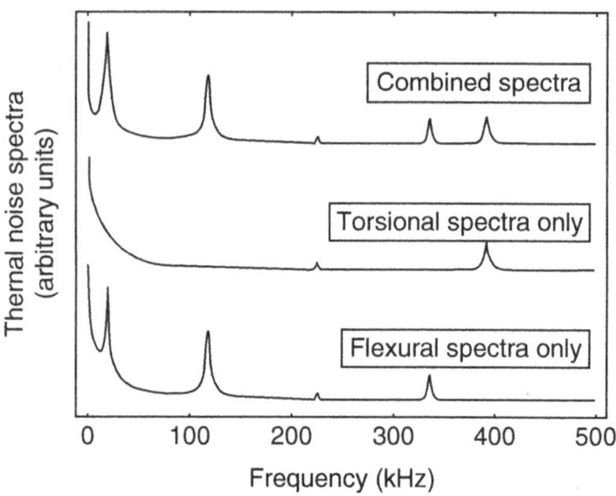


**Fig. 4.19** Thermal noise spectra due to flexural vibration, torsional vibration and a combination of the two [68]

Figure 4.19 shows the thermal noise spectra due to flexural vibration, torsional vibration and a combination of the two [68]. All three are shown to emphasize that data is usually a combination of the flexural and torsional vibration, and that the source of each peak should be analyzed with care. The peaks due to flexural and torsional vibration can be distinguished by picking the appropriate detector signals. Data from the middle plot should be used to obtain the torsional stiffness using this resonance technique.

An advantage of this technique is that it is relatively simple. However, the model relies on cantilever dimensions, which are sources of measurement uncertainty. In addition, it relies on knowledge of the Reynolds number of the fluid medium and the determination of the hydrodynamic function. It must be noted that the resonance method discussed here is based on the geometry of a rectangular cantilever beam. A solution to the hydrodynamic function is needed to extend this technique to triangular cantilevers. Alternatively, a wafer with both rectangular and triangular cantilevers present can be used, as described in Sect. 4.3.2.2.

### Calculation of Lateral Force from Torsional Stiffness

In the dynamic methods discussed in Sects. 4.4.2.1 and 4.4.2.2, the torsional stiffness $k_\phi$ (in N m) is obtained. Calculation of the lateral force from $k_\phi$ is not straightforward in this case and an additional step is necessary. Photodetector calibration is necessary in order to obtain the torsional sensitivity (in V/rad). Pettersson et al. [13] describes a procedure for the torsional calibration of the photodetector. Alternatively, if $k_\phi$ is converted into $k_{yT}$, the lateral sensitivity can be determined [70].

An example of a procedure which can be implemented in general is the "optical geometry" method proposed by Liu et al. [71]. They derived a relationship for the lateral force as the product of the half-width of the friction loop multiplied by a constant involving the stiffness parameter. This parameter is the product of $k_\phi$, $k_z$, the sensitivity of the input channel of the FFM and the tip length, divided by the product of the normal sensitivity of the force curve slope, the coefficient relating the sensitivities of the FFM and AFM photodiode pairs and a constant which relates the normal bending angle and the bending force acting on the cantilever tip. The normal bending is calculated with the knowledge of the cantilever's material properties and dimensions. These parameters are necessary in order to relate cantilever bending and torsion. The half-width of the friction loop and the sensitivity of the FFM input channel are related to the change in the photodiode signal resulting from torsion. This, in turn, is related to the change in photodiode signal from bending, the coefficient relating the sensitivities of the FFM and AFM and the cantilever's normal bending angle. As discussed earlier, the material properties and cantilever dimensions are sources of measurement uncertainties such that this method for calculating the lateral force from the torsional stiffness may not be accurate.

### *4.4.3 Discussion*

Table 4.2 lists the highlights for the lateral force and stiffness calibration methods discussed above. Quantitative measurement and calibration of the lateral force is not as straightforward as the corresponding methods for the normal force. This is because lateral motion involves bending or torsion of the cantilever during sliding in the parallel or perpendicular direction, relative to the long axis of the cantilever, respectively. Aside from classifying the methods discussed above as either static or dynamic, they can also be differentiated based on the measurable parameters that each model provides. The axial bending, wedge, and compliance hysteresis techniques all result in a value for the coefficient of friction. The lateral force can be easily calculated afterwards. Meanwhile, the lever, particle interaction, torsional added mass, and torsional resonance methods all give a value of the stiffness (torsional, or torsional and lateral). The stiffness can be used to calculate the lateral force, although it is not straightforward in the case of the torsional added mass and torsional resonance methods.

The distinction can also be made based on classifying them as "one-step" and "two-step" methods. Some methods, such as axial sliding and wedge, provide a direct means (one-step) for obtaining the lateral force. Meanwhile, other calibration techniques, such as the lever particle interaction, torsional added mass, and torsional resonance methods, consist of two steps. The first step is the determination of the torsional or lateral stiffness. The second step is the determination of the photodetector sensitivity (V/rad or V/m).

In evaluating the various techniques for measuring the lateral force of the cantilever, both the ease of implementation as well as the limitations of the respective models should be taken into consideration. The axial bending method is simple to implement, but the requirement of measuring the length of the cantilever and the tip introduces uncertainty into the resulting value of the coefficient of friction.

**Table 4.2** Summary of techniques for calibrating the lateral force and torsional stiffness

| Technique | Principle/key equation | Pros | Cons |
|---|---|---|---|
| *Static methods* | | | |
| Axial sliding[a] [9] | Height variation of piezo tube is monitored during sliding in the long axis of cantilever $\mu = \frac{F}{W} = \left[\frac{(\Delta H_1 + \Delta H_2)}{H_0}\right]\left(\frac{L}{4\ell}\right)$ | Simple; easy to implement | Requires measurement of cantilever length and tip height |
| Wedge method[a] [46] | Cantilever is scanned across calibration standard with two well-defined slopes $\mu + \frac{1}{\mu} = \frac{2\Delta_o'}{w_o' \sin 2\theta}$ | Does not require cantilever dimensions | Requires calibration standard; pull-off force value may cause measurement errors; computation method is not straightforward |
| Lever method [56] | Lever with glass fiber and sphere is attached to cantilever and angular deflection of cantilever is measured $k_\phi = \frac{k_z l^2}{\left(c_{z,L}/c_z - 1\right)}$ | – | Only practical for tipless cantilevers; requires attachment of lever to cantilever of interest, which is difficult and time consuming in practice |

| | | | |
|---|---|---|---|
| Cantilever with particle; Calibrated cantilever | perpendicular to cantilever with spherical tips (colloidal probe) and the twist of the cantilever is measured which is used to calculate $k_y$ and $k_\phi$ $k_{yT} = \frac{k_\phi}{l^2} = \frac{c_y^{ref} - c_y^0}{c_y^0} k_z^{ref} \left( \frac{D+h}{2D+h} \right)^2$ | | colloidal probes |
| Compliance hysteresis method[a] [61] | Tilt of cantilever and the hysteresis of the compliance lines during approach and withdrawal are measured | – | Requires accurate tip tilt angle value; computation method is not straightforward |
| Lateral electrical nanobalance [65] | Cantilever is calibrated by scanning tip in and out of the nanobalance platform | – | Requires fabricated nanobalance; signal is susceptible to nonlinearity and crosstalk |
| Lorentz force induction [66] 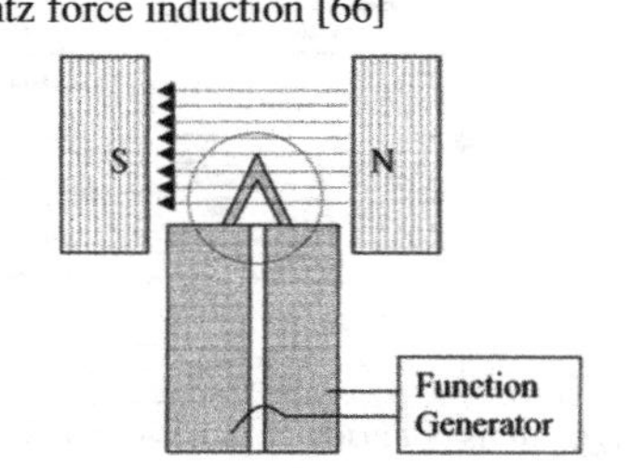  | Lorentz force is induced when an electric current is applied to cantilever in a magnetic field and the torsional resonance frequency is measured $k_\phi = \frac{1}{\phi} NISB \sin\theta$ | – | Requires magnetic field source, metal coating on cantilever, and knowledge of cantilever geometry |

(*continued*)

**Table 4.2** (continued)

| Technique | Principle/key equation | Pros | Cons |
|---|---|---|---|
| Diamagnetic levitation spring system [67] 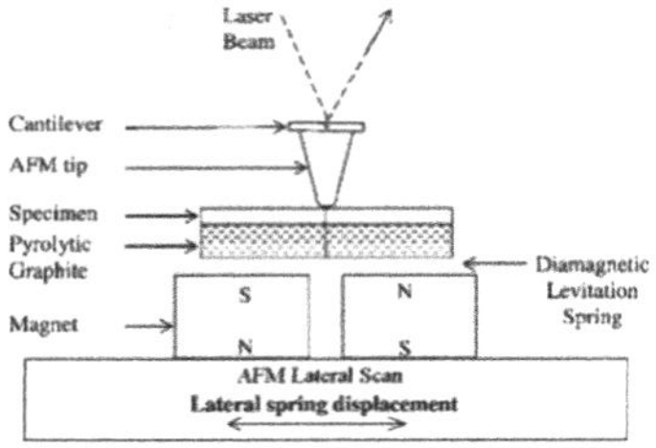 | Cantilever is placed in a set-up where specimen is on a pyrolytic graphite sheet and four magnets and the levitating graphite sheet is related to the lateral tip displacement | – | Additional instrumentation is required |
| *Dynamic methods* | | | |
| Torsional added mass method [68] 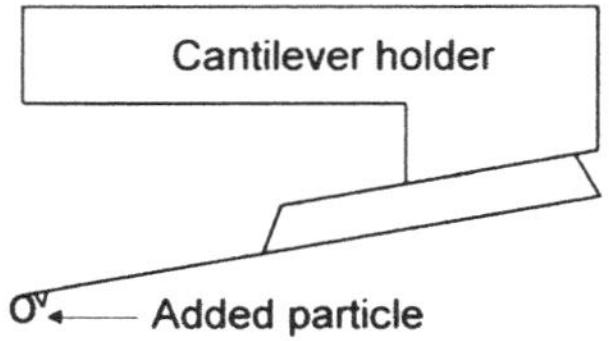 | A number of particles are individually attached to the end of the cantilever and the resonance frequency is measured $\frac{28\pi\rho_s r^5}{15} = \frac{k_\phi}{\omega_T^2} - J_e$ | – | Requires attachment of particles to cantilever; risk of cantilever damage; uncertainty in dimensions and errors in placement of added particle lead to significant error; conversion of torsional stiffness to lateral force is not straightforward |
| Torsional resonance method [68] 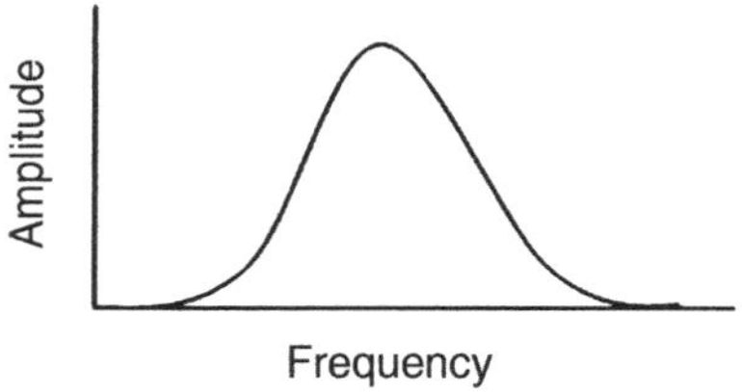 | The resonance frequency, quality factor and dimensions are measured. For a rectangular beam, $k_\phi = 0.1592\rho_f b^4 L Q_T \omega_T^2 \Gamma_i^T(\omega_T)$ | Relatively simple for torsional stiffness determination | Model relies on cantilever dimensions; requires Reynolds number of the fluid medium and calculation of the hydrodynamic function; conversion of torsional stiffness to lateral force is not straightforward |

The wedge method depends solely on the signals obtained during the friction experiment (on the sloped surfaces) and not on any cantilever dimension or property, which makes this technique desirable. Another benefit of this technique is that it can be used to ascertain the state of tip geometry [72]. However, it should be noted that the procedures described for performing the wedge method calibration have their own caveats. The use of a relatively steep calibration standard limits the applicability of the wedge method to sharp tips and colloidal probes with small radii of curvature. As pointed out earlier, inaccuracies in the pull-off force could affect the coefficient of friction obtained. Also, this method is not mathematically straightforward since careful consideration has to be made in selecting between the two values obtained in the quadratic equation solving for the coefficient of friction.

The lever and particle interaction methods are based on measuring the angular deflection of the cantilever and can measure both the torsional stiffness and lateral force. However, both the lever and particle interaction methods appear to be complicated to implement experimentally. The lever method is only practical for tipless cantilevers, while there is a risk of damaging the cantilever or dislodging the colloidal probe in the particle interaction method. The particle interaction method is also limited because it is only applicable to colloidal probes.

The compliance hysteresis method is performed in the longitudinal direction and yields a value of the coefficient of friction similar to the axial sliding and wedge methods. However, the model requires the tilt of the cantilever and its length, which may cause artifacts if measured incorrectly. Also, the computation method is not very straightforward.

The lateral electrical nanobalance (LEN), Lorentz force induction, and diamagnetic levitation spring methods are examples of novel calibration techniques that require the use of additional instrumentation in order to induce cantilever twisting and measure the resulting lateral displacement. It should be pointed out that there are other instrument-based methods that have been proposed in the literature. However, they were not included in the review, because either the measurement principle is similar to the ones discussed here, or the method is largely dependent on cantilever properties (such as mechanical properties of the cantilever and its dimensions). Two of the calibration procedures discussed in Sect. 4.4.1 under "Miscellaneous methods" are noteworthy due to their reported accuracy. Cumpson et al. [65] reported that in the LEN method, the lateral force uncertainty is as low as $\pm 7\%$, while the diamagnetic levitation method was reported to measure stiffness values with accuracy on the order of 0.1%. The Lorentz force induction method is of interest because it does not require the application of normal load, such that deflection is nominally zero, such that a pure torsional response is obtained. However, these potential advantages are outweighed by the requirement to acquire (or construct) and calibrate the measurement set-ups that will then be used to calibrate the cantilever.

The dynamic techniques (added mass and resonance) for determining the torsional stiffness have advantages and disadvantages similar to the normal stiffness methods that they were originally derived from. For the added mass technique, the main concern is the possibility of cantilever damage from the added sphere. In addition, the use of an inaccurate value for the sphere, as well as errors in placement of the sphere (e.g. off-axis), will lead to significant errors. For the resonance technique, an advantage is that the procedure enables the simultaneous determination of both the torsional and normal stiffness. However, the need for the dimensions of the cantilever (length and width) is a disadvantage as it adds some uncertainty in the measurement. Another disadvantage is that the lateral force cannot be determined easily from the torsional stiffness using these dynamic methods. An additional procedure to determine the sensitivity of the photodetector is necessary in order to complete the quantification of the lateral force.

Among all the techniques presented, the axial sliding, wedge and torsional resonance methods are preferred for calibrating lateral forces. These techniques are not limited to cantilever geometry and can be used on both integrated tips and colloidal probes. The first one (axial sliding) is recommended due to its relative simplicity, and the latter two, though not as convenient to implement, also have minimal susceptibility to cantilever damage. However, it should be pointed out that the wedge and torsional resonance methods have some susceptibility to measurement uncertainties, as described earlier.

It should be noted that one of the sources of error not discussed so far is the crosstalk (either optical or mechanical in nature) between the signals corresponding to lateral and normal cantilever deflection, as this is difficult to eliminate in general. Optical crosstalk takes place because of the rotational misalignment of the PSD. This causes an error in the topography and friction measurements, and is particularly pronounced for nanostructures with significant local variations in frictional properties [73]. Various approaches exist in the literature to analyze optical crosstalk [1, 2, 70]. Mechanical crosstalk is caused by the positional offset of the tip relative to the symmetry axis of the cantilever. The mechanical crosstalk artifact can be corrected for by using a set of relationships that account for the offset caused by the non-orthogonal alignment of the PSD in combination with the AFM tip being off-centered relative to the cantilever [74].

## 4.5 Outlook

Over the years, the use of the AFM has expanded beyond surface profiling to capabilities such as the nanomanipulation and the measurement of various physical properties. There is a need to calibrate the normal and lateral forces of AFM cantilevers in order to properly use AFM data in studies of tribological and mechanical properties, and as presented in this chapter, various techniques are available.

For calibrating the normal forces, the reference cantilever, resonance, and thermal noise methods are desirable. The use of a reference cantilever is recommended due to the simplicity of this method. The resonance and thermal noise methods are both relatively simple to implement, and the latter technique is regarded to provide accurate values of the normal stiffness as it is the least susceptible to measurement uncertainties. For lateral forces, the axial sliding, wedge and torsional resonance methods are desirable. The axial sliding technique is a simple technique. The wedge and torsional resonance methods are noteworthy as they possess minimal susceptibility to cantilever damage. However, there are other sources of measurement uncertainties involved with these techniques, most notably the effect of the pull-off force on the wedge method and the need for cantilever dimensions in the torsional resonance method.

There is a need to further understand the effect of optical and mechanical crosstalk, with the latter having been less studied so far. It would be of great benefit to experimentalists to have simple tools to deconvolute signal artifacts, thus ensuring accuracy in the measurement and analysis of normal and lateral forces. Lastly, it should be pointed out that most of the available calibration techniques are more applicable to the rectangular cantilever due to the simplicity of its geometry. Since in practice, triangular cantilevers are as widely-used as the rectangular cantilever, more geometry-independent experimental techniques should be developed.

## Appendix – Nomenclature

| ***Roman Symbols*** | |
|---|---|
| $A$ | Adhesive force |
| $a$ | Amplitude |
| $b$ | Cantilever width |
| $b_l$ | Width of the leg in a triangular cantilever |
| $c$ | Photodetector sensitivity |
| $d$ | Distance of the tip to the edge of the cantilever |
| $E$ | Young's modulus |
| $F$ | Friction force (lateral force) |
| $f$ | Frequency of the cantilever |
| $G$ | Shear modulus |
| $H$ | Piezo tube height in axial sliding method |
| $h$ | Cantilever thickness |
| $I$ | Area moment of inertia |
| $J$ | Mass moment of inertia |
| $\mathrm{k_B}$ | Boltzmann's constant |
| $k_x$ | Cantilever stiffness in the direction parallel to the longitudinal axis |
| $k_{yB}$ | Cantilever stiffness in the direction perpendicular to the longitudinal axis due to bending |
| $k_{yT}$ | Cantilever stiffness in the direction perpendicular to the longitudinal axis due to applied torque |
| $k_z$ | Cantilever stiffness in the normal direction |
| $k_\phi$ | Cantilever torsional stiffness |
| $L$ | Cantilever length |
| $l$ | Lever length |
| $\ell$ | Tip length |
| $M_T$ | Torsion moment |
| $m$ | Mass of the cantilever |
| $m^*$ | Effective mass of the cantilever |
| $m_s$ | Mass of added particle |

(*continued*)

| | |
|---|---|
| $P$ | Normal load component in axial sliding method, or applied load in the wedge method |
| $p$ | Area of the power spectrum in the thermal noise method |
| $Q$ | Quality factor |
| $r$ | Radius of added particle |
| $T$ | Temperature (in thermal tune method), or friction/horizontal force component (in axial sliding, wedge and compliance hysteresis methods) |
| $W$ | Normal load |
| $w$ | Half width of the friction loop in the wedge method |
| $z$ | Cantilever deflection |
| ***Greek Symbols*** | |
| $\alpha$ | One-half the included angle between the legs of a triangular cantilever |
| $\alpha_c, \beta_c$ | Lateral force calibration factors in the wedge method |
| $\gamma$ | Cantilever tilt relative to horizontal axis |
| $\Delta$ | Friction loop offset in the wedge method |
| $\delta_{\mathrm{I,II}}$ | Deflection of the cantilever in the parts I and II |
| $\varepsilon$ | Calibration factor in the lever method |
| $\eta$ | Viscosity of fluid medium |
| $\mu$ | Coefficient of friction |
| $\nu$ | Poisson's ratio |
| $\rho$ | Density |
| $\theta$ | Inclination of calibration standard in the wedge method |
| $\theta_{II}$ | Rotation of the legs of a triangular cantilever in the longitudinal direction |
| $\phi$ | Cantilever rotation from applied torque |
| $\omega$ | Angular frequency of the cantilever |
| $\Gamma$ | Hydrodynamic function in the resonance method |
| $\varsigma$ | Damping ratio |

## References


1. B. Bhushan, *Nanotribology and Nanomechanics – An Introduction*, 2nd edn. (Springer, Heidelberg, 2008)
2. B. Bhushan, *Springer Handbook of Nanotechnology*, 3rd edn. (Springer, Heidelberg, 2010)
3. B. Bhushan, K.J. Kwak, M. Palacio, Nanotribology and nanomechanics of AFM probe-based recording technology. J. Phys. Condens. Matter **20**, 365207 (2008)
4. G. Meyer, N.M. Amer, Simultaneous measurement of lateral and normal forces with an optical-beam-deflection atomic force microscope. Appl. Phys. Lett. **57**, 2089–2091 (1990)
5. T.R. Albrecht, S. Akamine, T.E. Carver, C.F. Quate, Microfabrication of cantilever styli for the atomic force microscope. J. Vac. Sci. Technol. A **8**, 3386–3396 (1990)
6. O. Wolter, T. Bayer, J. Greschner, Micromachined silicon sensors for scanning force microscopy. J. Vac. Sci. Technol. B **9**, 1353–1357 (1991)
7. B. Ohler, Application Note 94: Practical advice on the determination of cantilever spring constants (2007), http://www.veeco.com/library
8. J.E. Sader, I. Larson, P. Mulvaney, L.R. White, Method for the calibration of atomic force microscope cantilevers. Rev. Sci. Instrum. **66**, 3789–3798 (1995)
9. J. Ruan, B. Bhushan, Atomic-scale friction measurements using friction force microscopy: part I. General principles and new measurement techniques. ASME J. Tribol. **116**, 378–388 (1994)
10. J.M. Neumeister, W.A. Ducker, Lateral, normal and longitudinal spring constants of atomic force microscopy cantilevers. Rev. Sci. Instrum. **65**, 2527–2531 (1994)

11. C.A. Clifford, M.P. Seah, The determination of atomic force microscope cantilever spring constants via dimensional methods for nanomechanical analysis. Nanotechnology **16**, 1666–1680 (2005)
12. S.M. Cook, K.M. Lang, K.M. Chynoweth, M. Wigton, R.W. Simmonds, T.E. Schaffer, Practical implementation of dynamic methods for measuring atomic force microscope cantilever spring constants. Nanotechnology **17**, 2135–2145 (2006)
13. T. Pettersson, N. Nordgren, M.W. Rutland, A. Feiler, Comparison of different methods to calibrate torsional spring constant and photodetector for atomic force microscopy friction measurements in air and liquid. Rev. Sci. Instrum. **78**, 093702 (2007)
14. M.L.B. Palacio, B. Bhushan, Normal and lateral force calibration techniques for AFM cantilevers. Crit. Rev. Solid State Mater. Sci. **35**, 73–104 (2010); ibid, Erratum. **35**, 261 (2010)
15. D. Sarid, V. Elings, Review of scanning force microscopy. J. Vac. Sci. Technol. B **9**, 431–437 (1991)
16. S.P. Timoshenko, J.N. Goodier, *Theory of Elasticity*, 3rd edn. (McGraw-Hill, New York, 1970)
17. W.T. Thomson, M.D. Dahleh, *Theory of Vibration with Applications*, 5th edn. (Prentice Hall, Upper Saddle River, 1998)
18. J. Hutter, Comment on tilt of atomic force microscope cantilevers: effect on spring constant and adhesion measurements. Langmuir **21**, 2630–2632 (2005)
19. W.C. Young, R.G. Budynas, *Roark's Formulas for Stress and Strain*, 7th edn. (McGraw-Hill, New York, 2002)
20. H.J. Butt, P. Siedle, K. Seifert, K. Fendler, T. Seeger, E. Bamberg, A.L. Weisenhorn, K. Goldie, A. Engel, Scan speed limit in atomic force microscopy. J. Microsc. **169**, 75–84 (1993)
21. J.E. Sader, Parallel beam approximation for V-shaped atomic force microscope cantilevers. Rev. Sci. Instrum. **75**, 4583–4586 (1995)
22. Anonymous, Probes-Recommended Products, (Veeco Probes, Santa Barbara, 2008) http://www.veecoprobes.com
23. T.R. Albrecht, C.F. Quate, Atomic resolution imaging of a nonconductor by atomic force microscopy. J. Appl. Phys. **62**, 2599–2602 (1987)
24. J.E. Sader, Susceptibility of atomic force microscopy cantilevers to lateral forces. Rev. Sci. Instrum. **74**, 2438–2443 (2003)
25. J.E. Sader, R.C. Sader, Suitability of atomic force microscope cantilevers to lateral forces: experimental verification. Appl. Phys. Lett. **83**, 3195–3197 (2003)
26. M. Tortonese, M. Kirk, Characterization of application specific probes for SPMs. Proc. SPIE **3009**, 53–60 (1997)
27. T.J. Senden, W.A. Ducker, Experimental determination of spring constants in atomic force microscopy. Langmuir **10**, 1003–1004 (1994)
28. G.A. Shaw, J. Kramar, J. Pratt, SI-traceable spring constant calibration of microfabricated cantilevers for small force measurement. Exp. Mech. **47**, 143–151 (2007)
29. M.S. Kim, J.J. Choi, Y.K. Park, J.H. Kim, Atomic force microscope cantilever calibration device for quantified force metrology at micro- or nano-scale regime: the nano force calibrator (NFC). Metrologia **43**, 389–395 (2006)
30. J.R. Pratt, J.A. Kramar, D.B. Newell, D.T. Smith, Review of SI traceable force metrology for instrumented indentation and atomic force microscopy. Meas. Sci. Technol. **16**, 2129–2137 (2005)
31. M.S. Kim, J.J. Choi, J.H. Kim, Y.K. Park, Si-traceable determination of spring constants of various atomic force microscope cantilevers with a small uncertainty of 1%. Meas. Sci. Technol. **18**, 3351–3358 (2007)
32. P.J. Cumpson, J. Hedley, Accurate analytical measurements in the atomic force microscope: a microfabricated spring constant standard potentially traceable to the SI. Nanotechnology **14**, 1279–1288 (2003)

33. R. Leach, D. Chetwynd, L. Blunt, J. Haycocks, P. Harris, K. Jackson, S. Oldfield, S. Reilly, Recent advances in traceable nanoscale dimension and force metrology in the UK. Meas. Sci. Technol. **17**, 467–476 (2006)
34. I. Behrens, L. Doering, E. Peiner, Piezoresistive cantilever as portable micro force calibration standard. J. Micromech. Microeng. **13**, S171–S177 (2003)
35. V. Nesterov, Facility and methods for the measurement of micro and nano forces in the range below $10^{-5}$ N with a resolution of $10^{-12}$ N (development concept). Meas. Sci. Technol. **18**, 360–366 (2007)
36. J.P. Cleveland, S. Manne, D. Bocek, P.K. Hansma, A nondestructive method for determining the spring constant of cantilevers for scanning force microscopy. Rev. Sci. Instrum. **64**, 403–405 (1993)
37. J.E. Sader, Frequency response of cantilever beams immersed in various fluids with applications to the atomic force microscope. J. Appl. Phys. **84**, 64–76 (1998)
38. J.E. Sader, J.W.M. Chon, P. Mulvaney, Calibration of rectangular atomic force microscopy cantilevers. Rev. Sci. Instrum. **70**, 3967–3969 (1999)
39. J.E. Sader, J. Pacifico, C.P. Green, P. Mulvaney, General scaling law for stiffness measurement of small bodies with applications to the atomic force microscope. J. Appl. Phys. **97**, 124903 (2005)
40. B. Ohler, Cantilever spring constant calibration using laser Doppler vibrometry. Rev. Sci. Instrum. **78**, 063701 (2007)
41. J.L. Hutter, J. Bechhoefer, Calibration of atomic-force microscope tips. Rev. Sci. Instrum. **64**, 1868–1873 (1993)
42. H.J. Butt, M. Jaschke, Calculation of thermal noise in atomic force microscopy. Nanotechnology **6**, 1–7 (1995)
43. G.A. Matei, E.J. Thoreson, J.R. Pratt, D.B. Newell, N.A. Burnham, Precision and accuracy of thermal calibration of atomic force microscopy cantilevers. Rev. Sci. Instrum. **77**, 083703 (2006)
44. Y.L. Wang, X.Z. Zhao, F.Q. Zhou, Improved parallel scan method for nanofriction force measurement with atomic force microscopy. Rev. Sci. Instrum. **78**, 036107 (2007)
45. N.S. Tambe, Nanotribological investigations of materials, coatings and lubricants for nanotechnology applications at high sliding velocities, Ph.D. dissertation, The Ohio State University, 2005, Available from http://www.ohiolink.edu/etd/send-pdf.cgi?osu1109949835
46. D.F. Ogletree, R.W. Carpick, M. Salmeron, Calibration of frictional forces in atomic force microscopy. Rev. Sci. Instrum. **67**, 3298–3306 (1996)
47. J. Ruan, B. Bhushan, Atomic-scale and microscale friction of graphite and diamond using friction force microscopy. J. Appl. Phys. **76**, 5022–5035 (1994)
48. J. Ruan, B. Bhushan, Frictional behavior of highly oriented pyrolytic graphite. J. Appl. Phys. **76**, 8117–8120 (1994)
49. V.N. Koinkar, B. Bhushan, Effect of scan size and surface roughness on microscale friction measurements. J. Appl. Phys. **81**, 2472–2479 (1997)
50. S. Sundararajan, B. Bhushan, Topography-induced contributions to friction forces measured using an atomic force/friction force microscope. J. Appl. Phys. **88**, 4825–4831 (2000)
51. M. Varenberg, I. Etsion, G. Halperin, An improved wedge calibration method for lateral force in atomic force microscopy. Rev. Sci. Instrum. **74**, 3362–3367 (2003)
52. B. Bhushan, *Handbook of Micro/Nanotribology*, 2nd edn. (CRC Press, Boca Raton, 1999)
53. B. Bhushan, *Introduction to Tribology* (Wiley, New York, 2002)
54. E. Tocha, H. Schonherr, G.J. Vancso, Quantitative nanotribology by AFM: a novel universal calibration platform. Langmuir **22**, 2340–2350 (2006)
55. X. Ling, H.J. Butt, M. Kappl, Quantitative measurement of friction between single microspheres by friction force microscopy. Langmuir **23**, 8392–8399 (2007)
56. A. Feiler, P. Attard, I. Larson, Calibration of the torsional spring constant and the lateral photodiode response of frictional force microscopes. Rev. Sci. Instrum. **71**, 2746–2750 (2000)

57. G. Bogdanovic, A. Meurk, M.W. Rutland, Tip friction – torsional spring constant determination. Colloids Surf. B. Biointerfaces **19**, 397–405 (2000)
58. M.G. Reitsma, Lateral force calibration using a modified atomic force microscope cantilever. Rev. Sci. Instrum. **78**, 106102 (2007)
59. M.A.S. Quintanilla, D.T. Goddard, A calibration method for lateral forces for use with colloidal probe force microscopy cantilevers. Rev. Sci. Instrum. **79**, 023701 (2008)
60. S. Ecke, R. Raiteri, E. Bonaccurso, C. Reiner, H.J. Deiseroth, H.J. Butt, Measuring normal and friction forces acting on individual fine particles. Rev. Sci. Instrum. **72**, 4164–4170 (2001)
61. J. Stiernstedt, M.W. Rutland, P. Attard, A novel technique for the in situ calibration and measurement of friction with the atomic force microscope. Rev. Sci. Instrum. **76**, 083710 (2005)
62. P. Attard, A. Carambassis, M.W. Rutland, Dynamic surface force measurement. 2. Friction and the atomic force microscope. Langmuir **15**, 553–563 (1999)
63. J. Stiernstedt, M.W. Rutland, P. Attard, Erratum: A novel technique for the in situ calibration and measurement of friction with the atomic force microscope. Rev. Sci. Instrum. **77**, 019901 (2006)
64. P. Attard, Measurement and interpretation of elastic and viscoelastic properties with the atomic force microscope. J. Phys. Condens. Matter **19**, 473201 (2007)
65. P.J. Cumpson, J. Hedley, C.A. Clifford, Microelectromechanical device for lateral force calibration in the atomic force microscope: lateral electrical nanobalance. J. Vac. Sci. Technol. B **23**, 1992–1997 (2005)
66. S. Jeon, Y. Braiman, T. Thundat, Torsional spring constant obtained for an atomic force microscope cantilever. Appl. Phys. Lett. **84**, 1795–1797 (2004)
67. Q. Li, K.S. Kim, A. Rydberg, Lateral force calibration of an atomic force microscope with a diamagnetic levitation spring system. Rev. Sci. Instrum. **77**, 065105 (2006)
68. C.P. Green, H. Lioe, J.P. Cleveland, R. Proksch, P. Mulvaney, J.E. Sader, Normal and torsional spring constants of atomic force microscope cantilevers. Rev. Sci. Instrum. **75**, 1988–1996 (2004)
69. A.E.H. Love, *A Treatise on the Mathematical Theory of Elasticity* (Pergamon, London, 1959)
70. R.J. Cannara, M. Eglin, R.W. Carpick, Lateral force calibration in atomic force microscopy: a new lateral force calibration method and general guidelines for optimization. Rev. Sci. Instrum. **77**, 053701 (2006)
71. E. Liu, B. Blanpain, J.P. Celis, Calibration procedures for frictional measurements with a lateral force microscope. Wear **192**, 141–150 (1996)
72. R.G. Cain, M.G. Reitsma, S. Biggs, N.W. Page, Quantitative comparison of three calibration techniques for the lateral force microscope. Rev. Sci. Instrum. **72**, 3304–3312 (2001)
73. R. Piner, R.S. Ruoff, Cross talk between friction and height signals in atomic force microscopy. Rev. Sci. Instrum. **73**, 3392–3394 (2002)
74. D.B. Asay, S.H. Kim, Direct force balance method for atomic force microscopy lateral force calibration. Rev. Sci. Instrum. **77**, 043903 (2006)

# Chapter 5
# Noncontact Atomic Force Microscopy and Related Topics

**Franz J. Giessibl, Yasuhiro Sugawara, Seizo Morita, Hirotaka Hosoi, Kazuhisa Sueoka, Koichi Mukasa, Akira Sasahara, and Hiroshi Onishi**

**Abstract** Scanning probe microscopy (SPM) methods such as scanning tunneling microscopy (STM) and noncontact atomic force microscopy (NC-AFM) are the basic technologies for nanotechnology and also for future bottom-up processes. In Sect. 5.1, the principles of AFM such as its operating modes and the NC-AFM frequency-modulation method are fully explained. Then, in Sect. 5.2, applications of NC-AFM to semiconductors, which make clear its potential in terms of spatial resolution and function, are introduced. Next, in Sect. 5.3, applications of NC-AFM to insulators such as alkali halides, fluorides and transition-metal oxides are introduced. Lastly, in Sect. 5.4, applications of NC-AFM to molecules such as carboxylate ($RCOO^-$) with R = H, $CH_3$, $C(CH_3)_3$ and $CF_3$ are introduced. Thus, NC-AFM can observe atoms and molecules on various kinds of surfaces such as semiconductors, insulators and metal oxides with atomic or molecular resolution. These sections are essential to understand the state of the art and future possibilities for NC-AFM, which is the second generation of atom/molecule technology.

The scanning tunneling microscope (STM) is an atomic tool based on an electric method that measures the tunneling current between a conductive tip and a conductive surface. It can electrically observe individual atoms/molecules. It can characterize or analyze the electronic nature around surface atoms/molecules. In addition, it can manipulate individual atoms/molecules. Hence, the STM is the first generation of atom/molecule technology. On the other hand, the atomic force microscopy (AFM) is a unique atomic tool based on a mechanical method that can even deal with insulator surfaces. Since the invention of noncontact AFM (NC-AFM) in 1995, the NC-AFM and NC-AFM-based methods have rapidly developed into powerful surface tools on the atomic/molecular scales, because NC-AFM has the following characteristics: (1) it has true atomic resolution, (2) it can measure atomic force (so-called atomic force spectroscopy), (3) it can observe even insulators, and (4) it can measure mechanical responses such as elastic deformation. Thus, NC-AFM is the second generation of atom/molecule technology. Scanning probe microscopy (SPM) such as STM and NC-AFM is the basic technology for nanotechnology and also for future bottom-up processes.

B. Bhushan (ed.), *Nanotribology and Nanomechanics*,
DOI 10.1007/978-3-642-15283-2_5, © Springer-Verlag Berlin Heidelberg 2011

In Sect. 5.1, the principles of NC-AFM will be fully introduced. Then, in Sect. 5.2, applications to semiconductors will be presented. Next, in Sect. 5.3, applications to insulators will be described. And, in Sect. 5.4, applications to molecules will be introduced. These sections are essential to understanding the state of the art and future possibilities for NC-AFM.

## 5.1 Atomic Force Microscopy (AFM)

The atomic force microscope (AFM), invented by Binnig [1] and introduced in 1986 by Binnig et al. [2] is an offspring of the scanning tunneling microscope (STM) [3]. The STM is covered in several books and review articles, e.g. [4, 5, 6, 7, 8, 9]. Early in the development of STM it became evident that relatively strong forces act between a tip in close proximity to a sample. It was found that these forces could be put to good use in the atomic force microscope (AFM). Detailed information about the noncontact AFM can be found in [10, 11, 12].

### *5.1.1 Imaging Signal in AFM*

Figure 5.1 shows a sharp tip close to a sample. The potential energy between the tip and the sample $V_{ts}$ causes a $z$-component of the tip–sample force $F_{ts} = -\partial V_{ts}/\partial z$. Depending on the mode of operation, the AFM uses $F_{ts}$, or some entity derived from $F_{ts}$, as the imaging signal.

Unlike the tunneling current, which has a very strong distance dependence, $F_{ts}$ has long- and short-range contributions. We can classify the contributions by their range and strength. In vacuum, there are van-der-Waals, electrostatic and magnetic forces with a long range (up to 100 nm) and short-range chemical forces (fractions of nm).

The van-der-Waals interaction is caused by fluctuations in the electric dipole moment of atoms and their mutual polarization. For a spherical tip with radius $R$

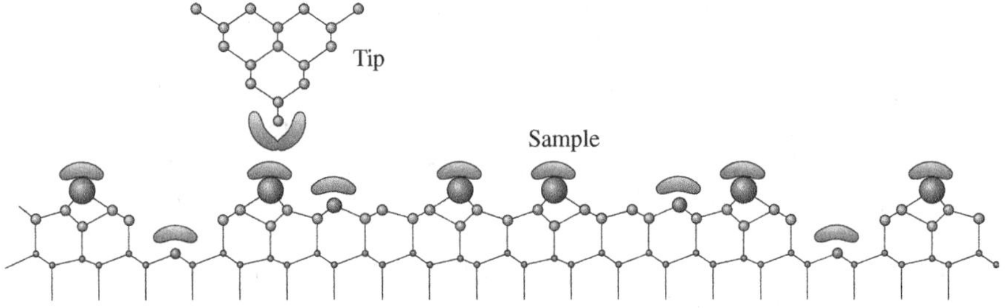


**Fig. 5.1** Schematic view of an AFM tip close to a sample

next to a flat surface ($z$ is the distance between the plane connecting the centers of the surface atoms and the center of the closest tip atom) the van-der-Waals potential is given by [13]

$$V_{\mathrm{vdW}} = -\frac{A_{\mathrm{H}}}{6z}. \tag{5.1}$$

The Hamaker constant $A_{\mathrm{H}}$ depends on the type of materials (atomic polarizability and density) of the tip and sample and is of the order of 1 eV for most solids [13].

When the tip and sample are both conductive and have an electrostatic potential difference $U \neq 0$, electrostatic forces are important. For a spherical tip with radius $R$, the force is given by [14]

$$F_{\mathrm{electrostatic}} = -\frac{\pi\varepsilon_0 R U^2}{z}. \tag{5.2}$$

Chemical forces are more complicated. Empirical model potentials for chemical bonds are the Morse potential (see e.g. [13])

$$V_{\mathrm{Morse}} = -E_{\mathrm{bond}}\left(2\mathrm{e}^{-\kappa(z-\sigma)} - \mathrm{e}^{-2\kappa(z-\sigma)}\right) \tag{5.3}$$

and the Lennard-Jones potential [13]

$$V_{\mathrm{Lennard-Jones}} = -E_{\mathrm{bond}}\left(2\frac{\sigma^6}{z^6} - \frac{\sigma^{12}}{z^{12}}\right). \tag{5.4}$$

These potentials describe a chemical bond with bonding energy $E_{\mathrm{bond}}$ and equilibrium distance $\sigma$. The Morse potential has an additional parameter: a decay length $\kappa$.

### 5.1.2 Experimental Measurement and Noise

Forces between the tip and sample are typically measured by recording the deflection of a cantilever beam that has a tip mounted on its end (Fig. 5.2). Today's microfabricated silicon cantilevers were first created in the group of Quate [15, 16, 17] and at IBM [18].

The cantilever is characterized by its spring constant $k$, eigenfrequency $f_0$ and quality factor $Q$.

For a rectangular cantilever with dimensions $w$, $t$ and $L$ (Fig. 5.2), the spring constant $k$ is given by [6]

$$k = \frac{E_{\mathrm{Y}} w t^3}{4L^3}, \tag{5.5}$$

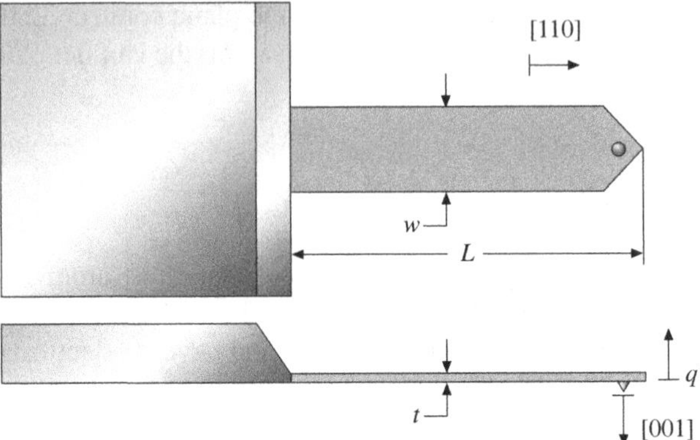


**Fig. 5.2** Top view and side view of a microfabricated silicon cantilever (schematic)

where $E_Y$ is the Young's modulus. The eigenfrequency $f_0$ is given by [6]

$$f_0 = 0.162 \frac{t}{L^2} \sqrt{\frac{E}{\rho}}, \tag{5.6}$$

where $\rho$ is the mass density of the cantilever material. The $Q$-factor depends on the damping mechanisms present in the cantilever. For micromachined cantilevers operated in air, $Q$ is typically a few hundred, while $Q$ can reach hundreds of thousands in vacuum.

In the first AFM, the deflection of the cantilever was measured with an STM; the back side of the cantilever was metalized, and a tunneling tip was brought close to it to measure the deflection [2]. Today's designs use optical (interferometer, beam-bounce) or electrical methods (piezoresistive, piezoelectric) to measure the cantilever deflection. A discussion of the various techniques can be found in [19], descriptions of piezoresistive detection schemes are found in [17, 20] and piezoelectric methods are explained in [21, 22, 5, 24].

The quality of the cantilever deflection measurement can be expressed in a schematic plot of the deflection noise density versus frequency as in Fig. 5.3.

The noise density has a $1/f$ dependence for low frequency and merges into a constant noise density (white noise) above the $1/f$ corner frequency.

### 5.1.3 Static AFM Operating Mode

In the static mode of operation, the force translates into a deflection $q' = F_{ts}/k$ of the cantilever, yielding images as maps of $z(x, y, F_{ts} = \text{const.})$. The noise level of the force measurement is then given by the cantilever's spring constant $k$ times the noise level of the deflection measurement. In this respect, a small value for $k$ increases force sensitivity. On the other hand, instabilities are more likely to occur with soft cantilevers (Sect. 5.1.1). Because the deflection of the cantilever should be

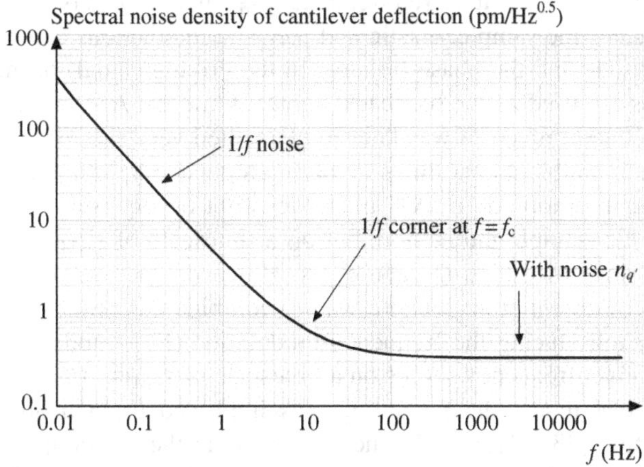


**Fig. 5.3** Schematic view of $1/f$ noise apparent in force detectors. Static AFMs operate in a frequency range from 0.01 Hz to a few hundred Hz, while dynamic AFMs operate at frequencies around 10 kHz to a few hundred kHz. The noise of the cantilever deflection sensor is characterized by the $1/f$ corner frequency $f_c$ and the constant deflection noise density $n_{q'}$ for the frequency range where white noise dominates

significantly larger than the deformation of the tip and sample, the cantilever should be much softer than the bonds between the bulk atoms in the tip and sample. Interatomic force constants in solids are in the range 10–100 N/m; in biological samples, they can be as small as 0.1 N/m. Thus, typical values for $k$ in the static mode are 0.01–5 N/m.

Even though it has been demonstrated that atomic resolution is possible with static AFM, the method can only be applied in certain cases. The detrimental effects of $1/f$-noise can be limited by working at low temperatures [25], where the coefficients of thermal expansion are very small or by building the AFM using a material with a low thermal-expansion coefficient [26]. The long-range attractive forces have to be canceled by immersing the tip and sample in a liquid [26] or by partly compensating the attractive force by pulling at the cantilever after jump-to-contact has occurred [27]. Jarvis et al. have canceled the long-range attractive force with an electromagnetic force applied to the cantilever [28]. Even with these restrictions, static AFM does not produce atomic resolution on reactive surfaces like silicon, as the chemical bonding of the AFM tip and sample poses an unsurmountable problem [29, 30].

### 5.1.4 Dynamic AFM Operating Mode

In the dynamic operation modes, the cantilever is deliberately vibrated. There are two basic methods of dynamic operation: amplitude-modulation (AM) and frequency-modulation (FM) operation. In AM-AFM [31], the actuator is driven by a fixed amplitude $A_{\mathrm{drive}}$ at a fixed frequency $f_{\mathrm{drive}}$ where $f_{\mathrm{drive}}$ is close to $f_0$. When the tip approaches the sample, elastic and inelastic interactions cause a change in both the amplitude and the phase (relative to the driving signal) of the cantilever. These changes are used as the feedback signal. While the AM mode was initially used in a noncontant mode, it was later implemented very successfully at a closer distance range in ambient conditions involving repulsive tip–sample interactions.

The change in amplitude in AM mode does not occur instantaneously with a change in the tip–sample interaction, but on a timescale of $\tau_{\mathrm{AM}} \approx 2Q/f_0$ and the AM mode is slow with high-$Q$ cantilevers. However, the use of high $Q$-factors reduces noise. *Albrecht* et al. found a way to combine the benefits of high $Q$ and high speed by introducing the frequency-modulation (FM) mode [32], where the change in the eigenfrequency settles on a timescale of $\tau_{\mathrm{FM}} \approx 1/f_0$.

Using the FM mode, the resolution was improved dramatically and finally atomic resolution [33, 34] was obtained by reducing the tip–sample distance and working in vacuum. For atomic studies in vacuum, the FM mode (Sect. 5.1.6) is now the preferred AFM technique. However, atomic resolution in vacuum can also be obtained with the AM mode, as demonstrated by Erlandsson et al. [35].

### 5.1.5 The Four Additional Challenges Faced by AFM

Some of the inherent AFM challenges are apparent by comparing the tunneling current and tip–sample force as a function of distance (Fig. 5.4).

The tunneling current is a monotonic function of the tip–sample distance and has a very sharp distance dependence. In contrast, the tip–sample force has long- and short-range components and is not monotonic.

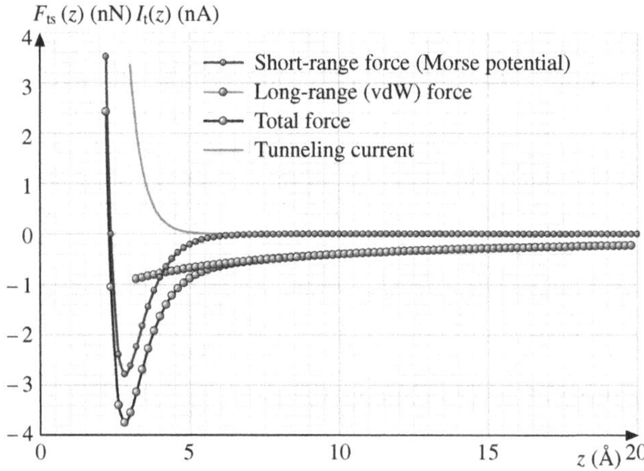


**Fig. 5.4** Plot of the tunneling current $I_t$ and force $F_{ts}$ (typical values) as a function of the distance $z$ between the front atom and surface atom layer

### Jump-to-Contact and Other Instabilities

If the tip is mounted on a soft cantilever, the initially attractive tip–sample forces can cause a sudden jump-to-contact when approaching the tip to the sample. This instability occurs in the quasistatic mode if [36, 37]

$$k < \max\left(-\frac{\partial^2 V_{ts}}{\partial z^2}\right) = k_{ts}^{max}. \tag{5.7}$$

Jump-to-contact can be avoided even for soft cantilevers by oscillating at a large enough amplitude $A$ [38]

$$kA > \max(-F_{ts}). \tag{5.8}$$

If hysteresis occurs in the $F_{ts}(z)$-relation, energy $\Delta E_{ts}$ needs to be supplied to the cantilever for each oscillation cycle. If this energy loss is large compared to the intrinsic energy loss of the cantilever, amplitude control can become difficult. An additional approximate criterion for $k$ and $A$ is then

$$\frac{kA^2}{2} \geq \frac{\Delta E_{ts} Q}{2\pi}. \tag{5.9}$$

### Contribution of Long-Range Forces

The force between the tip and sample is composed of many contributions: electrostatic, magnetic, van-der-Waals and chemical forces in vacuum. All of these force types except for the chemical forces have strong long-range components which conceal the atomic force components. For imaging by AFM with atomic resolution, it is desirable to filter out the long-range force contributions and only measure the force components which vary on the atomic scale. While there is no way to discriminate between long- and short-range forces in static AFM, it is possible to enhance the short-range contributions in dynamic AFM by proper choice of the oscillation amplitude $A$ of the cantilever.

### Noise in the Imaging Signal

Measuring the cantilever deflection is subject to noise, especially at low frequencies ($1/f$ noise). In static AFM, this noise is particularly problematic because of the approximate $1/f$ dependence. In dynamic AFM, the low-frequency noise is easily discriminated when using a bandpass filter with a center frequency around $f_0$.

### Nonmonotonic Imaging Signal

The tip–sample force is not monotonic. In general, the force is attractive for large distances and, upon decreasing the distance between tip and sample, the force turns repulsive (Fig. 5.4). Stable feedback is only possible on a monotonic subbranch of the force curve.

Frequency-modulation AFM helps to overcome challenges. The nonmonotonic imaging signal in AFM is a remaining complication for FM-AFM.

## *5.1.6 Frequency-Modulation AFM (FM-AFM)*

In FM-AFM, a cantilever with eigenfrequency $f_0$ and spring constant $k$ is subject to controlled positive feedback such that it oscillates with a constant amplitude $A$ [32], as shown in Fig. 5.5.

### Experimental Set-Up

The deflection signal is phase-shifted, routed through an automatic gain control circuit and fed back to the actuator. The frequency $f$ is a function of $f_0$, its quality factor $Q$, and the phase shift $\phi$ between the mechanical excitation generated at the actuator and the deflection of the cantilever. If $\phi = \pi/2$, the loop oscillates at $f = f_0$. Three physical observables can be recorded: (1) a change in the resonance frequency $\Delta f$, (2) the control signal of the automatic gain control unit as a measure of the tip–sample energy dissipation, and (3) an average tunneling current (for conducting cantilevers and tips).

### Applications

FM-AFM was introduced by Albrecht and coworkers in magnetic force microscopy [32]. The noise level and imaging speed was enhanced significantly compared to

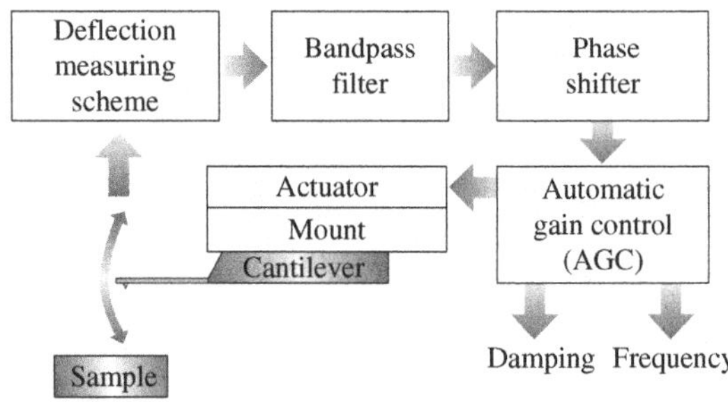


**Fig. 5.5** Block diagram of a frequency-modulation force sensor

amplitude-modulation techniques. Achieving atomic resolution on the Si(111)-(7×7) surface has been an important step in the development of the STM [39] and, in 1994, this surface was imaged by AFM with true atomic resolution for the first time [33] (Fig. 5.6).

The initial parameters which provided true atomic resolution (see caption of Fig. 5.6) were found empirically. Surprisingly, the amplitude necessary to obtain good results was very large compared to atomic dimensions. It turned out later that the amplitudes had to be so large to fulfill the stability criteria listed in Sect. 5.1.5. Cantilevers with $k \approx 2{,}000$ N/m can be operated with amplitudes in the Å-range [24].

### 5.1.7 Relation Between Frequency Shift and Forces

The cantilever (spring constant $k$, effective mass $m^*$) is a macroscopic object and its motion can be described by classical mechanics. Figure 5.7 shows the deflection $q'(t)$ of the tip of the cantilever: it oscillates with an amplitude $A$ at a distance $q(t)$ from a sample.

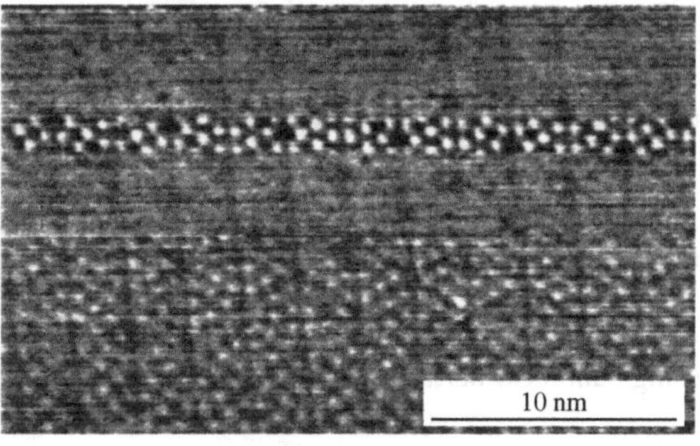


**Fig. 5.6** First AFM image of the Si(111)-(7×7) surface. Parameters: $k = 17$ Nm, $f_0 = 114$ kHz, $Q = 28000$, $A = 34$ nm, $\Delta f = -70$ Hz, $V_t = 0$ V

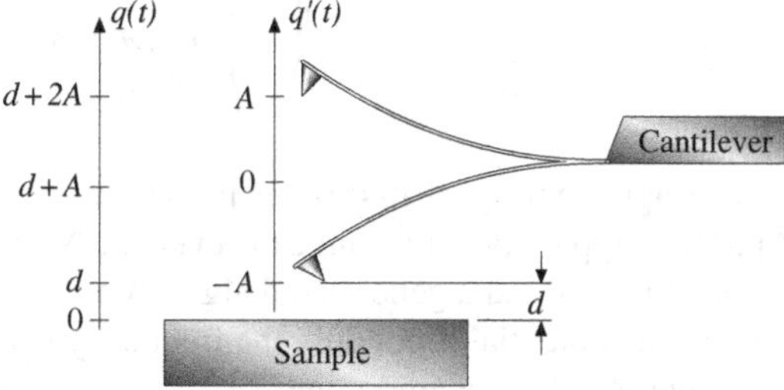


**Fig. 5.7** Schematic view of an oscillating cantilever and definition of geometric terms

### Generic Calculation

The Hamiltonian of the cantilever is

$$H = \frac{p^2}{2m^*} + \frac{kq'^2}{2} + V_{ts}(q) \tag{5.10}$$

where $p = m^* \mathrm{d}q'/\mathrm{d}t$. The unperturbed motion is given by

$$q'(t) = A\cos(2\pi f_0 t) \tag{5.11}$$

and the frequency is

$$f_0 = \frac{1}{2\pi}\sqrt{\frac{k}{m^*}}. \tag{5.12}$$

If the force gradient $k_{ts} = -\partial F_{ts}/\partial z = \partial^2 V_{ts}/\partial z^2$ is constant during the oscillation cycle, the calculation of the frequency shift is trivial

$$\Delta f = \frac{f_0}{2k} k_{ts}. \tag{5.13}$$

However, in classic FM-AFM $k_{ts}$ varies over orders of magnitude during one oscillation cycle and a perturbation approach, as shown below, has to be employed for the calculation of the frequency shift.

### Hamilton–Jacobi Method

The first derivation of the frequency shift in FM-AFM was achieved in 1997 [38] using canonical perturbation theory [40]. The result of this calculation is

$$\begin{aligned} \Delta f &= -\frac{f_0}{kA^2}\langle F_{ts} q'\rangle \\ &= -\frac{f_0}{kA^2}\int_0^{1/f_0} F_{ts}(d + A + q'(t))q'(t)\mathrm{d}t. \end{aligned} \tag{5.14}$$

The applicability of first-order perturbation theory is justified because, in FM-AFM, $E$ is typically in the range of several keV, while $V_{ts}$ is of the order of a few eV. *DÃ¼rig* [41] found a generalized algorithm that even allows one to reconstruct the tip–sample potential if not only the frequency shift, but the higher harmonics of the cantilever oscillation are known.

### A Descriptive Expression for Frequency Shifts as a Function of the Tip–Sample Forces

With integration by parts, the complicated expression (5.14) is transformed into a very simple expression that resembles (5.13) [42]

$$\Delta f = \frac{f_0}{2k} \int_{-A}^{A} k_{ts}(z - q') \frac{\sqrt{A^2 - q'^2}}{\frac{\pi}{2} k A^2} \mathrm{d}q'. \tag{5.15}$$

This expression is closely related to (5.13): the constant $k_{ts}$ is replaced by a weighted average, where the weight function $w(q', A)$ is a semicircle with radius $A$ divided by the area of the semicircle $\pi A^2/2$ (Fig. 5.8). For $A \to 0$, $w(q', A)$ is a representation of Dirac's delta function and the trivial zero-amplitude result of (5.13) is immediately recovered. The frequency shift results from a convolution between the tip–sample force gradient and weight function. This convolution can easily be reversed with a linear transformation and the tip–sample force can be recovered from the curve of frequency shift versus distance [42].

The dependence of the frequency shift on amplitude confirms an empirical conjecture: small amplitudes increase the sensitivity to short-range forces. Adjusting the amplitude in FM-AFM is comparable to tuning an optical spectrometer to a passing wavelength. When short-range interactions are to be probed, the amplitude should be in the range of the short-range forces. While using amplitudes in the Å-range has been elusive with conventional cantilevers because of the instability problems described in Sect. 5.1.5, cantilevers with a stiffness of the order

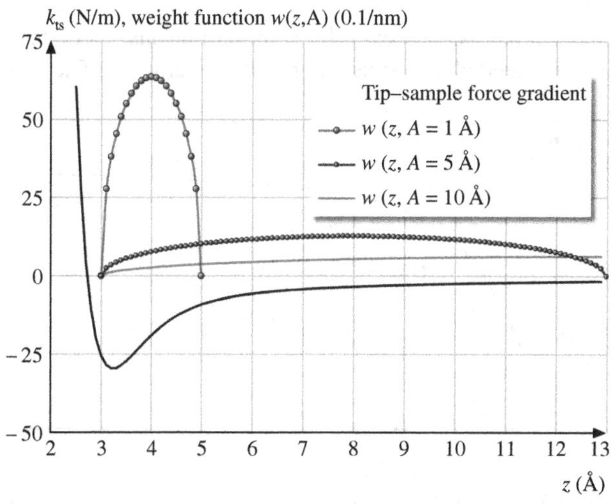


**Fig. 5.8** The tip–sample force gradient $k_{ts}$ and weight function for the calculation of the frequency shift

of 1,000 N/m like those introduced in [23] are well suited for small-amplitude operation.

### 5.1.8 Noise in Frequency Modulation AFM: Generic Calculation

The vertical noise in FM-AFM is given by the ratio between the noise in the imaging signal and the slope of the imaging signal with respect to $z$

$$\delta z = \frac{\delta \Delta f}{\left|\frac{\partial \Delta f}{\partial z}\right|}. \quad (5.16)$$

Figure 5.9 shows a typical curve of frequency shift versus distance. Because the distance between the tip and sample is measured indirectly through the frequency shift, it is clearly evident from Fig. 5.9 that the noise in the frequency measurement $\delta\Delta f$ translates into vertical noise $\delta z$ and is given by the ratio between $\delta\Delta f$ and the slope of the frequency shift curve $\Delta f(z)$ (5.16). Low vertical noise is obtained for a low-noise frequency measurement and a steep slope of the frequency-shift curve.

The frequency noise $\delta\Delta f$ is typically inversely proportional to the cantilever amplitude $A$ [32, 43]. The derivative of the frequency shift with distance is constant for $A \ll \lambda$ where $\lambda$ is the range of the tip–sample interaction and proportional to $A^{-1.5}$ for $A \gg \lambda$ [38]. Thus, minimal noise occurs if [44]

$$A_{\text{optimal}} \approx \lambda \quad (5.17)$$

for chemical forces, $\lambda \approx 1$ Å. However, for stability reasons, (Sect. 5.1.5) extremely stiff cantilevers are needed for small-amplitude operation. The excellent noise

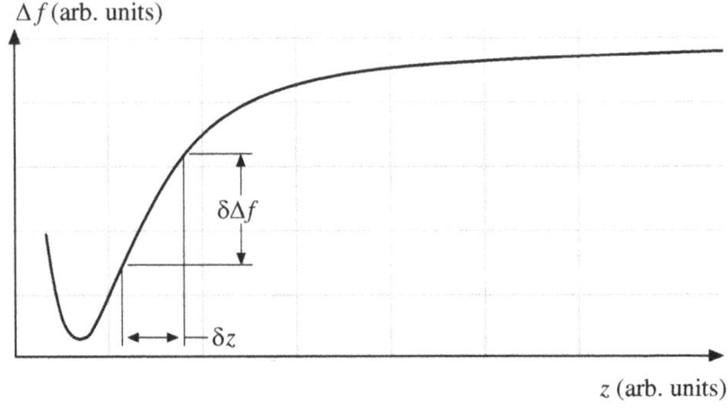


**Fig. 5.9** Plot of the frequency shift $\Delta f$ as a function of the tip–sample distance $z$. The noise in the tip–sample distance measurement is given by the noise of the frequency measurement $\delta\Delta f$ divided by the slope of the frequency shift curve

performance of the stiff cantilever and the small-amplitude technique has been verified experimentally [24].

### 5.1.9 Conclusion

Dynamic force microscopy, and in particular frequency-modulation atomic force microscopy has matured into a viable technique that allows true atomic resolution of conducting and insulating surfaces and spectroscopic measurements on individual atoms [10, 45]. Even true atomic resolution in lateral force microscopy is now possible [46]. Challenges remain in the chemical composition and structural arrangement of the AFM tip.

## 5.2 Applications to Semiconductors

For the first time, corner holes and adatoms on the Si(111)-(7 × 7) surface have been observed in very local areas by Giessible using pure noncontact AFM in ultrahigh vacuum (UHV) [33]. This was the breakthrough of true atomic-resolution imaging on a well-defined clean surface using the noncontact AFM. Since then, Si(111)-(7 × 7) [34, 35, 45, 47], InP(110) [48] and Si(100)-(2 × 1) [34] surfaces have been successively resolved with true atomic resolution. Furthermore, thermally induced motion of atoms or atomic-scale point defects on a InP(110) surface have been observed at room temperature [48]. In this section we will describe typical results of atomically resolved noncontact AFM imaging of semiconductor surfaces.

### 5.2.1 Si(111)-(7 × 7) Surface

Figure 5.10 shows the atomic-resolution images of the Si(111)-(7 × 7) surface [49]. Here, Fig. 5.10a (type I) was obtained using the Si tip without dangling, which is covered with an inert oxide layer. Figure 5.10b (type II) was obtained using the Si tip with a dangling bond, on which the Si atoms were deposited due the mechanical soft contact between the tip and the Si surface. The variable frequency shift mode was used. We can see not only adatoms and corner holes but also missing adatoms described by the dimer–adatom–stacking (DAS) fault model. We can see that the image contrast in Fig. 5.10b is clearly stronger than that in Fig. 5.10a.

Interestingly, by using the Si tip with a dangling bond, we observed contrast between inequivalent halves and between inequivalent adatoms of the 7 × 7 unit cell. Namely, as shown in Fig. 5.11a, the faulted halves (surrounded with a solid line) are brighter than the unfaulted halves (surrounded with a broken line). Here, the positions of the faulted and unfaulted halves were determined from the step

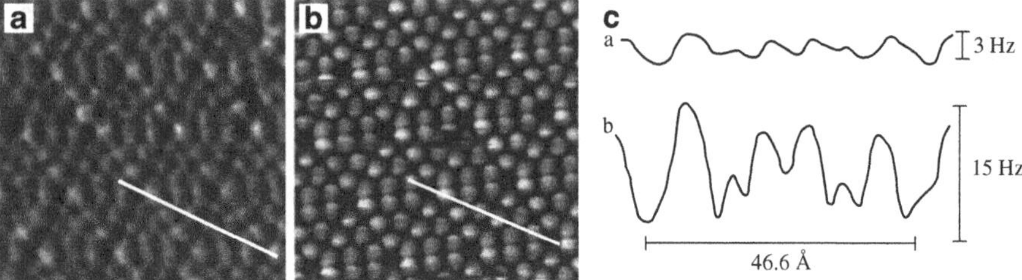


Fig. 5.10 Noncontact-mode AFM images of a Si(111)-(7×7) reconstructed surface obtained using the Si tips (**a**) without and (**b**) with a dangling bond. The scan area is 99 Å×99 Å. (**c**) The cross-sectional profiles along the long diagonal of the 7×7 unit cell indicated by the *white lines* in (**a**) and (**b**)

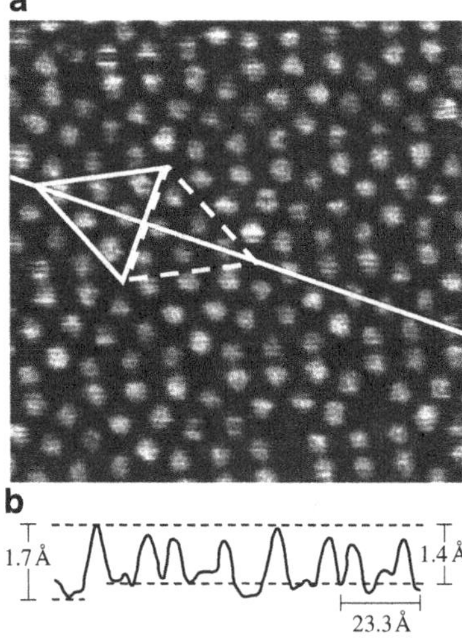


Fig. 5.11 (**a**) Noncontact mode AFM image with contrast of inequivalent adatoms and (**b**) a cross-sectional profile indicated by the *white line*. The halves of the 7×7 unit cell surrounded by the *solid line* and *broken line* correspond to the faulted and unfaulted halves, respectively. The scan area is 89 Å×89 Å

direction. From the cross-sectional profile along the long diagonal of the 7×7 unit cell in Fig. 5.11b, the heights of the corner adatoms are slightly higher than those of the adjacent center adatoms in the faulted and unfaulted halves of the unit cell. The measured corrugation are in the following decreasing order: Co-F $>$ Ce-F $>$ Co-U $>$ Ce-U, where Co-F and Ce-F indicate the corner and center adatoms in faulted halves, and Co-U and Ce-U indicate the corner and center adatoms in unfaulted halves, respectively. Averaging over several units, the corrugation height differences are estimated to be 0.25 Å, 0.15 Å and 0.05 Å for Co-F, Ce-F and Co-U, respectively, with respect to to Ce-U. This tendency, that the heights of the corner adatoms are higher than those of the center adatoms, is consistent with the experimental results using a silicon tip [47], although they could not determine the faulted and unfaulted halves of the unit cell in the measured AFM images.

However, this tendency is completely contrary to the experimental results using a tungsten tip [35]. This difference may originate from the difference between the tip materials, which seems to affect the interaction between the tip and the reactive sample surface. Another possibility is that the tip is in contact with the surface during the small fraction of the oscillating cycle in their experiments [35].

We consider that the contrast between inequivalent adatoms is not caused by tip artifacts for the following reasons: (1) each adatom, corner hole and defect was clearly observed, (2) the apparent heights of the adatoms are the same whether they are located adjacent to defects or not, and (3) the same contrast in several images for the different tips has been observed.

It should be noted that the corrugation amplitude of adatoms $\approx 1.4$ Å in Fig. 5.11b is higher than that of 0.8–1.0 Å obtained with the STM, although the depth of the corner holes obtained with noncontact AFM is almost the same as that observed with STM. Moreover, in noncontact-mode AFM images, the corrugation amplitude of adatoms was frequently larger than the depth of the corner holes. The origin of such large corrugation of adatoms may be due to the effect of the chemical interaction, but is not yet clear.

The atom positions, surface energies, dynamic properties and chemical reactivities on the Si(111)-(7×7) reconstructed surface have been extensively investigated theoretically and experimentally. From these investigations, the possible origins of the contrast between inequivalent adatoms in AFM images are the followings: the true atomic heights that correspond to the adatom core positions, the stiffness (spring constant) of interatomic bonding with the adatoms corresponding to the frequencies of the surface mode, the charge on the adatom, and the chemical reactivity of the adatoms. Table 5.1 summarizes the decreasing orders of the inequivalent adatoms for individual property. From Table 5.1, we can see that the calculated adatom heights and the stiffness of interatomic bonding cannot explain the AFM data, while the amount of charge of adatom and the chemical reactivity of adatoms can explain the our data. The contrast due to the amount of charge of adatom means that the AFM image is originated from the difference of the vdW or electrostatic physical interactions between the tip and the valence electrons at the adatoms. The contrast due to the chemical reactivity of adatoms means that the AFM image is originated from the difference of covalent bonding chemical interaction between the atoms at the tip apex and dangling bond of adatoms. Thus, we can see there are two possible interactions which explain the strong contrast between

**Table 5.1** Comparison between the adatom heights observed in an AFM image and the variety of properties for inequivalent adatoms

| | Decreasing order | Agreement |
|---|---|---|
| AFM image | Co-F > Ce-F > Co-U > Ce-U | – |
| Calculated height | Co-F > Co-U > Ce-F > Ce-U | × |
| Stiffness of interatomic bonding | Ce-U > Co-U > Ce-F > Co-F | × |
| Amount of charge of adatom | Co-F > Ce-F > Co-U > Ce-U | ○ |
| Calculated chemical reactivity | Faulted > unfaulted | ○ |
| Experimental chemical reactivity | Co-F > Ce-F > Co-U > Ce-U | ○ |

inequivalent adatoms of 7×7 unit cell observed using the Si tip with dangling bond.

The weak-contrast image in Fig. 5.10a is due to vdW and/or electrostatic force interactions. On the other hand, the strong-contrast images in Figs. 5.10b and 5.11a are due to a covalent bonding formation between the AFM tip with Si atoms and Si adatoms. These results indicate the capability of the noncontact-mode AFM to image the variation in chemical reactivity of Si adatoms. In the future, by controlling an atomic species at the tip apex, the study of chemical reactivity on an atomic scale will be possible using noncontact AFM.

### *5.2.2 Si(100)-(2×1) and Si(100)-(2×1):H Monohydride Surfaces*

In order to investigate the imaging mechanism of the noncontact AFM, a comparative study between a reactive surface and an insensitive surface using the same tip is very useful. Si(100)-(2×1):H monohydride surface is a Si(100)-(2×1) reconstructed surface that is terminated by a hydrogen atom. It does not reconstruct as metal is deposited on the semiconductor surface. The surface structure hardly changes. Thus, the Si(100)-(2×1):H monohydride surface is one of most useful surface for a model system to investigate the imaging mechanism, experimentally and theoretically. Furthermore, whether the interaction between a very small atom such as hydrogen and a tip apex is observable with noncontact AFM is interested. Here, we show noncontact AFM images measured on a Si(100)-(2×1) reconstructed surface with a dangling bond and on a Si(100)-(2×1):H monohydride surface on which the dangling bond is terminated by a hydrogen atom [50].

Figure 5.12a shows the atomic-resolution image of the Si(100)-(2×1) reconstructed surface. Pairs of bright spots arranged in rows with a 2×1 symmetry were observed with clear contrast. Missing pairs of bright spots were also observed, as indicated by arrows. Furthermore, the pairs of bright spots are shown by the white dashed arc and appear to be the stabilize-buckled asymmetric dimer structure. Furthermore, the distance between the pairs of bright spots is 3.2 ± 0.1 Å.

Figure 5.13a shows the atomic-resolution image of the Si(100)-(2×1):H monohydride surface. Pairs of bright spots arranged in rows were observed. Missing paired bright spots as well as those paired in rows and single bright spots were observed, as indicated by arrows. Furthermore, the distance between paired bright spots is 3.5 ± 0.1 Å. This distance of 3.5 ± 0.1 Å is 0.2 Å larger than that of the Si(100)-(2×1) reconstructed surface. Namely, it is found that the distance between bright spots increases in size due to the hydrogen termination.

The bright spots in Fig. 5.12 do not merely image the silicon-atom site, because the distance between the bright spots forming the dimer structure of Fig. 5.12a, 3.2 ± 0.1 Å, is lager than the distance between silicon atoms of every dimer structure model. (The maximum is the distance between the upper silicones in an

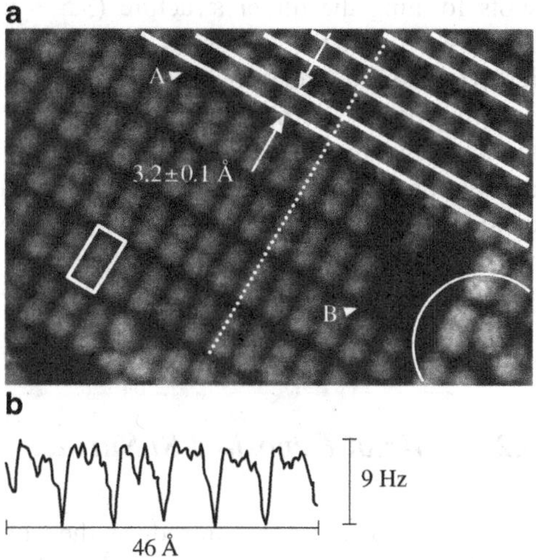

**Fig. 5.12** (**a**) Noncontact AFM image of a Si(001) (2×1) reconstructed surface. The scan area was 69×46 Å. One 2×1 unit cell is outlined with a *box*. *White rows* are superimposed to show the bright spots arrangement. The distance between the bright spots on the dimer row is 3.2 ± 0.1 Å. On the *white arc*, the alternative bright spots are shown. (**b**) Cross-sectional profile indicated by the *white dotted line*

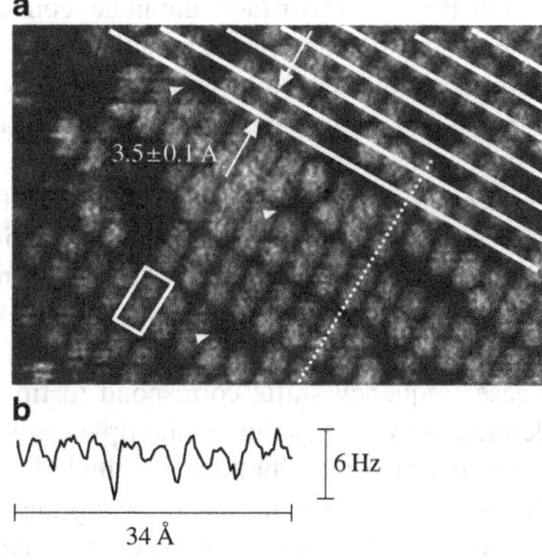


**Fig. 5.13** (**a**) Noncontact AFM image of Si(001)-(2×1):H surface. The scan area was 69×46 Å. One 2×1 unit cell is outlined with a *box*. *White rows* are superimposed to show the bright spots arrangement. The distance between the bright spots on the dimer row is 3.5 ± 0.1 Å. (**b**) Cross-sectional profile indicated by the *white dotted line*

asymmetric dimer structure 2.9 Å). This seems to be due to the contribution to the imaging of the chemical bonding interaction between the dangling bond from the apex of the silicon tip and the dangling bond on the Si(100)-(2×1) reconstructed surface. Namely, the chemical bonding interaction operates strongly, with strong direction dependence, between the dangling bond pointing out of the silicon dimer structure on the Si(100)-(2×1) reconstructed surface and the dangling bond pointing out of the apex of the silicon tip; a dimer structure is obtained with a larger separation than between silicones on the surface.

The bright spots in Fig. 5.13 seem to be located at hydrogen atom sites on the Si(100)-(2×1):H monohydride surface, because the distance between the bright spots forming the dimer structure (3.5 ± 0.1 Å) approximately agrees with the distance between the hydrogens, i.e., 3.52 Å. Thus, the noncontact AFM atomically resolved the individual hydrogen atoms on the topmost layer. On this surface, the dangling bond is terminated by a hydrogen atom, and the hydrogen atom on the topmost layer does not have chemical reactivity. Therefore, the interaction between the hydrogen atom on the topmost layer and the apex of the silicon tip does not contribute to the chemical bonding interaction with strong direction dependence as on the silicon surface, and the bright spots in the noncontact AFM image correspond to the hydrogen atom sites on the topmost layer.

### 5.2.3 *Metal Deposited Si Surface*

In this section, we will introduce the comparative study of force interactions between a Si tip and a metal-deposited Si surface, and between a metal adsorbed Si tip and a metal-deposited Si surface [51, 52]. As for the metal-deposited Si surface, Si(111)-$\sqrt{3} \times \sqrt{3}$-Ag (hereafter referred to as $\sqrt{3}$-Ag) surface was used.

For the $\sqrt{3}$-Ag surface, the honeycomb-chained trimer (HCT) model has been accepted as the appropriate model. As shown in Fig. 5.14, this structure contains a Si trimer in the second layer, 0.75 Å below the Ag trimer in the topmost layer. The topmost Ag atoms and lower Si atoms form covalent bonds. The interatomic distances between the nearest-neighbor Ag atoms forming the Ag trimer and between the lower Si atoms forming the Si trimer are 3.43 and 2.31 Å, respectively. The apexes of the Si trimers and Ag trimers face the $[11\bar{2}]$ direction and the direction tilted a little to the $[\bar{1}\bar{1}2]$ direction, respectively.

In Fig. 5.15, we show the noncontact AFM images measured using a normal Si tip at a frequency shift of (a) −37 Hz, (b) −43 Hz and (c) −51 Hz, respectively. These frequency shifts correspond to tip–sample distances of about 0–3 Å. We defined the zero position of the tip–sample distance, i.e., the contact point, as the point at which the vibration amplitude began to decrease. The rhombus indicates the $\sqrt{3} \times \sqrt{3}$ unit cell. When the tip approached the surface, the contrast of the noncontact AFM images become strong and the pattern changed remarkably. That is, by approaching the tip toward the sample surface, the hexagonal pattern, the trefoil-like pattern composed of three dark lines, and the triangle pattern can be observed sequentially. In Fig. 5.15a, the distance between the bright spots is 3.9 ± 0.2Å. In Fig. 5.15c, the distance between the bright spots is 3.0 ± 0.2 Å, and the direction of the apex of all the triangles composed of three bright spots is $[11\bar{2}]$.

In Fig. 5.16, we show the noncontact AFM images measured by using Ag-absorbed tip at a frequency shift of (a) −4.4 Hz, (b) −6.9 Hz and (c) −9.4 Hz, respectively. The tip–sample distances $Z$ are roughly estimated to be $Z = 1.9$, 0.6 and $\approx 0$ Å (in the noncontact region), respectively. When the tip approached the surface, the pattern of the noncontact AFM images did not change, although the

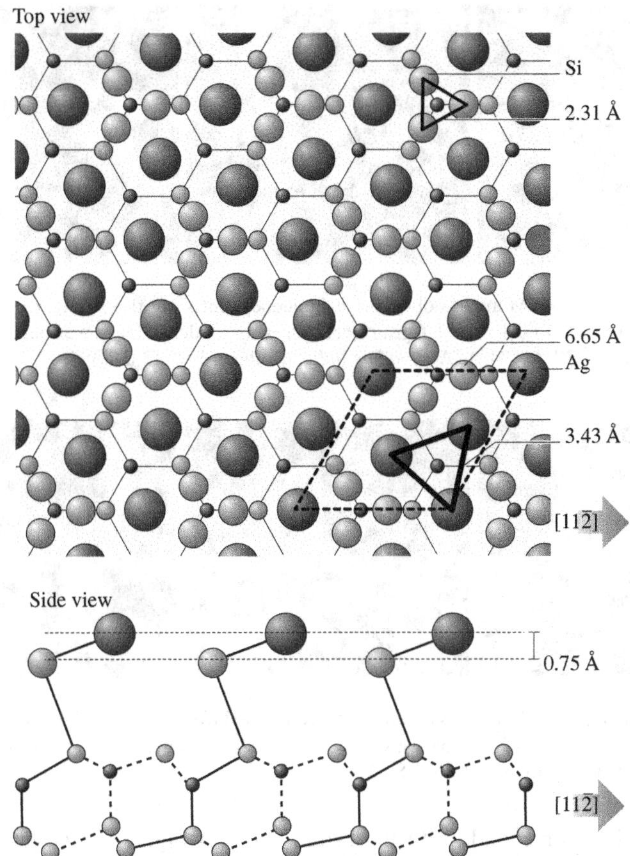


**Fig. 5.14** HCT model for the structure of the Si(111)-($\sqrt{3} \times \sqrt{3}$)-Ag surface. *Black closed circle*, *gray closed circle*, *open circle*, and *closed circle with horizontal line* indicate Ag atom at the topmost layer, Si atom at the second layer, Si atom at the third layer, and Si atom at the fourth layer, respectively. The *rhombus* indicates the $\sqrt{3} \times \sqrt{3}$ unit cell. The *thick, large, solid triangle* indicates an Ag trimer. The *thin, small, solid triangle* indicates a Si trimer

contrast become clearer. A triangle pattern can be observed. The distance between the bright spots is 3.5 ± 0.2 Å. The direction of the apex of all the triangles composed of three bright spots is tilted a little from the $[\bar{1}\bar{1}2]$ direction.

Thus, noncontact AFM images measured on Si(111)-($\sqrt{3} \times \sqrt{3}$)-Ag surface showed two types of distance dependence in the image patterns depending on the atom species on the apex of the tip.

By using the normal Si tip with a dangling bond, in Fig. 5.15a, the measured distance between the bright spot of 3.9 ± 0.2 Å agrees with the distance of 3.84 Å between the centers of the Ag trimers in the HCT model within the experimental error. Furthermore, the hexagonal pattern composed of six bright spots also agrees with the honeycomb structure of the Ag trimer in HCT model. So the most

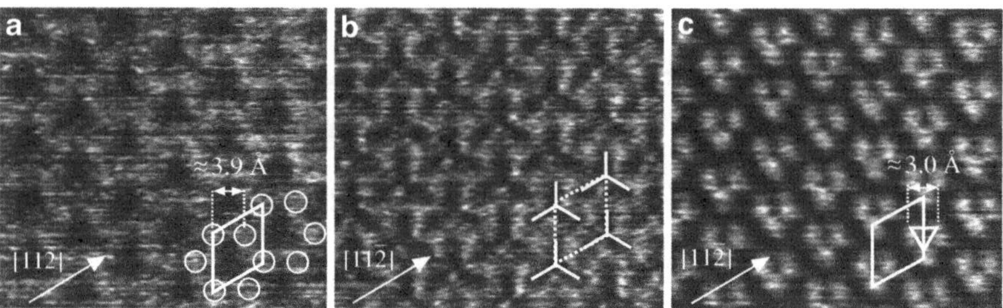


**Fig. 5.15** Noncontact AFM images obtained at frequency shifts of (**a**) −37 Hz, (**b**) −43 Hz, and (**c**) −51 Hz on a Si(111)-($\sqrt{3} \times \sqrt{3}$)-Ag surface. This distance dependence was obtained with a Si tip. The scan area is 38 Å×34 Å. A *rhombus* indicates the $\sqrt{3} \times \sqrt{3}$ unit cell

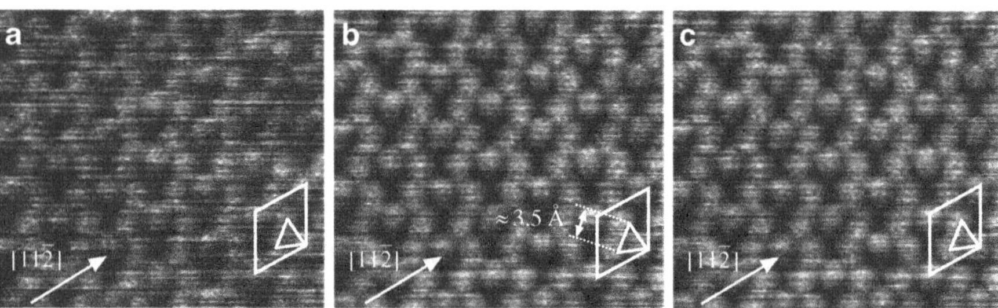


**Fig. 5.16** Noncontact AFM images obtained at frequency shifts of (**a**) −4.4 Hz, (**b**) −6.9 Hz, and (**c**) −9.4 Hz on a Si(111)-($\sqrt{3} \times \sqrt{3}$)-Ag surface. This distance dependence was obtained with the Ag-adsorbed tip. The scan area is 38 Å×34 Å

appropriate site corresponding to the bright spots in Fig. 5.15a is the site of the center of Ag trimers. In Fig. 5.15c, the measured distance of 3.0 ± 0.2 Å between the bright spots forming the triangle pattern agrees with neither the distance between the Si trimer of 2.31 Å nor the distance between the Ag trimer of 3.43 Å in the HCT model, while the direction of the apex of the triangles composed of three bright spots agrees with the $[11\bar{2}]$ direction of the apex of the Si trimer in the HCT model. So the most appropriate site corresponding to the bright spots in Fig. 5.15c is the intermediate site between the Si atoms and Ag atoms. On the other hand, by using the Ag-adsorbed tip, the measured distance between the bright spots of 3.5 ± 0.2 Å in Fig. 5.16 agrees with the distance of 3.43 Å between the nearest-neighbor Ag atoms forming the Ag trimer in the topmost layer in the HCT model within the experimental error. Furthermore, the direction of the apex of the triangles composed of three bright spots also agrees with the direction of the apex of the Ag trimer, i.e., tilted $[\bar{1}\bar{1}2]$, in the HCT model. So, the most appropriate site corresponding to the bright spots in Fig. 5.16 is the site of individual Ag atoms forming the Ag trimer in the topmost layer.

It should be noted that, by using the noncontact AFM with a Ag-adsorbed tip, for the first time, the individual Ag atom on the $\sqrt{3}$-Ag surface could be resolved in real space, although by using the noncontact AFM with an Si tip, it could not be resolved. So far, the $\sqrt{3}$-Ag surface has been observed by a scanning tunneling microscope (STM) with atomic resolution. However, the STM can also measure the local charge density of states near the Fermi level on the surface. From first-principle calculations, it was proven that unoccupied surface states are densely distributed around the center of the Ag trimer. As a result, bright contrast is obtained at the center of the Ag trimer with the STM.

Finally, we consider the origin of the atomic-resolution imaging of the individual Ag atoms on the $\sqrt{3}$-Ag surface. Here, we discuss the difference between the force interactions when using the Si tip and the Ag-adsorbed tip. As shown in Fig. 5.17a, when using the Si tip, there is a dangling bond pointing out of the topmost Si atom on the apex of the Si tip. As a result, the force interaction is dominated by physical bonding interactions, such as the Coulomb force, far from the surface and by chemical bonding interaction very close to the surface. Namely, if a reactive Si tip with a dangling bond approaches a surface, at distances far from the surface the Coulomb force acts between the electron localized on the dangling bond pointing out of the topmost Si atom on the apex of the tip, and the positive charge distributed around the center of the Ag trimer. At distances very close to the surface, the chemical bonding interaction will occur due to the onset of orbital hybridization between the dangling bond pointing out of the topmost Si atom on the apex of the Si tip and a Si–Ag covalent bond on the surface. Hence, the individual Ag atoms will not be resolved and the image pattern will change depending on the tip–sample distance. On the other hand, as shown in Fig. 5.17b, by using the Ag-adsorbed tip, the dangling bond localized out of topmost Si atom on the apex of the Si tip is terminated by the adsorbed Ag atom. As a result, even at very close tip–sample distances, the force interaction is dominated by physical bonding interactions such as the vdW force. Namely, if the Ag-adsorbed tip approaches the

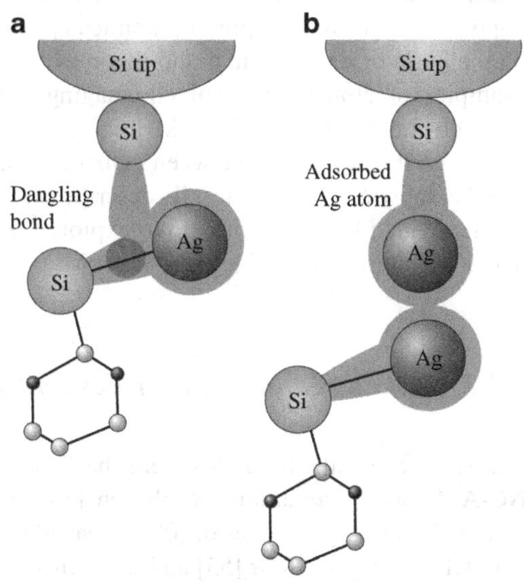


**Fig. 5.17** Schematic illustration of (**a**) the Si atom with dangling bond and (**b**) the Ag-adsorbed tip above the Si–Ag covalent bond on a Si (111)-($\sqrt{3} \times \sqrt{3}$)-Ag surface

surface, the vdW force acts between the Ag atom on the apex of the tip and the Ag or Si atom on the surface. Ag atoms in the topmost layer of the $\sqrt{3}$-Ag surface are located higher than the Si atoms in the lower layer. Hence, the individual Ag atoms (or their nearly true topography) will be resolved, and the image pattern will not change even at very small tip–sample distances. It should be emphasized that there is a possibility to identify or recognize atomic species on a sample surface using noncontact AFM if we can control the atomic species at the tip apex.

## 5.3 Applications to Insulators

Insulators such as alkali halides, fluorides, and metal oxides are key materials in many applications, including optics, microelectronics, catalysis, and so on. Surface properties are important in these technologies, but they are usually poorly understood. This is due to their low conductivity, which makes it difficult to investigate them using electron- and ion-based measurement techniques such as low-energy electron diffraction, ion-scattering spectroscopy, and scanning tunneling microscopy (STM). Surface imaging by noncontact atomic force microscopy (NC-AFM) does not require a sample surface with a high conductivity because NC-AFM detects a force between the tip on the cantilever and the surface of the sample. Since the first report of atomically resolved NC-AFM on a Si(111)-(7$\times$7) surface [33], several groups have succeeded in obtaining *true* atomic resolution images of insulators, including defects, and it has been shown that NC-AFM is a powerful new tool for atomic-scale surface investigation of insulators.

In this section we will describe typical results of atomically resolved NC-AFM imaging of insulators such as alkali halides, fluorides and metal oxides. For the alkali halides and fluorides, we will focus on contrast formation, which is the most important issue for interpreting atomically resolved images of binary compounds on the basis of experimental and theoretical results. For the metal oxides, typical examples of atomically resolved imaging will be exhibited and the difference between the STM and NC-AFM images will be demonstrated. Also, theoretical studies on the interaction between realistic Si tips and representative oxide surfaces will be shown. Finally, we will describe an antiferromagnetic NiO(001) surface imaged with a ferromagentic tip to explore the possibility of detecting short-range magnetic interactions using the NC-AFM.

### *5.3.1 Alkali Halides, Fluorides and Metal Oxides*

The surfaces of alkali halides were the first insulating materials to be imaged by NC-AFM with *true* atomic resolution [53]. To date, there have been reports on atomically resolved images of (001) cleaved surfaces for single-crystal NaF, RbBr, LiF, KI, NaCl, [54], KBr [55] and thin films of NaCl(001) on Cu(111) [56]. In this section we describe the contrast formation of alkali halides surfaces on the basis of experimental and theoretical results.

### Alkali Halides

In experiments on alkali halides, the symmetry of the observed topographic images indicates that the protrusions exhibit only one type of ions, either the positive or negatively charged ions. This leads to the conclusion that the atomic contrast is dominantly caused by electrostatic interactions between a charged atom at the apex of the tip and the surface ions, i.e. long-range forces between the macroscopic tip and the sample, such as the van der Waals force, are modulated by an alternating short-range electrostatic interaction with the surface ions. Theoretical work employing the atomistic simulation technique has revealed the mechanism for contrast formation on an ionic surface [57]. A significant part of the contrast is due to the displacement of ions in the force field, not only enhancing the atomic corrugations, but also contributing to the electrostatic potential by forming dipoles at the surface. The experimentally observed atomic corrugation height is determined by the interplay of the long- and short-range forces. In the case of NaCl, it has been experimentally demonstrated that a blunter tip produces a lager corrugation when the tip–sample distance is shorter [54]. This result shows that the increased long-range forces induced by a blunter tip allow for more stable imaging closer to the surface. The stronger electrostatic short-range interaction and lager ion displacement produce a more pronounced atomic corrugation. At steps and kinks on an NaCl thin film on Cu(111), the corrugation amplitude of atoms with low coordination number has been observed to increase by a factor of up to two more than that of atomically flat terraces [56]. The low coordination number of the ions results in an enhancement of the electrostatic potential over the site and an increase in the displacement induced by the interaction with the tip.

Theoretical study predicts that the image contrast depends on the chemical species at the apex of the tip. Bennewitz et al. [56] have performed the calculations using an MgO tip terminated by oxygen and an Mg ion. The magnitude of the atomic contrast for the Mg-terminated tip shows a slight increase in comparison with an oxygen-terminated tip. The atomic contrast with the oxygen-terminated tip is dominated by the attractive electrostatic interaction between the oxygen on the tip apex and the Na ion, but the Mg-terminated tip attractively interacts with the Cl ion. In other words, these results demonstrated that the species of the ion imaged as the bright protrusions depends on the polarity of the tip apex.

These theoretical results emphasized the importance of the atomic species at the tip apex for the alkali halide (001) surface, while it is not straightforward to define the nature of the tip apex experimentally because of the high symmetry of the surface structure. However, there are a few experiments exploring the possibilities to determine the polarity of the tip apex. Bennewitz et al. [58] studied imaging of surfaces of a mixed alkali halide crystal, which was designed to observe the chemically inhomogeneous surface. The mixed crystal is composed of 60% KCl and 40% KBr, with the Cl and Br ions interfused randomly in the crystal. The image of the cleaved $KCl_{0.6}Br_{0.4}(001)$ surface indicates that only one type of ion is imaged as protrusions, as if it were a pure alkali halide crystal. However, the amplitude of the atomic corrugation varies strongly between the positions of the ions imaged as depressions. This variation in the corrugations corresponds to the constituents of the crystal, i.e. the Cl and Br ions, and it is concluded that the tip apex is negatively charged. Moreover, the deep depressions can be assigned to Br ions by comparing the number with the relative density of anions. The difference between Cl and Br anions with different masses is enhanced in the damping signal measured simultaneously with the topographic image [59]. The damping is recorded as an increase in the excitation amplitude necessary to maintain the oscillation amplitude of the cantilever in the constant-amplitude mode [56]. Although the dissipation phenomena on an atomic scale are a subject under discussion, any dissipative interaction must generally induce energy losses in the cantilever oscillation [60, 61]. The measurement of energy dissipation has the potential to enable chemical discrimination on an atomic scale. Recently, a new procedure for species recognition on a alkali halide surface was proposed [62]. This method is based on a comparison between theoretical results and the site-specific measurement of frequency versus distance. The differences in the force curves measured at the typical sites, such as protrusion, depression, and their bridge position, are compared to the corresponding differences obtained from atomistic simulation. The polarity of the tip apex can be determined, leading to the identification of the surface species. This method is applicable to highly symmetric surfaces and is useful for determining the sign of the tip polarity.

## Fluorides

Fluorides are important materials for the progress of an atomic-scale-resolution NC-AFM imaging of insulators. There are reports in the literature of surface images for single-crystal $BaF_2$, $SrF_2$ [63], $CaF_2$ [64, 65, 66] and a CaF bilayer on Si(111) [67]. Surfaces of fluorite-type crystals are prepared by cleaving along the (111) planes. Their structure is more complex than the structure of alkali halides, which have a rock-salt structure. The complexity is of great interest for atomic-resolution imaging using NC-AFM and also for theoretical predictions of the interpretation of the atomic-scale contrast information.

The first atomically resolved images of a $CaF_2(111)$ surface were obtained in topographic mode [65], and the surface ions mostly appear as spherical caps. Barth et al. [68] have found that the $CaF_2(111)$ surface images obtained by using the constant-height mode, in which the frequency shift is recorded with a very low loop gain, can be categorized into two contrast patterns. In the first of these the ions appear as triangles and in the second they have the appearance of circles, similar to the contrast obtained in a topographic image. Theoretical studies demonstrated that these two different contrast patterns could be explained as a result of imaging with tips of different polarity [68, 69, 70]. When imaging with a positively charged (cation-terminated) tip, the triangular pattern appears. In this case, the contrast is dominated by the strong short-range electrostatic attraction between the positive tip and the negative F ions. The cross section along the [121] direction of the triangular image shows two maxima: one is a larger peak over the F(I) ions located in the topmost layer and the other is a smaller peak at the position of the F(III) ions in the third layer. The minima appear at the position of the Ca ions in the second layer. When imaging with a negatively charged (anion-terminated) tip, the spherical image contrast appears and the main periodicity is created by the Ca ions between the topmost and the third F ion layers. In the cross section along the [121] direction, the large maxima correspond to the Ca sites because of the strong attraction of the negative tip and the minima appear at the sites of maximum repulsion over the F(I) ions. At a position between two F ions, there are smaller maxima. This reflects the weaker repulsion over the F(III) ion sites compared to the protruding F(I) ion sites and a slower decay in the contrast on one side of the Ca ions.

The triangular pattern obtained with a positively charged tip appears at relatively large tip–sample distance, as shown in Fig. 5.18a. The cross section along the [121] direction, experiment results and theoretical studies both demonstrate the large-peak and small-shoulder characteristic for the triangular pattern image (Fig. 5.18d). When the tip approaches the surface more closely, the triangular pattern of the experimental images is more vivid (Fig. 5.18b), as predicted in the theoretical works. As the tip approaches, the amplitude of the shoulder increases until it is equal to that of the main peak, and this feature gives rise to the honeycomb pattern image, as shown in Fig. 5.18c. Moreover, theoretical results predict that the image contrast changes again when the tip apex is in close proximity to surface. Recently, Giessibl and Reichling [71] achieved atomic imaging in the repulsive region and proved experimentally the predicted change of the image contrast. As described here, there is good correspondence in the distance dependency of the image obtained by experimental and theoretical investigations.

From detailed theoretical analysis of the electrostatic potential [72], it was suggested that the change in displacement of the ions due to the proximity of the tip plays an important role in the formation of the image contrast. Such a drastic change in image contrast, depending on both the polarity of the terminated tip atom and on the tip–sample distance, is inherent to the fluoride (111) surface, and this image-contrast feature cannot be seen on the (001) surface of alkali halides with a simple crystal structure.

The results of careful experiments show another feature: that the cross sections taken along the three equivalent [121] directions do not yield identical results [68]. It is thought that this can be attributed to the asymmetry of the nanocluster at the tip apex, which leads to different interactions in the equivalent directions. A better understanding of the asymmetric image contrast may require more complicated modeling of the tip structure. In fact, it should be mentioned that perfect tips on an atomic scale can occasionally be obtained. These tips do yield identical results in forward and backward scanning, and cross sections in the three equivalent directions taken with this tip are almost identical [74].

The fluoride (111) surface is an excellent standard surface for calibrating tips on an atomic scale. The polarity of the tip-terminated atom can be determined from the image contrast pattern (spherical or triangular pattern). The irregularities in the tip structure can be detected, since the surface structure is highly symmetric. Therefore, once such a tip has been prepared, it can be used as a calibrated tip for imaging unknown surfaces.

The polarity and shape of the tip apex play an important role in interpreting NC-AFM images of alkali halide and fluorides surfaces. It is expected that the achievement of good correlation between experimental and theoretical studies will help to advance surface imaging of insulators by NC-AFM.

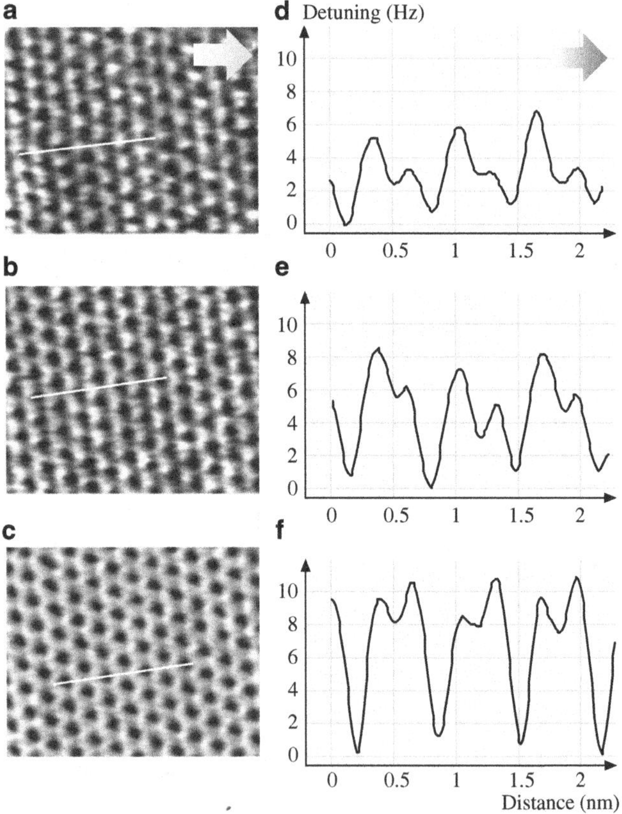


**Fig. 5.18** **(a)**–**(c)** $CaF_2(111)$ surface images obtained by using the constant-height mode. From **(a)** to **(c)** the frequency shift was lowered. The *white lines* represent the positions of the cross section. **(d)**–**(f)** The cross section extracted from the Fourier-filtered images of **(a)**–**(c)** (The *white* and *black arrows* represent the scanning direction. The images and the cross sections are from [68])

## Metal Oxides

Most of the metal oxides that have attracted strong interest for their technological importance are insulating. Therefore, in the case of atomically resolved imaging of metal oxide surfaces by STM, efforts to increase the conductivity of the sample are needed, such as, the introduction of anions or cations defects, doping with other atoms and surface observations during heating of the sample. However, in principle, NC-AFM provides the possibility of observing nonconductive metal oxides without these efforts. In cases where the conductivity of the metal oxides is high enough for a tunneling current to flow, it should be noted that most surface images obtained by NC-AFM and STM are not identical.

Since the first report of atomically resolved images on a $TiO_2(110)$ surface with oxygen point defects [75], they have also been reported on rutile $TiO_2(100)$ [76, 77, 78], anatase $TiO_2(001)$ thin film on $SrTiO_3(100)$ [79] and on $LaAO_3(001)$ [80], $SnO_2(110)$ [81], NiO(001) [82, 83], $SrTiO_3(100)$ [84], $CeO_2(111)$ [85] and $MoO_3(010)$ [86] surfaces. Also, Barth and Reichling have succeeded in obtaining atomically resolved NC-AFM images of a clean $\alpha$-$Al_2O_3(0001)$ surface [73] and of a UHV cleaved MgO (001) [87] surface, which are impossible to investigate using STM. In this section we describe typical results of the imaging of metal oxides by NC-AFM.

The $\alpha$-$Al_2O_3(0001)$ surface exists in several ordered phases that can reversibly be transformed into each other by thermal treatments and oxygen exposure. It is known that the high-temperature phase has a large $(\sqrt{31} \times \sqrt{31})R \pm 9°$ unit cell. However, the details of the atomic structure of this surface have not been revealed, and two models have been proposed. Barth and Reichling [73] have directly observed this reconstructed $\alpha$-$Al_2O_3(0001)$ surface by NC-AFM. They confirmed that the dominant contrast of the low-magnification image corresponds to a rhombic grid representing a unit cell of $(\sqrt{31} \times \sqrt{31})R + 9°$, as shown in Fig. 5.19a. Also, more details of the atomic structures were determined from the higher-magnification image

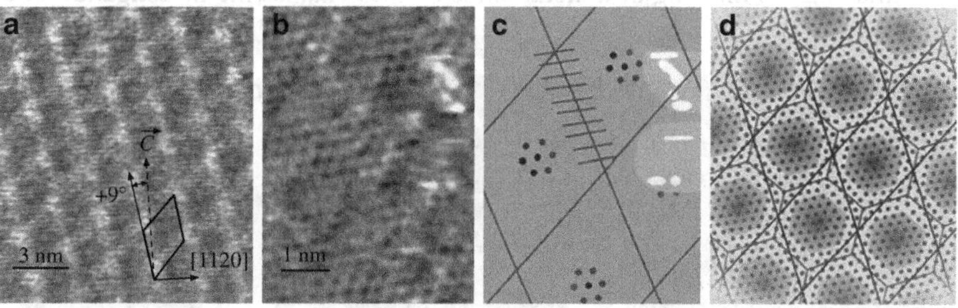


**Fig. 5.19** **(a)** Image of the high-temperature, reconstructed clean $\alpha$-$Al_2O_3$ surface obtained by using the constant-height mode. The *rhombus* represents the unit cell of the $(\sqrt{31} \times \sqrt{31})R + 9°$ reconstructed surface. **(b)** Higher-magnification image of **(a)**. Imaging was performed at a reduced tip–sample distance. **(c)** Schematic representation of the indicating regions of hexagonal order in the center of reconstructed rhombi. **(d)** Superposition of the hexagonal domain with reconstruction rhombi found by NC-AFM imaging. Atoms in the *gray shaded regions* are well ordered (The images and the schematic representations are from [73])

(Fig. 5.19b), which was taken at a reduced tip–sample distance. In this atomically resolved image, it was revealed that each side of the rhombus is intersected by ten atomic rows, and that a hexagonal arrangement of atoms exists in the center of the rhombi (Fig. 5.19c). This feature agrees with the proposed surface structure that predicts order in the center of the hexagonal surface domains and disorder at the domain boundaries. Their result is an excellent demonstration of the capabilities of the NC-AFM for the atomic-scale surface investigation of insulators.

The atomic structure of the $SrTiO_3(100)$-$(\sqrt{5} \times \sqrt{5})R26.6°$ surface, as well as that of $Al_2O_3(0001)$ can be determined on the basis of the results of NC-AFM imaging [84]. $SrTiO_3$ is one of the perovskite oxides, and its (100) surface exhibits the many different kinds of reconstructed structures. In the case of the $(\sqrt{5} \times \sqrt{5})R26°$ reconstruction, the oxygen vacancy–$Ti^{3+}$–oxygen model (where the terminated surface is $TiO_2$ and the observed spots are related to oxygen vacancies) was proposed from the results of STM imaging. As shown in Fig. 5.20, *Kubo* and Nozoye [84] have performed measurements using both STM and NC-AFM, and have found that the size of the bright spots as observed by NC-AFM is always smaller than that for STM measurement, and that the dark spots, which are not observed by STM, are arranged along the [001] and [010] directions in the NC-AFM image. A theoretical simulation of the NC-AFM image using first-principles calculations shows that the bright and dark spots correspond to Sr and oxygen atoms, respectively. It has been proposed that the structural model of the reconstructed surface consists of an ordered Sr adatom located at the oxygen fourfold site on the $TiO_2$-terminated layer (Fig. 5.20c).

Because STM images are related to the spatial distribution of the wave functions near the Fermi level, atoms without a local density of states near the Fermi level are generally invisible even on conductive materials. On the other hand, the NC-AFM image reflects the strength of the tip–sample interaction force originating from chemical, electrostatic and other interactions. Therefore, even STM and NC-AFM images obtained using an identical tip and sample may not be identical generally. The simultaneous imaging of a metal oxide surface enables the investigation of a

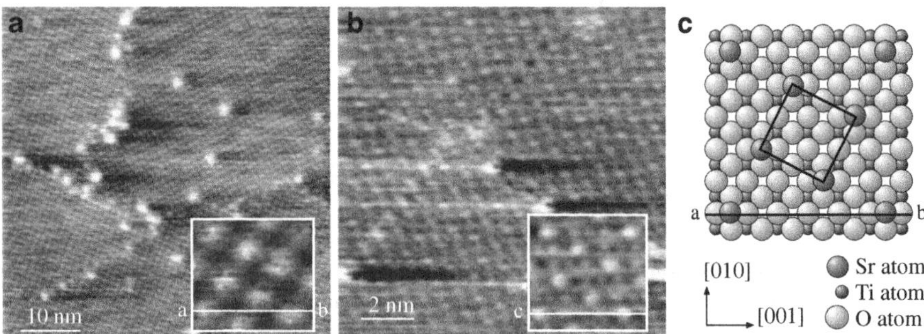


**Fig. 5.20** (**a**) STM and (**b**) NC-AFM images of a $SrTiO_3(100)$ surface. (**c**) A proposed model of the $SrTiO_3(100)$-$(\sqrt{5} \times \sqrt{5})R26.6°$ surface reconstruction. The images and the schematic representations are from [84]

more detailed surface structure. The images of a $TiO_2(110)$ surface simultaneously obtained with STM and NC-AFM [78] are a typical example. The STM image shows that the dangling-bond states at the tip apex overlap with the dangling bonds of the 3d states protruding from the Ti atom, while the NC-AFM primarily imaged the uppermost oxygen atom.

Recently, calculations of the interaction of a Si tip with metal oxides surfaces, such as $Al_2O_3(0001)$, $TiO_2(110)$, and MgO(001), were reported [88, 89]. Previous simulations of AFM imaging of alkali halides and fluorides assume that the tip would be oxides or contaminated and hence have been performed with a model of ionic oxide tips. In the case of imaging a metal oxide surface, pure Si tips are appropriate for a more realistic tip model because the tip is sputtered for cleaning in many experiments. The results of ab initio calculations for a Si tip with a dangling bond demonstrate that the balance between polarization of the tip and covalent bonding between the tip and the surface should determine the tip–surface force. The interaction force can be related to the nature of the surface electronic structure. For wide-gap insulators with a large valence-band offset that prevents significant electron-density transfer between the tip and the sample, the force is dominated by polarization of the tip. When the gap is narrow, the charge transfer increase and covalent bonding dominates the tip–sample interaction. The forces over anions (oxygen ions) in the surface are larger than over cations (metal ions), as they play a more significant role in charge transfer. This implies that a pure Si tip would always show the brightest contrast over the highest anions in the surface. In addition, Foster et al. [88] suggested the method of using applied voltage, which controls the charge transfer, during an AFM measurement to define the nature of tip apex.

The collaboration between experimental and theoretical studies has made great progress in interpreting the imaging mechanism for binary insulators surface and reveals that a well-defined tip with atomic resolution is preferable for imaging a surface. As described previously, a method for the evaluation of the nature of the tip has been developed. However, the most desirable solution would be the development of suitable techniques for well-defined tip preparation and a few attempts at controlled production of Si tips have been reported [24, 90, 91].

### 5.3.2 Atomically Resolved Imaging of a NiO(001) Surface

The transition metal oxides, such as NiO, CoO, and FeO, feature the simultaneous existence of an energy gap and unpaired electrons, which gives rise to a variety of magnetic property. Such magnetic insulators are widely used for the exchange biasing for magnetic and spintronic devices. NC-AFM enables direct surface imaging of magnetic insulators on an atomic scale. The forces detected by NC-AFM originate from several kinds of interaction between the surface and the tip, including magnetic interactions in some cases. Theoretical studies predict that short-range magnetic interactions such as the exchange interaction should enable the NC-AFM to image magnetic moments on an atomic scale. In this section, we will describe imaging of the antiferromagnetic NiO(001) surface using a ferromagnetic tip. Also, theoretical studies of the exchange force interaction between a magnetic tip and a sample will be described.

### Theoretical Studies of the Exchange Force

In the system of a magnetic tip and sample, the interaction detected by NC-AFM includes the short-range magnetic interaction in addition to the long-range magnetic dipole interaction. The energy of the short-range interaction depends on the electron spin states of the atoms on the apex of the tip and the sample surface, and the energy difference between spin alignments (parallel or antiparallel) is referred to as the exchange interaction energy. Therefore, the short-range magnetic interaction leads to the atomic-scale magnetic contrast, depending on the local energy difference between spin alignments.

In the past, extensive theoretical studies on the short-range magnetic interaction between a ferromagnetic tip and a ferromagnetic sample have been performed by a simple calculation [92], a tight-binding approximation [93] and first-principles calculations [94]. In the calculations performed by Nakamura et al. [94], three-atomic-layer Fe(001) films are used as a model for the tip and sample. The exchange force is defined as the difference between the forces in each spin configuration of the tip and sample (parallel and antiparallel). The result of this calculation demonstrates that the amplitude of the exchange force is measurable for AFM (about 0.1 nN). Also, they forecasted that the discrimination of the exchange force would enable direct imaging of the magnetic moments on an atomic scale. Foster and Shluger [95] have theoretically investigated the interaction between a spin-polarized H atom and a Ni atom on a NiO(001) surface. They demonstrated that the difference in magnitude in the exchange interaction between opposite-spin Ni ions in a NiO surface could be sufficient to be measured in a low-temperature NC-AFM experiment. Recently, first-principles calculation of the interaction of a ferromagnetic Fe tip with an NiO surface has demonstrated that it should be feasible to measure the difference in exchange force between opposite-spin Ni ions [96].

### Atomically Resolved Imaging Using Noncoated and Fe-Coated Si Tips

The detection of the exchange interaction is a challenging task for NC-AFM applications. An antiferromagnetic insulator NiO single crystal that has regularly aligned atom sites with alternating electron spin states is one of the best candidates to prove the feasibility of detecting the exchange force for the following reason. NiO has an antiferromagnetic $AF_2$ structure as the most stable below the Néel temperature of 525 K. This well-defined magnetic structure, in which Ni atoms on the (001) surface are arranged in a checkerboard pattern, leads to the simple interpretation of an image containing the atomic-scale contrast originating in the exchange force. In addition, a clean surface can easily be prepared by cleaving.

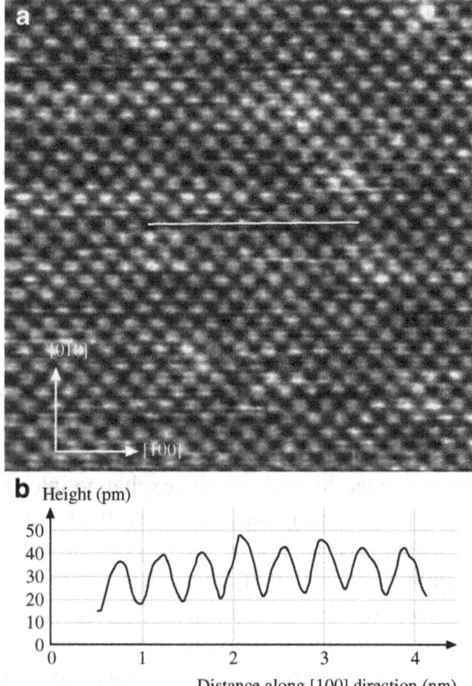


**Fig. 5.21** (**a**) Atomically resolved image obtained with an Fe-coated tip. (**b**) Shows the cross sections of the middle part in (**a**). Their corrugations are about 30 pm

Figure 5.21a shows an atomically resolved image of a NiO(001) surface with a ferromagnetic Fe-coated tip [97]. The bright protrusions correspond to atoms spaced about 0.42 nm apart, consistent with the expected periodic arrangement of the NiO(001) surface. The corrugation amplitude is typically 30 pm, which is comparable to the value previously reported [82, 83, 98, 99, 100], as shown in Fig. 5.21b. The atomic-resolution image (Fig. 5.21b), in which there is one maximum and one minimum within the unit cell, resembles that of the alkali halide (001) surface. The symmetry of the image reveals that only one type of atom appears to be at the maximum. From this image, it seems difficult to distinguish which of the atoms are observed as protrusions. The theoretical works indicate that a metal tip interacts strongly with the oxygen atoms on the MgO(001) surface [95]. From this result, it is presumed that the bright protrusions correspond to the oxygen atoms. However, it is still questionable which of the atoms are visible with a Fe-coated tip.

If the short-range magnetic interaction is included in the atomic image, the corrugation amplitude of the atoms should depend on the direction of the spin over the atom site. From the results of first-principles calculations [94], the contribution of the short-range magnetic interaction to the measured corrugation amplitude is expected to be about a few percent of the total interaction. Discrimination of such small perturbations is therefore needed. In order to reduce the noise, the corrugation amplitude was added on the basis of the periodicity of the NC-AFM image. In addition, the topographical asymmetry, which is the index characterizing the difference in atomic corrugation amplitude, has been defined [101]. The result shows that the value of the topographical asymmetry calculated from the image obtained with an Fe-coated Si tip depends on the direction of summing of the corrugation amplitude, and that the dependency corresponds to the antiferromagnetic spin ordering of the NiO(001) surface [101, 102]. Therefore, this result implies that the dependency of the topographical asymmetry originates in the short-range magnetic interaction. However, in some cases the topographic asymmetry with uncoated Si tips has a finite value [103]. The possibility that the asymmetry includes the influence of the structure of tip apex and of the relative orientation between the surface and tip cannot be excluded. In addition, it is suggested that the absence of unambiguous exchange contrast is due to the fact that surface ion instabilities occur at tip–sample distances that are small enough for a magnetic interact [100]. Another possibility is that the magnetic properties of the tips are not yet fully controlled because the topographic asymmetries obtained by Fe- and Ni-coated tips show no significant difference [103]. In any cases, a careful comparison is needed to evaluate the exchange interaction included in an atomic image.

From the aforementioned theoretical works, it is presumed that a metallic tip has the capability to image an oxygen atom as a bright protrusion. Recently, the magnetic properties of the NiO(001) surface were investigated by first-principles electronic-structure calculations [104]. It was shown that the surface oxygen has finite spin magnetic moment, which originates from symmetry breaking. We must take into account the possibility that a metal atom at the ferromagnetic tip apex may interact with a Ni atom on the second layer through a magnetic interaction mediated by the electrons in an oxygen atom on the surface.

The measurements presented here demonstrate the feasibility of imaging magnetic structures on an atomic scale by NC-AFM. In order to realize explicit detection of exchange force, further experiments and a theoretical study are required. In particular, the development of a tip with well-defined atomic structure and magnetic properties is essential for *exchange force microscopy*.

## 5.4 Applications to Molecules

In the future, it is expected that electronic, chemical, and medical devices will be downsized to the nanometer scale. To achieve this, visualizing and assembling individual molecular components is of fundamental importance. Topographic imaging of nonconductive materials, which is beyond the range of scanning tunneling microscopes, is a challenge for atomic force microscopy (AFM). Nanometer-sized domains of surfactants terminated with different functional groups have been identified by lateral force microscopy (LFM) [106] and by chemical force microscopy (CFM) [107] as extensions of AFM. At a higher resolution, a periodic array of molecules, Langmuir–Blodgett films [108] for example, was recognized by AFM. However, it remains difficult to visualize an isolated molecule, molecule vacancy, or the boundary of different periodic domains, with a microscope with the tip in contact.

### 5.4.1 Why Molecules and Which Molecules?

Access to individual molecules has not been a trivial task even for noncontact atomic force microscopy (NC-AFM). The force pulling the tip into the surface is less sensitive to the gap width ($r$), especially when chemically stable molecules cover the surface. The attractive potential between two stable molecules is shallow and exhibits $r^{-6}$ decay [13].

High-resolution topography of formate ($HCOO^-$) [109] was first reported in 1997 as a molecular adsorbate. The number of imaged molecules is now increasing because of the technological importance of molecular interfaces. To date, the following studies on molecular topography have been published: $C_{60}$ [105, 110], DNAs [111, 112], adenine and thymine [113], alkanethiols [113, 114], a perylene derivative (PTCDA) [115], a metal porphyrin (Cu-TBPP) [116], glycine sulfate [117], polypropylene [118], vinylidene fluoride [119], and a series of carboxylates ($RCOO^-$) [120, 121, 122, 123, 124, 125, 126]. Two of these are presented in Figs. 5.22 and 5.23 to demonstrate the current stage of achievement. The proceedings of the annual NC-AFM conference represent a convenient opportunity for us to update the list of molecules imaged.

### 5.4.2 Mechanism of Molecular Imaging

A systematic study of carboxylates ($RCOO^-$) with R = H, $CH_3$, $C(CH_3)_3$, C≡CH, and $CF_3$ revealed that the van der Waals force is responsible for the molecule-dependent microscope topography despite its long-range ($r^{-6}$) nature. Carboxylates adsorbed on the (110) surface of rutile $TiO_2$ have been extensively studied as a prototype for organic materials interfaced with an inorganic metal oxide [127].

**Fig. 5.22** The constant frequency-shift topography of domain boundaries on a $C_{60}$ multilayered film deposited on a Si(111) surface based on [105]. Image size: $35 \times 35\ nm^2$

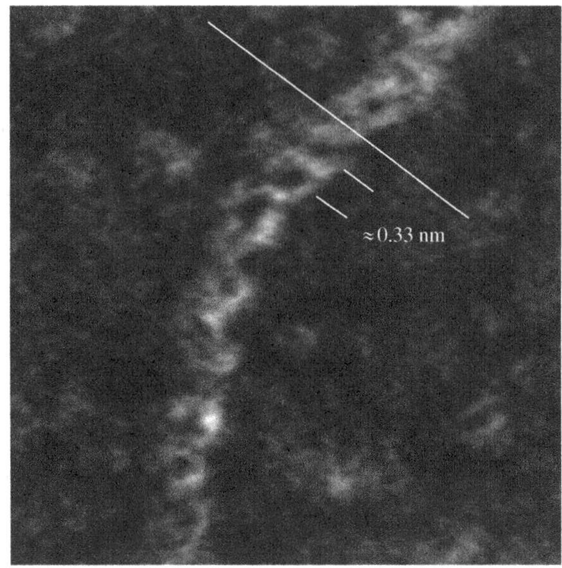


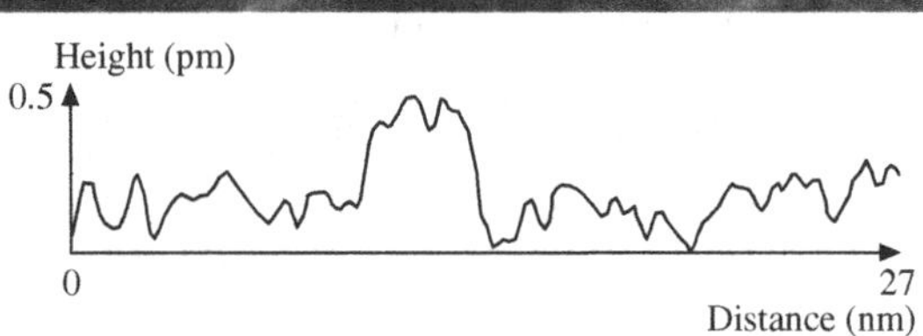


**Fig. 5.23** The constant frequency-shift topography of a DNA helix on a mica surface based on [111]. Image size: 43×43 $nm^2$. The image revealed features with a spacing of 3.3 nm, consistent with the helix turn of B-DNA

A carboxylic acid molecule (RCOOH) dissociates on this surface to a carboxylate ($RCOO^-$) and a proton ($H^+$) at room temperature, as illustrated in Fig. 5.24. The pair of negatively charged oxygen atoms in the $RCOO^-$ coordinate two positively charged Ti atoms on the surface. The adsorbed carboxylates create a long-range ordered monolayer. The lateral distances of the adsorbates in the ordered monolayer are regulated at 0.65 and 0.59 nm along the [110] and [001] directions. By scanning a mixed monolayer containing different carboxylates, the microscope topography of the terminal groups can be quantitatively compared while minimizing tip-dependent artifacts.

Figure 5.25 presents the observed constant frequency-shift topography of four carboxylates terminated by different alkyl groups. On the formate-covered surface of panel (a), individual formates (R = H) were resolved as protrusions of uniform brightness. The dark holes represent unoccupied surface sites. The cross section in the lower panel shows that the accuracy of the height measurement was 0.01 nm or better. Brighter particles appeared in the image when the formate monolayer was exposed to acetic acid ($CH_3COOH$) as shown in panel (b). Some formates were exchanged with acetates (R = $CH_3$) impinging from the gas phase [129]. Because the number of brighter spots increased with exposure time to acetic acid, the brighter particle was assigned to the acetate [121]. Twenty-nine acetates and

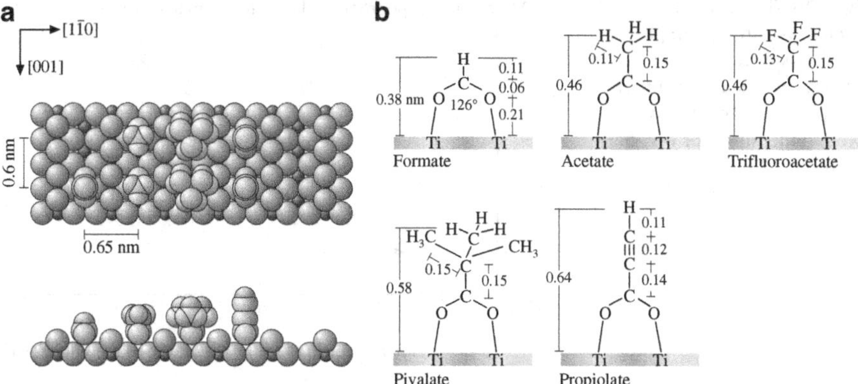


**Fig. 5.24** The carboxylates and $TiO_2$ substrate. (**a**) Top and side view of the ball model. *Small shaded* and *large shaded balls* represent Ti and O atoms in the substrate. Protons yielded in the dissociation reaction are not shown. (**b**) Atomic geometry of formate, acetate, pivalate, propiolate, and trifluoroacetate adsorbed on the $TiO_2(110)$ surface. The O–Ti distance and O–C–O angle of the formate were determined in the quantitative analysis using photoelectron diffraction [128]

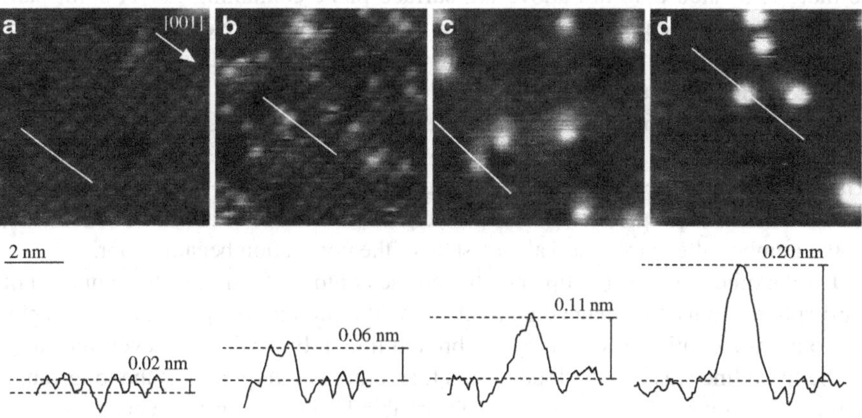


**Fig. 5.25** The constant frequency-shift topography of carboxylate monolayers prepared on the $TiO_2(110)$ surface based on [121,123,125]. Image size: $10 \times 10$ $nm^2$. (**a**) Pure formate monolayer; (**b**) formate–acetate mixed layer; (**c**) formate–pivalate mixed layer; (**d**) formate–propiolate mixed layer. Cross sections determined on the *lines* are shown in the *lower panel*

188 formates were identified in the topography. An isolated acetate and its surrounding formates exhibited an image height difference of 0.06 nm. Pivalate is terminated by bulky $R = (CH_3)_3$. Nine bright pivalates were surrounded by formates of ordinary brightness in the image of panel (c) [123]. The image height difference of an isolated pivalate over the formates was 0.11 nm. Propiolate with C≡CH is a needle-like adsorbate of single-atom diameter. That molecule exhibited in panel (d) a microscope topography 0.20 nm higher than that of the formate [125].

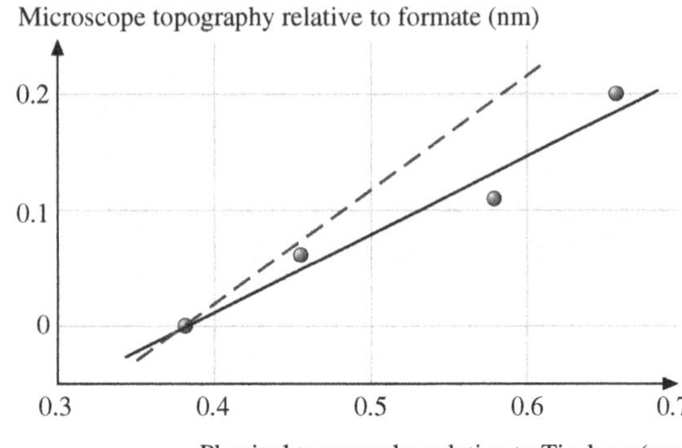


**Fig. 5.26** The constant frequency-shift topography of the alkyl-substituted carboxylates as a function of their physical topography given in the model of Fig. 5.3 based on [123]

The image topography of formate, acetate, pivalate, and propiolate followed the order of the size of the alkyl groups. Their physical topography can be assumed based on the C–C and C–H bond lengths in the corresponding RCOOH molecules in the gas phase [130], and is illustrated in Fig. 5.24. The top hydrogen atom of the formate is located 0.38 nm above the surface plane containing the Ti atom pair, while three equivalent hydrogen atoms of the acetate are more elevated at 0.46 nm. The uppermost H atoms in the pivalate are raised by 0.58 nm relative to the Ti plane. The H atom terminating the triple-bonded carbon chain in the propiolate is at 0.64 nm. Figure 5.26 summarizes the observed image heights relative to the formate, as a function of the physical height of the topmost H atoms given in the model. The straight line fitted the four observations [122]. When the horizontal axis was scaled with other properties (molecular weight, the number of atoms in a molecule, or the number of electrons in valence states), the correlation became poor.

On the other hand, if the tip apex traced the contour of a molecule composed of hard-sphere atoms, the image topography would reproduce the physical topography in a one-to-one ratio, as shown by the broken line in Fig. 5.26. However, the slope of the fitted line was 0.7. A slope of less than unity is interpreted as the long-range nature of the tip–molecule force. The observable frequency shift reflects the sum of the forces between the tip apex and individual molecules. When the tip passes above a tall molecule embedded in short molecules, it is pulled up to compensate for the increased force originating from the tall molecule. Forces between the lifted tip and the short molecules are reduced due to the increased tip–surface distance. Feedback regulation pushes down the probe to restore the lost forces.

This picture predicts that microscope topography is sensitive to the lateral distribution of the molecules, and that was in fact the case. Two-dimensionally clustered acetates exhibited enhanced image height over an isolated acetate [121]. The tip–molecule force therefore remained nonzero at distances over the lateral separation of the carboxylates on this surface (0.59–0.65 nm). Chemical bond interactions cannot be important across such a wide tip–molecule gap, whereas atom-scale images of Si(111)(7 $\times$ 7) are interpreted with the fractional formation of tip–surface chemical bonds [24, 45, 49]. Instead, the attractive component of the van der Waals force is probable responsible for the observed molecule-dependent topography. The absence of the tip–surface chemical bond is reasonable on the carboxylate-covered surface terminated with stable C–H bonds.

The attractive component of the van der Waals force contains electrostatic terms caused by permanent-dipole/permanent-dipole coupling, permanent-dipole/induced-dipole coupling, and induced-dipole/induced-dipole coupling (dispersion force). The four carboxylates examined are equivalent in terms of their permanent electric dipole, because the alkyl groups are nonpolar. The image contrast of one carboxylate relative to another is thus ascribed to the dispersion force and/or the force created by the coupling between the permanent dipole on the tip and the induced dipole on the molecule. If we further assume that the Si tip used exhibits the smallest permanent dipole, the dispersion force remains dominant to create the NC-AFM topography dependent on the nonpolar groups of atoms. A numerical simulation based on this assumption [125] successfully reproduced the propiolate topography of Fig. 5.25d. A calculation that does not include quantum chemical treatment is expected to work, unless the tip approaches the surface too closely, or the molecule possesses a dangling bond.

In addition to the contribution of the dispersion force, the permanent dipole moment of molecules may perturb the microscope topography through electrostatic coupling with the tip. Its possible role was demonstrated by imaging a fluorine-substituted acetate. The strongly polarized C–F bonds were expected to perturb the electrostatic field over the molecule. The constant frequency-shift topography of acetate ($R = CH_3$) and trifluoroacetate ($R = CF_3$) was indeed sensitive to the fluorine substitution. The acetate was observed to be 0.05 nm higher than the trifluoroacetate [122], although the F atoms in the trifluoroacetate as well as the H atoms in the acetate were lifted by 0.46 nm from the surface plane, as illustrated in Fig. 5.24.

### 5.4.3 Perspectives

The experimental results summarized in this section prove the feasibility of using NC-AFM to identify individual molecules. A systematic study on the constant frequency-shift topography of carboxylates with $R = CH_3$, $C(CH_3)_3$, $C{\equiv}CH$, and $CF_3$ has revealed the mechanism behind the high-resolution imaging of the chemically stable molecules. The dispersion force is primarily responsible for the molecule-dependent topography. The permanent dipole moment of the imaged molecule, if it exists, perturbs the topography through the electrostatic coupling with the tip. A tiny calculation containing empirical force fields works when simulating the microscope topography.

These results make us optimistic about analyzing physical and chemical properties of nanoscale supramolecular assemblies constructed on a solid surface. If the accuracy of topographic measurement is developed by one more order of magnitude, which is not an unrealistic target, it may be possible to identify structural isomers, chiral isomers, and conformational isomers of a molecule. Kelvin probe force microscopy (KPFM), an extension of NC-AFM, provides a nanoscale analysis of molecular electronic properties [118, 119]. Force spectroscopy with chemically modified tips seems promising for the detection of a selected chemical force. Operation in a liquid atmosphere [131] is required for the observation of biochemical materials in their natural environment.

## References


1. G. Binnig: Atomic force microscope, method for imaging surfaces with atomic resolution, US Patent 4724318 (1986).
2. G. Binnig, C.F. Quate, C. Gerber: Atomic force microscope, Phys. Rev. Lett. **56**, 930–933 (1986).
3. G. Binnig, H. Rohrer, C. Gerber, E. Weibel: Surface studies by scanning tunneling microscopy, Phys. Rev. Lett. **49**, 57–61 (1982).
4. G. Binnig, H. Rohrer: The scanning tunneling microscope, Sci. Am. **253**, 50–56 (1985).
5. G. Binnig, H. Rohrer: In touch with atoms, Rev. Mod. Phys. **71**, S320–S330 (1999).
6. C.J. Chen: *Introduction to Scanning Tunneling Microscopy* (Oxford University Press, Oxford, 1993).
7. H.-J. Güntherodt, R. Wiesendanger (eds.): *Scanning Tunneling Microscopy I–III* (Springer, Berlin, Heidelberg, 1991).
8. J.A. Stroscio, W.J. Kaiser (eds.): *Scanning Tunneling Microscopy* (Academic, Boston, 1993).
9. R. Wiesendanger: *Scanning Probe Microscopy and Spectroscopy: Methods and Applications* (Cambridge University Press, Cambridge, 1994).
10. S. Morita, R. Wiesendanger, E. Meyer (eds.): *Noncontact Atomic Force Microscopy* (Springer, Berlin, Heidelberg, 2002).
11. R. Garcia, R. Perez: Dynamic atomic force microscopy methods, Surf. Sci. Rep. **47**, 197–301 (2002).
12. F.J. Giessibl: Advances in atomic force microscopy, Rev. Mod. Phys. **75**, 949–983 (2003).
13. J. Israelachvili: *Intermolecular and Surface Forces*, 2nd edn. (Academic, London, 1991).
14. L. Olsson, N. Lin, V. Yakimov, R. Erlandsson: A method for in situ characterization of tip shape in AC-mode atomic force microscopy using electrostatic interaction, J. Appl. Phys. **84**, 4060–4064 (1998).
15. S. Akamine, R.C. Barrett, C.F. Quate: Improved atomic force microscopy images using cantilevers with sharp tips, Appl. Phys. Lett. **57**, 316–318 (1990).
16. T.R. Albrecht, S. Akamine, T.E. Carver, C.F. Quate: Microfabrication of cantilever styli for the atomic force microscope, J. Vac. Sci. Technol. A **8**, 3386–3396 (1990).
17. M. Tortonese, R.C. Barrett, C. Quate: Atomic resolution with an atomic force microscope using piezoresistive detection, Appl. Phys. Lett. **62**, 834–836 (1993).
18. O. Wolter, T. Bayer, J. Greschner: Micromachined silicon sensors for scanning force microscopy, J. Vac. Sci. Technol. **9**, 1353–1357 (1991).
19. D. Sarid: *Scanning Force Microscopy*, 2nd edn. (Oxford University Press, New York, 1994).
20. F.J. Giessibl, B.M. Trafas: Piezoresistive cantilevers utilized for scanning tunneling and scanning force microscope in ultrahigh vacuum, Rev. Sci. Instrum. **65**, 1923–1929 (1994).
21. P. Güthner, U.C. Fischer, K. Dransfeld: Scanning near-field acoustic microscopy, Appl. Phys. B **48**, 89–92 (1989).
22. K. Karrai, R.D. Grober: Piezoelectric tip–sample distance control for near field optical microscopes, Appl. Phys. Lett. **66**, 1842–1844 (1995).

23. F.J. Giessibl: High-speed force sensor for force microscopy and profilometry utilizing a quartz tuning fork, Appl. Phys. Lett. **73**, 3956–3958 (1998).
24. F.J. Giessibl, S. Hembacher, H. Bielefeldt, J. Mannhart: Subatomic features on the silicon (111)-(7×7) surface observed by atomic force microscopy, Science **289**, 422–425 (2000).
25. F. Giessibl, C. Gerber, G. Binnig: A low-temperature atomic force/scanning tunneling microscope for ultrahigh vacuum, J. Vac. Sci. Technol. B **9**, 984–988 (1991).
26. F. Ohnesorge, G. Binnig: True atomic resolution by atomic force microscopy through repulsive and attractive forces, Science **260**, 1451–1456 (1993).
27. F.J. Giessibl, G. Binnig: True atomic resolution on KBr with a low-temperature atomic force microscope in ultrahigh vacuum, Ultramicroscopy **42–44**, 281–286 (1992).
28. S.P. Jarvis, H. Yamada, H. Tokumoto, J.B. Pethica: Direct mechanical measurement of interatomic potentials, Nature **384**, 247–249 (1996).
29. L. Howald, R. Lüthi, E. Meyer, P. Güthner, H.-J. Güntherodt: Scanning force microscopy on the Si(111)7×7 surface reconstruction, Z. Phys. B **93**, 267–268 (1994).
30. L. Howald, R. Lüthi, E. Meyer, H.-J. Güntherodt: Atomic-force microscopy on the Si(111) 7×7 surface, Phys. Rev. B **51**, 5484–5487 (1995).
31. Y. Martin, C.C. Williams, H.K. Wickramasinghe: Atomic force microscope – force mapping and profiling on a sub 100-Å scale, J. Appl. Phys. **61**, 4723–4729 (1987).
32. T.R. Albrecht, P. Grütter, H.K. Horne, D. Rugar: Frequency modulation detection using high-$Q$ cantilevers for enhanced force microscope sensitivity, J. Appl. Phys. **69**, 668–673 (1991).
33. F.J. Giessibl: Atomic resolution of the silicon (111)-(7×7) surface by atomic force microscopy, Science **267**, 68–71 (1995).
34. S. Kitamura, M. Iwatsuki: Observation of silicon surfaces using ultrahigh-vacuum noncontact atomic force microscopy, Jpn. J. Appl. Phys. **35**, 668–L671 (1995).
35. R. Erlandsson, L. Olsson, P. Mårtensson: Inequivalent atoms and imaging mechanisms in AC-mode atomic-force microscopy of Si(111)7×7, Phys. Rev. B **54**, R8309–R8312 (1996).
36. N. Burnham, R.J. Colton: Measuring the nanomechanical and surface forces of materials using an atomic force microscope, J. Vac. Sci. Technol. A **7**, 2906–2913 (1989).
37. D. Tabor, R.H.S. Winterton: Direct measurement of normal and related van der Waals forces, Proc. R. Soc. Lond. A **312**, 435 (1969).
38. F.J. Giessibl: Forces and frequency shifts in atomic resolution dynamic force microscopy, Phys. Rev. B **56**, 16011–16015 (1997).
39. G. Binnig, H. Rohrer, C. Gerber, E. Weibel: 7×7 reconstruction on Si(111) resolved in real space, Phys. Rev. Lett. **50**, 120–123 (1983).
40. H. Goldstein: *Classical Mechanics* (Addison Wesley, Reading, 1980).
41. U. Dürig: Interaction sensing in dynamic force microscopy, New J. Phys. **2**, 5.1–5.12 (2000).
42. F.J. Giessibl: A direct method to calculate tip–sample forces from frequency shifts in frequency-modulation atomic force microscopy, Appl. Phys. Lett. **78**, 123–125 (2001).
43. U. Dürig, H.P. Steinauer, N. Blanc: Dynamic force microscopy by means of the phase-controlled oscillator method, J. Appl. Phys. **82**, 3641–3651 (1997).
44. F.J. Giessibl, H. Bielefeldt, S. Hembacher, J. Mannhart: Calculation of the optimal imaging parameters for frequency modulation atomic force microscopy, Appl. Surf. Sci. **140**, 352–357 (1999).
45. M.A. Lantz, H.J. Hug, R. Hoffmann, P.J.A. van Schendel, P. Kappenberger, S. Martin, A. Baratoff, H.-J. Güntherodt: Quantitative measurement of short-range chemical bonding forces, Science **291**, 2580–2583 (2001).
46. F.J. Giessibl, M. Herz, J. Mannhart: Friction traced to the single atom, Proc. Natl. Acad. Sci. USA **99**, 12006–12010 (2002).
47. N. Nakagiri, M. Suzuki, K. Oguchi, H. Sugimura: Site discrimination of adatoms in Si(111)-7×7 by noncontact atomic force microscopy, Surf. Sci. Lett. **373**, L329–L332 (1997).
48. Y. Sugawara, M. Ohta, H. Ueyama, S. Morita: Defect motion on an InP(110) surface observed with noncontact atomic force microscopy, Science **270**, 1646–1648 (1995).

49. T. Uchihashi, Y. Sugawara, T. Tsukamoto, M. Ohta, S. Morita: Role of a covalent bonding interaction in noncontact-mode atomic-force microscopy on Si(111)7 × 7, Phys. Rev. B **56**, 9834–9840 (1997).
50. K. Yokoyama, T. Ochi, A. Yoshimoto, Y. Sugawara, S. Morita: Atomic resolution imaging on Si(100)2 × 1 and Si(100)2 × 1-H surfaces using a non-contact atomic force microscope, Jpn. J. Appl. Phys. **39**, L113–L115 (2000).
51. Y. Sugawara, T. Minobe, S. Orisaka, T. Uchihashi, T. Tsukamoto, S. Morita: Non-contact AFM images measured on Si(111)3×3-Ag and Ag(111) surfaces, Surf. Interface Anal. **27**, 456–461 (1999).
52. K. Yokoyama, T. Ochi, Y. Sugawara, S. Morita: Atomically resolved Ag imaging on Si(111)3× 3-Ag surface with noncontact atomic force microscope, Phys. Rev. Lett. **83**, 5023–5026 (1999).
53. M. Bammerlin, R. Lüthi, E. Meyer, A. Baratoff, J. Lü, M. Guggisberg, C. Gerber, L. Howald, H.-J. Güntherodt: True atomic resolution on the surface of an insulator via ultrahigh vacuum dynamic force microscopy, Probe Microsc. J. **1**, 3–7 (1997).
54. M. Bammerlin, R. Lüthi, E. Meyer, A. Baratoff, J. Lü, M. Guggisberg, C. Loppacher, C. Gerber, H.-J. Güntherodt: Dynamic SFM with true atomic resolution on alkali halide surfaces, Appl. Phys. A **66**, S293–S294 (1998).
55. R. Hoffmann, M.A. Lantz, H.J. Hug, P.J.A. van Schendel, P. Kappenberger, S. Martin, A. Baratoff, H.-J. Güntherodt: Atomic resolution imaging and force versus distance measurements on KBr(001) using low temperature scanning force microscopy, Appl. Surf. Sci. **188**, 238–244 (2002).
56. R. Bennewitz, A.S. Foster, L.N. Kantotovich, M. Bammerlin, C. Loppacher, S. Schär, M. Guggisberg, E. Meyer, A.L. Shluger: Atomically resolved edges and kinks of NaCl islands on Cu(111): Experiment and theory, Phys. Rev. B **62**, 2074–2084 (2000).
57. A.I. Livshits, A.L. Shluger, A.L. Rohl, A.S. Foster: Model of noncontact scanning force microscopy on ionic surfaces, Phys. Rev. **59**, 2436–2448 (1999).
58. R. Bennewitz, O. Pfeiffer, S. Schär, V. Barwich, E. Meyer, L.N. Kantorovich: Atomic corrugation in nc-AFM of alkali halides, Appl. Surf. Sci. **188**, 232–237 (2002).
59. R. Bennewitz, S. Schär, E. Gnecco, O. Pfeiffer, M. Bammerlin, E. Meyer: Atomic structure of alkali halide surfaces, Appl. Phys. A **78**, 837–841 (2004).
60. M. Gauthier, L. Kantrovich, M. Tsukada: Theory of energy dissipation into surface viblationsed, in *Noncontact Atomic Force Microscopy*, ed. by S. Morita, R. Wiesendanger, E. Meyer (Springer, Berlin/Heidelberg, 2002) pp.371–394.
61. H.J. Hug, A. Baratoff: Measurement of dissipation induced by tip–sample interactions, in *Noncontact Atomic Force Microscopy*, ed. by S. Morita, R. Wiesendanger, E. Meyer (Springer, Berlin/Heidelberg, 2002) pp.395–431.
62. R. Hoffmann, L.N. Kantorovich, A. Baratoff, H.J. Hug, H.-J. Güntherodt: Sublattice identification in scanning force microscopy on alkali halide surfaces, Phys. Rev. B **92**, 146103–1–146103–4 (2004).
63. C. Barth, M. Reichling: Resolving ions and vacancies at step edges on insulating surfaces, Surf. Sci. **470**, L99–L103 (2000).
64. R. Bennewitz, M. Reichling, E. Matthias: Force microscopy of cleaved and electron-irradiated CaF_2(111) surfaces in ultra-high vacuum, Surf. Sci. **387**, 69–77 (1997).
65. M. Reichling, C. Barth: Scanning force imaging of atomic size defects on the $CaF_2$(111) surface, Phys. Rev. Lett. **83**, 768–771 (1999).
66. M. Reichling, M. Huisinga, S. Gogoll, C. Barth: Degradation of the $CaF_2$(111) surface by air exposure, Surf. Sci. **439**, 181–190 (1999).
67. A. Klust, T. Ohta, A.A. Bostwick, Q. Yu, F.S. Ohuchi, M.A. Olmstead: Atomically resolved imaging of a CaF bilayer on Si(111): Subsurface atoms and the image contrast in scanning force microscopy, Phys. Rev. B **69**, 035405–1–035405–5 (2004).
68. C. Barth, A.S. Foster, M. Reichling, A.L. Shluger: Contrast formation in atomic resolution scanning force microscopy of $CaF_2$(111): Experiment and theory, J. Phys. Condens. Matter **13**, 2061–2079 (2001).

69. A.S. Foster, C. Barth, A.L. Shulger, M. Reichling: Unambiguous interpretation of atomically resolved force microscopy images of an insulator, Phys. Rev. Lett. **86**, 2373–2376 (2001).
70. A.S. Foster, A.L. Rohl, A.L. Shluger: Imaging problems on insulators: What can be learnt from NC-AFM modeling on $CaF_2$?, Appl. Phys. A **72**, S31–S34 (2001).
71. F.J. Giessibl, M. Reichling: Investigating atomic details of the $CaF_2$(111) surface with a qPlus sensor, Nanotechnology **16**, S118–S124 (2005).
72. A.S. Foster, C. Barth, A.L. Shluger, R.M. Nieminen, M. Reichling: Role of tip structure and surface relaxation in atomic resolution dynamic force microscopy: $CaF_2$(111) as a reference surface, Phys. Rev. B **66**, 235417–1–235417–10 (2002).
73. C. Barth, M. Reichling: Imaging the atomic arrangements on the high-temperature reconstructed $\alpha$-$Al_2O_3$ surface, Nature **414**, 54–57 (2001).
74. M. Reichling, C. Barth: Atomically resolution imaging on fluorides, in *Noncontact Atomic Force Microscopy*, ed. by S. Morita, R. Wiesendanger, E. Meyer (Springer, Berlin/Heidelberg, 2002) pp. 109–123.
75. K. Fukui, H. Ohnishi, Y. Iwasawa: Atom-resolved image of the $TiO_2$(110) surface by noncontact atomic force microscopy, Phys. Rev. Lett. **79**, 4202–4205 (1997).
76. H. Raza, C.L. Pang, S.A. Haycock, G. Thornton: Non-contact atomic force microscopy imaging of $TiO_2$(100) surfaces, Appl. Surf. Sci. **140**, 271–275 (1999).
77. C.L. Pang, H. Raza, S.A. Haycock, G. Thornton: Imaging reconstructed $TiO_2$(100) surfaces with non-contact atomic force microscopy, Appl. Surf. Sci. **157**, 223–238 (2000).
78. M. Ashino, T. Uchihashi, K. Yokoyama, Y. Sugawara, S. Morita, M. Ishikawa: STM and atomic-resolution noncontact AFM of an oxygen-deficient $TiO_2$(110) surface, Phys. Rev. B **61**, 13955–13959 (2000).
79. R.E. Tanner, A. Sasahara, Y. Liang, E.I. Altmann, H. Onishi: Formic acid adsorption on anatase $TiO_2$(001)-($1 \times 4$) thin films studied by NC-AFM and STM, J. Phys. Chem. B **106**, 8211–8222 (2002).
80. A. Sasahara, T.C. Droubay, S.A. Chambers, H. Uetsuka, H. Onishi: Topography of anatase $TiO_2$ film synthesized on $LaAlO_3$(001), Nanotechnology **16**, S18–S21 (2005).
81. C.L. Pang, S.A. Haycock, H. Raza, P.J. Møller, G. Thornton: Structures of the $4 \times 1$ and $1 \times 2$ reconstructions of $SnO_2$(110), Phys. Rev. B **62**, R7775–R7778 (2000).
82. H. Hosoi, K. Sueoka, K. Hayakawa, K. Mukasa: Atomic resolved imaging of cleaved NiO (100) surfaces by NC-AFM, Appl. Surf. Sci. **157**, 218–221 (2000).
83. W. Allers, S. Langkat, R. Wiesendanger: Dynamic low-temperature scanning force microscopy on nickel oxide (001), Appl. Phys. A **72**, S27–S30 (2001).
84. T. Kubo, H. Nozoye: Surface Structure of SrTiO3(100)-($5\times5$)-*R* 26.6°, Phys. Rev. Lett. **86**, 1801–1804 (2001).
85. K. Fukui, Y. Namai, Y. Iwasawa: Imaging of surface oxygen atoms and their defect structures on $CeO_2$(111) by noncontact atomic force microscopy, Appl. Surf. Sci. **188**, 252–256 (2002).
86. S. Suzuki, Y. Ohminami, T. Tsutsumi, M.M. Shoaib, M. Ichikawa, K. Asakura: The first observation of an atomic scale noncontact AFM image of $MoO_3$(010), Chem. Lett. **32**, 1098–1099 (2003).
87. C. Barth, C.R. Henry: Atomic resolution imaging of the (001) surface of UHV cleaved MgO by dynamic scanning force microscopy, Phys. Rev. Lett. **91**, 196102–1–196102–4 (2003).
88. A.S. Foster, A.Y. Gal, J.M. Airaksinen, O.H. Pakarinen, Y.J. Lee, J.D. Gale, A.L. Shluger, R.M. Nieminen: Towards chemical identification in atomic-resolution noncontact AFM imaging with silicon tips, Phys. Rev. B **68**, 195420–1–195420–8 (2003).
89. A.S. Foster, A.Y. Gal, J.D. Gale, Y.J. Lee, R.M. Nieminen, A.L. Shluger: Interaction of silicon dangling bonds with insulating surfaces, Phys. Rev. Lett. **92**, 036101–1–036101–4 (2004).
90. T. Eguchi, Y. Hasegawa: High resolution atomic force microscopic imaging of the Si(111)-($7\times7$) surface: Contribution of short-range force to the images, Phys. Rev. Lett. **89**, 266105–1–266105–4 (2002).

91. T. Arai, M. Tomitori: A Si nanopillar grown on a Si tip by atomic force microscopy in ultrahigh vacuum for a high-quality scanning probe, Appl. Phys. Lett. **86**, 073110–1–073110–3 (2005).
92. K. Mukasa, H. Hasegawa, Y. Tazuke, K. Sueoka, M. Sasaki, K. Hayakawa: Exchange interaction between magnetic moments of ferromagnetic sample and tip: Possibility of atomic-resolution images of exchange interactions using exchange force microscopy, Jpn. J. Appl. Phys. **33**, 2692–2695 (1994).
93. H. Ness, F. Gautier: Theoretical study of the interaction between a magnetic nanotip and a magnetic surface, Phys. Rev. B **52**, 7352–7362 (1995).
94. K. Nakamura, H. Hasegawa, T. Oguchi, K. Sueoka, K. Hayakawa, K. Mukasa: First-principles calculation of the exchange interaction and the exchange force between magnetic Fe films, Phys. Rev. B **56**, 3218–3221 (1997).
95. A.S. Foster, A.L. Shluger: Spin-contrast in non-contact SFM on oxide surfaces: Theoretical modeling of NiO(001) surface, Surf. Sci. **490**, 211–219 (2001).
96. T. Oguchi, H. Momida: Electronic structure and magnetism of antiferromagnetic oxide surface – First-principles calculations, J. Surf. Sci. Soc. Jpn. **26**, 138–143 (2005).
97. H. Hosoi, M. Kimura, K. Sueoka, K. Hayakawa, K. Mukasa: Non-contact atomic force microscopy of an antiferromagnetic NiO(100) surface using a ferromagnetic tip, Appl. Phys. A **72**, S23–S26 (2001).
98. H. Hölscher, S.M. Langkat, A. Schwarz, R. Wiesendanger: Measurement of three-dimensional force fields with atomic resolution using dynamic force spectroscopy, Appl. Phys. Lett. **81**, 4428–4430 (2002).
99. S.M. Langkat, H. Hölscher, A. Schwarz, R. Wiesendanger: Determination of site specific interaction forces between an iron coated tip and the NiO(001) surface by force field spectroscopy, Surf. Sci. **527**, 12–20 (2003).
100. R. Hoffmann, M.A. Lantz, H.J. Hug, P.J.A. van Schendel, P. Kappenberger, S. Martin, A. Baratoff, H.-J. Güntherodt: Atomic resolution imaging and frequency versus distance measurement on NiO(001) using low-temperature scanning force microscopy, Phys. Rev. B **67**, 085402–1–085402–6 (2003).
101. H. Hosoi, K. Sueoka, K. Hayakawa, K. Mukasa: Atomically resolved imaging of a NiO(001) surface, in *Noncontact Atomic Force Microscopy*, ed. by S. Morita, R. Wiesendanger, E. Meyer (Springer, Berlin/Heidelberg, 2002) pp. 125–134.
102. K. Sueoka, A. Subagyo, H. Hosoi, K. Mukasa: Magnetic imaging with scanning force microscopy, Nanotechnology **15**, S691–S698 (2004).
103. H. Hosoi, K. Sueoka, K. Mukasa: Investigations on the topographic asymmetry of non-contact atomic force microscopy images of NiO(001) surface observed with a ferromagnetic tip, Nanotechnology **15**, 505–509 (2004).
104. H. Momida, T. Oguchi: First-principles studies of antiferromagnetic MnO and NiO surfaces, J. Phys. Soc. Jpn. **72**, 588–593 (2003).
105. K. Kobayashi, H. Yamada, T. Horiuchi, K. Matsushige: Structures and electrical properties of fullerene thin films on Si(111)-7 $\times$ 7 surface investigated by noncontact atomic force microscopy, Jpn. J. Appl. Phys. **39**, 3821–3829 (2000).
106. R.M. Overney, E. Meyer, J. Frommer, D. Brodbeck, R. Lüthi, L. Howald, H.-J. Güntherodt, M. Fujihira, H. Takano, Y. Gotoh: Friction measurements on phase-separated thin films with amodified atomic force microscope, Nature **359**, 133–135 (1992).
107. D. Frisbie, L.F. Rozsnyai, A. Noy, M.S. Wrighton, C.M. Lieber: Functional group imaging by chemical force microscopy, Science **265**, 2071–2074 (1994).
108. E. Meyer, L. Howald, R.M. Overney, H. Heinzelmann, J. Frommer, H.-J. Güntherodt, T. Wagner, H. Schier, S. Roth: Molecular-resolution images of Langmuir–Blodgett films using atomic force microscopy, Nature **349**, 398–400 (1992).
109. K. Fukui, H. Onishi, Y. Iwasawa: Imaging of individual formate ions adsorbed on $TiO_2$(110) surface by non-contact atomic force microscopy, Chem. Phys. Lett. **280**, 296–301 (1997).
110. K. Kobayashi, H. Yamada, T. Horiuchi, K. Matsushige: Investigations of $C_{60}$ molecules deposited on Si(111) by noncontact atomic force microscopy, Appl. Surf. Sci. **140**, 281–286 (1999).

111. T. Uchihashi, M. Tanigawa, M. Ashino, Y. Sugawara, K. Yokoyama, S. Morita, M. Ishikawa: Identification of B-form DNA in an ultrahigh vacuum by noncontact-mode atomic force microscopy, Langmuir **16**, 1349–1353 (2000).
112. Y. Maeda, T. Matsumoto, T. Kawai: Observation of single- and double-strand DNA using non-contact atomic force microscopy, Appl. Surf. Sci. **140**, 400–405 (1999).
113. T. Uchihashi, T. Ishida, M. Komiyama, M. Ashino, Y. Sugawara, W. Mizutani, K. Yokoyama, S. Morita, H. Tokumoto, M. Ishikawa: High-resolution imaging of organic monolayers using noncontact AFM, Appl. Surf. Sci. **157**, 244–250 (2000).
114. T. Fukuma, K. Kobayashi, T. Horiuchi, H. Yamada, K. Matsushige: Alkanethiol self-assembled monolayers on Au(111) surfaces investigated by non-contact AFM, Appl. Phys. A **72**, S109–S112 (2001).
115. B. Gotsmann, C. Schmidt, C. Seidel, H. Fuchs: Molecular resolution of an organic monolayer by dynamic AFM, Eur. Phys. J. B **4**, 267–268 (1998).
116. C. Loppacher, M. Bammerlin, M. Guggisberg, E. Meyer, H.-J. Güntherodt, R. Lüthi, R. Schlittler, J.K. Gimzewski: Forces with submolecular resolution between the probing tip and Cu-TBPP molecules on Cu(100) observed with a combined AFM/STM, Appl. Phys. A **72**, S105–S108 (2001).
117. L.M. Eng, M. Bammerlin, C. Loppacher, M. Guggisberg, R. Bennewitz, R. Lüthi, E. Meyer, H.-J. Güntherodt: Surface morphology, chemical contrast, and ferroelectric domains in TGS bulk single crystals differentiated with UHV non-contact force microscopy, Appl. Surf. Sci. **140**, 253–258 (1999).
118. S. Kitamura, K. Suzuki, M. Iwatsuki: High resolution imaging of contact potential difference using a novel ultrahigh vacuum non-contact atomic force microscope technique, Appl. Surf. Sci. **140**, 265–270 (1999).
119. H. Yamada, T. Fukuma, K. Umeda, K. Kobayashi, K. Matsushige: Local structures and electrical properties of organic molecular films investigated by non-contact atomic force microscopy, Appl. Surf. Sci. **188**, 391–398 (2000).
120. K. Fukui, Y. Iwasawa: Fluctuation of acetate ions in the $(2\times1)$-acetate overlayer on $TiO_2(110)$-$(1\times1)$ observed by noncontact atomic force microscopy, Surf. Sci. **464**, L719–L726 (2000).
121. A. Sasahara, H. Uetsuka, H. Onishi: Singlemolecule analysis by non-contact atomic force microscopy, J. Phys. Chem. B **105**, 1–4 (2001).
122. A. Sasahara, H. Uetsuka, H. Onishi: NC-AFM topography of HCOO and $CH_3COO$ molecules co-adsorbed on $TiO_2(110)$, Appl. Phys. A **72**, S101–S103 (2001).
123. A. Sasahara, H. Uetsuka, H. Onishi: Image topography of alkyl-substituted carboxylates observed by noncontact atomic force microscopy, Surf. Sci. **481**, L437–L442 (2001).
124. A. Sasahara, H. Uetsuka, H. Onishi: Noncontact atomic force microscope topography dependent on permanent dipole of individual molecules, Phys. Rev. B **64**, 121406 (2001).
125. A. Sasahara, H. Uetsuka, T. Ishibashi, H. Onishi: A needle-like organic molecule imaged by noncontact atomic force microscopy, Appl. Surf. Sci. **188**, 265–271 (2002).
126. H. Onishi, A. Sasahara, H. Uetsuka, T. Ishibashi: Molecule-dependent topography determined by noncontact atomic force microscopy: Carboxylates on $TiO_2(110)$, Appl. Surf. Sci. **188**, 257–264 (2002).
127. H. Onishi: Carboxylates adsorbed on $TiO_2(110)$, in *Chemistry of Nano-molecular Systems*, ed. by T. Nakamura (Springer, Berlin/Heidelberg, 2002) pp.75–89.
128. S. Thevuthasan, G.S. Herman, Y.J. Kim, S.A. Chambers, C.H.F. Peden, Z. Wang, R.X. Ynzunza, E.D. Tober, J. Morais, C.S. Fadley: The structure of formate on $TiO_2(110)$ by scanned-energy and scanned-angle photoelectron diffraction, Surf. Sci. **401**, 261–268 (1998).
129. H. Uetsuka, A. Sasahara, A. Yamakata, H. Onishi: Microscopic identification of a bimolecular reaction intermediate, J. Phys. Chem. B **106**, 11549–11552 (2002).
130. D.R. Lide: *Handbook of Chemistry and Physics*, 81st edn. (CRC, Boca Raton, 2000).
131. K. Kobayashi, H. Yamada, K. Matsushige: Dynamic force microscopy using FM detection in various environments, Appl. Surf. Sci. **188**, 430–434 (2002).

# Chapter 6
# Low-Temperature Scanning Probe Microscopy

**Markus Morgenstern, Alexander Schwarz, and Udo D. Schwarz**

**Abstract** This chapter is dedicated to scanning probe microscopy (SPM) operated at cryogenic temperatures, where the more fundamental aspects of phenomena important in the field of nanotechnology can be investigated with high sensitivity under well-defined conditions. In general, scanning probe techniques allow the measurement of physical properties down to the nanometer scale. Some techniques, such as scanning tunneling microscopy and scanning force microscopy, even go down to the atomic scale. Various properties are accessible. Most importantly, one can image the arrangement of atoms on conducting surfaces by scanning tunneling microscopy and on insulating substrates by scanning force microscopy. However, the arrangement of electrons (scanning tunneling spectroscopy), the force interaction between different atoms (scanning force spectroscopy), magnetic domains (magnetic force microscopy), the local capacitance (scanning capacitance microscopy), the local temperature (scanning thermo microscopy), and local light-induced excitations (scanning near-field microscopy) can also be measured with high spatial resolution. In addition, some techniques even allow the manipulation of atomic configurations.

Probably the most important advantage of the low-temperature operation of scanning probe techniques is that they lead to a significantly better signal-to-noise ratio than measuring at room temperature. This is why many researchers work below 100 K. However, there are also physical reasons to use low-temperature equipment. For example, the manipulation of atoms or scanning tunneling spectroscopy with high energy resolution can only be realized at low temperatures. Moreover, some physical effects such as superconductivity or the Kondo effect are restricted to low temperatures. Here, we describe the design criteria of low-temperature scanning probe equipment and summarize some of the most spectacular results achieved since the invention of the method about 30 years ago. We first focus on the scanning tunneling microscope, giving examples of atomic manipulation and the analysis of electronic properties in different material arrangements. Afterwards, we describe results obtained by scanning force microscopy, showing atomic-scale imaging on insulators, as well as force spectroscopy analysis. Finally, the magnetic force microscope, which images domain patterns in ferromagnets and vortex patterns in superconductors, is discussed. Although this list is far from

B. Bhushan (ed.), *Nanotribology and Nanomechanics*,
DOI 10.1007/978-3-642-15283-2_6, © Springer-Verlag Berlin Heidelberg 2011

complete, we feel that it gives an adequate impression of the fascinating possibilities of low-temperature scanning probe instruments.

In this chapter low temperatures are defined as lower than about 100 K and are normally achieved by cooling with liquid nitrogen or liquid helium. Applications in which SPMs are operated close to 0 °C are not covered in this chapter.

Nearly three decades ago, the first design of an experimental setup was presented where a sharp tip was systematically scanned over a sample surface in order to obtain local information on the tip–sample interaction down to the atomic scale. This original instrument used the tunneling current between a conducting tip and a conducting sample as a feedback signal and was thus named the *scanning tunneling microscope* [1]. Soon after this historic breakthrough, it became widely recognized that virtually any type of tip–sample interaction could be used to obtain local information on the sample by applying the same general principle, provided that the selected interaction was reasonably short-ranged. Thus, a whole variety of new methods has been introduced, which are denoted collectively as *scanning probe methods*. An overview is given, e.g., by Wiesendanger [2].

The various methods, especially the above mentioned scanning tunneling microscopy (STM) and scanning force microscopy (SFM) – which is often further classified into subdisciplines such as topography-reflecting atomic force microscopy (AFM), magnetic force microscopy (MFM) or electrostatic force microscopy (EFM) – have been established as standard methods for surface characterization on the nanometer scale. The reason is that they feature extremely high resolution (often down to the atomic scale for STM and AFM), despite a principally simple, compact, and comparatively inexpensive design.

A side-effect of the simple working principle and the compact design of many scanning probe microscopes (SPMs) is that they can be adapted to different environments such as air, all kinds of gaseous atmospheres, liquids or vacuum with reasonable effort. Another advantage is their ability to work within a wide temperature range. A microscope operation at higher temperatures is chosen to study surface diffusion, surface reactivity, surface reconstructions that only manifest at elevated temperatures, high-temperature phase transitions, or to simulate conditions as they occur, e.g., in engines, catalytic converters or reactors. Ultimately, the upper limit for the operation of an SPM is determined by the stability of the sample, but thermal drift, which limits the ability to move the tip in a controlled manner over the sample, as well as the depolarization temperature of the piezoelectric positioning elements might further restrict successful measurements.

On the other hand, low-temperature (LT) application of SPMs is much more widespread than operation at high temperatures. Essentially five reasons make researchers adapt their experimental setups to low-temperature compatibility. These are: (1) the reduced thermal drift, (2) lower noise levels, (3) enhanced stability of tip and sample, (4) the reduction in piezo hysteresis/creep, and (5) probably the most obvious, the fact that many physical effects are restricted to low temperature. Reasons 1–4 only apply unconditionally if the whole microscope body is kept at low temperature (typically in or attached to a bath cryostat, see Sect. 6.2). Setups in which only the sample is cooled may show considerably less favorable operating characteristics. As a result of 1–4, ultrahigh resolution and long-term stability can be achieved on a level that significantly exceeds what can be accomplished at room temperature even under the most favorable circumstances. Typical examples of effect 5 are superconductivity [3] and the Kondo effect [4].

## 6.1 Microscope Operation at Low Temperatures

Nevertheless, before we devote ourselves to a short overview of experimental LT-SPM work, we will take a closer look at the specifics of microscope operation at low temperatures, including a discussion of the corresponding instrumentation.

### *6.1.1 Drift*

Thermal drift originates from thermally activated movements of the individual atoms, which are reflected by the thermal expansion coefficient. At room temperature, typical values for solids are on the order of (1–50) $\times 10^{-6}$ $K^{-1}$. If the temperature could be kept precisely constant, any thermal drift would vanish, regardless of the absolute temperature of the system. The close coupling of the microscope to a large temperature bath that maintains a constant temperature ensures a significant reduction in thermal drift and allows for distortion-free long-term measurements. Microscopes that are efficiently attached to sufficiently large bath cryostats, therefore, show a one- to two-order-of-magnitude increase in thermal stability compared with nonstabilized setups operated at room temperature.

A second effect also helps suppress thermally induced drift of the probing tip relative to a specific location on the sample surface. The thermal expansion coefficients at liquid-helium temperatures are two or more orders of magnitude smaller than at room temperature. Consequently, the thermal drift during low-temperature operation decreases accordingly.

For some specific scanning probe methods, there may be additional ways in which a change in temperature can affect the quality of the data. In *frequency-modulation SFM* (FM-SFM), for example, the measurement principle relies on the accurate determination of the eigenfrequency of the cantilever, which is determined by its spring constant and its effective mass. However, the spring constant changes with temperature due to both thermal expansion (i.e., the resulting change in the cantilever dimensions) and the variation of the Young's modulus with temperature. Assuming drift rates of about 2 mK/min, as is typical for room-temperature measurements, this effect might have a significant influence on the obtained data.

### *6.1.2 Noise*

The theoretically achievable resolution in SPM often increases with decreasing temperature due to a decrease in thermally induced noise. An example is the thermal noise in SFM, which is proportional to the square root of the temperature [5, 6]. Lowering the temperature from $T = 300$ K to $T = 10$ K thus results in a reduction of the thermal frequency noise by more than a factor of five. Graphite, e.g., has been imaged with atomic resolution only at low temperatures due to its extremely low corrugation, which was below the room-temperature noise level [7, 8].

Another, even more striking, example is the spectroscopic resolution in *scanning tunneling spectroscopy* (STS). This depends linearly on the temperature [2] and is consequently reduced even more at LT than the thermal noise in AFM. This provides the opportunity to study structures or physical effects not accessible at room temperature such as spin and Landau levels in semiconductors [9].

Finally, it might be worth mentioning that the enhanced stiffness of most materials at low temperatures (increased Young's modulus) leads to a reduced coupling to external noise. Even though this effect is considered small [6], it should not be ignored.

### *6.1.3 Stability*

There are two major stability issues that considerably improve at low temperature. First, low temperatures close to the temperature of liquid helium inhibit most of the thermally activated diffusion processes. As a consequence, the sample surfaces show a significantly increased long-term stability, since defect motion or adatom diffusion is massively suppressed. Most strikingly, even single xenon atoms deposited on suitable substrates can be successfully imaged [10, 11] or even manipulated [12]. In the same way, low temperatures also stabilize the atomic configuration at the tip end by preventing sudden jumps of the most loosely bound, foremost tip atom(s). Secondly, the large cryostat that usually surrounds the microscope acts as an effective cryo-pump. Thus samples can be kept clean for several weeks, which is a multiple of the corresponding time at room temperature (about 3–4 h).

### *6.1.4 Piezo Relaxation and Hysteresis*

The last important benefit from low-temperature operation of SPMs is that artifacts from the response of the piezoelectric scanners are substantially reduced. After applying a voltage ramp to one electrode of a piezoelectric scanner, its immediate initial deflection, $l_0$, is followed by a much slower relaxation, $\Delta l$, with a logarithmic time dependence. This effect, known as piezo relaxation or *creep*, diminishes substantially at low temperatures, typically by a factor of ten or more.

As a consequence, piezo nonlinearities and piezo hysteresis decrease accordingly. Additional information is given by Hug et al. [13].

## 6.2 Instrumentation

The two main design criteria for all vacuum-based scanning probe microscope systems are: (1) to provide an efficient decoupling of the microscope from the vacuum system and other sources of external vibrations, and (2) to avoid most internal noise sources through the high mechanical rigidity of the microscope body itself. In vacuum systems designed for low-temperature applications, a significant degree of complexity is added, since, on the one hand, close thermal contact of the SPM and cryogen is necessary to ensure the (approximately) drift-free conditions described above, while, on the other hand, good vibration isolation (both from the outside world, as well as from the boiling or flowing cryogen) has to be maintained.

Plenty of microscope designs have been presented in the last 10–15 years, predominantly in the field of STM. Due to the variety of the different approaches, we will, somewhat arbitrarily, give two examples at different levels of complexity that might serve as illustrative model designs.

### *6.2.1 A Simple Design for a Variable-Temperature STM*

A simple design for a variable-temperature STM system is presented in Fig. 6.1; similar systems are also offered by Omicron (Germany) or Jeol (Japan). It should give an impression of what the minimum requirements are, if samples are to be investigated successfully at low temperatures. It features a single ultrahigh-vacuum (UHV) chamber that houses the microscope in its center. The general idea to keep the setup simple is that only the sample is cooled, by means of a flow cryostat that ends in the small liquid-nitrogen (LN) reservoir. This reservoir is connected to the sample holder with copper braids. The role of the copper braids is to attach the LN reservoir thermally to the sample located on the sample holder in an effective manner, while vibrations due to the flow of the cryogen should be blocked as much as possible. In this way, a sample temperature of about 100 K is reached. Alternatively, with liquid-helium operation, a base temperature of below 30 K can be achieved, while a heater that is integrated into the sample stage enables high-temperature operation up to 1,000 K.

A typical experiment would run as follows. First, the sample is brought into the system by placing it in the so-called *load-lock*. This small part of the chamber can be separated from the rest of the system by a valve, so that the main part of the system can remain under vacuum at all times (i.e., even if the load-lock is opened to introduce the sample). After vacuum is reestablished, the sample is transferred to the main chamber using the transfer arm. A linear-motion feedthrough enables the

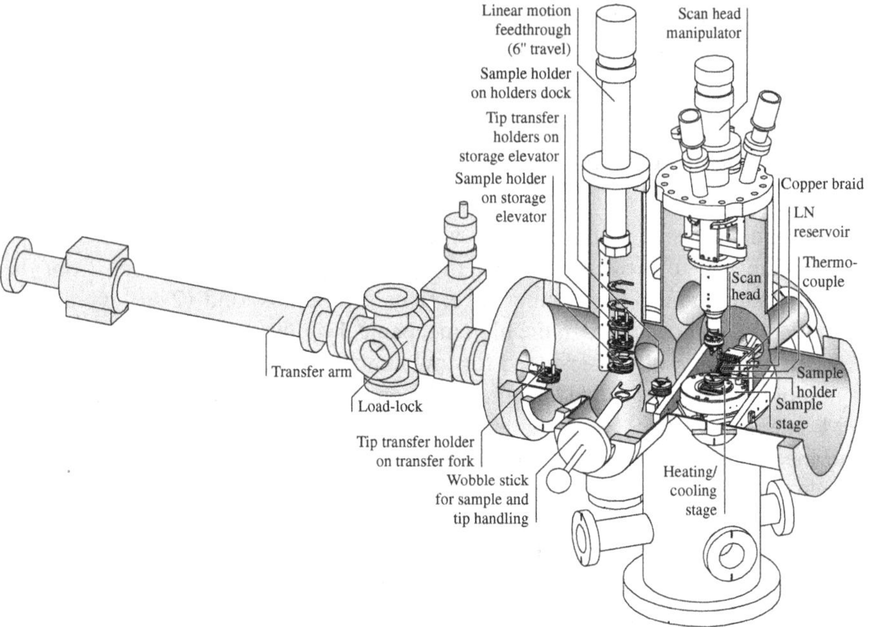


**Fig. 6.1** One-chamber UHV system with variable-temperature STM based on a flow cryostat design. (© RHK Technology, USA)

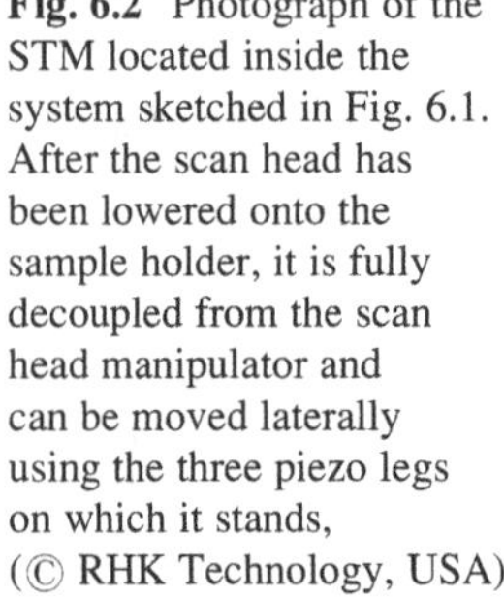

**Fig. 6.2** Photograph of the STM located inside the system sketched in Fig. 6.1. After the scan head has been lowered onto the sample holder, it is fully decoupled from the scan head manipulator and can be moved laterally using the three piezo legs on which it stands, (© RHK Technology, USA)

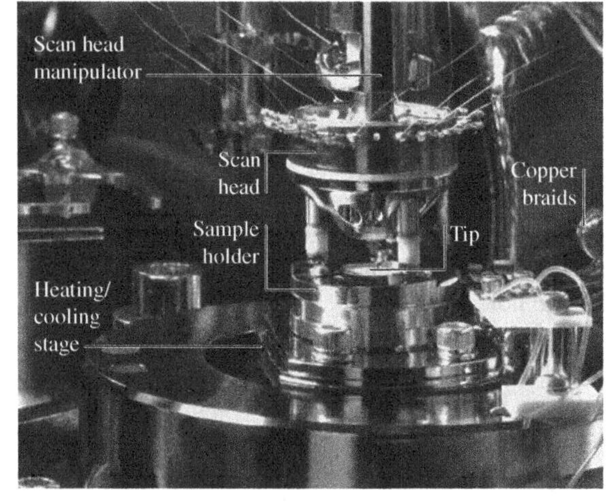


storage of sample holders or, alternatively, specialized holders that carry replacement tips for the STM. Extending the transfer arm further, the sample can be placed on the sample stage and subsequently cooled down to the desired temperature. The scan head, which carries the STM tip, is then lowered with the scan-head manipulator onto the sample holder (see Fig. 6.2). The special design of the scan head (see [14] for details) allows not only flexible positioning of the tip on any desired location on the sample surface but also compensates to a certain degree for the thermal drift that inevitably occurs in such a design due to temperature gradients.

In fact, thermal drift is often much more prominent in LT-SPM designs, where only the sample is cooled, than in room-temperature designs. Therefore, to benefit fully from the high-stability conditions described in the introduction, it is mandatory to keep the whole microscope at the exact same temperature. This is mostly realized by using bath cryostats, which add a certain degree of complexity.

### 6.2.2 A Low-Temperature SFM Based on a Bath Cryostat

As an example of an LT-SPM setup based on a bath cryostat, let us take a closer look at the LT-SFM system sketched in Fig. 6.3, which has been used to acquire the

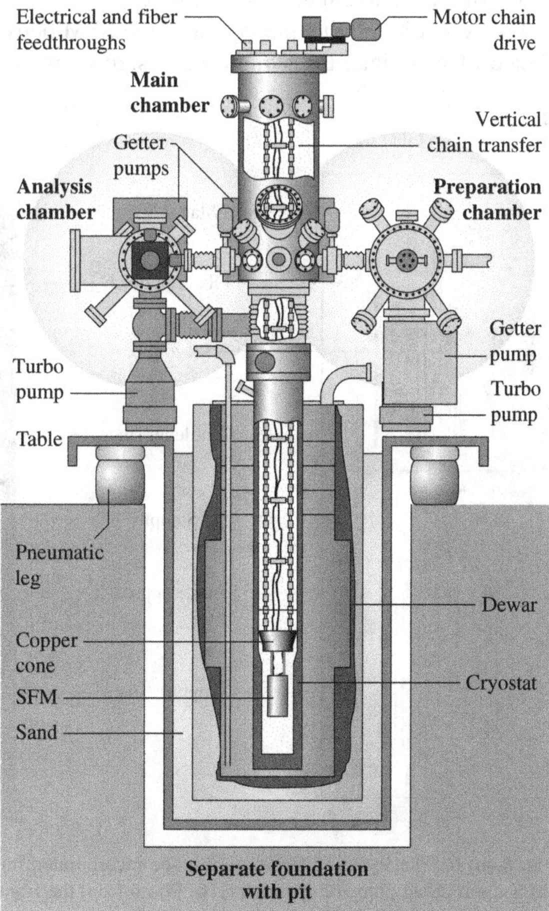


**Fig. 6.3** Three-chamber UHV and bath cryostat system for scanning force microscopy, front view

images on graphite, xenon, NiO, and InAs presented in Sect. 6.4. The force microscope is built into a UHV system that comprises three vacuum chambers: one for cantilever and sample preparation, which also serves as a transfer chamber, one for analysis purposes, and a main chamber that houses the microscope. A specially designed vertical transfer mechanism based on a double chain allows the lowering of the microscope into a UHV-compatible bath cryostat attached underneath the main chamber. To damp the system, it is mounted on a table carried by pneumatic damping legs, which, in turn, stand on a separate foundation to decouple it from building vibrations. The cryostat and dewar are separated from the rest of the UHV system by a bellow. In addition, the dewar is surrounded by sand for acoustic isolation.

In this design, tip and sample are exchanged at room temperature in the main chamber. After the transfer into the cryostat, the SFM can be cooled by either liquid nitrogen or liquid helium, reaching temperatures down to 10 K. An all-fiber interferometer as the detection mechanism for the cantilever deflection ensures high resolution, while simultaneously allowing the construction of a comparatively small, rigid, and symmetric microscope.

Figure 6.4 highlights the layout of the SFM body itself. Along with the careful choice of materials, the symmetric design eliminates most of the problems with

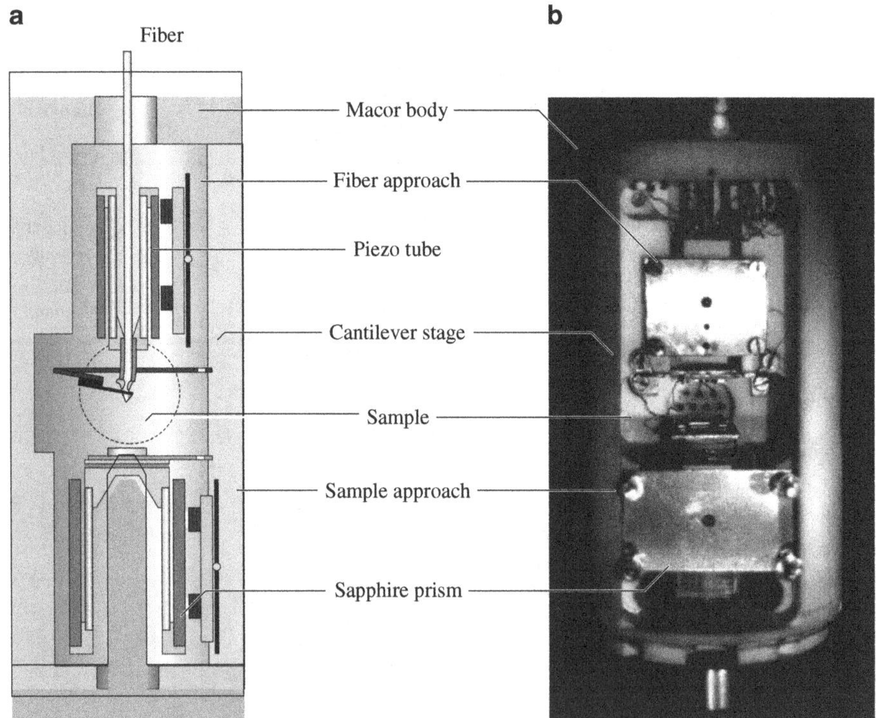


**Fig. 6.4a, b** The scanning force microscope incorporated into the system presented in Fig. 6.3. (**a**) Section along plane of symmetry. (**b**) Photo from the front

drift inside the microscope encountered when cooling or warming it up. The microscope body has an overall cylindrical shape with a height of 13 cm and a diameter of 6 cm and exact mirror symmetry along the cantilever axis. The main body is made of a single block of macor, a machinable glass ceramic, which ensures a rigid and stable design. For most of the metallic parts titanium was used, which has a temperature coefficient similar to macor. The controlled but stable accomplishment of movements, such as coarse approach and lateral positioning in other microscope designs, is a difficult task at low temperatures. The present design uses a special type of piezo motor that moves a sapphire prism (see the *fiber approach* and the *sample approach* labels in Fig. 6.4); it is described in detail in [15]. More information regarding this design is given in [16].

## 6.3 Scanning Tunneling Microscopy and Spectroscopy

In this section, we review some of the most important results achieved by LT-STM. After summarizing the results, placing emphasis on the necessity for LT equipment, we turn to the details of the different experiments and the physical meaning of the results obtained.

As described in Sect. 6.1, the LT equipment has basically three advantages for scanning tunneling microscopy (STM) and spectroscopy (STS): First, the instruments are much more stable with respect to thermal drift and coupling to external noise, allowing the establishment of new functionalities of the instrument. In particular, the LT-STM has been used to move atoms on a surface [12], cut molecules into pieces [17], reform bonds [18], charge individual atoms [19], and, consequently, establish new structures on the nanometer scale. Also, the detection of light resulting from tunneling into a particular molecule [20, 21], the visualization of thermally induced atomic movements [22], and the detection of hysteresis curves of individual atoms [23] require LT instrumentation.

Second, the spectroscopic resolution in STS depends linearly on temperature and is, therefore, considerably reduced at LT. This provides the opportunity to study physical effects inaccessible at room temperature. Examples are the resolution of spin and Landau levels in semiconductors [9], or the investigation of lifetime-broadening effects on the nanometer scale [24]. Also the imaging of distinct electronic wavefunctions in real space requires LT-STM [25]. More recently, vibrational levels, spin-flip excitations, and phonons have been detected with high spatial resolution at LT using the additional inelastic tunneling channel [26, 27, 28].

Third, many physical effects, in particular, effects guided by electronic correlations, are restricted to low temperature. Typical examples are superconductivity [3], the Kondo effect [4], and many of the electron phases found in semiconductors [29]. Here, LT-STM provides the possibility to study electronic effects on a local scale, and intensive work has been done in this field, the most elaborate with respect to high-temperature superconductivity [30, 31, 32].

## 6.3.1 Atomic Manipulation

Although manipulation of surfaces on the atomic scale can be achieved at room temperature [33, 34], only the use of LT-STM allows the placement of individual atoms at desired atomic positions [35]. The main reason is that rotation, diffusion or charge transfer of entities could be excited at higher temperature, making the intentionally produced configurations unstable.

The usual technique to manipulate atoms is to increase the current above a certain atom, which reduces the tip–atom distance, then to move the tip with the atom to a desired position, and finally to reduce the current again in order to decouple the atom and tip. The first demonstration of this technique was performed by Eigler and Schweizer [12], who used Xe atoms on a Ni(110) surface to write the three letters "IBM" (their employer) on the atomic scale (Fig. 6.5a).

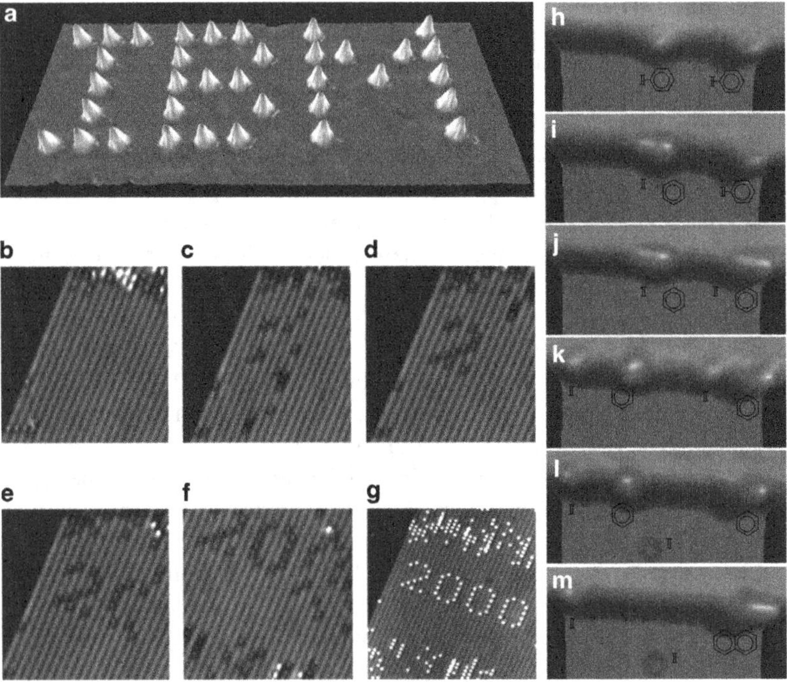


**Fig. 6.5** (**a**) STM image of single Xe atoms positioned on a Ni(110) surface in order to realize the letters "IBM" on the atomic scale (© D. Eigler, Almaden); (**b–f**) STM images recorded after different positioning processes of CO molecules on a Cu(110) surface; (**g**) final artwork greeting the new millennium on the atomic scale ((**b–g**) © G. Meyer, Zürich). (**h–m**) Synthesis of biphenyl from two iodobenzene molecules on Cu(111): First, iodine is abstracted from both molecules (**i, j**); then the iodine between the two phenyl groups is removed from the step (**k**), and finally one of the phenyls is slid along the Cu step (**l**) until it reacts with the other phenyl (**m**); the line drawings symbolize the actual status of the molecules ((**h–m**) © S. W. Hla and K. H. Rieder, Berlin)

Nowadays, many laboratories are able to move different kinds of atoms and molecules on different surfaces with high precision. An example featuring CO molecules on Cu(110) is shown in Fig. 6.5b–g. Even more complex structures than the "2,000", such as cascades of CO molecules that by mutual repulsive interaction mimic different kinds of logic gates, have been assembled and their functionality tested [36]. Although these devices are slow and restricted to low temperature, they nicely demonstrate the high degree of control achieved on the atomic scale.

The basic modes of controlled motion of atoms and molecules by the tip are pushing, pulling, and sliding. The selection of the particular mode depends on the tunneling current, i.e., the distance between tip and molecule, as well as on the particular molecule–substrate combination [37]. It has been shown experimentally that the potential landscape for the adsorbate movement is modified by the presence of the tip [38, 39] and that excitations induced by the tunneling current can trigger atomic or molecular motion [40, 41]. Other sources of motion are the electric field between tip and molecule or electromigration caused by the high current density [35]. The required lateral tip force for atomic motion has been measured for typical adsorbate–substrate combinations to be $\approx 0.1$ nN [42]. Other types of manipulation on the atomic scale are feasible. Some of them require a selective inelastic tunneling into vibrational or rotational modes of the molecules [43]. This leads to controlled desorption [44], diffusion [45], molecular rotation [46, 47], conformational change [48] or even controlled pick-up of molecules by the tip [18]. Dissociation can be achieved by voltage pulses [17] inducing local heating, even if the pulse is applied at distances of 100 nm away from the molecule [49]. Also, association of individual molecules [18, 50, 51, 52] can require voltage pulses in order to overcome local energy barriers. The process of controlled bond formation can even be used for doping of single $C_{60}$ molecules by up to four potassium atoms [53]. As an example of controlled manipulation, Fig. 6.5h–m shows the production of biphenyl from two iodobenzene molecules [54]. The iodine is abstracted by voltage pulses (Fig. 6.5i, j), then the iodine is moved to the terrace by the pulling mode (Fig. 6.5k, l), and finally the two phenyl parts are slid along the step edge until they are close enough to react (Fig. 6.5m). The chemical identification of the product is not deduced straightforwardly and partly requires detailed vibrational STM spectroscopy (see below and [55]).

Finally, also the charge state of a single atom or molecule can be manipulated, tested, and read out. A Au atom has been switched reversibly between two charge states using an insulating thin film as the substrate [19]. In addition, the carrier capture rate of a single impurity level within the bandgap of a semiconductor has been quantified [56], and the point conductance of a single atom has been measured and turned out to be a reproducible quantity [57]. These promising results might trigger a novel electronic field of manipulation of matter on the atomic scale, which is tightly related to the currently very popular field of molecular electronics.

### 6.3.2 Imaging Atomic Motion

Since individual manipulation processes last seconds to minutes, they probably cannot be used to manufacture large and repetitive structures. A possibility to construct such structures is self-assembled growth [58]. This partly relies on the temperature dependence of different diffusion processes on the surface. Detailed knowledge of the diffusion parameters is required, which can be deduced from sequences of STM images measured at temperatures close to the onset of the process [59]. Since many diffusion processes have their onset at LT, LT are partly required [22]. Consecutive images of so-called hexa-*tert*-butyl-decacyclene (HtBDC) molecules on Cu(110) recorded at $T = 194$ K are shown in Fig. 6.6a–c [60].

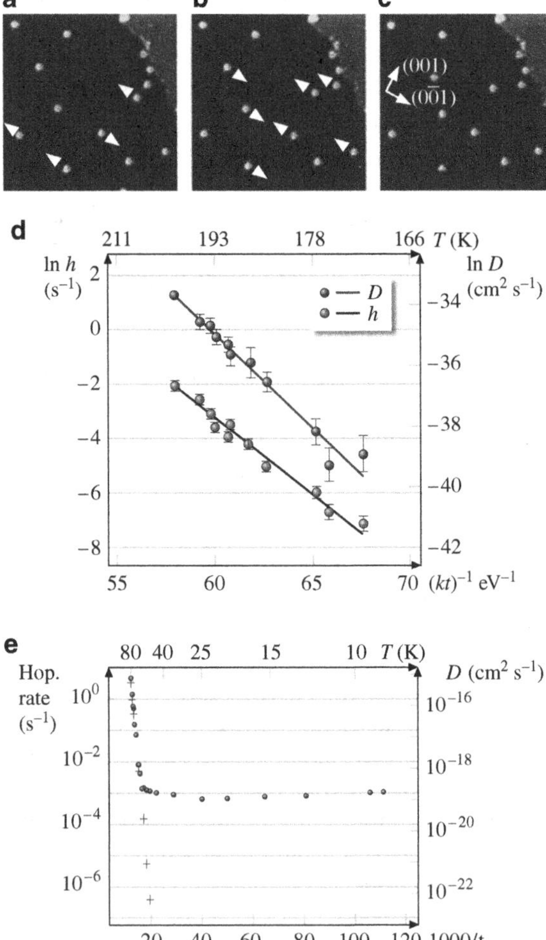


**Fig. 6.6** **(a–c)** Consecutive STM images of hexa-*tert*-butyl decacyclene molecules on Cu(110) imaged at $T = 194$ K; *arrows* indicate the direction of motion of the molecules between two images. **(d)** Arrhenius plot of the hopping rate *h* determined from images such as **(a–c)** as a function of inverse temperature (*grey symbols*); the *brown symbols* show the corresponding diffusion constant *D*; *lines* are fit results revealing an energy barrier of 570 meV for molecular diffusion (**(a–d)** © M. Schunack and F. Besenbacher, Aarhus). **(e)** Arrhenius plot for D (*crosses*) and H (*circles*) on Cu(001). The constant hopping rate of H below 65 K indicates a nonthermal diffusion process, probably tunneling (© W. Ho, Irvine)

As indicated by the arrows, the positions of the molecules change with time, implying diffusion. Diffusion parameters are obtained from Arrhenius plots of the determined hopping rate *h*, as shown in Fig. 6.6d. Of course, one must make sure that the diffusion process is not influenced by the presence of the tip, since it is known from manipulation experiments that the presence of the tip can move a molecule. However, particularly at low tunneling voltages, these conditions can be fulfilled.

Besides the determination of diffusion parameters, studies of the diffusion of individual molecules showed the importance of mutual interactions in diffusion, which can lead to concerted motion of several molecules [22], directional motion where smaller molecules carry larger ones [61] or, very interestingly, the influence of quantum tunneling [62]. The latter is deduced from the Arrhenius plot of hopping rates of H and D on Cu(001), as shown in Fig. 6.6e. The hopping rate of H levels off at about 65 K, while the hopping rate of the heavier D atom goes down to nearly zero, as expected from thermally induced hopping. Quantum tunneling has surprisingly also been found for vertical Sn displacements within a Sn adsorbate layer on Si(111) [63].

Other diffusion processes investigated by LT-STM include the movement of surface vacancies [64] or bulk interstitials close to the surface [65], the Brownian motion of vacancy islands [66] as well as laser-induced diffusion distinct from thermally excited diffusion [67].

### 6.3.3 *Detecting Light from Single Atoms and Molecules*

It had already been realized in 1988 that STM experiments are accompanied by light emission [68]. The fact that molecular resolution in the light intensity was achieved at LT (Fig. 6.7a, b) [20] raised the hope of performing quasi-optical experiments on the molecular scale. Meanwhile, it is clear that the basic emission process observed on metals is the decay of a local plasmon induced in the area around the tip by inelastic tunneling processes [69, 70]. Thus, the molecular resolution is basically a change in the plasmon environment, largely given by the increased height of the tip with respect to the surface above the molecule [71]. However, the electron can, in principle, also decay via single-particle excitations. Indeed, signatures of single-particle levels have been observed for a Na monolayer on Cu(111) [72] as well as for Ag adatom chains on NiAl(110) [21]. As shown in Fig. 6.7c, the peaks of differential photon yield $dY/dV$ as a function of applied bias $V$ are at identical voltages to the peaks in $dI/dV$ intensity. This is evidence that the density of states of the Ag adsorbates is responsible for the radiative decay. Photon emission spectra displaying much more details could be detected by depositing the adsorbates of interest on a thin insulating film [73, 74]. Figure 6.7d shows spectra of ZnEtiol deposited on a 0.5 nm-thick $Al_2O_3$ layer on NiAl(110). Importantly, the peaks within the light spectra do not shift with applied voltage, ruling out that they are due to a plasmon mode induced by the tip. As shown in Fig. 6.7e, the photon

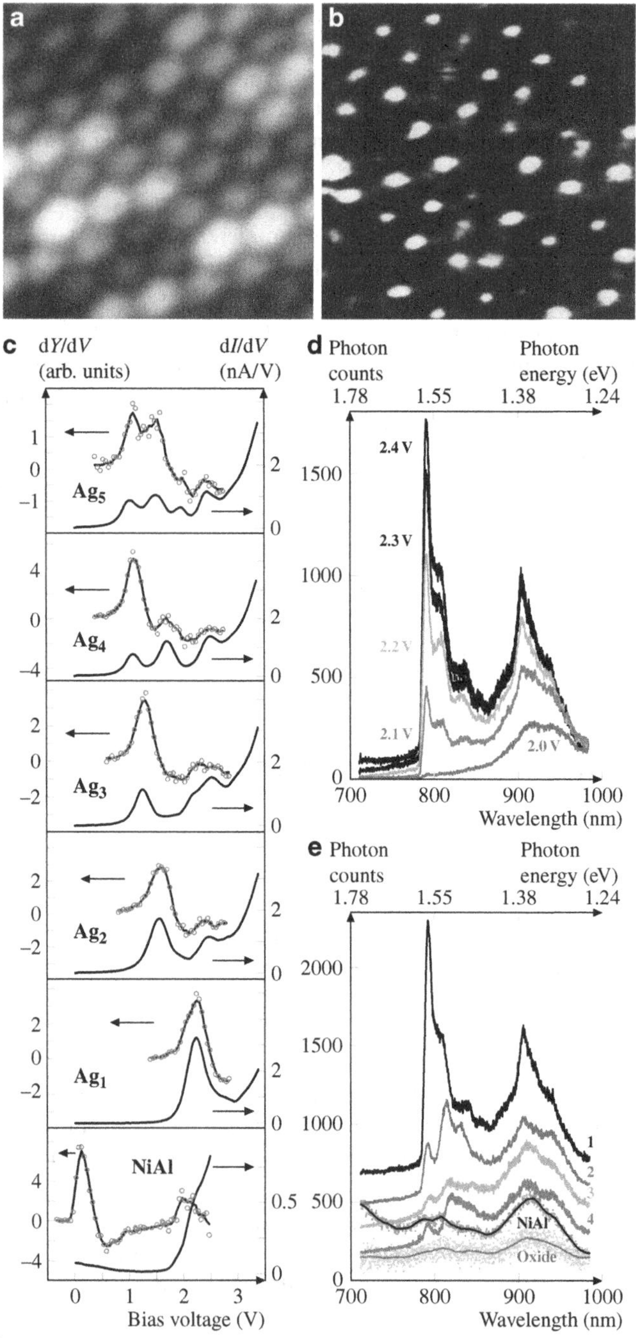


**Fig. 6.7** (**a**) STM image of $C_{60}$ molecules on Au(110) imaged at $T = 50$ K. (**b**) STM-induced photon intensity map of the same area; all photons from 1.5 to 2.8 eV contribute to the image,

spectra show distinct variations by changing the position within the molecule, demonstrating that atomically resolved maps of the excitation probability can be measured by STM.

Meanwhile, external laser light has also been coupled to the tunneling contact between the STM tip and a molecule deposited on an insulating film. A magnesium porphine molecule positioned below the tip could be charged reversibly either by increasing the voltage of the tip or by increasing the photon energy of the laser. This indicates selective absorption of light energy by the molecule leading to population of a novel charge level by tunneling electrons [75], a result that raises the hope that STM can probe photochemistry on the atomic scale. STM-induced light has also been detected from semiconductors [76], including heterostructures [77]. This light is again caused by single-particle relaxation of injected electrons, but without contrast on the atomic scale.

### 6.3.4 High-Resolution Spectroscopy

One of the most important modes of LT-STM is STS, which detects the differential conductivity $\mathrm{d}I/\mathrm{d}V$ as a function of the applied voltage $V$ and the position $(x, y)$. The $\mathrm{d}I/\mathrm{d}V$ signal is basically proportional to the local density of states (LDOS) of the sample, the sum over squared single-particle wavefunctions $\Psi_i$ [2]

$$\begin{aligned}\frac{\mathrm{d}I}{\mathrm{d}V}(V,x,y) &\propto LDOS(E,x,y)\\ &= \sum_{\Delta E} |\Psi_i(E,x,y)|^2,\end{aligned} \tag{6.1}$$

where $\Delta E$ is the energy resolution of the experiment. In simple terms, each state corresponds to a tunneling channel, if it is located between the Fermi levels ($E_\mathrm{F}$) of the tip and the sample. Thus, all states located in this energy interval contribute to $I$, while $\mathrm{d}I/\mathrm{d}V(V)$ detects only the states at the energy $E$ corresponding to $V$. The local intensity of each channel depends further on the LDOS of the state at the corresponding surface position and its decay length into vacuum. For s-like tip states, Tersoff and Hamann have shown that it is simply proportional to

**Fig. 6.7** (continued) tunneling voltage $V$ =−2.8 V (© R. Berndt, Kiel **(a, b)**). **(c)** Photon yield spectroscopy d$Y$/d$V$($V$) obtained above Ag chains ($Ag_n$) of different length consisting of $n$ atoms. For comparison, the differential conductivity d$I$/d$V$($V$) is also shown. The Ag chains are deposited on NiAl(110). The photon yield $Y$ is integrated over the spectral range from 750 to 775 nm. **(d)** Photon yield spectra $Y(E)$ measured at different tip voltages as indicated. The tip is positioned above a ZnEtiol molecule deposited on $Al_2O_3$/NiAl(110). Note that the *peaks* in $Y(E)$ do not shift with applied tip voltage; **(e)** $Y(E)$ spectra determined at different positions above the ZnEtiol molecule, $V$ =2.35 V, $I$ =0.5 nA. ((**c–e**) © W. Ho, Irvine)

the LDOS at the position of the tip [78]. Therefore, as long as the decay length is spatially constant, one measures the LDOS at the surface (6.1). Note that the contributing states are not only surface states, but also bulk states. However, surface states usually dominate if present. *Chen* has shown that higher orbital tip states lead to the so-called derivation rule [79]: $p_z$-type tip states detect $d(\mathrm{LDOS})/dz$, $d_{z^2}$-states detect $d^2(\mathrm{LDOS})/dz^2$, and so on. As long as the decay into vacuum is exponential and spatially constant, this leads only to an additional, spatially constant factor in $dI/dV$. Thus, it is still the LDOS that is measured (6.1). The requirement of a spatially constant decay is usually fulfilled on larger length scales, but not on the atomic scale [79]. There, states located close to the atoms show a stronger decay into vacuum than the less localized states in the interstitial region. This effect can lead to STS corrugations that are larger than the real LDOS corrugations [80].

The voltage dependence of $dI/dV$ is sensitive to a changing decay length with $V$, which increases with $V$. This influence can be reduced at higher $V$ by displaying $dI/dV/(I/V)$ [81]. Additionally, $dI/dV(V)$ curves might be influenced by possible structures in the DOS of the tip, which also contributes to the number of tunneling channels [82]. However, these structures can usually be identified, and only tips free of characteristic DOS structures are used for quantitative experiments.

Importantly, the energy resolution $\Delta E$ is largely determined by temperature. It is defined as the smallest energy distance of two $\delta$-peaks in the LDOS that can still be resolved as two individual peaks in $dI/dV(V)$ curves and is $\Delta E = 3.3\ k_B T$ [2]. The temperature dependence is nicely demonstrated in Fig. 6.8, where the tunneling gap of the superconductor Nb is measured at different temperatures [83]. The peaks at

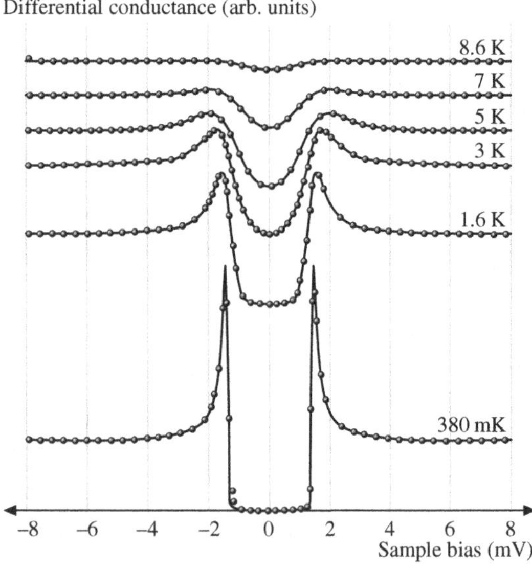


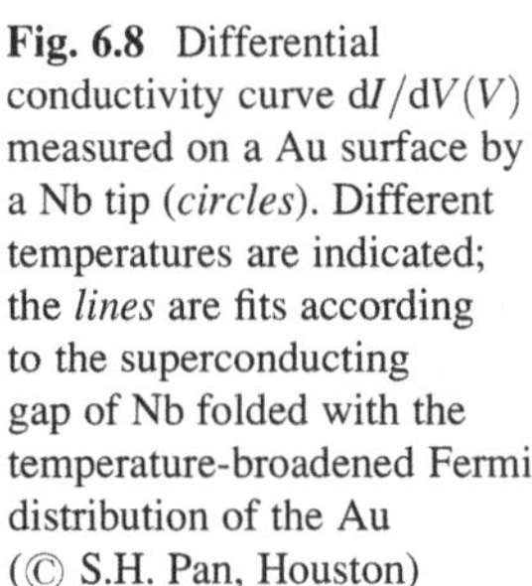
**Fig. 6.8** Differential conductivity curve $dI/dV(V)$ measured on a Au surface by a Nb tip (*circles*). Different temperatures are indicated; the *lines* are fits according to the superconducting gap of Nb folded with the temperature-broadened Fermi distribution of the Au (© S.H. Pan, Houston)

the rim of the gap get wider at temperatures well below the critical temperature of the superconductor ($T_c$ =9.2 K).

### Lifetime Broadening

Besides $\Delta E$, intrinsic properties of the sample lead to a broadening of spectroscopic features. Basically, the finite lifetime of the electron or hole in the corresponding state broadens its energetic width. Any kind of interaction such as electron–electron interaction can be responsible. Lifetime broadening has usually been measured by photoemission spectroscopy (PES), but it turned out that lifetimes of surface states on noble-metal surfaces determined by STS (Fig. 6.9a, b) are up to a factor of three larger than those measured by PES [84]. The reason is probably that defects broaden the PES spectrum. Defects are unavoidable in a spatially integrating technique such as PES, thus STS has the advantage of choosing a particularly clean area for lifetime measurements. The STS results can be successfully compared with theory, highlighting the dominating influence of intraband transitions for the surface-state lifetime on Au(111) and Cu(111), at least close to the onset of the surface band [24].

With respect to band electrons, the analysis of the width of the band onset on $dI/dV(V)$ curves has the disadvantage of being restricted to the onset energy. Another method circumvents this problem by measuring the decay of standing electron waves scattered from a step edge as a function of energy [85]. Figure 6.9c,d shows the resulting oscillating $dI/dV$ signal measured for two different energies. To deduce the coherence length $L_\Phi$, which is inversely proportional to the lifetime $\tau_\Phi$, one has to consider that the finite energy resolution $\Delta E$ in the experiment also leads to a decay of the standing wave away from the step edge. The dotted fit line using $L_\Phi = \infty$ indicates this effect and, more importantly, shows a discrepancy from the measured curve. Only including a finite coherence length of 6.2 nm results in good agreement, which in turn determines $L_\Phi$ and thus $\tau_\Phi$, as displayed in Fig. 6.9c. The found $1/E^2$ dependence of $\tau_\Phi$ points to a dominating influence of electron–electron interactions at higher energies in the surface band.

### Landau and Spin Levels

Moreover, the increased energy resolution at LT allows the resolution of electronic states that are not resolvable at room temperature (RT); for example, Landau and spin quantization appearing in a magnetic field $B$ have been probed on InAs(110) [9, 86]. The corresponding quantization energies are given by $E_{\mathrm{Landau}} = \hbar eB/m_{\mathrm{eff}}$ and $E_{\mathrm{spin}} = g\mu B$. Thus InAs is a good choice, since it exhibits a low effective mass $m_{\mathrm{eff}}/m_e = 0.023$ and a high $g$-factor of 14 in the bulk conduction band. The values in metals are $m_{\mathrm{eff}}/m_e \approx 1$ and $g \approx 2$, resulting in energy splittings of only 1.25 and

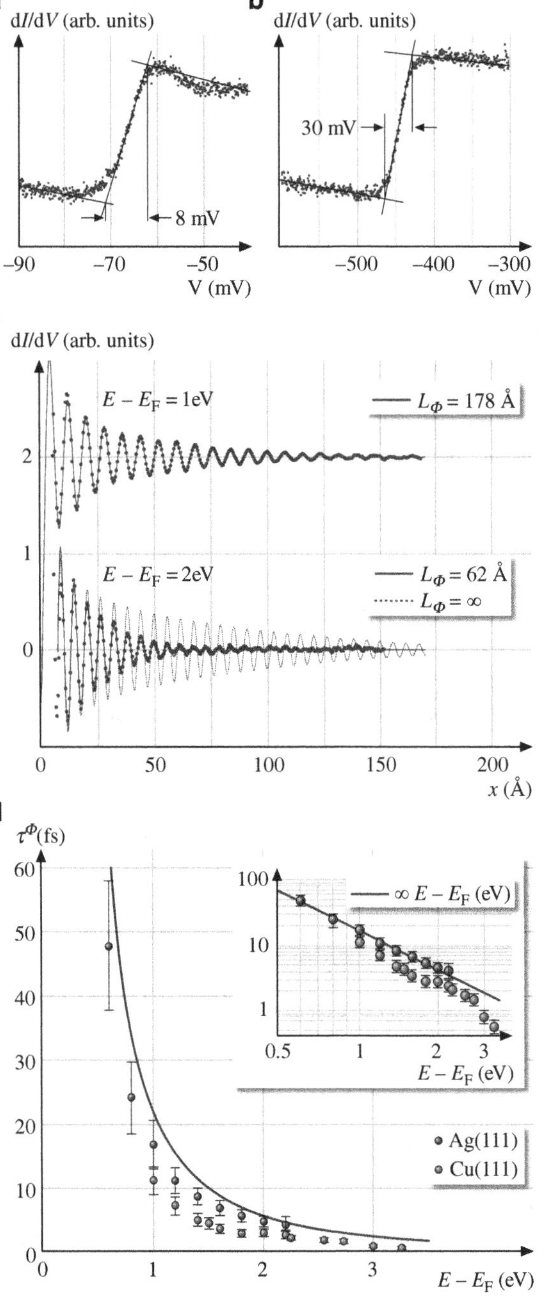


**Fig. 6.9** **(a, b)** Spatially averaged $\mathrm{d}I/\mathrm{d}V(V)$ curves of Ag(111) and Cu(111); both surfaces exhibit a surface state with parabolic dispersion, starting at −65 and −430 meV, respectively. The *lines* are drawn to determine the energetic width of the onset of these surface bands (**(a, b)** © R. Berndt, Kiel); **(c)** $\mathrm{d}I/\mathrm{d}V$ intensity as a function of position away from a step edge of Cu(111) measured at the voltages ($E - E_\mathrm{F}$), as indicated (*points*); the *lines* are fits assuming standing electron waves with a phase coherence length $L_\Phi$ as marked; **(d)** resulting phase coherence time as a function of energy for Ag(111) and Cu(111). *Inset* shows the same data on a double-logarithmic scale, evidencing the $E^{-2}$ dependence (*line*) (**(c, d)** © H. Brune, Lausanne)

1.2 meV at $B = 10$ T. This is obviously lower than the typical lifetime broadenings discussed in the previous section and also close to $\Delta E = 1.1$ meV achievable at $T = 4$ K.

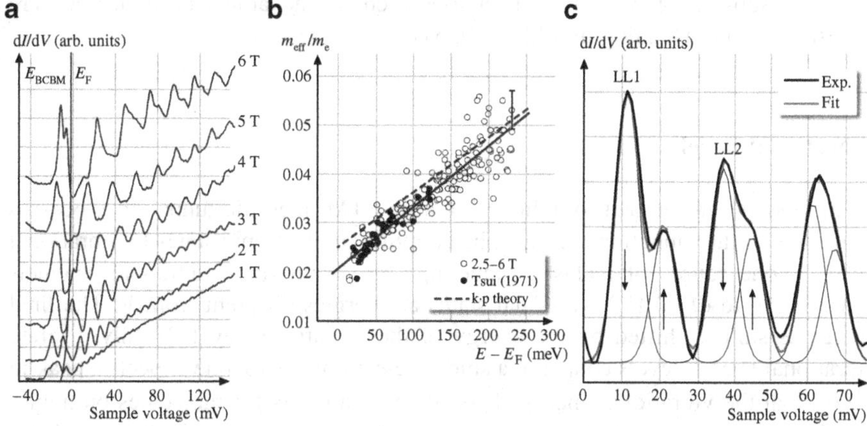


**Fig. 6.10** (**a**) $dI/dV$ curves of $n$-InAs(110) at different magnetic fields, as indicated; $E_{BCBM}$ marks the bulk conduction band minimum; oscillations above $E_{BCBM}$ are caused by Landau quantization; the *double peaks* at $B = 6$ T are caused by spin quantization. (**b**) Effective-mass data deduced from the distance of adjacent Landau peaks $\Delta E$ according to $\Delta E = heB/m_{eff}$ (*open symbols*); *filled symbols* are data from planar tunnel junctions (Tsui), the *solid line* is a mean-square fit of the data and the *dashed line* is the expected effective mass of InAs according to $\boldsymbol{k} \cdot \boldsymbol{p}$ theory. (**c**) Magnification of a $dI/dV$ curve at $B = 6$ T, exhibiting spin splitting; the Gaussian curves marked by *arrows* are the fitted spin levels

Fortunately, the electron density in doped semiconductors is much lower, and thus the lifetime increases significantly. Figure 6.10a shows a set of spectroscopy curves obtained on InAs(110) in different magnetic fields [9]. Above $E_F$, oscillations with increasing intensity and energy distance are observed. They show the separation expected from Landau quantization. In turn, they can be used to deduce $m_{eff}$ from the peak separation (Fig. 6.10b). An increase of $m_{eff}$ with increasing $E$ has been found, as expected from theory. Also, at high fields, spin quantization is observed (Fig. 6.10c). It is larger than expected from the bare $g$-factor due to contributions from exchange enhancement [87].

### Atomic Energy Levels

Another opportunity at LT is to study electronic states and resonances of single adatoms. A complicated resonance is the Kondo resonance described below. A simpler resonance is a surface state bound at the adatom potential. It appears as a spatially localized peak below the onset of the extended surface state (Fig. 6.9a) [88, 89]. A similar resonance caused by a mixing of bulk states of the NiAl(110) substrate with atomic Au levels has been used to detect exchange splitting in Au dimers as a function of interatomic distance [90]. Single magnetic adatoms on the same surface also exhibit a double-peak resonance, but here due to the influence of spin-split d-levels of the adsorbate [91]. Atomic and molecular states decoupled from the substrate have finally been observed, if the atoms or molecules are deposited on an insulating thin film [19, 51].

### Vibrational Levels

As discussed with respect to light emission in STM, inelastic tunneling processes contribute to the tunneling current. The coupling of electronic states to vibrational levels is one source of inelastic tunneling [26]. It provides additional channels contributing to $dI/dV(V)$ with final states at energies different from $V$. The final energy is simply shifted by the energy of the vibrational level. If only discrete vibrational energy levels couple to a smooth electronic DOS, one expects a peak in $d^2I/dV^2$ at the vibrational energy. This situation appears for molecules on noble-metal surfaces. As usual, the isotope effect can be used to verify the vibrational origin of the peak. First indications of vibrational levels have been found for $H_2O$ and $D_2O$ on $TiO_2$ [92], and completely convincing work has been performed for $C_2H_2$ and $C_2D_2$ on Cu(001) [26] (Fig. 6.11a). The technique has been used to identify individual molecules on the surface by their characteristic vibrational levels [55]. Moreover, the orientation of complexes with respect to the surface can be determined to a certain extent, since the vibrational excitation depends on the position of the tunneling current within the molecule. Finally, the excitation of certain molecular levels can induce such corresponding motions as hopping [45], rotation [47] (Fig. 6.11b–e) or desorption [44], leading to additional possibilities for manipulation on the atomic scale.

In turn, the manipulation efficiency as a function of applied voltage can be used to identify vibrational energies within the molecule, even if they are not detectable directly by $d^2I/dV^2$ spectroscopy [93]. Multiple vibronic excitations are found by positioning the molecule on an insulating film, leading to the observation of equidistant peaks in $d^2I/dV^2(V)$ [94].

### Other Inelastic Excitations

The tunneling current can not only couple to vibrational modes of molecules, but also to other degrees of freedom. It has been shown that phonon modes can be observed in carbon nanotubes [95, 96], on graphite [97], and on metal surfaces [28]. One finds distinct dependencies of excitation probability on the position of the STM tip with respect to the investigated structure.

First indications for the $d^2I/dV^2(V)$-based detection of extended magnons [98] and plasmons [97] have also been published.

The inelastic tunneling current has, moreover, been used to study single spin-flip excitations in magnetic field for atoms and atomic assemblies deposited on a thin insulator. The excitation probability was high enough to observe the spin flip even as a step in $dI/dV$ instead of as a peak in $d^2I/dV^2$. Figure 6.12a shows the $dI/dV$

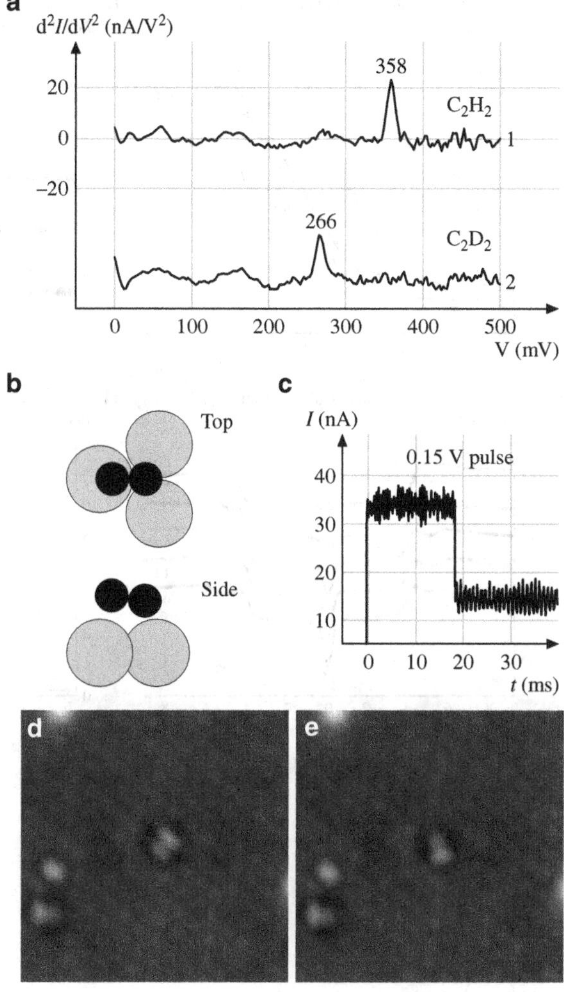


**Fig. 6.11** **(a)** $d^2I/dV^2$ curves taken above a $C_2H_2$ and a $C_2D_2$ molecule on Cu(100); the *peaks* correspond to the C–H or C–D stretch-mode energy of the molecule, respectively. **(b)** Sketch of $O_2$ molecule on Pt(111). **(c)** Tunneling current above an $O_2$ molecule on Pt(111) during a voltage pulse of 0.15 V; the jump in current indicates rotation of the molecule. **(d,e)** STM images of an $O_2$ molecule on Pt(111) ($V$ =0.05 V), prior to and after rotation induced by a voltage pulse to 0.15 V ((**a–e**) © W. Ho, Irvine)

curves recorded above a single Mn atom on $Al_2O_3$/NiAl(110) at different $B$ fields. The linear shift of the step with $B$ field is obvious, and the step voltage can be fitted by $eV = g\mu_B B$ with $\mu_B$ being the Bohr magneton and a reasonable $g$-factor of $g \approx 2$ [27]. Figure 6.12b shows the d$I$/d$V$ spectra obtained on a Mn dimer embedded in CuN/Cu(100). A step is already visible at $B$ =0 T, splitting into three steps at higher field. This result can be explained straightforwardly, as sketched in the inset, by a singlet–triplet transition of the combined two spins (coupled by an exchange energy of $J \approx 6$ meV). Investigating longer chains revealed an even–odd asymmetry, i.e., chains consisting of 2, 4, 6, ... atoms exhibit a singlet–triplet transition, while chains of 1, 3, 5, ... atoms exhibit a transition from $S = 5/2$ to $S = 3/2$. This indicates antiferromagnetic coupling within the chain [99]. Figure 6.12c shows

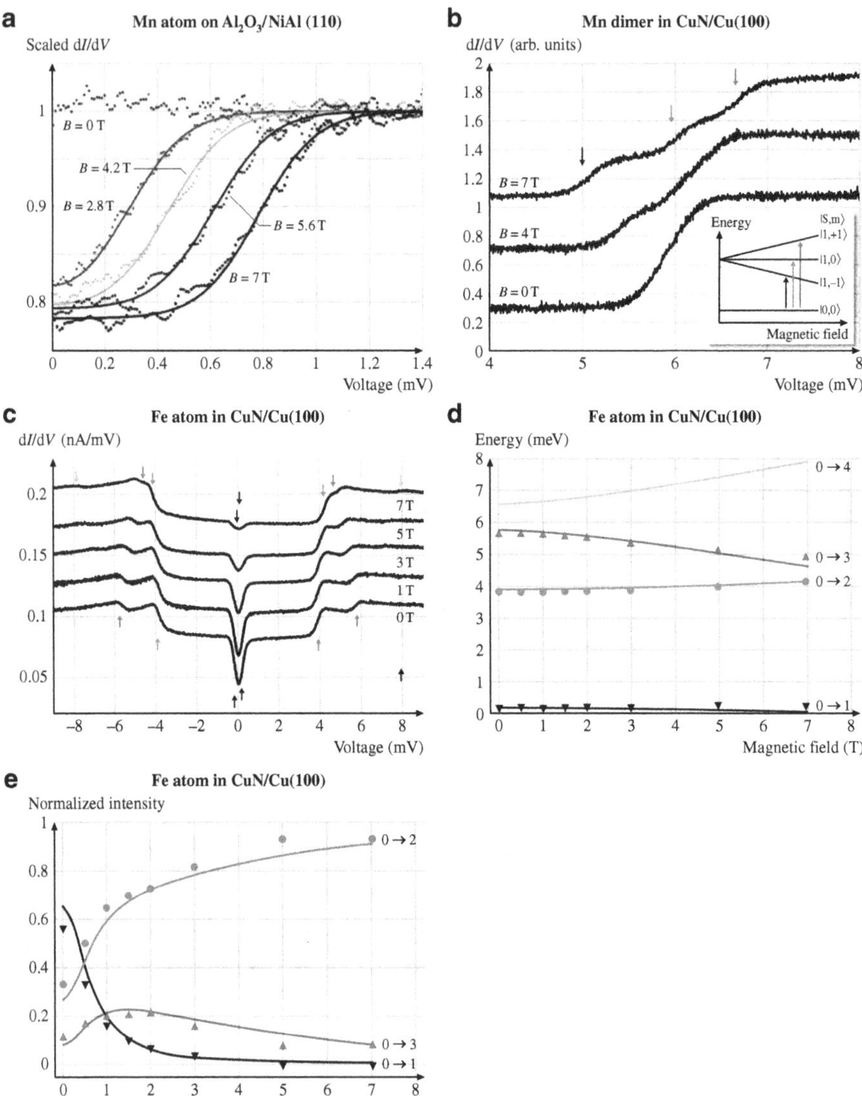


**Fig. 6.12** **(a)** d*I*/d*V* curves taken above a single Mn atom deposited on $Al_2O_3$/NiAl(110) at different magnetic fields as indicated; **(b)** d*I*/d*V* curves taken above a Mn dimer deposited onto CuN/Cu(100) at different magnetic fields as indicated; *inset* shows the three possible spin-flip transitions between singlet and triplet; **(c)** d*I*/d*V* curves taken above a single Fe atom deposited onto CuN/Cu(100) at different magnetic fields as indicated; transitions are marked by *arrows*; **(d,e)** energy and intensity of the steps in d*I*/d*V* measured with magnetic field along the direction of the N rows of the CuN surface (*symbols*) in comparison with calculated results (*lines*) (© A. Heinrich, C. F. Hirjibehedin, Almaden)

spectra of a single Fe atom embedded within CuN. The spectrum reveals several steps already at $B = 0$ T, showing that different spin orientations $S_z$ must exhibit different energies due to magnetic anisotropy. In order to determine the anisotropy, the step energies and intensities were measured at different magnetic fields applied in three different directions. Amazingly, the results could be fitted completely by a single model with an out-of-plane anisotropy of $D = -1.55$ meV and an in-plane anisotropy of $E = 0.31$ meV. Therefore, one has to assume five different spin states of the Fe being mixtures of the five $S_z$ states of a total Fe spin of $|S| = 2$. The excellent fit is shown for energy and intensity of a particular $B$-field direction in Fig. 6.12d, e [100].

The different experiments of inelastic tunneling demonstrate that details of atomic excitations in a solid environment can be probed by LT-STM, even if they are not of primary electronic origin. This might be a highly productive method in the near future. A complementary novel approach to inelastic effects might be the recently developed radiofrequency STM [101], which could give access to low-energy excitations, such as GHz spin-wave modes in nanostructures, which are not resolvable by $\mathrm{d}^2I/\mathrm{d}V^2$ at LT.

### Kondo Resonance

A rather intricate interaction effect is the Kondo effect. It results from a second-order scattering process between itinerate states and a localized state [102]. The two states exchange some degree of freedom back and forth, leading to a divergence of the scattering probability at the Fermi level of the itinerate state. Due to the divergence, the effect strongly modifies sample properties. For example, it leads to an unexpected increase in resistance with decreasing temperature for metals containing magnetic impurities [4]. Here, the exchanged degree of freedom is the spin. A spectroscopic signature of the Kondo effect is a narrow peak in the DOS at the Fermi level, continuously disappearing above a characteristic temperature (the Kondo temperature). STS provides the opportunity to study this effect on the local scale [103, 104].

Figure 6.13a–d shows an example of Co clusters deposited on a carbon nanotube [105]. While only a small dip at the Fermi level, probably caused by curvature influences on the $\pi$-orbitals, is observed without Co (Fig. 6.13b) [106], a strong peak is found around a Co cluster deposited on top of the tube (Co cluster is marked in Fig. 6.13a). The peak is slightly shifted with respect to $V = 0$ mV due to the so-called Fano resonance [107], which results from interference of the tunneling processes into the localized Co level and the itinerant nanotube levels. The resonance disappears within several nanometers of the cluster, as shown in Fig. 6.13d.

The Kondo effect has also been detected for different magnetic atoms deposited on noble-metal surfaces [103, 104]. There, it disappears at about 1 nm from the magnetic impurity, and the effect of the Fano resonance is more pronounced, contributing to dips in $\mathrm{d}I/\mathrm{d}V(V)$ curves instead of peaks. Detailed investigations show that the d-level occupation of the adsorbate [108] as well as the surface charge

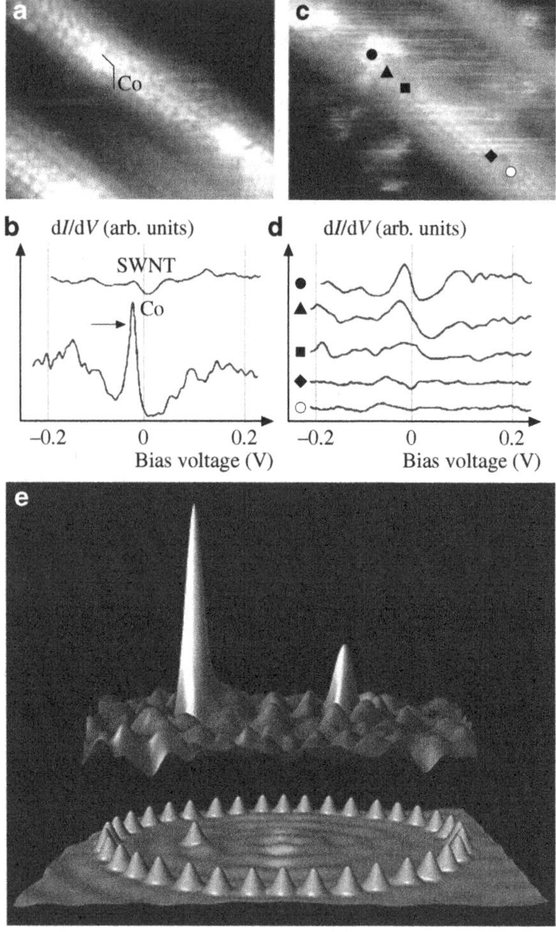


**Fig. 6.13** (**a**) STM image of a Co cluster on a single-wall carbon nanotube (SWNT). (**b**) d$I$/d$V$ curves taken directly above the Co cluster (Co) and far away from the Co cluster (SWNT); the *arrow* marks the Kondo peak. (**c**) STM image of another Co cluster on a SWNT with *symbols* marking the positions where the d$I$/d$V$ curves displayed in (**d**) are taken. (**d**) d$I$/d$V$ curves taken at the positions marked in (**c**) ((**a–d**) © C. Lieber, Cambridge). (**e**) *Lower part:* STM image of a quantum corral of elliptic shape made from Co atoms on Cu(111); one Co atom is placed at one of the foci of the ellipse. *Upper part:* map of the strength of the Kondo signal in the corral; note that there is also a Kondo signal at the focus that is not covered by a Co atom ((**e**) © D. Eigler, Almaden)

density [109, 110] matter for the Kondo temperature. Exchange interaction between adsorbates tunable by their mutual distance can be used to tune the Kondo temperature [111] or even to destroy the Kondo resonance completely [112]. Meanwhile, magnetic molecules have also been shown to exhibit Kondo resonances. This increases the tunability of the Kondo effect, e.g., by the selection of adequate ligands surrounding the localized spins [113, 114], by distant association of other molecules [115] or by conformational changes within the molecule [116].

A fascinating experiment has been performed by Manoharan et al. [117], who used manipulation to form an elliptic cage for the surface states of Cu(111) (Fig. 6.13e, bottom). This cage was constructed to have a quantized level at $E_{\mathrm{F}}$. Then, a cobalt atom was placed in one focus of the elliptic cage, producing a Kondo resonance. Surprisingly, the same resonance reappeared in the opposite focus, but not away from the focus (Fig. 6.13e, top). This shows amazingly that complex local effects such as the Kondo resonance can be wave-guided to remote points.

### 6.3.5 Imaging Electronic Wavefunctions

#### Bloch Waves

Since STS measures the sum of squared wavefunctions (6.1), it is an obvious task to measure the local appearance of the most simple wavefunctions in solids, namely Bloch waves. The atomically periodic part of the Bloch wave is always measured if atomic resolution is achieved (inset of Fig. 6.15a). However, the long-range wavy part requires the presence of scatterers. The electron wave impinges on the scatterer and is reflected, leading to self-interference. In other words, the phase of the Bloch wave becomes fixed by the scatterer.

Such self-interference patterns were first found on graphite(0001) [118] and later on noble-metal surfaces, where adsorbates or step edges scatter the surface states (Fig. 6.14a) [25]. Fourier transforms of the real-space images reveal the $k$-space distribution of the corresponding states [119], which may include additional contributions besides the surface state [120]. Using particular geometries such as so-called quantum corrals, the Bloch waves can be confined (Fig. 6.14b). Depending on the geometry of the corral, the result state looks rather complex, but it can usually be reproduced by simple calculations involving single-particle states only [121].

Meanwhile, Bloch waves in semiconductors scattered at charged dopants (Fig. 6.14c, d) [122], Bloch states confined in semiconducting or organic quantum dots (Fig. 6.14e–g) [123, 124, 125], and quantum wells [126], as well as Bloch waves confined in short-cut carbon nanotubes (Fig. 6.14h, i) [127, 128] have been visualized. In special nanostructures, it was even possible to extract the phase of the wavefunction by using the mathematically known transformation matrices of so-called isospectral structures, i.e., geometrically different structures exhibiting exactly the same spatially averaged density of states. The resulting wavefunctions $\Psi(x)$ are shown in Fig. 6.14j [129].

More localized structures, where a Bloch wave description is not appropriate, have been imaged, too. Examples are the highest occupied molecular orbital (HOMO) and lowest unoccupied molecular orbital (LUMO) of pentacene molecules deposited on NaCl/Cu(100) (Fig. 6.14k, l) [51], the different molecular states of $C_{60}$ on Ag(110) (Fig. 6.3m–o) [130], the anisotropic states of Mn acceptors in a semiconducting host [131, 132], and the hybridized states developing within short monoatomic Au chains, which develop particular states at the end of the chains [133, 134]. Using pairs of remote Mn acceptors, even symmetric and antisymmetric pair wavefunctions have been imaged in real space [135].

The central requirements for a detailed imaging of wavefunctions are LT for an appropriate energetic distinction of an individual state, adequate decoupling of the state from the substrate in order to decrease lifetime-induced broadening effects, and, partly, the selection of a system with an increased Bohr radius in order to increase the spatial extension of details above the lateral resolution of STM, thereby improving, e.g., the visibility of bonding and antibonding pair states within a dimer [135].

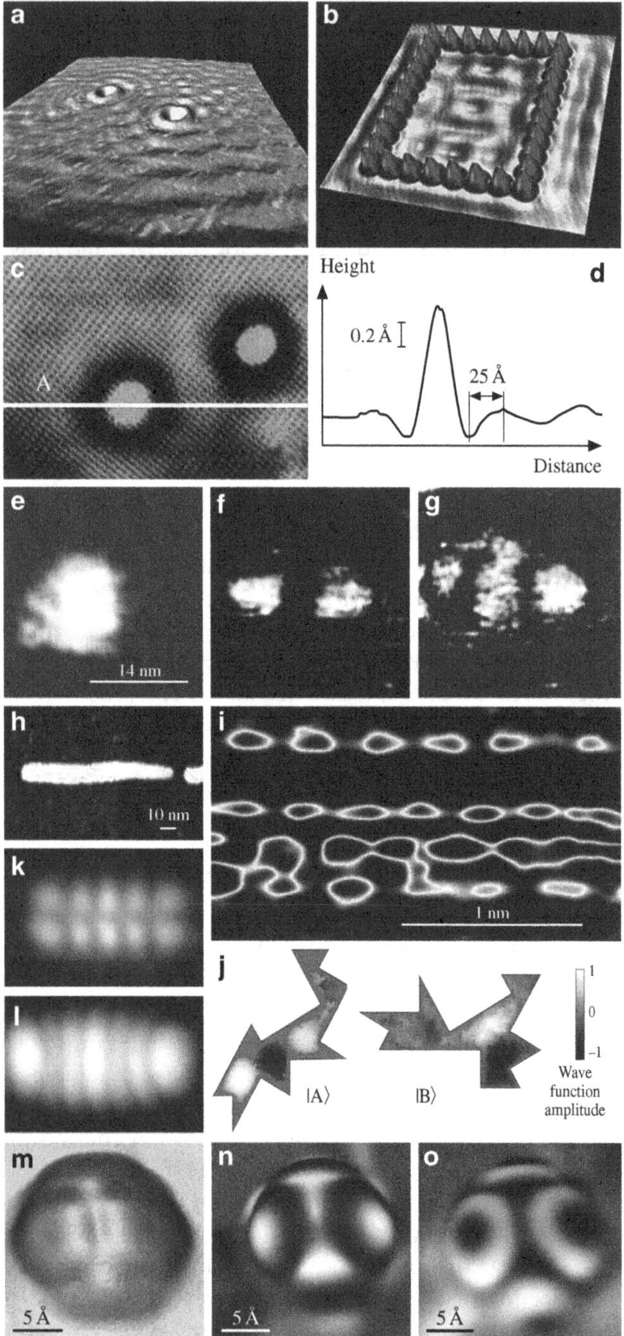


**Fig. 6.14** (**a**) Low-voltage STM image of Cu(111) including two defect atoms; the waves are electronic Bloch waves scattered at the defects; (**b**) low-voltage STM image of a rectangular

## Wavefunctions in Disordered Systems

More complex wavefunctions result from interactions. A nice playground to study such interactions is doped semiconductors. The reduced electron density with respect to metals increases the importance of electron interactions with potential disorder and other electrons. Applying a magnetic field quenches the kinetic energy, thus enhancing the importance of interactions. A dramatic effect can be observed on InAs(110), where three-dimensional (3-D) bulk states are measured. While the usual scattering states around individual dopants are observed at $B = 0$ T (Fig. 6.15a) [136], stripe structures are found at high magnetic field (Fig. 6.15b) [137]. They run along equipotential lines of the disorder potential. This can be understood by recalling that the electron tries to move in a cyclotron circle, which becomes a cycloid path along an equipotential line within an inhomogeneous electrostatic potential [138].

The same effect has been found in two-dimensional (2-D) electron systems (2-DES) of InAs at the same large $B$-field (Fig. 6.15d) [139]. However the scattering states at $B = 0$ T are much more complex in 2-D (Fig. 6.15c) [140]. The reason is the tendency of a 2-DES to exhibit closed scattering paths [141]. Consequently, the self-interference does not result from scattering at individual scatterers, but from complicated self-interference paths involving many scatterers. Nevertheless, the wavefunction pattern can be reproduced by including these effects within the calculations.

Reducing the dimensionality to one dimension (1-D) leads again to complicated self-interference patterns due to the interaction of the electrons with several impurities [142, 143]. For InAs, they can be reproduced by single-particle

**Fig. 6.14** (continued) quantum corral made from single atoms on Cu(111); the pattern inside the corral is the confined state of the corral close to $E_F$; (© D. Eigler, Almaden **(a, b)**); **(c)** STM image of GaAs(110) around a Si donor, $V = -2.5$ V; the line scan along A, shown in **(d)**, exhibits an additional oscillation around the donor caused by a standing Bloch wave; the grid-like pattern corresponds to the atomic corrugation of the Bloch wave (© H. van Kempen, Nijmegen **(c, d)**); **(e–g)** d$I$/d$V$ images of a self-assembled InAs quantum dot deposited on GaAs(100) and measured at different $V$ (**(e)** 1.05 V, **(f)** 1.39 V, **(g)** 1.60 V). The *images* show the squared wavefunctions confined within the quantum dot, which exhibit zero, one, and two nodal lines with increasing energy. **(h)** STM image of a short-cut carbon nanotube; **(i)** greyscale plot of the d$I$/d$V$ intensity inside the short-cut nanotube as a function of position ($x$-axis) and tunneling voltage ($y$-axis); four wavy patterns of different wavelength are visible in the voltage range from − 0.1 to 0.15 V (© C. Dekker, Delft **(h, i)**); **(j)** two reconstructed wavefunctions confined in so-called isospectral corrals made of CO molecules on Cu(111). Note that $\Psi(x)$ instead of $|\Psi(x)|^2$ is displayed, exhibiting positive and negative values. This is possible since the transplantation matrix transforming one isospectral wavefunction into another is known (© H. Manoharan, Stanford **(j)**); **(k, l)** STM images of a pentacene molecule deposited on NaCl/Cu(100) and measured with a pentacene molecule at the apex of the tip at $V = -2.5$ V (**(k)**, HOMO = highest occupied molecular orbital) and $V = 2.5$ V (**(l)**, LUMO = lowest unoccupied molecular orbital) (© J. Repp, Regensburg **(k, l)**); **(m)** STM image of a $C_{60}$ molecule deposited on Ag(100), $V = 2.0$ V; **(n, o)** d$I$/d$V$ images of the same molecule at $V = 0.4$ V **(n)**, 1.6 V **(n)** (**(m–o)**) © M. Crommie, Berkeley

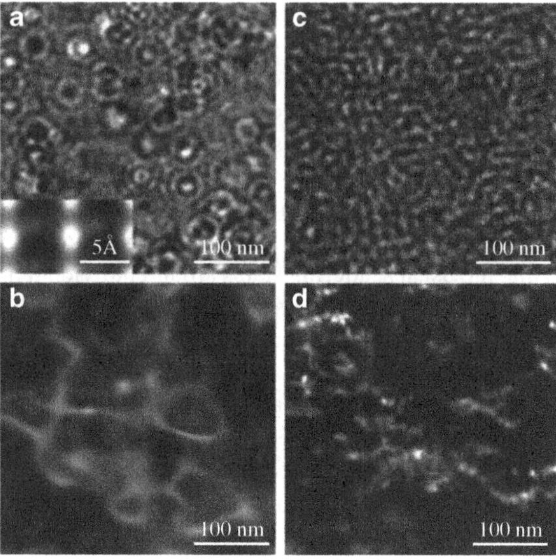


**Fig. 6.15** (**a**) $dI/dV$ image of InAs(110) at $V = 50$ mV, $B = 0$ T; circular wave patterns corresponding to standing Bloch waves around each sulphur donor are visible; *inset* shows a magnification revealing the atomically periodic part of the Bloch wave. (**b**) Same as (**a**), but at $B = 6$ T; the stripe structures are drift states. (**c**) $dI/dV$ image of a 2-D electron system on InAs(110) induced by the deposition of Fe, $B = 0$ T. (**d**) Same as (**c**) but at $B = 6$ T; note that the contrast in (**a**) is increased by a factor of ten with respect to (**b–d**)

calculations. However, experiments imaging self-interference patterns close to the end of a C-nanotube are interpreted as indications of spin-charge separation, a genuine property of 1-D electrons not feasible within the single-particle description [144].

## Charge Density Waves, Jahn–Teller Distortion

Another interaction modifying the LDOS is the electron–phonon interaction. Phonons scatter electrons between different Fermi points. If the wavevectors connecting Fermi points exhibit a preferential orientation, a so-called Peierls instability occurs [145]. The corresponding phonon energy goes to zero, the atoms are slightly displaced with the periodicity of the corresponding wavevector, and a charge density wave (CDW) with the same periodicity appears. Essentially, the CDW increases the overlap of the electronic states with the phonon by phase-fixing with respect to the atomic lattice. The Peierls transition naturally occurs in one-dimensional (1-D) systems, where only two Fermi points are present and hence preferential orientation is pathological. It can also occur in 2-D systems if large parts of the Fermi line run in parallel.

STS studies of CDWs are numerous (e.g., [146, 147]). Examples of a 1-D CDW on a quasi-1-D bulk material and of a 2-D CDW are shown in Fig. 6.16a–d and Fig. 6.16e–h, respectively [148, 149]. In contrast to usual scattering states, where LDOS corrugations are only found close to the scatterer, the corrugations

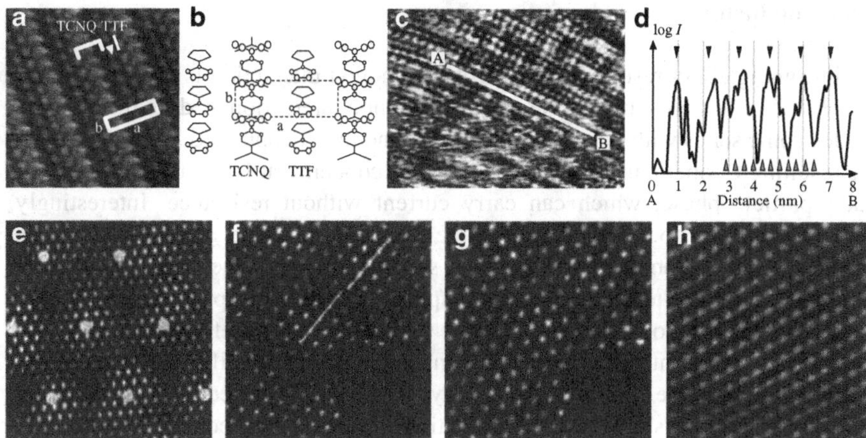


**Fig. 6.16** **(a)** STM image of the *ab*-plane of the organic quasi-1-D conductor tetrathiafulvalene tetracyanoquinodimethane (TTF-TCNQ), $T = 300$ K; while the TCNQ chains are conducting, the TTF chains are insulating. **(b)** Stick-and-ball model of the *ab*-plane of TTF-TCNQ. **(c)** STM image taken at $T = 61$ K; the additional modulation due to the Peierls transition is visible in the profile along line AB shown in **(d)**; the *brown triangles* mark the atomic periodicity and the *black triangles* the expected CDW periodicity (**(a–d)** © M. Kageshima, Kanagawa). **(e–h)** Low-voltage STM images of the two-dimensional CDW system 1 T-$TaS_2$ at $T = 242$ K **(e)**, 298 K **(f)**, 349 K **(g)**, and 357 K **(h)**. A long-range, hexagonal modulation is visible besides the atomic spots; its periodicity is highlighted by *large white dots* in **(e)**; the additional modulation obviously weakens with increasing $T$, but is still apparent in **(f)** and **(g)**, as evidenced in the lower-magnification images in the *insets* (**(e–h)** © C. Lieber, Cambridge)

of CDWs are continuous across the surface. Heating the substrate toward the transition temperature leads to a melting of the CDW lattice, as shown in Fig. 6.16f–h.

CDWs have also been found on monolayers of adsorbates such as a monolayer of Pb on Ge(111) [150]. These authors performed a nice temperature-dependent study revealing that the CDW is nucleated by scattering states around defects, as one might expect [151]. Some of the transitions have been interpreted as more complex Mott–Hubbard transitions caused primarily by electron–electron interactions [152]. One-dimensional systems have also been prepared on surfaces showing Peierls transitions [153, 154]. Finally, the energy gap occurring at the transition has been studied by measuring $\mathrm{d}I/\mathrm{d}V(V)$ curves [155].

A more local crystallographic distortion due to electron–lattice interactions is the Jahn–Teller effect. Here, symmetry breaking by elastic deformation can lead to the lifting of degeneracies close to the Fermi level. This results in an energy gain due to the lowering of the energy of the occupied levels. By tuning the Fermi level of an adsorbate layer to a degeneracy via doping, such a Jahn–Teller deformation has been induced on a surface and visualized by STM [156].

## Superconductors

An intriguing effect resulting from electron–phonon interaction is superconductivity. Here, the attractive part of the electron–phonon interaction leads to the coupling of electronic states with opposite wavevector and mostly opposite spin [157]. Since the resulting Cooper pairs are bosons, they can condense at LT, forming a coherent many-particle phase, which can carry current without resistance. Interestingly, defect scattering does not influence the condensate if the coupling along the Fermi surface is homogeneous (s-wave superconductor). The reason is that the symmetry of the scattering of the two components of a Cooper pair effectively leads to a scattering from one Cooper pair state to another without affecting the condensate. This is different if the scatterer is magnetic, since the different spin components of the pair are scattered differently, leading to an effective pair breaking, which is visible as a single-particle excitation within the superconducting gap. On a local scale, this effect was first demonstrated by putting Mn, Gd, and Ag atoms on a Nb(110) surface [158]. While the nonmagnetic Ag does not modify the gap shown in Fig. 6.17a, it is modified in an asymmetric fashion close to Mn or Gd adsorbates, as shown in Fig. 6.17b. The asymmetry of the additional intensity is caused by the breaking of the particle–hole symmetry due to the exchange interaction between the localized Mn state and the itinerate Nb states.

Another important local effect is caused by the relatively large coherence length of the condensate. At a material interface, the condensate wavefunction cannot stop abruptly, but overlaps into the surrounding material (proximity effect). Consequently, a superconducting gap can be measured in areas of nonsuperconducting material. Several studies have shown this effect on the local scale using metals and doped semiconductors as surrounding materials [159, 160].

While the classical type I superconductors are ideal diamagnets, the so-called type II superconductors can contain magnetic flux. The flux forms vortices, each containing one flux quantum. These vortices are accompanied by the disappearance of the superconducting gap and, therefore, can be probed by STS [161]. LDOS maps measured inside the gap lead to bright features in the area of the vortex core. Importantly, the length scale of these features is different from the length scale of the magnetic flux due to the difference between the London penetration depth and the electronic coherence length. Thus, STS probes a different property of the vortex than the usual magnetic imaging techniques (see Sect. 6.4.4). Surprisingly, first measurements of the vortices on $NbSe_2$ revealed vortices shaped as a sixfold star [162] (Fig. 6.17c). With increasing voltage inside the gap, the orientation of the star rotates by 30° (Fig. 6.17d, e). The shape of these stars could finally be reproduced by theory, assuming an anisotropic pairing of electrons in the superconductor (Fig. 6.17f–h) [163]. Additionally, bound states inside the vortex core, which result from confinement by the surrounding superconducting material, are found [162]. Further experiments investigated the arrangement of the vortex lattice, including transitions between hexagonal and quadratic lattices [164], the influence of pinning centers [165], and the vortex motion induced by current [166].

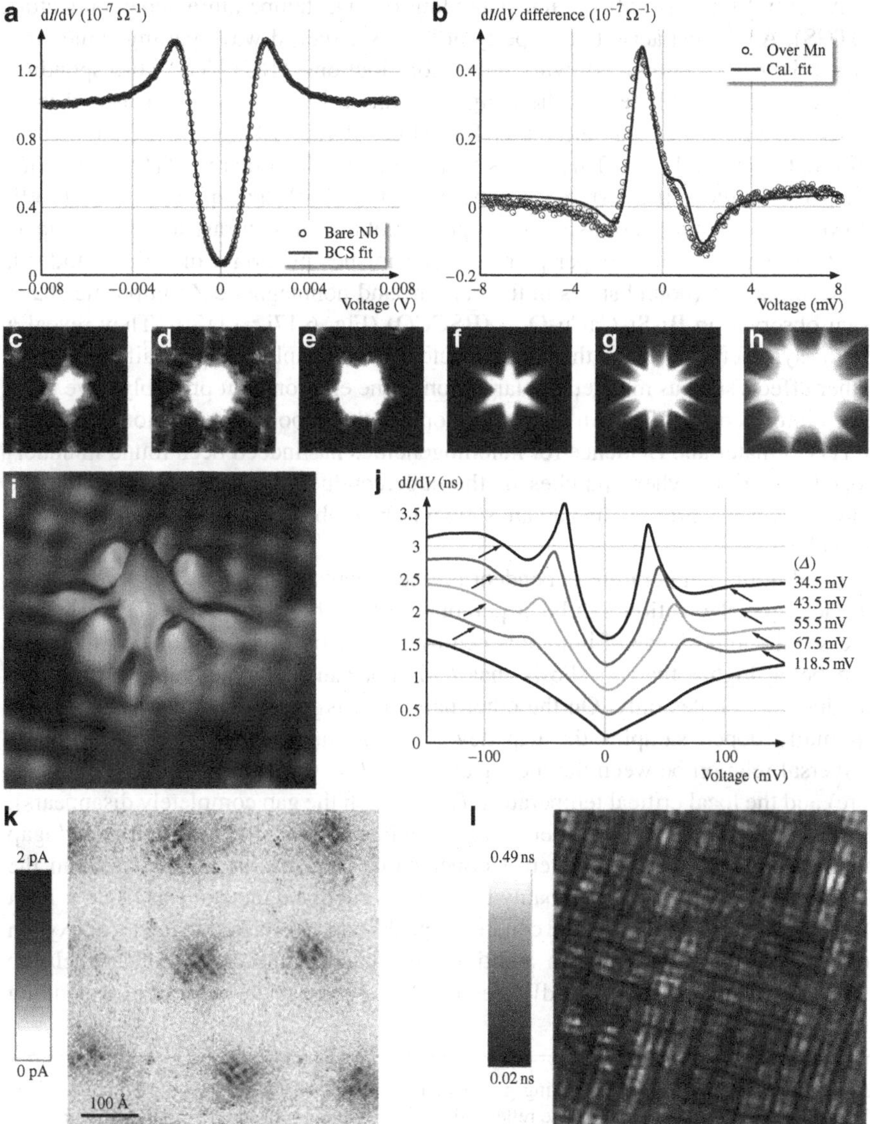


**Fig. 6.17** **(a)** $dI/dV$ curve of Nb(110) at $T = 3.8$ K (*symbols*) in comparison with a BCS fit of the superconducting gap of Nb (*line*). **(b)** Difference between the $dI/dV$ curve taken directly above a Mn atom on Nb(110) and the $dI/dV$ curve taken above clean Nb(110) (*symbols*) in comparison with a fit using the Bogulubov–de Gennes equations (*line*) ((© D. Eigler, Almaden **(a, b)**). **(c–e)** $dI/dV$ images of a vortex core in the type II superconductor 2H-$NbSe_2$ at 0 mV **(c)**, 0.24 mV **(d)**, and 0.48 mV **(e)** ((**c–e**) © H. F. Hess). **(f–h)** Corresponding calculated LDOS images within the Eilenberger framework ((**f–h**) © K. Machida, Okayama). **(i)** Overlap of an STM image at $V = -100$ mV (background 2-D image) and a $dI/dV$ image at $V = 0$ mV (overlapped 3-D image) of optimally doped $Bi_2Sr_2CaCu_2O_{8+\delta}$ containing 0.6% Zn impurities. The STM image shows the atomic structure of the cleavage plane, while the $dI/dV$ image shows a bound state within the

A central topic is still the understanding of high-temperature superconductors (HTCS). An almost accepted property of HTCS is their d-wave pairing symmetry, which is partly combined with other contributions [167]. The corresponding $k$-dependent gap (where $k$ is the reciprocal lattice vector) can be measured indirectly by STS using a Fourier transformation of the LDOS($x$, $y$) determined at different energies [168]. This shows that LDOS modulations in HTCS are dominated by simple self-interference patterns of the Bloch-like quasiparticles [169]. However, scattering can also lead to pair breaking (in contrast to s-wave superconductors), since the Cooper-pair density vanishes in certain directions. Indeed, scattering states (bound states in the gap) around nonmagnetic Zn impurities have been observed in $Bi_2Sr_2CaCu_2O_{8+\delta}$ (BSCCO) (Fig. 6.17i, j) [170]. They reveal a d-like symmetry, but not the one expected from simple Cooper-pair scattering. Other effects such as magnetic polarization in the environment probably have to be taken into account [171]. An interesting topic is the importance of inhomogeneities in HTCS materials. Evidence for inhomogeneities has indeed been found in underdoped materials, where puddles of the superconducting phase identified by the coherence peaks around the gap are shown to be embedded in nonsuperconducting areas [30].

In addition, temperature-dependent measurements of the gap size development at each spatial position exhibit a percolation-type behavior above $T_c$ [32]. This stresses the importance of inhomogeneities, but the observed percolation temperature being higher than $T_c$ shows that $T_c$ is not caused by percolation of superconducting puddles only. On the other hand, it was found that for overdoped and optimally doped samples the gap develops continuously across $T_c$, showing a universal relation between the local gap size $\Delta(T = 0)$ (measured at low temperature) and the local critical temperature $T_p$ (at which the gap completely disappears): $2\Delta(T = 0)/k_B T_p \approx 8$. The latter result is evidence that the so-called pseudogap phase is a phase with incoherent Cooper pairs. The results are less clear in the underdoped region, where probably two gaps complicate the analysis. Below $T_c$, it turns out that the strength of the coherence peak is anticorrelated to the local oxygen acceptor density [169] and, in addition, correlated to the energy of an inelastic phonon excitation peak in d$I$/d$V$ spectra [31]. Figure 6.17j shows corresponding

---

**Fig. 6.17** (continued) superconducting gap, which is located around a single Zn impurity. The fourfold symmetry of the bound state reflects the d-like symmetry of the superconducting pairing function; **(j)** d$I$/d$V$ spectra of $Bi_2Sr_2CaCu_2O_{8+\delta}$ measured at different positions of the surface at $T = 4.2$ K; the phonon peaks are marked by *arrows*, and the determined local gap size $\Delta$ is indicated; note that the strength of the phonon peak increases with the strength of the coherence peaks surrounding the gap; **(k)** LDOS in the vortex core of slightly overdoped $Bi_2Sr_2CaCu_2O_{8+\delta}$, $B = 5$ T; the d$I$/d$V$ image taken at $B = 5$ T is integrated over $V = 1$–$12$ mV, and the corresponding d$I$/d$V$ image at $B = 0$ T is subtracted to highlight the LDOS induced by the magnetic field. The checkerboard pattern within the seven vortex cores exhibits a periodicity, which is fourfold with respect to the atomic lattice shown in **(i)** and is thus assumed to be a CDW; **(l)** STM image of cleaved $Ca_{1.9}Na_{0.1}CuO_2Cl_2$ at $T = 0.1$ K, i.e., within the superconducting phase of the material; a checkerboard pattern with fourfold periodicity is visible on top of the atomic resolution (© S. Davis, Cornell and S. Uchida, Tokyo **(i–l)**)

spectra taken at different positions, where the coherence peaks and the nearby phonon peaks marked by arrows are clearly visible. The phonon origin of the peak has been proven by the isotope effect, similar to Fig. 6.11a. The strong intensity of the phonon side-peak as well as the correlation of its strength with the coherence peak intensity points towards an important role of electron–phonon coupling for the pairing mechanism. However, since the gap size does not scale with the strength of the phonon peak [172], other contributions must be involved too.

Of course, vortices have also been investigated for HTCS [173]. Bound states are found, but at energies that are in disagreement with simple models, assuming a Bardeen–Cooper–Schrieffer (BCS)-like d-wave superconductor [174, 175]. Theory predicts, instead, that the bound states are magnetic-field-induced spin density waves, stressing the competition between antiferromagnetic order and superconductivity in HTCS materials [176]. Since the spin density wave is accompanied by a charge density wave of half wavelength, it can be probed by STS [177]. Indeed, a checkerboard pattern of the right periodicity has been found in and around vortex cores in BSCCO (Fig. 6.17k). Similar checkerboards, which do not show any $E(k)$ dispersion, have also been found in the underdoped pseudogap phase at temperatures higher than the superconducting transition temperature [178] or at dopant densities lower than the critical doping [179]. Depending on the sample, the patterns can be either homogeneous or inhomogeneous and exhibit slightly different periodicities. However, the fact that the pattern persists within the superconducting phase as shown in Fig. 6.17l, at least for Na-CCOC, indicates that the corresponding phase can coexist with superconductivity. This raises the question of whether spin density waves are the central opponent to HTCS. Interestingly, a checkerboard pattern of similar periodicity, but without long-range order, is also found, if one displays the particle–hole asymmetry of $dI/dV(V)$ intensity in underdoped samples at low temperature [180]. Since the observed asymmetry is known to be caused by the lifting of the correlation gap with doping, the checkerboard pattern might be directly linked to the corresponding localized holes in the CuO planes appearing at low doping. Although a comprehensive model for HTCS materials is still lacking, STS contributes significantly to disentangling this puzzle.

Even more complex superconductors are based on heavy fermions, where superconductivity is known to coexist with ferromagnetism. First attempts to obtain information about these materials by STM have been made using very low temperature (190 mK). They exhibit indeed spatial fluctuations of the superconducting gap [181, 182]. However, the key issue for these materials is still the preparation of high-quality surfaces similar to the HTCS materials, where cleavage was extremely advantageous to obtain high-quality data.

Notice that all the measurements described above have probed the superconducting phase only indirectly by measuring the quasiparticle LDOS. The superconducting condensate itself could principally also be probed directly using Cooper-pair tunneling between a superconducting tip and a superconducting sample. A proof of principle of this detection scheme has indeed been given at low tunneling resistance ($R \approx 50\ \mathrm{k\Omega}$) [183], but meaningful spatially resolved data are still lacking.

**Complex Systems (Manganites)**

Complex phase diagrams are not restricted to HTCS materials (cuprates). They exist with similar complexity for other doped oxides such as manganites. Only a few studies of these materials have been performed by STS. Some of them show the inhomogeneous evolution of metallic and insulating phases across a metal–insulator transition [184, 185]. Within layered materials, such a phase separation has been found to be absent [186]. This experiment performed on LaSrMnO revealed, in addition, a peculiar atomic structure, which appears only locally. It has been attributed to the observation of a local polaron bound to a defect. Since inhomogeneities seem to be crucial also in these materials, a local method such as STS might continue to be important for the understanding of their complex properties.

## *6.3.6 Imaging Spin Polarization: Nanomagnetism*

Conventional STS couples to the LDOS, i.e., the charge distribution of the electronic states. Since electrons also have spin, it is desirable to also probe the spin distribution of the states. This can be achieved by spin-polarized STM (SP-STM) using a tunneling tip covered by a ferromagnetic material [187]. The coating acts as a spin filter or, more precisely, the tunneling current depends on the relative angle $\alpha_{ij}$ between the spins of the tip and the sample according to cos $(\alpha_{ij})$. Consequently, a particular tip is not sensitive to spin orientations of the sample perpendicular to the spin orientation of the tip. Different tips have to be prepared to detect different spin orientations. Moreover, the stray magnetic field of the tip can perturb the spin orientation of the sample. To avoid this, a technique using antiferromagnetic Cr as a tip coating material has been developed [188]. This avoids stray fields, but still provides a preferential spin orientation of the few atoms at the tip apex that dominate the tunneling current. Depending on the thickness of the Cr coating, spin orientations perpendicular or parallel to the sample surface, implying corresponding sensitivities to the spin directions of the sample, are achieved.

SP-STM has been used to image the evolution of magnetic domains with increasing $B$ field (Fig. 6.18a–d) [189], the antiferromagnetic order of a Mn monolayer on W(110) [190], as well as of a Fe monolayer on W(100) (Fig. 6.18e) [191], and the out-of-plane orientation of a magnetic vortex core in the center of a nanomagnet exhibiting the flux closure configuration [192].

In addition, more complex atomic spin structures showing chiral or noncollinear arrangements have been identified [193, 194, 195]. Even the spin orientation of a single adatom could be detected, if the adatom is placed either directly on a ferromagnetic island [196] or close to a ferromagnetic stripe [23]. In the latter case, hysteresis curves of the ferromagnetic adatoms could be measured, as shown in Fig. 6.18f–h. It was found that the adatoms couple either ferromagnetically (Fig. 6.18g) or antiferromagnetically (Fig. 6.18h) to the close-by magnetic stripe;

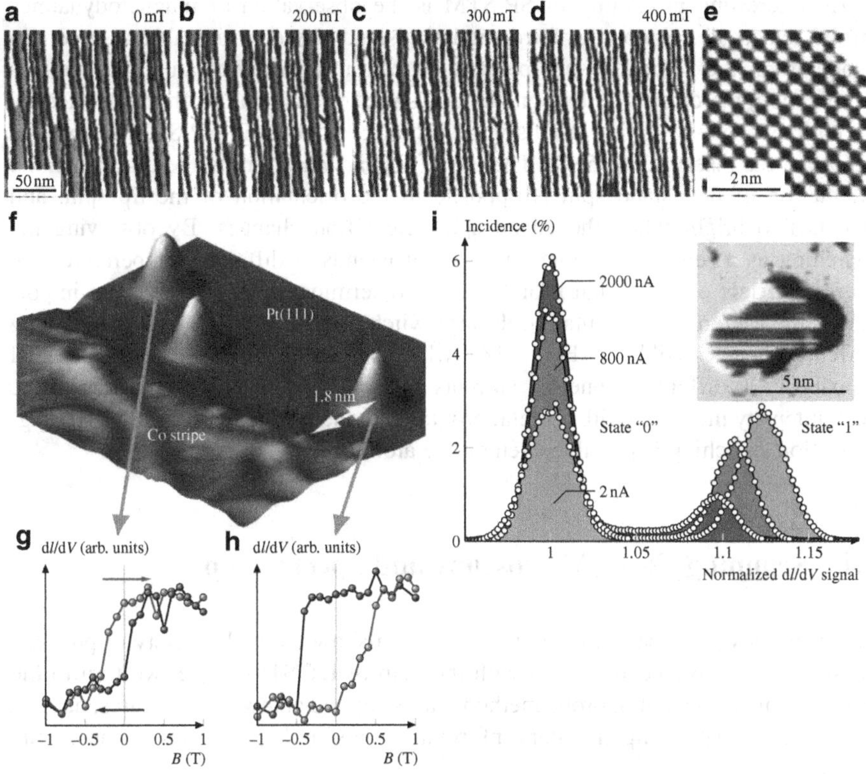


**Fig. 6.18** **(a–d)** Spin-polarized STM images of 1.65 monolayers of Fe deposited on a stepped W (110) surface measured at different $B$ fields, as indicated. Double-layer and monolayer Fe stripes are formed on the W substrate; only the double-layer stripes exhibit magnetic contrast with an out-of-plane sensitive tip, as used here. *White* and *grey areas* correspond to different domains. Note that more white areas appear with increasing field (© M. Bode, Argonne **(a–d)**). **(e)** STM image of an antiferromagnetic Fe monolayer on W(001) exhibiting a checkerboard pattern of spin-down (*dark*) and spin-up (*bright*) atoms (© A. Kubetzka, Hamburg); **(f)** STM image of a Pt(111) surface with a Co stripe deposited at the Pt edge as marked. Single Co atoms, visible as *three hills*, are deposited subsequently on the surface at $T = 25$ K; **(g, h)** d$I$/d$V$($B$) curves obtained above the Co atoms marked in **(f)** using a spin-polarized tip at $V = 0.3$ V. The *colors* mark the sweeping direction of the $B$ field. Obviously the resulting contrast is hysteretic with $B$ and opposite for the two Co atoms. This indicates a different sign of ferromagnetic coupling to the Co stripe. (© J. Wiebe, Hamburg **(f–h)**); **(i)** observed incidences of differential conductivities above a single monolayer Fe island on W(110) with a spin-polarized tip. The *three curves* are recorded at different tunneling currents and the increasing asymmetry shows a preferential spin direction with increasing spin-polarized current. *Inset*: d$I$/d$V$ image of the Fe island at $T = 56$ K showing the irregular change of d$I$/d$V$ intensity (© S. Krause, Hamburg)

i.e., the hysteresis is either in phase or out of phase with the hysteresis of the stripe. This behavior, depending on adatom–stripe distance in an oscillating fashion, directly visualizes the famous Ruderman–Kittel–Kasuya–Yoshida (RKKY) interaction [23].

An interesting possibility of SP-STM is the observation of magnetodynamics on the nanoscale. Nanoscale ferromagnetic islands become unstable at a certain temperature, the so-called superparamagnetic transition temperature. Above this temperature, the direction of magnetization switches back and forth due to thermal excitations. This switching results in a stripe-like contrast in SP-STM images, as visible in the inset of Fig. 6.18i. The island appears dark during the time when the orientation of the island spin is opposite to the orientation of the tip spin, and switches to bright when the island spin orientation changes. By observing the switching as a function of time on different islands at different temperatures the energy barriers of individual islands can be determined [197]. Even more importantly, the preferential orientation during switching can be tuned by the tunneling current. This is visible in Fig. 6.18i, which shows the measured orientational probability at different tunneling currents [198]. The observed asymmetry in the peak intensity increases with current, providing evidence that current-induced magnetization switching is possible even on the atomic scale.

## 6.4 Scanning Force Microscopy and Spectroscopy

The examples discussed in the previous section show the wide variety of physical questions that have been tackled with the help of LT-STM. Here, we turn to the other prominent scanning probe method that is applied at low temperatures, namely SFM, which gives complementary information on sample properties on the atomic scale.

The ability to detect *forces* sensitively with spatial resolution down to the atomic scale is of great interest, since force is one of the most fundamental quantities in physics. Mechanical force probes usually consist of a cantilever with a tip at its free end that is brought close to the sample surface. The cantilever can be mounted parallel or perpendicular to the surface (general aspects of force probe designs are described in Chap. 3). Basically, two methods exist to detect forces with cantilever-based probes: the *static* and the *dynamic* mode (see Chap. 2). They can be used to generate a laterally resolved image (*microscopy* mode) or determine its distance dependence (*spectroscopy* mode). One can argue about this terminology, since spectroscopy is usually related to energies and not to distance dependencies. Nevertheless, we will use it throughout the text, because it avoids lengthy paraphrases and is established in this sense throughout the literature.

In the static mode, a force that acts on the tip bends the cantilever. By measuring its deflection $\Delta z$ the tip–sample force $F_{\text{ts}}$ can be directly calculated from Hooke's law: $F_{\text{ts}} = c_z \Delta z$, where $c_z$ denotes the spring constant of the cantilever. In the various dynamic modes, the cantilever is oscillated with amplitude $A$ at or near its eigenfrequency $f_0$, but in some applications also off-resonance. At ambient pressures or in liquids, amplitude modulation (AM-SFM) is used to detect amplitude changes or the phase shift between the driving force and cantilever oscillation. In vacuum, the frequency shift $\Delta f$ of the cantilever due to a tip–sample interaction is measured by the frequency-modulation technique (FM-SFM). The nomenclature is not standardized. Terms such as tapping mode or intermittent contact mode are used instead of AM-SFM, and NC-AFM (noncontact atomic force microscopy) or DFM (dynamic force microscopy) instead of FM-SFM or FM-AFM. However, all these modes are *dynamic*, i.e., they involve an oscillating cantilever and can be used in the noncontact, as well as in the contact, regime. Therefore, we believe that the best and most consistent way is to distinguish them by their different detection schemes. Converting the measured quantity (amplitude, phase or frequency shift) into a physically meaningful quantity, e.g., the tip–sample interaction force $F_{\mathrm{ts}}$ or the force gradient $\partial F_{\mathrm{ts}}/\partial_z$, is not always straightforward and requires an analysis of the equation of motion of the oscillating tip (see Chaps. 5 and 7).

Whatever method is used, the resolution of a cantilever-based force detection is fundamentally limited by its intrinsic *thermomechanical* noise. If the cantilever is in thermal equilibrium at a temperature $T$, the equipartition theorem predicts a thermally induced *root-mean-square* (RMS) motion of the cantilever in the $z$ direction of $z_{\mathrm{RMS}} = (k_{\mathrm{B}}T/c_{\mathrm{eff}})^{1/2}$, where $k_{\mathrm{B}}$ is the Boltzmann constant and $c_{\mathrm{eff}} = c_z + \partial F_{\mathrm{ts}}/\partial_z$. Note that usually $\mathrm{d}F_{\mathrm{ts}}/\mathrm{d}_z \gg c_z$ in the contact mode and $\mathrm{d}F_{\mathrm{ts}}/\mathrm{d}_z < c_z$ in the noncontact mode. Evidently, this fundamentally limits the force resolution in the static mode, particularly if operated in the noncontact mode. Of course, the same is true for the different dynamic modes, because the thermal energy $k_{\mathrm{B}}T$ excites the eigenfrequency $f_0$ of the cantilever. Thermal noise is *white* noise, i.e., its spectral density is flat. However, if the cantilever transfer function is taken into account, one can see that the thermal energy mainly excites $f_0$. This explains the term "thermo" in thermomechanical noise, but what is the "mechanical" part?

A more detailed analysis reveals that the thermally induced cantilever motion is given by

$$z_{\mathrm{RMS}} = \sqrt{\frac{2k_{\mathrm{B}}TB}{\pi c_z f_0 Q}}, \tag{6.2}$$

where $B$ is the measurement bandwidth and $Q$ is the quality factor of the cantilever. Analogous expressions can be obtained for all quantities measured in dynamic modes, because the deflection noise translates, e.g., into frequency noise [5]. Note that $f_0$ and $c_z$ are correlated with each other via $2\pi f_0 = (c_z/m_{\mathrm{eff}})^{1/2}$, where the effective mass $m_{\mathrm{eff}}$ depends on the geometry, density, and elasticity of the material. The $Q$-factor of the cantilever is related to the external damping of the cantilever motion in a medium and to the intrinsic damping within the material. This is the "mechanical" part of the fundamental cantilever noise.

It is possible to operate a low-temperature force microscope directly immersed in the cryogen [199, 200] or in the cooling gas [201], whereby the cooling is simple and very effective. However, it is evident from (6.2) that the smallest fundamental noise is achievable in vacuum, where the $Q$-factors are more than 100 times larger than in air, and at low temperatures.

The best force resolution up to now, which is better than $1 \times 10^{-18}$ N/Hz$^{1/2}$, has been achieved by Mamin et al. [202] in vacuum at a temperature below 300 mK. Due to the reduced thermal noise and the lower thermal drift, which results in a higher stability of the tip–sample gap and a better signal-to-noise ratio, the highest resolution is possible at low temperatures in ultrahigh vacuum with FM-SFM. A vertical RMS noise below 2 pm [203, 204] and a force resolution below 1 aN [202] have been reported.

Besides the reduced noise, the application of force detection at low temperatures is motivated by the increased stability and the possibility to observe phenomena that appear below a certain critical temperature $T_c$, as outlined on page 664. The experiments, which have been performed at low temperatures until now, were motivated by at least one of these reasons and can be roughly divided into four groups:

(i) Atomic-scale imaging
(ii) Force spectroscopy
(iii) Investigation of quantum phenomena by measuring electrostatic forces
(iv) Utilizing magnetic probes to study ferromagnets, superconductors, and single spins

In the following, we describe some exemplary results.

### *6.4.1 Atomic-Scale Imaging*

In a simplified picture, the dimensions of the tip end and its distance to the surface limit the lateral resolution of force microscopy, since it is a near-field technique. Consequently, atomic resolution requires a stable single atom at the tip apex that has to be brought within a distance of some tenths of a nanometer of an atomically flat surface. The latter condition can only be fulfilled in the dynamic mode, where the additional restoring force $c_z A$ at the lower turnaround point prevents the jump-to-contact. As described in Chap. 5, by preventing the so-called jump-to-contact, *true* atomic resolution is nowadays routinely obtained in vacuum by FM-AFM. The nature of the short-range tip–sample interaction during imaging with atomic resolution has been studied experimentally as well as theoretically. Si(111)-(7 × 7) was the first surface on which true atomic resolution was achieved [205], and several studies have been performed at low temperatures on this well-known material [206, 207, 208]. First-principles simulations performed on semiconductors with a silicon tip revealed that *chemical* interactions, i.e., a significant charge redistribution between the dangling bonds of the tip and sample, dominate the atomic-scale contrast [209, 210, 211]. On V–III semiconductors, it was found that only one atomic species, the group V atoms, is imaged as protrusions with a silicon tip [210, 211]. Furthermore, these simulations revealed that the sample, as well as the tip atoms, are noticeably displaced from their equilibrium position due to the interaction forces. At low temperatures, both aspects could be observed with silicon tips on indium arsenide [203, 212]. On weakly interacting surfaces the short-range interatomic van der Waals force has been believed responsible for the atomic-scale contrast [213, 214, 215].

### Chemical Sensitivity of Force Microscopy

The (110) surface of the III–V semiconductor indium arsenide exhibits both atomic species in the top layer (see Fig. 6.19a). Therefore, this sample is well suited to study the chemical sensitivity of force microscopy [203]. In Fig. 6.19b, the usually observed atomic-scale contrast on InAs(110) is displayed. As predicted, the arsenic atoms, which are shifted by 80 pm above the indium layer due to the $(1 \times 1)$ relaxation, are imaged as protrusions. While this general appearance was similar for most tips, two other distinctively different contrasts were also observed: a second protrusion (Fig. 6.19c) and a sharp depression (Fig. 6.19d). The arrangement of these two features corresponds well to the zigzag configuration of the indium and arsenic atoms along the $[1\bar{1}0]$ direction. A sound explanation would be as follows: the contrast usually obtained with one feature per surface unit cell corresponds to a silicon-terminated tip, as predicted by simulations. A different atomic species at the tip apex, however, can result in a very different charge redistribution. Since the atomic-scale contrast is due to a chemical interaction, the two other contrasts would then correspond to a tip that has been accidentally contaminated with sample material (an arsenic- or indium-terminated tip apex). Nevertheless, this explanation has not yet been verified by simulations for this material.

### Tip-Induced Atomic Relaxation

Schwarz et al. [203] were able to visualize directly the predicted tip-induced relaxation during atomic-scale imaging near a point defect. Figure 6.20 shows

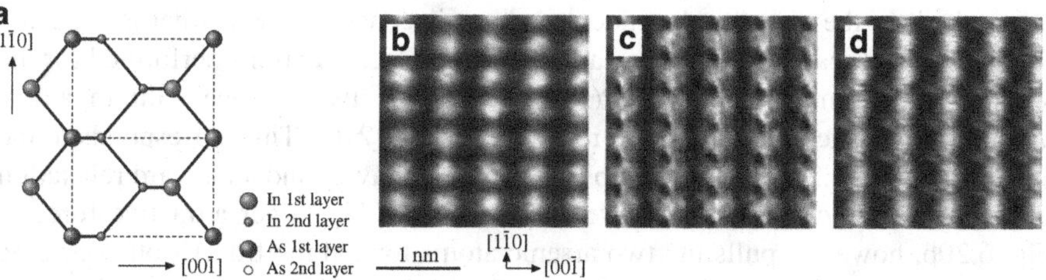


**Fig. 6.19a–d** The structure of InAs(110) as seen from above **(a)** and three FM-AFM images of this surface obtained with different tips at 14 K **(b–d)**. In **(b)**, only the arsenic atoms are imaged as protrusions, as predicted for a silicon tip. The two features in **(c)** and **(d)** corresponds to the zigzag arrangement of the indium and arsenic atoms. Since force microscopy is sensitive to short-range chemical forces, the appearance of the indium atoms can be associated with a chemically different tip apex

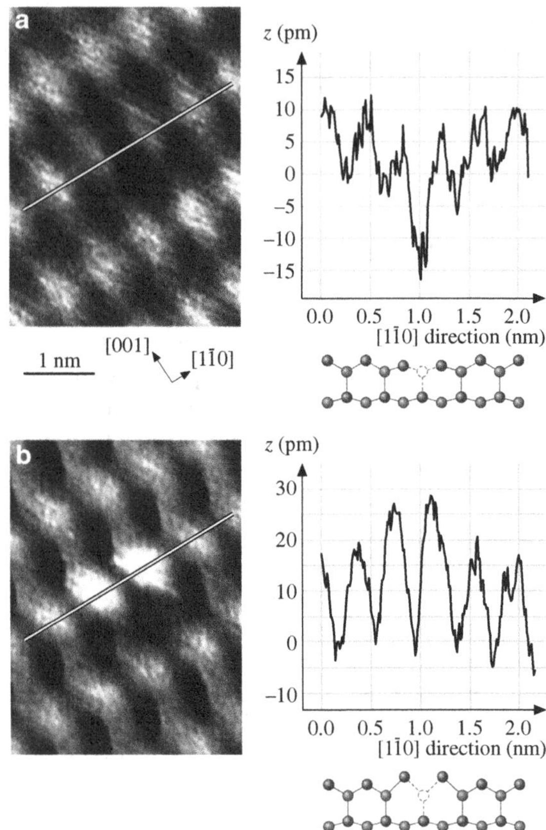


**Fig. 6.20a, b** Two FM-AFM images of the identical indium-site point defect (presumably an indium vacancy) recorded at 14 K. If the tip is relatively far away, the theoretically predicted inward relaxation of two arsenic atoms adjacent to an indium vacancy is visible **(a)**. At a closer tip–sample distance **(b)**, the two arsenic atoms are pulled farther toward the tip compared with the other arsenic atoms, since they have only two instead of three bonds

two FM-AFM images of the same point defect recorded with different constant frequency shifts on InAs(110), i.e., the tip was closer to the surface in Fig. 6.20b compared with Fig. 6.20a. The arsenic atoms are imaged as protrusions with the silicon tip used. From the symmetry of the defect, an indium-site defect can be inferred, since the distance-dependent contrast is consistent with what is expected for an indium vacancy. This expectation is based on calculations performed for the similar III–V semiconductor GaP(110), where the two surface gallium atoms around a P-vacancy were found to relax downward [216]. This corresponds to the situation in Fig. 6.20a, where the tip is relatively far away and an inward relaxation of the two arsenic atoms is observed. The considerably larger attractive force in Fig. 6.20b, however, pulls the two arsenic atoms toward the tip. All other arsenic atoms are also pulled, but they are less displaced, because they have three bonds to the bulk, while the two arsenic atoms in the neighborhood of an indium vacancy have only two bonds. This direct experimental proof of the presence of tip-induced relaxations is also relevant for STM measurements, because the tip–sample distances are similar during atomic-resolution imaging. Moreover, the result demonstrates that FM-AFM can probe elastic properties on an atomic level.

## Imaging of Weakly Interacting van der Waals Surfaces

For weakly interacting van der Waals surfaces, much smaller atomic corrugation amplitudes are expected compared with strongly interacting surfaces of semiconductors. A typical example is graphite, a layered material, where the carbon atoms are covalently bonded and arranged in a honeycomb structure within the (0001) plane. Individual graphene layers stick together by van der Waals forces. Due to the *ABA* stacking, three distinctive sites exist on the (0001) surface: carbon atoms with (*A*-type) and without (*B*-type) neighbor in the next graphite layer and the *hollow site* (*H*-site) in the hexagon center. In static contact force microscopy as well as in STM the contrast usually exhibits a trigonal symmetry with a periodicity of 246 pm, where *A*- and *B*-site carbon atoms could not be distinguished. However, in high-resolution FM-AFM images acquired at low temperatures, a large maximum and two different minima have been resolved, as demonstrated by the profiles along the three equivalent [1-100] directions in Fig. 6.21a. A simulation using the Lennard-Jones (LJ) potential, given by the short-range interatomic van der Waals force, reproduced these three features very well (dotted line). Therefore, the large maximum could be assigned to the *H*-site, while the two different minima represent *A*- and *B*-type carbon atoms [214].

Compared with graphite, the carbon atoms in a single-walled carbon nanotube (SWNT), which consists of a single rolled-up graphene layer, are indistinguishable. For the first time Ashino et al. [215] successfully imaged the curved surface of a SWNT with atomic resolution. Note that, for geometric reasons, atomic resolution is only achieved on the top (see Fig. 6.21b). Indeed, as shown in Fig. 6.21b, all profiles between two hollow sites across two neighboring carbon atoms are symmetric [217]. Particularly, curves 1 and 2 exhibit two minima of equal depth, as predicted by theory (cf., dotted line). The assumption used in the simulation (dotted lines in the profiles of Fig. 6.21) that interatomic van der Waals forces are responsible for the atomic-scale contrast has been supported by a quantitative evaluation of force spectroscopy data obtained on SWNTs [215].

Interestingly, the image contrast on graphite and SWNTs is inverted with respect to the arrangement of the atoms, i.e., the minima correspond to the positions of the carbon atoms. This can be related to the small carbon–carbon distance of only 142 pm, which is in fact the smallest interatomic distance that has been resolved with FM-AFM so far. The van der Waals radius of the front tip atom, (e.g., 210 pm for silicon) has a radius that is significantly larger than the intercarbon distance. Therefore, next-nearest-neighbor interactions become important and result in contrast inversion [217].

While experiments on graphite and SWNTs basically take advantage of the increased stability and signal-to-noise ratio at low temperatures, solid xenon (melting temperature $T_{\mathrm{m}} = 161$ K) can only be observed at sufficient low temperatures [8]. In addition, xenon is a pure van der Waals crystal and, since it is an insulator, FM-AFM is the only real-space method available today that allows the study of solid xenon on the atomic scale.

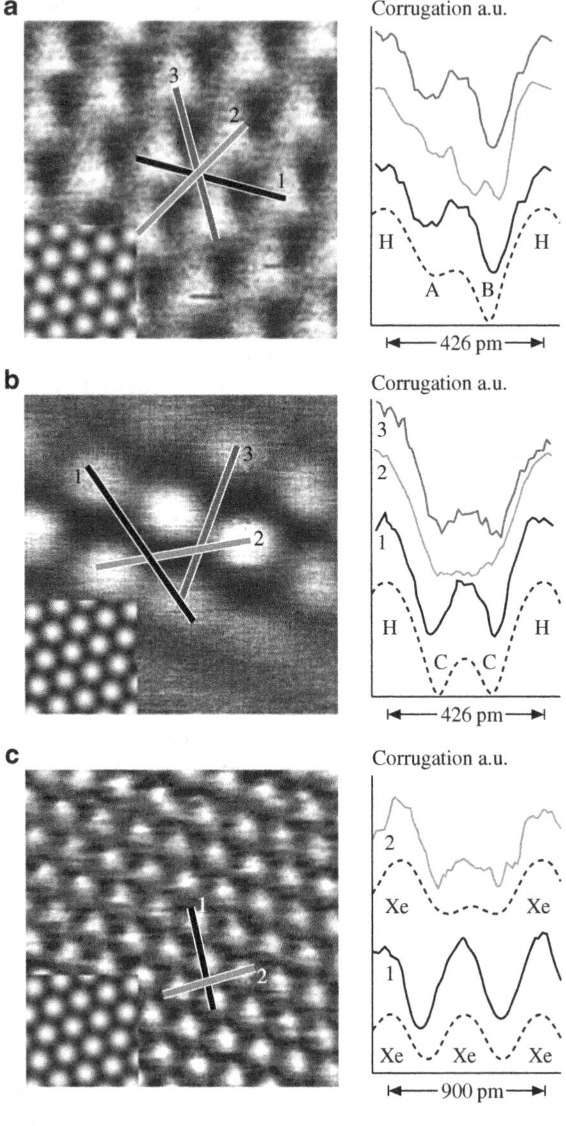


**Fig. 6.21a–c** FM-AFM images of (**a**) graphite(0001), (**b**) a single-walled carbon nanotube (SWNT), and (**c**) Xe (111) recorded at 22 K. On the right side, line sections taken from the experimental data (*solid lines*) are compared with simulations (*dotted lines*). *A*- and *B*-type carbon atoms, as well as the hollow site (*H*-site) on graphite can be distinguished, but are imaged with inverted contrast, i.e., the carbon sites are displayed as minima. Such an inversion does not occur on Xe(111)

Allers et al. [8] adsorbed a well-ordered xenon film on cold graphite(0001) ($T < 55$ K) and studied it subsequently at 22 K by FM-AFM (Fig. 6.21c). The sixfold symmetry and the distance between the protrusions corresponds well with the nearest-neighbor distance in the close-packed (111) plane of bulk xenon, which crystallizes in a face-centered cubic structure. A comparison between experiment and simulation confirmed that the protrusions correspond to the position of the xenon atoms [214]. However, the simulated corrugation amplitudes do not fit as well as for graphite (see sections in Fig. 6.21c). A possible reason is that tip-induced relaxations, which were not considered in the simulations, are more important for this pure van der Waals crystal xenon than they are for graphite, because in-plane graphite exhibits strong covalent bonds. Nevertheless, the results demonstrated for the first time that a weakly bonded van der Waals crystal could be imaged nondestructively on the atomic scale. Note that on Xe(111) no contrast inversion exists, presumably because the separation between Xe sites is about 450 pm, i.e., twice as large as the van der Waals radius of a silicon atom at the tip end.

### Atomic Resolution Using Small Oscillation Amplitudes

All the examples above described used spring constants and amplitudes on the order of 40 N/m and 10 nm, respectively, to obtain atomic resolution. However, Giessibl et al. [218] pointed out that the optimal amplitude should be on the order of the characteristic decay length $\lambda$ of the relevant tip–sample interaction. For short-range interactions, which are responsible for the atomic-scale contrast, $\lambda$ is on the order of 0.1 nm. On the other hand, stable imaging without a jump-to-contact is only possible as long as the restoring force $c_z A$ at the lower turnaround point of each cycle is larger than the maximal attractive tip–sample force. Therefore, reducing the desired amplitude by a factor of 100 requires a 100 times larger spring constant. Indeed, Hembacher et al. [219] could demonstrate atomic resolution with small amplitudes (about 0.25 nm) and large spring constants (about 1,800 N/m) utilizing a qPlus sensor [220]. Figure 6.22 shows a constant-height image of graphite recorded at 4.9 K within the repulsive regime. Note that, compared with Fig. 6.21a, b, the contrast is inverted, i.e., the carbon atoms appear as maxima. This is expected, because the imaging interaction is switched from attractive to repulsive regime [213, 217].

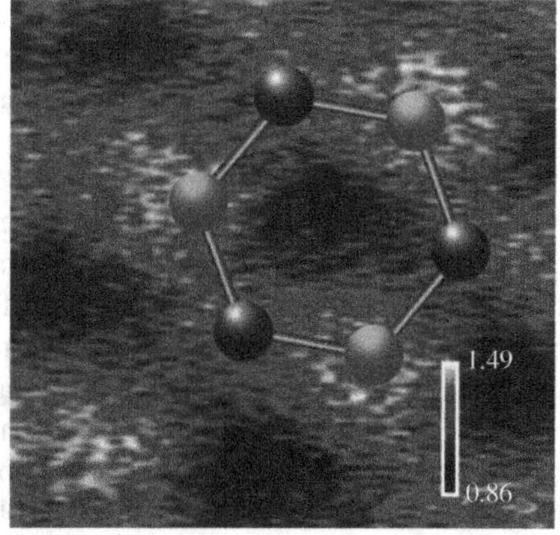


**Fig. 6.22** Constant-height FM-AFM image of graphite (0001) recorded at 4.9 K using a small amplitude ($A = 0.25$ nm) and a large spring constant ($c_z = 1,800$ N/m). As in Fig. 6.20a, *A*- and *B*-site carbon atoms can be distinguished. However, they appear as maxima, because imaging has been performed in the repulsive regime (© F. J. Giessibl [219])

### 6.4.2 Force Spectroscopy

A wealth of information about the nature of the tip–sample interaction can be obtained by measuring its distance dependence. This is usually done by recording the measured quantity (deflection, frequency shift, amplitude change, phase shift) and applying an appropriate voltage ramp to the $z$-electrode of the scanner piezo, while the $z$-feedback is switched off. According to (6.2), low temperatures and high $Q$-factors (vacuum) considerably increase the force resolution. In the static mode, long-range forces and contact forces can be examined. Force measurements at small tip–sample distances are inhibited by the *jump-to-contact* phenomenon: If the force gradient $\partial F_{\mathrm{ts}}/\partial_z$ becomes larger than the spring constant $c_z$, the cantilever cannot resist the attractive tip–sample forces and the tip snaps onto the surface. Sufficiently large spring constants prevent this effect, but reduce the force resolution. In the dynamic modes, the jump-to-contact can be avoided due to the additional restoring force ($c_zA$) at the lower turnaround point. The highest sensitivity can be achieved in vacuum by using the FM technique, i.e., by recording $\Delta f(z)$ curves. An alternative FM spectroscopy method, the recording of $\Delta f(A)$ curves, has been suggested by Hölscher et al. [221]. Note that, if the amplitude is much larger than the characteristic decay length of the tip–sample force, the frequency shift cannot simply be converted into force gradients by using $\partial F_{\mathrm{ts}}/\partial_z = 2c_z\Delta f/f_0$ [222]. Several methods have been published to convert $\Delta f(z)$ data into the tip–sample potential $V_{\mathrm{ts}}(z)$ and tip–sample force $F_{\mathrm{ts}}(z)$ [223, 224, 225, 226].

#### Measurement of Interatomic Forces at Specific Atomic Sites

FM force spectroscopy has been successfully used to measure and determine quantitatively the short-range chemical force between the foremost tip atom and specific surface atoms [177, 227, 228]. Figure 6.23 displays an example for the quantitative determination of the short-range force. Figure 6.23a shows two $\Delta f(z)$ curves measured with a silicon tip above a corner hole and above an adatom. Their position is indicated by arrows in the inset, which displays the atomically resolved Si(111)-(7 $\times$ 7) surface. The two curves differ from each other only for small tip–sample distances, because the long-range forces do not contribute to the atomic-scale contrast. The low, thermally induced lateral drift and the high stability at low temperatures were required to precisely address the two specific sites. To extract the short-range force, the long-range van der Waals and/or electrostatic forces can be subtracted from the total force. The black curve in Fig. 6.23b has been reconstructed from the $\Delta f(z)$ curve recorded above an adatom and represents the total force. After removing the long-range contribution from the data, the much steeper brown line is obtained, which corresponds to the short-range force between the adatom and the atom at the tip apex. The measured maximum attractive force ($-2.1$ nN) agrees well with that obtained from first-principles calculations ($-2.25$ nN).

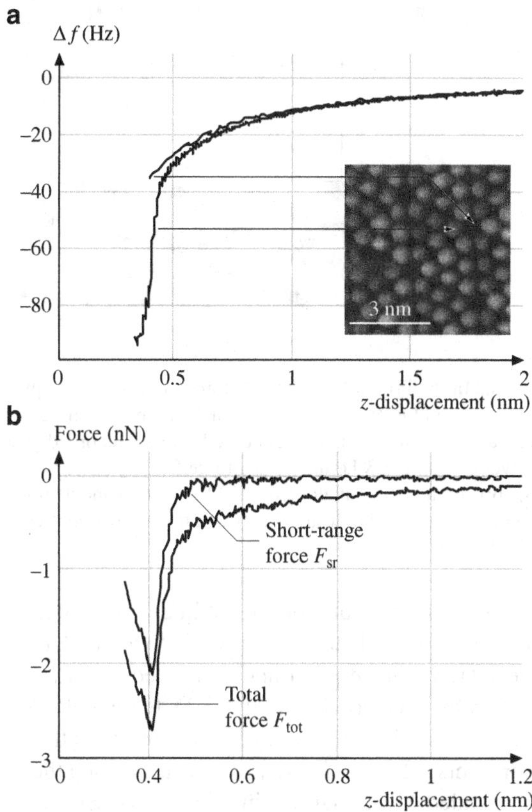


**Fig. 6.23a, b** FM force spectroscopy on specific atomic sites at 7.2 K. In (**a**), an FM-SFM image of the Si(111)-(7 × 7) surface is displayed together with two $\Delta f(z)$ curves, which have been recorded at the positions indicated by the *arrows*, i.e., above the corner hole (*black*) and above an adatom (*brown*). In (**b**), the total force above an adatom (*black line*) has been recovered from the $\Delta f(z)$ curve. After subtraction of the long-range part, the short-range force can be determined (*brown line*) (Courtesy of H. J. Hug; cf. [227])

By determining the maximal attractive short-range force between tip apex atom and surface atom as a fingerprint Sugimoto et al. [229] were able to utilize force spectroscopy data for chemical identification. They demonstrated this concept using a Si tip and a surface with Sn, Pb, and Si adatoms located at equivalent lattice sites on a Si(111) substrate. Since the experiment was performed at room temperature, the signal-to-noise-ratio had to be increased by averaging about 100 curves at every atom species, which required an appropriate atom tracking scheme [230].

### Three-Dimensional Force Field Spectroscopy

Further progress with the FM technique has been made by Hölscher et al. [231]. They acquired a complete 3-D force field on NiO(001) with atomic resolution (*3-D force field spectroscopy*). In Fig. 6.24, the atomically resolved FM-AFM image of NiO(001) is shown together with the coordinate system used and the tip to illustrate the measurement principle. NiO(001) crystallizes in the rock-salt structure. The distance between the protrusions corresponds to the lattice constant of 417 pm,

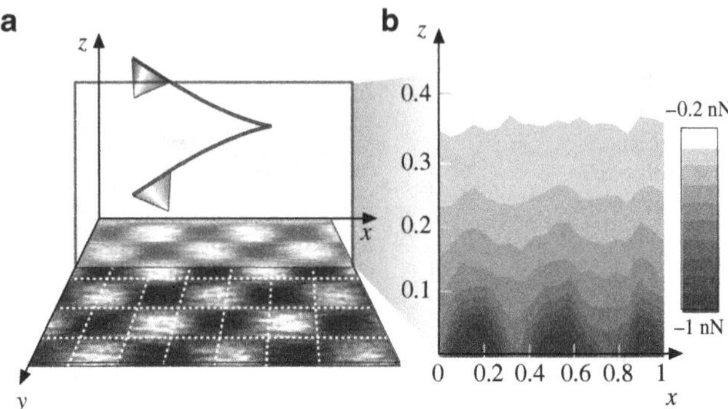


**Fig. 6.24a, b** Principle of the 3-D force field spectroscopy method (**a**) and a 2-D cut through the 3-D force field $F_{ts}(x, y, z)$ recorded at 14 K (**b**). At all 32 × 32 image points of the 1 nm × 1 nm scan area on NiO(001), a $\Delta f(z)$ curve has been recorded. The $\Delta f(x, y, z)$ data set obtained is then converted into the 3-D tip–sample force field $F_{ts}(x, y, z)$. The *shaded slice* $F_{ts}(x, y = \text{const}, z)$ in (**a**) corresponds to a cut along the [100] direction and demonstrates that atomic resolution has been obtained, because the distance between the protrusions corresponds well to the lattice constant of nickel oxide

i.e., only one type of atom (most likely the oxygen) is imaged as a protrusion. In an area of 1 nm ×1 nm, 32 × 32 individual $\Delta f(z)$ curves have been recorded at every $(x, y)$ image point and converted into $F_{ts}(z)$ curves. The $\Delta f(x, y, z)$ data set is thereby converted into the 3-D force field $F_{ts}(x, y, z)$. Figure 6.24, where a specific $x$–$z$-plane is displayed, demonstrates that atomic resolution is achieved. It represents a 2-D cut $F_{ts}(x, y = \text{const}, z)$ along the [100] direction (corresponding to the shaded slice marked in Fig. 6.24). Since a large number of curves have been recorded, Langkat et al. [228] could evaluate the whole data set by standard statistical means to extract the long- and short-range forces. A possible future application of 3-D force field spectroscopy could be to map the short-range forces of complex molecules with functionalized tips in order to resolve locally their chemical reactivity. A first step in this direction has been accomplished on SWNTs. Its structural unit, a hexagonal carbon ring, is common to all aromatic molecules. Like the constant frequency-shift image of an SWNT shown in Fig. 6.21b the force map shows clear differences between hollow sites and carbon sites [215]. Analyzing site-specific individual force curves extracted from the 3-D data revealed a maximum attractive force of $\approx -0.106$ nN above $H$-sites and $\approx -0.075$ nN above carbon sites. Since the attraction is one order of magnitude weaker than on Si(111)-(7 × 7) (Fig. 6.23b), it has been inferred that the short-range interatomic van der Waals force and not a chemical force is responsible for atomic-scale contrast formation on such nonreactive surfaces. It is worth mentioning that 3-D force field spectroscopy data have been acquired at room temperature as well [232, 233].

Apart from calculating the vertical tip–sample force Schwarz et al. [234] demonstrated that it is also possible to obtain the lateral tip–sample force from 3-D data sets. First, the tip–sample potential $V_{ts}(x, y, z)$ has to be determined. Then

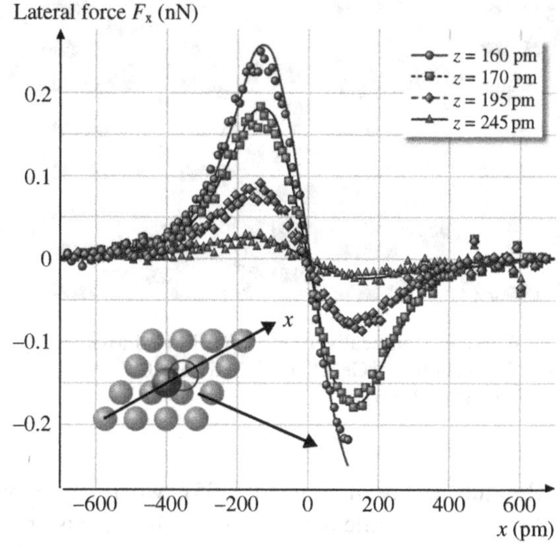


**Fig. 6.25** Lateral force curves recorded at constant tip–sample separation $z$. At the lowest separation a discontinuity appears, which marks the jump of the Co atom form one site to the next as indicated in the *inset* (©. M. Ternes [42]).

the lateral force components can be calculated by taking the derivative with respect to the $x$- and $y$-coordinate, respectively. This technique has been employed to determine the lateral force needed to move an atom sideways by Ternes et al. [42]. They recorded constant-height data at different tip–sample distances to obtain first $V_{ts}(x, z)$ and then the lateral force $F_x(x, z) = \mathrm{d}/\mathrm{d}x V_{ts}(x, z)$. Four curves of the whole data set are displayed in Fig. 6.25. The discontinuity of the lateral force at the lowest adjusted tip–sample distance ($z = 160$ pm) indicates the jump of the Co atom from one hollow site to the next on Pt(111), cf., inset. It takes place at about 210 pN.

### 6.4.3 *Atomic Manipulation*

Nowadays, atomic-scale manipulation is routinely performed using an STM tip (see Sect. 6.3.1). In most of these experiments an adsorbate is dragged with the tip by using an attractive force between the foremost tip apex atoms and the adsorbate. By adjusting a large or a small tip–surface distance via the tunneling resistance, it is possible to switch between imaging and manipulation. Recently, it has been demonstrated that controlled manipulation of individual atoms is also possible in the dynamic mode of atomic force microscopy, i.e., FM-AFM. Vertical manipulation was demonstrated by pressing the tip in a controlled manner into a Si(111)-(7 × 7) surface [236]. The strong repulsion leads to the removal of the selected silicon atom. The process could be traced by recording the frequency shift and the damping signal during the approach. For lateral manipulation a *rubbing*

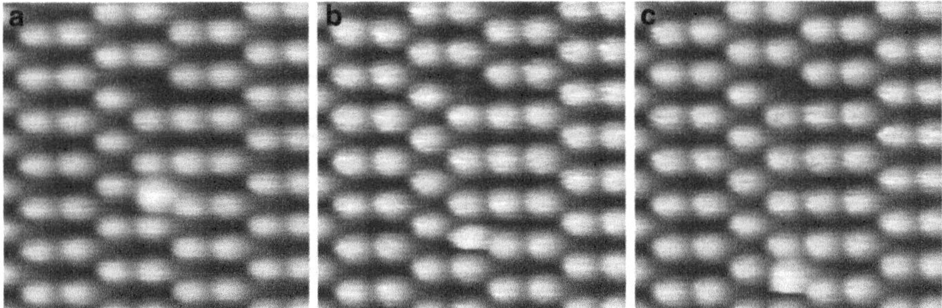


**Fig. 6.26a–c** Consecutively recorded FM-AFM images showing the tip-induced manipulation of a Ge adatom on Ge(111)-c(2 × 8) at 80 K. Scanning was performed from bottom to top (© N. Oyabu [235])

technique has been utilized [235], where the slow scan axis is halted above a selected atom, while the tip–surface distance is gradually reduced until the selected atom hops to a new stable position. Figure 6.26 shows a Ge adatom on Ge(111)-c (2 × 8) that was moved during scanning in two steps from its original position (Fig. 6.26a) to its final position (Fig. 6.26c). In fact, manipulation by FM-AFM is reproducible and fast enough to write nanostructures in a bottom-up process with single atoms [237].

## 6.4.4 Electrostatic Force Microscopy

Electrostatic forces are readily detectable by a force microscope, because the tip and sample can be regarded as two electrodes of a capacitor. If they are electrically connected via their back sides and have different work functions, electrons will flow between the tip and sample until their Fermi levels are equalized. As a result, an electric field, and consequently an attractive electrostatic force, exists between them at zero bias. This *contact potential difference* can be balanced by applying an appropriate bias voltage. It has been demonstrated that individual doping atoms in semiconducting materials can be detected by electrostatic interactions due to the local variation of the surface potential around them [238, 239].

### Detection of Edge Channels in the Quantum Hall Regime

At low temperatures, electrostatic force microscopy has been used to measure the electrostatic potential in the quantum Hall regime of a *two-dimensional electron gas* (2-DEG) buried in epitaxially grown GaAs/AlGaAs heterostructures [240, 241, 242, 243]. In the 2-DEG, electrons can move freely in the $x$–$y$-plane, but they cannot move in $z$-direction. Electrical transport properties of a 2-DEG are

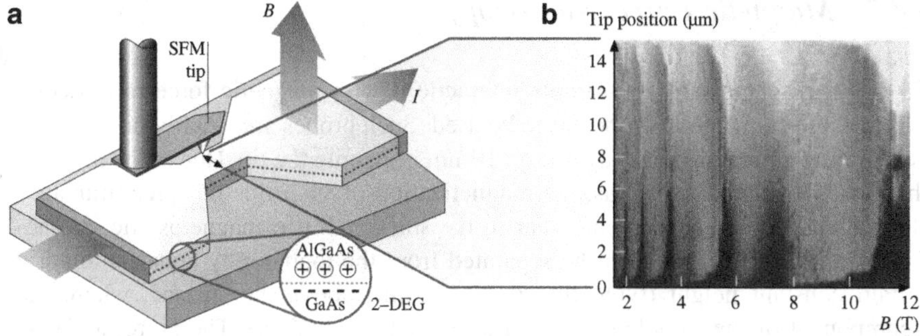


**Fig. 6.27a, b** Configuration of the Hall bar within a low-temperature ($T < 1$ K) force microscope **(a)** and profiles (*y*-axis) at different magnetic field (*x*-axis) of the electrostatic potential across a 14-µm-wide Hall bar in the quantum Hall regime **(b)**. The external magnetic field is oriented perpendicular to the 2-DEG, which is buried below the surface. *Bright* and *dark regions* reflect the characteristic changes of the electrostatic potential across the Hall bar at different magnetic fields and can be explained by the existence of the theoretically predicted edge channels (© E. Ahlswede [242])

very different compared with normal metallic conduction. Particularly, the Hall resistance $R_H = h/ne^2$ (where $h$ represents Planck's constant, $e$ is the electron charge, and $n = 1, 2, \ldots$) is quantized in the quantum Hall regime, i.e., at sufficiently low temperatures ($T < 4$ K) and high magnetic fields (up to 20 T). Under these conditions, theoretical calculations predict the existence of *edge channels* in a Hall bar. A Hall bar is a strip conductor that is contacted in a specific way to allow longitudinal and transversal transport measurements in a perpendicular magnetic field. The current is not evenly distributed over the cross-section of the bar, but passes mainly along rather thin paths close to the edges. This prediction has been verified by measuring profiles of the electrostatic potential across a Hall bar in different perpendicular external magnetic fields [240, 241, 242].

Figure 6.27a shows the experimental setup used to observe these edge channels on top of a Hall bar with a force microscope. The tip is positioned above the surface of a Hall bar under which the 2-DEG is buried. The direction of the magnetic field is oriented perpendicular to the 2-DEG. Note that, although the 2-DEG is located several tens of nanometers below the surface, its influence on the electrostatic surface potential can be detected. In Fig. 6.27b, the results of scans perpendicular to the Hall bar are plotted against the magnitude of the external magnetic field. The value of the electrostatic potential is grey-coded in arbitrary units. In certain field ranges, the potential changes linearly across the Hall bar, while in other field ranges the potential drop is confined to the edges of the Hall bar. The predicted edge channels can explain this behavior. The periodicity of the phenomenon is related to the filling factor $\nu$, i.e., the number of Landau levels that are filled with electrons (Sect. 6.3.4). Its value depends on $1/B$ and is proportional to the electron concentration $n_e$ in the 2-DEG ($\nu = n_e h/eB$, where $h$ represents Planck's constant and $e$ the electron charge).

### 6.4.5 Magnetic Force Microscopy

To detect magnetostatic tip–sample interactions with magnetic force microscopy (MFM), a ferromagnetic probe has to be used. Such probes are readily prepared by evaporating a thin magnetic layer, e.g., 10 nm iron, onto the tip. Due to the in-plane shape anisotropy of thin films, the magnetization of such tips lies predominantly along the tip axis, i.e., perpendicular to the surface. Since magnetostatic interactions are long range, they can be separated from the topography by scanning at a certain constant height (typically around 20 nm) above the surface, where the $z$-component of the sample stray field is probed (Fig. 6.28a). Therefore, MFM is always operated in noncontact mode. The signal from the cantilever is directly recorded while the $z$-feedback is switched off. MFM can be operated in the static mode or in the dynamic modes (AM-MFM at ambient pressures and FM-MFM in vacuum). A lateral resolution below 50 nm can be routinely obtained.

#### Observation of Domain Patterns

MFM is widely used to visualize domain patterns of ferromagnetic materials. At low temperatures, Moloni et al. [244] observed the domain structure of magnetite below its Verwey transition temperature ($T_V = 122$ K), but most of the work has

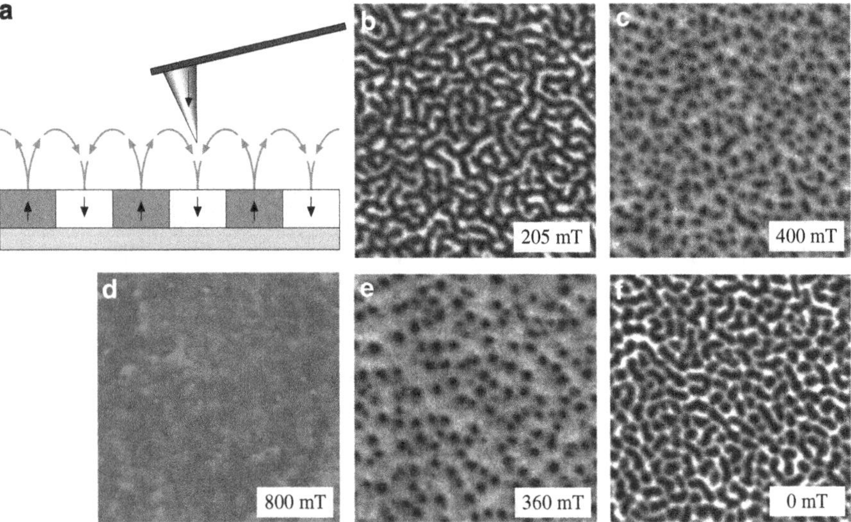


**Fig. 6.28a–f** Principle of MFM operation **(a)** and field-dependent domain structure of a ferromagnetic thin film **(b–f)** recorded at 5.2 K with FM-MFM. All images were recorded on the same 4 μm × 4 μm scan area. The $La_{0.7}Ca_{0.3}MnO_3/LaAlO_3$ system exhibits a substrate-induced out-of-plane anisotropy. *Bright* and *dark areas* are visible and correspond to attractive and repulsive magnetostatic interactions, respectively. The series shows how the domain pattern evolves along the major hysteresis loop, i.e., from zero field to saturation at 600 mT and back to zero field

concentrated on thin films of $La_{1-x}Ca_xMnO_3$ [245, 246, 247]. Below $T_V$, the conductivity decreases by two orders of magnitude and a small structural distortion is observed. The domain structure of this mixed-valence manganite is of great interest, because its resistivity strongly depends on the external magnetic field, i.e., it exhibits a large colossal-magnetoresistive effect. To investigate the field dependence of the domain patterns under ambient conditions, electromagnets have to be used. They can cause severe thermal drift problems due to Joule heating of the coils by large currents. Flux densities on the order of 100 mT can be achieved. In contrast, much larger flux densities (more than 10 T) can be rather easily produced by implementing a superconducting magnet in low-temperature setups. Using such a design, Liebmann et al. [247] recorded the domain structure along the major hysteresis loop of $La_{0.7}Ca_{0.3}MnO_3$ epitaxially grown on $LaAlO_3$ (Fig. 6.28b–f). The film geometry (with thickness of 100 nm) favors an in-plane magnetization, but the lattice mismatch with the substrate induces an out-of-plane anisotropy. Thereby, an irregular pattern of strip domains appears at zero field. If the external magnetic field is increased, the domains with antiparallel orientation shrink and finally disappear in saturation (Fig. 6.28b, c). The residual contrast in saturation (Fig. 6.28d) reflects topographic features. If the field is decreased after saturation (Fig. 6.28e, f), cylindrical domains first nucleate and then start to grow. At zero field, the maze-type domain pattern has evolved again. Such data sets can be used to analyze domain nucleation and the domain growth mode. Moreover, due to the negligible drift, domain structure and surface morphology can be directly compared, because every MFM can be used as a regular topography-imaging force microscope.

### Detection of Individual Vortices in Superconductors

Numerous low-temperature MFM experiments have been performed on superconductors [248, 249, 250, 251, 252, 253, 254, 255]. Some basic features of superconductors have been mentioned already in Sect. 6.3.5. The main difference of STM/STS compared to MFM is its high sensitivity to the electronic properties of the surface. Therefore, careful sample preparation is a prerequisite. This is not so important for MFM experiments, since the tip is scanned at a certain distance above the surface.

Superconductors can be divided into two classes with respect to their behavior in an external magnetic field. For type I superconductors, all magnetic flux is entirely excluded below their critical temperature $T_c$ (Meissner effect), while for type II superconductors, cylindrical inclusions (*vortices*) of normal material exist in a superconducting matrix (*vortex* state). The radius of the vortex *core*, where the Cooper-pair density decreases to zero, is on the order of the coherence length $\xi$. Since the superconducting gap vanishes in the core, they can be detected by STS (see Sect. 6.3.5). Additionally, each vortex contains one magnetic quantum flux $\Phi = h/2e$ (where $h$ represents Planck's constant and $e$ the electron charge). Circular supercurrents around the core screen the magnetic field associated with a vortex; their radius is given by the London penetration depth $\lambda$ of the material.

This magnetic field of the vortices can be detected by MFM. Investigations have been performed on the two most popular copper oxide high-$T_c$ superconductors, $YBa_2Cu_3O_7$ [248, 249, 251] and $Bi_2Sr_2CaCu_2O_8$ [249, 255], on the only elemental conventional type II superconductor Nb [252, 253] and on the layered compound crystal $NbSe_2$ [250, 252].

Most often, vortices have been generated by cooling the sample from the normal state to below $T_c$ in an external magnetic field. After such a *field-cooling* procedure, the most energetically favorable vortex arrangement is a regular triangular Abrikosov lattice. Volodin et al. [250] were able to observe such an Abrikosov lattice on $NbSe_2$. The intervortex distance $d$ is related to the external field during $B$ cool down via $d = (4/3)^{1/4}(\Phi/B)^{1/2}$. Another way to introduce vortices into a type II superconductor is vortex penetration from the edge by applying a magnetic field at temperatures below $T_c$. According to the Bean model, a vortex density gradient exists under such conditions within the superconducting material. Pi et al. [255] slowly increased the external magnetic field until the vortex front approaching from the edge reached the scanning area.

If the vortex configuration is dominated by the *pinning* of vortices at randomly distributed structural defects, no Abrikosov lattice emerges. The influence of pinning centers can be studied easily by MFM, because every MFM can be used to scan the topography in its AFM mode. This has been done for natural growth defects by Moser et al. [251] on $YBa_2Cu_3O_7$ and for $YBa_2Cu_3O_7$ and niobium thin films, respectively, by Volodin et al. [254]. Roseman and Grütter [256] investigated the formation of vortices in the presence of an artificial structure on niobium films, while Pi et al. [255] produced columnar defects by heavy-ion bombardment in a $Bi_2Sr_2CaCu_2O_8$ single crystal to study the strong pinning at these defects.

Figure 6.29 demonstrates that MFM is sensitive to the polarity of vortices. In Fig. 6.29a, six vortices have been produced in a niobium film by field cooling in +0.5 mT. The external magnetic field and tip magnetization are parallel, and

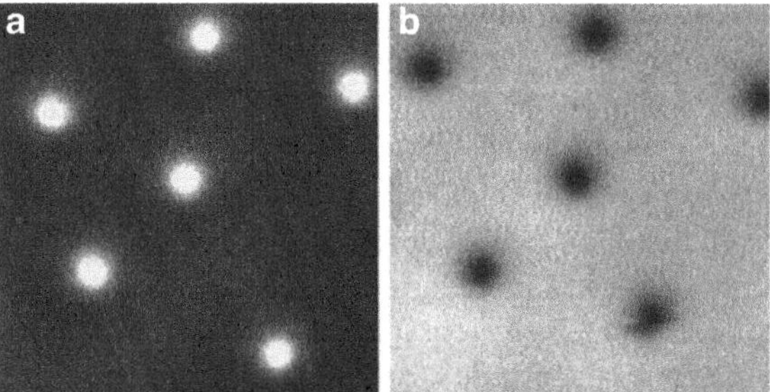


**Fig. 6.29a, b** Two 5 μm ×5 μm FM-MFM images of vortices in a niobium thin film after field-cooling at 0.5 mT (**a**) and −0.5 mT (**b**), respectively. Since the external magnetic field was parallel in (**a**) and antiparallel in (**b**) with respect to the tip magnetization, the vortices exhibit opposite contrast. Strong pinning dominates the position of the vortices, since they appear at identical locations in (**a**) and (**b**) and are not arranged in a regular Abrikosov lattice (© P. Grütter [253])

therefore the tip–vortex interaction is attractive (bright contrast). To remove the vortices, the niobium was heated above $T_c$ ($\approx 9$ K). Thereafter, vortices of opposite polarity were produced by field-cooling in $-0.5$ mT, which appear dark in Fig. 6.29b. The vortices are probably bound to strong pinning sites, because the vortex positions are identical in both images of Fig. 6.29. By imaging the vortices at different scanning heights, Roseman et al. [253] tried to extract values for the London penetration depth from the scan-height dependence of their profiles. While good qualitative agreement with theoretical predictions has been found, the absolute values do not agree with published literature values. The disagreement was attributed to the convolution between the tip and vortex stray fields. Better values might be obtained with calibrated tips.

### *6.4.6 Magnetic Exchange Force Microscopy*

The resolution of MFM is limited to the nanometer range, because the long-range magnetostatic tip–sample interaction is not localized between individual surface atoms and the foremost tip apex atom [257]. As early as 1991 Wiesendanger et al. [258] proposed that the short-range magnetic exchange interaction could be utilized to image the configuration of magnetic moments with atomic resolution. For the suggested test system NiO(001), an antiferromagnetic insulator, Momida and Oguchi [259] provided density-functional calculations. They found a magnetic exchange force between the magnetic moments of a single iron atom (the tip) and nickel surface atoms of more than 0.1 nN at tip–sample distances below 0.5 nm. Recently, Kaiser et al. [260] were able to prove the feasibility of magnetic exchange force microscopy (MExFM) on NiO(001). The superexchange between neighboring {111} planes via bridging oxygen atoms results in a row-wise antiferromagnetic configuration of magnetic moments on the (001) surface. Hence the magnetic surface unit cell is twice as large as the chemical surface unit cell. Figure 6.30a shows the atomic-scale contrast due to a pure chemical interaction. Maxima and minima correspond to the oxygen and nickel atoms, respectively. Their arrangement represents the $(1 \times 1)$ surface unit cell. Figure 6.30b exhibits an additional modulation on chemically and structurally equivalent rows of nickel atoms (the minima). The structure corresponds to the $(2 \times 1)$ magnetic surface unit cell. Since the spin-carrying nickel 3d states are highly localized, the magnetic contrast only becomes significant at very small tip–sample distances. More recently Schmidt et al. [261] were able to perform MExFM with much better signal-to-noise ratio on an itinerant metallic system: the antiferromagnetic iron monolayer on W(001). Density-functional theory performed with a realistic tip model indicated significant relaxations of tip and sample atoms during imaging. Moreover, a comparison between simulation and experimental data revealed complex interplay between chemical and magnetic interaction, which results in the observed atomic-scale contrast.

Even more ambitious is the proposed detection of individual nuclear spins by magnetic resonance force microscopy (MRFM) using a magnetic tip [263, 264].

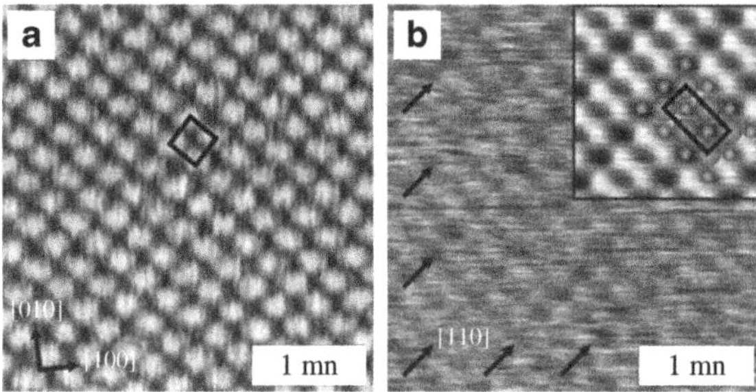


**Fig. 6.30** (**a**) Pure chemical contrast on NiO(001) obtained with AFM using a nonmagnetic tip. Oxygen and nickel atoms are represented as maxima and minima, respectively, forming the (1 × 1) surface unit cell (*black square*). *Arrows* indicate the main crystallographic directions. (**b**) Additional modulation on neighboring nickel rows along the [110] direction (see *arrows*) due to the magnetic exchange interaction obtained with MExFM using a magnetic tip. The (2 × 1) structure (*black rectangle in the inset*) represents the magnetic surface unit cell. The *inset* is tiled together from the averaged magnetic unit cell calculated from the raw data, whereby the signal-to-noise ratio is significantly increased

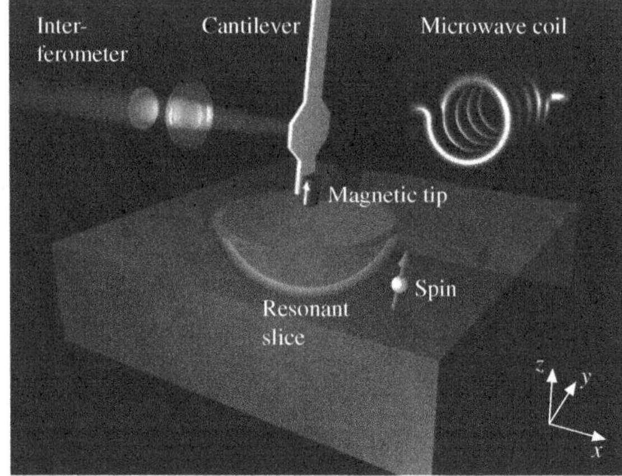


**Fig. 6.31** MRFM setup. The cantilever with the magnetic tip oscillates parallel to the surface. Only electron spins within a hemispherical slice, where the stray field of the tip plus the external field matches the condition for magnetic resonance, can contribute to the MRFM signal due to cyclic spin inversion (© D. Rugar [262])

Conventionally, nuclear spins are investigated by nuclear magnetic resonance (NMR), a spectroscopic technique to obtain microscopic chemical and physical information about molecules. An important application of NMR for medical diagnostics of the inside of humans is magnetic resonance imaging (MRI). This tomographic imaging technique uses the NMR signal from thin slices through the body to reconstruct its three-dimensional structure. Currently, at least $10^{12}$ nuclear spins must be present in a given volume to obtain a significant MRI signal. The ultimate goal of MRFM is to combine aspects of force microscopy with MRI to achieve true 3-D imaging with atomic resolution and elemental selectivity.

The experimental setup is sketched in Fig. 6.31. An oscillating cantilever with a magnetic tip at its end points toward the surface. The spherical resonant slice within the sample represents those points where the stray field from the tip and the external field match the condition for magnetic resonance. The cyclic spin flip causes a slight shift of the cantilever frequency due to the magnetic force exerted by the spin on the tip. Since the forces are extremely small, very low temperatures are required.

To date, no individual nuclear spins have been detected by MRFM. However, the design of ultrasensitive cantilevers has made considerable progress, and the detection of forces below $1\times10^{-18}$ N has been achieved [202]. Therefore, it has become possible to perform nuclear magnetic resonance [265], and ferromagnetic resonance [266] experiments of spin ensembles with micrometer resolution. Moreover, in $SiO_2$ the magnetic moment of a single electron, which is three orders of magnitude larger than the nuclear magnetic moment, could be detected [262] using the setup shown in Fig. 6.31 at 1.6 K. This major breakthrough demonstrates the capability of force microscopy to detect single spins.

## References


1. G. Binnig, H. Rohrer, C. Gerber, E. Weibel: Surface studies by scanning tunneling microscopy, Phys. Rev. Lett. **49**, 57–61 (1982)
2. R. Wiesendanger: *Scanning Probe Microscopy and Spectroscopy* (Cambridge Univ. Press, Cambridge, 1994)
3. M. Tinkham: *Introduction to Superconductivity* (McGraw–Hill, New York, 1996)
4. J. Kondo: Theory of dilute magnetic alloys, Solid State Phys. **23**, 183–281 (1969)
5. T.R. Albrecht, P. Grütter, H.K. Horne, D. Rugar: Frequency modulation detection using high-$Q$ cantilevers for enhanced force microscope sensitivity, J. Appl. Phys. **69**, 668–673 (1991)
6. F.J. Giessibl, H. Bielefeld, S. Hembacher, J. Mannhart: Calculation of the optimal imaging parameters for frequency modulation atomic force microscopy, Appl. Surf. Sci. **140**, 352–357 (1999)
7. W. Allers, A. Schwarz, U.D. Schwarz, R. Wiesendanger: Dynamic scanning force microscopy at low temperatures on a van der Waals surface: Graphite(0001), Appl. Surf. Sci. **140**, 247–252 (1999)
8. W. Allers, A. Schwarz, U.D. Schwarz, R. Wiesendanger: Dynamic scanning force microscopy at low temperatures on a noble-gas crystal: Atomic resolution on the xenon(111) surface, Europhys. Lett. **48**, 276–279 (1999)
9. M. Morgenstern, D. Haude, V. Gudmundsson, C. Wittneven, R. Dombrowski, R. Wiesendanger: Origin of Landau oscillations observed in scanning tunneling spectroscopy on $n$-InAs(110), Phys. Rev. B **62**, 7257–7263 (2000)
10. D.M. Eigler, P.S. Weiss, E.K. Schweizer, N.D. Lang: Imaging Xe with a low-temperature scanning tunneling microscope, Phys. Rev. Lett. **66**, 1189–1192 (1991)
11. P.S. Weiss, D.M. Eigler: Site dependence of the apparent shape of a molecule in scanning tunneling micoscope images: Benzene on Pt{111}, Phys. Rev. Lett. **71**, 3139–3142 (1992)
12. D.M. Eigler, E.K. Schweizer: Positioning single atoms with a scanning tunneling microscope, Nature **344**, 524–526 (1990)
13. H. Hug, B. Stiefel, P.J.A. van Schendel, A. Moser, S. Martin, H.-J. Güntherodt: A low temperature ultrahigh vacuum scanning force microscope, Rev. Sci. Instrum. **70**, 3627–3640 (1999)
14. S. Behler, M.K. Rose, D.F. Ogletree, F. Salmeron: Method to characterize the vibrational response of a beetle type scanning tunneling microscope, Rev. Sci. Instrum. **68**, 124–128 (1997)
15. C. Wittneven, R. Dombrowski, S.H. Pan, R. Wiesendanger: A low-temperature ultrahigh-vacuum scanning tunneling microscope with rotatable magnetic field, Rev. Sci. Instrum. **68**, 3806–3810 (1997)

16. W. Allers, A. Schwarz, U.D. Schwarz, R. Wiesendanger: A scanning force microscope with atomic resolution in ultrahigh vacuum and at low temperatures, Rev. Sci. Instrum. **69**, 221–225 (1998)
17. G. Dujardin, R.E. Walkup, P. Avouris: Dissociation of individual molecules with electrons from the tip of a scanning tunneling microscope, Science **255**, 1232–1235 (1992)
18. H.J. Lee, W. Ho: Single-bond formation and characterization with a scanning tunneling microscope, Science **286**, 1719–1722 (1999)
19. J. Repp, G. Meyer, F.E. Olsson, M. Persson: Controlling the charge state of individual gold adatoms, Science **305**, 493–495 (2004)
20. R. Berndt, R. Gaisch, J.K. Gimzewski, B. Reihl, R.R. Schlittler, W.D. Schneider, M. Tschudy: Photon emission at molecular resolution induced by a scanning tunneling microscope, Science **262**, 1425–1427 (1993)
21. G.V. Nazin, X.H. Qui, W. Ho: Atomic engineering of photon emission with a scanning tunneling microscopy, Phys. Rev. Lett. **90**, 216110–1–216110–4 (2003)
22. B.G. Briner, M. Doering, H.P. Rust, A.M. Bradshaw: Microscopic diffusion enhanced by adsorbate interaction, Science **278**, 257–260 (1997)
23. F. Meier, L. Zhou, J. Wiebe, R. Wiesendanger: Revealing magnetic interactions from single-atom magnetization curves, Science **320**, 82–86 (2008)
24. J. Kliewer, R. Berndt, E.V. Chulkov, V.M. Silkin, P.M. Echenique, S. Crampin: Dimensionality effects in the lifetime of surface states, Science **288**, 1399–1401 (2000)
25. M.F. Crommie, C.P. Lutz, D.M. Eigler: Imaging standing waves in a two-dimensional electron gas, Nature **363**, 524–527 (1993)
26. B.C. Stipe, M.A. Rezaei, W. Ho: Single-molecule vibrational spectroscopy and microscopy, Science **280**, 1732–1735 (1998)
27. A.J. Heinrich, J.A. Gupta, C.P. Lutz, D. Eigler: Single-atom spin-flip spectroscopy, Science **306**, 466–469 (2004)
28. H. Gawronski, M. Mehlhorn, K. Morgenstern: Imaging phonon excitation with atomic resolution, Science **319**, 930–933 (2008)
29. C.W.J. Beenakker, H. van Houten: Quantum transport in semiconductor nanostructures, Solid State Phys. **44**, 1–228 (1991)
30. K.M. Lang, V. Madhavan, J.E. Hoffman, E.W. Hudson, H. Eisaki, S. Uchida, J.C. Davis: Imaging the granular structure of high-$T_c$ superconductivity in underdoped $Bi_2Sr_2CaCu_2O_{8+\delta}$, Nature **415**, 412–416 (2002)
31. J. Lee, K. Fujita, K. McElroy, J.A. Slezak, M. Wang, Y. Aiura, H. Bando, M. Ishikado, T. Masui, J.X. Zhu, A.V. Balatsky, H. Eisaki, S. Uchida, J.C. Davis: Interplay of electron-lattice interactions and superconductivity in $Bi_2Sr_2CaCu_2O_{8+\delta}$, Nature **442**, 546–550 (2006)
32. K.K. Gomes, A.N. Pasupathy, A. Pushp, S. Ono, Y. Ando, A. Yazdani: Visualizing pair formation on the atomic scale in the high-$T_c$ superconductor $Bi_2Sr_2CaCu_2O_{8+\delta}$, Nature **447**, 569–572 (2007)
33. R.S. Becker, J.A. Golovchenko, B.S. Swartzentruber: Atomic-scale surface modifications using a tunneling microscope, Nature **325**, 419–421 (1987)
34. P.G. Piva, G.A. DiLabio, J.L. Pitters, J. Zikovsky, M. Rezeq, S. Dogel, W.A. Hofer, R.A. Wolkow: Field regulation of single-molecule conductivity by a charged surface atom, Nature **435**, 658–661 (2005)
35. J.A. Stroscio, D.M. Eigler: Atomic and molecular manipulation with the scanning tunneling microscope, Science **254**, 1319–1326 (1991)
36. A.J. Heinrich, J.A. Gupta, C.P. Lutz, D. Eigler: Molecule cascades, Science **298**, 1381–1387 (2002)
37. L. Bartels, G. Meyer, K.H. Rieder: Basic steps of lateral manipulation of single atoms and diatomic clusters with a scanning tunneling microscope, Phys. Rev. Lett. **79**, 697–700 (1997)
38. J.A. Stroscio, R.J. Celotta: Controlling the dynamics of a single atom in lateral atom manipulation, Science **306**, 242–247 (2004)

39. J.J. Schulz, R. Koch, K.H. Rieder: New mechanism for single atom manipulation, Phys. Rev. Lett. **84**, 4597–4600 (2000)
40. J.A. Stroscio, F. Tavazza, J.N. Crain, R.J. Celotta, A.M. Chaka: Electronically induced atom motion in engineered $CoCu_n$ nanostructures, Science **313**, 948–951 (2006)
41. M. Lastapis, M. Martin, D. Riedel, L. Hellner, G. Comtet, G. Dujardin: Picometer-scale electronic control of molecular dynamics inside a single molecule, Science **308**, 1000–1003 (2005)
42. M. Ternes, C.P. Lutz, C.F. Hirjibehedin, F.J. Giessibl, A.J. Heinrich: The force needed to move an atom on a surface, Science **319**, 1066–1069 (2008)
43. J.I. Pascual, N. Lorente, Z. Song, H. Conrad, H.P. Rust: Selectivity in vibrationally mediated single-molecule chemistry, Nature **423**, 525–528 (2003)
44. T.C. Shen, C. Wang, G.C. Abeln, J.R. Tucker, J.W. Lyding, P. Avouris, R.E. Walkup: Atomic-scale desorption through electronic and vibrational excitation mechanisms, Science **268**, 1590–1592 (1995)
45. T. Komeda, Y. Kim, M. Kawai, B.N.J. Persson, H. Ueba: Lateral hopping of molecules induced by excitations of internal vibration mode, Science **295**, 2055–2058 (2002)
46. Y.W. Mo: Reversible rotation of antimony dimers on the silicon(001) surface with a scanning tunneling microscope, Science **261**, 886–888 (1993)
47. B.C. Stipe, M.A. Rezaei, W. Ho: Inducing and viewing the rotational motion of a single molecule, Science **279**, 1907–1909 (1998)
48. P. Liljeroth, J. Repp, G. Meyer: Current-induced hydrogen tautomerization and conductance switching of napthalocyanine molecules, Science **317**, 1203–1206 (2007)
49. P. Maksymovych, D.B. Dougherty, X.-Y. Zhu, J.T. Yates Jr.: Nonlocal dissociative chemistry of adsorbed molecules induced by localized electron injection into metal surfaces, Phys. Rev. Lett. **99**, 016101–1–016101–4 (2007)
50. G.V. Nazin, X.H. Qiu, W. Ho: Visualization and spectroscopy of a metal-molecule-metal bridge, Science **302**, 77–81 (2003)
51. J. Repp, G. Meyer, S. Paavilainen, F.E. Olsson, M. Persson: Imaging bond formation between a gold atom and pentacene on an insulating surface, Science **312**, 1196–1199 (2006)
52. S. Katano, Y. Kim, M. Hori, M. Trenary, M. Kawai: Reversible control of hydrogenation of a single molecule, Science **316**, 1883–1886 (2007)
53. R. Yamachika, M. Grobis, A. Wachowiak, M.F. Crommie: Controlled atomic doping of a single $C_60$ molecule, Science **304**, 281–284 (2004)
54. S.W. Hla, L. Bartels, G. Meyer, K.H. Rieder: Inducing all steps of a chemical reaction with the scanning tunneling microscope tip: Towards single molecule engineering, Phys. Rev. Lett. **85**, 2777–2780 (2000)
55. Y. Kim, T. Komeda, M. Kawai: Single-molecule reaction and characterization by vibrational excitation, Phys. Rev. Lett. **89**, 126104–1–126104–4 (2002)
56. M. Berthe, R. Stiufiuc, B. Grandidier, D. Deresmes, C. Delerue, D. Stievenard: Probing the carrier capture rate of a single quantum level, Science **319**, 436–438 (2008)
57. N. Neel, J. Kröger, L. Limot, K. Palotas, W.A. Hofer, R. Berndt: Conductance and Kondo effect in a controlled single-atom contact, Phys. Rev. Lett. **98**, 016801–1–016801–4 (2007)
58. F. Rosei, M. Schunack, P. Jiang, A. Gourdon, E. Laegsgaard, I. Stensgaard, C. Joachim, F. Besenbacher: Organic molecules acting as templates on metal surfaces, Science **296**, 328–331 (2002)
59. E. Ganz, S.K. Theiss, I.S. Hwang, J. Golovchenko: Direct measurement of diffusion by hot tunneling microscopy: Activations energy, anisotropy, and long jumps, Phys. Rev. Lett. **68**, 1567–1570 (1992)
60. M. Schunack, T.R. Linderoth, F. Rosei, E. Laegsgaard, I. Stensgaard, F. Besenbacher: Long jumps in the surface diffusion of large molecules, Phys. Rev. Lett. **88**, 156102–1–156102–4 (2002)
61. K.L. Wong, G. Pawin, K.Y. Kwon, X. Lin, T. Jiao, U. Solanki, R.H.J. Fawcett, L. Bartels, S. Stolbov, T.S. Rahman: A molecule carrier, Science **315**, 1391–1393 (2007)

62. L.J. Lauhon, W. Ho: Direct observation of the quantum tunneling of single hydrogen atoms with a scanning tunneling microscope, Phys. Rev. Lett. **85**, 4566–4569 (2000)
63. F. Ronci, S. Colonna, A. Cricenti, G. LeLay: Evidence of Sn adatoms quantum tunneling at the $\alpha$-Sn/Si(111) surface, Phys. Rev. Lett. **99**, 166103–1–166103–4 (2007)
64. N. Kitamura, M. Lagally, M.B. Webb: Real-time observation of vacancy diffusion on Si (001)-(2 $\times$ 1) by scanning tunneling microscopy, Phys. Rev. Lett. **71**, 2082–2085 (1993)
65. M. Morgenstern, T. Michely, G. Comsa: Onset of interstitial diffusion determined by scanning tunneling microscopy, Phys. Rev. Lett. **79**, 1305–1308 (1997)
66. K. Morgenstern, G. Rosenfeld, B. Poelsema, G. Comsa: Brownian motion of vacancy islands on Ag(111), Phys. Rev. Lett. **74**, 2058–2061 (1995)
67. L. Bartels, F. Wang, D. Möller, E. Knoesel, T.F. Heinz: Real-space observation of molecular motion induced by femtosecond laser pulses, Science **305**, 648–651 (2004)
68. B. Reihl, J.H. Coombs, J.K. Gimzewski: Local inverse photoemission with the scanning tunneling microscope, Surf. Sci. **211/212**, 156–164 (1989)
69. R. Berndt, J.K. Gimzewski, P. Johansson: Inelastic tunneling excitation of tip-induced plasmon modes on noble-metal surfaces, Phys. Rev. Lett. **67**, 3796–3799 (1991)
70. P. Johansson, R. Monreal, P. Apell: Theory for light emission from a scanning tunneling microscope, Phys. Rev. B **42**, 9210–9213 (1990)
71. J. Aizpurua, G. Hoffmann, S.P. Apell, R. Berndt: Electromagnetic coupling on an atomic scale, Phys. Rev. Lett. **89**, 156803–1–156803–4 (2002)
72. G. Hoffmann, J. Kliewer, R. Berndt: Luminescence from metallic quantum wells in a scanning tunneling microscope, Phys. Rev. Lett. **78**, 176803–1–176803–4 (2001)
73. X.H. Qiu, G.V. Nazin, W. Ho: Vibrationally resolved fluorescence excited with submolecular precission, Science **299**, 542–546 (2003)
74. E. Cavar, M.C. Blüm, M. Pivetta, F. Patthey, M. Chergui, W.D. Schneider: Fluorescence and phosphorescence from individual $C_60$ molecules excited by local electron tunneling, Phys. Rev. Lett. **95**, 196102–1–196102–4 (2005)
75. S.W. Wu, N. Ogawa, W. Ho: Atomic-scale coupling of photons to single-molecule junctions, Science **312**, 1362–1365 (2006)
76. A. Downes, M.E. Welland: Photon emission from Si(111)-(7 $\times$ 7) induced by scanning tunneling microscopy: atomic scale and material contrast, Phys. Rev. Lett. **81**, 1857–1860 (1998)
77. M. Kemerink, K. Sauthoff, P.M. Koenraad, J.W. Geritsen, H. van Kempen, J.H. Wolter: Optical detection of ballistic electrons injected by a scanning–tunneling microscope, Phys. Rev. Lett. **86**, 2404–2407 (2001)
78. J. Tersoff, D.R. Hamann: Theory and application for the scanning tunneling microscope, Phys. Rev. Lett. **50**, 1998–2001 (1983)
79. C.J. Chen: *Introduction to Scanning Tunneling Microscopy* (Oxford University Press, Oxford, 1993)
80. J. Winterlin, J. Wiechers, H. Brune, T. Gritsch, H. Hofer, R.J. Behm: Atomic-resolution imaging of close-packed metal surfaces by scanning tunneling microscopy, Phys. Rev. Lett. **62**, 59–62 (1989)
81. J.A. Stroscio, R.M. Feenstra, A.P. Fein: Electronic structure of the Si(111) 2 $\times$ 1 surface by scanning-tunneling microscopy, Phys. Rev. Lett. **57**, 2579–2582 (1986)
82. A.L. Vázquez de Parga, O.S. Hernan, R. Miranda, A. Levy Yeyati, N. Mingo, A. Martín-Rodero, F. Flores: Electron resonances in sharp tips and their role in tunneling spectroscopy, Phys. Rev. Lett. **80**, 357–360 (1998)
83. S.H. Pan, E.W. Hudson, J.C. Davis: Vacuum tunneling of superconducting quasiparticles from atomically sharp scanning tunneling microscope tips, Appl. Phys. Lett. **73**, 2992–2994 (1998)
84. J.T. Li, W.D. Schneider, R. Berndt, O.R. Bryant, S. Crampin: Surface-state lifetime measured by scanning tunneling spectroscopy, Phys. Rev. Lett. **81**, 4464–4467 (1998)
85. L. Bürgi, O. Jeandupeux, H. Brune, K. Kern: Probing hot-electron dynamics with a cold scanning tunneling microscope, Phys. Rev. Lett. **82**, 4516–4519 (1999)

86. J.W.G. Wildoer, C.J.P.M. Harmans, H. van Kempen: Observation of Landau levels at the InAs(110) surface by scanning tunneling spectroscopy, Phys. Rev. B **55**, R16013–R16016 (1997)
87. M. Morgenstern, V. Gudmundsson, C. Wittneven, R. Dombrowski, R. Wiesendanger: Nonlocality of the exchange interaction probed by scanning tunneling spectroscopy, Phys. Rev. B **63**, 201301(R)-1–201301(R)-4 (2001)
88. F.E. Olsson, M. Persson, A.G. Borisov, J.P. Gauyacq, J. Lagoute, S. Fölsch: Localization of the Cu(111) surface state by single Cu adatoms, Phys. Rev. Lett. **93**, 206803–1–206803–4 (2004)
89. L. Limot, E. Pehlke, J. Kröger, R. Berndt: Surface-state localization at adatoms, Phys. Rev. Lett. **94**, 036805–1–036805–4 (2005)
90. N. Nilius, T.M. Wallis, M. Persson, W. Ho: Distance dependence of the interaction between single atoms: Gold dimers on NiAl(110), Phys. Rev. Lett. **90**, 196103–1–196103–4 (2003)
91. H.J. Lee, W. Ho, M. Persson: Spin splitting of s and p states in single atoms and magnetic coupling in dimers on a surface, Phys. Rev. Lett. **92**, 186802–1–186802–4 (2004)
92. M.V. Grishin, F.I. Dalidchik, S.A. Kovalevskii, N.N. Kolchenko, B.R. Shub: Isotope effect in the vibrational spectra of water measured in experiments with a scanning tunneling microscope, JETP Lett. **66**, 37–40 (1997)
93. Y. Sainoo, Y. Kim, T. Okawa, T. Komeda, H. Shigekawa, M. Kawai: Excitation of molecular vibrational modes with inelastic scanning tunneling microscopy: Examination through action spectra of *cis*-2-butene on Pd(110), Phys. Rev. Lett. **95**, 246102–1–246102–4 (2005)
94. X.H. Qiu, G.V. Nazin, W. Ho: Vibronic states in single molecule electron transport, Phys. Rev. Lett. **92**, 206102–1–206102–4 (2004)
95. L. Vitali, M. Burghard, M.A. Schneider, L. Liu, S.Y. Wu, C.S. Jayanthi, K. Kern: Phonon spectromicroscopy of carbon nanostructures with atomic resolution, Phys. Rev. Lett. **93**, 136103–1–136103–4 (2004)
96. B.J. LeRoy, S.G. Lemay, J. Kong, C. Dekker: Electrical generation and absorption of phonons in carbon nanotubes, Nature **432**, 371–374 (2004)
97. L. Vitali, M.A. Schneider, K. Kern, L. Wirtz, A. Rubio: Phonon and plasmon excitation in inelastic scanning tunneling spectroscopy of graphite, Phys. Rev. B **69**, 121414–1–121414–4 (2004)
98. T. Balashov, A.F. Takacz, W. Wulfhekel, J. Kirschner: Magnon excitation with spin-polarized scanning tunneling microscopy, Phys. Rev. Lett. **97**, 187201–1–187201–4 (2006)
99. C.F. Hirjibehedin, C.P. Lutz, A.J. Heinrich: Spin coupling in engineered atomic structures, Science **312**, 1021–1024 (2006)
100. C.F. Hirjibehedin, C.Y. Lin, A.F. Otte, M. Ternes, C.P. Lutz, B.A. Jones, A.J. Heinrich: Large magnetic anisotropy of a single atomic spin embedded in a surface molecular network, Science **317**, 1199–1203 (2007)
101. U. Kemiktarak, T. Ndukum, K.C. Schwab, K.L. Ekinci: Radio-frequency scanning tunneling microscopy, Nature **450**, 85–89 (2007)
102. A. Hewson: *From the Kondo Effect to Heavy Fermions* (Cambridge Univ. Press, Cambridge, 1993)
103. V. Madhavan, W. Chen, T. Jamneala, M.F. Crommie, N.S. Wingreen: Tunneling into a single magnetic atom: Spectroscopic evidence of the Kondo resonance, Science **280**, 567–569 (1998)
104. J. Li, W.D. Schneider, R. Berndt, B. Delley: Kondo scattering observed at a single magnetic impurity, Phys. Rev. Lett. **80**, 2893–2896 (1998)
105. T.W. Odom, J.L. Huang, C.L. Cheung, C.M. Lieber: Magnetic clusters on single-walled carbon nanotubes: the Kondo effect in a one-dimensional host, Science **290**, 1549–1552 (2000)
106. M. Ouyang, J.L. Huang, C.L. Cheung, C.M. Lieber: Energy gaps in metallic single-walled carbon nanotubes, Science **292**, 702–705 (2001)
107. U. Fano: Effects of configuration interaction on intensities and phase shifts, Phys. Rev. **124**, 1866–1878 (1961)

108. P. Wahl, L. Diekhöner, M.A. Schneider, L. Vitali, G. Wittich, K. Kern: Kondo temperature of magnetic impurities at surfaces, Phys. Rev. Lett. **93**, 176603–1–176603–4 (2004)
109. Y.S. Fu, S.H. Ji, X. Chen, X.C. Ma, R. Wu, C.C. Wang, W.H. Duan, X.H. Qiu, B. Sun, P. Zhang, J.F. Jia, Q.K. Xue: Manipulating the Kondo resonance through quantum size effects, Phys. Rev. Lett. **99**, 256601–1–256601–4 (2007)
110. J. Henzl, K. Morgenstern: Contribution of the surface state to the observation of the surface Kondo resonance, Phys. Rev. Lett. **98**, 266601–1–266601–4 (2007)
111. P. Wahl, P. Simon, L. Diekhöner, V.S. Stepanyuk, P. Bruno, M.A. Schneider, K. Kern: Exchange interaction between single magnetic adatoms, Phys. Rev. Lett. **98**, 056601–1–056601–4 (2007)
112. T. Jamneala, V. Madhavan, M.F. Crommie: Kondo response of a single antiferromagnetic chromium trimer, Phys. Rev. Lett. **87**, 256804–1–256804–4 (2001)
113. P. Wahl, L. Diekhöner, G. Wittich, L. Vitali, M.A. Schneider, K. Kern: Kondo effect of molecular complexes at surfaces: Ligand control of the local spin coupling, Phys. Rev. Lett. **95**, 166601–1–166601–4 (2005)
114. A. Zhao, Q. Li, L. Chen, H. Xiang, W. Wang, S. Pan, B. Wang, X. Xiao, J. Yang, J.G. Hou, Q. Zhu: Controlling the Kondo effect of an adsorbed magnetic ion through its chemical bonding, Science **309**, 1542–1544 (2005)
115. V. Iancu, A. Deshpande, S.W. Hla: Manipulation of the Kondo effect via two-dimensional molecular assembly, Phys. Rev. Lett. **97**, 266603–1–266603–4 (2006)
116. L. Gao, W. Ji, Y.B. Hu, Z.H. Cheng, Z.T. Deng, Q. Liu, N. Jiang, X. Lin, W. Guo, S.X. Du, W.A. Hofer, X.C. Xie, H.J. Gao: Site-specific Kondo effect at ambient temperatures in iron-based molecules, Phys. Rev. Lett. **99**, 106402–1–106402–4 (2007)
117. H.C. Manoharan, C.P. Lutz, D.M. Eigler: Quantum mirages formed by coherent projection of electronic structure, Nature **403**, 512–515 (2000)
118. H.A. Mizes, J.S. Foster: Long-range electronic perturbations caused by defects using scanning tunneling microscopy, Science **244**, 559–562 (1989)
119. P.T. Sprunger, L. Petersen, E.W. Plummer, E. Laegsgaard, F. Besenbacher: Giant Friedel oscillations on beryllium(0001) surface, Science **275**, 1764–1767 (1997)
120. P. Hofmann, B.G. Briner, M. Doering, H.P. Rust, E.W. Plummer, A.M. Bradshaw: Anisotropic two-dimensional Friedel oscillations, Phys. Rev. Lett. **79**, 265–268 (1997)
121. E.J. Heller, M.F. Crommie, C.P. Lutz, D.M. Eigler: Scattering and adsorption of surface electron waves in quantum corrals, Nature **369**, 464–466 (1994)
122. M.C.M.M. van der Wielen, A.J.A. van Roij, H. van Kempen: Direct observation of Friedel oscillations around incorporated Si_Ga dopants in GaAs by low-temperature scanning tunneling microscopy, Phys. Rev. Lett. **76**, 1075–1078 (1996)
123. O. Millo, D. Katz, Y.W. Cao, U. Banin: Imaging and spectroscopy of artificial-atom states in core/shell nanocrystal quantum dots, Phys. Rev. Lett. **86**, 5751–5754 (2001)
124. T. Maltezopoulos, A. Bolz, C. Meyer, C. Heyn, W. Hansen, M. Morgenstern, R. Wiesendanger: Wave-function mapping of InAs qunatum dots by scanning tunneling spectroscopy, Phys. Rev. Lett. **91**, 196804–1–196804–4 (2003)
125. R. Temirov, S. Soubatch, A. Luican, F.S. Tautz: Free-electron-like dispersion in an organic monolayer film on a metal substrate, Nature **444**, 350–353 (2006)
126. K. Suzuki, K. Kanisawa, C. Janer, S. Perraud, K. Takashina, T. Fujisawa, Y. Hirayama: Spatial imaging of two-dimensional electronic states in semiconductor quantum wells, Phys. Rev. Lett. **98**, 136802–1–136802–4 (2007)
127. L.C. Venema, J.W.G. Wildoer, J.W. Janssen, S.J. Tans, L.J.T. Tuinstra, L.P. Kouwenhoven, C. Dekker: Imaging electron wave functions of quantized energy levels in carbon nanotubes, Nature **283**, 52–55 (1999)
128. S.G. Lemay, J.W. Jannsen, M. van den Hout, M. Mooij, M.J. Bronikowski, P.A. Willis, R.E. Smalley, L.P. Kouwenhoven, C. Dekker: Two-dimensional imaging of electronic wavefunctions in carbon nanotubes, Nature **412**, 617–620 (2001)

129. C.R. Moon, L.S. Matos, B.K. Foster, G. Zeltzer, W. Ko, H.C. Manoharan: Quantum phase extraction in isospectral electronic nanostructures, Science **319**, 782–787 (2008)
130. X. Lu, M. Grobis, K.H. Khoo, S.G. Louie, M.F. Crommie: Spatially mapping the spectral density of a single $C_60$ molecule, Phys. Rev. Lett. **90**, 096802–1–096802–4 (2003)
131. A.M. Yakunin, A.Y. Silov, P.M. Koenraad, J.H. Wolter, W. van Roy, J. de Boeck, J.M. Tang, M.E. Flatte: Spatial structure of an individual Mn acceptor in GaAs, Phys. Rev. Lett. **92**, 216806–1–216806–4 (200405)
132. F. Marczinowski, J. Wiebe, J.M. Tang, M.E. Flatte, F. Meier, M. Morgenstern, R. Wiesendanger: Local electronic structure near Mn acceptors in InAs: Surface-induced symmetry breaking and coupling to host states, Phys. Rev. Lett. **99**, 157202–1–157202–4 (2007)
133. N. Nilius, T.M. Wallis, W. Ho: Development of one-dimensional band structure in artificial gold chains, Science **297**, 1853–1856 (2002)
134. J.N. Crain, D.T. Pierce: End states in one-dimensional atom chains, Science **307**, 703–706 (2005)
135. D. Kitchen, A. Richardella, J.M. Tang, M.E. Flatte, A. Yazdani: Atom-by-atom substitution of Mn in GaAs and visualization of their hole-mediated interactions, Nature **442**, 436–439 (2006)
136. C. Wittneven, R. Dombrowski, M. Morgenstern, R. Wiesendanger: Scattering states of ionized dopants probed by low temperature scanning tunneling spectroscopy, Phys. Rev. Lett. **81**, 5616–5619 (1998)
137. D. Haude, M. Morgenstern, I. Meinel, R. Wiesendanger: Local density of states of a three-dimensional conductor in the extreme quantum limit, Phys. Rev. Lett. **86**, 1582–1585 (2001)
138. R. Joynt, R.E. Prange: Conditions for the quantum Hall effect, Phys. Rev. B **29**, 3303–3317 (1984)
139. M. Morgenstern, J. Klijn, R. Wiesendanger: Real space observation of drift states in a two-dimensional electron system at high magnetic fields, Phys. Rev. Lett. **90**, 056804–1–056804–4 (2003)
140. M. Morgenstern, J. Klijn, C. Meyer, M. Getzlaff, R. Adelung, R.A. Römer, K. Rossnagel, L. Kipp, M. Skibowski, R. Wiesendanger: Direct comparison between potential landscape and local density of states in a disordered two-dimensional electron system, Phys. Rev. Lett. **89**, 136806–1–136806–4 (2002)
141. E. Abrahams, P.W. Anderson, D.C. Licciardello, T.V. Ramakrishnan: Scaling theory of localization: absence of quantum diffusion in two dimensions, Phys. Rev. Lett. **42**, 673–676 (1979)
142. C. Meyer, J. Klijn, M. Morgenstern, R. Wiesendanger: Direct measurement of the local density of states of a disordered one-dimensional conductor, Phys. Rev. Lett. **91**, 076803–1–076803–4 (2003)
143. N. Oncel, A. van Houselt, J. Huijben, A.S. Hallbäck, O. Gurlu, H.J.W. Zandvliet, B. Poelsema: Quantum confinement between self-organized Pt nanowires on Ge(001), Phys. Rev. Lett. **95**, 116801–1–116801–4 (2005)
144. J. Lee, S. Eggert, H. Kim, S.J. Kahng, H. Shinohara, Y. Kuk: Real space imaging of one-dimensional standing waves: Direct evidence for a Luttinger liquid, Phys. Rev. Lett. **93**, 166403–1–166403–4 (2004)
145. R.E. Peierls: *Quantum Theory of Solids* (Clarendon, Oxford, 1955)
146. C.G. Slough, W.W. McNairy, R.V. Coleman, B. Drake, P.K. Hansma: Charge-density waves studied with the use of a scanning tunneling microscope, Phys. Rev. B **34**, 994–1005 (1986)
147. X.L. Wu, C.M. Lieber: Hexagonal domain-like charge-density wave of TaS_2 determined by scanning tunneling microscopy, Science **243**, 1703–1705 (1989)
148. T. Nishiguchi, M. Kageshima, N. Ara-Kato, A. Kawazu: Behaviour of charge density waves in a one-dimensional organic conductor visualized by scanning tunneling microscopy, Phys. Rev. Lett. **81**, 3187–3190 (1998)
149. X.L. Wu, C.M. Lieber: Direct observation of growth and melting of the hexagonal-domain charge-density-wave phase in 1 T-$TaS_2$ by scanning tunneling microscopy, Phys. Rev. Lett. **64**, 1150–1153 (1990)

150. J.M. Carpinelli, H.H. Weitering, E.W. Plummer, R. Stumpf: Direct observation of a surface charge density wave, Nature **381**, 398–400 (1996)
151. H.H. Weitering, J.M. Carpinelli, A.V. Melechenko, J. Zhang, M. Bartkowiak, E.W. Plummer: Defect-mediated condensation of a charge density wave, Science **285**, 2107–2110 (1999)
152. S. Modesti, L. Petaccia, G. Ceballos, I. Vobornik, G. Panaccione, G. Rossi, L. Ottaviano, R. Larciprete, S. Lizzit, A. Goldoni: Insulating ground state of Sn/Si(111)-(3×3)R30°, Phys. Rev. Lett. **98**, 126401–1–126401–4 (2007)
153. H.W. Yeom, S. Takeda, E. Rotenberg, I. Matsuda, K. Horikoshi, J. Schäfer, C.M. Lee, S.D. Kevan, T. Ohta, T. Nagao, S. Hasegawa: Instability and charge density wave of metallic quantum chains on a silicon surface, Phys. Rev. Lett. **82**, 4898–4901 (1999)
154. K. Swamy, A. Menzel, R. Beer, E. Bertel: Charge-density waves in self-assembled halogen-bridged metal chains, Phys. Rev. Lett. **86**, 1299–1302 (2001)
155. J.J. Kim, W. Yamaguchi, T. Hasegawa, K. Kitazawa: Observation of Mott localization gap using low temperature scanning tunneling spectroscopy in commensurate 1 T-$TaSe_2$, Phys. Rev. Lett. **73**, 2103–2106 (1994)
156. A. Wachowiak, R. Yamachika, K.H. Khoo, Y. Wang, M. Grobis, D.H. Lee, S.G. Louie, M.F. Crommie: Visualization of the molecular Jahn–Teller effect in an insulating $K_4C_{60}$ monolayer, Science **310**, 468–470 (2005)
157. J. Bardeen, L.N. Cooper, J.R. Schrieffer: Theory of superconductivity, Phys. Rev. **108**, 1175–1204 (1957)
158. A. Yazdani, B.A. Jones, C.P. Lutz, M.F. Crommie, D.M. Eigler: Probing the local effects of magnetic impurities on superconductivity, Science **275**, 1767–1770 (1997)
159. S.H. Tessmer, M.B. Tarlie, D.J. van Harlingen, D.L. Maslov, P.M. Goldbart: Probing the superconducting proximity effect in $NbSe_2$ by scanning tunneling micrsocopy, Phys. Rev. Lett. **77**, 924–927 (1996)
160. K. Inoue, H. Takayanagi: Local tunneling spectroscopy of Nb/InAs/Nb superconducting proximity system with a scanning tunneling microscope, Phys. Rev. B **43**, 6214–6215 (1991)
161. H.F. Hess, R.B. Robinson, R.C. Dynes, J.M. Valles, J.V. Waszczak: Scanning-tunneling-microscope observation of the Abrikosov flux lattice and the density of states near and inside a fluxoid, Phys. Rev. Lett. **62**, 214–217 (1989)
162. H.F. Hess, R.B. Robinson, J.V. Waszczak: Vortex-core structure observed with a scanning tunneling microscope, Phys. Rev. Lett. **64**, 2711–2714 (1990)
163. N. Hayashi, M. Ichioka, K. Machida: Star-shaped local density of states around vortices in a type-II superconductor, Phys. Rev. Lett. **77**, 4074–4077 (1996)
164. H. Sakata, M. Oosawa, K. Matsuba, N. Nishida: Imaging of vortex lattice transition in $YNi_2B_2C$ by scanning tunneling spectroscopy, Phys. Rev. Lett. **84**, 1583–1586 (2000)
165. S. Behler, S.H. Pan, P. Jess, A. Baratoff, H.-J. Güntherodt, F. Levy, G. Wirth, J. Wiesner: Vortex pinning in ion-irradiated $NbSe_2$ studied by scanning tunneling microscopy, Phys. Rev. Lett. **72**, 1750–1753 (1994)
166. R. Berthe, U. Hartmann, C. Heiden: Influence of a transport current on the Abrikosov flux lattice observed with a low-temperature scanning tunneling microscope, Ultramicroscopy **42-44**, 696–698 (1992)
167. N.C. Yeh, C.T. Chen, G. Hammerl, J. Mannhart, A. Schmehl, C.W. Schneider, R.R. Schulz, S. Tajima, K. Yoshida, D. Garrigus, M. Strasik: Evidence of doping-dependent pairing symmetry in cuprate superconductors, Phys. Rev. Lett. **87**, 087003–1–087003–4 (2001)
168. K. McElroy, R.W. Simmonds, J.E. Hoffman, D.H. Lee, J. Orenstein, H. Eisaki, S. Uchida, J.C. Davis: Relating atomic-scale electronic phenomena to wave-like quasiparticle states in superconducting $Bi_2Sr_2CaCu_2O_{8+\delta}$, Nature **422**, 592–596 (2003)
169. K. McElroy, J. Lee, J.A. Slezak, D.H. Lee, H. Eisaki, S. Uchida, J.C. Davis: Atomic-scale sources and mechanism of nanoscale electronic disorder in $Bi_2Sr_2CaCu_2O_{8+\delta}$, Science **309**, 1048–1052 (2005)

170. S.H. Pan, E.W. Hudson, K.M. Lang, H. Eisaki, S. Uchida, J.C. Davis: Imaging the effects of individual zinc impurity atoms on superconductivity in $Bi_2Sr_2CaCu_2O_{8+\delta}$, Nature **403**, 746–750 (2000)
171. A. Polkovnikov, S. Sachdev, M. Vojta: Impurity in a d-wave superconductor: Kondo effect and STM spectra, Phys. Rev. Lett. **86**, 296–299 (2001)
172. A.N. Pasupathy, A. Pushp, K.K. Gomes, C.V. Parker, J. Wen, Z. Xu, G. Gu, S. Ono, Y. Ando, A. Yazdani: Electronic origin of the inhomogeneous pairing interaction in the high-$T_c$ superconductor $Bi_2Sr_2CaCu_2O_{8+\delta}$, Science **320**, 196–201 (2008)
173. I. Maggio-Aprile, C. Renner, E. Erb, E. Walker, Ø. Fischer: Direct vortex lattice imaging and tunneling spectroscopy of flux lines on $YBa_2Cu_3O_{7-\delta}$, Phys. Rev. Lett. **75**, 2754–2757 (1995)
174. C. Renner, B. Revaz, K. Kadowaki, I. Maggio-Aprile, Ø. Fischer: Observation of the low temperature pseudogap in the vortex cores of $Bi_2Sr_2CaCu_2O_{8+\delta}$, Phys. Rev. Lett. **80**, 3606–3609 (1998)
175. S.H. Pan, E.W. Hudson, A.K. Gupta, K.W. Ng, H. Eisaki, S. Uchida, J.C. Davis: STM studies of the electronic structure of vortex cores in $Bi_2Sr_2CaCu_2O_{8+\delta}$, Phys. Rev. Lett. **85**, 1536–1539 (2000)
176. D.P. Arovas, A.J. Berlinsky, C. Kallin, S.C. Zhang: Superconducting vortex with antiferromagnetic core, Phys. Rev. Lett. **79**, 2871–2874 (1997)
177. J.E. Hoffmann, E.W. Hudson, K.M. Lang, V. Madhavan, H. Eisaki, S. Uchida, J.C. Davis: A four unit cell periodic pattern of quasi-particle states surrounding vortex cores in $Bi_2Sr_2CaCu_2O_{8+\delta}$, Science **295**, 466–469 (2002)
178. M. Vershinin, S. Misra, S. Ono, Y. Abe, Y. Ando, A. Yazdani: Local ordering in the pseudogap state of the high-$T_c$ superconductor $Bi_2Sr_2CaCu_2O_{8+\delta}$, Science **303**, 1995–1998 (2004)
179. T. Hanaguri, C. Lupien, Y. Kohsaka, D.H. Lee, M. Azuma, M. Takano, H. Takagi, J.C. Davis: A 'checkerborad' electronic crystal state in lightly hole-doped $Ca_{2-x}Na_xCuO_2Cl_2$, Nature **430**, 1001–1005 (2004)
180. Y. Kohsaka, C. Taylor, K. Fujita, A. Schmidt, C. Lupien, T. Hanaguri, M. Azuma, M. Takano, H. Eisaki, H. Takagi, S. Uchida, J.C. Davis: An intrinsic bond-centered electronic glass with unidirectional domains in underdoped cuprates, Science **315**, 1380–1385 (2007)
181. M. Crespo, H. Suderow, S. Vieira, S. Bud'ko, P.C. Canfield: Local superconducting density of states of ErNi_2B_2C, Phys. Rev. Lett. **96**, 027003–1–027003–4 (2006)
182. H. Suderow, S. Vieira, J.D. Strand, S. Bud'ko, P.C. Canfield: Very-low-temperature tunneling spectroscopy in the heavy-fermion superconductor PrOs_4Sb_12, Phys. Rev. B **69**, 060504–1–060504–4 (2004)
183. O. Naaman, W. Teizer, R.C. Dynes: Fluctuation dominated Josephson tunneling with a scanning tunneling microscope, Phys. Rev. Lett. **87**, 097004–1–097004–4 (2001)
184. M. Fäth, S. Freisem, A.A. Menovsky, Y. Tomioka, J. Aaarts, J.A. Mydosh: Spatially inhomogeneous metal–insulator transition in doped manganites, Science **285**, 1540–1542 (1999)
185. C. Renner, G. Aeppli, B.G. Kim, Y.A. Soh, S.W. Cheong: Atomic-scale images of charge ordering in a mixed-valence manganite, Nature **416**, 518–521 (2000)
186. H.M. Ronnov, C. Renner, G. Aeppli, T. Kimura, Y. Tokura: Polarons and confinement of electronic motion to two dimensions in a layered manganite, Nature **440**, 1025–1028 (2006)
187. M. Bode, M. Getzlaff, R. Wiesendanger: Spin-polarized vacuum tunneling into the exchange-split surface state of Gd(0001), Phys. Rev. Lett. **81**, 4256–4259 (1998)
188. A. Kubetzka, M. Bode, O. Pietzsch, R. Wiesendanger: Spin-polarized scanning tunneling microscopy with antiferromagnetic probe tips, Phys. Rev. Lett. **88**, 057201–1–057201–4 (2002)
189. O. Pietzsch, A. Kubetzka, M. Bode, R. Wiesendanger: Observation of magnetic hysteresis at the nanometer scale by spin-polarized scanning tunneling spectroscopy, Science **292**, 2053–2056 (2001)

190. S. Heinze, M. Bode, A. Kubetzka, O. Pietzsch, X. Xie, S. Blügel, R. Wiesendanger: Real-space imaging of two-dimensional antiferromagnetism on the atomic scale, Science **288**, 1805–1808 (2000)
191. A. Kubetzka, P. Ferriani, M. Bode, S. Heinze, G. Bihlmayer, K. von Bergmann, O. Pietzsch, S. Blügel, R. Wiesendanger: Revealing antiferromagnetic order of the Fe monolayer on W (001): Spin-polarized scanning tunneling microscopy and first-principles calculations, Phys. Rev. Lett. **94**, 087204–1–087204–4 (2005)
192. A. Wachowiak, J. Wiebe, M. Bode, O. Pietzsch, M. Morgenstern, R. Wiesendanger: Internal spin-structure of magnetic vortex cores observed by spin-polarized scanning tunneling microscopy, Science **298**, 577–580 (2002)
193. M. Bode, M. Heide, K. von Bergmann, P. Ferriani, S. Heinze, G. Bihlmeyer, A. Kubetzka, O. Pietzsch, S. Blügel, R. Wiesendanger: Chiral magnetic order at surfaces driven by inversion asymmetry, Nature **447**, 190–193 (2007)
194. K. von Bergmann, S. Heinze, M. Bode, E.Y. Vedmedenko, G. Bihlmayer, S. Blügel, R. Wiesendanger: Observation of a complex nanoscale magnetic structure in a hexagonal Fe monolayer, Phys. Rev. Lett. **96**, 167203–1–167203–4 (2006)
195. C.L. Gao, U. Schlickum, W. Wulfhekel, J. Kirschner: Mapping the surface spin structure of large unit cells: Reconstructed Mn films on Fe(001), Phys. Rev. Lett. **98**, 107203–1–107203–4 (2007)
196. Y. Yayon, V.W. Brar, L. Senapati, S.C. Erwin, M.F. Crommie: Observing spin polarization of individual magnetic adatoms, Phys. Rev. Lett. **99**, 067202–1–067202–4 (2007)
197. M. Bode, O. Pietzsch, A. Kubetzka, R. Wiesendanger: Shape-dependent thermal switching behavior of superparamagnetic nanoislands, Phys. Rev. Lett. **92**, 067201–1–067201–4 (2004)
198. S. Krause, L. Berbil-Bautista, G. Herzog, M. Bode, R. Wiesendanger: Current-induced magnetization switching with a spin-polarized scanning tunneling microscope, Science **317**, 1537–1540 (2007)
199. M.D. Kirk, T.R. Albrecht, C.F. Quate: Low-temperature atomic force microscopy, Rev. Sci. Instrum. **59**, 833–835 (1988)
200. D. Pelekhov, J. Becker, J.G. Nunes: Atomic force microscope for operation in high magnetic fields at millikelvin temperatures, Rev. Sci. Instrum. **70**, 114–120 (1999)
201. J. Mou, Y. Jie, Z. Shao: An optical detection low temperature atomic force microscope at ambient pressure for biological research, Rev. Sci. Instrum. **64**, 1483–1488 (1993)
202. H.J. Mamin, D. Rugar: Sub-attonewton force detection at millikelvin temperatures, Appl. Phys. Lett. **79**, 3358–3360 (2001)
203. A. Schwarz, W. Allers, U.D. Schwarz, R. Wiesendanger: Dynamic mode scanning force microscopy of $n$-InAs(110)-(1 $\times$ 1) at low temperatures, Phys. Rev. B **61**, 2837–2845 (2000)
204. W. Allers, S. Langkat, R. Wiesendanger: Dynamic low-temperature scanning force microscopy on nickel oxide(001), Appl. Phys. A **72**, S27–S30 (2001)
205. F.J. Giessibl: Atomic resolution of the silicon(111)-(7 $\times$ 7) surface by atomic force microscopy, Science **267**, 68–71 (1995)
206. M.A. Lantz, H.J. Hug, P.J.A. van Schendel, R. Hoffmann, S. Martin, A. Baratoff, A. Abdurixit, H.-J. Güntherodt: Low temperature scanning force microscopy of the Si (111)-(7 $\times$ 7) surface, Phys. Rev. Lett. **84**, 2642–2645 (2000)
207. K. Suzuki, H. Iwatsuki, S. Kitamura, C.B. Mooney: Development of low temperature ultrahigh vacuum force microscope/scanning tunneling microscope, Jpn. J. Appl. Phys. **39**, 3750–3752 (2000)
208. N. Suehira, Y. Sugawara, S. Morita: Artifact and fact of Si(111)-(7 $\times$ 7) surface images observed with a low temperature noncontact atomic force microscope (LT-NC-AFM), Jpn. J. Appl. Phys. **40**, 292–294 (2001)
209. R. Pérez, M.C. Payne, I. Štich, K. Terakura: Role of covalent tip–surface interactions in noncontact atomic force microscopy on reactive surfaces, Phys. Rev. Lett. **78**, 678–681 (1997)

210. S.H. Ke, T. Uda, R. Pérez, I. Štich, K. Terakura: First principles investigation of tip–surface interaction on GaAs(110): Implication for atomic force and tunneling microscopies, Phys. Rev. B **60**, 11631–11638 (1999)
211. J. Tobik, I. Štich, R. Pérez, K. Terakura: Simulation of tip–surface interactions in atomic force microscopy of an InP(110) surface with a Si tip, Phys. Rev. B **60**, 11639–11644 (1999)
212. A. Schwarz, W. Allers, U.D. Schwarz, R. Wiesendanger: Simultaneous imaging of the In and As sublattice on InAs(110)-(1 $\times$ 1) with dynamic scanning force microscopy, Appl. Surf. Sci. **140**, 293–297 (1999)
213. H. Hölscher, W. Allers, U.D. Schwarz, A. Schwarz, R. Wiesendanger: Interpretation of 'true atomic resolution' images of graphite (0001) in noncontact atomic force microscopy, Phys. Rev. B **62**, 6967–6970 (2000)
214. H. Hölscher, W. Allers, U.D. Schwarz, A. Schwarz, R. Wiesendanger: Simulation of NC-AFM images of xenon(111), Appl. Phys. A **72**, S35–S38 (2001)
215. M. Ashino, A. Schwarz, T. Behnke, R. Wiesendanger: Atomic-resolution dynamic force microscopy and spectroscopy of a single-walled carbon nanotube: characterization of interatomic van der Waals forces, Phys. Rev. Lett. **93**, 136101–1–136101–4 (2004)
216. G. Schwarz, A. Kley, J. Neugebauer, M. Scheffler: Electronic and structural properties of vacancies on and below the GaP(110) surface, Phys. Rev. B **58**, 1392–1499 (1998)
217. M. Ashino, A. Schwarz, H. Hölscher, U.D. Schwarz, R. Wiesendanger: Interpretation of the atomic scale contrast obtained on graphite and single-walled carbon nanotubes in the dynamic mode of atomic force microscopy, Nanotechnology **16**, 134–137 (2005)
218. F.J. Giessibl, H. Bielefeldt, S. Hembacher, J. Mannhart: Calculation of the optimal imaging parameters for frequency modulation atomic force microscopy, Appl. Surf. Sci. **140**, 352–357 (1999)
219. S. Hembacher, F.J. Giessibl, J. Mannhart, C.F. Quate: Local spectroscopy and atomic imaging of tunneling current, forces, and dissipation on graphite, Phys. Rev. Lett. **94**, 056101–1–056101–4 (2005)
220. F.J. Giessibl: High-speed force sensor for force microscopy and profilometry utilizing a quartz tuning fork, Appl. Phys. Lett. **73**, 3956–3958 (1998)
221. H. Hölscher, W. Allers, U.D. Schwarz, A. Schwarz, R. Wiesendanger: Determination of tip–sample interaction potentials by dynamic force spectroscopy, Phys. Rev. Lett. **83**, 4780–4783 (1999)
222. H. Hölscher, U.D. Schwarz, R. Wiesendanger: Calculation of the frequency shift in dynamic force microscopy, Appl. Surf. Sci. **140**, 344–351 (1999)
223. B. Gotsman, B. Anczykowski, C. Seidel, H. Fuchs: Determination of tip–sample interaction forces from measured dynamic force spectroscopy curves, Appl. Surf. Sci. **140**, 314–319 (1999)
224. U. Dürig: Extracting interaction forces and complementary observables in dynamic probe microscopy, Appl. Phys. Lett. **76**, 1203–1205 (2000)
225. F.J. Giessibl: A direct method to calculate tip–sample forces from frequency shifts in frequency-modulation atomic force microscopy, Appl. Phys. Lett. **78**, 123–125 (2001)
226. J.E. Sader, S.P. Jarvis: Accurate formulas for interaction force and energy in frequency modulation force spectroscopy, Appl. Phys. Lett. **84**, 1801–1803 (2004)
227. M.A. Lantz, H.J. Hug, R. Hoffmann, P.J.A. van Schendel, P. Kappenberger, S. Martin, A. Baratoff, H.-J. Güntherodt: Quantitative measurement of short-range chemical bonding forces, Science **291**, 2580–2583 (2001)
228. S.M. Langkat, H. Hölscher, A. Schwarz, R. Wiesendanger: Determination of site specific forces between an iron coated tip and the NiO(001) surface by force field spectroscopy, Surf. Sci. **527**, 12–20 (2002)
229. Y. Sugimoto, P. Pou, M. Abe, P. Jelinek, R. Pérez, S. Morita, O. Custance: Chemical identification of individual surface atoms by atomic force microscopy, Nature **446**, 64–67 (2007)
230. M. Abe, Y. Sugimoto, O. Custance, S. Morita: Room-temperature reproducible spatial force spectroscopy using atom-tracking technique, Appl. Phys. Lett. **87**, 173503 (2005)

231. H. Hölscher, S.M. Langkat, A. Schwarz, R. Wiesendanger: Measurement of three-dimensional force fields with atomic resolution using dynamic force spectroscopy, Appl. Phys. Lett. **81**, 4428–4430 (2002)
232. A. Schirmeisen, D. Weiner, H. Fuchs: Single atom contact mechanics: From atomic scale energy barrier to mechanical relaxation hysteresis, Phys. Rev. Lett. **97**, 136101 (2006)
233. M. Abe, Y. Sugimoto, T. Namikawa, K. Morita, N. Oyabu, S. Morita: Drift-compensated data acquisition performed at room temperature with frequency modulation atomic force microscopy, Appl. Phys. Lett. **90**, 203103 (2007)
234. A. Schwarz, H. Hölscher, S.M. Langkat, R. Wiesendanger: Three-dimensional force field spectroscopy, AIP Conf. Proc. **696**, 68 (2003)
235. N. Oyabu, Y. Sugimoto, M. Abe, O. Custance, S. Morita: Lateral manipulation of single atoms at semiconductor surfaces using atomic force microscopy, Nanotechnology **16**, 112–117 (2005)
236. N. Oyabu, O. Custance, I. Yi, Y. Sugawara, S. Morita: Mechanical vertical manipulation of selected single atoms by soft nanoindentation using near contact atomic force microscopy, Phys. Rev. Lett. **90**, 176102–1–176102–4 (2004)
237. Y. Sugimoto, M. Abe, S. Hirayama, N. Oyabu, O. Custance, S. Morita: Atom inlays performed at room temperature using atomic force microscopy, Nat. Mater. **4**, 156–160 (2005)
238. C. Sommerhalter, T.W. Matthes, T. Glatzel, A. Jäger-Waldau, M.C. Lux-Steiner: High-sensitivity quantitative Kelvin probe microscopy by noncontact ultra-high-vacuum atomic force microscopy, Appl. Phys. Lett. **75**, 286–288 (1999)
239. A. Schwarz, W. Allers, U.D. Schwarz, R. Wiesendanger: Dynamic mode scanning force microscopy of $n$-InAs(110)-(1 $\times$ 1) at low temperatures, Phys. Rev. B **62**, 13617–13622 (2000)
240. K.L. McCormick, M.T. Woodside, M. Huang, M. Wu, P.L. McEuen, C. Duruoz, J.S. Harris: Scanned potential microscopy of edge and bulk currents in the quantum Hall regime, Phys. Rev. B **59**, 4656–4657 (1999)
241. P. Weitz, E. Ahlswede, J. Weis, K. von Klitzing, K. Eberl: Hall-potential investigations under quantum Hall conditions using scanning force microscopy, Physica E **6**, 247–250 (2000)
242. E. Ahlswede, P. Weitz, J. Weis, K. von Klitzing, K. Eberl: Hall potential profiles in the quantum Hall regime measured by a scanning force microscope, Physica B **298**, 562–566 (2001)
243. M.T. Woodside, C. Vale, P.L. McEuen, C. Kadow, K.D. Maranowski, A.C. Gossard: Imaging interedge-state scattering centers in the quantum Hall regime, Phys. Rev. B **64**, 041310–1–041310–4 (2001)
244. K. Moloni, B.M. Moskowitz, E.D. Dahlberg: Domain structures in single crystal magnetite below the Verwey transition as observed with a low-temperature magnetic force microscope, Geophys. Res. Lett. **23**, 2851–2854 (1996)
245. Q. Lu, C.C. Chen, A. de Lozanne: Observation of magnetic domain behavior in colossal magnetoresistive materials with a magnetic force microscope, Science **276**, 2006–2008 (1997)
246. G. Xiao, J.H. Ross, A. Parasiris, K.D.D. Rathnayaka, D.G. Naugle: Low-temperature MFM studies of CMR manganites, Physica C **341–348**, 769–770 (2000)
247. M. Liebmann, U. Kaiser, A. Schwarz, R. Wiesendanger, U.H. Pi, T.W. Noh, Z.G. Khim, D.W. Kim: Domain nucleation and growth of $La_0.7Ca_0.3MnO_{3-\delta}/LaAlO_3$ films studied by low temperature MFM, J. Appl. Phys. **93**, 8319–8321 (2003)
248. A. Moser, H.J. Hug, I. Parashikov, B. Stiefel, O. Fritz, H. Thomas, A. Baratoff, H.J. Güntherodt, P. Chaudhari: Observation of single vortices condensed into a vortex-glass phase by magnetic force microscopy, Phys. Rev. Lett. **74**, 1847–1850 (1995)
249. C.W. Yuan, Z. Zheng, A.L. de Lozanne, M. Tortonese, D.A. Rudman, J.N. Eckstein: Vortex images in thin films of YBa_2Cu_3O_7-$x$ and Bi_2Sr_2Ca_1Cu_2O_8-$x$ obtained by low-temperature magnetic force microscopy, J. Vac. Sci. Technol. B **14**, 1210–1213 (1996)
250. A. Volodin, K. Temst, C. van Haesendonck, Y. Bruynseraede: Observation of the Abrikosov vortex lattice in $NbSe_2$ with magnetic force microscopy, Appl. Phys. Lett. **73**, 1134–1136 (1998)
251. A. Moser, H.J. Hug, B. Stiefel, H.J. Güntherodt: Low temperature magnetic force microscopy on $YBa_2Cu_3O_{7-\delta}$ thin films, J. Magn. Magn. Mater. **190**, 114–123 (1998)

252. A. Volodin, K. Temst, C. van Haesendonck, Y. Bruynseraede: Imaging of vortices in conventional superconductors by magnetic force microscopy images, Physica C **332**, 156–159 (2000)
253. M. Roseman, P. Grütter: Estimating the magnetic penetration depth using constant-height magnetic force microscopy images of vortices, New J. Phys. **3**, 24.1–24.8 (2001)
254. A. Volodin, K. Temst, C. van Haesendonck, Y. Bruynseraede, M.I. Montero, I.K. Schuller: Magnetic force microscopy of vortices in thin niobium films: Correlation between the vortex distribution and the thickness-dependent film morphology, Europhys. Lett. **58**, 582–588 (2002)
255. U.H. Pi, T.W. Noh, Z.G. Khim, U. Kaiser, M. Liebmann, A. Schwarz, R. Wiesendanger: Vortex dynamics in $Bi_2Sr_2CaCu_2O_8$ single crystal with low density columnar defects studied by magnetic force microscopy, J. Low Temp. Phys. **131**, 993–1002 (2003)
256. M. Roseman, P. Grütter, A. Badia, V. Metlushko: Flux lattice imaging of a patterned niobium thin film, J. Appl. Phys. **89**, 6787–6789 (2001)
257. A. Schwarz, R. Wiesendanger: Magnetic sensitive force microscopy, Nano Today **3**, 28–39 (2008)
258. R. Wiesendanger, D. Bürgler, G. Tarrach, A. Wadas, D. Brodbeck, H.J. Güntherodt, G. Güntherodt, R.J. Gambio, R. Ruf: Vacuum tunneling of spin-polarized electrons detected by scanning tunneling microscopy, J. Vac. Sci. Technol. B **9**, 519–524 (1991)
259. H. Momida, T. Oguchi: First-principles study on exchange force image of NiO(001) surface using a ferromagnetic Fe probe, Surf. Sci. **590**, 42–50 (2005)
260. U. Kaiser, A. Schwarz, R. Wiesendanger: Magnetic exchange force microscopy with atomic resolution, Nature **446**, 522–525 (2007)
261. R. Schmidt, C. Lazo, H. Hölscher, U.H. Pi, V. Caciuc, A. Schwarz, R. Wiesendanger, S. Heinze: Probing the magnetic exchange forces of iron on the atomic scale, Nano Lett. **9**, 200–204 (2008)
262. D. Rugar, R. Budakian, H.J. Mamin, B.W. Chui: Single spin detection by magnetic resonance force microscopy, Nature **430**, 329–332 (2004)
263. J.A. Sidles, J.L. Garbini, G.P. Drobny: The theory of oscillator-coupled magnetic resonance with potential applications to molecular imaging, Rev. Sci. Instrum. **63**, 3881–3899 (1992)
264. J.A. Sidles, J.L. Garbini, K.J. Bruland, D. Rugar, O. Züger, S. Hoen, C.S. Yannoni: Magnetic resonance force microscopy, Rev. Mod. Phys. **67**, 249–265 (1995)
265. D. Rugar, O. Züger, S. Hoen, C.S. Yannoni, H.M. Vieth, R.D. Kendrick: Force detection of nuclear magnetic resonance, Science **264**, 1560–1563 (1994)
266. Z. Zhang, P.C. Hammel, P.E. Wigen: Observation of ferromagnetic resonance in a microscopic sample using magnetic resonance force microscopy, Appl. Phys. Lett. **68**, 2005–2007 (1996)

# Chapter 7
# Dynamic Modes of Atomic Force Microscopy

**André Schirmeisen, Boris Anczykowski, Hendrik Hölscher, and Harald Fuchs**

**Abstract** This chapter presents an introduction to the concept of the dynamic operational modes of the atomic force microscope (dynamic AFM). While the static (or contact-mode) AFM is a widespread technique to obtain nanometer-resolution images on a wide variety of surfaces, true atomic-resolution imaging is routinely observed only in the dynamic mode. We will explain the jump-to-contact phenomenon encountered in static AFM and present the dynamic operational mode as a solution to avoid this effect. The dynamic force microscope is modeled as a harmonic oscillator to gain a basic understanding of the underlying physics in this mode.

On closer inspection, the dynamic AFM comprises a whole family of operational modes. A systematic overview of the different modes typically found in force microscopy is presented with special attention paid to the distinct features of each mode. Two modes of operation dominate the application of dynamic AFM. First, the amplitude modulation mode (also called tapping mode) is shown to exhibit an instability, which separates the purely attractive force interaction regime from the attractive–repulsive regime. Second, the self-excitation mode is derived and its experimental realization is outlined. While the tapping mode is primarily used for imaging in air and liquid, the self-excitation mode is typically used under ultrahigh vacuum (UHV) conditions for atomic-resolution imaging. In particular, we explain the influence of different forces on spectroscopy curves obtained in dynamic force microscopy. A quantitative link between the experimental spectroscopy curves and the interaction forces is established.

Force microscopy in air suffers from small quality factors of the force sensor (i.e., the cantilever beam), which are shown to limit the resolution. Also, the above-mentioned instability in the amplitude modulation mode often hinders imaging of soft and fragile samples. A combination of the amplitude modulation with the self-excitation mode is shown to increase the quality, or $Q$-factor, and extend the regime of stable operation. This so-called $Q$-control module allows one to increase as well as decrease the $Q$-factor. Apart from the advantages of dynamic force microscopy as a nondestructive, high-resolution imaging method, it can also be used to obtain information about energy-dissipation phenomena at the nanometer scale. This measurement channel can provide crucial information on electric and magnetic surface properties. Even atomic-resolution imaging has been obtained in the dissipation mode.

B. Bhushan (ed.), *Nanotribology and Nanomechanics*,
DOI 10.1007/978-3-642-15283-2_7, 

Therefore, in the last section, the quantitative relation between the experimental measurement channels and the dissipated power is derived.

## 7.1 Motivation – Measurement of a Single Atomic Bond

The direct measurement of the force interaction between two distinct molecules has been a challenge for scientists for many years now. The fundamental forces responsible for the solid state of matter can be directly investigated, ultimately between defined single molecules. However, it has not been until 2001 that the chemical forces could be quantitatively measured for a single atomic bond [1]. How can we reliably measure forces that may be as small as one billionth of 1 N? How can we identify one single pair of atoms as the source of the force interaction?

The same mechanical principle that is used to measure the gravitational force exerted by your body weight (e.g., with the scale in your bathroom) can be employed to measure the forces between single atoms. A spring with a defined elasticity is compressed by an arbitrary force (e.g., your weight). The compression $\Delta z$ of the spring (with spring constant $k$) is a direct measure of the force $F$ exerted, which in the regime of elastic deformation obeys Hooke's law

$$F = k\Delta z. \tag{7.1}$$

The only difference with regard to your bathroom scale is the sensitivity of the measurement. Typically springs with a stiffness of 0.1–10 N/m are used, which will be deflected by 0.1–10 nm upon application of an interatomic force of some nN. Experimentally, a laser deflection technique is used to measure the movement of the spring. The spring is a bendable cantilever microfabricated from a silicon wafer. If a sufficiently sharp tip, usually directly attached to the cantilever, is approached toward a surface within some nanometers, we can measure the interaction forces through changes in the deflected laser beam. This is a static measurement and is hence called *static AFM*. Alternatively, the cantilever can be excited to vibrate at its resonant frequency. Under the influence of tip–sample forces the resonant frequency (and consequently also the amplitude and phase) of the cantilever will change and serve as measurement parameters. This approach is called *dynamic AFM*. Due to the multitude of possible operational modes, expressions such as noncontact mode, intermittent contact mode, tapping mode, frequency modulation (FM) mode, amplitude-modulation (AM) mode, self-excitation mode, constant-excitation mode, or constant-amplitude mode are found in the literature, which will be systematically categorized in the following paragraphs.

In fact, the first AFMs were operated in dynamic mode. In 1986, Binnig et al. presented the concept of the atomic force microscope [2]. The deflection of the cantilever with the tip was measured with subangstrom precision by an additional scanning tunneling microscope (STM). While the cantilever was externally oscillated close to its resonant frequency, the amplitude and phase of the oscillation were measured. If the tip is approached toward the surface, the oscillation parameters, amplitude and phase, are influenced by the tip–surface interaction, and can therefore be used as feedback channels. Typically, a certain setpoint for the amplitude is defined, and the feedback loop will adjust the tip–sample distance such that the amplitude remains constant. The control parameter is recorded as a function of the lateral position of the tip with respect to the sample, and the scanned image essentially represents the surface topography.

What then is the difference between the static and dynamic modes of operation for the AFM? Static deflection AFM directly gives the interaction force between tip and sample using (7.1). In the dynamic mode, we find that the resonant frequency, amplitude, and phase of the oscillation change as a consequence of the interaction forces (and also dissipative processes, as discussed in the final section).

In order to obtain a basic understanding of the underlying physics, it is instructive to consider a highly simplified case. Assume that the vibration amplitude is small compared with the range of force interaction. Since van der Waals forces range over typical distances of 10 nm, the vibration amplitude should be less than 1 nm. Furthermore, we require that the force gradient $\partial F_{ts}/\partial z$ does not vary significantly over one oscillation cycle. We can view the AFM setup as a coupling of two springs (Fig. 7.1). Whereas the cantilever is represented by a spring with spring constant $k$, the force interaction between the tip and the surface can be modeled by a second spring. The derivative of the force with respect to the tip–sample distance is the force gradient and represents the spring constant $k_{ts}$ of the interaction spring. This spring constant $k_{ts}$ is constant only with respect to one oscillation cycle, but varies with the average tip–sample distance as the probe is approached to the sample. The two springs are effectively coupled in parallel, since sample and tip support are rigidly connected for a given value of $z_0$. Therefore, we can write for the total spring constant of the AFM system

$$k_{total} = k + k_{ts} = k - \frac{\partial F_{ts}}{\partial z}. \quad (7.2)$$

From the simple harmonic oscillator (neglecting any damping effects) we find that the resonant frequency $\omega$ of the system is shifted by $\Delta\omega$ from the free resonant frequency $\omega_0$ due to the force interaction

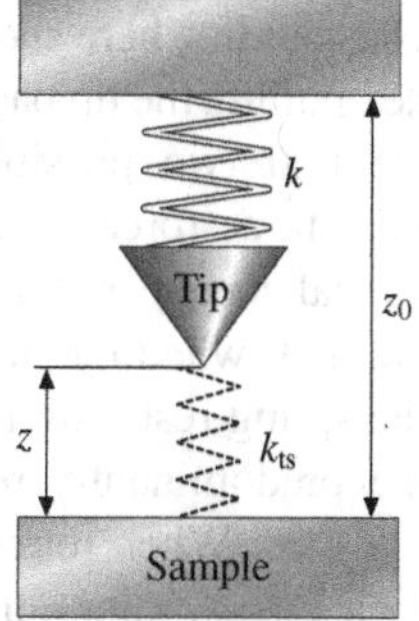


**Fig. 7.1** Model of the AFM tip while experiencing tip–sample forces. The tip is attached to a cantilever with spring constant $k$, and the force interaction is modeled by a spring with a stiffness equal to the force gradient. Note that the force interaction spring is not constant, but depends on the tip–sample distance $z$

$$\omega^2 = (\omega_0 + \Delta\omega)^2 = \frac{k_{\text{total}}}{m^*} = \frac{\left(k + \frac{\partial F_{\text{ts}}}{\partial z}\right)}{m^*}. \tag{7.3}$$

Here $m^*$ represents the effective mass of the cantilever. A detailed analysis of how $m^*$ is related to the geometry and total mass of the cantilever can be found in the literature [3]. In the approximation that $\Delta\omega$ is much smaller than $\omega_0$, we can write

$$\frac{\Delta\omega}{\omega_0} \cong -\frac{1}{2k}\frac{\partial F_{\text{ts}}}{\partial z}. \tag{7.4}$$

Therefore, we find that the frequency shift of the cantilever resonance is proportional to the force gradient of the tip–sample interaction.

Although the above consideration is based on a highly simplified model, it shows qualitatively that in dynamic force microscopy we will find that the oscillation frequency depends on the force gradient, whereas static force microscopy measures the force itself. In principle, we can calculate the force curve from the force gradient and vice versa (neglecting a constant offset). It seems, therefore, that the two methods are equivalent, and our choice will depend on whether we can measure the beam deflection or the frequency shift with better precision at the cost of technical effort.

However, we have neglected one important issue for the operation of the AFM thus far: the mechanical stability of the measurement. In static AFM, the tip is slowly approached toward the surface. The force between the tip and the surface will always be counteracted by the restoring force of the cantilever. Figure 7.2 shows a typical force–distance curve. Upon approach of the tip toward the sample, the negative attractive forces, representing van der Waals or chemical interaction forces, increase until a maximum is reached. This turnaround point is due to the onset of repulsive forces caused by Coulomb repulsion, which will start to dominate upon further approach. The spring constant of the cantilever is represented by the slope of the straight line. The position of the $z$-transducer (typically a piezoelectric element), which moves the probe, is at the intersection of the line with the horizontal axis. The position of the tip, shifted from the probe's base due to the lever bending, can be found at the intersection of the cantilever line with the force curve. Hence, the total force is zero, i.e., the cantilever is in its equilibrium position (note that the spring constant line here shows attractive forces, although in reality the forces are repulsive, i.e., pulling the tip back from the surface). As soon as position A in Fig. 7.2 is reached, we find two possible intersection points, and upon further approach there are even three force equilibrium points. However, between points A and B the tip is at a local energy minimum and, therefore, will still follow the force curve. However, at point B, when the adhesion force upon further approach would become larger than the spring restoring force, the tip will suddenly jump to point C. We can then probe the predominantly repulsive force interaction by further reducing the tip–sample distance. When retracting the tip, we will pass point C, because the tip is still in a local energy minimum. Only at position D will the tip jump suddenly to point A again,

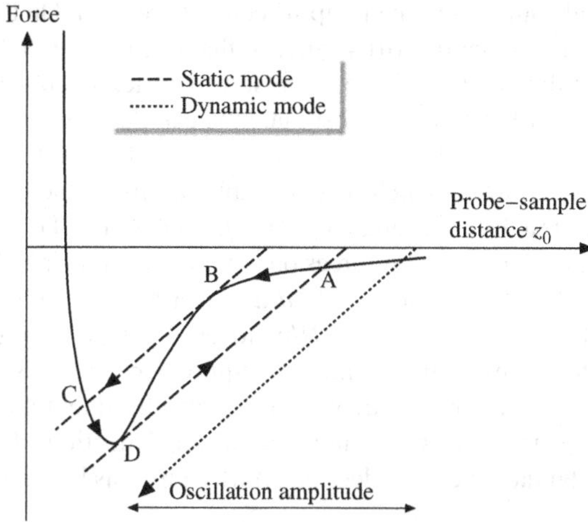


**Fig. 7.2** Force–distance curve of a typical tip–sample interaction. In static-mode AFM the tip would follow the force curve until point B is reached. If the slope of the force curve becomes larger than the spring constant of the cantilever (*dashed line*) the tip will suddenly jump to position C. Upon retraction a different path will be followed along D and A again. In dynamic AFM the cantilever oscillates with amplitude. Although the equilibrium position of the oscillation is far from the surface, the tip will experience the maximum attractive force at point D during some parts of the oscillation cycle. However, the total force is always pointing away from the surface, therefore avoiding an instability

since the restoring force now exceeds the adhesion. From Fig. 7.2 we can see that the sudden instability will happen at exactly the point where the slope of the adhesion force exceeds the slope of the spring constant. Therefore, if the negative force gradient of the tip–sample interaction will at any point exceed the spring constant, a mechanical instability occurs. Mathematically speaking, we demand that for a stable measurement

$$-\left.\frac{\partial F_{\text{ts}}}{\partial z}\right|_z < k, \qquad \text{for all points } z. \tag{7.5}$$

This mechanical instability is often referred to as the *jump-to-contact* phenomenon.

Looking at Fig. 7.2, we realize that large parts of the force curve cannot be measured if the jump-to-contact phenomenon occurs. We will not be able to measure the point at which the attractive forces reach their maximum, representing the temporary chemical bonding of the tip and the surface atoms. Secondly, the sudden instability, the jump-to-contact, will often cause the tip to change the very last tip or surface atoms. A smooth, careful approach needed to measure the full force curve does not seem feasible. Our goal of measuring the chemical interaction forces of two single molecules may become impossible.

There are several solutions to the jump-to-contact problem: On the one hand, we can simply choose a sufficiently stiff spring, so that (7.5) is fulfilled at all points of the force curve. On the other hand, we can resort to a trick to enhance the counteracting force of the cantilever: We can oscillate the cantilever with large amplitude, thereby making it virtually stiffer at the point of strong force interaction.

Consider the first solution, which seems simpler at first glance. Chemical bonding forces extend over a distance range of about 0.1 nm. Typical binding energies of a couple of eV will lead to adhesion forces on the order of some nN. Force gradients will, therefore, reach values of some 10 N/m. A spring for stable force measurements will have to be as stiff as 100 N/m to ensure that no instability occurs (a safety factor of ten seems to be a minimum requirement, since usually one cannot be sure a priori that only one atom will dominate the interaction). In order to measure the nN interaction force, a static cantilever deflection of 0.01 nm has to be detected. With standard beam deflection AFM setups this becomes a challenging task.

This problem was solved by using an in situ optical interferometer measuring the beam deflection at liquid-nitrogen temperature in a UHV environment [4, 5]. In order to ensure that the force gradients are smaller than the lever spring constant (50 N/m), the tips were fabricated to terminate in only three atoms, thereby minimizing the total force interaction. The field ion microscope (FIM) is a tool which allows scanning probe microscopy (SPM) tips to be engineered down to atomic dimensions. This technique not only allows imaging of the tip apex with atomic precision, but also can be used to manipulate the tip atoms by field evaporation [6], as shown in Fig. 7.3. Atomic interaction forces were measured with subnanonewton precision, revealing force curves of only a few atoms interacting without mechanical hysteresis. However, the technical effort to achieve this type of measurement is considerable, and most researchers today have resorted to the second solution.

The alternative solution can be visualized in Fig. 7.2. The straight, dashed line now represents the force values of the oscillating cantilever, with amplitude $A$

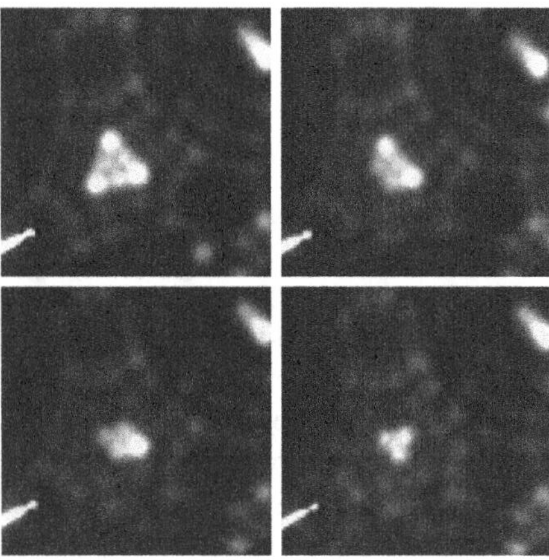

**Fig. 7.3** Manipulation of the apex atoms of an AFM tip using field ion microscopy (FIM). *Images* were acquired at a tip bias of 4.5 kV. The last six atoms of the tip can be inspected in this example. Field evaporation to remove single atoms is performed by increasing the bias voltage for a short time to 5.2 kV. Each of the outer three atoms can be consecutively removed, eventually leaving a trimer tip apex

assuming Hooke's law is valid. This is the tensile force of the cantilever spring pulling the tip away from the sample. The restoring force of the cantilever is at all points stronger than the adhesion force. For example, the total force at point D is still pointing away from the sample, although the spring has the same stiffness as before. Mathematically speaking, the measurement is stable as long as the cantilever spring force $F_{cb} = kA$ is larger than the attractive tip–sample force $F_{ts}$ [7]. In the static mode we would already experience an instability at that point. However, in the dynamic mode, the spring is preloaded with a force stronger than the attractive tip–sample force. The equilibrium point of the oscillation is still far away from the point of closest contact of the tip and surface atoms. The total force curve can now be probed by varying the equilibrium point of the oscillation, i.e., by adjusting the $z$-piezo.

The diagram also shows that the oscillation amplitude has to be quite large if fairly soft cantilevers are to be used. With lever spring constants of 10 N/m, the amplitude must be at least 1 nm to ensure that forces of 1 nN can be reliably measured. In practical applications, amplitudes of 10–100 nm are used to stay on the safe side. This means that the oscillation amplitude is much larger than the force interaction range. The above simplification, that the force gradient remains constant within one oscillation cycle, does not hold anymore. Measurement stability is gained at the cost of a simple quantitative analysis of the experiments. In fact, dynamic AFM was first used to obtain atomic resolution images of clean surfaces [8], and it took another 6 years [1] before quantitative measurements of single bond forces were obtained.

The technical realization of dynamic-mode AFMs is based on the same key components as a static AFM setup. The most common principle is the method of laser deflection sensing (Fig. 7.4). A laser beam is focused on the back side of a

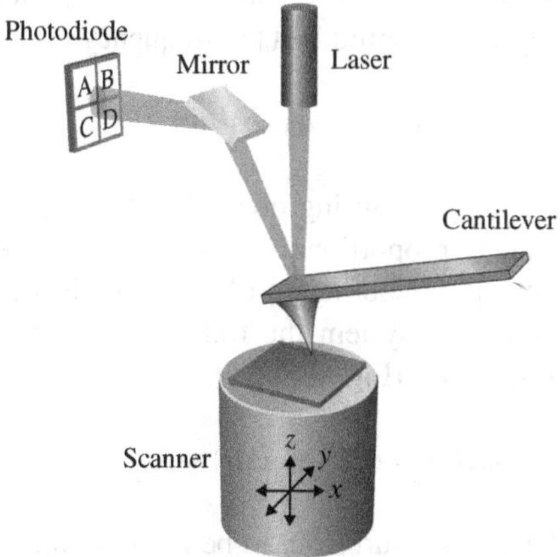


**Fig. 7.4** Representation of an AFM setup with the laser beam deflection method. Cantilever and tip are microfabricated from silicon wafers. A laser beam is deflected from the back side of the cantilever and again focused on a photosensitive diode via an adjustable mirror. The diode is segmented into four quadrants, which allows measurement of vertical and torsional bending of the cantilever (artwork by D. Ebeling rendered with POV-Ray)

microfabricated cantilever. The reflected laser spot is detected with a position-sensitive diode (PSD). This photodiode is sectioned into two parts that are read out separately (usually even a four-quadrant diode is used to detect torsional movements of the cantilever for lateral friction measurements). With the cantilever at equilibrium, the spot is adjusted such that the two sections show the same intensity. If the cantilever bends up or down, the spot moves, and the difference signal between the upper and lower sections is a measure of the bending.

In order to enhance sensitivity, several groups have adopted an interferometer system to measure the cantilever deflection. A thorough comparison of different measurement methods with analysis of sensitivity and noise level is given in reference [3].

The cantilever is mounted on a device that allows the beam to be oscillated. Typically a piezo element directly underneath the cantilever beam serves this purpose. The reflected laser beam is analyzed for oscillation amplitude, frequency, and phase difference. Depending on the mode of operation, a feedback mechanism will adjust oscillation parameters and/or tip–sample distance during the scanning. The setup can be operated in air, UHV, and even fluids. This allows measurement of a wide range of surface properties from atomic-resolution imaging [8] up to studying biological processes in liquid [9,10].

## 7.2 Harmonic Oscillator: a Model System for Dynamic AFM

The oscillating cantilever has three degrees of freedom: the amplitude, the frequency, and the phase difference between excitation and oscillation. Let us consider the damped driven harmonic oscillator. The cantilever is mounted on a piezoelectric element that is oscillating with amplitude $A_\mathrm{d}$ at frequency $\omega$

$$z_\mathrm{d}(t) = A_\mathrm{d}\cos(\omega t). \tag{7.6}$$

We assume that the cantilever spring obeys Hooke's law. Secondly, we introduce a friction force that is proportional to the speed of the cantilever motion, whereas $\alpha$ denotes the damping coefficient (Amontons' law). With Newton's first law we find for the oscillating system the following equation of motion for the position $z(t)$ of the cantilever tip (Fig. 7.1)

$$m\ddot{z}(t) = \alpha\dot{z}(t) - kz(t) - kz_\mathrm{d}(t). \tag{7.7}$$

We define $\omega_0^2 = k/m^*$, which turns out to be the resonant frequency of the free (undamped, i.e., $\alpha = 0$) oscillating beam. We further define the dimensionless quality factor $Q = m^*\omega_0/\alpha$, antiproportional to the damping coefficient. The quality factor describes the number of oscillation cycles after which the damped oscillation amplitude decays to 1/e of the initial amplitude with no external excitation ($A_\mathrm{d} = 0$). After some basic math, this results in

$$\ddot{z}(t) + \frac{\omega_0}{Q}\dot{z}(t) + \omega_0^2 z(t) = A_d \omega_0^2 \cos(\omega t). \tag{7.8}$$

The solution is a linear combination of two regimes [11]. Starting from rest and switching on the piezo excitation at $t = 0$, the amplitude will increase from zero to the final magnitude and reach a steady state, where the amplitude, phase, and frequency of the oscillation stay constant over time. The steady-state solution $z_1(t)$ is reached after $2Q$ oscillation cycles and follows the external excitation with amplitude $A_0$ and phase difference $\varphi$

$$z_1(t) = A_0 \cos(\omega t + \varphi). \tag{7.9}$$

The oscillation amplitude in the transient regime during the first $2Q$ cycles is

$$z_2(t) = A_t e^{\frac{-\omega_0 t}{2Q}} \sin(\omega_0 t + \phi_t). \tag{7.10}$$

We emphasize the important fact that the exponential term causes $z_2(t)$ to decrease exponentially with time constant $\tau$

$$\tau = \frac{2Q}{\omega_0}. \tag{7.11}$$

In vacuum conditions, only the internal dissipation due to bending of the cantilever is present, and $Q$ reaches values of 10,000 at typical resonant frequencies of 100,000 Hz. These values result in a relatively long transient regime of $\tau$ 30 ms, which limits the possible operational modes for dynamic AFM (for a detailed analysis see Albrecht et al. [11]). Changes in the measured amplitude, which reflect a change of atomic forces, will have a time lag of 30 ms, which is very slow considering one wants to scan a 200 × 200 point image within a few minutes. In air, however, viscous damping due to air friction dominates and $Q$ drops to less than 1,000, resulting in a time constant below the millisecond level. This response time is fast enough to use the amplitude as a measurement parameter.

If we evaluate the steady-state solution $z_1(t)$ in the differential equation, we find the following well-known solution for amplitude and phase of the oscillation as a function of the excitation frequency $\omega$:

$$A_0 = \frac{A_d Q \omega_0^2}{\sqrt{\omega^2 \omega_0^2 + Q^2 \left(\omega_0^2 - \omega^2\right)^2}}, \tag{7.12}$$

$$\varphi = \arctan\left(\frac{\omega \omega_0}{Q(\omega_0^2 - \omega^2)}\right) \tag{7.13}$$

Amplitude and phase diagrams are depicted in Fig. 7.5. As can be seen from (7.12), the amplitude will reach its maximum at a frequency different from $\omega_0$ if $Q$

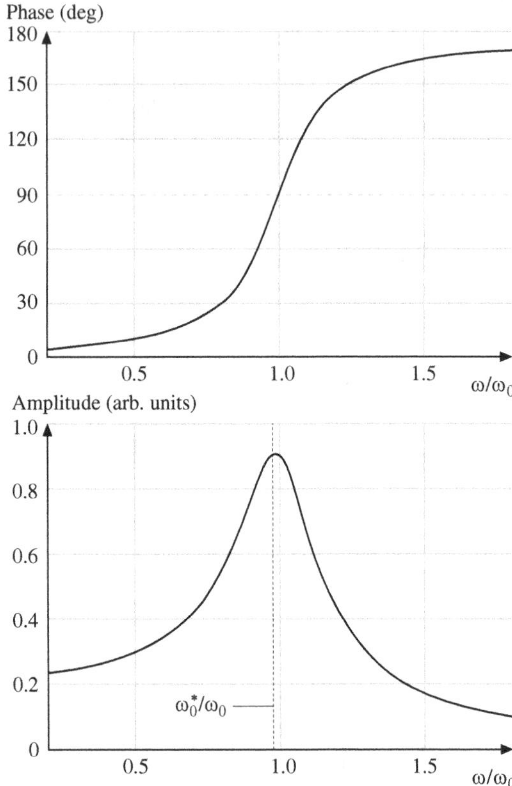


**Fig. 7.5** Curves of amplitude and phase versus excitation frequency for the damped harmonic oscillator, with a quality factor of $Q = 4$

has a finite value. The damping term of the harmonic oscillator causes the resonant frequency to shift from $\omega_0$ to $\omega_0{}^*$

$$\omega_0^* = \omega_0\sqrt{1 - \frac{1}{2Q^2}}. \tag{7.14}$$

The shift is negligible for $Q$-factors of 100 and above, which is the case for most applications in vacuum or air. However, for measurements in liquids, $Q$ can be smaller than 10 and $\omega_0$ differs significantly from $\omega_0{}^*$. As we will discuss later, it is also possible to enhance $Q$ by using a special excitation method called $Q$-control.

In the case that the excitation frequency is equal to the resonant frequency of the undamped cantilever $\omega = \omega_0$, we find the useful relation

$$A_0 = Q\,A_\mathrm{d}, \qquad \text{for} \quad \omega = \omega_0. \tag{7.15}$$

Since $\omega_0{}^* \approx \omega_0$ for most cases, we find that (7.15) holds true for exciting the cantilever at its resonance. From a similar argument, the phase becomes approximately 90° for the resonance case. We also see that, in order to reach vibration amplitudes of some 10 nm, the excitation only has to be as small as 1 pm, for typical cantilevers operated in vacuum.

So far we have not considered an additional force term, describing the interaction between the probing tip and the sample. For typical, large vibration amplitudes of 10–100 nm the tip experiences a whole range of force interactions during one single oscillation cycle, rather than one defined tip–sample force. How this problem can be attacked will be shown in the next paragraphs.

## 7.3 Dynamic AFM Operational Modes

While the quantitative interpretation of force curves in contact AFM is straightforward using (7.1), we explained in the previous paragraphs that its application to assess short-range attractive interatomic forces is rather limited. The dynamic mode of operation seems to open a viable direction toward achieving this task. However interpretation of the measurements generally appears to be more difficult. Different operational modes are employed in dynamic AFM, and the following paragraphs are intended to distinguish these modes and categorize them in a systematic way.

The oscillation trajectory of a dynamically driven cantilever is determined by three parameters: the amplitude, the phase, and the frequency. Tip–sample interactions can influence all three parameters, in the following, termed the internal parameters. The oscillation is driven externally, with excitation amplitude $A_d$ and excitation frequency $\omega$. These variables will be referred to as the external parameters. The external parameters are set by the experimentalist, whereas the internal parameters are measured and contain the crucial information about the force interaction. In scanning probe applications, it is common to control the probe–surface distance $z_0$ in order to keep an internal parameter constant (i.e., the tunneling current in STM or the beam deflection in contact AFM), which represents a certain tip–sample interaction. In $z$-spectroscopy mode, the distance is varied in a certain range, and the change of the internal parameters is measured as a fingerprint of the tip–sample interactions.

In dynamic AFM the situation is rather complex. Any of the internal parameters can be used for feedback of the tip–sample distance $z_0$. However, we already realized that, in general, the tip–sample forces could only be fully assessed by measuring all three parameters. Therefore, dynamic AFM images are difficult to interpret. A solution to this problem is to establish additional feedback loops, which keep the internal parameters constant by adjusting the external variables. In the simplest setup, the excitation frequency and the excitation amplitude are set to predefined values. This is the so-called amplitude-modulation (AM) mode or tapping mode. As stated before, in principle, any of the internal parameters can be used for feedback to the tip–sample distance – in AM mode the amplitude signal is used. A certain amplitude (smaller than the free oscillation amplitude) at a frequency close to the resonance of the cantilever is chosen, the tip is approached toward the surface under investigation, and the approach is stopped as soon as the setpoint amplitude is reached. The oscillation phase is usually recorded during the scan; however, the shift of the resonant frequency of the cantilever cannot be directly accessed, since this degree of freedom is blocked by the external excitation at a fixed frequency. It turns out that this mode is simple to operate from a technical perspective. Therefore, it is one of the most commonly used modes in dynamic AFM operated in air, and even in liquid. The strength of this mode is the easy and reliable high-resolution imaging of a large variety of surfaces.

It is interesting to discuss the AM mode in the situation that the external excitation frequency is much lower than the resonant frequency [12,13]. This results in a quasistatic measurement, although a dynamic oscillation force is applied, and therefore this mode can be viewed as a hybrid between static and dynamic AFM. Unfortunately, it has the drawbacks of the static mode, namely that stiff spring constants must be used and therefore the sensitivity of the deflection measurement must be very good, typically employing a high-resolution interferometer. Still, it has the advantage of the static measurement in terms of quantitative interpretation, since in the regime of small amplitudes ($<0.1$ nm) direct interpretation of the experiments is possible. In particular, the force gradient at tip–sample distance $z_0$ is given by the change of the amplitude $A$ and the phase angle $\varphi$

$$\left.\frac{\partial F_{\text{ts}}}{\partial z}\right|_{z_0} = k\left(1 - \frac{A_0}{A}\cos\phi\right). \tag{7.16}$$

In effect, the modulated AFM technique can profit from an enhanced sensitivity due to the use of lock-in techniques, which allows the measurement of the amplitude and phase of the oscillation signal with high precision.

As stated before, the internal parameters can be fed back to the external excitation variables. One of the most useful applications in this direction is the self-excitation system. Here the resonant frequency of the cantilever is detected and selected again as the excitation frequency. In a typical setup, the cantilever is self-oscillated with a phase shift of 90° by feeding back the detector signal to the excitation piezo. In this way the cantilever is always excited at its actual resonance [14]. Tip–sample interaction forces then only influence the resonant frequency, but do not change the two other parameters of the oscillation (amplitude and phase). Therefore, it is sufficient to measure the frequency shift induced by the tip–sample interaction. Since the phase remains at a fixed value, the oscillating system is much better defined than before, and the degrees of freedom for the oscillation are reduced. To even reduce the last degree of freedom an additional feedback loop can be incorporated to keep the oscillation amplitude $A$ constant by varying the excitation amplitude $A_{\text{d}}$. Now, all internal parameters have a fixed relation to the external excitation variables, the system is well defined, and all parameters can be assessed during the measurement.

In the following section we want to discuss the two most popular operational modes, tapping mode and self-excitation mode, in more detail.

### 7.3.1 *Amplitude-Modulation/Tapping-Mode AFM*

In tapping mode, or AM-AFM, the cantilever is excited externally at a constant frequency close to its resonance. Oscillation amplitude and phase during approach of tip and sample serve as the experimental observation channels. Figure 7.6 shows a diagram of a typical tapping-mode AFM setup. The oscillation of the cantilever is detected with the photodiode, whose output signal is analyzed with a lock-in amplifier to obtain amplitude and phase information. The amplitude is then compared with the setpoint, and the resulting difference or error signal is fed into the proportional–integral–differential (PID) controller, which adjusts the *z*-piezo, i.e., the probe–sample distance, accordingly. The external modulation unit supplies the signal for the excitation piezo, and at the same time the oscillation signal serves as the reference for the lock-in amplifier. As shown by the following applications the tapping mode is typically used to measure surface topography and other material parameters on the nanometer scale. The tapping mode is mostly used in ambient conditions and in liquids.

High-resolution imaging has been extensively performed in the area of materials science. Due to its technical relevance the investigation of polymers has been the focus of many studies (see, e.g., a recent review about AFM imaging on polymers by Magonov [16]). In Fig. 7.7 the topography of a diblock copolymer ($BC_{0.26}$-$3A_{0.53}F_8H_{10}$) at different magnifications is shown [15]. On the large scan (Fig. 7.7a) the large-scale structure of the microphase-separated polystyrene (PS) cylinders (within a polyisoprene (PI) matrix) lying parallel to the substrate can be seen. In the high-resolution image (Fig. 7.7b) a surface substructure of regular domes can be seen, which were found to be related to the cooling process during the polymer preparation.

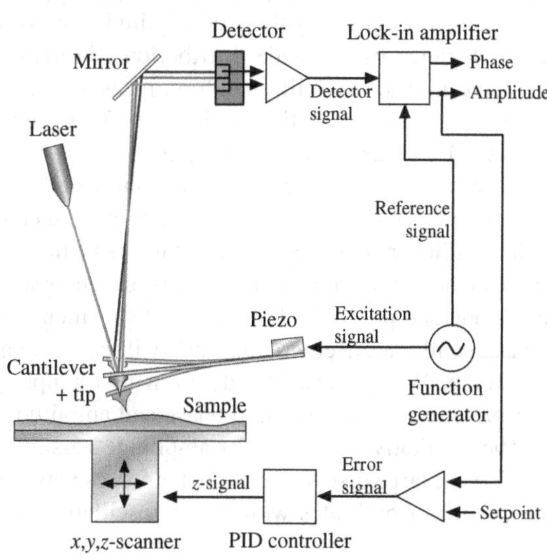


**Fig. 7.6** Setup of a dynamic force microscope operated in the AM or tapping mode. A laser beam is deflected by the back side of the cantilever and the deflection is detected by a split photodiode. The excitation frequency is chosen externally with a modulation unit, which drives the excitation piezo. A lock-in amplifier analyzes the phase and amplitude of the cantilever oscillation. The amplitude is used as the feedback signal for the probe–sample distance control

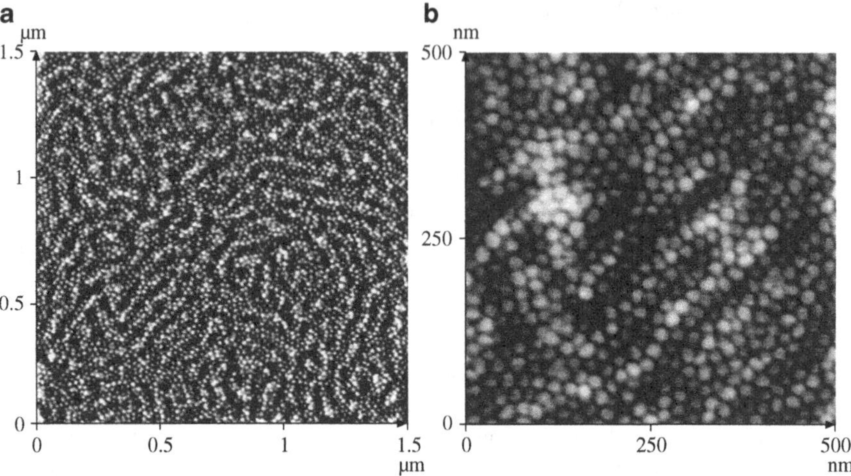


**Fig. 7.7** Tapping-mode images of $BC_{0.26}$–$3A_{0.53}F_8H_{10}$ at (**a**) low resolution and (**b**) high resolution. The height scale is 10 nm (After [15], © American Chemical Society, 2001)

Imaging in liquids opens up the possibility of the investigation of biological samples in their natural environment. For example Möller et al. [17] have obtained high-resolution images of the topography of hexagonally packed intermediate (HPI) layer of *Deinococcus radiodurans* with tapping-mode AFM. Another interesting example is the imaging of DNA in liquid, as shown in Fig. 7.8. Jiao et al. [10] measured the time evolution of a single DNA strand interacting with a molecule as shown by a sequence of images acquired in liquid over a time period of several minutes.

For a quantitative interpretation of tip–sample forces one has to consider that during one oscillation cycle with amplitudes of 10–100 nm the tip–sample interaction will range over a wide distribution of forces, including attractive as well as repulsive forces. We will, therefore, measure a convolution of the force–distance curve with the oscillation trajectory. This complicates the interpretation of AM-AFM measurements appreciably.

At the same time, the resonant frequency of the cantilever will change due to the appearing force gradients, as could already be seen in the simplified model in (7.4). If the cantilever is excited exactly at its resonant frequency before, it will be excited off resonance after interaction forces are encountered. This, in turn, changes the amplitude and phase in (7.12) and (7.13), which serve as the measurement signals. Consequently, a different amplitude will cause a change in the encountered effective force. We can see already from this simple *gedanken* experiment that the interpretation of the measured phase and amplitude curves is not straightforward.

The qualitative behavior for amplitude versus $z_0$-position curves is depicted in Fig. 7.9. At large distances, where the forces between tip and sample are negligible, the cantilever oscillates with its free oscillation amplitude. Upon approach of the

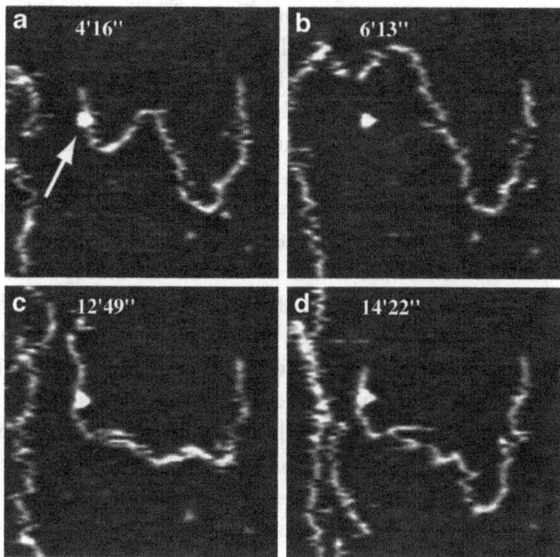


**Fig. 7.8** Dynamic p53–DNA interactions observed by time-lapse tapping-mode AFM imaging in solution. Both p53 protein and DNA were weakly adsorbed to a mica surface by balancing the buffer conditions. **(a)** A p53 protein molecule (*arrow*) was bound to a DNA fragment. The protein **(b)** dissociated from and then **(c)** reassociated with the DNA fragment. **(d)** A downward movement of the DNA with respect to the protein occurred, constituting a *sliding* event whereby the protein changes its position on the DNA. Image size: 620 nm. Grey scale (height) range: 4 nm. Time units: min, s. (© T. Schäffer, University of Münster)

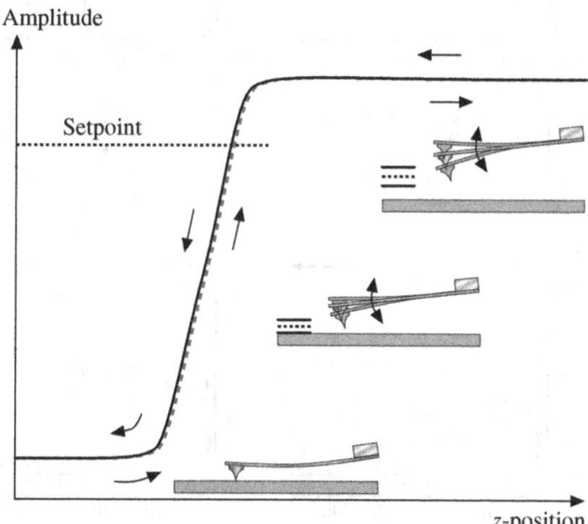


**Fig. 7.9** Simplified model showing the oscillation amplitude in tapping-mode AFM for various probe–sample distances

probe toward the surface the interaction forces cause the amplitude to change, typically resulting in an amplitude that gets smaller with continuously decreased tip–sample distance. This is expected, since the force–distance curve will eventually reach the repulsive part and the tip is hindered from indenting further into the sample, resulting in smaller oscillation amplitudes.

However, in order to gain some qualitative insight into the complex relationship between forces and oscillation parameters, we resort to numerical simulations. Anczykowski et al. [18,19] have calculated the oscillation trajectory of the cantilever under the influence of a given force model. van der Waals interactions were considered the only effective attractive forces, and the total interaction resembled a Lennard–Jones-type potential. Mechanical relaxations of the tip and sample surface were treated in the limits of continuum theory with the numerical Muller–Yushchenko–Derjaguin/Burgess–Hughes–White MYD/BHW [20,21] approach, which allows the simulations to be compared with experiments.

The cantilever trajectory was analyzed by numerically solving the differential equation (7.7) extended by the tip–sample force. The results of the simulation for the amplitude and phase of the tip oscillation as a function of $z$-position of the probe are presented in Fig. 7.10. One has to keep in mind that the $z$-position of the probe is not equivalent to the real tip–sample distance at equilibrium position, since the cantilever might bend statically due to the interaction forces. The behavior of the cantilever can be subdivided into three different regimes.

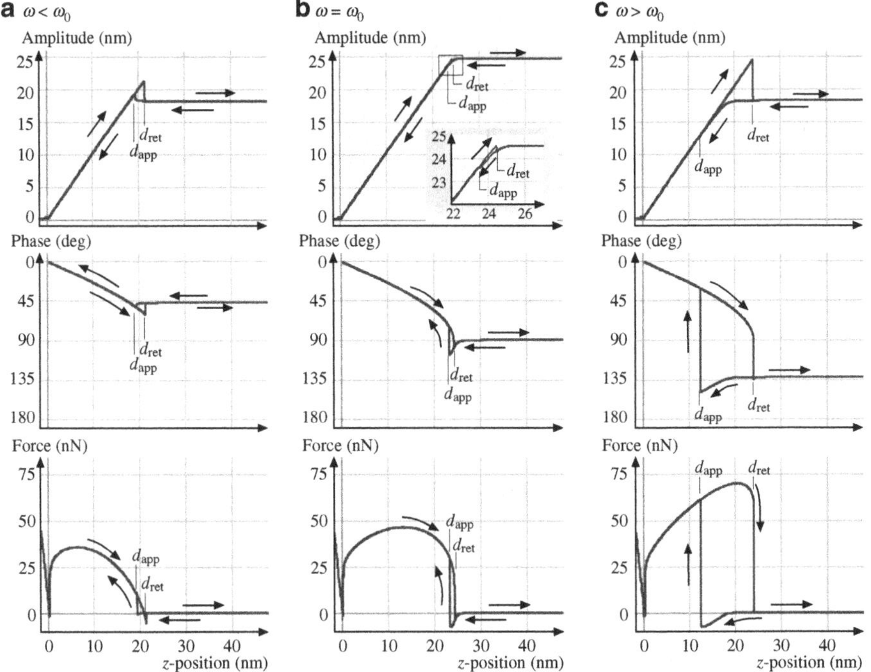


**Fig. 7.10** Amplitude and phase diagrams with excitation frequency: (**a**) below, (**b**) exactly at, and (**c**) above the resonant frequency for tapping-mode AFM from numerical simulations. Additionally, the *bottom diagrams* show the interaction forces at the point of closest tip–sample distance, i.e., the lower turnaround point of the oscillation

We distinguish the cases in which the beam is oscillated below its resonant frequency $\omega_0$, exactly at $\omega_0$, and above $\omega_0$. In the following, we will refer to $\omega_0$ as the resonant frequency, although the correct resonant frequency is $\omega_0^*$ if taking into account the finite $Q$-value.

Clearly, Fig. 7.10 exhibits more features than were anticipated from the initial, simple arguments. Amplitude and phase seem to change rather abruptly at certain points when the $z_0$-position is decreased. Additionally, we find hysteresis between approach and retraction.

As an example, let us start by discussing the discontinuous features in the AFM spectroscopy curves of the first case, where the excitation frequency is smaller than $\omega_0$. Consider the oscillation amplitude as a function of excitation frequency in Fig. 7.5. Upon approach of probe and sample, attractive forces will lower the effective resonant frequency of the oscillator. Therefore, the excitation frequency will now be closer to the resonant frequency, causing the vibration amplitude to increase. This, in turn, reduces the tip–sample distance, which again gives rise to a stronger attractive force. The system becomes unstable until the point $z_0 = d_{app}$ is reached, where repulsive forces stop the self-enhancing instability. This can be clearly observed in Fig. 7.10a. Large parts of the force–distance curve cannot be measured due to this instability.

In the second case, where the excitation equals the free resonant frequency, only a small discontinuity is observed upon reduction of the $z$-position. Here, a shift of the resonant frequency toward smaller values, induced by the attractive force interaction, will reduce the oscillation amplitude. The distance between tip and sample is, therefore, reduced as well, and the self-amplifying effect with the sudden instability does not occur as long as repulsive forces are not encountered. However, at closer tip–sample distances, repulsive forces will cause the resonant frequency to shift again toward higher values, increasing the amplitude with decreasing tip–sample distance. Therefore, a self-enhancing instability will also occur in this case, but at the crossover from purely attractive forces to the regime where repulsive forces occur. Correspondingly, a small kink in the amplitude curve can be observed in Fig. 7.10b. An even clearer indication of this effect is manifested by the sudden change in the phase signal at $d_{app}$.

In the last case, with $\omega > \omega_0$, the effect of amplitude reduction due to the resonant frequency shift is even larger. Again, we find no instability in the amplitude signal during approach in the attractive force regime. However, as soon as the repulsive force regime is reached, the instability occurs due to the induced positive frequency shift. Consequently, a large jump in the phase curve from values smaller than 90° to values larger than 90° is observed. The small change in the amplitude curve is not resolved in the simulated curves in Fig. 7.10c; however, it can be clearly seen in the experimental curves.

Figure 7.11 depicts the corresponding experimental amplitude and phase curves. The measurements were performed in air with a Si cantilever approaching a silicon wafer, with a cantilever resonant frequency of 299.95 kHz. Qualitatively, all prominent features of the simulated curves can also be found in the experimental

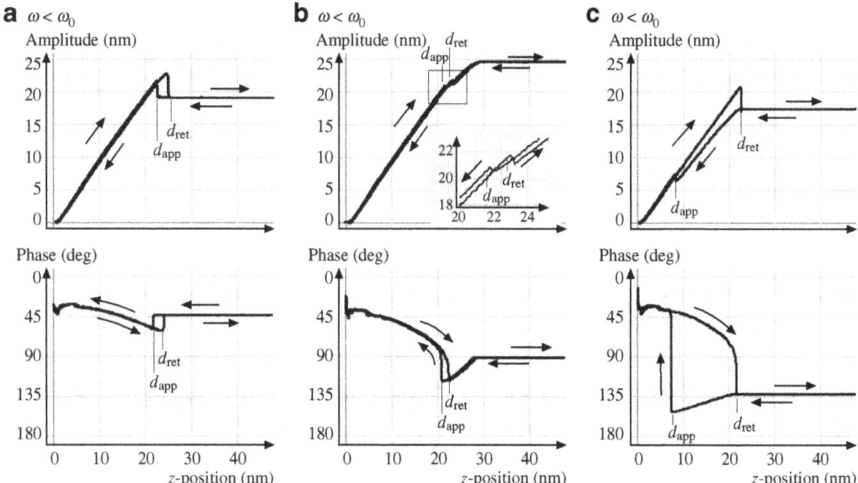


**Fig. 7.11** Amplitude and phase diagrams with excitation frequency: (**a**) below, (**b**) exactly at, and (**c**) above the resonant frequency for tapping-mode AFM from experiments with a Si cantilever on a Si wafer in air

data sets. Hence, the above model seems to capture the important factors necessary for an appropriate description of the experimental situation.

However, what is the reason for this unexpected behavior? We have to turn to the numerical simulations again, where we have access to all physical parameters, in order to understand the underlying processes. The lower part of Fig. 7.10 also shows the interaction force between the tip and the sample at the point of closest approach, i.e., the sample-sided turnaround point of the oscillation. We see that exactly at the points of the discontinuities the total interaction force changes from the net-attractive regime to the attractive–repulsive regime, also termed the intermittent contact regime. The term net-attractive is used to emphasize that the total force is attractive, despite the fact that some minor contributions might still originate from repulsive forces. As soon as a minimum distance is reached, the tip also starts to experience repulsive forces, which completely changes the oscillation behavior. In other words, the dynamic system switches between two oscillatory states.

Directly related to this fact is the second phenomenon: the hysteresis effect. We find separate curves for the approach of the probe toward the surface and the retraction. This seems to be somewhat counterintuitive, since the tip is constantly approaching and retracting from the surface and the average values of amplitude and phase should be independent of the direction of the average tip–sample distance movement. Hysteresis between approach and retraction within one oscillation due to dissipative processes should directly influence amplitude and phase. However, no dissipation models were included in the simulation. In this case, the hysteresis in Fig. 7.11 is due to the fact that the oscillation jumps into different modes; the system exhibits bistability. This effect is often observed in oscillators under the influence of nonlinear forces (e.g., [22]).

For the interpretation of these effects it is helpful to look at Fig. 7.12, which shows the behavior of the simulated tip trajectory and the force during one oscillation cycle over time. The data is shown for the $z$-positions where hysteresis is observed, while Fig. 7.12a was taken during the approach and Fig. 7.12b during the retraction. Excitation was in resonance, where the amplitude shows small hysteresis. Also note that the amplitude is almost exactly the same in Fig. 7.12a,b. We see that the oscillation at the same $z$-position exhibits two different modes: Whereas in Fig. 7.12a the experienced force is net-attractive, in Fig. 7.12b the tip is exposed to attractive and repulsive interactions. Experimental and simulated data show that the change between the net-attractive and intermittent contact mode takes place at different $z$-positions ($d_{app}$ and $d_{ret}$) for approach and retraction. Between $d_{app}$ and $d_{ret}$ the system is in a bistable mode. Depending on the history of the measurement, e.g., whether the position $d_{app}$ during the approach (or $d_{ret}$ during retraction) has been reached, the system flips to the other oscillation mode. While the amplitude might not be influenced strongly, the phase is a clear indicator of the mode switch. On the other hand, if point $d_{app}$ is never reached during the approach, the system will stay in the net-attractive regime and no hysteresis is observed, i.e., the system remains stable.

In conclusion, we find that, although a qualitative interpretation of the interaction forces is possible from the amplitude and phase curves, they do not give direct quantitative knowledge of tip–sample force interactions. However, it is a very useful tool for imaging nanometer-sized structures in a wide variety of setups, in air or even in liquid. We find that two distinct modes exist for the externally excited oscillation – the net-attractive and the intermittent contact mode – which describe

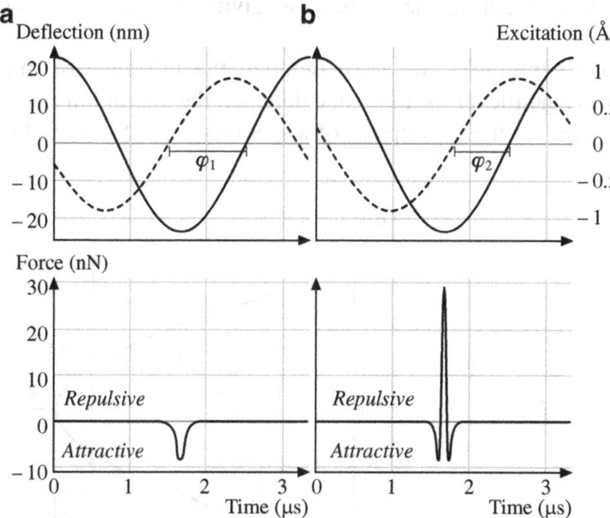


**Fig. 7.12** Simulation of the tapping-mode cantilever oscillation in the **(a)** net-attractive and **(b)** the intermittent contact regime. The *dashed line* represents the excitation amplitude and the *solid line* is the oscillation amplitude

what kind of forces govern the tip–sample interaction. The phase can be used as an indicator of the current mode of the system.

In particular, it can be easily seen that, if the free resonant frequency of the cantilever is higher than the excitation frequency, the system cannot stay in the net-attractive regime due to a self-enhancing instability. Since in many applications involving soft and delicate biological samples strong repulsive forces should be avoided, the tapping-mode AFM should be operated at frequencies equal to or above the free resonant frequency [23]. Even then, statistical changes of tip–sample forces during the scan might induce a sudden jump into the intermittent contact mode, and the previously explained hysteresis will tend to keep the system in this mode. It is, therefore, of great importance to tune the oscillation parameters in such a way that the AFM stays in the net-attractive regime [24]. A concept that achieves this task is the $Q$-control system, which will be discussed in some detail in the forthcoming paragraphs.

A last word concerning the overlap of simulation and experimental data: Whereas the qualitative agreement down to the detailed shape of hysteresis and instabilities is rather striking, we still find some quantitative discrepancies between the positions of the instabilities $d_{\mathrm{app}}$ and $d_{\mathrm{ret}}$. This is probably due to the simplified force model, which only takes into account van der Waals and repulsive forces. Especially at ambient conditions, an omnipresent water meniscus between tip and sample will give rise to much stronger attractive and also dissipative forces than considered in the model. A very interesting feature is that the simulated phase curves in the intermittent contact regime tend to have a steeper slope in the simulation than in the experiments (Fig. 7.13). We will show later that this effect is a fingerprint of an effect that had not been included in the above simulation at all: dissipative processes during the oscillation, giving rise to an additional loss of oscillation energy.

In the above paragraphs, we have outlined the influence of the tip–sample interaction on the cantilever oscillation, calculated the maximum tip–sample interaction forces based on the assumption of a specific model force, and subsequently

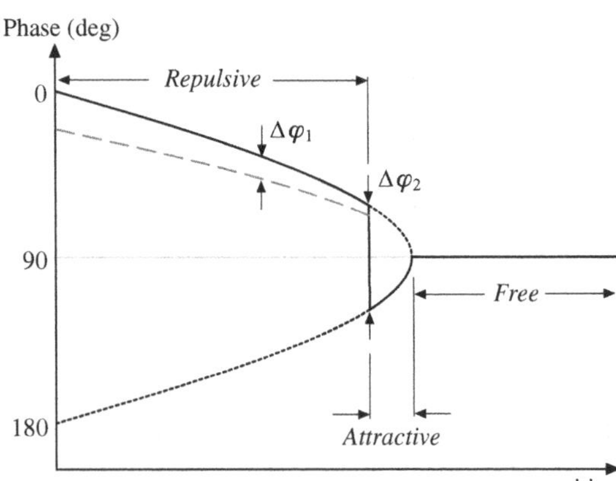


**Fig. 7.13** Phase shift in tapping mode as a function of tip–sample distance

discussed possible routes for image optimization. However, in practical imaging, the tip–sample interaction is not known a priori. In contrast, the ability to measure the continuous tip–sample interaction force as a function of both the tip–sample distance as well as the lateral location (e.g., in order to identify different bond strengths on chemically inhomogeneous surfaces) would add a tool of great value to the force-microscopist's toolbox.

Surprisingly, despite the more than 15 years during which the amplitude-modulation technique has been used, it was only recently that two solutions to this inversion problem have been suggested [25,26]. As already discussed in the previous paragraphs, conventional force–distance curves suffer from a *jump-to-contact* due to attractive surface forces. As a result, the most interesting range of the tip–sample force, the last few nanometers above the surface, is left out, and conventional force–distance curves thus mainly serve to determine adhesion forces.

As shown by Hölscher [25] the tip–sample force can be calculated with the help of the integral equation

$$F_{\mathrm{ts}}(D) = -\frac{\partial}{\partial D}\int_{D}^{D+2A} \frac{\kappa(z)}{\sqrt{z-D}}\,\mathrm{d}z, \tag{7.17a}$$

where

$$\kappa = \frac{kA^{3/2}}{\sqrt{2}}\left(\frac{A_{\mathrm{d}}\cos\varphi}{A} - \frac{\omega_0^2-\omega^2}{\omega^2}\right). \tag{7.17b}$$

It is now straightforward to recover the tip–sample force using (17a,b) from a *spectroscopy experiment*, i.e., an experiment where the amplitude and the phase are continuously measured as a function of the actual nearest tip–sample distance $D = z_0 - A$ at a fixed location above the sample surface. With this input, one first calculates $\kappa$ as a function of $D$. In a second step, the tip–sample force is computed by solving the integral in (7.17a) numerically.

A verification of the algorithm is shown in Fig. 7.14, which presents computer simulations of the method by calculating numerical solutions of the equation of motion. Figure 7.14a, b shows the resulting curves of amplitude and phase versus distance during approach, respectively. The subsequent reconstruction of the tip–sample interaction based on the data provided by the curves of amplitude and phase versus distance is presented in Fig. 7.14d. The assumed tip–sample force and energy dissipation are plotted by solid lines, while the reconstructed data is indicated by symbols; the excellent agreement demonstrates the reliability of the method. Nonetheless, it is important to recognize that the often observed instability in the curves of amplitude and phase versus distance affects the reconstruction of the tip–sample force. If such an instability occurs, experimentally accessible $\kappa(D)$ values will feature a *gap* at a specific range of tip–sample distances $D$. This is illustrated in Fig. 7.14c, where the gap is indicated by an arrow and the question

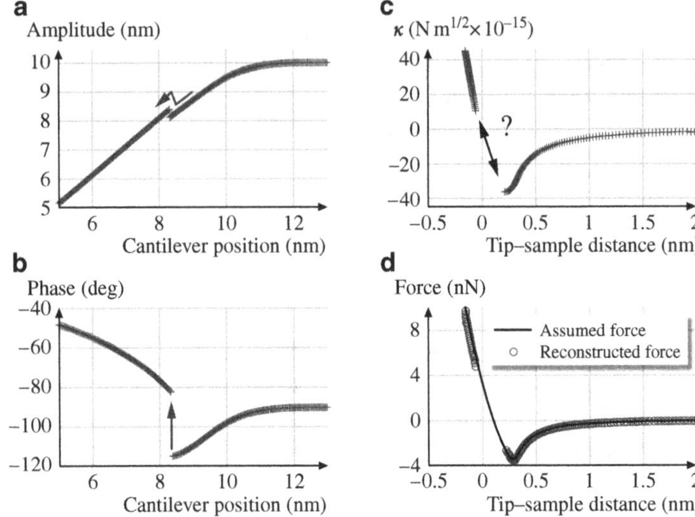


**Fig. 7.14** Numerical verification of the proposed force spectroscopy method for the tapping mode. The numerically calculated curves of amplitude **(a)** and phase **(b)** versus distance during the approach towards the sample surface. Both curves reveal the typical instability resulting also in a gap for the $\kappa$-curve. The tip–sample interaction force **(d)** can be recalculated from this data set by the application of (17a). As discussed in the text, the integration over the gap has to be handled with care

mark. Since calculation of the integral (17a) requires knowledge of all $\kappa$ values within the oscillation range, one might be tempted to extrapolate the missing values in the gap. This could be a workable solution if, as in our example, the accessible $\kappa$ values appear smooth and, in particular, the lower turning point of the $\kappa(D)$ values is clearly visible. In most realistic cases, however, the curves are unlikely to look as smooth as in our simulation and/or the lower turning point might not be reached, and we thus advise utmost caution in applying any extrapolation for missing data points.

### 7.3.2 Self-Excitation Modes

Despite the wide range of technical applications of the AM mode of dynamic AFM, it has been found unsuitable for measurements in an environment extremely useful for scientific research: vacuum or ultrahigh vacuum (UHV) with pressures reaching $10^{-10}$ mbar. The STM has already shown how much insight can be gained from experiments under those conditions.

Consider (7.11) from the previous section. The time constant $\tau$ for the amplitude to adjust to a different tip–sample force scales with $1/Q$. In vacuum applications, the $Q$-factor of the cantilever is on the order of 10,000, which means that $\tau$ is in

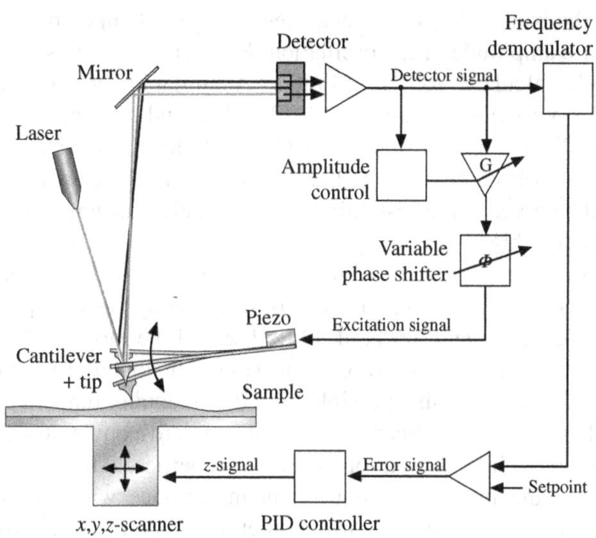


**Fig. 7.15** Dynamic AFM operated in the self-excitation mode, where the oscillation signal is directly fed back to the excitation piezo. The detector signal is amplified with the variable gain $G$ and phase-shifted by phase $\phi$. The frequency demodulator detects the frequency shift due to tip–sample interactions, which serves as the control signal for the probe–sample distance

the range of some 10 ms. This time constant is clearly too long for a scan of at least $100 \times 100$ data points. On the other hand, the resonant frequency of the system will react instantaneously to tip–sample forces. This has led Albrecht et al. [11] to use a modified excitation scheme.

The system is always oscillated at its resonant frequency. This is achieved by feeding back the oscillation signal from the cantilever into the excitation piezo element. Figure 7.15 pictures the method in a block diagram. The signal from the PSD is phase-shifted by 90° (and, therefore, always exciting in resonance) and used as the excitation signal of the cantilever. An additional feedback loop adjusts the excitation amplitude in such a way that the oscillation amplitude remains constant. This ensures that the tip–sample distance is not influenced by changes in the oscillation amplitude. The only degree of freedom that the oscillation system still has that can react to the tip–sample forces is the change of the resonant frequency. This shift of the frequency is detected and used as the setpoint signal for surface scans. Therefore, this mode is also called the frequency-modulation (FM) mode.

Let us take a look at the sensitivity of the dynamic AFM. If electronic noise, laser noise, and thermal drift can be neglected, the main noise contribution will come from thermal excitations of the cantilever. A detailed analysis of a dynamic system yields for the minimum detectable force gradient the relation [11]

$$\left.\frac{\partial F}{\partial z}\right|_{\min} = \sqrt{\frac{4kk_{\mathrm{B}}TB}{\omega_0 Q\langle z_{\mathrm{osc}}^2\rangle}}. \tag{7.18}$$

Here, $B$ is the bandwidth of the measurement, $T$ the temperature, and $\langle z_{\text{osc}}^2 \rangle$ is the mean-square amplitude of the oscillation. Please note that this sensitivity limit was deliberately calculated for the FM mode. A similar analysis of the AM mode, however, yields virtually the same result [27]. We find that the minimum detectable force gradient, i.e., the measurement sensitivity, is inversely proportional to the square root of the $Q$-factor of the cantilever. This means that it should be possible to achieve very high-resolution imaging under vacuum conditions where the $Q$-factor is very high.

A breakthrough in high-resolution AFM imaging was the atomic resolution imaging of the Si(111)-(7 $\times$ 7) surface reconstruction by Giessibl [8] under UHV conditions. Moreover, Sugawara et al. [28] observed the motion of single atomic defects on InP with true atomic resolution. However, imaging on conducting or semiconducting surfaces is also possible with the scanning tunneling microscope (STM) and these first noncontact atomic force microscopy (NC-AFM) images provided little new information on surface properties. The true potential of NC-AFM lies in the imaging of nonconducting surface with atomic precision, which was first demonstrated by Bammerlin et al. [29] on NaCl. A long-standing question about the surface reconstruction of the technological relevant material aluminum oxide could be answered by Barth and Reichling [30], who imaged the atomic structure of the high-temperature phase of $\alpha$-$Al_2O_3$(0001).

The high-resolution capabilities of noncontact atomic force microscopy are nicely demonstrated by the images shown in Fig. 7.16. Allers et al. [31] imaged steps and defects on the insulator nickel oxide with atomic resolution. Recently, Kaiser et al. [32] succeeded in imaging the antiferromagnetic structure of NiO (001). Nowadays, true atomic resolution is routinely obtained by various research groups (for an overview, see [33,34,35,36]).

However, we are concerned with measuring atomic force potentials of a single pair of molecules. Clearly, FM-mode AFM will allow us to identify single atoms, and with sufficient care we will be able to ensure that only one atom from the tip contributes to the total force interaction. Can we, therefore, fill in the last piece of information and find a quantitative relation between the oscillation parameters and the force?

A good insight into the cantilever dynamics can be drawn from the tip potential displayed in Fig. 7.17 [37]. If the cantilever is far away from the sample surface, the tip moves in a symmetric parabolic potential (dotted line), and the oscillation is harmonic. In such a case, the tip motion is sinusoidal and the resonant frequency is determined by the eigenfrequency $f_0$ of the cantilever. If the cantilever approaches the sample surface, the potential is changed, given by an effective potential $V_{\text{eff}}$ (solid line) which is the sum of the parabolic potential and the tip–sample interaction potential $V_{\text{ts}}$ (dashed line). This effective potential differs from the original parabolic potential and shows an asymmetric shape. As a result the oscillation becomes inharmonic, and the resonant frequency of the cantilever depends on the oscillation amplitude.

Gotsmann and Fuchs [38] investigated this relation with a numerical simulation. During each oscillation cycle the tip experiences a whole range of forces. For each

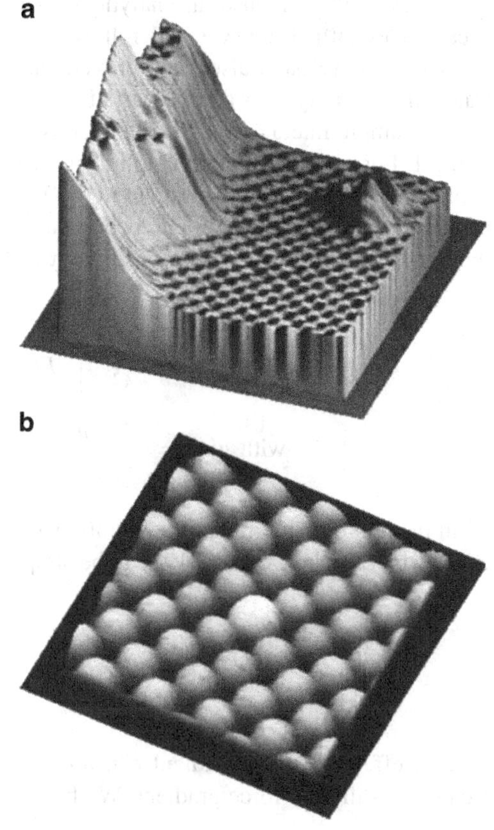


**Fig. 7.16** Imaging of a NiO (001) sample surface with a noncontact AFM. (**a**) Surface step and an atomic defect. The lateral distance between two atoms is 4.17 Å. (**b**) A dopant atom is imaged as a light protrusion about 0.1 Å higher as the other atoms. (© of W. Allers, S. Langkat, University of Hamburg)

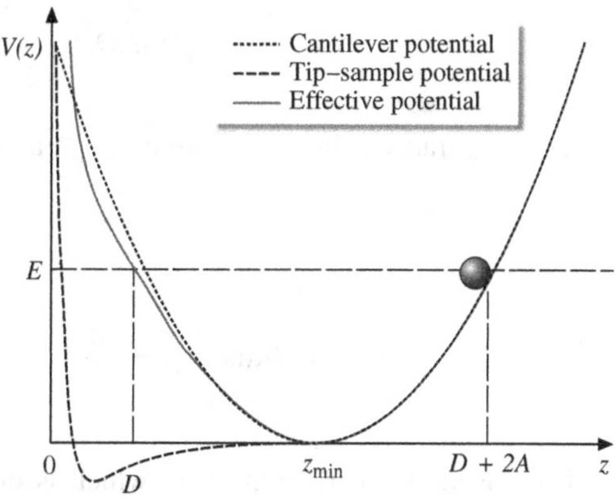


**Fig. 7.17** The frequency shift in dynamic force microscopy is caused by the tip–sample interaction potential (*dashed line*), which alters the harmonic cantilever potential (*dotted line*). Therefore, the tip moves in an anharmonic and asymmetric effective potential (*solid line*). Here $z_{min}$ is the minimum position of the effective potential (After [37])

step during the approach the differential equation for the whole oscillation loop (including also the feedback system) was evaluated and finally the quantitative relation between force and frequency shift was revealed.

However, there is also an analytical relationship, if some approximations are accepted [39,40]. Here, we will follow the route as indicated by [40], although alternative ways have also been proven successful. Consider the tip oscillation trajectory reaching over a large part of force gradient curve in Fig. 7.2. We model the tip–sample interaction as a spring constant of stiffness $k_{\mathrm{ts}}(z) = \partial F/\partial z|_{z0}$ as in Fig. 7.1. For small oscillation amplitudes we already found that the frequency shift is proportional to the force gradient in (7.4). For large amplitudes, we can calculate an effective force gradient $k_{\mathrm{eff}}$ as a convolution of the force and the fraction of time that the tip spends between the positions $x$ and $x + \mathrm{d}x$

$$\begin{aligned} k_{\mathrm{eff}}(z) &= \frac{2}{\pi A^2} \int_{z}^{z+2A} F(x) g\left(\frac{x-z}{A} - 1\right) \mathrm{d}x, \\ \text{with } g(u) &= -\frac{u}{\sqrt{1-u^2}}. \end{aligned} \tag{7.19}$$

In the approximation that the vibration amplitude is much larger than the range of the tip–sample forces, (7.19) can be simplified to

$$k_{\mathrm{eff}}(z) = \frac{\sqrt{2}}{\pi} A^{3/2} \int_{z}^{\infty} \frac{F(x)}{\sqrt{x-z}} \mathrm{d}x. \tag{7.20}$$

This effective force gradient can now be used in (7.4), the relation between frequency shift and force gradient. We find

$$\Delta f = \frac{f_0}{\sqrt{2}\pi k A^{3/2}} \int_{z}^{\infty} \frac{F(x)}{\sqrt{x-z}} \mathrm{d}x. \tag{7.21}$$

If we separate the integral from other parameters, we can define

$$\begin{aligned} \Delta f &= \frac{f_0}{k A^{3/2}} \gamma(z), \\ \text{with } \gamma(z) &= \frac{1}{\sqrt{2}\pi} \int_{z}^{\infty} \frac{F(x)}{\sqrt{x-z}} \mathrm{d}x. \end{aligned} \tag{7.22}$$

This means we can define $\gamma(z)$, which is only dependent on the shape of the force curve $F(z)$ but independent of the external parameters of the oscillation. The function $\gamma(z)$ is also referred to as the *normalized frequency shift* [7], a very useful parameter, which allows us to compare measurements independent of resonant frequency, amplitude, and spring constant of the cantilever.

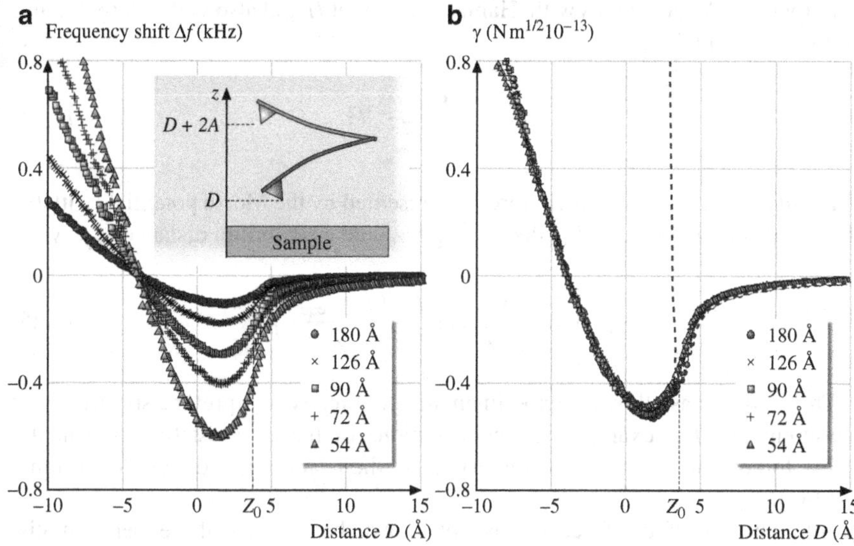


**Fig. 7.18** **(a)** Frequency-shift curves for different oscillation amplitudes for a silicon tip on a graphite surface in UHV, **(b)** $\gamma$ curves calculated from the $\Delta f$ curves in **(a)** (After [41], © The American Physical Society)

The dependence of the frequency shift on the vibration amplitude is an especially useful relation, since this parameter can be easily varied during one experiment. A nice example is depicted in Fig. 7.18, where frequency shift curves for different amplitudes were found to collapse into one curve in the $\gamma(z)$-diagram [41].

This relationship has been nicely exploited for the calibration of the vibration amplitude by Guggisberg [42], which is a problem often encountered in dynamic AFM operation and worthy of discussion. One approaches tip and sample and records curves of frequency shift versus distance, which show a reproducible shape. Then, the $z$-feedback is disabled, and several curves with different amplitudes are acquired. The amplitudes are typically chosen by adjusting the amplitude setpoint in volts. One has to take care that drift in the $z$-direction is negligible. An analysis of the corresponding $\gamma(z)$-curves will show the same curves (as in Fig. 7.18), but the curves will be shifted in the horizontal axis. These shifts correspond to the change in amplitude, allowing one to correlate the voltage values with the $z$-distances.

For the often encountered force contributions from electrostatic, van der Waals, and chemical binding forces the frequency shift has been calculated from the force laws. In the approximation that the tip radius $R$ is larger than the tip–sample distance $z$, an electrostatic potential $V$ will yield a normalized frequency shift of [43]

$$\gamma(z) = \frac{\pi \varepsilon_0 R V^2}{\sqrt{2}} z^{-1/2}. \tag{7.23}$$

For van der Waals forces with Hamaker constant $H$ and also with $R$ larger than $z$ we find accordingly

$$\gamma(z) = \frac{HR}{12\sqrt{2}} z^{-3/2}. \tag{7.24}$$

Finally, short-range chemical forces represented by the Morse potential (with the parameters binding energy $U_0$, decay length $\lambda$, and equilibrium distance $z_{\text{equ}}$) yield

$$\gamma(z) = \frac{U_0\sqrt{2}}{\sqrt{\pi\lambda}} \exp\left(-\frac{(z - z_{\text{equ}})}{\lambda}\right). \tag{7.25}$$

These equations allow the experimentalist to directly interpret the spectroscopic measurements. For example, the contributions of the electrostatic and van der Waals forces can be easily distinguished by their slope in a double-logarithmic plot [43].

Alternatively, if the force law is not known beforehand, the experimentalist wants to analyze the experimental frequency-shift data curves and extract the force or energy potential curves. We therefore have to invert the integral in (7.21) to find the tip–sample interaction potential $V_{\text{ts}}$ from the $\gamma(z)$-curves [40]

$$V_{\text{ts}}(z) = \sqrt{2} \int_z^\infty \frac{\gamma(x)}{\sqrt{x - z}} \mathrm{d}x. \tag{7.26}$$

Using this method, quantitative force curves were extracted from $\Delta f$ spectroscopy measurements on different, atomically resolved sites of the Si(111)-(7 $\times$ 7) reconstruction [1]. Comparison with theoretical molecular dynamics (MD) simulations showed good quantitative agreement with theory and confirmed the assumption that force interactions were governed by a single atom at the tip apex. Our initially formulated goal seems to be achieved: With FM-AFM we have found a powerful method that allows us to measure the chemical bond formation of single molecules. The last uncertainty, the exact shape and identity of the tip apex atom, can possibly be resolved by employing the FIM technique to characterize the tip surface in combination with FM-AFM.

All the above equations are only valid in the approximation that the oscillation amplitudes are much larger than the distance range of the encountered forces. However, for amplitudes of, e.g., 10 nm and long-range forces such as electrostatic interactions this approximation is no longer valid. Several approaches have been proposed by different authors to solve this issue [44,45,46]. The matrix method [45,47] uses the fact that in a real experiment the frequency shift curve is not continuous, but rather a set of discrete values acquired at equidistant points. Therefore the integral in (18) can be substituted by a sum and the equation can be rewritten as a linear equation system, which in return can be easily inverted by appropriate matrix operations. This *matrix method* is a very simple and general method for the AFM user to extract force curves from experimental frequency-shift curves without the restrictions of the large-amplitude approximation.

The concept of dynamic force spectroscopy can be also extended to three-dimensional (3-D) force spectroscopy by mapping the complete force field above the sample surface [48]. Figure 7.19a shows a schematic of the measurement principle. Curves of frequency shift versus distance are recorded on a matrix of points perpendicular to the sample surface. From this frequency shift data the complete three-dimensional force field between tip and sample can be recovered with atomic resolution. Figure 7.19b shows a cut through the force field as a two-dimensional map. The 3-D force technique has been applied also to a NaCl(100) surface, where not only conservative but also the dissipative tip–sample interaction could be measured in full space [49]. On the one hand, the forces were measured in the attractive as well as repulsive regime, allowing for the determination of the local minima in the corresponding potential energy curves in Fig. 7.20. This information is directly related to the atomic energy barriers responsible for a multitude of

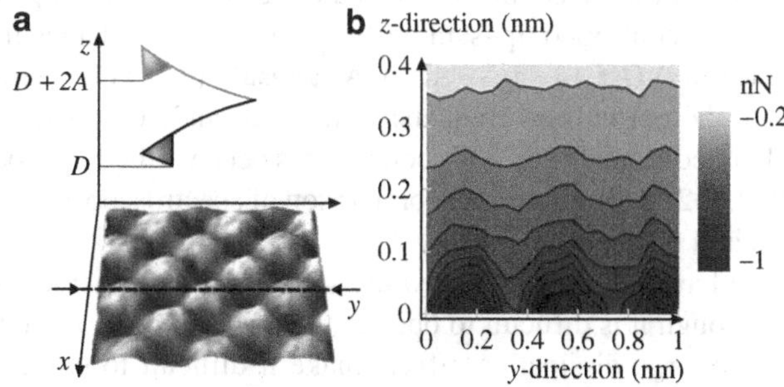


**Fig. 7.19** **(a)** Principle of 3-D force spectroscopy. The cantilever oscillates near the sample surface and measure the frequency shift in an *xyz*-box. The three-dimensional surface shows the topography of the sample (image size 1 ×1 nm$^2$) obtained immediately before the recording of the spectroscopy field. **(b)** The reconstructed force field of NiO(001) shows atomic resolution. The data are taken along the *line* shown in **(a)**

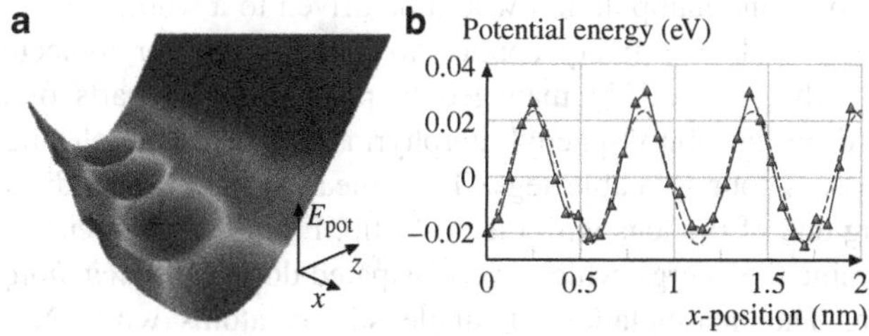


**Fig. 7.20** **(a)** Three-dimensional representation of the interaction energy map determine from 3-D force spectroscopy experiments on a NaCl(100) crystal surface. The *circular depressions* represent the local energy minima. **(b)** Potential energy profile obtained from **(a)** by collecting the minimum-energy values along the *x*-axis. This curve thus directly reveals the potential energy barrier of $\Delta E$ =48 meV separating the local energy minima

dynamic phenomena in surface science, such as diffusion, faceting, and crystalline growth. The direct comparison of conservative with the simultaneously acquired dissipative processes furthermore allowed the determination of atomic-scale mechanical relaxation processes.

In this context it is worth pointing out a slightly different dynamic AFM method. While in the typical FM-AFM setup the oscillation amplitude is controlled to stay constant by a dedicated feedback circuit, one could simply keep the excitation amplitude constant; this has been termed the *constant-excitation* (CE) mode, as opposed to the *constant-amplitude* (CA) mode. It is expected that this mode will be gentler to the surface, because any dissipative interaction will reduce the amplitude and therefore prevent a further reduction of the effective tip–sample distance. This mode has been employed to image soft biological molecules such as DNA or thiols in UHV [50].

At first glance, quantitative interpretation of the obtained frequency spectra seems more complicated, since the amplitude as well as the tip–sample distance is altered during the measurement. However, it was found by Hölscher et al. [51] that for the CE mode in the large-amplitude approximation the distance and the amplitude channel can be decoupled by calculating the effective tip–sample distance from the piezo-controlled tip–sample distance $z_0$ and the change in the amplitude with distance $A(z)$:$D(z_0) = z_0 - A(z_0)$. As a result, (7.22) can then be directly used to calculate the normalized frequency shift $\gamma(D)$ and consequently the force curve can be obtained from (7.26). This concept has been verified in experiments by Schirmeisen et al. [52] through a direct comparison of spectroscopy curves acquired in the CE mode and CA mode.

Until now, we have always associated the self-excitation scheme with vacuum applications. Although it is difficult to operate the FM-AFM in constant-amplitude mode in air, since large dissipative effects make it difficult to ensure a constant amplitude, it is indeed possible to use the constant-excitation FM-AFM in air or even in liquid [51,53,54]. Interestingly, a low-budget construction set (employing a tuning-fork force sensor) for a CE-mode dynamic AFM setup has been published on the internet (http://sxm4.uni-muenster.de).

If it is possible to measure atomic-scale forces with the NC-AFM, it should vice versa also be possible to exert forces with similar precision. In fact, the new and exciting field of nanomanipulation would be driven to a whole new dimension if defined forces could be reliably applied to single atoms or molecules. In this respect, Loppacher et al. [55] managed to push different parts of an isolated Cu-tetra-3,5 di-tertiary-butyl-phenyl porphyrin (Cu-TBBP) molecule, which is known to possess four rotatable legs. They measured the force–distance curves while pushing one of the legs with the AFM tip. From the force curves they were able to determine the energy which was dissipated during the *switching* process of the molecule. The manipulation of single silicon atoms with NC-AFM was demonstrated by Oyabu et al. [56], who removed single atoms from a Si(111)-$7 \times 7$ surface with the AFM tip and could subsequently deposit atoms from the tip on the surface again. This technique was further improved by Sugimoto et al. [57], who wrote artificial atomic structures with single Sn atoms. The possibility to exert and measure forces simultaneously during single atom or molecule manipulation is an exciting new application of high-resolution NC-AFM experiments.

## 7.4 $Q$-Control

We have already discussed the virtues of a high $Q$ value for high-sensitivity measurements: The minimum detectable force gradient was inversely proportional to the square root of $Q$. In vacuum, $Q$ mainly represents the internal dissipation of the cantilever during oscillation, an internal damping factor. Low damping is obtained by using high-quality cantilevers, which are cut (or etched) from defect-free, single-crystal silicon wafers. Under ambient or liquid conditions, the quality factor is dominated by dissipative interactions between the cantilever and the surrounding medium, and $Q$ values can be as low as 100 for air or even 5 in liquid. Still, we ask if it is somehow possible to compensate for the damping effect by exciting the cantilever in a sophisticated way.

It turns out that the shape of the resonance curves in Fig. 7.5 can be influenced toward higher (or lower) $Q$ values by an amplitude feedback loop. In principle, there are several mechanisms to couple the amplitude signal back to the cantilever, e.g., by the photothermal effect [58] or capacitive forces [59]. Figure. 7.21 shows a method in which the amplitude feedback is mediated directly by the excitation piezo [60]. This has the advantage that no additional mechanical setups are necessary.

The working principle of the feedback loop can be understood by analyzing the equation of motion of the modified dynamic system

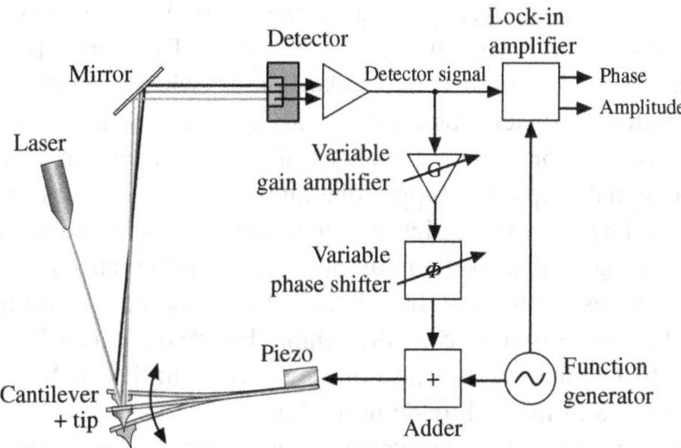


**Fig. 7.21** Schematic diagram of a $Q$-control feedback circuit with an externally driven dynamic AFM. The tapping-mode setup is in effect extended by an additional feedback loop

$$m^*\ddot{z}(t) + \alpha\dot{z}(t) + kz(t) - F_{ts}[z_0 + z(t)] \\ = F_{ext}\cos(\omega t) + G\,e^{i\phi}z(t). \tag{7.27}$$

This ansatz takes into account the feedback of the detector signal through a phase shifter, amplifier, and adder as an additional force, which is linked to the cantilever deflection $z(t)$ through the gain $G$ and the phase shift $e^{i\phi}$. We assume that the oscillation can be described by a harmonic oscillation trajectory. With a phase shift of $\phi = \pm\pi/2$ we find

$$e^{\pm i\pi/2}z(t) = \pm\frac{1}{\omega}\dot{z}(t). \tag{7.28}$$

This means, that the additional feedback force signal $Ge^{i\varphi}z(t)$ is proportional to the velocity of the cantilever, just like the damping term in the equation of motion. We can define an effective damping constant $\alpha_{eff}$, which combines the two terms

$$m^*\ddot{z}(t) + \alpha_{eff}\dot{z}(t) + kz(t) - F_{ts}[z_0 + z(t)] \\ = F_{ext}\cos(\omega t), \\ \text{with } \alpha_{eff} = \alpha \mp \frac{1}{\omega}G, \text{ for } \phi = \pm\frac{\pi}{2}. \tag{7.29}$$

Equation 7.28 shows that the damping of the oscillator can be enhanced or weakened by choosing $\phi = +\frac{\pi}{2}$ or $\phi = -\frac{\pi}{2}$, respectively. The feedback loop therefore allows us to vary the effective quality factor $Q_{eff} = m\omega_0/\alpha_{eff}$ of the complete dynamic system. Hence, this system was termed $Q$-control. Figure 7.22 shows experimental data regarding the effect of $Q$-control on the amplitude and phase as a function of the external excitation frequency [60]. In this example, $Q$-control was able to increase the $Q$-value by a factor of >40.

The effect of improved image contrast is demonstrated in Fig. 7.23. Here, a computer hard disk was analyzed with a magnetic tip in tapping mode, where the magnetic contrast is observed in the phase image. The upper part shows the magnetic data structures recorded in standard mode, whereas in the lower part of the image $Q$-control feedback was activated, giving rise to an improved signal, i.e., magnetic contrast. A more detailed analysis of measurements on a magnetic tape shows that the signal amplitude (upper diagrams in Fig. 7.24) was increased by a factor of 12.4 by the $Q$-control feedback. The lower image shows a noise analysis of the signal, indicating an improvement of the signal-to-noise ratio by a factor of 2.3.

It might be interesting to note that $Q$-control can also be applied in FM mode, which might be counterintuitive at first sight. However, it has been shown by Ebeling et al. [61,62] that the increase of the $Q$-factor in liquids helps to increase the imaging features of the FM mode in liquids.

The diagrams represent measurements in air with an AFM operated in AM mode. Only then can we make a distinction between excitation and vibration frequency, since in the FM mode these two frequencies are equal by definition.

Although the relation between sensitivity and $Q$-factor in (7.17a) is the same for AM and FM mode, it must be critically investigated to see whether the enhanced quality factor by $Q$-control can be inserted into the equation for FM-mode AFM. In vacuum applications, $Q$ is already very high, which makes it unnecessary to operate an additional $Q$-control module.

As stated before, we can also use $Q$-control to enhance the damping in the oscillating system. This would decrease the sensitivity of the system. However, on the other hand, the response time of the amplitude change is decreased as well. For tapping-mode applications, where high-speed scanning is the goal, $Q$-control was able to reduce the scan speed limiting relaxation time [63].

A large quality factor $Q$ does not only have the virtue of increasing the force sensitivity of the instrument. It also has the advantage of increasing the parameter space of stable AFM operation in AM-mode AFM. Consider the resonance curve of Fig. 7.5. When approaching the tip toward the surface there are two competing mechanisms: On the one hand, we bring the tip closer to the sample, which results in an increase in attractive forces (Fig. 7.2). On the other hand, for the case $\omega > \omega_0$, the resonant frequency of the cantilever is shifted toward smaller values due to the

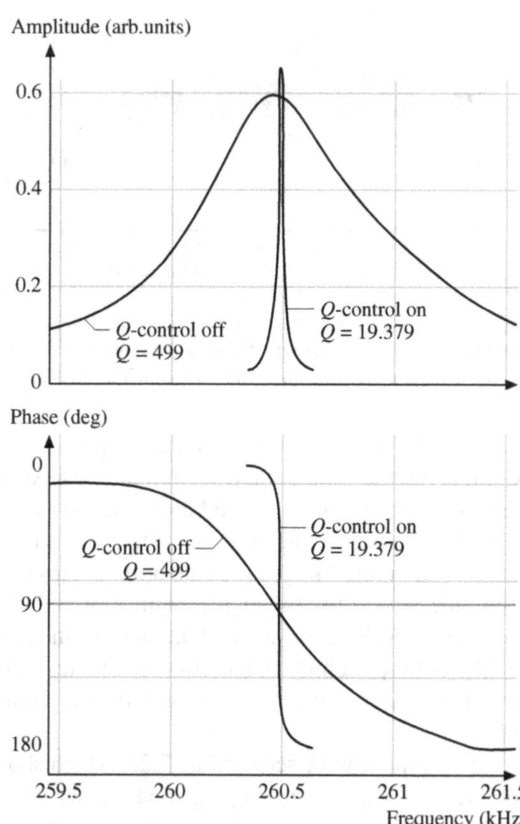


**Fig. 7.22** Amplitude and phase diagrams measured in air with a Si cantilever far away from the sample. The quality factor can be increased from 450 to 20000 by using the $Q$-control feedback method

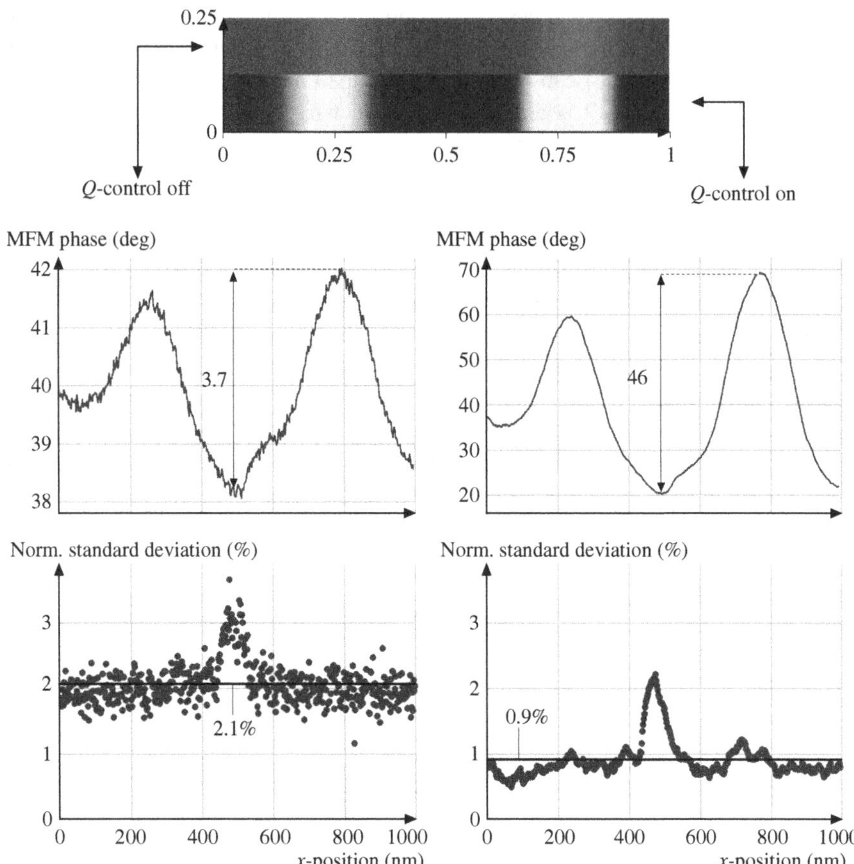


**Fig. 7.23** Enhancement of the contrast in the phase channel due to $Q$-control on a magnetic hard disk measured with a magnetic tip in tapping-mode AFM in air. Scan size 5 ×5 μm, phase range 10 (www.nanoanalytics.com) (MFM – magnetic force microscopy)

attractive forces, which causes the amplitude to become smaller. This is the desirable regime, where stable operation of the AFM is possible in the net-attractive regime. However, as explained before, below a certain tip–sample separation $d_{\mathrm{app}}$, the system switches suddenly into intermittent contact mode, where surface modifications are more likely due to the onset of strong repulsive forces. The steeper the amplitude curve, the larger the regime of stable, net-attractive AFM operation. Looking at Fig. 7.22 we find that the slope of the amplitude curve is governed by the quality factor $Q$. A high $Q$, therefore, facilitates stable operation of the AM-AFM in the net-attractive regime. (A more detailed discussion about this topic can be found in [64]).

An example can be seen in Fig. 7.25, which shows a surface scan of an ultrathin organic film acquired in tapping mode under ambient conditions. First, the inner square was scanned without the $Q$ enhancement, and then a wider surface area was

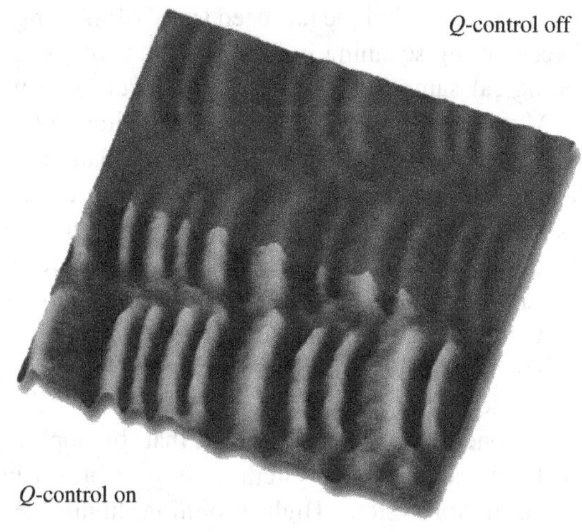


**Fig. 7.24** Signal-to-noise analysis with a magnetic tip in tapping-mode AFM on a magnetic tape sample with $Q$-control

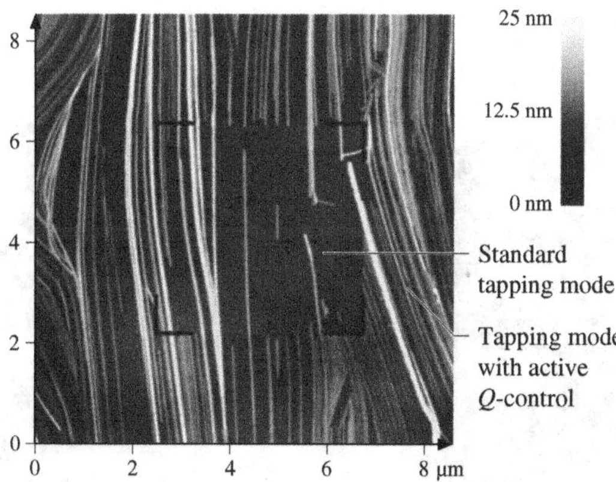


**Fig. 7.25** Imaging of a delicate organic surface with $Q$-control. Sample was a Langmuir–Blodgett film (ethyl-2,3-dihydroxyoctadecanoate) on a mica substrate. The topographical image clearly shows that the highly sensitive sample surface can only be imaged nondestructively with active $Q$-control, whereas the periodic repulsive contact with the probe in standard operation without $Q$-control leads to significant modification or destruction of the surface structure (© L. Chi and coworkers, University of Münster)

scanned with applied $Q$-control. The high quality factor provides a larger parameter space for operating the AFM in the net-attractive regime, allowing good resolution of the delicate organic surface structure. Without the $Q$-control the surface structures are deformed and even destroyed due to the strong repulsive tip–sample interactions [65,66,67]. This also allowed imaging of DNA structures without predominantly depressing the soft material during imaging. It was then possible to observe a DNA diameter close to the theoretical value with the $Q$-control feedback [68].

The same technique has been successfully employed to minimize the interaction forces during scanning in liquids. This is of special relevance for imaging delicate biological samples in environments such as water or buffer solution. When the AFM probe is submerged in a liquid medium, the oscillation of the AFM cantilever is strongly affected by hydrodynamic damping. This typically leads to quality factors $<10$ and accordingly to a loss in force sensitivity. However, the $Q$-control technique allows the effective quality factor to be increased by about three orders of magnitude in liquids. Figure 26 shows results of scanning DNA structures on a mica substrate under buffer solution [69]. Comparison of the topographic data obtained in standard tapping mode and under $Q$-control, in particular the difference in the observed DNA height, indicates that the imaging forces were successfully reduced by employing $Q$-control.

In conclusion, we have shown that, by applying an additional feedback circuit to the dynamic AFM system, it is possible to influence the quality factor $Q$ of the oscillator system. High-resolution, high-speed, or low-force scanning is then possible.

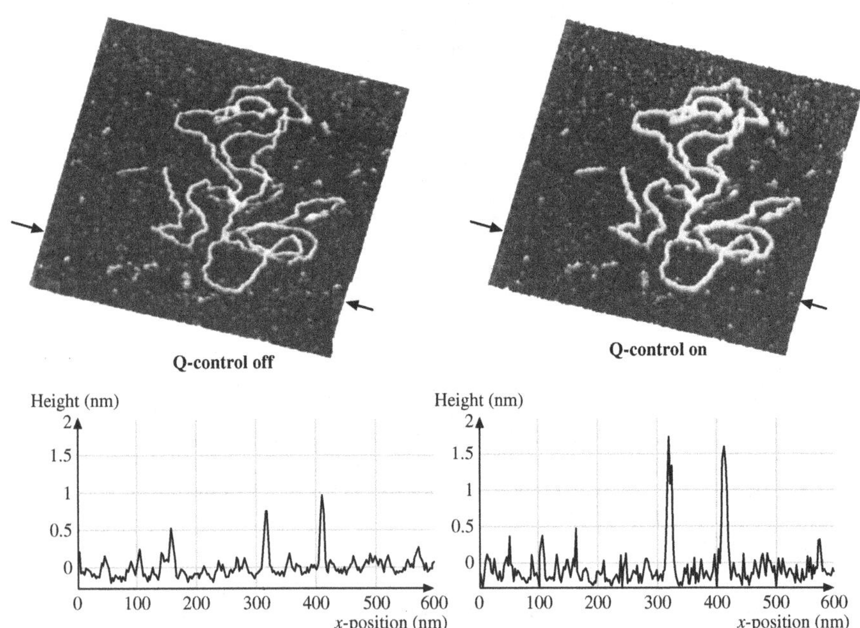


**Fig. 7.26** AFM images of DNA on mica scanned in buffer solution ($600 \times 600$ nm$^2$). Each scan line was scanned twice – in standard tapping mode during the first scan of the line (*left data*) and with $Q$-control being activated by a trigger signal during the subsequent scan of the same line (*right data*). This interleave technique allows direct comparison of the results of the two modes obtained on the same surface area while minimizing drift effects. Cross-sections of the topographic data reveal that the observed DNA height is significantly higher in the case of imaging under $Q$-control (© D. Ebeling, University of Münster)

## 7.5 Dissipation Processes Measured with Dynamic AFM

Dynamic AFM methods have proven their great potential for imaging surface structures at the nanoscale, and we have also discussed methods that allow the assessment of forces between distinct single molecules. However, there is another physical mechanism that can be analyzed with the dynamic mode and has been mentioned in some previous paragraphs: energy dissipation.

In Fig. 7.12 we have already shown an example where the phase signal in tapping mode cannot be explained by conservative forces alone; dissipative processes must also play a role. In constant-amplitude FM mode, where the quantitative interpretation of experiments has proven to be less difficult, an intuitive distinction between conservative and dissipative tip–sample interaction is possible. We have shown the correlation between forces and frequency shifts of the oscillating system, but we have neglected one experimental input channel. The excitation amplitude, which is necessary to keep the oscillation amplitude constant, is a direct indication of the energy dissipated during one oscillation cycle. Dürig [70] and Hölscher et al. [71] have shown that, in self-excitation mode (with an excitation–oscillation phase difference of 90°), conservative and dissipative interactions can be strictly separated. Part of this energy is dissipated in the cantilever itself; another part is due to external viscous forces in the surrounding medium. However, more interestingly, some energy is dissipated at the tip–sample junction. This mechanism is the focus of the following paragraphs.

In contrast to conservative forces acting at the tip–sample junction, which at least in vacuum can be understood in terms of van der Waals, electrostatic, and chemical interactions, the dissipative processes are poorly understood. Stowe et al. [72] have shown that, if a voltage potential is applied between tip and sample, charges are induced in the sample surface, which will follow the tip motion (in their setup the oscillation was parallel to the surface). Due to the finite resistance of the sample material, energy will be dissipated during the charge movement. This effect has been exploited to image the doping level of semiconductors. Energy dissipation has also been observed in imaging magnetic materials. Liu and Grütter [73] found that energy dissipation due to magnetic interactions was enhanced at the boundaries of magnetic domains, which was attributed to domain wall oscillations. Even a simple system such as two clean metal surfaces which are moved in close proximity can give rise to frictional forces. Stipe et al. [74] have measured the energy dissipation due to fluctuating electromagnetic fields between two closely spaced gold surfaces, which was later interpreted by Volokitin and Persson [75] in terms of *van der Waals friction*.

However, also in the absence of external electromagnetic fields, energy dissipation was observed in close proximity of tip and sample, within 1 nm. Clearly, mechanical surface relaxations must give rise to energy losses. One could model the AFM tip as a small hammer, hitting the surface at high frequency, possibly resulting in phonon excitations. From a continuum-mechanics point of view, we assume that the mechanical relaxation of the surface is not only governed by elastic responses.

Viscoelastic effects of soft surfaces will also render a significant contribution to energy dissipation. The whole area of phase imaging in tapping mode is concerned with those effects [76,77,78,79].

In the atomistic view, the last tip atom can be envisaged to change position while experiencing the tip–sample force field. A strictly reversible change of position would not result in a loss of energy. Still, it has been pointed out by Sasaki and Tsukada [80] that a change in atom position would result in a change in the force interaction itself. Therefore, it is possible that the tip atom changes position at different tip–surface distances during approach and retraction, effectively causing atomic-scale hysteresis to develop. Hoffmann et al. [13] and Hembacher et al. [81] have measured the short-range energy dissipation for different combinations of tip and surface materials in UHV. For atomic-resolution experiments at low temperatures on graphite [81] it was found that the energy dissipation is a step-like function. A similar shape of dissipation curves was found in a theoretical analysis by Kantorovich and Trevethan [82], where the energy dissipation was directly associated with atomic instabilities at the sample surface.

The dissipation channel has also been used to image surfaces with atomic resolution [83]. Instead of feeding back the distance on the frequency shift, the excitation amplitude in FM mode has been used as the control signal. The Si(111)-$(7 \times 7)$ reconstruction was successfully imaged in this mode. The step edges of monatomic NaCl islands on single-crystalline copper have also rendered atomic-resolution contrast in the dissipation channel [84]. The dissipation processes discussed so far are mostly in the configuration in which the tip is oscillated perpendicular to the surface. Friction is usually referred to as the energy loss due to lateral movement of solid bodies in contact. It is interesting to note in this context that Israelachivili [85] has pointed out a quantitative relationship between lateral and vertical (with respect to the surface) dissipation. He states that the hysteresis in vertical force–distance curves should equal the energy loss in lateral friction. An experimental confirmation of this conjecture at the molecular level is still lacking.

Physical interpretation of energy-dissipation processes at the atomic scale seems to be a daunting task at this point. Notwithstanding, we can find a quantitative relation between the energy loss per oscillation cycle and the experimental parameters in dynamic AFM, as will be shown in the following section.

In static AFM it was found that permanent changes of the sample surface by indentations can cause hysteresis between approach and retraction. The area between the approach and retraction curves in a force–distance diagram represents the lost or dissipated energy caused by the irreversible change of the surface structure. In dynamic-mode AFM, the oscillation parameters such as amplitude, frequency, and phase must contain the information about the dissipated energy per cycle. So far, we have resorted to a treatment of the equation of motion of the cantilever vibration in order to find a quantitative correlation between forces and the experimental parameters. For the dissipation it is useful to treat the system from the energy-conservation point of view.

Assuming that a dynamic system is in equilibrium, the average energy input must equal the average energy output or dissipation. Applying this rule to an AFM running in dynamic mode means that the average power fed into the cantilever oscillation by an external driver, denoted by $\overline{P}_{\text{in}}$, must equal the average power dissipated by the motion of the cantilever beam $\overline{P}_0$ and by tip–sample interaction $\overline{P}_{\text{tip}}$

$$\bar{P}_{\text{in}} = \bar{P}_0 + \bar{P}_{\text{tip}}. \tag{7.30}$$

The term $\overline{P}_{\text{tip}}$ is what we are interested in, since it gives us a direct physical quantity to characterize the tip–sample interaction. Therefore, we have first to calculate and then measure the two other terms in (7.30) in order to determine the power dissipated when the tip periodically probes the sample surface. This requires an appropriate rheological model to describe the dynamic system. Although there are investigations in which the complete flexural motion of the cantilever beam has been considered [86], a simplified model, comprising a spring and two dashpots (Fig. 7.27), represents a good approximation in this case [87].

The spring, characterized by the constant $k$ according to Hooke's law, represents the only channel through which power $P_{\text{in}}$ can be delivered to the oscillating tip $z(t)$ by the external driver $z_{\text{d}}(t)$. Therefore, the instantaneous power fed into the dynamic system is equal to the force exerted by the driver times the velocity of the driver (the force which is necessary to move the base side of the dashpot can be neglected, since this power is directly dissipated and therefore does not contribute to the power delivered to the oscillating tip)

$$P_{\text{in}}(t) = F_{\text{d}}(t)\dot{z}_{\text{d}}(t) = k[z(t) - z_{\text{d}}(t)]\dot{z}_{\text{d}}(t). \tag{7.31}$$

Assuming a sinusoidal steady-state response and that the base of the cantilever is driven sinusoidally (7.6) with amplitude $A_{\text{d}}$ and frequency $\omega$, the deflection from equilibrium of the end of the cantilever follows (7.9), where $A$ and $0 \leq \phi \leq \pi$ are the oscillation amplitude and phase shift, respectively. This allows us to calculate the average power input per oscillation cycle by integrating (7.30) over one period $T = 2\pi/\omega$

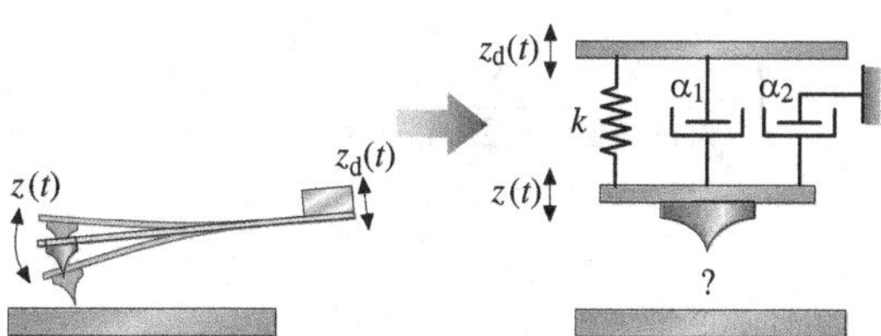


**Fig. 7.27** Rheological models applied to describe the dynamic AFM system, comprising the oscillating cantilever and tip interacting with the sample surface. The movement of the cantilever base and the tip is denoted by $z_{\text{d}}(t)$ and $z(t)$, respectively. The cantilever is characterized by the spring constant $k$ and the damping constant $\alpha$. In a first approach, damping is broken into two pieces $\alpha_1$ and $\alpha_2$: first, intrinsic damping caused by the movement of the cantilever's tip relative to its base, and second, damping related to the movement of the cantilever body in a surrounding medium, e.g., air damping

$$\bar{P}_{\text{in}} = \frac{1}{T}\int_0^T P_{\text{in}}(t)\text{d}t = \frac{1}{2}k\omega A_{\text{d}}A\sin\phi. \tag{7.32}$$

This contains the familiar result that the maximum power is delivered to an oscillator when the response is 90° out of phase with the drive.

The simplified rheological model as depicted in Fig. 7.27 exhibits two major contributions to the damping term $\overline{P}_0$. Both are related to the motion of the cantilever body and assumed to be well modeled by viscous damping with coefficients $\alpha_1$ and $\alpha_2$. The dominant damping mechanism in UHV conditions is intrinsic damping, caused by the deflection of the cantilever beam, i.e., the motion of the tip relative to the cantilever base. Therefore the instantaneous power dissipated by such a mechanism is given by

$$P_{01}(t) = |F_{01}(t)\dot{z}(t)| = |\alpha_1[\dot{z}(t) - \dot{z}_{\text{d}}(t)]\dot{z}(t)|. \tag{7.33}$$

Note that the absolute value has to be calculated, since all dissipated power is *lost* and therefore cannot be returned to the dynamic system.

However, when running an AFM in ambient conditions an additional damping mechanism has to be considered. Damping due to the motion of the cantilever body in the surrounding medium, e.g., air damping, is in most cases the dominant effect. The corresponding instantaneous power dissipation is given by

$$P_{02}(t) = |F_{02}(t)\dot{z}(t)| = \alpha_2\dot{z}^2(t). \tag{7.34}$$

In order to calculate the average power dissipation, (7.33) and (7.34) have to be integrated over one complete oscillation cycle. This procedure yields

$$\begin{aligned}\bar{P}_{01} &= \frac{1}{T}\int_0^T P_{01}(t)\text{d}t \\ &= \frac{1}{\pi}\alpha_1\omega^2 A\Big[(A - A_{\text{d}}\cos\varphi)\arcsin \\ &\times \left(\frac{A - A_{\text{d}}\cos\varphi}{\sqrt{A^2 + A_{\text{d}}^2 - 2AA_{\text{d}}\cos\varphi}}\right) + A_{\text{d}}\sin\ \varphi\Big]\end{aligned} \tag{7.35}$$

and

$$\bar{P}_{02} = \frac{1}{T}\int_0^T P_{02}(t)\text{d}t = \frac{1}{2}\alpha_2\omega^2 A^2. \tag{7.36}$$

Considering the fact that commonly used cantilevers exhibit a quality factor of at least several hundreds (in UHV even several tens of thousands), we can assume that the oscillation amplitude is significantly larger than the drive amplitude when the dynamic system is driven at or near its resonance frequency: $A \gg A_d$. Therefore (7.34) can be simplified in first-order approximation to an expression similar to (35). Combining the two equations yields the total average power dissipated by the oscillating cantilever

$$\bar{P}_0 = \frac{1}{2}\alpha\omega^2 A^2, \text{ with } \alpha = \alpha_1 + \alpha_2, \tag{7.37}$$

where $\alpha$ denotes the overall effective damping constant.

We can now solve (7.30) for the power dissipation localized to the small interaction volume of the probing tip with the sample surface, represented by the question mark in Fig. 7.27. Furthermore by expressing the damping constant $\alpha$ in terms of experimentally accessible quantities such as the spring constant $k$, the quality factor $Q$, and the natural resonant frequency $\omega_0$ of the free oscillating cantilever, $\alpha = k/Q\omega_0$, we obtain

$$\begin{aligned}\bar{P}_{\text{tip}} &= \bar{P}_{\text{in}} - \bar{P}_0 \\ &= \frac{1}{2}\frac{k\omega}{Q}\left(Q_{\text{cant}} A_d A \sin\varphi - A^2 \frac{\omega}{\omega_0}\right).\end{aligned} \tag{7.38}$$

Note that so far no assumptions have been made on how the AFM is operated, except that the motion of the oscillating cantilever has to remain sinusoidal to a good approximation. Therefore (7.38) is applicable to a variety of different dynamic AFM modes.

For example, in FM-mode AFM the oscillation frequency $\omega$ changes due to tip–sample interaction while at the same time the oscillation amplitude $A$ is kept constant by adjusting the drive amplitude $A_d$. By measuring these quantities, one can apply (38) to determine the average power dissipation related to tip–sample interaction. In spectroscopy applications usually $A_d(z)$ is not measured directly, but a signal $G(z)$ proportional to $A_d(z)$ is acquired, representing the gain factor applied to the excitation piezo. With the help of (15) we can write

$$A_d(z) = \frac{A_0 G(z)}{Q G_0}. \tag{7.39}$$

where $A_0$ and $G_0$ are the amplitude and gain at large tip–sample distances where the tip–sample interactions are negligible.

Now let us consider the tapping-mode AFM. In this case the cantilever is driven at a fixed frequency and with constant drive amplitude, while the oscillation amplitude and phase shift may change when the probing tip interacts with the sample surface. Assuming that the oscillation frequency is chosen to be $\omega_0$, (37) can be further simplified again by employing (15) for the free oscillation amplitude $A_0$. This calculation yields

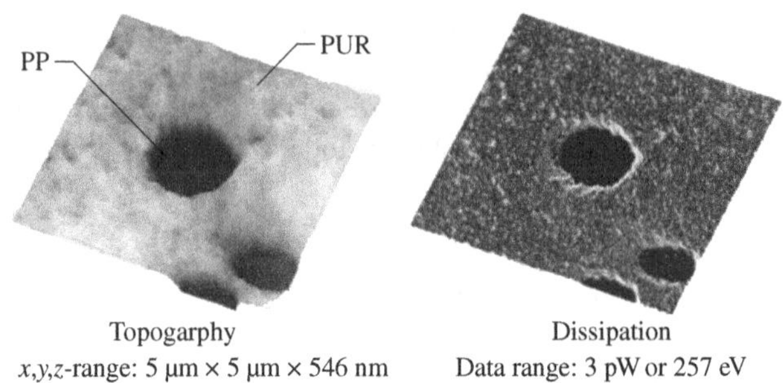


**Fig. 7.28** Topography and phase image in tapping-mode AFM of a polymer blend composed of polypropylene (PP) particles embedded in a polyurethane (PUR) matrix. The dissipation image shows a strong contrast between the harder PP (little dissipation, *dark*) and the softer PUR (large dissipation, *bright*) surface

$$\bar{P}_{\text{tip}} = \frac{1}{2} \frac{k\omega_0}{Q_{\text{cant}}} (A_0 A \sin\varphi - A^2). \tag{7.40}$$

Equation (40) implies that, if the oscillation amplitude $A$ is kept constant by a feedback loop, as is commonly done in tapping mode, simultaneously acquired phase data can be interpreted in terms of energy dissipation [77,79,88,89]. When analyzing such phase images [90,91,92] one has also to consider the fact that the phase may also change due to the transition from net-attractive ($\varphi > 90°$) to intermittent contact ($\varphi < 90°$) interaction between the tip and the sample [19,60,93,94]. For example, consider the phase shift in tapping mode as a function of $z$-position (Fig. 7.12). If phase measurements are performed close to the point where the oscillation switches from the net-attractive to the intermittent contact regime, a large contrast in the phase channel is observed. However, this contrast is not due to dissipative processes. Only a variation of the phase signal within the intermittent contact regime will give information about the tip–sample dissipative processes.

An example of dissipation measurement is depicted in Fig. 7.28. The surface of a polymer blend was imaged in air, simultaneously acquiring the topography and dissipation. The dissipation on the softer polyurethane matrix is significantly larger than on the embedded, mechanically stiffer polypropylene particles.

## 7.6 Conclusions

Dynamic force microscopy is a powerful tool, which is capable of imaging surfaces with atomic precision. It also allows us to look at surface dynamics and can operate in vacuum, air or even liquid. However, the oscillating cantilever system introduces a level of complexity which disallows straightforward interpretation of acquired images. An exception is the self-excitation mode, where tip–sample forces can be successfully extracted from spectroscopic experiments. However, not only conservative forces can be investigated with dynamic AFM; energy dissipation also influences the cantilever oscillation and can therefore serve as a new information channel.

Open questions are still concerned with the exact geometric and chemical identity of the probing tip, which significantly influences the imaging and spectroscopic results. Using predefined tips such as single-walled nanotubes or using atomic-resolution techniques such as field ion microscopy to image the tip itself are possible approaches addressing this issue.

## References

1. M.A. Lantz, H.J. Hug, R. Hoffmann, P.J.A. van Schendel, P. Kappenberger, S. Martin, A. Baratoff, H.-J. Güntherodt: Quantitative measurement of short-range chemical bonding forces, Science **291**, 2580–2583 (2001)
2. G. Binnig, C.F. Quate, C. Gerber: Atomic force microscope, Phys. Rev. Lett. **56**, 930–933 (1986)
3. O. Marti: AFM Instrumentation and Tips, in *Handbook of Micro/Nanotribology*, 2nd edn., ed. by B. Bushan (CRC, Boca Raton, 1999) pp. 81–144
4. G. Cross, A. Schirmeisen, A. Stalder, P. Grütter, M. Tschudy, U. Dürig: Adhesion interaction between atomically defined tip and sample, Phys. Rev. Lett. **80**, 4685–4688 (1998)
5. A. Schirmeisen, G. Cross, A. Stalder, P. Grütter, U. Dürig: Metallic adhesion and tunneling at the atomic scale, New J. Phys. **2**, 1–29 (2000)
6. A. Schirmeisen: Metallic adhesion and tunneling at the atomic scale. Ph.D. Thesis, McGill University, Montréal, 1999 pp. 29–38
7. F.J. Giessibl: Forces and frequency shifts in atomicresolution dynamic-force microscopy, Phys. Rev. B **56**, 16010–16015 (1997)
8. F.J. Giessibl: Atomic resolution of the silicon (111)-(7 × 7) surface by atomic force microscopy, Science **267**, 68–71 (1995)
9. M. Bezanilla, B. Drake, E. Nudler, M. Kashlev, P.K. Hansma, H.G. Hansma: Motion and enzymatic degradation of DNA in the atomic forcemicroscope, Biophys. J. **67**, 2454–2459 (1994)
10. Y. Jiao, D.I. Cherny, G. Heim, T.M. Jovin, T.E. Schäffer: Dynamic interactions of p53 with DNA in solution by time-lapse atomic force microscopy, J. Mol. Biol. **314**, 233–243 (2001)
11. T.R. Albrecht, P. Grütter, D. Horne, D. Rugar: Frequency modulation detection using high-$Q$ cantilevers for enhanced force microscopy sensitivity, J. Appl. Phys. **69**, 668–673 (1991)
12. S.P. Jarvis, M.A. Lantz, U. Dürig, H. Tokumoto: Off resonance ac mode force spectroscopy and imaging with an atomic force microscope, Appl. Surf. Sci. **140**, 309–313 (1999)
13. P.M. Hoffmann, S. Jeffery, J.B. Pethica, H.Ö. Özer, A. Oral: Energy dissipation in atomic force microscopy and atomic loss processes, Phys. Rev. Lett. **87**, 265502–265505 (2001)
14. H. Hölscher, B. Gotsmann, W. Allers, U.D. Schwarz, H. Fuchs, R. Wiesendanger: Comment on damping mechanism in dynamic force microscopy, Phys. Rev. Lett. **88**, 019601 (2002)
15. E. Sivaniah, J. Genzer, G.H. Fredrickson, E.J. Kramer, M. Xiang, X. Li, C. Ober, S. Magonov: Periodic surface topology of three-arm semifluorinated alkane monodendron diblock copolymers, Langmuir **17**, 4342–4346 (2001)
16. S.N. Magonov: Visualization of polymer structures with atomic force microscopy, in *Applied Scanning Probe Methods*, ed. by H. Fuchs, M. Hosaka, B. Bhushan (Springer, Berlin, 2004) pp. 207–250

17. C. Möller, M. Allen, V. Elings, A. Engel, D.J. Müller: Tapping-mode atomic force microscopy produces faithful high-resolution images of protein surfaces, Biophys. J. **77**, 1150–1158 (1999)
18. B. Anczykowski, D. Krüger, H. Fuchs: Cantilever dynamics in quasinoncontact force microscopy: Spectroscopic aspects, Phys. Rev. B **53**, 15485–15488 (1996)
19. B. Anczykowski, D. Krüger, K.L. Babcock, H. Fuchs: Basic properties of dynamic force spectroscopy with the scanning force microscope in experiment and simulation, Ultramicroscopy **66**, 251–259 (1996)
20. V.M. Muller, V.S. Yushchenko, B.V. Derjaguin: On the influence of molecular forces on the deformation of an elastic sphere and its sticking to a rigid plane, J. Colloid Interface Sci. **77**, 91–101 (1980)
21. B.D. Hughes, L.R. White: 'Soft' contact problems in linear elasticity, Q. J. Mech. Appl. Math. **32**, 445–471 (1979)
22. P. Gleyzes, P.K. Kuo, A.C. Boccara: Bistable behavior of a vibrating tip near a solid surface, Appl. Phys. Lett. **58**, 2989–2991 (1991)
23. A. San Paulo, R. Garcia: High-resolution imaging of antibodies by tapping-mode atomic force microscopy: Attractive and repulsive tip–sample interaction regimes, Biophys. J. **78**, 1599–1605 (2000)
24. D. Krüger, B. Anczykowski, H. Fuchs: Physical properties of dynamic force microscopies in contact and noncontact operation, Ann. Phys. **6**, 341–363 (1997)
25. H. Hölscher: Quantitative measurement of tip–sample interactions in amplitude modulation atomic force microscopy, Appl. Phys. Lett. **89**, 123109 (2006)
26. M. Lee, W. Jhe: General theory of amplitude-modulation atomic force microscopy, Phys. Rev. Lett. **97**, 036104 (2006)
27. Y. Martin, C.C. Williams, H.K. Wickramasinghe: Atomic force microscope – force mapping and profiling on a sub 100-Å scale, J. Appl. Phys. **61**, 4723–4729 (1987)
28. Y. Sugawara, M. Otha, H. Ueyama, S. Morita: Defect motion on an InP(110) surface observed with noncontact atomic force microscopy, Science **270**, 1646–1648 (1995)
29. M. Bammerlin, R. Lüthi, E. Meyer, A. Baratoff, J. Lue, M. Guggisberg, C. Gerber, L. Howald, H.-J. Güntherodt: True atomic resolution on the surface of an insulator via ultrahigh vacuum dynamic dynamic force microscopy, Probe Microsc. **1**, 3–9 (1996)
30. C. Barth, M. Reichling: Imaging the atomic arrangement on the high-temperature reconstructed $\alpha$-$Al_2O_3$(0001) Surface, Nature **414**, 54–57 (2001)
31. W. Allers, S. Langkat, R. Wiesendanger: Dynamic low-temperature scanning force microscopy on nickel oxyde(001), Appl. Phys. A [Suppl.] **72**, S27–S30 (2001)
32. U. Kaiser, A. Schwarz, R. Wiesendanger: Magnetic exchange force microscopy with atomic resolution, Nature **446**, 522–525 (2007)
33. S. Morita, R. Wiesendanger, E. Meyer: *Noncontact Atomic Force Microscopy* (Springer, Berlin, Heidelberg, 2002)
34. R. García, R. Pérez: Dynamic atomic force microscopy methods, Surf. Sci. Rep. **47**, 197–301 (2002)
35. F.J. Giessibl: Advances in atomic force microscopy, Rev. Mod. Phys. **75**, 949–983 (2003)
36. H. Hölscher, A. Schirmeisen: Dynamic force microscopy and spectroscopy, Adv. Imaging Electron Phys. **135**, 41–101 (2005)
37. H. Hölscher, U.D. Schwarz, R. Wiesendanger: Calculation of the frequency shift in dynamic force microscopy, Appl. Surf. Sci. **140**, 344–351 (1999)
38. B. Gotsmann, H. Fuchs: Dynamic force spectroscopy of conservative and dissipative forces in an Al- Au(111) tip–sample system, Phys. Rev. Lett. **86**, 2597–2600 (2001)
39. H. Hölscher, W. Allers, U.D. Schwarz, A. Schwarz, R. Wiesendanger: Determination of tip–sample interaction potentials by dynamic force spectroscopy, Phys. Rev. Lett. **83**, 4780–4783 (1999)
40. U. Dürig: Relations between interaction force and frequency shift in large-amplitude dynamic force microscopy, Appl. Phys. Lett. **75**, 433–435 (1999)

41. H. Hölscher, A. Schwarz, W. Allers, U.D. Schwarz, R. Wiesendanger: Quantitative analysis of dynamicforce-spectroscopy data on graphite (0001) in the contact and noncontact regime, Phys. Rev. B **61**, 12678–12681 (2000)
42. M. Guggisberg: Lokale Messung von atomaren Kräften. Ph.D. Thesis, University of Basel, Basel, 2000, pp. 9–11, in German
43. M. Guggisberg, M. Bammerlin, E. Meyer, H.-J. Güntherodt: Separation of interactions by noncontact force microscopy, Phys. Rev. B **61**, 11151–11155 (2000)
44. U. Dürig: Extracting interaction forces and complementary observables in dynamic probemicroscopy, Appl. Phys. Lett. **76**, 1203–1205 (2000)
45. F.J. Giessibl: A direct method to calculate tip– Sample forces from frequency shifts in frequencymodulation atomic force microscopy, Appl. Phys. Lett. **78**, 123–125 (2001)
46. J.E. Sader, S.P. Jarvis: Accurate formulas for interaction force and energy in frequency modulation force spectroscopy, Appl. Phys. Lett. **84**, 1801–1803 (2004)
47. O. Pfeiffer: Quantitative dynamische Kraft- und Dissipationsmikroskopie auf molekularer Skala. Ph.D. Thesis, Universität Basel, Basel, 2004, in German
48. H. Hölscher, S.M. Langkat, A. Schwarz, R. Wiesendanger: Measurement of three-dimensional force fields with atomic resolution using dynamic force spectroscopy, Appl. Phys. Lett. **81**, 4428 (2002)
49. A. Schirmeisen, D. Weiner, H. Fuchs: Single-atom contact mechanics: From atomic scale energy barrier to mechanical relaxation hysteresis, Phys. Rev. Lett. **97**, 136101 (2006)
50. T. Uchihashi, T. Ishida, M. Komiyama, M. Ashino, Y. Sugawara, W. Mizutani, K. Yokoyama, S. Morita, H. Tokumoto, M. Ishikawa: High-resolution imaging of organic monolayers using noncontact AFM, Appl. Surf. Sci. **157**, 244–250 (2000)
51. H. Hölscher, B. Gotsmann, A. Schirmeisen: Dynamic force spectroscopy using the frequency modulation technique with constant excitation, Phys. Rev. B **68**, 153401-1–153401-4 (2003)
52. A. Schirmeisen, H. Hölscher, B. Anczykowski, D. Weiner, M.M. Schäfer, H. Fuchs: Dynamic force spectroscopy using the constant-excitation and constant-amplitude modes, Nanotechnology **16**, 13–17 (2005)
53. T. Uchihashi, M.J. Higgins, S. Yasuda, S.P. Jarvis, S. Akita, Y. Nakayama, J.E. Sader: Quantitative force measurements in liquid using frequency modulation atomic force microscopy, Appl. Phys. Lett. **85**, 3575 (2004)
54. J.-E. Schmutz, H. Hölscher, D. Ebeling, M.M. Schäfer, B. Ancyzkowski: Mapping the tip–sample interactions on DPPC and DNA by dynamic force spectroscopy under ambient conditions, Ultramicroscopy **107**, 875–881 (2007)
55. C. Loppacher, M. Guggisberg, O. Pfeiffer, E. Meyer, M. Bammerlin, R. Lüthi, R. Schlittler, J.K. Gimzewski, H. Tang, C. Joachim: Direct determination of the energy required to operate a single molecule switch, Phys. Rev. Lett. **90**, 066107-1–066107-4 (2003)
56. N. Oyabu, O. Custance, I. Yi, Y. Sugawara, S. Morita: Mechanical vertical manipulation of selected single atoms by soft nanoindentation using near contact atomic force microscopy, Phys. Rev. Lett. **90**, 176102 (2003)
57. Y. Sugimoto, M. Abe, S. Hirayama, N. Oyabu, O. Custance, S. Morita: Atom inlays performed at room temperature using atomic force microscopy, Nat. Mater. **4**, 156–159 (2005)
58. J. Mertz, O. Marti, J. Mlynek: Regulation of a microcantilever response by force feedback, Appl. Phys. Lett. **62**, 2344–2346 (1993)
59. D. Rugar, P. Grütter: Mechanical parametric amplification and thermomechanical noise squeezing, Phys. Rev. Lett. **67**, 699–702 (1991)
60. B. Anczykowski, J.P. Cleveland, D. Krüger, V.B. Elings, H. Fuchs: Analysis of the interaction mechanisms in dynamic mode SFM by means of experimental data and computer simulation, Appl. Phys. A **66**, 885 (1998)
61. D. Ebeling, H. Hölscher, B. Anczykowski: Increasing the Q-factor in the constant-excitation mode of frequency-modulation atomic force microscopy in liquid, Appl. Phys. Lett. **89**, 203511 (2006)

62. D. Ebeling, H. Hölscher: Analysis of the constant-excitation mode in frequency-modulation atomic force microscopy with active Q-control applied in ambient conditions and liquids, J. Appl. Phys. **102**, 114310 (2007)
63. T. Sulchek, G.G. Yaralioglu, C.F. Quate, S.C. Minne: Characterization and optimisation of scan speed for tapping-mode atomic force microscopy, Rev. Sci. Instrum. **73**, 2928–2936 (2002)
64. H. Hölscher, U.D. Schwarz: Theory of amplitude modulation atomic force microscopy with and without Q-control, Int. J. Nonlinear Mech. **42**, 608–625 (2007)
65. L.F. Chi, S. Jacobi, B. Anczykowski, M. Overs, H.-J. Schäfer, H. Fuchs: Supermolecular periodic structures in monolayers, Adv. Mater. **12**, 25–30 (2000)
66. S. Gao, L.F. Chi, S. Lenhert, B. Anczykowski, C. Niemeyer, M. Adler, H. Fuchs: High-quality mapping of DNA–protein complexes by dynamic scanning forcemicroscopy, ChemPhysChem **6**, 384–388 (2001)
67. B. Zou, M. Wang, D. Qiu, X. Zhang, L.F. Chi, H. Fuchs: Confined supramolecular nanostructures of mesogen-bearing amphiphiles, Chem. Commun. **9**, 1008–1009 (2002)
68. B. Pignataro, L.F. Chi, S. Gao, B. Anczykowski, C. Niemeyer, M. Adler, H. Fuchs: Dynamic scanning force microscopy study of self-assembled DNA– protein nanostructures, Appl. Phys. A **74**, 447–452 (2002)
69. D. Ebeling, H. Hölscher, H. Fuchs, B. Anczykowski, U.D. Schwarz: Imaging of biomaterials in liquids: A comparison between conventional and $Q$-controlled amplitude modulation ('tapping mode') atomic force microscopy, Nanotechnology **17**, S221–S226 (2005)
70. U. Dürig: Interaction sensing in dynamic force microscopy, New J. Phys. **2**, 1–5 (2000)
71. H. Hölscher, B. Gotsmann, W. Allers, U.D. Schwarz, H. Fuchs, R. Wiesendanger: Measurement of conservative and dissipative tip–sample interaction forces with a dynamic force microscope using the frequency modulation technique, Phys. Rev. B **64**, 075402 (2001)
72. T.D. Stowe, T.W. Kenny, D.J. Thomson, D. Rugar: Silicon dopant imaging by dissipation force microscopy, Appl. Phys. Lett. **75**, 2785–2787 (1999)
73. Y. Liu, P. Grütter: Magnetic dissipation force microscopy studies of magnetic materials, J. Appl. Phys. **83**, 7333–7338 (1998)
74. B.C. Stipe, H.J. Mamin, T.D. Stowe, T.W. Kenny, D. Rugar: Noncontact friction and force fluctuations between closely spaced bodies, Phys. Rev. Lett. **87**, 96801-1–96801-4 (2001)
75. A.I. Volokitin, B.N.J. Persson: Resonant photon tunneling enhancement of the van der Waals friction, Phys. Rev. Lett. **91**, 106101-1–106101-4 (2003)
76. J. Tamayo, R. Garcia: Effects of elastic and inelastic interactions on phase contrast images in tappingmode scanning force microscopy, Appl. Phys. Lett. **71**, 2394–2396 (1997)
77. J.P. Cleveland, B. Anczykowski, A.E. Schmid, V.B. Elings: Energy dissipation in tapping-mode atomic force microscopy, Appl. Phys. Lett. **72**, 2613–2615 (1998)
78. R. García, J. Tamayo, M. Calleja, F. García: Phase contrast in tapping-mode scanning force microscopy, Appl. Phys. A **66**, S309–S312 (1998)
79. B. Anczykowski, B. Gotsmann, H. Fuchs, J.P. Cleveland, V.B. Elings: How to measure energy dissipation in dynamic mode atomic force microscopy, Appl. Surf. Sci. **140**, 376–382 (1999)
80. N. Sasaki, M. Tsukada: Effect of microscopic nonconservative process on noncontact atomic force microscopy, Jpn. J. Appl. Phys. **39**, 1334 (2000)
81. S. Hembacher, F.J. Giessibl, J. Mannhart, C.F. Quate: Local spectroscopy and atomic imaging of tunneling current, forces, and dissipation on graphite, Phys. Rev. Lett. **94**, 056101-1–056101-4 (2005)
82. L.N. Kantorovich, T. Trevethan: General theory of microscopic dynamical response in surface probe microscopy: From imaging to dissipation, Phys. Rev. Lett. **93**, 236102-1–236102-4 (2004)
83. R. Lüthi, E. Meyer, M. Bammerlin, A. Baratoff, L. Howald, C. Gerber, H.-J. Güntherodt: Ultrahigh vacuum atomic force microscopy: True atomic resolution, Surf. Rev. Lett. **4**, 1025–1029 (1997)

84. R. Bennewitz, A.S. Foster, L.N. Kantorovich, M. Bammerlin, C. Loppacher, S. Schär, M. Guggisberg, E. Meyer, A.L. Shluger: Atomically resolved edges and kinks of NaCl islands on Cu(111): Experiment and theory, Phys. Rev. B **62**, 2074–2084 (2000)
85. J. Israelachvili: *Intermolecular and Surface Forces* (Academic, London, 1992)
86. U. Rabe, J. Turner, W. Arnold: Analysis of the highfrequency response of atomic force microscope cantilevers, Appl. Phys. A **66**, 277 (1998)
87. T.R. Rodriguez, R. García: Tip motion in amplitude modulation (tapping-mode) atomic-force microscopy: Comparison between continuous and point-mass models, Appl. Phys. Lett. **80**, 1646–1648 (2002)
88. J. Tamayo, R. García: Relationship between phase shift and energy dissipation in tapping-mode scanning force microscopy, Appl. Phys. Lett. **73**, 2926–2928 (1998)
89. R. García, J. Tamayo, A. San Paulo: Phase contrast and surface energy hysteresis in tapping mode scanning force microcopy, Surf. Interface Anal. **27**(5/6), 312–316 (1999)
90. S.N. Magonov, V.B. Elings, M.H. Whangbo: Phase imaging and stiffness in tapping-mode atomic force microscopy, Surf. Sci. **375**, 385–391 (1997)
91. J.P. Pickering, G.J. Vancso: Apparent contrast reversal in tapping mode atomic force microscope images on films of polystyrene-b-polyisoprene-b-polystyrene, Polym. Bull. **40**, 549–554 (1998)
92. X. Chen, S.L. McGurk, M.C. Davies, C.J. Roberts, K.M. Shakesheff, S.J.B. Tendler, P.M. Williams, J. Davies, A.C. Dwakes, A. Domb: Chemical and morphological analysis of surface enrichment in a biodegradable polymer blend by phase-detection imaging atomic force microscopy, Macromolecules **31**, 2278–2283 (1998)
93. A. Kühle, A.H. Sørensen, J. Bohr: Role of attractive forces in tapping tip force microscopy, J. Appl. Phys. **81**, 6562–6569 (1997)
94. A. Kühle, A.H. Sørensen, J.B. Zandbergen, J. Bohr: Contrast artifacts in tapping tip atomic force microscopy, Appl. Phys. A **66**, 329–332 (1998)

# Chapter 8
# Molecular Recognition Force Microscopy: From Molecular Bonds to Complex Energy Landscapes

**Peter Hinterdorfer, Andreas Ebner, Hermann Gruber, Ruti Kapon, and Ziv Reich**

**Abstract** Atomic force microscopy (AFM), developed in the late 1980s to explore atomic details on hard material surfaces, has evolved into a method capable of imaging fine structural details of biological samples. Its particular advantage in biology is that measurements can be carried out in aqueous and physiological environments, which opens the possibility to study the dynamics of biological processes in vivo. The additional potential of the AFM to measure ultralow forces at high lateral resolution has paved the way for measuring inter- and intramolecular forces of biomolecules on the single-molecule level. Molecular recognition studies using AFM open the possibility to detect specific ligand–receptor interaction forces and to observe molecular recognition of a single ligand–receptor pair. Applications include biotin–avidin, antibody–antigen, nitrilotriacetate (NTA)–hexahistidine 6, and cellular proteins, either isolated or in cell membranes.

The general strategy is to bind ligands to AFM tips and receptors to probe surfaces (or vice versa). In a force–distance cycle, the tip is first approached towards the surface, whereupon a single receptor–ligand complex is formed due to the specific ligand receptor recognition. During subsequent tip–surface retraction a temporarily increasing force is exerted on the ligand–receptor connection, thus reducing its lifetime until the interaction bond breaks at a critical (unbinding) force. Such experiments allow for estimation of affinity, rate constants, and structural data of the binding pocket. Comparing them with values obtained from ensemble-average techniques and binding energies is of particular interest. The dependences of unbinding force on the rate of load increase exerted on the receptor–ligand bond reveal details of the molecular dynamics of the recognition process and energy landscapes. Similar experimental strategies have also been used for studying intramolecular force properties of polymers and unfolding–refolding kinetics of filamentous proteins. Recognition imaging, developed by combing dynamic force microscopy with force spectroscopy, allows for localization of receptor sites on surfaces with nanometer positional accuracy.

B. Bhushan (ed.), *Nanotribology and Nanomechanics*,
DOI 10.1007/978-3-642-15283-2_8, 

## Abbreviations

| | |
|---|---|
| AFM | atomic force microscope |
| AFM | atomic force microscopy |
| BFP | biomembrane force probe |
| BSA | bovine serum albumin |
| DC | direct-current |
| DFM | dynamic force microscopy |
| DFS | dynamic force spectroscopy |
| DNA | deoxyribonucleic acid |
| DTSSP | 3,3′-dithio-bis(sulfosuccinimidylproprionate) |
| FS | force spectroscopy |
| GDP | guanosine diphosphate |
| GTP | guanosine triphosphate |
| HUVEC | human umbilical venous endothelial cell |
| ICAM-1 | intercellular adhesion molecules 1 |
| ICAM-2 | intercellular adhesion molecules 2 |
| IgG | immunoglobulin G |
| LFA-1 | leukocyte function-associated antigen-1 |
| MRFM | magnetic resonance force microscopy |
| MRFM | molecular recognition force microscopy |
| NHS | *N*-hydroxysuccinimidyl |
| NTA | nitrilotriacetate |
| OT | optical tweezers |
| PDP | 2-pyridyldithiopropionyl |
| PDP | pyridyldithiopropionate |
| PEG | polyethylene glycol |
| PSGL-1 | P-selectin glycoprotein ligand-1 |
| RGD | arginine–glycine–aspartic |
| SATP | (*S*-acetylthio)propionate |
| SFA | surface forces apparatus |
| SFD | shear flow detachment |
| TREC | topography and recognition |

Molecular recognition plays a pivotal role in nature. Signaling cascades, enzymatic activity, genome replication and transcription, cohesion of cellular structures, interaction of antigens and antibodies, and metabolic pathways all rely critically on specific recognition. In fact, every process which requires molecules to interact with each other in a specific manner requires that they be able to recognize each other.

Molecular recognition studies emphasize specific interactions between receptors and their cognitive ligands. Despite a growing body of literature on the structure and function of receptor–ligand complexes, it is still not possible to predict reaction kinetics or energetics for any given complex formation, even when the structures are known. Additional insights, in particular into the molecular dynamics within the complex during the association and dissociation process, are needed. The high-end strategy is to probe the energy landscape that underlies the interactions between molecules whose structures are known with atomic resolution.

Receptor–ligand complexes are usually formed by a few, noncovalent weak interactions between contacting chemical groups in complementary determining regions, supported by framework residues providing structurally conserved scaffolding. Both the complementary determining regions and the framework have a considerable amount of plasticity and flexibility, allowing for conformational movements during association and dissociation. In addition to knowledge about structure, energies, and kinetic constants, information about these movements is required to understand the recognition process. Deeper insight into the nature of these movements as well as the spatiotemporal action of the many weak interactions, in particular the cooperativity of bond formation, is the key to understanding receptor–ligand recognition.

For this, experiments at the single-molecule level, and on time scales typical for receptor–ligand complex formation and dissociation, are required. The methodology described in this chapter for investigating molecular dynamics of receptor–ligand interactions, molecular recognition force microscopy (MRFM) [1, 2, 3], is based on atomic force microscope (AFM) technology [4]. The ability of the AFM [4] to measure ultralow forces at high lateral resolution together with its unique capability to operate in an aqueous and physiological environment opens the possibility of studying biological recognition processes in vivo. The interaction between a receptor and a ligand complex is studied by exerting a force on the complex and following the dissociation process over time. Dynamic aspects of recognition are addressed in force spectroscopy (FS) experiments, where distinct force–loading rate profiles are used to provide insight into the energy landscape underlying the reaction. It is also possible to investigate the force–time behavior to unravel changes of conformation which occur during the dissociation process. It will be shown that MRFM is a versatile tool to explore kinetic and structural details of receptor–ligand recognition.

## 8.1 Ligand Tip Chemistry

In MRFM experiments, the binding of ligands immobilized on AFM tips to surface-bound receptors (or vice versa) is studied by applying a force to the receptor–ligand complex. The force reduces the lifetime of the bond, ultimately leading to its disassociation. The distribution of forces at which rupture occurs, and its dependence on parameters such as loading rate and temperature, can be used to provide insight into the interaction. This type of setup requires careful AFM tip sensor design, including tight attachment of the ligands to the tip surface. In the first pioneering demonstrations of single-molecule recognition force measurements [1, 2], strong physical adsorption of bovine serum albumin (BSA) was used to

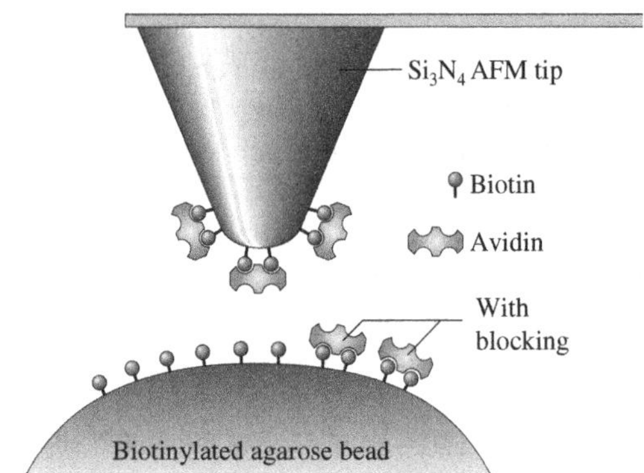


**Fig. 8.1** Avidin-functionalized AFM tip. A dense layer of biotinylated BSA was adsorbed onto the tip and subsequently saturated with avidin. The biotinylated agarose bead opposing the tip also contained a high surface density of reactive sites. These were partly blocked with avidin to achieve single-molecule binding events (After [2])

directly coat the tip [2] or a glass bead glued to it [1]. This physisorbed protein layer then served as a matrix for biochemical modifications with chemically active ligands (Fig. 8.1). In spite of the large number of probe molecules on the tip ($10^3$–$10^4$ /$nm^2$) the low fraction of properly oriented molecules, or internal blocks of most reactive sites (Fig. 8.1), allowed measurement of single receptor–ligand unbinding forces. Nevertheless, parallel breakage of multiple bonds was predominately observed with this configuration.

To measure interactions between isolated receptor–ligand pairs, strictly defined conditions need to be fulfilled. Covalently coupling ligands to gold-coated tip surfaces via freely accessible SH groups guarantees sufficiently stable attachment because these bonds are about ten times stronger than typical ligand–receptor interactions [5]. This chemistry has been used to detect the forces between complementary deoxyribonucleic acid (DNA) strands [1] as well as between isolated nucleotides [6]. Self-assembled monolayers of dithio-bis(succinimidylundecanoate) were formed to enable covalent coupling of biomolecules via amines [7] and were used to study the binding strength between cell adhesion proteoglycans [8] and between biotin-directed immunoglobulin G (IgG) antibodies and biotin [9]. Vectorial orientation of Fab molecules on gold tips was achieved by site-directed chemical binding via their SH groups [10], without the need for additional linkers. To this end, antibodies were digested with papain and subsequently purified to generate Fab fragments with freely accessible SH groups in the hinge region.

Gold surfaces exhibit a unique and selective affinity for thiols, although the adhesion strength of the resulting bonds is comparatively weak [5]. Since all commercially available AFM tips are etched from silicon nitride or silicon oxide material, deposition of a gold layer onto the tip surface is required prior to using this chemistry. Therefore, designing a sensor with covalent attachments of biomolecules to the silicon surface may be more straightforward. Amine functionalization procedures, a strategy widely used in surface biochemistry, were applied using ethanolamine [3, 11] and various silanization methods [12, 13, 14, 15], as a first step in thoroughly developed surface anchoring protocols suitable for single-molecule experiments. Since the amine surface density determines, to a large extent, the number of ligands on the tip which can specifically bind to the receptors on the surface, it has to be sufficiently low to guarantee single-molecular recognition events [3, 11]. Typically, these densities are kept between 200 and 500 molecules/$\mu m^2$, which for AFM tips with radii of $\approx$5–20 nm, amounts to about one molecule per effective tip area. A striking example of a minimally ligated tip was given by *Wong* et al. [16], who derivatized a few carboxyl groups present at the open end of carbon nanotubes attached to the tips of gold-coated Si cantilevers.

In a number of laboratories, a distensible and flexible linker was used to distance the ligand molecule from the tip surface (e.g., [3, 13]) (Fig. 8.2). At a given low number of spacer molecules per tip, the ligand can freely orient and diffuse within a certain volume, provided by the length of the tether, to achieve unconstrained binding to its receptor. The unbinding process occurs with little torque and the ligand molecule escapes the danger of being squeezed between the tip and the surface. This approach also opens the possibility of site-directed coupling for a defined orientation of the ligand relative to the receptor at receptor–ligand unbinding. As a cross-linking element, polyethylene glycol (PEG), a water-soluble nontoxic polymer with a wide range of applications in surface technology and clinical research, was often used [17]. PEG is known to prevent surface adsorption of proteins and lipid structures and therefore appears ideally suited for this purpose. Glutaraldehyde [12] and DNA [1] were also successfully applied as molecular spacers in recognition force studies. Cross-linker lengths, ideally arriving at a good compromise between high tip molecule mobility and narrow lateral resolution of the target recognition site, varied from 2 to 100 nm.

For coupling to the tip surface and to the ligand, the cross-linker typically carries two different functional ends, e.g., an amine reactive *N*-hydroxysuccinimidyl (NHS) group on one end, and a thiol reactive 2-pyridyldithiopropionyl group (PDP) [18, 19] on the other (Fig. 8.2). This sulfur chemistry is highly advantageous,

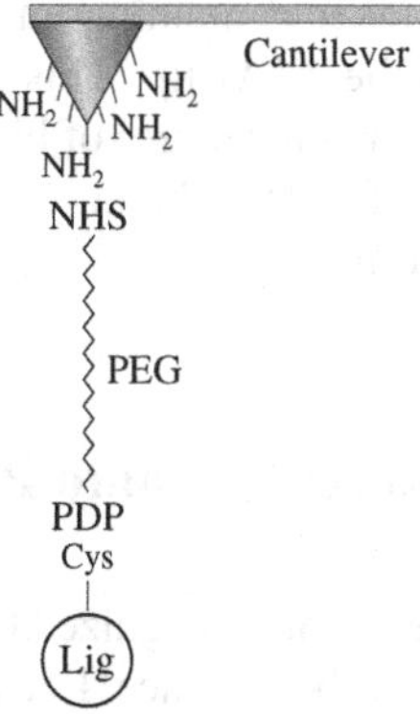


**Fig. 8.2** Linkage of ligands to AFM tips. Ligands were covalently coupled to AFM tips via a heterobifunctional polyethylene glycol (PEG) derivative of 8 nm length. Silicon tips were first functionalized with ethanolamine ($NH_2$–$C_2H_4OH \cdot HCl$). Then, the *N*-hydroxy-succinimide (NHS)-end of the PEG linker was covalently bound to amines on the tip surface before ligands were attached to the pyridyldithiopropionate (PDP) end via a free thiol or cysteine

since it is very reactive and readily enables site-directed coupling. However, free thiols are hardly available on native ligands and must often be added by chemical derivatization.

Different strategies have been used to achieve this goal. Lysine residues were derivatized with the short heterobifunctional linker *N*-succinimidyl-3-(*S*-acetylthio) propionate (SATP) [18]. Subsequent deprotection with $NH_2OH$ led to reactive SH groups. Alternatively, lysins can be directly coupled via aldehyde groups [15]. The direct coupling of proteins via an NHS–PEG–aldehyde linker allows binding via lysine groups without prederivatization. Nevertheless, since both ends are reactive against amino groups, loop formation can occur between adjacent $NH_2$ groups on the tip. The probability for this side-effect is significantly lowered (1) by the much higher amino reactivity of the NHS ester in comparison with the aldehyde function and (2) by high linker concentration. Another disadvantage of the latter two methods is that it does not allow for site-specific coupling of the cross-linker, since lysine residues are quite abundant. Several protocols are commercially available (Pierce, Rockford, IL) to generate active antibody fragments with free cysteines. Half-antibodies are produced by cleaving the two disulfide bonds in the central region of the heavy chain using 2-mercaptoethylamine HCl [20], and Fab fragments are generated from papain digestion [10]. The most elegant methods are to introduce a cysteine into the primary sequence of proteins or to append a thiol group to the end of a DNA strand [21], allowing for well-defined sequence-specific coupling of the ligand to the cross-linker.

An attractive alternative for covalent coupling is provided by the widely used nitrilotriacetate (NTA)-$His_6$ system. The strength of binding in this system, which is routinely used in chromatographic and biosensor matrices, is significantly larger than that between most ligand–receptor pairs [22–24]. For receptor–ligand interactions with very high unbinding force, NTA can be substituted with a recently developed Tris-NTA linker [25, 26]. Since a $His_6$ tag can be readily introduced in recombinant proteins, a cross-linker containing an (Tris-)NTA residue is ideally suited for coupling proteins to the AFM tip. This generic, site-specific coupling strategy also allows rigorous and ready control of binding specificity by using $Ni^{2+}$ as a molecular switch of the NTA–$His_6$ bond. A detailed description of actual coupling strategies can by found in [26].

## 8.2 Immobilization of Receptors onto Probe Surfaces

To enable force detection, the receptors recognized by the ligand-functionalized tip need to be firmly attached to the probed surface. Loose association will unavoidably lead to pull-off of the receptors from the surface by the tip-immobilized ligands, precluding detection of the interaction force.

Freshly cleaved muscovite mica is a perfectly pure and atomically flat surface and, therefore, ideally suited for MRFM studies. The strong negative charge of mica also accomplishes very tight electrostatic binding of various biomolecules; for example, lysozyme [20] and avidin [27] strongly adhere to mica at pH $< 8$. In such cases, simple adsorption of the receptors from solution is sufficient, since attachment is strong enough to withstand pulling. Nucleic acids can also be firmly bound to mica through mediatory divalent cations such as $Zn^{2+}$, $Ni^{2+}$ or $Mg^{2+}$ [28]. The strongly acidic sarcoplasmic domain of the skeletal muscle calcium release channel (RYR1) was likewise absorbed to mica via $Ca^{2+}$ bridges [29]. By carefully optimizing buffer conditions, similar strategies were used to deposit protein crystals and bacterial layers onto mica in defined orientations [30, 31].

The use of nonspecific electrostatic-mediated binding is however quite limited and generally offers no means to orient the molecules over the surface in a desirable direction. Immobilization by covalent attachment must therefore be frequently explored. When glass, silicon or mica are used as probe surfaces, immobilization is essentially the same as described above for tip functionalization. The number of reactive SiOH groups of the chemically relatively inert mica can be optionally increased by water plasma treatment [32]. As with tips, cross-linkers are also often used to provide receptors with motional freedom and to prevent surface-induced protein denaturation [3]. Immobilization can be controlled, to some extent, by using photoactivatable cross-linkers such as *N*-5-azido-2-nitrobenzoyl-oxysuccinimide [33].

A major limitation of silicon chemistry is that it does not allow for high surface densities, i.e., $>1{,}000/\mu m^2$. By comparison, the surface density of a monolayer of streptavidin is about 60,000 molecules/$\mu m^2$ and that of a phospholipid monolayer may exceed $10^6$ molecules/$\mu m^2$. The latter high density is also achievable by chemisorption of alkanethiols to gold. Tightly bound functionalized alkanethiol monolayers formed on ultraflat gold surfaces provide excellent probes for AFM [9] and readily allow for covalent and noncovalent attachment of biomolecules [9, 34] (Fig. 8.3).

Kada et al. [35] reported on a new strategy to immobilize proteins on gold surfaces using phosphatidyl choline or phosphatidyl ethanolamine analogues containing dithiophospholipids at their hydrophobic tail. Phosphatidyl ethanolamine, which is chemically reactive, was derivatized with a long-chain biotin for molecular recognition of streptavidin molecules in an initial study [35]. These self-assembled phospholipid monolayers closely mimic the cell surface and minimize nonspecific adsorption. Additionally, they can be spread as insoluble monolayers at an air–water interface. Thereby, the ratio of functionalized thiolipids to host lipids accurately defines the surface density of bioreactive sites in the monolayer. Subsequent transfer onto gold substrates leads to covalent, and hence tight, attachment of the monolayer.

MRFM has also been used to study the interactions between ligands and cell surface receptors in situ, on fixed or unfixed cells. In these studies, it was found that the immobilization of cells strongly depends on cell type. Adherent cells are readily usable for MRFM whereas cells that grow in suspension need to be adsorbed onto the probe surface. Various protocols for tight immobilization of cells over a surface are available. For adherent cells, the easiest approach is to grow the cells directly on glass or other surfaces suitable for MRFM [36]. Firm immobilization of

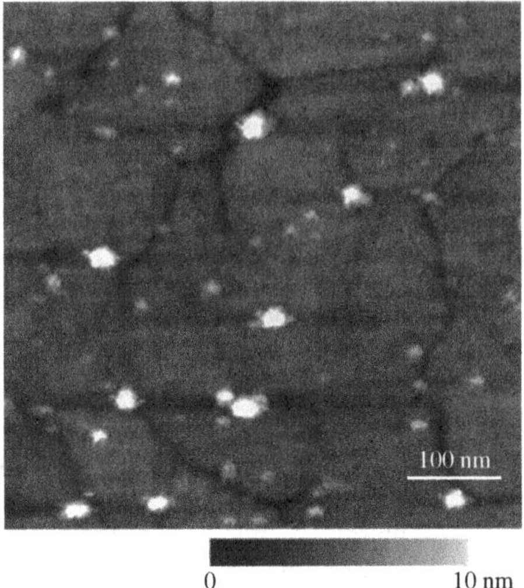


**Fig. 8.3** AFM image of hisRNAP molecules specifically bound to nickel-NTA domains on a functionalized gold surface. Alkanethiols terminated with ethylene glycol groups to resist unspecific protein adsorption served as a host matrix and were doped with 10% nickel-NTA alkanthiols. The sample was prepared to achieve full monolayer coverage. Ten individual hisRNAP molecules can be clearly visualized bound to the surface. The more abundant, smaller, lower features are NTA islands with no bound molecules. The underlying morphology of the gold can also be distinguished (After [34])

non- and weakly adhering cells can be achieved by various adhesive coatings such as Cell-Tak [37], gelatin, or polylysine. Hydrophobic surfaces such as gold or carbon are also very useful to immobilize nonadherent cells or membranes [38]. Covalent attachment of cells to surfaces can be accomplished by cross-linkers that carry reactive groups, such as those used for immobilization of molecules [37]. Alternatively, one can use cross-linkers carrying a fatty-acid moiety that can penetrate into the lipid bilayer of the cell membrane. Such linkers provide sufficiently strong fixation without interference with membrane proteins [37].

## 8.3 Single-Molecule Recognition Force Detection

Measurements of interaction forces traditionally rely on ensemble techniques such as shear flow detachment (SFD) [39] and the surface force apparatus (SFA) [40]. In SFD, receptors are fixed to a surface to which ligands carried by beads or presented on the cell surface bind specifically. The surface-bound particles are then subjected to a fluid shear stress that disrupts the ligand–receptor bonds. However, the force acting between single molecular pairs can only be estimated because the net force applied to the particles can only be approximated and the number of bonds per particle is unknown.

SFA measures the forces between two surfaces to which different interacting molecules are attached, using a cantilever spring as force probe and interferometry for detection. The technique, which has a distance resolution of ≈1Å, allows the measurement of adhesive and compressive forces and rapid transient effects to be followed in real time. However, the force sensitivity of the technique (≈10 nN) does not allow for single-molecule measurements of noncovalent interaction forces.

The biomembrane force probe (BFP) technique uses pressurized membrane capsules rather than mechanical springs as a force transducer (Fig. 8.4; see, for example, [41]). To form the transducer, a red blood cell or a lipid bilayer vesicle is pressurized into the tip of a glass micropipette. The spring constant of the capsule can then be varied over several orders of magnitude by suction. This simple but highly effective configuration enables the measurement of forces ranging from 0.1–1,000 pN with a force resolution of about 1 pN, allowing probing of single-molecular bonds.

In optical tweezers (OT), small dielectric particles (beads) are manipulated by electromagnetic traps [42, 43]. Three-dimensional light intensity gradients of a focused laser beam are used to pull or push particles with nanometer positional accuracy. Using this technique, forces in the range of $10^{-13}$–$10^{-10}$ N can be measured accurately. Optical tweezers have been used extensively to measure the force-generating properties of various molecular motors at the single-molecule level [44–46] and to obtain force–extension profiles of single DNA [47] or protein [48] molecules. Defined, force-controlled twisting of DNA using rotating magnetically manipulated particles gave further insights into DNA's viscoelastic properties [49, 50].

AFM has successfully been used to measure the interaction forces between various single-molecular pairs [1, 2, 3]. In these measurements, one of the binding partners is immobilized onto a tip mounted at the end of a flexible cantilever that functions as a force transducer and the other is immobilized over a hard surface such as mica or glass. The tip is initially brought to, and subsequently retracted from the surface, and the interaction (unbinding) force is measured by following the

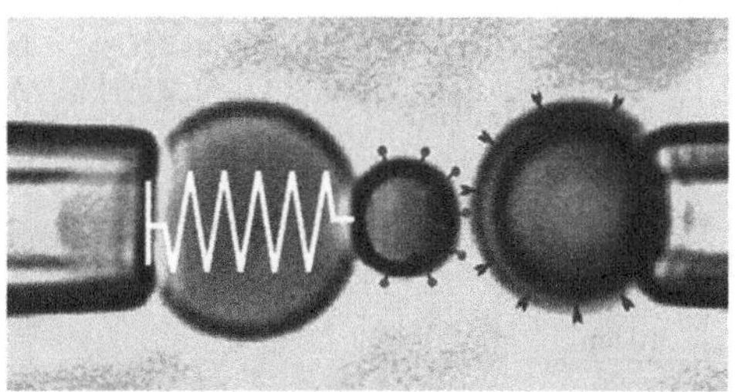

**Fig. 8.4** Experimental setup of the biomembrane force probe (BFP). The spring in the BFP is a pressurized membrane capsule. Its spring constant is set by membrane tension, which is controlled by micropipette suction. The BFP tip is formed by a glass microbead with diameter of 1–2 μm chemically glued to the membrane. The BFP (*on the left*) was kept stationary and the test surface, formed by another microbead (*on the right*), was translated to or from contact with the BFP tip by precision piezoelectric control (After [41])

cantilever deflection, which is monitored by measuring the reflection of a laser beam focused on the back of the cantilever using a split photodiode. Approach and retract traces obtained from the unbinding of a single-molecular pair is shown in Fig. 8.5 [3]. In this experiment, the binding partners were immobilized onto their respective surfaces through a distensible PEG tether.

Cantilever deflection, $\Delta x$, relates directly to the force $F$ acting on it through Hook's law $F = k\Delta x$, where $k$ is the spring constant of the cantilever. During most of the approach phase (trace, and points 1–5), when the tip and the surface are sufficiently far away from each other (1–4), cantilever deflection remains zero because the molecules are still unbound from each other. Upon contact (4) the cantilever bends upwards (4–5) due to a repulsive force that increases linearly as the tip is pushed further into the surface. If the cycle was futile, and no binding had occurred, retraction of the tip from the surface (retrace, 5–7) will lead to a gradual relaxation of the cantilever to its rest position (5–4). In such cases, the retract curve will look very much like the approach curve. On the other hand, if binding had occurred, the cantilever will bend downwards as the cantilever is retracted from the surface (retrace, 4–7). Since the receptor and ligand were tethered to the surfaces through flexible cross-linkers, the shape of the attractive force–distance profile is nonlinear, in contrast to the profile obtained during contact (4–7). The exact shape of the retract curve depends on the elastic properties of the cross-linker used for

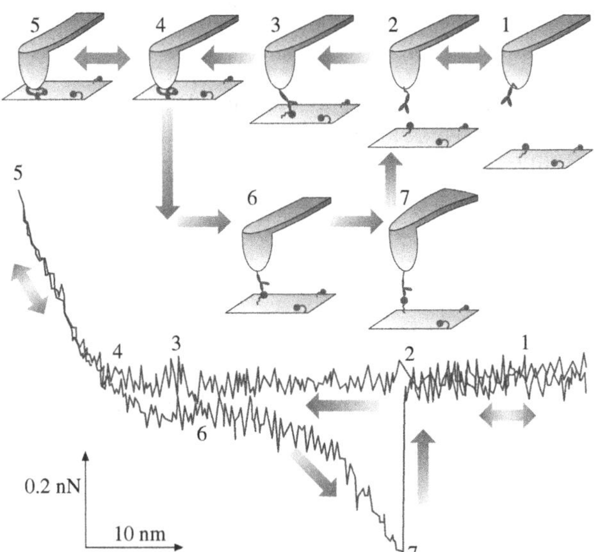


**Fig. 8.5** Single-molecule recognition event detected with AFM: a force–distance cycle, measured with an amplitude of 100 nm at a sweep frequency of 1 Hz, for an antibody–antigen pair in PBS. Binding of the antibody immobilized on the tip to the antigen on the surface, which occurs during the approach (*trace points 1–5*), results in a parabolic retract force curve (*points 6–7*) reflecting the extension of the distensible cross-linker antibody–antigen connection. The force increases until unbinding occurs at a force of 268 pN (*points 7 to 2*) (After [3])

immobilization [17, 51] and exhibits parabolic-like characteristics, reflecting an increase of the spring constant of the cross-linker during extension. The downward bending of the retracting cantilever continues until the ramping force reaches a critical value that dissociates the ligand–receptor complex (unbinding force, 7). Unbinding of the complex is indicated by a sharp spike in the retract curve that reflects an abrupt recoil of the cantilever to its rest position. Specificity of binding is usually demonstrated by block experiments in which free ligands are added to mask receptor sites over the surface.

The force resolution of the AFM, $\Delta F = (k_B T k)^{1/2}$, is limited by the thermal noise of the cantilever which, in turn, is determined by its spring constant. A way to reduce thermal fluctuations of cantilevers without changing their stiffness or lowering the temperature is to increase the apparent damping constant. Applying an actively controlled external dissipative force to cantilevers to achieve such an increase, *Liang* et al. [52] reported a 3.4-fold decrease in thermal noise amplitude. The smallest forces that can be detected with commercially available cantilevers are in the range of a few piconewtons. Decreasing cantilever dimensions enables the range of detectable forces to be pushed to smaller forces since small cantilevers have lower coefficients of viscous damping [53]. Such miniaturized cantilevers also have much higher resonance frequencies than conventional cantilevers and, therefore, allow for faster measurements.

The atomic force microscope (AFM) [4] is the force-measuring method with the smallest sensor and therefore provides the highest lateral resolution. Radii of commercially available AFM tips vary between 2 and 50 nm. In contrast, the particles used for force sensing in SFD, BFP, and OT are in the 1–10 μm range, and the surfaces used in SFA exceed millimeter extensions. The small apex of the AFM tip allows visualization of single biomolecules with molecular to submolecular resolution [28, 30, 31].

Besides the detection of intermolecular forces, the AFM also shows great potential for measuring forces acting within molecules. In these experiments, the molecule is clamped between the tip and the surface and its viscoelastic properties are studied by force–distance cycles.

## 8.4 Principles of Molecular Recognition Force Spectroscopy

Molecular recognition is mediated by a multitude of noncovalent interactions whose energy is only slightly higher than that of thermal energy. Due to the power-law dependence of these interactions on distance, the attractive forces between noncovalently interacting molecules are extremely short-ranged. A close geometrical and chemical fit within the binding interface is therefore a prerequisite for productive association. The weak bonds that govern molecular cohesion are believed to be formed in a spatially and temporarily correlated fashion. Protein binding often involves structural rearrangements that can be either localized or global. These rearrangements often bear functional significance by modulating the activity of the interactants. Signaling pathways, enzyme activity, and the activation and inactivation of genes all depend on conformational changes induced in proteins by ligand binding.

The strength of binding is usually given by the binding energy $E_b$, which amounts to the free energy difference between the bound and the free state, and which can readily be determined by ensemble measurements. $E_b$ determines the ratio of bound complexes [RL] to the product of free reactants [R][L] at equilibrium and is related to the equilibrium dissociation constant $K_D$ through $E_b = -RT \ln(K_D)$, where $R$ is the gas constant. $K_D$ itself is related to the empirical association ($k_{on}$) and dissociation ($k_{off}$) rate constants through $K_D = k_{off}/k_{on}$. In order to obtain an estimate for the interaction force $f$, from the binding energy $E_b$, the depth of the binding pocket may be used as a characteristic length scale $l$. Using typical values of $E_b = 20k_BT$ and $l = 0.5$ nm, an order-of-magnitude estimate of $f(= E_b/l) \approx 170$ pN is obtained for the binding strength of a single-molecular pair. Classical mechanics describes bond strength as the gradient in energy along the direction of separation. Unbinding therefore occurs when the applied force exceeds the steepest gradient in energy. This purely mechanical description of molecular bonds, however, does not provide insights into the microscopic determinants of bond formation and rupture.

Noncovalent bonds have limited lifetimes and will therefore break even in the absence of external force on characteristic time scales needed for spontaneous dissociation $\left(\tau(0) = k_{off}^{-1}\right)$. When pulled faster than $\tau(0)$, however, bonds will resist detachment. Notably, the unbinding force may approach and even exceed the adiabatic limit given by the steepest energy gradient of the interaction potential, if rupture occurs in less time than needed for diffusive relaxation ($10^{-10}$–$10^{-9}$ s for biomolecules in viscous aqueous medium) and friction effects become dominant [55]. Therefore, unbinding forces do not resemble unitary values and the dynamics of the experiment critically affects the measured bond strengths. On the time scale of AFM experiments (milliseconds to seconds), thermal impulses govern the unbinding process. In the thermal activation model, the lifetime of a molecular complex in solution is described by a Boltzmann ansatz, $\tau(0) = \tau_{osc} \exp[E_b/(k_BT)]$ [56], where $\tau_{osc}$ is the inverse of the natural oscillation frequency and $E_b$ is the height of the energy barrier for dissociation. This gives a simple Arrhenius dependency of dissociation rate on barrier height.

A force acting on a complex deforms the interaction free energy landscape and lowers barriers for dissociation (Fig. 8.6). As a result of the latter, bond lifetime is shortened. The lifetime $\tau(f)$ of a bond loaded with a constant force $f$ is given by $\tau(f) = \tau_{osc} \exp[(E_b - x_\beta f)/(k_BT)]$ [56], where $x_\beta$ marks the thermally averaged projection of the energy barrier along the direction of the force. A detailed analysis of the relation between bond strength and lifetime was performed by Evans and Ritchie [57], using Kramers' theory for overdamped kinetics. For a sharp barrier, the lifetime $\tau(f)$ of a bond subjected to a constant force $f$ relates to its characteristic lifetime $\tau(0)$ according to $\tau(f) = \tau(0) \exp[-x_\beta f/(k_BT)]$ [3]. However, in most pulling experiments the applied force is not constant. Rather, it increases in a complex, nonlinear manner, which depends on the pulling velocity, the spring constant

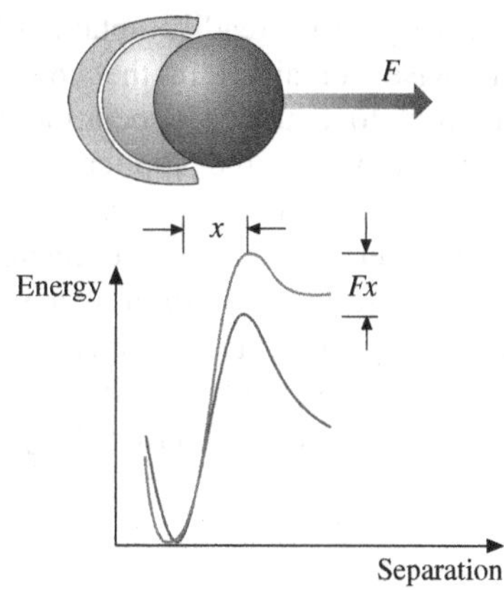


**Fig. 8.6** Dissociation over a single sharp energy barrier. Under a constant force, the barrier is lowered by the applied force *F*. This gives rise to a characteristic length scale $x_\beta$ that is interpreted as the distance of the energy barrier from the energy minimum along the projection of the force (After [54])

of the cantilever, and the force–distance profile of the molecular complex. Nevertheless, contributions arising from thermal activation manifest themselves mostly near the point of detachment. Therefore, the change of force with time or the loading rate $r$ ($= \mathrm{d}f/\mathrm{d}t$) can be derived from the product of the pulling velocity and the effective spring constant at the end of the force curve, just before unbinding occurs.

The dependence of the rupture force on the loading rate (force spectrum) in the thermally activated regime was first derived by Evans and Ritchie [57] and described further by Strunz et al. [54]. Forced dissociation of receptor–ligand complexes using AFM or BFP can often be regarded as an irreversible process because the molecules are kept away from each other after unbinding occurs (rebinding can be safely neglected when measurements are made with soft springs). Rupture itself is a stochastic process, and the likelihood of bond survival is expressed in the master equation as a time-dependent probability $N(t)$ to be in the bound state under a steady ramp of force, namely $\mathrm{d}N(t)/\mathrm{d}t = -\,k_{\mathrm{off}}(rt)N(t)$ [54]. This results in a distribution of unbinding forces $P(F)$ parameterized by the loading rate [54, 57, 58]. The most probable force for unbinding $f^*$, given by the maximum of the distribution, relates to the loading rate through $f^* = f_\beta \ln\left(rk_{off}^{-1}/f_\beta\right)$, where the force scale $f_\beta$ is set by the ratio of thermal energy to $x_\beta$ [54, 57]. Thus, the unbinding force scales linearly with the logarithm of the loading rate. For a single barrier, this would give rise to a simple, linear force spectrum $f^*$ versus log ($r$). In cases where the escape path is traversed by several barriers, the curve will follow a sequence of linear regimes, each marking a particular barrier [41, 57, 58]. Transition from one regime to the other is associated with an abrupt change of slope determined by the characteristic barrier length scale and signifying that a crossover between barriers has occurred.

Dynamic force spectroscopy (DFS) exploits the dependence of bond strength on the loading rate to obtain detailed insights into intra- and intermolecular interactions. By measuring bond strength over a broad range of loading rates, length scales and relative heights of energy barriers traversing the free energy surface can be readily obtained. The lifetime of a bond at any given force is likewise contained in the complete force distribution [3]. Finally, one may attempt to extract dissociation rate constants by extrapolation to zero force [59]. However, the application of force acts to select the dissociation path. Since the kinetics of reactions is pathway dependent, such a selection implies that kinetic parameters extracted from force–probe experiments may differ from those obtained from assays conducted in the absence of external force. For extremely fast complexation/decomplexation kinetics the forces can be independent of the loading rate, indicating that the experiments were carried out under thermodynamic equilibrium [60].

## 8.5 Recognition Force Spectroscopy: From Isolated Molecules to Biological Membranes

### *8.5.1 Forces, Energies, and Kinetic Rates*

Conducted at fixed loading rates, pioneering measurements of interaction forces provided single points in continuous spectra of bond strengths [41]. Not unexpectedly, the first interaction studied was that between biotin and its extremely high-affinity receptors avidin [2] and streptavidin [1]. The unbinding forces measured for these interactions were 250–300 pN and 160 pN, for streptavidin and avidin, respectively. During this initial phase it was also revealed that different unbinding forces can be obtained for the same pulling velocity if the spring constant of the cantilever is varied [1], consistent with the aforementioned dependency of bond strength on the loading rate. The interaction force between several biotin analogues and avidin or streptavidin [61] and between biotin and a set of streptavidin mutants [62] was investigated and found to generally correlate with the equilibrium binding enthalpy and the enthalpic activation barrier. No correlation with the equilibrium free energy of binding or the activation free energy barrier to dissociation was observed, suggesting that internal energies rather than entropic contributions were probed by the force measurements [62].

In another pioneering study, Lee et al. [21] measured the forces between complementary 20-base DNA strands covalently attached to a spherical probe and surface. The interaction forces fell into three different distributions amounting to the rupture of duplexes consisting of 12, 16, and 20 base pairs. The average rupture force per base pair was $\approx$70 pN. When a long, single-stranded DNA was analyzed, both intra- and interchain forces were observed, the former probing the elastic properties of the molecule. Hydrogen bonds between nucleotides have been probed for all 16 combinations of the four DNA bases [6]. Directional hydrogen-bonding interactions were measured only when complementary bases were present on tip and probe surface, indicating that AFM can be used to follow specific pairing of DNA strands.

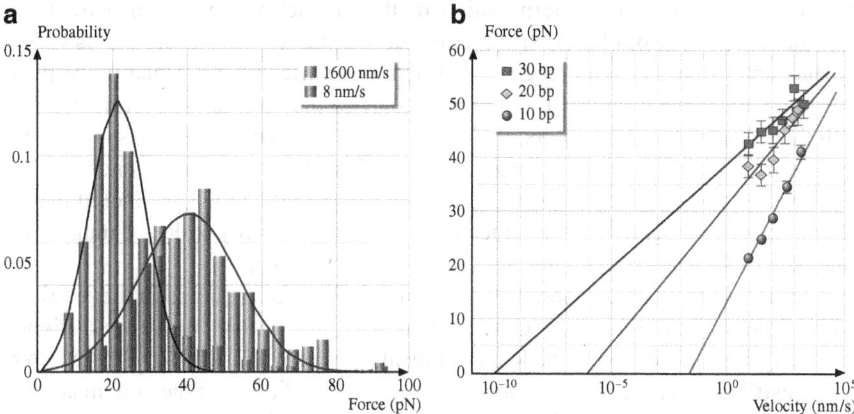


**Fig. 8.7** Dependence of the unbinding force between DNA single-strand duplexes on the retract velocity. In addition to the expected logarithmic behavior on the loading rate, the unbinding force scales with the length of the strands, increasing from the 10- to 20- to 30-base-pair duplexes (After [14])

Strunz et al. [14] measured the forces required to separate individual double-stranded DNA molecules of 10, 20, and 30 base pairs (Fig. 8.7). The parameters describing the energy landscape, i.e., the distance from the energy barrier to the minimum energy along the separation path and the logarithm of the thermal dissociation rate, were found to be proportional to the number of base pairs of the DNA duplex. Such scaling suggests that unbinding proceeds in a highly cooperative manner characterized by one length scale and one time scale. Studying the dependence of rupture forces on temperature, it was proposed by Schumakovitch et al. [63] that entropic contributions play an important role in the unbinding of complementary DNA strands [63].

Prevalent as it is, molecular recognition has mostly been discussed in the context of interactions between antibodies and antigens. To maximize motional freedom and to overcome problems associated with misorientation and steric hindrance, antibodies and antigens were immobilized onto the AFM tip and probe surface via flexible molecular spacers [3, 9, 12, 13]. By optimizing the antibody density over the AFM tip [3, 11], the interaction between individual antibody–antigen pairs could be examined. Binding of antigen to the two Fab fragments of the antibody was shown to occur independently and with equal probability. Single antibody–antigen recognition events were also recorded with tip-bound antigens interacting with intact antibodies [9,12] or with single-chain Fv fragments [13]. The latter study also showed that an Fv mutant whose affinity to the antigen was attenuated by about 10-fold dissociated from the antigen under applied forces that were 20% lower than those required to unbind the wild-type (Fv) antibody.

Besides measurements of interaction forces, single-molecule force spectroscopy also allows estimation of association and dissociation rate constants, notwithstanding the concern stated above [3, 11, 23, 59, 64, 65], and measurement of structural parameters of the binding pocket [3, 11, 14, 64, 65]. Quantification of the association rate constant $k_{on}$ requires determination of the interaction time needed for half-maximal probability of binding ($t_{1/2}$). This can be obtained from experiments where the encounter time between receptor and ligand is varied over a broad range [64]. Given that the concentration of ligand molecules on the tip available for interaction with the surface-bound receptors $c_{eff}$ is known, the association rate constant can be derived from $k_{on} = t_{0.5}{}^{-1} c_{eff}{}^{-1}$. Determination of the effective ligand concentration requires knowledge of the effective volume $V_{eff}$ explored by the tip-tethered ligand which, in turn, depends on the tether length. Therefore, only order-of-magnitude estimates of $k_{on}$ can be obtained from such measurements [64].

Additional information about the unbinding process is contained in the distributions of the unbinding forces. Concomitant with the shift of maxima to higher unbinding forces, increasing the loading rate also leads to an increase in the width $\sigma$ of the distributions [23, 41], indicating that at lower loading rates the system adjusts closer to equilibrium. The lifetime $\tau(f)$ of a bond under an applied force was estimated by the time the cantilever spends in the force window spanned by the standard deviation of the most probable force for unbinding [3]. In the case of $Ni^{2+}$-$His_6$, the lifetime of the complex decreased from 17 to 2.5 ms when the force was increased from 150 to 194 pN [23]. The data fit well to Bell's model, confirming the predicted exponential dependence of bond lifetime on the applied force, and yielded an estimated lifetime at zero force of about 15 s. A more direct measurement of $\tau$ is afforded by force-clamp experiments in which the applied force is kept constant by a feedback loop. This configuration was first adapted for use with AFM by Oberhauser et al. [66], who employed it to study the force dependence of the unfolding probability of the I27 and I28 modules of cardiac titin as well as of the full-length protein [66].

However, as discussed above, in most experiments the applied force is not constant but varies with time, and the measured bond strength depends on the loading rate [55, 57, 67]. In accordance with this, experimentally measured unbinding forces do not assume unitary values but rather vary with both pulling velocity [59, 64] and cantilever spring constant [1]. The slopes of the force versus loading rate curves contain information about the length scale $x_\beta$ of prominent energy barriers along the force-driven dissociation pathway, which may be related to the depth of the binding pocket of the interaction [64]. The predicted logarithmic dependence of the unbinding force on the loading rate holds well when the barriers are stationary with force, as confirmed by a large number of unbinding and unfolding experiments [14, 23, 41, 59, 64, 65, 68]. However, if the position of the transition state is expected to vary along the reaction coordinate with the force, as for example when the curvature at the top of the barrier is small, the strict logarithmic dependence gives way to more complex forms.

Schleirf and Rief [69] used a Kramers diffusion model to calculate the probability force distributions when the barrier cannot be assumed to be stationary. Notably, although the position of the transition state predicted by the Bell model was smaller than that predicted by the Kramer analysis by 6Å, the most probable unfolding forces showed an almost perfect logarithmic dependence on the pulling velocity, indicating that great care should be taken before the linear theory of DFS is applied.

Failure to fit force distributions at high loading rates using a Bell model was also reported by Neuert et al. [70] for the interaction of digoxigenin and antidigoxigenin. Poor matches were observed in the crossover region between the two linear regimes of the force spectrum as well. Klafter et al. also suggested a method to analyze force spectra which also does not assume stationarity of the energy barrier [71]. In their treatment, they find a $(\ln r)^{2/3}$ dependence of the mean force of dissociation, where $r$ is the loading rate. They also find the distribution of unbinding forces to be asymmetric, as indeed observed many times. Evstigneev and Reimann [72] suggest that the practice of fitting this asymmetric distribution with a Gaussian one in order to extract the mean rupture leads to the latter's overestimation and consequently to an overestimate of the force-free dissociation rate. They suggest an optimized statistical data analysis which overcomes this limitation by combining data at many pulling rates into a single distribution of the probability of rupture versus force.

The force spectra may also be used to derive the dissociation rate constant $k_{off}$ by extrapolation to zero force [59, 64, 65]. As mentioned above, values derived in this manner may differ from those obtained from bulk measurements because only a subset of dissociation pathways defined by the force is sampled. Nevertheless, a simple correlation between unbinding forces and thermal dissociation rates was obtained for a set consisting of nine different Fv fragments constructed from point mutations of three unrelated antifluorescein antibodies [65, 70]. This correlation, which implies a close similarity between the force- and thermally driven pathways explored during dissociation, was probably due to the highly rigid nature of the interaction, which proceeds in a lock-and-key fashion. The force spectra obtained for the different constructs exhibited a single linear regime, indicating that in all cases unbinding was governed by a single prominent energy barrier (Fig. 8.8).

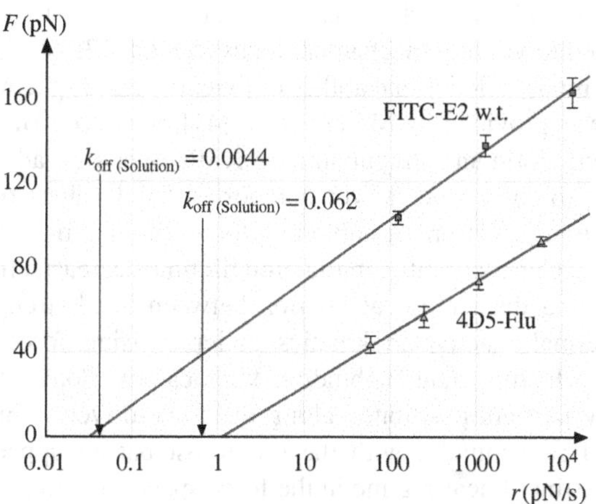


**Fig. 8.8** The dependence of the unbinding force on the loading rate for two antifluorescein antibodies. For both FITC-E2 w.t. and 4D5-Flu a strictly single-exponential dependence was found in the range accessed, indicating that only a single energy barrier was probed. The same energy barrier dominates dissociation without forces applied because extrapolation to zero force matches kinetic off-rates determined in solution (indicated by the *arrow*) (After [65])

Interestingly, the position of the energy barrier along the forced-dissociation pathway was found to be proportional to the height of the barrier and, thus, most likely includes contributions arising from elastic stretching of the antibodies during the unbinding process.

A good correspondence between dissociation rates derived from mechanical unbinding experiments and from bulk assays was also reported by Neuert et al. [70]. In this case, the experimental system consisted of digoxigenin and its specific antibody. This pair is used as a noncovalent coupler in various applications, including forced-unbinding experiments. The force spectra obtained for the complex suggested that the unbinding path is traversed by two activation energy barriers located at $x_\beta = 0.35$ nm and $x_\beta = 1.15$ nm. Linear fit of the low-force regime revealed a dissociation rate at zero force of 0.015 $s^{-1}$, in close agreement with the 0.023 $s^{-1}$ value obtained from bulk measurements made on antidigoxigenin Fv fragments.

## 8.5.2 Complex Bonds and Energy Landscapes

The energy landscapes that describe proteins are generally not smooth. Rather, they are traversed by multiple energy barriers of various heights that render them highly corrugated or rugged. All these barriers affect the kinetics and conformational dynamics of proteins and any one of them may govern interaction lifetime and strength on certain time scales. Dynamic force spectroscopy provides an excellent tool to detect energy barriers which are difficult or impossible to detect by conventional, near-equilibrium assays and to probe the free energy surface of proteins and protein complexes. It also provides a natural means to study interactions which are normally subjected to varying mechanical loads [59, 64, 73, 74, 75].

A beautiful demonstration of the ability of dynamic force spectroscopy to reveal hidden barriers was provided by Merkel et al. [41], who used BFP to probe bond formation between biotin and streptavidin or avidin over a broad range of loading rates. In contrast to early studies which reported fixed values of bond strength [61, 62], a continuous spectrum of unbinding forces ranging from 5 to 170 pN was obtained (Fig. 8.9). Concomitantly, interaction lifetime decreased from about 1 min to 0.001 s, revealing the reciprocal relation between bond strength and lifetime expected for thermally activated kinetics under a rising force. Most notably, depending on the loading rate, unbinding kinetics was dominated by different activation energy barriers positioned along the force-driven unbinding pathway. Barriers emerged sequentially, with the outermost barrier appearing first, each giving rise to a distinct linear regime in the force spectrum. Going from one linear regime to the next was associated with an abrupt change in slope, indicating that a crossover between an outer to (more) inner barrier had occurred. The position of two of the three barriers identified in the force spectra was consistent with the location of prominent transition states revealed by molecular dynamics simulations [55, 67]. However, as was mentioned earlier, unbinding is not necessarily confined

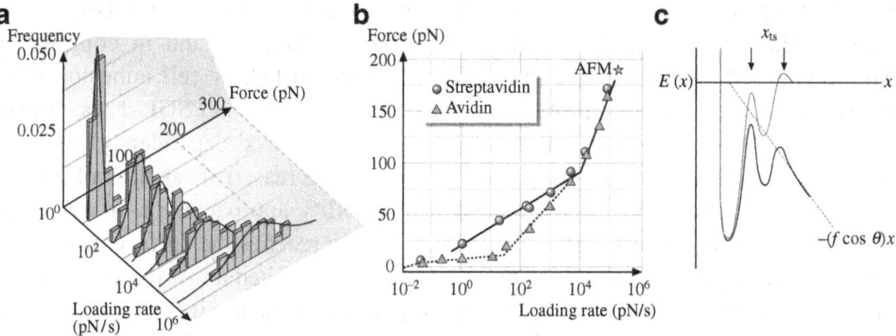


**Fig. 8.9** Unbinding force distributions and energy landscape of a complex molecular bond. **(a)** Force histograms of single biotin–streptavidin bonds recorded at different loading rates. The shift in peak location and the increase in width with increasing loading rate is clearly demonstrated. **(b)** Dynamic force spectra for biotin–streptavidin (*circles*) and biotin–avidin (*triangles*). The slopes of the linear regimes mark distinct activation barriers along the direction of force. **(c)** Conceptual energy landscape traversed along a reaction coordinate under force. The external force $f$ adds a mechanical potential that tilts the energy landscape and lowers the barriers. The inner barrier starts to dominate when the outer has fallen below it due to the applied force (After [41])

to a single, well-defined path, and may take different routes even when directed by an external force. Molecular dynamics simulations of force-driven unbinding of an antibody–antigen complex characterized by a highly flexible binding pocket revealed a large heterogeneity of enforced dissociation pathways [76].

The rolling of leukocytes on activated endothelium is a first step in the emergence of leukocytes out of the blood stream into sites of inflammation. This rolling, which occurs under hydrodynamic shear forces, is mediated by selectins, a family of extended, calcium-dependent lectin receptors present on the surface of endothelial cells. To fulfill their function, selectins and their ligands exhibit a unique combination of mechanical properties: they associate rapidly and avidly and can tether cells over very long distances by their long, extensible structure. In addition, complexes formed between selectins and their ligands can withstand high tensile forces and dissociate in a controllable manner, which allows them to maintain rolling without being pulled out of the cell membrane.

Fritz et al. [59] used dynamic force spectroscopy to study the interaction between P-selectin and its leukocyte-expressed surface ligand P-selectin glycoprotein ligand-1 (PSGL-1). Modeling both intermolecular and intramolecular forces, as well as adhesion probability, they were able to obtain detailed information on rupture forces, elasticity, and the kinetics of the interaction. Complexes were able to withstand forces up to 165 pN and exhibited a chain-like elasticity with a molecular spring constant of 5.3 pN/nm and a persistence length of 0.35 nm. Rupture forces and the lifetime of the complexes exhibited the predicted logarithmic dependence on the loading rate.

An important characteristics of the interaction between P-selectin and PSGL-1, which is highly relevant to the biological function of the complex, was found by investigating the dependence of the adhesion probability between the two molecules on the velocity of the AFM probe. Counterintuitively and in contrast to experiments with avidin–biotin [61], antibody–antigen [3], or cell adhesion proteoglycans [8], the adhesion probability between P-selectin and PSGL-1 was found to *increase* with increasing velocities [59]. This unexpected dependency explains the increase in leukocyte tethering probability with increased shear flow observed in rolling experiments. Since the adhesion probability approached 1, it was concluded that binding occurs instantaneously as the tip reaches the surface and, thus, proceeds with a very fast on-rate. The complex also exhibited a fast forced off-rate. Such a fast-on/fast-off kinetics is probably important for the ability of leukocytes to bind and detach rapidly from the endothelial cell surface. Likewise, the long contour length of the complex together with its high elasticity reduces the mechanical loading on the complex upon binding and allows leukocyte rolling even at high shear rates.

Evans et al. [73] used BPF to study the interaction between PSGL-1 and another member of the selectin family, L-selectin. The force spectra, obtained over a range of loading rates extending from 10 to 100000 pN/s, revealed two prominent energy barriers along the unbinding pathway: an outer barrier, probably constituted by an array of hydrogen bonds, that impeded dissociation under slow detachment, and an inner, $Ca^{2+}$-dependent barrier that dominated dissociation under rapid detachment. The observed hierarchy of inner and outer activation barriers was proposed to be important for multibond recruitment during selectin-mediated function.

Using force-clamp AFM [66], bond lifetimes were directly measured in dependence on a constantly applied force. For this, lifetime–force relations of P-selectin complexed to two forms of P-selectin glycoprotein ligand 1 (PSGL-1) and to G1, a blocking monoclonal antibody against P-selectin, respectively, were determined [75]. Both monomeric (sPSGL-1) and dimeric PSGL-1 exhibited a biphasic relationship between lifetime and force in their interaction to P-selectin (Fig. 8.10a,b). The bond lifetimes initially increased, indicating the presence of catch bonds. After reaching a maximum, the lifetimes decreased with force, indicating a catch bond. In

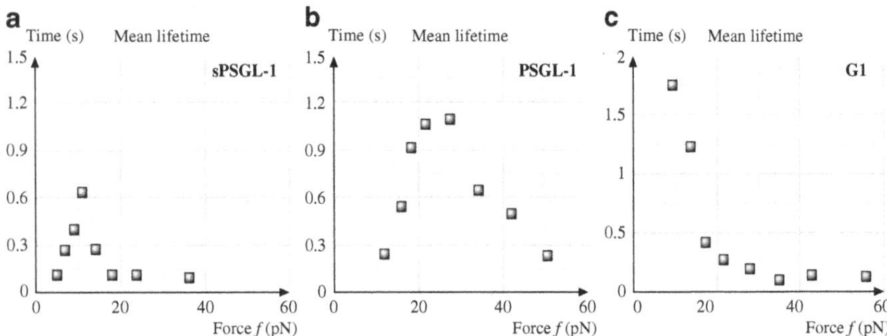


**Fig. 8.10** Lifetimes of bonds of single-molecular complexes, depending on a constantly applied force. **(a)** sPSGL-1/P-selectin: catch bond and slip bond. **(b)** PSGL-1/P-selectin: catch bond and slip bond. **(c)** G1/P-selectin: slip bond only (After [75])

contrast, the P-selectin/G1 bond lifetimes decreased exponentially with force (Fig. 8.10c), displaying typical slip bond characteristics that are well described by the single-energy-barrier Bell model. The curves of lifetime against force for the two forms of PSGL1-1 had similar biphasic shapes (Fig. 8.10a,b), but the PSGL-1 curve (Fig. 8.10b) was shifted relative to the sPSGL-1 curve (Fig. 8.10a), approximately doubling the force and the lifetime. These data suggest that sPSGL-1 forms monomeric bonds with P-selectin, whereas PSGL-1 forms dimeric bonds with P-selectin. In agreement with the studies describes above, it was concluded that the use of force-induced switching from catch to slip bonds might be physiologically relevant for the tethering and rolling process of leukocytes on selectins [75].

Baumgartner et al. [64] used AFM to probe specific trans-interaction forces and conformational changes of recombinant vascular endothelial (VE)-cadherin strand dimers. VE-cadherins are cell-surface proteins that mediate the adhesion of cells in the vascular endothelium through $Ca^{2+}$-dependent homophilic interactions of their N-terminal extracellular domains. Acting as such they play an important role in the regulation of intercellular adhesion and communication in the inner surface of blood vessels. Unlike selectin-mediated adhesion, association between trans-interacting VE dimers was slow and independent of probe velocity, and complexes were ruptured at relatively low forces. These differences were attributed to the fact that, as opposed to selectins, cadherins mediate adhesion between resting cells. Mechanical stress on the junctions is thus less intense and high-affinity binding is not required to establish and maintain intercellular adhesion. Determination of $Ca^{2+}$ dependency of recognition events between tip- and surface-bound VE-cadherins revealed a surprisingly high $K_D$ (1.15 mM), which is very close to the free extracellular $Ca^{2+}$ concentration in the body. Binding also revealed a strong dependence on calcium concentration, giving rise to an unusually high Hill coefficient of $\approx$5. This steep dependency suggests that local changes of free extracellular $Ca^{2+}$ in the narrow intercellular space may facilitate rapid remodeling of intercellular adhesion and permeability.

Odorico et al. [77] used DFS to explore the energy landscape underlying the interaction between a chelated uranyl compound and a monoclonal antibody raised against the uranyl-dicarboxy-phenanthroline complex. To isolate contributions of the uranyl moiety to the binding interaction, measurements were performed with and without the ion in the chelating ligand. In the presence of uranyl, the force spectra contained two linear regimes, suggesting the presence of at least two major energy barriers along the unbinding pathway. To relate the experimental data to molecular events, the authors constructed a model with a variable fragment of the antibody and used computational graphics to dock the chelated uranyl ion into the binding pocket. The analysis suggested that the inner barrier ($x_\beta = 0.5$Å) reflects the rupture of coordination bonds between the uranium atom and an Asp residue, whereas the outer barrier ($x_\beta = 3.9$Å) amounts to the detachment of the entire ligand from the Ab binding site.

Nevo et al. [78, 79] used single-molecule force spectroscopy to discriminate between alternative mechanisms of protein activation (Fig. 8.11). The activation of proteins by other proteins, protein domains or small ligands is a central process in biology, e.g., in signalling pathways and enzyme activity. Moreover, activation and

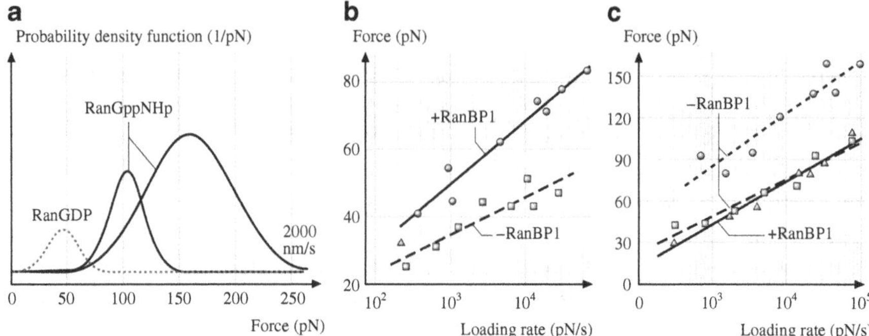


**Fig. 8.11** Protein activation revealed by force spectroscopy. Ran and importin $\beta$ (imp$\beta$) were immobilized onto the AFM cantilevered tip and mica, respectively, and the interaction force was measured at different loading rates in the absence or presence of RanBP1, which was added as a mobile substrate to the solution in the AFM liquid cell. Unbinding force distributions obtained for imp$\beta$–Ran complexes at pulling velocity of 2,000 nm/s. Association of imp$\beta$ with Ran loaded with GDP (**a**) or with nonhydrolyzable GTP analogue (GppNHp) (**b**) gives rise to uni- or bimodal force distributions, respectively, reflecting the presence of one or two bound states. (**b–c**) Force spectra obtained for complexes of imp$\beta$ with RanGDP or with RanGppNHp, in the absence (*dashed lines*) or presence (*solid lines*) of RanBP1. The results indicate that activation of imp$\beta$–RanGDP and imp–RanGTP complexes by RanBP1 proceeds through induced-fit and dynamic population-shift mechanisms, respectively (see text for details) (After [78,79])

deactivation of genes both depend on the switching of proteins between alternative functional states. Two general mechanisms have been proposed. The induced-fit model assigns changes in protein activity to conformational changes triggered by effector binding. The population-shift model, on the other hand, ascribes these changes to a redistribution of *preexisting* conformational isomers. According to this model, also known as the preequilibrium or conformational selection model, protein structure is regarded as an ensemble of conformations existing in equilibrium. The ligand binds to one of these conformations, i.e., the one to which it is most complementary, thus shifting the equilibrium in favor of this conformation. Discrimination between the two models of activation requires that the distribution of conformational isomers in the ensemble is known. Such information, however, is very hard to obtain from conventional bulk methods because of ensemble averaging.

Using AFM, Nevo and coworkers measured the unbinding forces of two related protein complexes in the absence or presence of a common effector. The complexes consisted of the nuclear transport receptor importin $\beta$(imp$\beta$) and the small GTPase Ran. The difference between them was the nucleotide-bound state of Ran, which was either guanosine diphosphate (GDP) or guanosine-5'-triphosphate (GTP). The effector molecule was the Ran-binding protein RanBP1. Loaded with GDP, Ran associated weakly with imp$\beta$ to form a single bound state characterized by unimodal distributions of small unbinding forces (Fig. 8.11a, dotted line). Addition of Ran BP1 resulted in a marked shift of the distribution to higher unbinding forces (Fig. 8.11b, dashed to solid line). These results were interpreted to be consistent with an induced-fit mechanism where binding of RanBP1 induces a conformational change in the complex, which, in turn, strengthens the interaction between imp$\beta$ and Ran(GDP). In contrast, association of RanGTP with imp$\beta$ was found to lead to alternative bound states of relatively low and high adhesion strength represented by partially overlapping force distributions (Fig. 8.11a, solid line). When RanBP1 was added to the solution, the higher-strength population, which predominated the ensemble in the absence of the effector (Fig. 8.11c, dashed lines), was diminished, and the lower-strength conformation became correspondingly more populated (Fig. 8.11c, solid line). The means of the distributions, however, remain unchanged, indicating that the strength of the interaction in the two states of the complex had not been altered by the effector. These data fit a dynamic population-shift mechanism in which RanBP1 binds selectively to the lower-strength conformation of RanGTP–imp$\beta$, changing the properties and function of the complex by shifting the equilibrium between its two states.

The complex between imp$\beta$ and RanGTP was also used in studies aimed to measure the energy landscape roughness of proteins. The roughness of the energy landscapes that describe proteins has numerous effects on their folding and binding as well as on their behavior at equilibrium, since undulations in the free energy surface can attenuate diffusion dramatically. Thus, to understand how proteins fold, bind, and function, one needs to know not only the energy of their initial and final states, but also the roughness of the energy surface that connects them. However, for a long time, knowledge of protein energy landscape roughness came solely from theory and simulations of small model proteins.

Adopting Zwanzig's theory of diffusion in rough potentials [80], Hyeon and Thirumalai [81] proposed that the energy landscape roughness of proteins can be measured from single-molecule mechanical unfolding experiments conducted at different temperatures. In particular, their simulations showed that at a constant loading rate the most probable force for unfolding increases because of roughness that acts to attenuate diffusion. Because this effect is temperature dependent, an overall energy scale of roughness, $\varepsilon$, can be derived from plots of force versus loading rate acquired at two arbitrary temperatures. Extending this theory to the case of unbinding, and performing single-molecule force spectroscopy measurements, Nevo et al. [82] extracted the overall energy scale of roughness $\varepsilon$ for RanGTP–imp$\beta$. The results yielded $\varepsilon > 5k_{\mathrm{B}}T$, indicating a bumpy energy surface, which is consistent with the unusually high structural flexibility of imp$\beta$ and its ability to interact with different, structurally distinct ligands in a highly specific manner. This mechanistic principle may also be applicable to other proteins whose function demands highly specific and regulated interactions with multiple ligands.

More recently, the same type of analysis using three temperatures and pulling speeds in the range of 100 to 38,000 nm/s, was applied to derive $\varepsilon$ for the well-studied streptavidin–biotin interaction [83]. Analysis of the Bell parameters revealed considerable widening of the inner barrier for the transition with temperature, reflecting perhaps a softening of the dominant hydrogen-bond network that stabilizes the ground state of the complex. In contrast, the position of the outer barrier did not change significantly upon increase of the temperature. Estimations of $\varepsilon$ were made at four different forces, 75, 90, 135, and 156 pN, with the first two forces belonging to the first linear loading regime of the force spectrum (outer barrier) and the last two to the second (inner barrier). The values obtained were consistent *within each of the two regimes*, averaging at 7.5 and $\approx 5.5k_BT$ along the outer and inner barriers of the transition, respectively. The difference was attributed to contributions from the intermediate state of the reaction, which is suppressed (along with the outer barrier) at high loading rates. The origin of roughness was attributed to competition of solvent water molecules with some of the hydrogen bonds that stabilize the complex and to the aforementioned 3–4 loop of streptavidin, which is highly flexible and, therefore, may induce the formation of multiple conformational substates in the complex. It was also proposed by the authors that the large roughness detected in the energy landscape of streptavidin–biotin is a significant contributor to the unusually slow dissociation kinetics of the complex and may account for the discrepancies in the unbinding forces measured for this pair.

### *8.5.3 Live Cells and Membranes*

Thus far, there have been only a few attempts to apply recognition force spectroscopy to cells. In one of the early studies, Lehenkari and Horton [84] measured the unbinding forces between integrin receptors present on the surface of intact cells and several RGD-containing (Arg–Gly–Asp) ligands. The unbinding forces measured were found to be cell and amino acid sequence specific, and sensitive to pH and the divalent cation composition of the cellular culture medium. In contrast to short linear RGD hexapeptides, larger peptides and proteins containing the RGD sequence showed different binding affinities, demonstrating that the context of the RGD motif within a protein has a considerable influence upon its interaction with the receptor. In another study, Chen and Moy [85] used AFM to measure the adhesive strength between concanavalin A (Con A) coupled to an AFM tip and Con A receptors on the surface of NIH3T3 fibroblasts. Cross-linking of receptors with either glutaraldehyde or 3, 3′-dithio-bis(sulfosuccinimidylproprionate) (DTSSP) led to an increase in adhesion that was attributed to enhanced cooperativity among adhesion complexes. The results support the notion that receptor cross-linking can increase adhesion strength by creating a shift towards cooperative binding of receptors. Pfister et al. [86] investigated the surface localization of HSP60 on stressed and unstressed human umbilical venous endothelial cells (HUVECs). By detecting specific single-molecule binding events between the monoclonal antibody AbII-13 tethered to AFM tips and HSP60 molecules on cells, clear evidence was found for the occurrence of HSP60 on the surface of stressed HUVECs, but not on unstressed HUVECs.

The sidedness and accessibility of protein epitopes of the $Na^{2+}$/D-glucose cotransporter 1 (SGLT1) was probed in intact brush border membranes by a tip-bound antibody directed against an amino acid sequence close to the glucose binding site [38]. Binding of glucose and transmembrane transport altered both the binding probability and the most probable unbinding force, suggesting changes in the orientation and conformation of the transporter. These studies were extended to live SGLT1-transfected CHO cells [87]. Using AFM tips carrying the substrate 1-$\beta$-thio-D-glucose, direct evidence could be obtained that, in the presence of sodium, a sugar binding site appears on the SGLT1 surface. It was shown that this binding site accepts the sugar residue of the glucoside phlorizin, free D-glucose and D-galactose, but not free L-glucose. The data indicate the importance of stereoselectivity for sugar binding and transport.

Studies on the interaction between leukocyte function-associated antigen-1 (LFA-1) and its cognate ligand, intercellular adhesion molecules 1 and 2 (ICAM-1 and ICAM-2), which play a crucial role in leukocyte adhesion, revealed two prominent barriers [74, 88]. The experimental system consisted of LFA-1-expressing Jurkat T-cells attached to the end of the AFM cantilever and surface-immobilized ICAM-1 or -2. For both ICAM-1 and ICAM-2, the force spectra exhibited fast and slow loading regimes, amounting to a sharp, inner energy barrier ($x_\beta \approx 0.56$Å and 1.5Å, for complexes formed with ICAM-1 and ICAM-2) and a shallow, outer barrier ($x_\beta \approx 3.6$Å and 4.9Å), respectively. Addition of $Mg^{2+}$ led to an increase of the rupture forces measured in the slow loading regime, indicating an increment of the outer barrier in the presence of the divalent cation. Comparison between the force spectra obtained for the complexes formed between LFA-1 and ICAM-1 or ICAM-2 indicated that, in the fast loading regime, the rupture of LFA-1–ICAM-1 depends more steeply on the loading rate than that of LFA-1–ICAM-2. The difference in dynamic strength between the two interactions was attributed to the presence of wider barriers in the LFA-1–ICAM-2 complex, which render the interaction more receptive to the applied load. The enhanced sensitivity of complexes with ICAM-2 to pulling forces was proposed to be important for the ability of ICAM-2 to carry out routine immune surveillance, which might otherwise be impeded due to frequent adhesion events.

## 8.6 Recognition Imaging

Besides measuring interaction strengths, locating binding sites over biological surfaces such as cells or membranes is of great interest. To achieve this goal, force detection must be combined with high-resolution imaging.

Ludwig et al. [89] used chemical force microscopy to image a streptavidin pattern with a biotinylated tip. An approach–retract cycle was performed at each point of a raster, and topography, adhesion, and sample elasticity were extracted from the local force ramps. This strategy was also used to map binding sites on cells [90, 91] and to differentiate between red blood cells of different blood groups (A and 0) using AFM tips functionalized with a group A-specific lectin [92].

Identification and localization of single antigenic sites was achieved by recording force signals during the scanning of an AFM tip coated with antibodies along a single line across a surface immobilized with a low density of antigens [3, 11]. Using this method, antigens could be localized over the surface with positional accuracy of 1.5 nm. A similar configuration used by Willemsen et al. [93] enabled the simultaneous acquisition of height and adhesion-force images with near molecular resolution.

The aforementioned strategies of force mapping either lack high lateral resolution [89] and/or are much slower [3, 11, 93] than conventional topographic imaging since the frequency of the force-sensing retract–approach cycles is limited by hydrodynamic damping. In addition, the ligand needs to be detached from the receptor in each retract–approach cycle, necessitating large working amplitudes (50 nm). Therefore, the surface-bound receptor is inaccessible to the tip-immobilized ligand on the tip during most of the time of the experiment. This problem, however, should be overcome with the use of small cantilevers [53], which should increase the speed for force mapping because the hydrodynamic forces are significantly reduced and the resonance frequency is higher than that of commercially available cantilevers. Short cantilevers were recently applied to follow the association and dissociation of individual chaperonin proteins, GroES to GroEL, in real time using dynamic force microscopy topography imaging [94].

An imaging method for mapping antigenic sites on surfaces was developed [20] by combining molecular recognition force spectroscopy [3] with dynamic force microscopy (DFM) [28, 96]. In DFM, the AFM tip is oscillated across a surface and the amplitude reduction arising from tip–surface interactions is held constant by a feedback loop that lifts or lowers the tip according to the detected amplitude signal. Since the tip contacts the surface only intermittently, this technique provides very gentle tip–surface interactions and the specific interaction of the antibody on the tip with the antigen on the surface can be used to localize antigenic sites for recording recognition images. The AFM tip is magnetically coated and oscillated by an alternating magnetic field at very small amplitudes while being scanned along the surface. Since the oscillation frequency is more than a hundred times faster than typical frequencies in conventional force mapping, the data acquisition rate is much higher. This method was recently extended to yield fast, simultaneous acquisition of two independent maps, i.e., a topography image and a lateral map of recognition sites, recorded with nm resolution at experimental times equivalent to normal AFM imaging [95, 97, 98].

Topography and recognition images were simultaneously obtained (TREC imaging) using a special electronic circuit (PicoTrec, Agilent, Chandler, AZ) (Fig. 8.12a). Maxima ($U_{\mathrm{up}}$) and minima ($U_{\mathrm{down}}$) of each sinusoidal cantilever deflection period were depicted in a peak detector, filtered, and amplified. Direct-current (DC) offset signals were used to compensate for the thermal drifts of the cantilever. $U_{\mathrm{up}}$ and $U_{\mathrm{down}}$ were fed into the AFM controller, with $U_{\mathrm{down}}$ driving the feedback loop to record the height (i.e., topography) image and $U_{\mathrm{up}}$ providing the data for constructing the recognition image (Fig. 8.12a). Since we used cantilevers with low $Q$-factor ($\approx$1 in liquid) driven at frequencies below resonance, the two types of information were independent. In this way, topography and recognition image were recorded simultaneously and independently.

The circuit was applied to mica containing singly distributed avidin molecules using a biotinylated AFM tip [95]. The sample was imaged with an antibody-containing tip, yielding the topography (Fig. 8.12b, left image) and the recognition

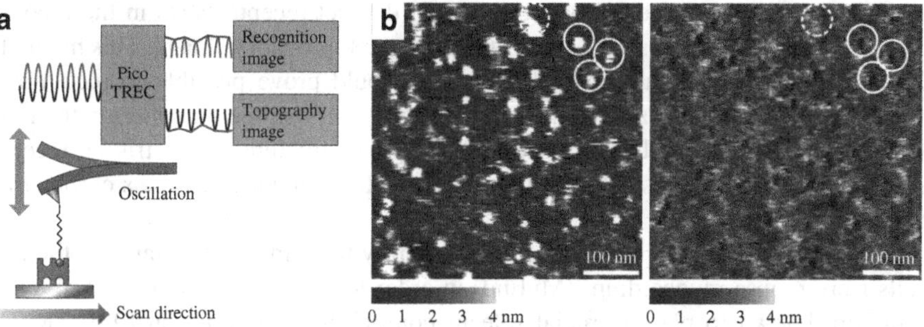


**Fig. 8.12** Simultaneous topography and recognition (TREC) imaging. (**a**) Principle: the cantilever oscillation is split into lower and upper parts, resulting in simultaneously acquired topography and recognition images. (**b**) Avidin was electrostatically adsorbed to mica and imaged with a biotin-tethered tip. Good correlation between topography (*left image, bright spots*) and recognition (*right image, dark spots*) was found (*solid circles*). Topographical spots without recognition denote structures lacking specific interaction (*dashed circle*). Scan size was 500 nm (After [95])

image (Fig. 8.12b, right image) at the same time. The tip oscillation amplitude (5 nm) was chosen to be slightly smaller than the extended cross-linker length (8 nm), so that both the antibody remained bound while passing a binding site and the reduction of the upwards deflection was of sufficient significance compared with the thermal noise. Since the spring constant of the polymeric cross-linker increases nonlinearly with the tip–surface distance (Fig. 8.5), the binding force is only sensed close to full extension of the cross-linker (given at the maxima of the oscillation period). Therefore, the recognition signals were well separated from the topographic signals arising from the surface, in both space ($\Delta z \approx 5$ nm) and time (half-oscillation period $\approx$0.1 ms).

The bright dots with 2–3 nm height and 15–20 nm diameter visible in the topography image (Fig. 8.12b, left image) represent single avidin molecules stably adsorbed onto the flat mica surface. The recognition image shows black dots at positions of avidin molecules (Fig. 8.12b, right image) because the oscillation maxima are lowered due to the physical avid–biotin connection established during recognition. That the lateral positions of the avidin molecules obtained in the topography image are spatially correlated with the recognition signals of the recognition image is indicated by solid circles in the images (Fig. 8.12). Recognition between the antibody on the tip and the avidin on the surface took place for almost all avidin molecules, most likely because avidin contains four biotin binding sites, two on either side. Thus, one would assume to have always binding epitopes oriented away from the mica surface and accessible to the biotinylated tip, resulting in a high binding efficiency. Structures observed in the topography image and not detected in the recognition image were very rare (dotted circle in Fig. 8.12b).

It is important to note that topography and recognition images were recorded at speeds typical for standard AFM imaging and were therefore considerably faster than conventional force mapping. With this methodology, topography and recognition images can be obtained at the same time and distinct receptor sites in the recognition image can be assigned to structures from the topography image. This method is applicable to any ligand, and therefore it should prove possible to recognize many types of proteins or protein layers and carry out epitope mapping on the nm scale on membranes, cells, and complex biological structures. In a striking recent example, histone proteins H3 were identified and localized in a complex chromatin preparation [98].

Recently, TREC imaging was applied to gently fixed microvascular endothelial cells from mouse myocardium (MyEnd) in order to visualize binding sites of VE-cadherin, known to play a crucial role in homophilic cell adhesion [99]. TREC images were acquired with AFM tips coated with a recombinant VE-cadherin. The recognition images revealed prominent, irregularly shaped *dark* spots (domains) with size from 30 to 250 nm. The domains enriched in VE-cadherins molecules were found to be collocated with the cytoskeleton filaments supporting the anchorage of VE-cadherins to F-actin. Compared with conventional techniques such as immunochemistry or single-molecule optical microscopy, TREC represents an alternative method to quickly obtain the local distribution of receptors on cell surface with unprecedented lateral resolution of several nanometers.

## 8.7 Concluding Remarks

Atomic force microscopy has evolved to become an imaging method that can yield the finest structural details on live, biological samples in their native, aqueous environment under ambient conditions. Due to its high lateral resolution and sensitive force detection capability, it is now possible to measure molecular forces of biomolecules on the single-molecule level. Well beyond the proof-of-principle stage of the pioneering experiments, AFM has now developed into a high-end analysis method for exploring kinetic and structural details of interactions underlying protein folding and molecular recognition. The information obtained from force spectroscopy, being on a single-molecule level, includes physical parameters not accessible by other methods. In particular, it opens up new perspectives to explore the dynamics of biological processes and interactions.

## References

1. G.U. Lee, D.A. Kidwell, R.J. Colton, Sensing discrete streptavidin-biotin interactions with atomic force microscopy. Langmuir **10**, 354–357 (1994).
2. E.L. Florin, V.T. Moy, H.E. Gaub, Adhesion forces between individual ligand receptor pairs. Science **264**, 415–417 (1994).
3. P. Hinterdorfer, W. Baumgartner, H.J. Gruber, K. Schilcher, H. Schindler, Detection and localization of individual antibody-antigen recognition events by atomic force microscopy. Proc. Natl. Acad. Sci. USA **93**, 3477–3481 (1996).

4. G. Binnig, C.F. Quate, C. Gerber, Atomic force microscope. Phys. Rev. Lett. **56**, 930–933 (1986).
5. M. Grandbois, W. Dettmann, M. Benoit, H.E. Gaub, How strong is a covalent bond. Science **283**, 1727–1730 (1999).
6. T. Boland, B.D. Ratner, Direct measurement of hydrogen bonding in DNA nucleotide bases by atomic force microscopy. Proc. Natl. Acad. Sci. USA **92**, 5297–5301 (1995).
7. P. Wagner, M. Hegner, P. Kernen, F. Zaugg, G. Semenza, Covalent immobilization of native biomolecules onto Au(111) via N-hydroxysuccinimide ester functionalized self assembled monolayers for scanning probe microscopy. Biophys. J. **70**, 2052–2066 (1996).
8. U. Dammer, O. Popescu, P. Wagner, D. Anselmetti, H.-J. Güntherodt, G.M. Misevic, Binding strength between cell adhesion proteoglycans measured by atomic force microscopy. Science **267**, 1173–1175 (1995).
9. U. Dammer, M. Hegner, D. Anselmetti, P. Wagner, M. Dreier, W. Huber, H.-J. Güntherodt, Specific antigen/antibody interactions measured by force microscopy. Biophys. J. **70**, 2437–2441 (1996).
10. Y. Harada, M. Kuroda, A. Ishida, Specific and quantized antibody-antigen interaction by atomic force microscopy. Langmuir **16**, 708–715 (2000).
11. P. Hinterdorfer, K. Schilcher, W. Baumgartner, H.J. Gruber, H. Schindler, A mechanistic study of the dissociation of individual antibody-antigen pairs by atomic force microscopy. Nanobiology **4**, 39–50 (1998).
12. S. Allen, X. Chen, J. Davies, M.C. Davies, A.C. Dawkes, J.C. Edwards, C.J. Roberts, J. Sefton, S.J.B. Tendler, P.M. Williams, Spatial mapping of specific molecular recognition sites by atomic force microscopy. Biochemistry **36**, 7457–7463 (1997).
13. R. Ros, F. Schwesinger, D. Anselmetti, M. Kubon, R. Schäfer, A. Plückthun, L. Tiefenauer, Antigen binding forces of individually addressed single-chain Fv antibody molecules. Proc. Natl. Acad. Sci. USA **95**, 7402–7405 (1998).
14. T. Strunz, K. Oroszlan, R. Schäfer, H.-J. Güntherodt, Dynamic force spectroscopy of single DNA molecules. Proc. Natl. Acad. Sci. USA **96**, 11277–11282 (1999).
15. A. Ebner, P. Hinterdorfer, H.J. Gruber, Comparison of different aminofunctionalization strategies for attachment of single antibodies to AFM cantilevers. Ultramicroscopy **107**, 922–927 (2007).
16. S.S. Wong, E. Joselevich, A.T. Woolley, C.L. Cheung, C.M. Lieber, Covalently functionalyzed nanotubes as nanometre-sized probes in chemistry and biology. Nature **394**, 52–55 (1998).
17. P. Hinterdorfer, F. Kienberger, A. Raab, H.J. Gruber, W. Baumgartner, G. Kada, C. Riener, S. Wielert-Badt, C. Borken, H. Schindler, Poly(ethylene glycol): An ideal spacer for molecular recognition force microscopy/spectroscopy. Single Mol. **1**, 99–103 (2000).
18. T. Haselgrübler, A. Amerstorfer, H. Schindler, H.J. Gruber, Synthesis and applications of a new poly(ethylene glycol) derivative for the crosslinking of amines with thiols. Bioconjug. Chem. **6**, 242–248 (1995).
19. A.S.M. Kamruzzahan, A. Ebner, L. Wildling, F. Kienberger, C.K. Riener, C.D. Hahn, P.D. Pollheimer, P. Winklehner, M. Holzl, B. Lackner, D.M. Schorkl, P. Hinterdorfer, H.J. Gruber, Antibody linking to atomic force microscope tips via disulfide bond formation. Bioconjug. Chem. **17**(6), 1473–1481 (2006).
20. A. Raab, W. Han, D. Badt, S.J. Smith-Gill, S.M. Lindsay, H. Schindler, P. Hinterdorfer, Antibody recognition imaging by force microscopy. Nat. Biotech. **17**, 902–905 (1999).
21. G.U. Lee, A.C. Chrisey, J.C. Colton, Direct measurement of the forces between complementary strands of DNA. Science **266**, 771–773 (1994).
22. M. Conti, G. Falini, B. Samori, How strong is the coordination bond between a histidine tag and Ni-nitriloacetate? An experiment of mechanochemistry on single molecules. Angew. Chem. **112**, 221–224 (2000).
23. F. Kienberger, G. Kada, H.J. Gruber, V.P. Pastushenko, C. Riener, M. Trieb, H.-G. Knaus, H. Schindler, P. Hinterdorfer, Recognition force spectroscopy studies of the NTA-His6 bond. Single Mol. **1**, 59–65 (2000).

24. L. Schmitt, M. Ludwig, H.E. Gaub, R. Tampe, A metal-chelating microscopy tip as a new toolbox for single-molecule experiments by atomic force microscopy. Biophys. J. **78**, 3275–3285 (2000).
25. S. Lata, A. Reichel, R. Brock, R. Tampe, J. Piehler, High-affinity adaptors for switchable recognition of histidine-tagged proteins. J. Am. Chem. Soc. **127**, 10205–10215 (2005).
26. A. Ebner, L. Wildling, R. Zhu, C. Rankl, T. Haselgrübler, P. Hinterdorfer, H.J. Gruber, Functionalization of probe tips and supports for single molecule force microscopy. Top. Curr. Chem. **285**, 29–76 (2008).
27. C. Yuan, A. Chen, P. Kolb, V.T. Moy, Energy landscape of avidin-biotin complexes measured burey atomic force microscopy. Biochemistry **39**, 10219–10223 (2000).
28. W. Han, S.M. Lindsay, M. Dlakic, R.E. Harrington, Kinked DNA. Nature **386**, 563 (1997).
29. G. Kada, L. Blaney, L.H. Jeyakumar, F. Kienberger, V.P. Pastushenko, S. Fleischer, H. Schindler, F.A. Lai, P. Hinterdorfer, Recognition force microscopy/spectroscopy of ion channels: Applications to the skeletal muscle $Ca^{2+}$ release channel (RYR1). Ultramicroscopy **86**, 129–137 (2001).
30. D.J. Müller, W. Baumeister, A. Engel, Controlled unzipping of a bacterial surface layer atomic force microscopy. Proc. Natl. Acad. Sci. USA **96**, 13170–13174 (1999).
31. F. Oesterhelt, D. Oesterhelt, M. Pfeiffer, A. Engle, H.E. Gaub, D.J. Müller, Unfolding pathways of individual bacteriorhodopsins. Science **288**, 143–146 (2000).
32. E. Kiss, C.-G. Gölander, Chemical derivatization of muscovite mica surfaces. Colloids Surf. **49**, 335–342 (1990).
33. S. Karrasch, M. Dolder, F. Schabert, J. Ramsden, A. Engel, Covalent binding of biological samples to solid supports for scanning probe microscopy in buffer solution. Biophys. J. **65**, 2437–2446 (1993).
34. N.H. Thomson, B.L. Smith, N. Almqvist, L. Schmitt, M. Kashlev, E.T. Kool, P.K. Hansma, Oriented, active *escherichia coli* RNA polymerase: An atomic force microscopy study. Biophys. J. **76**, 1024–1033 (1999).
35. G. Kada, C.K. Riener, P. Hinterdorfer, F. Kienberger, C.M. Stroh, H.J. Gruber, Dithiophospholipids for biospecific immobilization of proteins on gold surfaces. Single Mol. **3**, 119–125 (2002).
36. C. LeGrimellec, E. Lesniewska, M.C. Giocondi, E. Finot, V. Vie, J.P. Goudonnet, Imaging of the surface of living cells by low-force contact-mode atomic force microscopy. Biophys. J. **75**(2), 695–703 (1998).
37. K. Schilcher, P. Hinterdorfer, H.J. Gruber, H. Schindler, A non-invasive method for the tight anchoring of cells for scanning force microscopy. Cell. Biol. Int. **21**, 769–778 (1997).
38. S. Wielert-Badt, P. Hinterdorfer, H.J. Gruber, J.-T. Lin, D. Badt, H. Schindler, R.K.-H. Kinne, Single molecule recognition of protein binding epitopes in brush border membranes by force microscopy. Biophys. J. **82**, 2767–2774 (2002).
39. P. Bongrand, C. Capo, J.-L. Mege, A.-M. Benoliel, Use of hydrodynamic flows to study cell adhesion, In *Physical Basis of Cell Adhesion*, ed. by P. Bongrand (CRC Press, Boca Raton, 1988) pp. 125–156.
40. J.N. Israelachvili, *Intermolecular and Surface Forces*, 2nd edn. (Academic, New York, 1991).
41. R. Merkel, P. Nassoy, A. Leung, K. Ritchie, E. Evans, Energy landscapes of receptor-ligand bonds explored by dynamic force spectroscopy. Nature **397**, 50–53 (1999).
42. A. Askin, Optical trapping and manipulation of neutral particles using lasers. Proc. Natl. Acad. Sci. USA **94**, 4853–4860 (1997).
43. K.C. Neuman, S.M. Block, Optical trapping. Rev. Sci. Instrum. **75**, 2787–2809 (2004).
44. K. Svoboda, C.F. Schmidt, B.J. Schnapp, S.M. Block, Direct observation of kinesin stepping by optical trapping interferometry. Nature **365**, 721–727 (1993).
45. S.M. Block, C.L. Asbury, J.W. Shaevitz, M.J. Lang, Probing the kinesin reaction cycle with a 2D optical force clamp. Proc. Natl. Acad. Sci. USA **100**, 2351–2356 (2003).
46. A.E.M. Clemen, M. Vilfan, J. Jaud, J. Zhang, M. Barmann, M. Rief, Force-dependent stepping kinetics of myosin-V. Biophys. J. **88**, 4402–4410 (2005).

47. S. Smith, Y. Cui, C. Bustamante, Overstretching B-DNA: The elastic response of individual double-stranded and single-stranded DNA molecules. Science **271**, 795–799 (1996).
48. M.S.Z. Kellermayer, S.B. Smith, H.L. Granzier, C. Bustamante, Folding-unfolding transitions in single titin molecules characterized with laser tweezers. Sience **276**, 1112–1216 (1997).
49. T.R. Strick, J.F. Allemend, D. Bensimon, A. Bensimon, V. Croquette, The elasticity of a single supercoiled DNA molecule. Biophys. J. **271**, 1835–1837 (1996).
50. T. Lionnet, D. Joubaud, R. Lavery, D. Bensimon, V. Croquette, Wringing out DNA. Phys. Rev. Lett. **96**, 178102 (2006).
51. F. Kienberger, V.P. Pastushenko, G. Kada, H.J. Gruber, C. Riener, H. Schindler, P. Hinterdorfer, Static and dynamical properties of single poly(ethylene glycol) molecules investigated by force spectroscopy. Single Mol. **1**, 123–128 (2000).
52. S. Liang, D. Medich, D.M. Czajkowsky, S. Sheng, J.-Y. Yuan, Z. Shao, Thermal noise reduction of mechanical oscillators by actively controlled external dissipative forces. Ultramicroscopy **84**, 119–125 (2000).
53. M.B. Viani, T.E. Schäffer, A. Chand, M. Rief, H.E. Gaub, P.K. Hansma, Small cantilevers for force spectroscopy of single molecules. J. Appl. Phys. **86**, 2258–2262 (1999).
54. T. Strunz, K. Oroszlan, I. Schumakovitch, H.-J. Güntherodt, M. Hegner, Model energy landscapes and the force-induced dissociation of ligand-receptor bonds. Biophys. J. **79**, 1206–1212 (2000).
55. H. Grubmüller, B. Heymann, P. Tavan, Ligand binding: Molecular mechanics calculation of the streptavidin-biotin rupture force. Science **271**, 997–999 (1996).
56. G.I. Bell, Models for the specific adhesion of cells to cells. Science **200**, 618–627 (1978).
57. E. Evans, K. Ritchie, Dynamic strength of molecular adhesion bonds. Biophys. J. **72**, 1541–1555 (1997).
58. E. Evans, K. Ritchie, Strength of a weak bondconnecting flexible polymer chains. Biophys. J. **76**, 2439–2447 (1999).
59. J. Fritz, A.G. Katopidis, F. Kolbinger, D. Anselmetti, Force-mediated kinetics of single P-selectin/ligand complexes observed by atomic force microscopy. Proc. Natl. Acad. Sci. USA **95**, 12283–12288 (1998).
60. T. Auletta, M.R. de Jong, A. Mulder, F.C.J.M. van Veggel, J. Huskens, D.N. Reinhoudt, S. Zou, S. Zapotocny, H. Schönherr, G.J. Vancso, L. Kuipers, $\beta$-cyclodextrin host-guest complexes probed under thermodynamic equilibrium: Thermodynamics and force spectroscopy. J. Am. Chem. Soc. **126**, 1577–1584 (2004).
61. V.T. Moy, E.-L. Florin, H.E. Gaub, Adhesive forces between ligand and receptor measured by AFM. Science **266**, 257–259 (1994).
62. A. Chilkoti, T. Boland, B. Ratner, P.S. Stayton, The relationship between ligand-binding thermodynamics and protein-ligand interaction forces measured by atomic force microscopy. Biophys. J. **69**, 2125–2130 (1995).
63. I. Schumakovitch, W. Grange, T. Strunz, P. Bertoncini, H.-J. Güntherodt, M. Hegner, Temperature dependence of unbinding forces between complementary DNA strands. Biophys. J. **82**, 517–521 (2002).
64. W. Baumgartner, P. Hinterdorfer, W. Ness, A. Raab, D. Vestweber, H. Schindler, D. Drenckhahn, Cadherin interaction probed by atomic force microscopy. Proc. Natl. Acad. Sci. USA **8**, 4005–4010 (2000).
65. F. Schwesinger, R. Ros, T. Strunz, D. Anselmetti, H.-J. Güntherodt, A. Honegger, L. Jermutus, L. Tiefenauer, A. Plückthun, Unbinding forces of single antibody-antigen complexes correlate with their thermal dissociation rates. Proc. Natl. Acad. Sci. USA **29**, 9972–9977 (2000).
66. A.F. Oberhauser, P.K. Hansma, M. Carrion-Vazquez, J.M. Fernandez, Stepwise unfolding of titin under force-clamp atomic force microscopy. Proc. Natl. Acad. Sci. USA **16**, 468–472 (2000).
67. S. Izraelev, S. Stepaniants, M. Balsera, Y. Oono, K. Schulten, Molecular dynamics study of unbinding of the avidin-biotin complex. Biophys. J. **72**, 1568–1581 (1997).

68. M. Rief, F. Oesterhelt, B. Heyman, H.E. Gaub, Single molecule force spectroscopy on polysaccharides by atomic force microscopy. Science **275**, 1295–1297 (1997).
69. M. Schlierf, M. Rief, Single-molecule unfolding force distributions reveal a funnel-shaped energy landscape. Biophys. J. **90**, L33 (2006).
70. G. Neuert, C. Albrecht, E. Pamir, H.D. Gaub, Dynamic force spectroscopy of the digoxigenin-antibody complex. FEBS Letters **580**, 505–509 (2006).
71. O.K. Dudko, A.E. Filippov, J. Klafter, M. Urback, Beyond the conventional description of dynamic force spectroscopy of adhesion bonds. Proc. Natl. Acad. Sci. USA **100**, 11378–11381 (2003).
72. M. Evstigneev, P. Reimann, Dynamic force spectroscopy: Optimized data analysis. Phys. Rev. E **68**, 045103(R) (2003).
73. E. Evans, E. Leung, D. Hammer, S. Simon, Chemically distinct transition states govern rapid dissociation of single L-selectin bonds under force. Proc. Natl. Acad. Sci. USA **98**, 3784–3789 (2001).
74. X. Zhang, E. Woijcikiewicz, V.T. Moy, Force spectroscopy of the leukocyte function-associated antigen-1/intercellular adhesion molecule-1 interaction. Biophys. J. **83**, 2270–2279 (2002).
75. B.T. Marshall, M. Long, J.W. Piper, T. Yago, R.P. McEver, Z. Zhu, Direct observation of catch bonds involving cell adhesion molecules. Nature **423**, 190–193 (2003).
76. B. Heymann, H. Grubmüller, Molecular dynamics force probe simulations of antibody/antigen unbinding: Entropic control and non additivity of unbinding forces. Biophys. J. **81**, 1295–1313 (2001).
77. M. Odorico, J.M. Teulon, T. Bessou, C. Vidaud, L. Bellanger, S.W. Chen, E. Quemeneur, P. Parot, J.L. Pellequer, Energy landscape of chelated uranyl: Antibody interactions by dynamic force spectroscopy. Biophys. J. **93**, 645 (2007).
78. R. Nevo, C. Stroh, F. Kienberger, D. Kaftan, V. Brumfeld, M. Elbaum, Z. Reich, P. Hinterdorfer, A molecular switch between two bound states in the RanGTP-importin$\beta$1 interaction. Nat. Struct. Mol. Biol. **10**, 553–557 (2003).
79. R. Nevo, V. Brumfeld, M. Elbaum, P. Hinterdorfer, Z. Reich, Direct discrimination between models of protein activation by single-molecule force measurements. Biophys. J. **87**, 2630–2634 (2004).
80. R. Zwanzig, Diffusion in a rough potential. Proc. Natl. Acad. Sci. USA **85**, 2029–2030 (1988).
81. C.B. Hyeon, D. Thirumalai, Can energy landscape roughness of proteins and RNA be measured by using mechanical unfolding experiments? Proc. Natl. Acad. Sci. USA **100**, 10249–10253 (2003).
82. R. Nevo, V. Brumfeld, R. Kapon, P. Hinterdorfer, Z. Reich, Direct measurement of protein energy landscape roughness. EMBO Reports **6**, 482–486 (2005).
83. F. Rico, V.T. Moy, Energy landscape roughness of the streptavidin-biotin interaction. J. Mol. Recognit. **20**, 495–501 (2007).
84. P.P. Lehenkari, M.A. Horton, Single integrin molecule adhesion forces in intact cells measured by atomic force microscopy. Biochem. Biophys. Res. Commun. **259**, 645–650 (1999).
85. A. Chen, V.T. Moy, Cross-linking of cell surface receptors enhances cooperativity of molecular adhesion. Biophys. J. **78**, 2814–2820 (2000).
86. G. Pfister, C.M. Stroh, H. Perschinka, M. Kind, M. Knoflach, P. Hinterdorfer, G. Wick, Detection of HSP60 on the membrane surface of stressed human endothelial cells by atomic force and confocal microscopy. J. Cell Sci. **118**, 1587–1594 (2005).
87. T. Puntheeranurak, L. Wildling, H.J. Gruber, R.K.H. Kinne, P. Hinterdorfer, Ligands on the string: single molecule studies on the interaction of antibodies and substrates with the surface of the $Na^{+}$-glucose cotransporter SGLT1 in living cells. J. Cell Sci. **119**, 2960–2967 (2006).
88. E.P. Wojcikiewicz, M.H. Abdulreda, X. Zhang, V.T. Moy, Force spectroscopy of LFA-1 and its ligands, ICAM-1 and ICAM-2. Biomacromolecules **7**, 3188 (2006).
89. M. Ludwig, W. Dettmann, H.E. Gaub, Atomic force microscopy imaging contrast based on molecuar recognition. Biophys. J. **72**, 445–448 (1997).

90. P.P. Lehenkari, G.T. Charras, G.T. Nykänen, M.A. Horton, Adapting force microscopy for cell biology. Ultramicroscopy **82**, 289–295 (2000).
91. N. Almqvist, R. Bhatia, G. Primbs, N. Desai, S. Banerjee, R. Lal, Elasticity and adhesion force mapping reveals real-time clustering of growth factor receptors and associated changes in local cellular rheological properties. Biophys. J. **86**, 1753–1762 (2004).
92. M. Grandbois, M. Beyer, M. Rief, H. Clausen-Schaumann, H.E. Gaub, Affinity imaging of red blood cells using an atomic force microscope. J. Histochem. Cytochem. **48**, 719–724 (2000).
93. O.H. Willemsen, M.M.E. Snel, K.O. van der Werf, B.G. de Grooth, J. Greve, P. Hinterdorfer, H.J. Gruber, H. Schindler, Y. van Kyook, C.G. Figdor, Simultaneous height and adhesion imaging of antibody antigen interactions by atomic force microscopy. Biophys. J. **57**, 2220–2228 (1998).
94. B.V. Viani, L.I. Pietrasanta, J.B. Thompson, A. Chand, I.C. Gebeshuber, J.H. Kindt, M. Richter, H.G. Hansma, P.K. Hansma, Probing protein-protein interactions in real time. Nat. Struct. Biol. **7**, 644–647 (2000).
95. A. Ebner, F. Kienberger, G. Kada, C.M. Stroh, M. Geretschläger, A.S.M. Kamruzzahan, L. Wildling, W.T. Johnson, B. Ashcroft, J. Nelson, S.M. Lindsay, H.J. Gruber, P. Hinterdorfer, Localization of single avidin biotin interactions using simultaneous topography and molecular recognition imaging. ChemPhysChem **6**, 897–900 (2005).
96. W. Han, S.M. Lindsay, T. Jing, A magnetically driven oscillating probe microscope for operation in liquid. Appl. Phys. Lett. **69**, 1–3 (1996).
97. C.M. Stroh, A. Ebner, M. Geretschläger, G. Freudenthaler, F. Kienberger, A.S.M. Kamruzzahan, S.J. Smith-Gill, H.J. Gruber, P. Hinterdorfer, Simultaneous topography and recognition imaging using force microscopy. Biophys. J. **87**, 1981–1990 (2004).
98. C. Stroh, H. Wang, R. Bash, B. Ashcroft, J. Nelson, H.J. Gruber, D. Lohr, S.M. Lindsay, P. Hinterdorfer, Single-molecule recognition imaging microscope. Proc. Natl. Acad. Sci. USA **101**, 12503–12507 (2004).
99. L. Chtcheglova, J. Waschke, L. Wildling, D. Drenckhahn, P. Hinterdorfer, Nano-scale dynamic recognition imaging on vascular endothelial cells. Biophys. J. **9**(3), L11–L13 (2007).

# Part II
# Nanomechanics

# Chapter 9
# Nanomechanical Properties of Solid Surfaces and Thin Films

**Adrian B. Mann**

**Abstract** Instrumentation for the testing of mechanical properties on the submicron scale has developed enormously in recent years. This has enabled the mechanical behavior of surfaces, thin films, and coatings to be studied with unprecedented accuracy. In this chapter, the various techniques available for studying nanomechanical properties are reviewed with particular emphasis on nanoindentation. The standard methods for analyzing the raw data obtained using these techniques are described, along with the main sources of error. These include residual stresses, environmental effects, elastic anisotropy, and substrate effects. The methods that have been developed for extracting thin-film mechanical properties from the often convoluted mix of film and substrate properties measured by nanoindentation are discussed. Interpreting the data is frequently difficult, as residual stresses can modify the contact geometry and, hence, invalidate the standard analysis routines. Work hardening in the deformed region can also result in variations in mechanical behavior with indentation depth. A further unavoidable complication stems from the ratio of film to substrate mechanical properties and the depth of indentation in comparison to film thickness. Even very shallow indentations may be influenced by substrate properties if the film is hard and very elastic but the substrate is compliant. Under these circumstances nonstandard methods of analysis must be used. For multilayered systems many different mechanisms affect the nanomechanical behavior, including Orowan strengthening, Hall–Petch behavior, image force effects, coherency and thermal stresses, and composition modulation.

The application of nanoindentation to the study of phase transformations in semiconductors, fracture in brittle materials, and mechanical properties in biological materials are described. Recent developments such as the testing of viscoelasticity using nanoindentation methods are likely to be particularly important in future studies of polymers and biological materials. The importance of using a range of complementary methods such as electron microscopy, in situ AFM imaging, acoustic monitoring, and electrical contact measurements is emphasized. These are especially important on the nanoscale because so many different physical and chemical processes can affect the measured mechanical properties.

B. Bhushan (ed.), *Nanotribology and Nanomechanics*,
DOI 10.1007/978-3-642-15283-2_9, 

## 9.1 Introduction

When two bodies come into contact their surfaces experience the first and usually largest mechanical loads. Hence, characterizing and understanding the mechanical properties of surfaces is of paramount importance in a wide range of engineering applications. Obvious examples of where surface mechanical properties are important are in wear-resistant coatings on reciprocating surfaces and hard coatings for machine tool bits. This chapter details the current methods for measuring the mechanical properties of surfaces and highlights some of the key experimental results that have been obtained.

The experimental technique that is highlighted in this chapter is nanoindentation. This is for the simple reason that it is now recognized as the preferred method for testing thin film and surface mechanical properties. Despite this recognition, there are still many pitfalls for the unwary researcher when performing nanoindentation tests. The commercial instruments that are currently available all have attractive, user-friendly software, which makes the performance and analysis of nanoindentation tests easy. Hidden within the software, however, are a myriad of assumptions regarding the tests that are being performed and the material that is being examined. Unless the researcher is aware of these, there is a real danger that the results obtained will say more about the analysis routines than they do about the material being tested.

## 9.2 Instrumentation

The instruments used to examine nanomechanical properties of surfaces and thin films can be split into those based on point probes and those complimentary methods that can be used separately or in conjunction with point probes. The complimentary methods include a wide variety of techniques ranging from optical tests such as micro-Raman spectroscopy to high-energy diffraction studies using X-rays, neutrons, or electrons to mechanical tests such as bulge or blister testing.

Point-probe methods have developed from two historically different methodologies, namely, scanning probe microscopy [1] and microindentation [2]. The two converge at a length scale between 10–1,000 nm. Point-probe mechanical tests in this range are often referred to as nanoindentation.

### 9.2.1 *AFM and Scanning Probe Microscopy*

Atomic force microscopy (AFM) and other scanning probe microscopies are covered in detail elsewhere in this volume, but it is worth briefly highlighting the main features in order to demonstrate the similarities to nanoindentation. There are now a myriad of different variants on the basic scanning probe microscope. All use piezoelectric stacks to move either a probe tip or the sample with subnanometer precision in the lateral and vertical planes. The probe itself can be as simple as a tungsten wire electrochemically polished to give a single atom at the tip, or as complex as an AFM tip that is bio-active with, for instance, antigens attached. A range of scanning probes have been developed with the intention of measuring specific physical properties such as magnetism and heat capacity.

To measure mechanical properties with an AFM, the standard configuration is a hard probe tip (such as silicon nitride or diamond) mounted on a cantilever (see Fig. 9.1). The elastic deflection of the cantilever is monitored either directly or via a feedback mechanism to measure the forces acting on the probe. In general, the forces experienced by the probe tip split into attractive or repulsive forces. As the tip approaches the surface, it experiences intermolecular forces that are attractive, although they can be repulsive under certain circumstances [3]. Once in contact with the surface the tip usually experiences a combination of attractive intermolecular forces and repulsive elastic forces. Two schools of thought exist regarding the attractive forces when the tip is in contact with the surface. The first is often referred to as the DMT or Bradley model. It holds that attractive forces only act outside the region of contact [4–7]. The second theory, usually called the JKR model, assumes that all the forces experienced by the tip, whether attractive or repulsive, act in the region of contact [8]. Most real nanoscale contacts lie somewhere between these two theoretical extremes.

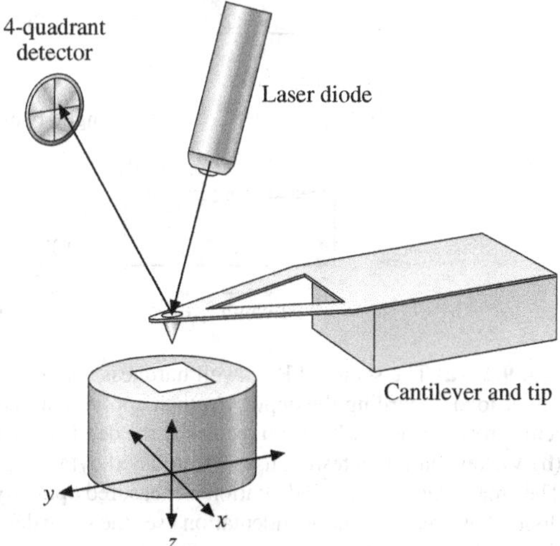


**Fig. 9.1** Diagram of a commercial AFM. The AFM tip is mounted on a compliant cantilever, and a laser light is reflected off the back of the cantilever onto a position-sensitive detector (four-quadrant detector). Any movements of the cantilever beam cause a deflection of the laser light that the detector senses. The sample is moved using piezo-electric stack, and forces are calculated from the cantilever's stiffness and the measured deflection

### 9.2.2 Nanoindentation

The fundamental difference between AFM and nanoindentation is that during a nanoindentation experiment an external load is applied to the indenter tip. This load enables the tip to be pushed into the sample, creating a nanoscale impression on the surface, otherwise referred to as a nanoindentation or nanoindent.

Conventional indentation or microindentation tests involve pushing a hard tip of known geometry into the sample surface using a fixed peak load. The area of indentation that is created is then measured, and the mechanical properties of the sample, in particular its hardness, is calculated from the peak load and the indentation area. Various types of indentation testing are used in measuring hardness, including Rockwell, Vickers, and Knoop tests. The geometries and definitions of hardness used in these tests are shown by Fig. 9.2.

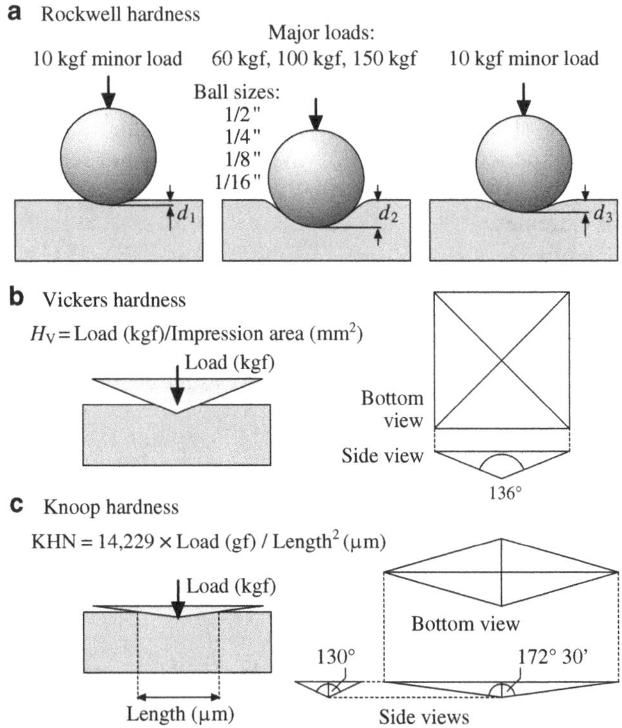


**Fig. 9.2** (**a**) The standard Rockwell hardness test involves pushing a ball into the sample with a minor load, recording the depth, $d_1$, then applying a major load and recording the depth, $d_2$, then returning to the lower load and recording the depth, $d_3$. Using the depths, the hardness is calculated. (**b**) Vickers hardness testing uses a four-sided pyramid pushed into the sample with a known load. The area of the resulting indentation is measured optically and the hardness calculated as the load is divided by area. (**c**) Knoop indentation uses the same definition of hardness as the Vickers test, load divided by area, but the indenter geometry has one long diagonal and one short diagonal

When indentations are performed on the nanoscale there is a basic problem in measuring the size of the indents. Standard optical techniques cannot easily be used to image anything smaller than a micron, while electron microscopy is simply impractical due to the time involved in finding and imaging small indents. To overcome these difficulties, nanoindentation methods have been developed that continuously record the load, displacement, time, and contact stiffness throughout the indentation process. This type of continuously recording indentation testing was originally developed in the former Soviet Union [9–13] as an extension of micro-indentation tests. It was applied to nanoscale indentation testing in the early 1980s [14, 15], hence, giving rise to the field of nanoindentation testing.

In general, nanoindentation instruments include a loading system that may be electrostatic, electromagnetic, or mechanical, along with a displacement measuring system that may be capacitive or optical. Schematics of several commercial nanoindentation instruments are shown in Fig. 9.3a–c.

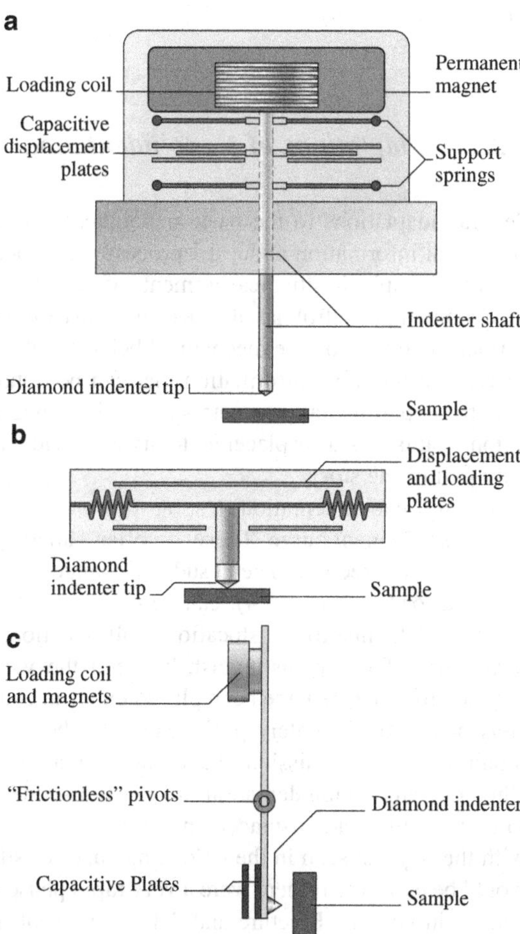


**Fig. 9.3** Schematics of three commercial nanoindentation devices made by (**a**) MTS Nanoinstruments, Oak Ridge, Tennessee, (**b**) Hysitron Inc., Minneapolis, Minnesota, (**c**) Micro Materials Limited, Wrexham, UK. Instruments and use electromagnetic loading, while uses electrostatic loads

Among the many advantages of nanoindentation over conventional microindentation testing is the ability to measure the elastic, as well as the plastic properties of the test sample. The elastic modulus is obtained from the contact stiffness ($S$) using the following equation that appears to be valid for all elastic contacts [16, 17]:

$$S = \frac{2}{\sqrt{\pi}} E_r \sqrt{A} . \quad (9.1)$$

$A$ is the contact area and $E_r$ is the reduced modulus of the tip and sample as given by:

$$\frac{1}{E_r} = \frac{\left(1 - \nu_t^2\right)}{E_t} + \frac{\left(1 - \nu_s^2\right)}{E_s} , \quad (9.2)$$

where $E_t$, $\nu_t$ and $E_s$, $\nu_s$ are the elastic modulus and Poissons ratio of the tip and sample, respectively.

### 9.2.3 Adaptations of Nanoindentation

Several adaptations to the basic nanoindentation setup have been used to obtain additional information about the processes that occur during nanoindentation testing, for example, in situ measurements of acoustic emissions and contact resistance. Environmental control has also been used to examine the effects of temperature and surface chemistry on the mechanical behavior of nanocontacts. In general, it is fair to say that the more information that can be obtained and the greater the control over the experimental parameters the easier it will be to understand the nanoindentation results. Load-displacement curves provide a lot of information, but they are only part of the story.

During nanoindentation testing discontinuities are frequently seen in the load–displacement curve. These are often called “pop-ins” or “pop-outs”, depending on their direction. These sudden changes in the indenter displacement, at a constant load (see Fig. 9.4), can be caused by a wide range of events, including fracture, delamination, dislocation multiplication, or nucleation and phase transformations. To help distinguish between the various sources of discontinuities, acoustic transducers have been placed either in contact with the sample or immediately behind the indenter tip. For example, the results of nanoindentation tests that monitor acoustic emissions have shown that the phase transformations seen in silicon during nanoindentation are not the sudden events that they would appear to be from the load–displacement curve. There is no acoustic emission associated with the pop-out seen in the unloading curve of silicon [18]. An acoustic emission would be expected if there were a very rapid phase transformation causing a sudden change in volume. Fracture and delamination of films, however, give very strong

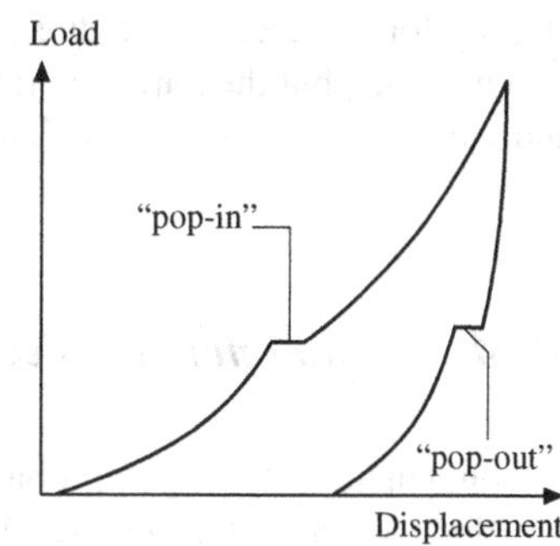


**Fig. 9.4** Sketch of a load/displacement curve showing a pop-in and a pop-out

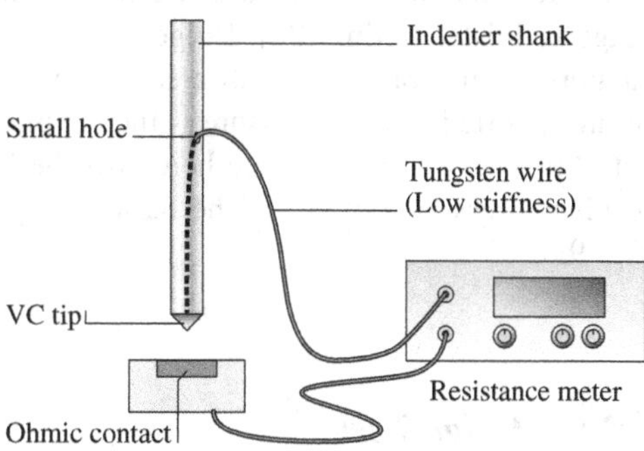


**Fig. 9.5** Schematic of the basic setup for making contact resistance measurements during nanoindentation testing

acoustic signals [19], but the exact form of the signal appears to be more closely related to the sample geometry than to the event [20].

Additional information about the nature of the deformed region under the nanoindentation can be obtained by performing in situ measurements of contact resistance. The basic setup for this type of testing is shown in Fig. 9.5. An electrically conductive tip is needed to study contact resistance. Consequently, a conventional diamond tip is of limited use. Elastic, hard, and metallically conductive materials such as vanadium carbide can be used as substitutes for diamond [21, 22], or a thin conductive film (e.g., Ag) can be deposited on the diamond's surface (such a film is easily transferred to the indented surface so great care must be taken if multiple indents are performed). Measurements of contact resistance have been most useful for examining phase transformations in semiconductors [21, 22] and the dielectric breakdown of oxide films under mechanical loading [23].

One factor that is all too frequently neglected during nanoindentation testing is the effect of the experimental environment. Two obvious ways in which the environment can affect the results of nanoindentation tests are increases in temperature, which give elevated creep rates, and condensation of water vapor, which modifies the tip-sample interactions. Both of these environmental effects have been shown to significantly affect the measured mechanical properties and the modes of deformation that occur during nanoindentation [24–27]. Other environmental effects, for instance, those due to photoplasticity or hydrogen ion absorption, are also possible, but they are generally less troublesome than temperature fluctuations and variations in atmospheric humidity.

### 9.2.4 Complimentary Techniques

Nanoindentation testing is probably the most important technique for characterizing the mechanical properties of thin films and surfaces, but there are many alternative or additional techniques that can be used. One of the most important alternative methods for measuring the mechanical properties of thin films uses bulge or blister testing [28]. Bulge tests are performed on thin films mounted on supporting substrates. A small area of the substrate is removed to give a window of unsupported film. A pressure is then applied to one side of the window causing it to bulge. By measuring the height of the bulge, the stress-strain curve and the residual stress are obtained. The basic configuration for bulge testing is shown in Fig. 9.6.

### 9.2.5 Bulge Tests

The original bulge tests used circular windows because they are easier to analyze mathematically, but now square and rectangular windows have become common [29]. These geometries tend to be easier to fabricate. Unfortunately, there are several sources of errors in bulge testing that can potentially lead to large errors in the measured mechanical properties. These errors at one time led to the belief that multilayer films can show a "super modulus" effect, where the elastic modulus

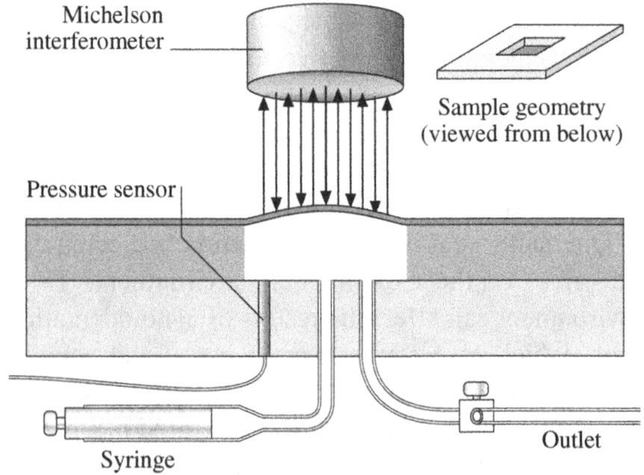


**Fig. 9.6** Schematic of the basic setup for bulge testing. The sample is prepared so that it is a thin membrane, and then a pressure is applied to the back of the membrane to make it bulge upwards. The height of the bulge is measured using an interferometer

of the multilayer is several times that of its constituent layers [30]. It is now accepted that any enhancement to the elastic modulus in multilayer films is small, on the order of 15% [31]. The main sources of error stem from compressive stresses in the film (tensile stress is not a problem), small variations in the dimensions of the window, and uncertainty in the exact height of the bulge. Despite these difficulties, one advantage of bulge testing over nanoindentation testing is that the stress state is biaxial, so that only properties in the plane of the film are measured. In contrast, nanoindentation testing measures a combination of in-plane and out-of-plane properties.

### 9.2.6 Acoustic Methods

Acoustic and ultrasonic techniques have been used for many years to study the elastic properties of materials. Essentially, these techniques take advantage of the fact that the velocity of sound in a material is dependent on the inter-atomic or inter-molecular forces in the material. These, of course, are directly related to the material's elastic constants. In fact, any nonlinearity of inter-atomic forces enables slight variations in acoustic signals to be used as a measure of residual stress.

An acoustic method ideally suited to studying surfaces is scanning acoustic microscopy (SAM) [32]. There are also several other techniques that have been used to study surface films and multilayers, but we will first consider SAM in detail. In a SAM, a lens made of sapphire is used to bring acoustic waves to focus via a coupling fluid on the surface. A small piezoelectric transducer at the top of the lens generates the acoustic signal. The same transducer can be used to detect the signal when the SAM is used in reflection mode. The use of a transducer as both generator and detector, a common imaging mode, necessitates the use of a pulsed rather than a continuous acoustic signal. Continuous waves can be used if phase changes are used to build up the image. The transducer lens generates two types of acoustic waves in the material: longitudinal and shear. The ability of a solid to sustain both types of wave (liquids can only sustain longitudinal waves) gives rise to a third type of acoustic wave called a Rayleigh, or surface, wave. These waves are generated as a result of superposition of the shear and longitudinal waves with a common phase velocity. The stresses and displacements associated with a Rayleigh wave are only of significance to a depth of $\approx 0.6$ Rayleigh wavelengths below the solid surface. Hence, using SAM to examine Rayleigh waves in a material is a true surface characterization technique.

Using a SAM in reflection mode gives an image where the contrast is directly related to the Rayleigh wave velocity, which is in turn a function of the material's elastic constants. The resolution of the image depends on the frequency of the transducer used, i.e., for a 2 GHz signal a resolution better than 1 μm is achievable. The contrast in the image results from the interference of two different waves in the coupling fluid. Rayleigh waves that are excited in the surface "leak" into the coupling fluid and interfere with the acoustic signal that is directly reflected back from the surface. It is usually assumed that the properties of the coupling fluid are well characterized. The interference of the two waves gives a characteristic $V(z)$ curve, as illustrated by Fig. 9.7, where $z$ is the separation between the lens and the surface. Analyzing the periodicity of the $V(z)$ curves provides information on the Rayleigh wave velocity. As with other acoustic waves, the Rayleigh velocity is related to the elastic constants of the material. When using the SAM for a material's characterization, the lens is usually held in a fixed position on the surface. By using a lens designed specifically to give a line-focus beam, rather than the standard spherical lens, it is possible to use SAM to look at anisotropy in the wave velocity [33] and hence in elastic properties by producing waves with a specific direction.

One advantage of using SAM in conjunction with nanoindentation to characterize a surface is that the measurements obtained with the two methods have a slightly different dependence on the test material's elastic properties, $E_s$ and $v_s$ (the elastic modulus and Poisson's ratio). As a result, it is possible to use SAM and nanoindentation combined to find both $E_s$ and $v_s$, as illustrated by Fig. 9.8 [34]. This is not possible when using only one of the techniques alone.

In addition to measuring surface properties, SAM has been used to study thin films on a surface. However, the Rayleigh wave velocity can be dependent on a complex mix of the film and substrate properties. Other acoustic methods have been utilized to study freestanding films. A freestanding film can be regarded as a plate, and, therefore, it is possible to excite Lamb waves in the film. Using a pulsed laser to generate the waves and a heterodyne interferometer to detect the arrival of the Lamb wave, it is possible to measure the flexural modulus of the film [35]. This has been successfully demonstrated for multilayer films with a total thickness $<10$ μm. In the plate configuration, due to the nonlinearity of elastic properties, it is also possible to measure stress. This has been demonstrated for horizontally polarized shear waves in plates [36], but thin plates require very high frequency transducers or laser sources.

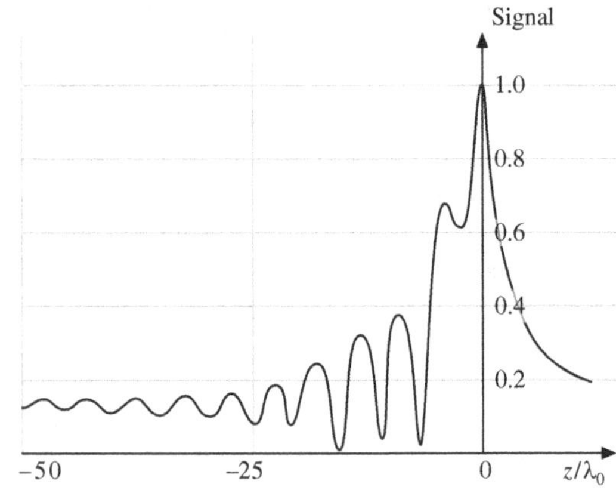


**Fig. 9.7** A typical $V(z)$ curve obtained with a SAM when testing fused silica

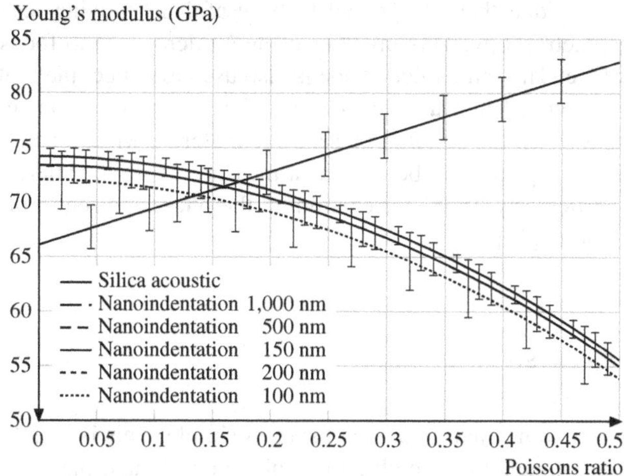


**Fig. 9.8** Because SAM and nanoindentation have different dependencies on Young's modulus, $E$, and Poissons ratio, $\nu$, it is possible to use the two techniques in combination to find $E$ and $\nu$ [34]. On the graph, the intersection of the curves gives $E$ and $\nu$

## 9.2.7 Imaging Methods

When measuring the mechanical properties of a surface or thin film using nanoindentation, it is not always easy to visualize what is happening. In many instances there is a risk that the mechanical data can be completely misinterpreted if the geometry of the test is not as expected. To expedite the correct interpretation of the mechanical data, it is generally worthwhile to use optical, electron, or atomic force microscopy to image the nanoindentations. Obviously, optical techniques are only of use for larger indentations, but they will often reveal the presence of median or lateral cracks [37]. Electron microscopy and AFM, however, can be used to examine even the smallest nanoindentations. The principle problem with these microscopy techniques is the difficulty in finding the nanoindentations. It is usually necessary to make large, "marker" indentations in the vicinity of the nanoindentations to be examined in order to find them [38].

It is possible to see features such as extrusions with a scanning electron microscope (SEM) [39], as well as pile-up and sink-in around the nanoindents, though AFM is generally better for this. Transmission electron microscopy (TEM) is useful for examining what has happened subsurface, for instance, the indentation induced dislocations in a metal [40] or the phases present under a nanoindent in silicon [21]. However, with TEM there is the added difficulty of sample preparation and the associated risk of observing artifacts. Recently, there has been considerable interest in the use of focused ion beams to cut cross sections through nanoindents [41]. When used in conjunction with SEM or TEM this provides an excellent means to see what has happened in the subsurface region.

One other technique that has proved to be useful in studying nanoindents is micro-Raman spectroscopy. This involves using a microscope to focus a laser on the sample surface. The same microscope is also used to collect the scattered laser light, which is then fed into a spectroscope. The Raman peaks in the spectrum provide information on the bonding present in a material, while small shifts in the wave number of the peaks can be used as a measure of strain. Micro-Raman has proven to be particularly useful for examining the phases present around nanoindentations in silicon [42].

## 9.3 Data Analysis

The analysis of nanoindentation data is far from simple. This is mostly due to the lack of effective models that are able to combine elastic and plastic deformation under a contact. However, provided certain precautions are taken, the models for perfectly elastic deformation and ideal plastic materials can be used in the analysis of nanoindentation data. For this reason, it is worth briefly reviewing the models for perfect contacts.

### *9.3.1 Elastic Contacts*

The theoretical modeling of elastic contacts can be traced back many years, at least to the late nineteenth century and the work of Hertz (1882) [43] and Boussinesq (1885) [44]. These models, which are still widely used today, consider two axisymmetric curved surfaces in contact over an elliptical region (see Fig. 9.9). The contact

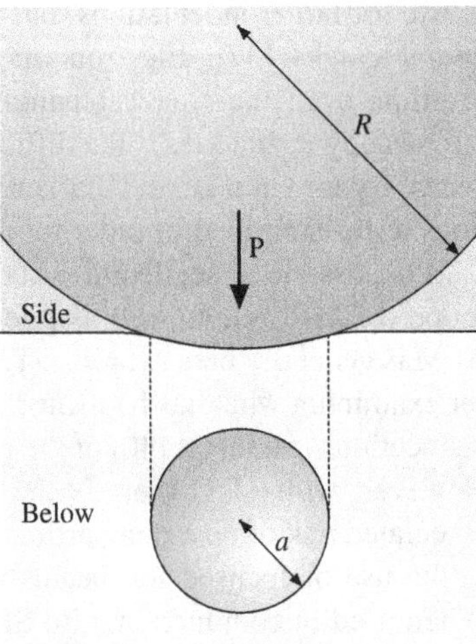


**Fig. 9.9** Hertzian contact of a sphere, radius $R$, on a semi-infinite, flat surface. The contact in this case is a circular region of radius $a$

region is taken to be small in comparison to the radius of curvature of the contacting surfaces, which are treated as elastic half-spaces. For an elastic sphere, radius $R$, in contact with a flat, elastic half-space, the contact region will be circular and the Hertz model gives the following relationships:

$$a = \sqrt[3]{\frac{3PR}{4E_r}}, \tag{9.3}$$

$$\delta = \sqrt[3]{\frac{9P^2}{16RE_r^2}}, \tag{9.4}$$

$$P_0 = \sqrt[3]{\frac{6PE_r^2}{\pi^3 R^2}}, \tag{9.5}$$

where $a$ is the radius of the contact region, $E_r$ is given by (9.2), $\delta$ is the displacement of the sphere into the surface, $P$ is the applied load and $P_0$ is the maximum pressure under the contact (in this case at the center of the contact).

The work of Hertz and Boussinesq was extended by Love [45, 46] and later by Sneddon [47], who simplified the analysis using Hankel transforms. Love showed how Boussinesq's model could be used for a flat-ended cylinder and a conical indenter, while Sneddon produced a generalized relationship for any rigid axisymmetric punch pushed into an elastic half-space. Sneddon applied his new analysis to punches of various shapes and derived the following relationships between the applied load, $P$, and displacement, $\delta$, into the elastic half-space for, respectively, a flat-ended cylinder, a cone of semi-vertical angle $\phi$, and a parabola of revolution where $a^2 = 2k\delta$:

$$P = \frac{4\mu a\delta}{1-v}, \tag{9.6}$$

$$P = \frac{4\mu \cot\phi}{\pi(1-v)}\delta^2, \tag{9.7}$$

$$P = \frac{8\mu}{3(1-v)}\left(2k\delta^3\right)^{1/2}, \tag{9.8}$$

where $\mu$ and $v$ are the shear modulus and Poisson's ratio of the elastic half-space, respectively.

The key point to note about (9.6–9.8) is that they all have the same basic form, namely:

$$P = \alpha\delta^m, \tag{9.9}$$

where $\alpha$ and $m$ are constants for each geometry.

Equation (9.9) and the relationships developed by Hertz and his successors, (9.3–9.8), form the foundation for much of the current nanoindentation data analysis routines.

### 9.3.2 Indentation of Ideal Plastic Materials

Plastic deformation during indentation testing is not easy to model. However, the indentation response of ideal plastic metals was considered by Tabor in his classic text, "The Hardness of Metals" [48]. An ideal plastic material (or more accurately an ideal elastic-plastic material) has a linear stress-strain curve until it reaches its elastic limit and then yields plastically at a yield stress,$Y_0$, that remains constant even after deformation has commenced. In a 2-D problem, the yielding occurs because the Huber-Mises [49] criterion has been reached. In other words, the maximum shear stress acting on the material is around $1.15Y_0/2$.

First, we consider a 2-D flat punch pushed into an ideal plastic material. By using the method of slip lines it is found that the mean pressure, $P_m$, across the end of the punch is related to the yield stress by:

$$P_m = 3Y_0\,. \tag{9.10}$$

If the Tresca criterion [50] is used, then $P_m$ is closer to $2.6Y_0$. In general, for both 2-D and three-dimensional punches pushed into ideal plastic materials, full plasticity across the entire contact region can be expected when $P_m = 2.6$ to $3.0Y_0$. However, significant deviations from this range can be seen if, for instance, the material undergoes work-hardening during indentation, or the material is a ceramic, or there is friction between the indenter and the surface.

The apparently straightforward relationship between $P_m$ and $Y_0$ makes the mean pressure a very useful quantity to measure. In fact, $P_m$ is very similar to the Vickers hardness, $H_V$, of a material:

$$H_V = 0.927P_m\,. \tag{9.11}$$

During nanoindentation testing it is the convention to take the mean pressure as the nanohardness. Thus, the "nanohardness", $H$, is defined as the peak load, $P$, applied during a nanoindentation divided by the projected area, $A$, of the nanoindentation in the plane of the surface, hence:

$$H = \frac{P}{A}\,. \tag{9.12}$$

### 9.3.3 Adhesive Contacts

During microindentation testing and even most nanoindentation testing the effects of intermolecular and surface forces can be neglected. Very small nanoindentations, however, can be influenced by the effects of intermolecular forces between the sample and the tip. These adhesive effects are most readily seen when testing soft polymers, but there is some evidence that forces between the tip and sample may be important in even relatively strong materials [51, 52].

Contact adhesion is usually described by either the JKR or DMT model, as discussed earlier in this chapter. Both the models consider totally elastic spherical contacts under the influence of attractive surface forces. The JKR model considers the surface forces in terms of the associated surface energy, whereas the DMT model considers the effects of adding van der Waals forces to the Hertzian contact model. The differences between the two models are illustrated by Fig. 9.10.

For nanoindentation tests conducted in air the condensation of water vapor at the tip-sample interface usually determines the size of the adhesive force acting during unloading. The effects of water vapor on a single nanoasperity contact have been studied using force-controlled AFM techniques [53] and, more recently, nanoindentation methods [26]. Unsurprisingly, it has also been found that water vapor can affect the deformation of surfaces during nanoindentation testing [27].

In addition to water vapor, other surface adsorbates can cause dramatic changes in the nanoscale mechanical behavior. For instance, oxygen on a clean metal surface can cause an increase in the apparent strength of the metal [54]. These effects are likely to be related to, firstly, changes in the surface and intermolecular forces acting between the tip and the sample and, secondly, changes in the

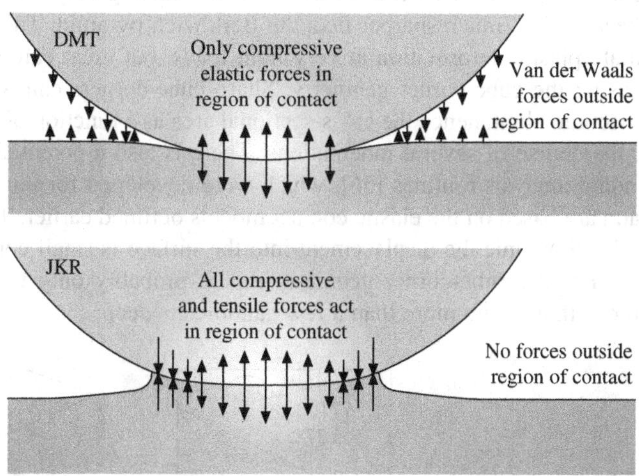


**Fig. 9.10** The contact geometry for the DMT and JKR models for adhesive contact. Both models are based on the Hertzian model. In the DMT, model van der Waals forces outside the region of contact introduce an additional load in the Hertz model. But for the JKR model, it is assumed that tensile, as well as compressive stresses can be sustained within the region of contact

mechanical stability of surface nanoasperities and ledges. Adsorbates can help stabilize atomic-scale variations in surface morphology, thereby making defect generation at the surface more difficult.

### 9.3.4 Indenter Geometry

All of the indenter geometries considered up to this point have been axisymmetric, largely because they are easier to deal with theoretically. Unfortunately, fabricating axisymmetric nanoindentation tips is extremely difficult, because shaping a hard tip on the scale of a few nanometers is virtually impossible. Despite these problems, there has been considerable effort put into the use of spherical nanoindentation tips [55]. This clearly demonstrates that the spherical geometry can be useful at larger indentation depths.

Because of the problems associated with creating axisymmetric nanoindentation tips, pyramidal indenter geometries have now become standard during nanoindentation testing. The most common geometries are the three-sided Berkovich pyramid and cube-corner (see Fig. 9.11). The Berkovich pyramid is based on the four-sided Vickers pyramid, the opposite sides of which make an 136° angle. For both the Vickers and Berkovich pyramids the cross-sectional area of the pyramid's base, $A$, is related to the pyramid's height, $D$, by:

$$A = 24.5D^2. \tag{9.13}$$

The cube-corner geometry is now widely used for making very small nanoindentations, because it is much sharper than the Berkovich pyramid. This makes it easier to initiate plastic deformation at very light loads, but great care should be taken when using the cube-corner geometry. Sharp cube-corners can wear down quickly and become blunt, hence the cross-sectional area as a function of depth can change over the course of several indentations. There is also a potential problem with the standard analysis routines [56], which were developed for much blunter geometries and are based on the elastic contact models outlined earlier. The elastic contact models all assume the displacement into the surface is small compared to the tip radius. For the cube-corner geometry this is probably only the case for nanoindentations that are no more than a few nanometers deep.

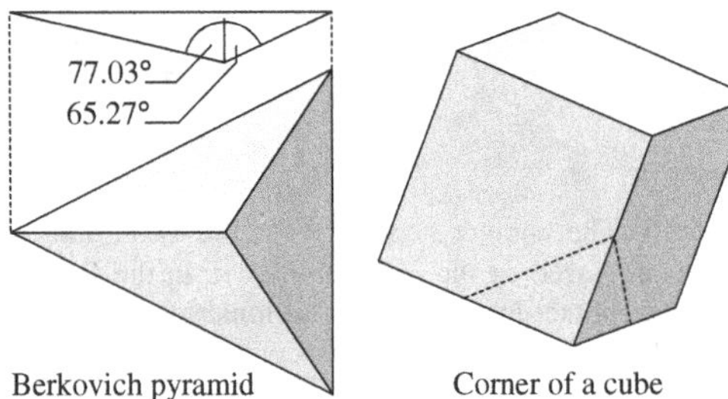


**Fig. 9.11** The ideal geometry for the three-sided Berkovich pyramid and cube corner tips

### 9.3.5 Analyzing Load/Displacement Curves

The load/displacement curves obtained during nanoindentation testing are deceptively simple. Most newcomers to the area will see the curves as being somewhat akin to the stress/strain curves obtained during tensile testing. There is also a real temptation just to use the values of hardness, $H$, and elastic modulus, $E$, obtained from standard analysis software packages as the "true" values. This may be the case in many instances, but for very shallow nanoindents and tests on thin films the geometry of the contact can differ significantly from the geometry assumed in the analysis routines. Consequently, experimentalists should think very carefully about the test itself before concluding that the values of $H$ and $E$ are correct.

The basic shape of a load/displacement curve can reveal a great deal about the type of material being tested. Figure 9.12 shows some examples of ideal curves for materials with different elastic moduli and yield stresses. Discontinuities in the load/displacement curve can also provide information on such processes as fracture, dislocation nucleation, and phase transformations. Initially, though, we will consider ideal situations such as those illustrated by Fig. 9.12.

The loading section of the load/displacement curve approximates a parabola [57] whose width depends on a combination of the material's elastic and plastic properties. The unloading curve, however, has been shown to follow a more general relationship [56] of the form:

$$P = \alpha(\delta - \delta_i)^m, \tag{9.14}$$

where $\delta$ is the total displacement and $\delta_i$ is the intercept of the unloading curve with the displacement axis shown in Fig. 9.13.

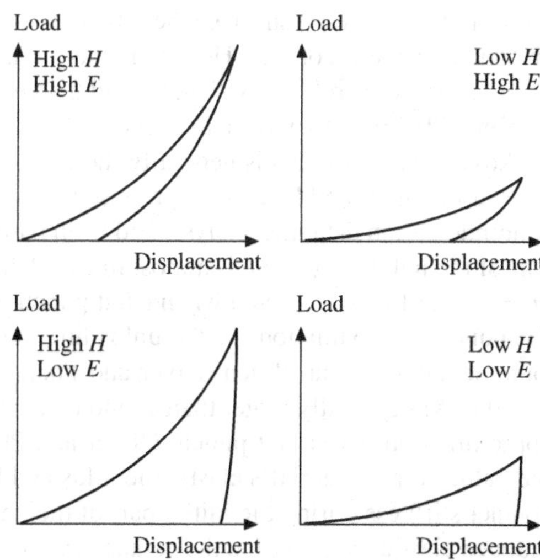


**Fig. 9.12** Examples of load/displacement curves for idealized materials with a range of hardness and elastic properties

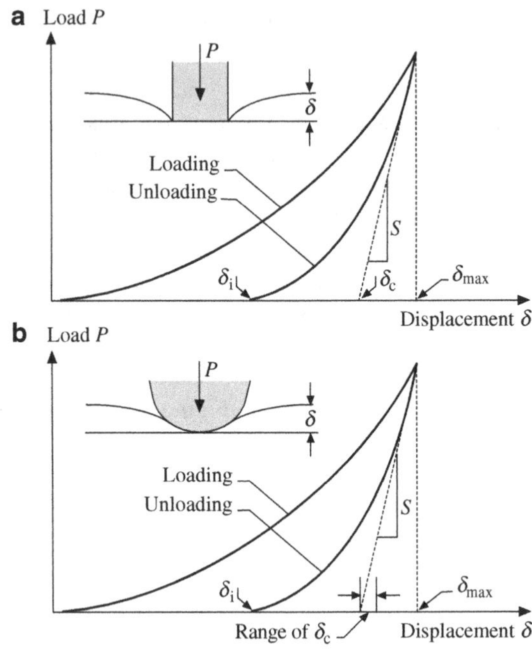


**Fig. 9.13** Analysis of the load/displacement curve gives the contact stiffness, $S$, and the contact depth, $\delta_c$. These can then be used to find the hardness, $H$, and elastic, or Young's modulus, $E$. (**a**) The first method of analysis [58–60] assumed the unloading curve could be approximated by a flat punch on an elastic half-space. (**b**) A more refined analysis [56] uses a paraboloid on an elastic half-space

Equation (9.14) is essentially the same as (9.9) but with the origin displaced. Since (9.9) is obtained by considering purely elastic deformation, it follows that the unloading curve is exhibiting purely elastic behavior. Since the shape of the unloading curve is determined by the elastic recovery of the indented region, it is not entirely surprising that its shape resembles that found for purely elastic deformation. What is fortuitous is that the elastic analysis used for an elastic half-space seems to be valid for a surface where there is a plastically formed indentation crater present under the contact. However, the validity of this analysis may only hold when the crater is relatively shallow and the geometry of the surface does not differ significantly from that of a flat, elastic half-space. For nanoindentations with a Berkovich pyramid, this is generally the case.

Before Oliver and Pharr [56] proposed their now standard method for analyzing nanoindentation data, the analysis had been based on the observation that the initial part of the unloading curve is almost linear. A linear unloading curve, equivalent to $m = 1$ in (9.14), is expected when a flat punch is used on an elastic half-space. The flat punch approximation for the unloading curve was used in [58–60] to analyze nanoindentation data. When Oliver and Pharr looked at a range of materials they found $m$ was typically larger than 1, and that $m = 1.5$, or a paraboloid, was a better approximation than a flat punch. Oliver and Pharr used (9.1) and (9.12) to obtain the values for a material's elastic modulus and hardness. Equation (9.1) relates the contact stiffness during the initial part of the unloading curve (see Fig. 9.13) to the reduced elastic modulus and the contact area at the peak load. Equation (9.12) gives the hardness as the peak load divided by the contact area. It is immediately obvious that the key to measuring the mechanical properties of a material is knowing the contact area at the peak load. This is the single most important factor in analyzing nanoindentation data. Most mistakes in the analysis come from incorrect assumptions about the contact area.

To find the contact area, a function relating the contact area, $A_c$, to the contact depth, $\delta_c$, is needed. For a perfect Berkovich pyramid this would be the same as (9.13). But since making a perfect nanoindenter tip is impossible, an expanded equation is used:

$$A_c(\delta_c) = 24.5\delta_c^2 + \sum_{j=1}^{7} C_j \sqrt[2^j]{\delta_c}\,, \tag{9.15}$$

where $C_j$ are calibration constants of the tip.

There is a crucial step in the analysis before $A_c$ can be calculated, namely, finding $\delta_c$. The contact depth is not the same as the indentation depth, because the surface around the indentation will be elastically deflected during loading, as illustrated by Fig. 9.14. Sneddon's analysis [47] provides a way to calculate the deflection of the surface at the edge of an axisymmetric contact. Subtracting the deflection from the total indentation depth at peak load gives the contact depth. For a paraboloid, as used by Oliver and Pharr [56] in their analysis, the elastic deflection at the edge of the contact is given by:

$$\delta_s = \varepsilon\frac{P}{S} = 0.75\frac{P}{S}\,, \tag{9.16}$$

where $S$ is the contact stiffness and $P$ the peak load. The constant $\varepsilon$ is 0.75 for a paraboloid, but ranges between 0.72 (conic indenter) and 1 (flat punch). Figure 9.15 shows how the contact depth depends on the value of $\varepsilon$. The contact depth at the peak load is, therefore:

$$\delta_c = \delta - \delta_s\,. \tag{9.17}$$

Using the load/displacement data from the unloading curve and (9.1), (9.2), (9.12), (9.14–9.17), the hardness and reduced elastic modulus for the test sample can be calculated. To find the elastic modulus of the sample, $E_s$, it is also necessary to know Poisson's ratio, $\nu_s$, for the sample, as well as the elastic modulus, $E_t$, and

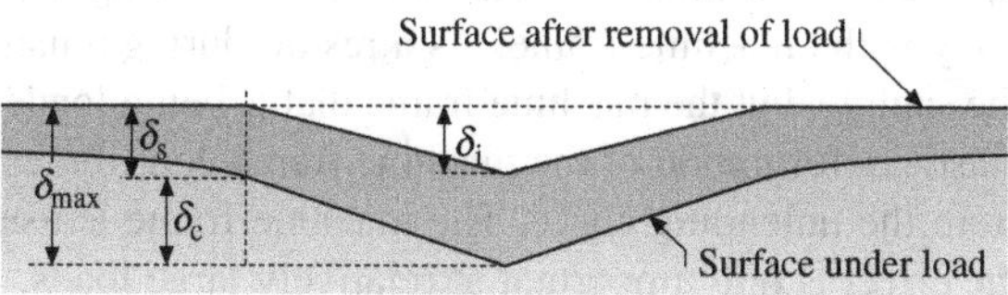


**Fig. 9.14** Profile of surface under load and unloaded showing how $\delta_c$ compares to $\delta_i$ and $\delta_{max}$

**Fig. 9.15** Load-displacement curve showing how $\delta_c$ varies with ε

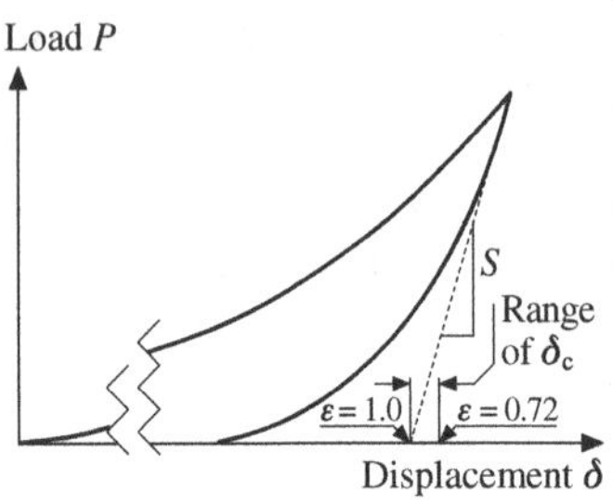


Poisson's ratio, $\nu_t$, of the indenter tip. For diamond these are 1,141 and 0.07 GPa, respectively.

There also remains the issue of calibrating the tip shape, or finding the values for $C_j$ in (9.15). Knowing the exact expansion of $A_c(\delta_c)$ is vital if the values for $E_s$ and $H$ are to be accurate. Several methods for calibrating the tip shape have been used, including imaging the tip with an electron microscope, measuring the size of nanoindentations using SEM or TEM of negative replicas, and using scanning probes to examine either the tip itself or the nanoindentations made with the tip. There are strengths and weaknesses to each of these methods. In general, however, the accuracy and usefulness of the methods depends largely on how patient and rigorous the experimentalist is in performing the calibration.

Because of the experimental difficulties and time involved in calibrating the tip shape by these methods, Oliver and Pharr [56] developed a method for calibration based on standard specimens. With a standard specimen that is mechanically isotropic and has a known $E$ and $H$ that does not vary with indentation depth, it should be possible to perform nanoindentations to a range of depths, and then use the analysis routines in reverse to deduce the tip area function, $A_c(\delta_c)$. In other words, if you perform a nanoindentation test, you can find the contact stiffness, $S$, at the peak load, $P$, and the contact depth, $\delta_c$, from the unloading curve. Then if you know $E$ a priori, (9.1) can be used to calculate the contact area, $A$, and, hence, you have a value for $A_c$ at a depth $\delta_c$. Repeating this procedure for a range of depths will give a numerical version of the function $A_c(\delta_c)$. Then, it is simply a case of fitting (9.15) to the numerical data. If the hardness, $H$, is known and not a function of depth, and the calibration specimen was fully plastic during testing, then essentially the same approach could be used but based on (9.12). Situations where a constant $H$ is used to calibrate the tip are extremely rare.

In addition to the tip shape function, the machine compliance must be calibrated. Basic Newtonian mechanics tells us that for the tip to be pushed into a surface the tip must be pushing off of another body. During nanoindentation testing the other body is the machine frame. As a result, during a nanoindentation test it is not just the sample, but the machine frame that is being loaded. Consequently, a very small elastic deformation of the machine frame contributes to the total stiffness obtained from the unloading curve. The machine frame is usually very stiff, $>10^6$ N/m, so the effect is only important at relatively large loads.

To calibrate the machine frame stiffness or compliance, large nanoindentations are made in a soft material such as aluminum with a known, isotropic elastic modulus. For very deep nanoindentations made with a Berkovich pyramid, the contact area, $A_c(\delta_c)$, can be reasonably approximated to $24.5\delta_c^2$, thus (9.1) can be used to find the expected contact stiffness for the material. Any difference between the expected value of $S$ and the value measured from the unloading curve will be due to the compliance of the machine frame. Performing a number of deep nanoindentations enables an accurate value for the machine frame compliance to be obtained.

Currently, because of its ready availability and predictable mechanical properties, the most popular calibration material is fused silica ($E = 72$ GPa, $\nu = 0.17$), though aluminum is still used occasionally.

### 9.3.6 Modifications to the Analysis

Since the development of the analysis routines in the early 1990s, it has become apparent that the standard analysis of nanoindentation data is not applicable in all situations, usually because errors occur in the calculated contact depth or contact area. Pharr et al. [61–64] have used finite element modeling (FEM) to help understand and overcome the limitations of the standard analysis. Two important sources of errors have been identified in this way. The first is residual stress at the sample surface. The second is the change in the shape of nanoindents after elastic recovery.

The effect of residual stresses at a surface on the indentation properties has been the subject of debate for many years [65–67]. The perceived effect was that compressive stresses increased hardness, while tensile stresses decreased hardness. Using FEM it is possible to model a pointed nanoindenter being pushed into a model material that is in residual tension or compression. An FEM model of nanoindentation into aluminum alloy 8009 [61] has confirmed earlier experimental observations [68] indicating that the contact area calculated from the unloading curve is incorrect if there are residual stresses. In the FEM model of an aluminum alloy the mechanical behavior of the material is modeled using a stress-strain curve, which resembles that of an elastic-perfectly-plastic metal with a flow stress of 425.6 MPa. Yielding starts at 353.1 MPa and includes a small amount of work hardening. The FEM model was used to find the contact area directly and using the simulated unloading curve in conjunction with Oliver and Pharr's method. The results as a function of residual stress are illustrated in Fig. 9.16. Note that the differences between the two measured contact areas lead to miscalculations of $E$ and $H$.

Errors in the calculated contact area stem from incorrect assumptions about the pile-up and sink-in at the edge of the contact, as illustrated by Fig. 9.17. The Oliver and Pharr analysis assumes the geometry of the sample surface is the same as that given by Sneddon [47] in his analytical model for the indentation of elastic surfaces. Clearly, for materials where there is significant plastic deformation, it is possible

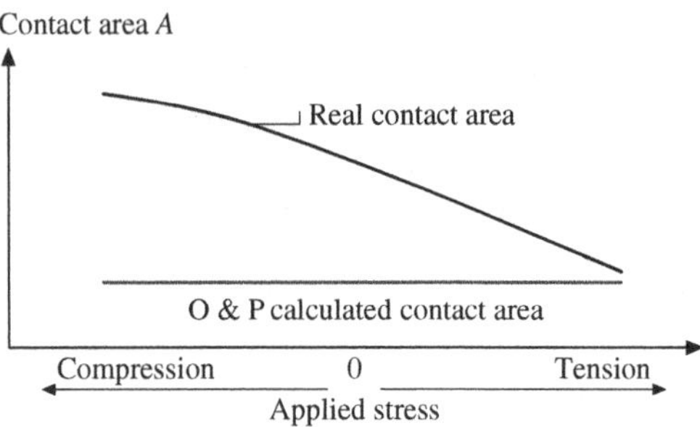


**Fig. 9.16** When a surface is in a state of stress there is a significant difference between the contact area calculated using the Oliver and Pharr method and the actual contact area [61]. For an aluminum alloy this can lead to significant errors in the calculated hardness and elastic modulus

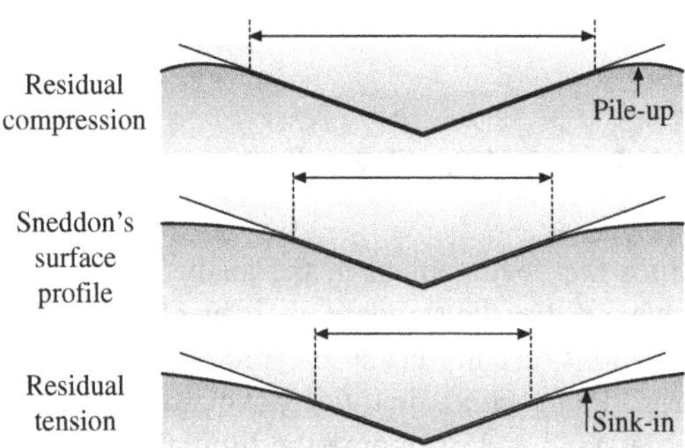


**Fig. 9.17** Pile-up and sink-in are affected by residual stresses, and, hence, errors are introduced into standard Oliver and Pharr analysis

that there will be large deviations from the surface geometry found using Sneddon's elastic model. In reality, the error in the contact area depends on how much the geometry of the test sample surface differs from that of the calibration material (typically fused silica). It is possible that a test sample, even without a residual stress, will have a different surface geometry and, hence, contact area at a given depth, when compared to the calibration material. This is often seen for thin films on a substrate (e.g., Tsui et al. [69, 70]). Residual stresses increase the likelihood that the contact area calculated using Oliver and Pharr's method will be incorrect.

The issue of sink-in and pile-up is always a factor in nanoindentation testing. However, there is still no effective way to deal with these phenomena other than reverting to imaging of the indentations to identify the true contact area. Even this is difficult, as the edge of an indentation is not easy to identify using AFM or electron microscopy. One approach that has been used [71] with some success is measuring the ratio $E_r^2/H$, rather than $E_r$ and $H$ separately. Because $E_r$ is proportional to $1/\sqrt{A}$ and $H$ is proportional to $1/A$, $E_r^2/H$ should be independent of $A$ and, hence, unaffected by pile-up or sink-in. While this does not provide quantitative values for mechanical properties, it does provide a way to identify any variations in mechanical properties with indentation depth or between similar samples with different residual stresses.

Another source of error in the Oliver and Pharr analysis is due to incorrect assumptions about the nanoindentation geometry after unloading [63]. Once again,

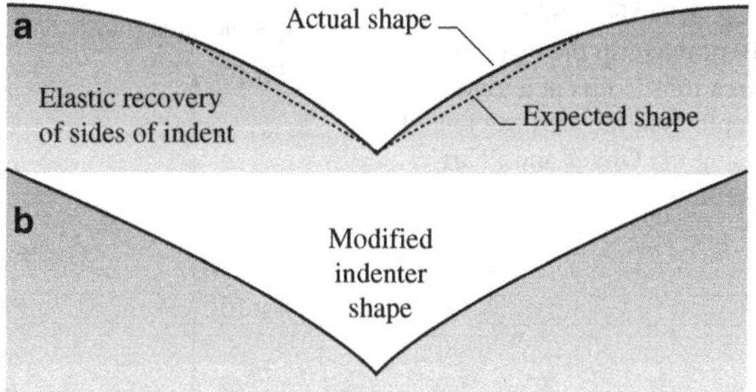


**Fig. 9.18** (**a**) Hay et al. [63] found from experiments and FEM simulations that the actual shape of an indentation after unloading is not as expected. (**b**) They introduced a $\gamma$ term to correct for this effect. This assumes the indenter has slightly concave sides

this is due to differences between the test sample and the calibration material. The exact shape of an unloaded nanoindentation on a material exhibiting elastic recovery is not simply an impression of the tip shape; rather, there is some elastic recovery of the nanoindentation sides giving them a slightly convex shape (see Fig. 9.18). The shape actually depends on Poisson's ratio, so the standard Oliver and Pharr analysis will only be valid for a material where $\nu = 0.17$, the value for fused silica, assuming it is used for the calibration.

To deal with the variations in the recovered nanoindentation shape, it has been suggested [63] that a modified nanoindenter geometry with a slightly concave side be used in the analysis (see Fig. 9.18). This requires a modification to (9.1):

$$S = \gamma 2E_{\mathrm{r}}\sqrt{\frac{A}{\pi}}, \tag{9.18}$$

where $\gamma$ is a correction term dependent on the tip geometry. For a Berkovich pyramid the best value is:

$$\gamma = \frac{\frac{\pi}{4} + 0.15483073\cot\Phi\left(\frac{(1-2\nu_{\mathrm{s}})}{4(1-\nu_{\mathrm{s}})}\right)}{\left[\frac{\pi}{2} - 0.83119312\cot\Phi\left(\frac{(1-2\nu_{\mathrm{s}})}{4(1-\nu_{\mathrm{s}})}\right)\right]^2}, \tag{9.19}$$

where $\Phi = 70.32°$. For a cube corner the correction can be even larger and $\gamma$ is given by:

$$\gamma = 1 + \left(\frac{(1-2\nu_{\mathrm{s}})}{4(1-\nu_{\mathrm{s}})\tan\Phi}\right), \tag{9.20}$$

where $\Phi = 42.28°$. Figure 9.19 shows how the modified contact area varies with depth for a real diamond Berkovich pyramid.

The validity of the $\gamma$-modified geometry is questionable from the perspective of contact mechanics since it relies on assuming an incorrect geometry for the nanoindenter tip to correct for an error in the geometry of the nanoindentation impression. The values for $E$ and $H$ obtained using the $\gamma$-modification are, however,

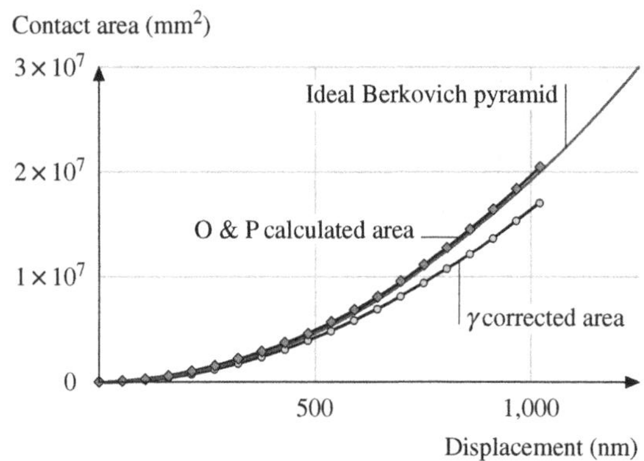


**Fig. 9.19** For a real Berkovich tip the $\gamma$ corrected area [63] is less at a given depth than the area calculated using the Oliver and Pharr method

good and can be significantly different to the values obtained with the standard Oliver and Pharr analysis.

### *9.3.7 Alternative Methods of Analysis*

All of the preceding discussion on the analysis of nanoindentation curves has focused on the unloading curve, virtually ignoring the loading curve data. This is for the simple reason that the unloading curve can in many cases be regarded as purely elastic, whereas the shape of the loading curve is determined by a complex mix of elastic and plastic properties.

It is clear that there is substantially more data in the loading curve if it can be extracted. Page et al. [57, 72] have explored the possibility of curve fitting to the loading data using a combination of elastic and plastic properties. By a combination of analysis and empirical fitting to experimental data, it was suggested that the loading curve is of the following form:

$$P = E\left(\psi\sqrt{\frac{H}{E}} + \phi\sqrt{\frac{E}{H}}\right)^{-2}\delta^2, \tag{9.21}$$

where $\psi$ and $\phi$ are determined experimentally to be 0.930 and 0.194, respectively. For homogenous samples this equation gives a linear relationship between $P$ and $\delta^2$. Coatings, thin film systems, and samples that strain-harden can give significant deviations from linearity. Analysis of the loading curve has yet to gain popularity as a standard method for examining nanoindentation data, but it should certainly be regarded as a prime area for further investigation.

Another alternative method of analysis is based on the work involved in making an indentation. In essence, the nanoindentation curve is a plot of force against distance indicating integration under the loading curve will give the total work of indentation, or the sum of the elastic strain energy and the plastic work of indentation. Integrating under the unloading curve should give only the elastic strain energy. Thus, the work involved in both elastic and plastic deformation during nanoindentation can be found. Cheng and Cheng [73] combined measurements of the work of indentation with a dimensional analysis that deals with the effects of scaling in a material that work-hardens to estimate $H/E_r$. They subsequently evaluated $H$ and $E$ using the Oliver and Pharr approach to find the contact area.

### 9.3.8 Measuring Contact Stiffness

As discussed earlier, it is possible to add a small AC load on top of the DC load used during nanoindentation testing, providing a way to measure the contact stiffness throughout the entire loading and unloading cycle [74, 75]. The AC load is typically at a frequency of ≈60 Hz and creates a dynamic system, with the sample acting as a spring with stiffness $S$ (the contact stiffness), and the nanoindentation system acting as a series of springs and dampers. Figure 9.20 illustrates how the small AC load is added to the DC load. Figure 9.21 shows how the resulting dynamic system can be modeled. An analysis of the dynamic system gives the following relationships for $S$ based on the amplitude of the AC displacement oscillation and the phase difference between the AC load and displacement signals:

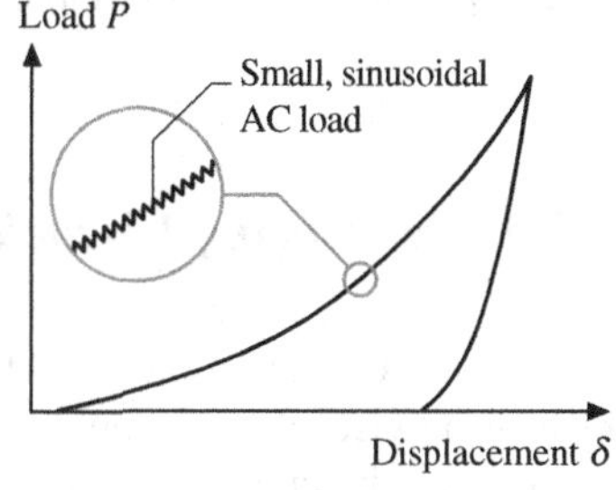


**Fig. 9.20** A small AC load can be added to the DC load. This enables the contact stiffness, $S$, to be calculated throughout the indentation cycle

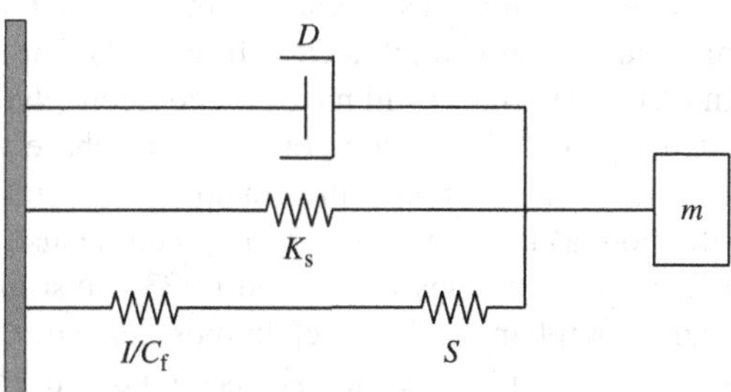


**Fig. 9.21** The dynamic model used in the analysis of the AC response of a nanoindentation device

$$\left|\frac{P_{\text{os}}}{\delta(\omega)}\right| = \sqrt{\left[\left(S^{-1}+C_{\text{f}}\right)^{-1}+K_{\text{s}}-m\omega^2\right]^2+\omega^2D^2}\,, \tag{9.22}$$

$$\tan(\chi) = \frac{\omega D}{\left(S^{-1}+C_{\text{f}}\right)^{-1}+K_{\text{s}}-m\omega^2}\,, \tag{9.23}$$

where $C_{\text{f}}$ is the load frame compliance (the reciprocal of the load frame stiffness), $K_{\text{s}}$ is the stiffness of the support springs (typically in the region of 50–100 N/m), $D$ is the damping coefficient, $P_{\text{os}}$ is the magnitude of the load oscillation, $\delta(\omega)$ is the magnitude of the displacement oscillation, $\omega$ is the oscillation frequency, $m$ is the mass of the indenter, and $\chi$ is the phase angle between the force and the displacement.

In order to find $S$ using either (9.22) or (9.23), it is necessary to calibrate the dynamic response of the system when the tip is not in contact with a sample ($S^{-1} = 0$). This calibration combined with the standard DC calibrations will provide the values for all of the constants in the two equations. All that needs to be measured in order to obtain $S$ is either $\delta(\omega)$ or $\chi$, both of which are measured by the lock-in amplifier used to generate the AC signal. Since the $S$ obtained is the same as the $S$ in (9.1), it follows that the Oliver and Pharr analysis can be applied to obtain $E_{\text{r}}$ and $H$ throughout the entire nanoindentation cycle.

The dynamic analysis detailed here was developed for the MTS Nanoindenter™ (Oakridge, Tennessee), but a similar analysis has been applied to other commercial instruments such as the Hysitron Triboscope™ (Minneapolis, Minnesota) [76]. For all instruments, an AC oscillation is used in addition to the DC voltage, and a dynamic model is used to analyze the response.

### *9.3.9 Measuring Viscoelasticity*

Using an AC oscillation in addition to the DC load introduces the possibility of measuring viscoelastic properties during nanoindentation testing. This has recently been the subject of considerable interest with researchers looking at the loss modulus, storage modulus, and loss tangent of various polymeric materials [25, 77]. Recording the displacement response to the AC force oscillation enables the complex modulus (including the loss and storage modulus) to be found. If the modulus is complex, it is clear from (9.1) that the stiffness also becomes complex. In fact, the stiffness will have two components: $S'$, the component in phase with the AC force and $S''$, the component out of phase with the AC force.

The dynamic model illustrated in Fig. 9.21 is no longer appropriate for this situation, as the contact on the test sample also includes a damping term, shown in Fig. 9.22. Equations (9.22) and (9.23) must also be revised. Neglecting the load frame compliance, $C_{\text{f}}$, which in most real situations is negligible, (9.22) and (9.23) when the sample damping, $D_{\text{s}}$, is included become:

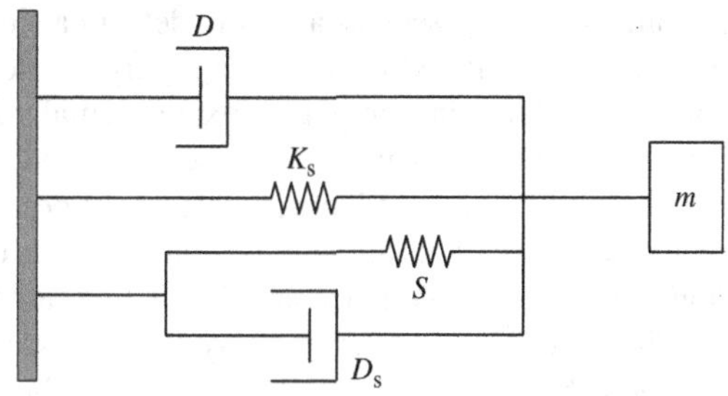


**Fig. 9.22** The simplified dynamic model used when the sample is viscoelastic. It is assumed that the load frame compliance is negligible

$$\left|\frac{P_{os}}{\delta(\omega)}\right| = \sqrt{\{S + K_s - m\omega^2\}^2 + \omega^2(D + D_s)^2}, \tag{9.24}$$

$$\tan(\chi) = \frac{\omega(D + D_s)}{S + K_s - m\omega^2}. \tag{9.25}$$

In order to find the loss modulus and storage modulus, (9.1) is used to relate $S'$ (storage component) and $S''$ (loss component) to the complex modulus.

This method for measuring viscoelastic properties using nanoindentation has now been proven in principal, but has still only been applied to a very small range of polymers and remains an area of future growth.

## 9.4 Modes of Deformation

As described earlier, the analysis of nanoindentation data is based firmly on the results of elastic continuum mechanics. In reality, this idealized, purely elastic situation rarely occurs. For very shallow contacts on metals with thin surface films such as oxides, carbon layers, or organic layers [78, 79], the contact can initially be very similar to that modeled by Hertz and, later, Sneddon. It is very important to realize that this in itself does not constitute proof that the contact is purely elastic, because in many cases a small number of defects are present. These may be preexisting defects that move in the strain-field generated beneath the contact. Alternatively, defects can be generated either when the contact is first made or during the initial loading [52, 80]. When defects such as short lengths of dislocation are present the curves may still appear to be elastic even though inelastic processes like dislocation glide and cross-slip are taking place.

### *9.4.1 Defect Nucleation*

Nucleation of defects during nanoindentation testing has been the subject of many experimental [81, 82] and theoretical studies [83, 84]. This is probably because nanoindentation is seen as a way to deform a small, defect-free volume of material to its elastic limit and beyond in a highly controlled geometry. There are, unfortunately, problems in comparing experimental results with theoretical predictions, largely because the kinetic processes involved in defect nucleation are difficult to model. Simulations conducted at 0 K do not permit kinetic processes, and molecular dynamics simulations are too fast (nanoseconds or picoseconds). Real nanoindentation experiments take place at ≈293 K and last for seconds or even minutes.

Kinetic effects appear in many forms, for instance, during the initial contact between the indenter tip and the surface when defects can be generated by the combined action of the impact velocity and surface forces [51]. A second example of a kinetic effect occurs during hold cycles at large loads when what appears to be an elastic contact can suddenly exhibit a large discontinuity in the displacement data [80]. Figure 9.23 shows how these kinetic effects can affect the nanoindentation data and the apparent yield point load.

During the initial formation of a contact, the deformation of surface asperities [51] and ledges [85] can create either point defects or short lengths of dislocation line. During the subsequent loading, the defects can help in the nucleation and multiplication of dislocations. The large strains present in the region surrounding the contact, coupled with the existence of defects generated on contact, can result in the extremely rapid multiplication of dislocations and, hence, pronounced discontinuities in the load-displacement curve. It is important to realize that the discontinuities are due to the rapid multiplication of dislocations, which may or may not occur at the same time that the first dislocation is nucleated. Dislocations may have been present for some time with the discontinuity only occurring when the existing defects are configured appropriately, as a Frank–Read source, for instance. Even under large strains, the time taken for a dislocation source to form from preexisting defects may be long. It is, therefore, not surprising that large discontinuities can be seen during hold cycles or unloading.

The generation of defects at the surface and the initiation of yielding is a complex process that is extremely dependent on surface asperities and surface forces.

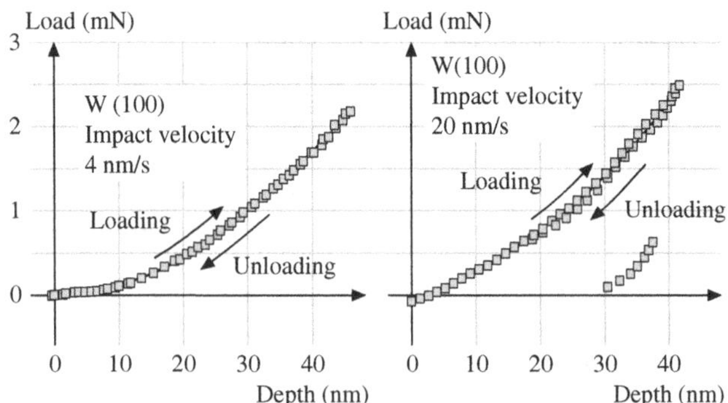


**Fig. 9.23** Load-displacement curves for W(100) showing how changes in the impact velocity can cause a transition from perfectly elastic behavior to yielding during unloading

These, in turn, are closely related to the surface chemistry. It is not only the magnitude of surface forces, but also their range in comparison to the height of surface asperities that determines whether defects are generated on contact. Small changes in the surface chemistry or the velocity of the indenter tip when it first contacts the surface, can cause a transition from a situation in which defects are generated on contact to one where the contact is purely elastic [52].

When the generation of defects during the initial contact is avoided and the deformed region under the contact is truly defect free, then the yielding of the sample should occur at the yield stress of a perfect crystal lattice. The load at which plastic deformation commences under these circumstances becomes very reproducible [86]. Unfortunately, nanoindenter tips on the near-atomic scale are not perfectly smooth or axisymmetric. As a result, accurately measuring the yield stress is very difficult. In fact, a slight rotation in the plane of the surface of either the sample or the tip can give a substantial change in the observed yield point load. Coating the surface in a cushioning self-assembled monolayer [87] can alleviate some of these variations, but it also introduces a large uncertainty in the contact area. Surface oxide layers, which may be several nanometers thick, have also been found to enhance the elastic behavior seen for very shallow nanoindentations on metallic surfaces [78]. Removal of the oxide has been shown to alleviate the initial elastic response.

While nanoindentation testing is ideal for examining the mechanical properties of defect-free volumes and looking at the generation of defects in perfect crystal lattices, it should be clear from the preceding discussion that great care must be taken in examining how the surface properties and the loading rate affect the results, particularly when comparisons are being made to theoretical models for defect generation.

### *9.4.2 Variations with Depth*

Ideal elastic-plastic behavior, as described by Tabor [48], can be seen during indentation testing, provided the sample has been work-hardened so that the flow stress is a constant. However, it is often the case that the mechanical properties appear to change as the load (or depth) is increased. This apparent change can be a result of several processes, including work-hardening during the test. This is a particularly important effect for soft metals like copper. These metals usually have a high hardness at shallow depths, but it decreases asymptotically with increasing indentation depth to a hardness value that may be less than half that observed at shallow depths. This type of behavior is due to the increasing density of geometrically necessary dislocations at shallow depths [88]. Hence the effects of work-hardening are most pronounced at shallow depths. For hard materials the effect is less obvious.

Work-hardening is one of the factors that contribute to the so-called indentation size effect (ISE), whereby at shallow indentation depths the material appears to be harder. The ISE has been widely observed during microindentation testing, with at least part of the effect appearing to result from the increased difficulty in optically measuring the area of an indentation when it is small. During nanoindentation testing the ISE can also be observed, but it is often due to the tip area function, $A_c(\delta_c)$, being incorrectly calibrated. However, there are physical reasons other than work-hardening for expecting an increase in mechanical strength in small volumes. As described in the previous section, small volumes of crystalline materials can have either no defects or only a small number of defects present, making plastic yielding more difficult. Also, because of dislocation line tension, the shear stress required to make a dislocation bow out increases as the radius of the bow decreases. Thus, the shear stress needed to make a dislocation bow out in a small volume is greater than it is in a large volume. These physical reasons for small volumes appearing stronger than large volumes are particularly important in thin film systems, as will be discussed later. Note, however, that these physical reasons for increased hardness do not apply for an amorphous material such as fused silica, which partially explains its value as a calibration material.

### 9.4.3 *Anisotropic Materials*

The analysis methods detailed earlier were concerned primarily with the interpretation of data from nanoindentations in isotropic materials where the elastic modulus is assumed to be either independent of direction or a polycrystalline average of a material's elastic constants. Many crystalline materials exhibit considerable anisotropy in their elastic constants, hence, these analysis techniques may not always be appropriate. The theoretical problem of a rigid indenter pressed into an elastic, anisotropic half-space has been considered by Vlassak and Nix [89]. Their aim was to identify the feasibility of interpreting data from a depth-sensing indentation apparatus for samples with elastic constants that are anisotropic. Nanoindentation experiments [90] have shown the validity of the elastic analysis for crystalline zinc, copper, and beta-brass. The observed indentation modulus for zinc, as predicted, varied by as much as a factor of 2 between different orientations. The variations in the observed hardness values for the same materials were smaller, with a maximum variation with orientation of 20% detected in zinc. While these variations are clearly detectable with nanoindentation techniques, the variations are small in comparison to the actual anisotropy of the test material's elastic properties. This is because the indentation modulus is a weighted average of the stiffness in all directions.

At this time the effects of anisotropy on the hardness measured using nanoindentation have not been fully explored. For materials with many active slip planes it is likely that the small anisotropy observed by Vlassak and Nix is correct once plastic flow has been initiated. It is possible, however, that for defect-free crystalline specimens with a limited number of active slip planes that very shallow nanoindentations may show a much larger anisotropy in the observed hardness and initial yield point load.

### 9.4.4 Fracture and Delamination

Indentation testing has been widely used to study fracture in brittle materials [91], but the lower loads and smaller deformation regions of nanoindentation tests make it harder to initiate cracks and, hence, less useful as a way to evaluate fracture toughness. To overcome these problems the cube corner geometry, which generates larger shear stresses than the Berkovich pyramid, has been used with nanoindentation testing to study fracture [92]. These studies have had mixed success, because the cube corner geometry blunts very quickly when used on hard materials. In many cases, brittle materials are very hard.

Depth sensing indentation is better suited to studying delamination of thin films. Recent work extends the research conducted by Marshall et al. [93, 94], who examined the deformation of residually stressed films by indentation. A schematic of their analysis is given by Fig. 9.24. Their indentations were several microns deep, but the basic analysis is valid for nanoindentations. The analysis has been extended to multilayers [95], which is important since it enables a quantitative assessment of adhesion energy when an additional stressed film has been deposited on top of the film and substrate of interest. The additional film limits the plastic deformation of the film of interest and also applies extra stress that aids in the delamination. After indentation, the area of the delaminated film is measured optically or with an AFM to assess the extent of the delamination. This measurement, coupled with the

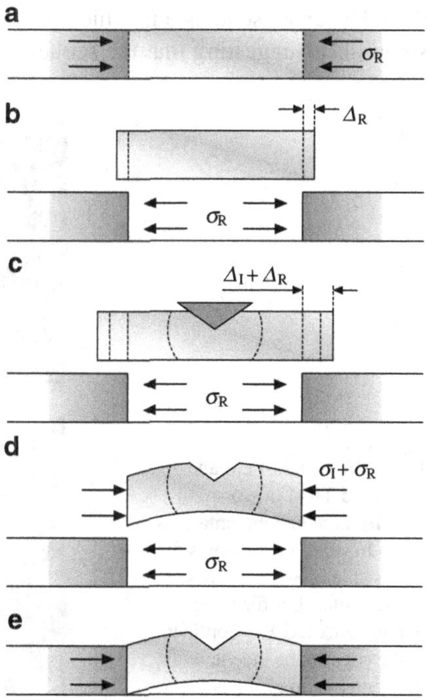

**Fig. 9.24** To model delamination Kriese et al. [95] adapted the model developed by Marshall and Evans [93]. The model considers a segment of removed stressed film that is allowed to expand and then indented, thereby expanding it further. Replacing the segment in its original position requires an additional stress, and the segment bulges upwards

load-displacement data, enables quantitative assessment of the adhesion energy to be made for metals [96] and polymers [97].

### 9.4.5 Phase Transformations

The pressure applied to the surface of a material during indentation testing can be very high. Equation (9.10) indicates that the pressure during plastic yielding is about three times the yield stress. For many materials, high hydrostatic pressures can cause phase transformations, and provided the transformation pressure is less than the pressure required to cause plastic yielding, it is possible during indentation testing to induce a phase transformation. This was first reported for silicon [98], but it has also been speculated [99] that many other materials may show the same effects. Most studies still focus on silicon because of its enormous technological importance, although there is some evidence that germanium also undergoes a phase transformation during the nanoindentation testing [100].

Recent results [21, 22, 41, 101, 102] indicate that there are actually multiple phase transformations during the nanoindentation of silicon. TEM of nanoindentations in diamond cubic silicon have shown the presence of amorphous-Si and the body-centered cubic BC-8 phase (see Fig. 9.25). Micro-Raman spectroscopy has indicated the presence of a further phase, the rhombhedral R-8 (see Fig. 9.26). For many nanoindentations on silicon there is a characteristic discontinuity in the unloading curve (see Fig. 9.27), which seems to correlate with a phase transformation. The exact sequence in which the phases form is still highly controversial with some [42], suggesting that the sequence during loading and unloading is:

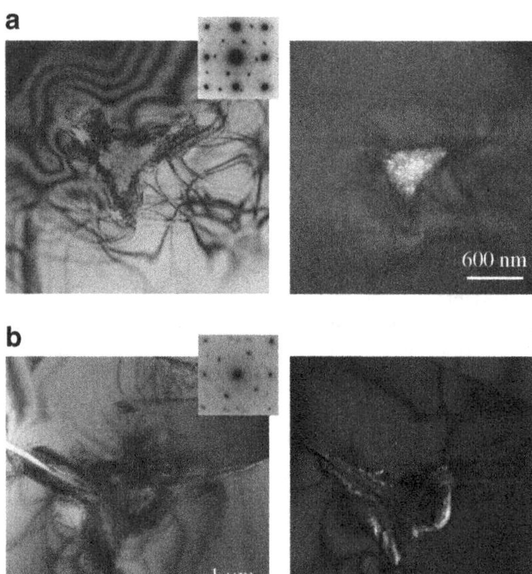


**Fig. 9.25** Bright-field and dark-field TEM of (**a**) small and (**b**) large nanoindentations in Si. In small nanoindents the metastable phase BC-8 is seen in the center, but for large nanoindents BC-8 is confined to the edge of the indent, while the center is amorphous

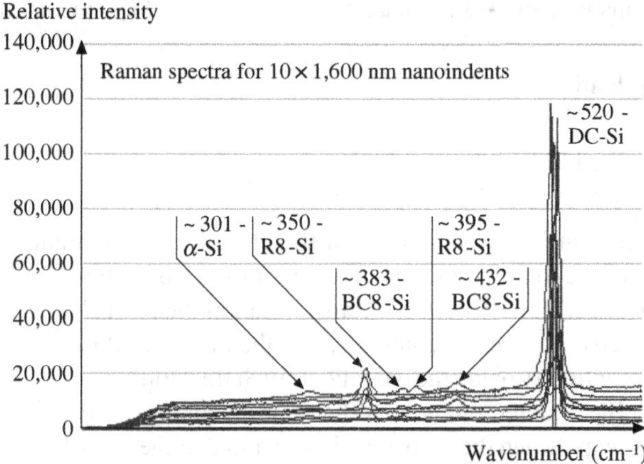


**Fig. 9.26** Micro-Raman generally shows the BC-8 and R-8 Si phases that are at the edge of the nanoindents, but the amorphous phase in the center is not easy to detect, as it is often subsurface and the Raman peak is broad

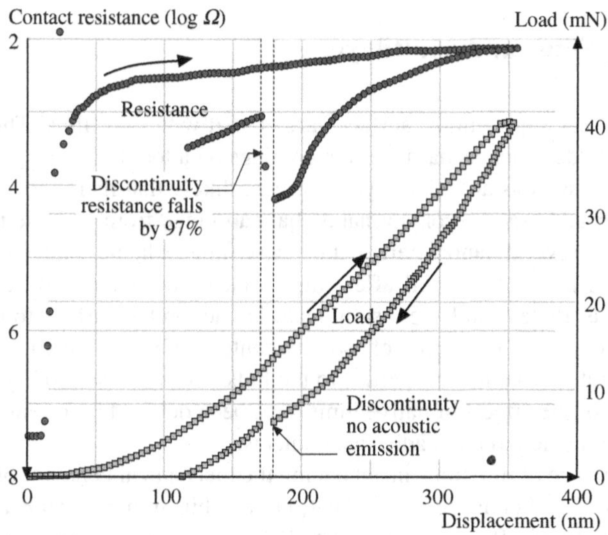


**Fig. 9.27** Nanoindentation curves for deep indents on Si show a discontinuity during unloading and simultaneously a large drop in contact resistance

**Increasing load** →
Diamond cubic Si → β-Sn Si

← **Decreasing load**
BC-8 Si and R-8 Si ← β-Sn Si

Other groups [21, 22] suggest that the above sequence is only valid for shallow nanoindentations that do not exhibit an unloading discontinuity. For large nanoindentations that show an unloading discontinuity, they suggest the sequence will be:

**Increasing load →**
Diamond cubic Si → $\beta$-Sn Si

**← Decreasing load**
$\alpha$-Si ← BC-8 Si and R-8 Si ← $\beta$-Sn Si

The disagreement is over the origin of the unloading discontinuity. Mann et al. [21] suggest it is due to the formation of amorphous silicon, while Gogotsi et al. [42] believe it is the $\beta$-Sn Si to BC-8 or R-8 transformation. Mann et al. argue that the high contact resistance before the discontinuity and the low contact resistance afterwards rule out the discontinuity being the metallic $\beta$-Sn Si transforming to the more resistive BC-8 or R-8. The counterargument is that amorphous Si is only seen with micro-Raman spectroscopy when the unloading is very rapid or there is a large nanoindentation with no unloading discontinuity. The importance of unloading rate and cracking in determining the phases present are further complications. The controversy will remain until in situ characterization of the phases present is undertaken.

## 9.5 Thin Films and Multilayers

In almost all real applications, surfaces are coated with thin films. These may be intentionally added such as hard carbide coatings on a tool bit, or they may simply be native films such as an oxide layer. It is also likely that there will be adsorbed films of water and organic contaminants that can range from a single molecule in thickness up to several nanometers. All of these films, whether native or intentionally placed on the surface, will affect the surface's mechanical behavior on the nanoscale. Adsorbates can have a significant impact on the surface forces [3] and, hence, the geometry and stability of asperity contacts. Oxide films can have dramatically different mechanical properties to the bulk and will also modify the surface forces. Some of the effects of native films have been detailed in the earlier sections on dislocation nucleation and adhesive contacts.

The importance of thin films in enhancing the mechanical behavior of surfaces is illustrated by the abundance of publications on thin film mechanical properties (see for instance Nix [88] or Cammarata [31] or Was and Foecke [103]). In the following sections, the mechanical properties of films intentionally deposited on the surface will be discussed.

### 9.5.1 *Thin Films*

Measuring the mechanical properties of a single thin surface film has always been difficult. Any measurement performed on the whole sample will inevitably be dominated by the bulk substrate. Nanoindentation, since it looks at the mechanical properties of a very small region close to the surface, offers a possible solution to the problem of measuring thin film mechanical properties. However, there are certain inherent problems in using nanoindentation testing to examine the properties of thin films. The problems stem in part from the presence of an interface between the film and substrate. The quality of the interface can be affected by many variables, resulting in a range of effects on the apparent elastic and plastic properties of the film. In particular, when the deformation region around the indent approaches the interface, the indentation curve may exhibit features due to the thin film, the bulk, the interface, or a combination of all three. As a direct consequence of these complications, models for thin-film behavior must attempt to take into account not only the properties of the film and substrate, but also the interface between them.

If, initially, the effect of the interface is neglected, it is possible to divide thin-coated systems into a number of categories that depend on the values of $E$ (elastic modulus) and $Y$ (the yield stress) of the film and substrate. These categories are typically [104, 105]:

1. Coatings with high $E$ and high $Y$, substrates with high $E$ and high $Y$
2. Coatings with high $E$ and high $Y$, substrates with high or low $E$ and low $Y$
3. Coatings with high or low $E$ and low $Y$, substrates with high $E$ and high $Y$
4. Coatings with high or low $E$ and low $Y$, substrates with high or low $E$ and low $Y$.

The reasons for splitting thin film systems into these different categories have been amply demonstrated experimentally by Whitehead and Page [104, 105] and theoretically by Fabes et al. [106]. Essentially, hard, elastic materials (high $E$ and $Y$) will possess smaller plastic zones than soft, inelastic (low $E$ and $Y$) materials. Thus, when different combinations of materials are used as film and substrate, the overall plastic zone will differ significantly. In some cases, the plasticity is confined to the film, and in other cases, it is in both the film and substrate, as shown by Fig. 9.28.

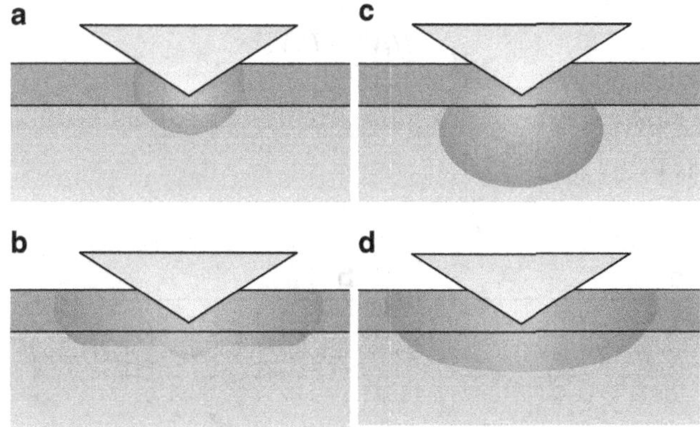


**Fig. 9.28** Variations in the plastic zone for indents on films and substrates of different properties. (**a**) Film and substrate have high $E$ and $Y$, (**b**) film has a high $E$ and $Y$, substrate has a high or low $E$ and low $Y$, (**c**) film has a high or low $E$ and a low $Y$, and substrate has a high $E$ and $Y$, (**d**) film has a high or low $E$ and a low $Y$, and substrate has a high or low $E$ and a low $Y$

If the standard nanoindentation analysis routines are to be used, it is essential that the plastic zone and the elastic strain field are both confined to the film and do not reach the substrate. Clearly, this is difficult to achieve unless extremely shallow nanoindentations are used. There is an often quoted 10% rule, that says nanoindents in a film must have a depth of less than 10% of the film's thickness if only the film properties are to be measured. This has no real validity [107]. There are film/substrate combinations for which 10% is very conservative, while for other combinations even 5% may be too deep. The effect of the substrate for different combinations of film and substrate properties has been studied using FEM [108], which has shown that the maximum nanoindentation depth to measure film only properties decreases in moving from soft on hard to hard on soft combinations. For a very soft film on a hard substrate, nanoindentations of 50% of the film thickness are alright, but this drops to <10% for a hard film on a soft substrate. For a very strong film on a soft substrate, the surface film behaves like an elastic membrane or a bending plate.

Theoretical analysis of thin-film mechanical behavior is difficult. One theoretical approach that has been adopted uses the volumes of plastically deformed material in the film and substrate to predict the overall hardness of the system. However, it should be noted that this method is only really appropriate for soft coatings and indentation depths below the thickness of the coating (see cases c and d of Fig. 9.28), otherwise the behavior will be closer to that detailed later and shown by Fig. 9.29.

The technique of combining the mechanical properties of the film and substrate to evaluate the overall hardness of the system is generally referred to as the rule of mixtures. It stems from work by Burnett et al. [109–111] and Sargent [112], who derived a weighted average to relate the "composite" hardness ($H$) to the volumes of plastically deformed material in the film ($V_f$) and substrate ($V_s$) and their respective values of hardness, $H_f$ and $H_s$. Thus,

$$H = \frac{H_f V_f + H_s V_s}{V_{total}}, \tag{9.26}$$

where $V_{total}$ is $V_f + V_s$.

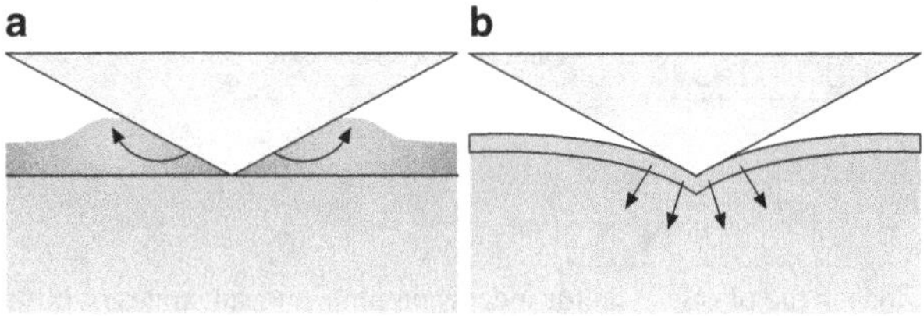


**Fig. 9.29** Two different modes of deformation during nanoindentation of films. In (**a**) materials move upwards and outwards, while in (**b**) the film acts like a membrane and the substrate deforms

Equation (9.26) was further developed by Burnett and Page [109] to take into account the indentation size effect. They replaced $H_s$ with $K\delta_c^{n-2}$, where $K$ and $n$ are experimentally determined constants dependent on the indenter and sample, and $\delta_c$ is the contact depth. This expression is derived directly from Meyer's law for spherical indentations, which gives the relationship $P = Kd^n$ between load, $P$, and the indentation dimension, $d$. Burnett and Page also employed a further refinement to enable the theory to fit experimental results from a specific sample, ion-implanted silicon. This particular modification essentially took into account the different sizes of the plastic zones in the two materials by multiplying $H_s$ by a dimensionless factor $(V_s/V_{total})$. While this seems to be a sensible approach, it is mostly empirical, and the physical justification for using this particular factor is not entirely clear. Later, Burnett and Rickerby [110, 111] took this idea further and tried to generalize the equations to take into account all of the possible scenarios. Thus, the following equations were suggested:

$$H = \frac{H_f(\Omega^3)V_f + H_s V_s}{V_{total}}, \tag{9.27}$$

$$H = \frac{H_f V_f + H_s(\Omega^3)V_s}{V_{total}}. \tag{9.28}$$

The first of these, (9.27), deals with the case of a soft film on a hard substrate, and the second, (9.28), with a hard film on a soft substrate. The $\Omega$ term expresses the variation of the total plastic zone from the ideal hemispherical shape. This was taken still further by Bull and Rickerby [113], who derived an approximation for $\Omega$ based on the film and substrate zone radii being related to their respective hardness and elastic modulii [114, 115]. Hence:

$$\Omega = (E_f H_s / E_s H_f)^l, \tag{9.29}$$

where $l$ is determined empirically. $E_f$ and $H_f$ and $E_s$ and $H_s$ are the elastic modulus and hardness of the film and substrate, respectively.

Experimental data [116] indicate that the effect of the substrate on the elastic modulus of the film can be quite different than the effect on hardness, due to the zones of the elastic and plastic strain fields being different sizes.

Chechechin et al. [117] have recently studied the behavior of $Al_2O_3$ films of various thicknesses on different substrates. Their results indicate that many of the models correctly predict the transition between the properties of the film and those of the substrate, but do not always fit the observed hardness against depth curves. This group have also studied the pop-in behavior of $Al_2O_3$ films [118] and have attempted to model the range of loads and depths at which they occur via a Weibull-type distribution, as utilized in fracture analysis.

A point raised by Burnett and Rickerby should be emphasized. They state that there are two very distinct modes of deformation during nanoindentation testing. The first, referred to as Tabor's [48] model for low $Y/E$ materials, involves the buildup of material at the side of the indenter through movement of material at slip lines. The second, for materials with large $Y/E$ does not result in surface pile-up. The displaced material is then accommodated by radial displacements [115]. The point is that a thin, strong, and well-bonded surface film can cause a substrate that would normally deform by Tabor's method to behave more like a material with high $Y/E$ (see Fig. 9.29). It should be noted that this only applies as long as the film does not fail.

In recent theoretical and experimental studies the importance of material pile-up and sink-in has been investigated extensively. As discussed in an earlier section, pile-up can be increased by residual compressive stresses, but even in the absence of residual stresses pile-up can introduce a significant error in the calculated contact area. This is most pronounced in materials that do not work-harden [62]. For these materials using the Oliver and Pharr method fails to account for the pile-up and results in a large error in the values for $E$ and $H$. For thin films Tsui et al. have used a focused ion beam to section through Knoop indentations in both soft films on hard substrates [69] and hard films on soft substrates [70]. The soft films, as expected, exhibit pile-up, while the hard film acts more like a membrane and the indentation exhibits sink-in with most of the plasticity in the substrate. Thus, there are three clearly identifiable factors affecting the pile-up and sink-in around nanoindents during testing of thin films:

1. Residual stresses
2. Degree of work-hardening
3. Ratio of film and substrate mechanical properties.

The bonding or adhesion between the film and substrate could also be added to this list. And it should not be forgotten that the depth of the nanoindentation relative to the film thickness also affects pile-up. For a very deep nanoindentation into a thin, soft film on a hard substrate pile-up is reduced, due to the combined constraints on the film of the tip and substrate [119]. Due to all of these complications, using nanoindentation to study thin film mechanical properties is fraught with danger. Many unprepared researchers have misguidedly taken the values of $E$ and $H$ obtained during nanoindentation testing to be absolute values only to find out later that the values contain significant errors.

Many of the problems associated with nanoindentation testing are related to incorrectly calculating the contact area, $A$. The Joslin and Oliver method [71] is one way that $A$ can be removed from the calculations. This approach has been used with some success to look at strained epitaxial II/VI semiconductor films [120], but there is evidence that the lattice mismatch in these films can cause dramatic changes in the mechanical properties of the films [121]. This may be due to image forces and the film/substrate interface acting as a barrier to dislocation motion. Recently, it has been shown that using films and substrates with known matching elastic moduli, it is possible to use the assumption of constant elastic modulus with depth to evaluate $H$ [122]. In effect, this is using (9.1) to evaluate $A$ from the contact stiffness data, and then substituting the value for $A$ into (9.12). The value of $E$ is measured independently, for instance, using acoustic techniques.

### 9.5.2 Multilayers

Multilayered materials with individual layers that are a micron or less in thickness, sometimes referred to as superlattices, can exhibit substantial enhancements in hardness or strength. This should be distinguished from the super modulus effect discussed earlier, which has been shown to be largely an artifact. The enhancements in hardness can be as much as 100% when compared to the value expected from the rule of mixtures, which is essentially a weighted average of the hardness for the constituents of the two layers [123]. Table 9.1 shows how the properties of isostructural multilayers can show a substantial increase in hardness over that for fully interdiffused layers. The table also shows how there can be a substantial enhancement in hardness for non-isostructural multilayers compared to the values for the same materials when they are homogeneous.

There are many factors that contribute to enhanced hardness in multilayers. These can be summarized as [103]:

1. Hall-Petch behavior
2. Orowan strengthening
3. Image effects
4. Coherency and thermal stresses
5. Composition modulation

Hall-Petch behavior is related to dislocations piling-up at grain boundaries. (Note that pile-up is used to describe two distinct effects: One is material building up at the side of an indentation, the other is an accumulation of dislocations on a slip-plane.) The dislocation pile-up at grain boundaries impedes the motion of dislocations. For materials with a fine grain structure there are many grain boundaries, and, hence,

**Table 9.1** Results for some experimental studies of multilayer hardness

| Study | Multilayer | Maximum hardness and multilayer repeat length | Reference hardness value | Range of hardness values for multilayers |
|---|---|---|---|---|
| *Isostructural* Knoop hardness [124] | Cu/Ni | 524 at 11.6 nm | 284 (interdiffused) | 295–524 |
| *Non-isostructural* nanoindentation [125] | Mo/NbN | 33 GPa at 2 nm | NbN – 17 GPa<br>Mo – 2.7 GPa<br>Wo – 7 GPa | 12–33 GPa |
| | W/NbN | 29 GPa at 3 nm | (individual layer materials) | 23–29 GPa |

dislocations find it hard to move. In polycrystalline multilayers, it is often the case that the size of the grains within a layer scales with the layer thickness so that reducing the layer thickness reduces the grain size. Thus, the Hall-Petch relationship (below) should be applicable to polycrystalline multilayer films with the grain size, $d_g$, replaced by the layer thickness.

$$Y = Y_0 + k_{HP}\, d_g^{-0.5}, \tag{9.30}$$

where $Y$ is the enhanced yield stress, $Y_0$ is the yield stress for a single crystal, and $k_{HP}$ is a constant.

There is an ongoing argument about whether Hall-Petch behavior really takes place in nanostructured multilayers. The basic model assumes many dislocations are present in the pile-up, but such large dislocation pile-ups are not seen in small grains [126] and are unlikely to be present in multilayers. As a direct consequence, studies have found a range of values, between 0 and $-1$, for the exponent in (9.30), rather than the $-0.5$ predicted for Hall-Petch behavior.

Orowan strengthening is due to dislocations in layered materials being effectively pinned at the interfaces. As a result, the dislocations are forced to bow out along the layers. In narrow films, dislocations are pinned at both the top and bottom interfaces of a layer and bow out parallel to the plane of the interface [127, 128]. Forcing a dislocation to bow out in a layered material requires an increase in the applied shear stress beyond that required to bow out a dislocation in a homogeneous sample. This additional shear stress would be expected to increase as the film thickness is reduced.

Image effects were suggested by Koehler [129] as a possible source of enhanced yield stress in multilayered materials. If two metals, A and B, are used to make a laminate and one of them, A, has a high dislocation line energy, but the other, B, has a low dislocation line energy, then there will be an increased resistance to dislocation motion due to image forces. However, if the individual layers are thick enough that there may be a dislocation source present within the layer, then dislocations could pile-up at the interface. This will create a local stress concentration point and the enhancement to the strength will be very limited. If the layers are thin enough that there will be no dislocation source present, the enhanced mechanical strength may be substantial. In Koehler's model only nearest neighbor layers were taken to contribute to the image forces. However, this was extended to include more layers [130] without substantial changes in the results. The consequence on image effects of reducing the thickness of the individual layers in a multilayer is that it prevents dislocation sources from being active within the layer.

For many multilayer systems there is an increase in strength as the bilayer repeat length is reduced, but there is often a critical repeat length (e.g., 3 nm for the W/NbN multilayer of Table 9.1) below which the strength falls. One explanation for the fall in strength involves the effects of coherency and thermal stresses on dislocation energy. Unlike image effects where the energy of dislocations are a maximum or minimum in the center of layers, the energy maxima and minima are at the interfaces for coherency stresses. Combining the effects of varying moduli and coherency stresses shows that the dependence of strength on layer thickness has a peak near the repeat period where coherency strains begin decreasing [131].

Another source of deviations in behavior at very small repeat periods is the imperfect nature of interfaces. With the exception of atomically perfect epitaxial films, interfaces are generally not atomically flat and there is some interdiffusion. For the Cu/Ni film of Table 9.1, the effects of interdiffusion on hardness were examined [124] by annealing the multilayers. The results were in agreement with a model by Krzanowski [132] that predicted the variations in hardness would be proportional to the amplitude of the composition modulation.

It is interesting to note that the explanations for enhanced mechanical properties in multilayered materials are all based on dislocation mechanisms. So it would seem natural to assume that multilayered materials that do not contain dislocations will show no enhanced hardness over their rule of mixtures values. This has been verified by studies on amorphous metal multilayers [133], which shows that the hardness of the multilayers, firstly, lies between that of the two individual materials and, secondly, has almost no variation with repeat period.

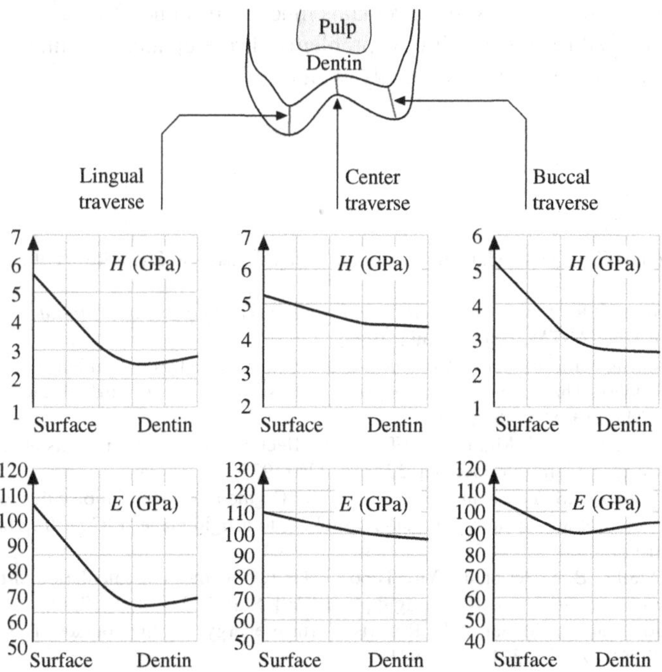


**Fig. 9.30** Variations in $E$ and $H$ across human dental enamel. The sample is an upper second molar cut in cross section from the lingual to the buccal side. Nanoindentations are performed across the surface to examine how the mechanical properties vary

## 9.6 Developing Areas

Over the past 20–30 years, the driving force for studying nanomechanical behavior of surfaces and thin films has been largely, though not exclusively, the microelectronics industry. The importance of electronics to the modern world is only likely to grow in the foreseeable future, but other technological areas may overtake microelectronics as the driving force for research, including the broad fields of biomaterials and nanotechnology. In several places in this chapter, a number of developing areas have been mentioned. These include nanoscale measurements of viscoelasticity and the study of environmental effects (temperature and surface chemistry) on nanomechanical properties. Both of these topics will be vital in the study of biological systems and, as a result, will be increasingly important from a research point of view.

We still have a relatively rudimentary understanding of the nanomechanics of complex biological systems such as bone cells (osteoblasts and osteoclasts) and skin cells (fibroblasts), or even, for that matter, simpler biological structures such as dental enamel. For example, Fig. 9.30 shows how the mechanical properties of dental enamel can vary within a single tooth [134]. But this is still a relatively large-scale measure of mechanical behavior. The prismatic structure of enamel means that there are variations in mechanical properties on a range of scales from millimeters down to nanometers.

In terms of data analysis, there remains much to be done. If an analysis method can be developed that deals with the problems of pile-up and sink-in, the utility of nanoindentation testing will be greatly enhanced.

## References

1. G. Binnig, C.F. Quate, C. Gerber, Atomic force microscope. Phys. Rev. Lett. **56**, 930–933 (1986)
2. P.J. Blau, B.R. Lawn (eds.), *Microindentation Techniques in Materials Science and Engineering* (ASTM, Pennsylvania, 1986)
3. J.N. Israelachvili, *Intermolecular and Surface Forces* (Academic, London, 1992)
4. R.S. Bradley, The cohesive force between solid surfaces and the surface energy of solids. Philos. Mag. **13**, 853–862 (1932)
5. B.V. Derjaguin, V.M. Muller, YuP Toporov, Effect of contact deformations on the adhesion of particles. J. Coll. Interface Sci. **53**, 314–326 (1975)
6. V.M. Muller, V.S. Yuschenko, B.V. Derjaguin, On the influence of molecular forces on the deformation of an elastic sphere and its sticking to a rigid plane. J. Coll. Interface Sci. **77**, 91–101 (1980)
7. V.M. Muller, B.V. Derjaguin, YuP Toporov, On two methods of calculation of the force of sticking of an elastic sphere to a rigid plane. Coll. Surf. **7**, 251–259 (1983)
8. K.L. Johnson, K. Kendal, A.D. Roberts, Surface energy and the contact of elastic solids. Proc. R. Soc. A **324**, 301–320 (1971)
9. A.P. Ternovskii, V.P. Alekhin, MKh Shorshorov, M.M. Khrushchov, V.N. Skvortsov, Zavod Lab. **39**, 1242 (1973)
10. S.I. Bulychev, V.P. Alekhin, MKh Shorshorov, A.P. Ternovskii, G.D. Shnyrev, Determining Young's modulus from the indenter penetration diagram. Zavod Lab. **41**, 1137 (1975)

11. S.I. Bulychev, V.P. Alekhin, MKh Shorshorov, A.P. Ternovskii, Mechanical properties of materials studied from kinetic diagrams of load versus depth of impression during microimpression. Prob. Prochn. **9**, 79 (1976)
12. S.I. Bulychev, V.P. Alekhin, Zavod Lab. **53**, 76 (1987)
13. MKh Shorshorov, S.I. Bulychev, V.P. Alekhin, Sov. Phys. Doklady **26**, 769 (1982)
14. J.B. Pethica, Microhardness tests with penetration depths less than ion implanted layer thickness, in *Ion Implantation into Metals*, ed. by V. Ashworth, W. Grant, R. Procter (Pergamon, Oxford, 1982), p. 147
15. D. Newey, M.A. Wilkens, H.M. Pollock, An ultra-low-load penetration hardness tester. J. Phys. E Sci. Instrum. **15**, 119 (1982)
16. D. Kendall, D. Tabor, An ultrasonic study of the area of contact between stationary and sliding surfaces. Proc. R. Soc. A **323**, 321–340 (1971)
17. G.M. Pharr, W.C. Oliver, F.R. Brotzen, On the generality of the relationship among contact stiffness, contact area and elastic-modulus during indentation. J. Mater. Res. **7**, 613 (1992)
18. T.P. Weihs, C.W. Lawrence, B. Derby, C.B. Scruby, J.B. Pethica, Acoustic emissions during indentation tests. MRS Symp. Proc. **239**, 361–366 (1992)
19. D.F. Bahr, J.W. Hoehn, N.R. Moody, W.W. Gerberich, Adhesion and acoustic emission analysis of failures in nitride films with 14 metal interlayer. Acta Mater. **45**, 5163 (1997)
20. D.F. Bahr, W.W. Gerberich, Relationships between acoustic emission signals and physical phenomena during indentation. J. Mat. Res. **13**, 1065 (1998)
21. A.B. Mann, D. van Heerden, J.B. Pethica, T.P. Weihs, Size-dependent phase transformations during point-loading of silicon. J. Mater. Res. **15**, 1754 (2000)
22. A.B. Mann, D. van Heerden, J.B. Pethica, P. Bowes, T.P. Weihs, Contact resistance and phase transformations during nanoindentation of silicon. Philos. Mag. A **82**, 1921 (2002)
23. S. Jeffery, C.J. Sofield, J.B. Pethica, The influence of mechanical stress on the dielectric breakdown field strength of $SiO_2$ films. Appl. Phys. Lett. **73**, 172 (1998)
24. B.N. Lucas, W.C. Oliver, Indentation power-law creep of high-purity indium. Metall. Trans. A **30**, 601 (1999)
25. S.A. Syed Asif, Time dependent micro deformation of materials. Ph.D. Thesis, (Oxford University, Oxford 1997)
26. S.A. Syed Asif, R.J. Colton, K.J. Wahl, Nanoscale surface mechanical property measurements: force modulation techniques applied to nanoindentation, in *Interfacial Properties on the Submicron Scale*, ed. by J. Frommer, R. Overney (ACS Books, Washington, 2000)
27. A.B. Mann, J.B. Pethica, Nanoindentation studies in a liquid environment. Langmuir **12**, 4583 (1996)
28. J.W. Beams, Mechanical properties of thin films of gold and silver, in *Structure and Properties of Thin Films*, ed. by C.A. Neugebauer, J.B. Newkirk, D.A. Vermilyea (Wiley, New York, 1959), pp. 183–192
29. J.J. Vlassak, W.D. Nix, A new bulge test technique for the determination of Youngs modulus and Poissons ratio of thin-films. J. Mater. Res. **7**, 3242 (1992)
30. W.M.C. Yang, T. Tsakalakos, J.E. Hilliard, Enhanced elastic modulus in composition modulated gold-nickel and copper-palladium foils. J. Appl. Phys. **48**, 876 (1977)
31. R.C. Cammarata, Mechanical properties of nanocomposite thin-films. Thin Solid Films **240**, 82 (1994)
32. G.A.D. Briggs, *Acoustic Microscopy* (Clarendon, Oxford, 1992)
33. J. Kushibiki, N. Chubachi, Material characterization by line-focus-beam acoustic microscope. IEEE Trans. Sonics Ultrasonics **32**, 189–212 (1985)
34. M.J. Bamber, K.E. Cooke, A.B. Mann, B. Derby, Accurate determination of Young's modulus and Poisson's ratio of thin films by a combination of acoustic microscopy and nanoindentation. Thin Solid Films **398**, 299–305 (2001)
35. S.E. Bobbin, R.C. Cammarata, J.W. Wagner, Determination of the flexural modulus of thin-films from measurement of the 1st arrival of the symmetrical Lamb wave. Appl. Phys. Lett. **59**, 1544–1546 (1991)

36. R.B. King, C.M. Fortunko, Determination of in plane residual-stress state in plates using horizontally polarized shear waves. J. Appl. Phys. **54**, 3027–3035 (1983)
37. R.F. Cook, G.M. Pharr, Direct observation and analysis of indentation cracking in glasses and ceramics. J. Am. Ceram. Soc. **73**, 787–817 (1990)
38. T.F. Page, W.C. Oliver, C.J. McHargue, The deformation-behavior of ceramic crystals subjected to very low load (nano)indentations. J. Mater. Res. **7**, 450–473 (1992)
39. G.M. Pharr, W.C. Oliver, D.S. Harding, New evidence for a pressure-induced phase-transformation during the indentation of silicon. J. Mater. Res. **6**, 1129–1130 (1991)
40. C.F. Robertson, M.C. Fivel, The study of submicron indent-induced plastic deformation. J. Mater. Res. **14**, 2251–2258 (1999)
41. J.E. Bradby, J.S. Williams, J. Wong-Leung, M.V. Swain, P. Munroe, Transmission electron microscopy observation of deformation microstructure under spherical indentation in silicon. Appl. Phys. Lett. **77**, 3749–3751 (2000)
42. Y.G. Gogotsi, V. Domnich, S.N. Dub, A. Kailer, K.G. Nickel, Cyclic nanoindentation and Raman microspectroscopy study of phase transformations in semiconductors. J. Mater. Res. **15**, 871–879 (2000)
43. H. Hertz, Über die Berührung fester elastischer Körper. J. Reine Angew. Math. **92**, 156–171 (1882)
44. J. Boussinesq, *Application des potentiels à l'étude de l'équilibre et du mouvement des solides élastiques* (Blanchard, Paris, 1885); reprint (1996)
45. A.E.H. Love, The stress produced in a semi-infinite solid by pressure on part of the boundary. Philos. Trans. R. Soc. **228**, 377–420 (1929)
46. A.E.H. Love, Boussinesq's problem for a rigid cone. Quarter. J. Math. **10**, 161 (1939)
47. I.N. Sneddon, The relationship between load and penetration in the axisymmetric Boussinesq problem for a punch of arbitrary profile. Int. J. Eng. Sci. **3**, 47–57 (1965)
48. D. Tabor, *Hardness of Metals* (Oxford University Press, Oxford, 1951)
49. R. von Mises, Mechanik der festen Körper in plastisch deformablen Zustand. Goettinger Nachr. Math. Phys. **K1**, 582–592 (1913)
50. H. Tresca, Sur l'ecoulement des corps solids soumis s fortes pression. Compt. Rend. **59**, 754 (1864)
51. A.B. Mann, J.B. Pethica, The role of atomic-size asperities in the mechanical deformation of nanocontacts. Appl. Phys. Lett. **69**, 907–909 (1996)
52. A.B. Mann, J.B. Pethica, The effect of tip momentum on the contact stiffness and yielding during nanoindentation testing. Philos. Mag. A **79**, 577–592 (1999)
53. S.P. Jarvis, Atomic force microscopy and tip-surface interactions. Ph.D. Thesis (Oxford University, Oxford 1993)
54. J.B. Pethica, D. Tabor, Contact of characterised metal surfaces at very low loads: Deformation and adhesion. Surf. Sci. **89**, 182 (1979)
55. J.S. Field, M.V. Swain, Determining the mechanical-properties of small volumes of materials from submicrometer spherical indentations. J. Mater. Res. **10**, 101–112 (1995)
56. W.C. Oliver, G.M. Pharr, An improved technique for determining hardness and elastic-modulus using load and displacement sensing indentation experiments. J. Mater. Res. **7**, 1564–1583 (1992)
57. S.V. Hainsworth, H.W. Chandler, T.F. Page, Analysis of nanoindentation load-displacement loading curves. J. Mater. Res. **11**, 1987–1995 (1996)
58. J.L. Loubet, J.M. Georges, O. Marchesini, G. Meille, Vickers indentation curves of magnesium oxide (MgO). Mech. Eng. **105**, 91–92 (1983)
59. J.L. Loubet, J.M. Georges, O. Marchesini, G. Meille, Vickers indentation curves of magnesium oxide (MgO). J. Tribol. Trans. ASME **106**, 43–48 (1984)
60. M.F. Doerner, W.D. Nix, A method for interpreting the data from depth sensing indentation experiments. J. Mater. Res. **1**, 601–609 (1986)
61. A. Bolshakov, W.C. Oliver, G.M. Pharr, Influences of stress on the measurement of mechanical properties using nanoindentation. 2. Finite element simulations. J. Mater. Res. **11**, 760–768 (1996)

62. A. Bolshakov, G.M. Pharr, Influences of pileup on the measurement of mechanical properties by load and depth sensing instruments. J. Mater. Res. **13**, 1049–1058 (1998)
63. J.C. Hay, A. Bolshakov, G.M. Pharr, A critical examination of the fundamental relations used in the analysis of nanoindentation data. J. Mater. Res. **14**, 2296–2305 (1999)
64. G.M. Pharr, T.Y. Tsui, A. Bolshakov, W.C. Oliver, Effects of residual-stress on the measurement of hardness and elastic-modulus using nanoindentation. MRS Symp. Proc. **338**, 127–134 (1994)
65. T.R. Simes, S.G. Mellor, D.A. Hills, A note on the influence of residual-stress on measured hardness. J. Strain Anal. Eng. Des. **19**, 135–137 (1984)
66. W.R. Lafontaine, B. Yost, C.Y. Li, Effect of residual-stress and adhesion on the hardness of copper-films deposited on silicon. J. Mater. Res. **5**, 776–783 (1990)
67. W.R. Lafontaine, C.A. Paszkiet, M.A. Korhonen, C.Y. Li, Residual stress measurements of thin aluminum metallizations by continuous indentation and X-ray stress measurement techniques. J. Mater. Res. **6**, 2084–2090 (1991)
68. T.Y. Tsui, W.C. Oliver, G.M. Pharr, Influences of stress on the measurement of mechanical properties using nanoindentation. 1. Experimental studies in an aluminum alloy. J. Mater. Res. **11**, 752–759 (1996)
69. T.Y. Tsui, J. Vlassak, W.D. Nix, Indentation plastic displacement field: Part I. The case of soft films on hard substrates. J. Mater. Res. **14**, 2196–2203 (1999)
70. T.Y. Tsui, J. Vlassak, W.D. Nix, Indentation plastic displacement field: Part II. The case of hard films on soft substrates. J. Mater. Res. **14**, 2204–2209 (1999)
71. D.L. Joslin, W.C. Oliver, A new method for analyzing data from continuous depth-sensing microindentation tests. J. Mater. Res. **5**, 123–126 (1990)
72. M.R. McGurk, T.F. Page, Using the P-delta(2) analysis to deconvolute the nanoindentation response of hard-coated systems. J. Mater. Res. **14**, 2283–2295 (1999)
73. Y.T. Cheng, C.M. Cheng, Relationships between hardness, elastic modulus, and the work of indentation. Appl. Phys. Lett. **73**, 614–616 (1998)
74. J.B. Pethica, W.C. Oliver, Mechanical properties of nanometer volumes of material: Use of the elastic response of small area indentations. MRS Symp. Proc. **130**, 13–23 (1989)
75. W.C. Oliver, J.B. Pethica, Method for continuous determination of the elastic stiffness of contact between two bodies, United States Patent Number 4,848,141, (1989)
76. S.A.S. Asif, K.J. Wahl, R.J. Colton, Nanoindentation and contact stiffness measurement using force modulation with a capacitive load-displacement transducer. Rev. Sci. Instrum. **70**, 2408–2413 (1999)
77. J.L. Loubet, W.C. Oliver, B.N. Lucas, Measurement of the loss tangent of low-density polyethylene with a nanoindentation technique. J. Mater. Res. **15**, 1195–1198 (2000)
78. W.W. Gerberich, J.C. Nelson, E.T. Lilleodden, P. Anderson, J.T. Wyrobek, Indentation induced dislocation nucleation: The initial yield point. Acta Mater. **44**, 3585–3598 (1996)
79. J.D. Kiely, J.E. Houston, Nanomechanical properties of Au(111), (001), and (110) surfaces. Phys. Rev. B **57**, 12588–12594 (1998)
80. D.F. Bahr, D.E. Wilson, D.A. Crowson, Energy considerations regarding yield points during indentation. J. Mater. Res. **14**, 2269–2275 (1999)
81. D.E. Kramer, K.B. Yoder, W.W. Gerberich, Surface constrained plasticity: Oxide rupture and the yield point process. Philos. Mag. A **81**, 2033–2058 (2001)
82. S.G. Corcoran, R.J. Colton, E.T. Lilleodden, W.W. Gerberich, Anomalous plastic deformation at surfaces: Nanoindentation of gold single crystals. Phys. Rev. B **55**, 16057–16060 (1997)
83. E.B. Tadmor, R. Miller, R. Phillips, M. Ortiz, Nanoindentation and incipient plasticity. J. Mater. Res. **14**, 2233–2250 (1999)
84. J.A. Zimmerman, C.L. Kelchner, P.A. Klein, J.C. Hamilton, S.M. Foiles, Surface step effects on nanoindentation, Phys. Rev. Lett. **87**, article 165507 (1–4) (2001)
85. J.D. Kiely, R.Q. Hwang, J.E. Houston, Effect of surface steps on the plastic threshold in nanoindentation. Phys. Rev. Lett. **81**, 4424–4427 (1998)

86. A.B. Mann, P.C. Searson, J.B. Pethica, T.P. Weihs, The relationship between near-surface mechanical properties, loading rate and surface chemistry. Mater. Res. Soc. Symp. Proc. **505**, 307–318 (1998)
87. R.C. Thomas, J.E. Houston, T.A. Michalske, R.M. Crooks, The mechanical response of gold substrates passivated by self-assembling monolayer films. Science **259**, 1883–1885 (1993)
88. W.D. Nix, Elastic and plastic properties of thin films on substrates: Nanoindentation techniques. Mater. Sci. Eng. A **234**, 37–44 (1997)
89. J.J. Vlassak, W.D. Nix, Indentation modulus of elastically anisotropic half-spaces. Philos. Mag. A **67**, 1045–1056 (1993)
90. J.J. Vlassak, W.D. Nix, Measuring the elastic properties of anisotropic materials by means of indentation experiments. J. Mech. Phys. Solids **42**, 1223–1245 (1994)
91. B.R. Lawn, *Fracture of Brittle Solids* (Cambridge University Press, Cambridge, 1993)
92. G.M. Pharr, Measurement of mechanical properties by ultra-low load indentation. Mater. Sci. Eng. A **253**, 151–159 (1998)
93. D.B. Marshall, A.G. Evans, Measurement of adherence of residually stressed thin-films by indentation. 1. Mechanics of interface delamination. J. Appl. Phys. **56**, 2632–2638 (1984)
94. C. Rossington, A.G. Evans, D.B. Marshall, B.T. Khuriyakub, Measurement of adherence of residually stressed thin-films by indentation. 2. Experiments with ZnO/Si. J. Appl. Phys. **56**, 2639–2644 (1984)
95. M.D. Kriese, W.W. Gerberich, N.R. Moody, Quantitative adhesion measures of multilayer films: Part I. Indentation mechanics. J. Mater. Res. **14**, 3007–3018 (1999)
96. M.D. Kriese, W.W. Gerberich, N.R. Moody, Quantitative adhesion measures of multilayer films: Part II. Indentation of W/Cu, W/W, Cr/W. J. Mater. Res. **14**, 3019–3026 (1999)
97. M. Li, C.B. Carter, M.A. Hillmyer, W.W. Gerberich, Adhesion of polymer-inorganic interfaces by nanoindentation. J. Mater. Res. **16**, 3378–3388 (2001)
98. D.R. Clarke, M.C. Kroll, P.D. Kirchner, R.F. Cook, B.J. Hockey, Amorphization and conductivity of silicon and germanium induced by indentation. Phys. Rev. Lett. **60**, 2156–2159 (1988)
99. J.J. Gilman, Insulator-metal transitions at microindentation. J. Mater. Res. **7**, 535–538 (1992)
100. G.M. Pharr, W.C. Oliver, R.F. Cook, P.D. Kirchner, M.C. Kroll, T.R. Dinger, D.R. Clarke, Electrical-resistance of metallic contacts on silicon and germanium during indentation. J. Mater. Res. **7**, 961–972 (1992)
101. A. Kailer, Y.G. Gogotsi, K.G. Nickel, Phase transformations of silicon caused by contact loading. J. Appl. Phys. **81**, 3057–3063 (1997)
102. J.E. Bradby, J.S. Williams, J. Wong-Leung, M.V. Swain, P. Munroe, Mechanical deformation in silicon by micro-indentation. J. Mater. Res. **16**, 1500–1507 (2000)
103. G.S. Was, T. Foecke, Deformation and fracture in microlaminates. Thin Solid Films **286**, 1–31 (1996)
104. A.J. Whitehead, T.F. Page, Nanoindentation studies of thin-film coated systems. Thin Solid Films **220**, 277–283 (1992)
105. A.J. Whitehead, T.F. Page, Nanoindentation studies of thin-coated systems. NATO ASI Ser. E **233**, 481–488 (1993)
106. B.D. Fabes, W.C. Oliver, R.A. McKee, F.J. Walker, The determination of film hardness from the composite response of film and substrate to nanometer scale indentations. J. Mater. Res. **7**, 3056–3064 (1992)
107. T.F. Page, S.V. Hainsworth, Using nanoindentation techniques for the characterization of coated systems – a critique. Surface Coat. Technol. **61**, 201–208 (1993)
108. X. Chen, J.J. Vlassak, Numerical study on the measurement of thin film mechanical properties by means of nanoindentation. J. Mater. Res. **16**, 2974–2982 (2001)
109. P.J. Burnett, T.F. Page, Surface softening in silicon by ion-implantation. J. Mater. Sci. **19**, 845–860 (1984)
110. P.J. Burnett, D.S. Rickerby, The mechanical-properties of wear resistant coatings. 1. Modeling of hardness behavior. Thin Solid Films **148**, 41–50 (1987)

111. P.J. Burnett, D.S. Rickerby, The mechanical-properties of wear resistant coatings. 2. Experimental studies and interpretation of hardness. Thin Solid Films **148**, 51–65 (1987)
112. P.M. Sargent, A better way to present results from a least-squares fit to experimental-data – an example from microhardness testing. J. Test. Eval. **14**, 122–127 (1986)
113. S.J. Bull, D.S. Rickerby, Evaluation of coatings. Br. Ceram. Trans. J. **88**, 177–183 (1989)
114. B.R. Lawn, A.G. Evans, D.B. Marshall, Elastic/plastic indentation damage in ceramics: the median/radial crack system. J. Am. Ceram. Soc. **63**, 574–581 (1980)
115. R. Hill, *The Mathematical Theory of Plasticity* (Clarendon, Oxford, 1950)
116. W.C. Oliver, C.J. McHargue, S.J. Zinkle, Thin-film characterization using a mechanical-properties microprobe. Thin Solid Films **153**, 185–196 (1987)
117. N.G. Chechechin, J. Bottiger, J.P. Krog, Nanoindentation of amorphous aluminum oxide films. 1. Influence of the substrate on the plastic properties. Thin Solid Films **261**, 219–227 (1995)
118. N.G. Chechechin, J. Bottiger, J.P. Krog, Nanoindentation of amorphous aluminum oxide films. 2. Critical parameters for the breakthrough and a membrane effect in thin hard films on soft substrates. Thin Solid Films **261**, 228–235 (1995)
119. D.E. Kramer, A.A. Volinsky, N.R. Moody, W.W. Gerberich, Substrate effects on indentation plastic zone development in thin soft films. J. Mater. Res. **16**, 3150–3157 (2001)
120. A.B. Mann, Nanomechanical measurements: Surface and environmental effects. Ph.D. Thesis (Oxford University Press, Oxford 1995)
121. A.B. Mann, J.B. Pethica, W.D. Nix, S. Tomiya, Nanoindentation of epitaxial films: A study of pop-in events. Mater. Res. Soc. Symp. Proc. **356**, 271–276 (1995)
122. R. Saha, W.D. Nix, Effects of the substrate on the determination of thin film mechanical properties by nanoindentation. Acta Mater. **50**, 23–38 (2002)
123. S.A. Barnett, Deposition and mechanical properties of superlattice thin films, in *Physics of Thin Films*, ed. by M.H. Francombe, J.L. Vossen (Academic, New York, 1993)
124. R.R. Oberle, R.C. Cammarata, Dependence of hardness on modulation amplitude in electro-deposited Cu-Ni compositionally modulated thin-films. Scripta Metall. **32**, 583–588 (1995)
125. A. Madan, Y.Y. Wang, S.A. Barnett, C. Engstrom, H. Ljungcrantz, L. Hultman, M. Grimsditch, Enhanced mechanical hardness in epitaxial nonisostructural Mo/NbN and W/NbN superlattices. J. Appl. Phys. **84**, 776–785 (1998)
126. R. Venkatraman, J.C. Bravman, Separation of film thickness and grain-boundary strengthening effects in Al thin-films on Si. J. Mater. Res. **7**, 2040–2048 (1992)
127. J.D. Embury, J.P. Hirth, On dislocation storage and the mechanical response of fine-scale microstructures. Acta Mater. **42**, 2051–2056 (1994)
128. D.J. Srolovitz, S.M. Yalisove, J.C. Bilello, Design of multiscalar metallic multilayer composites for high-strength, high toughness, and low CTE mismatch. Metall. Trans. A **26**, 1805–1813 (1995)
129. J.S. Koehler, Attempt to design a strong solid. Phys. Rev. B **2**, 547–551 (1970)
130. S.V. Kamat, J.P. Hirth, B. Carnahan, Image forces on screw dislocations in multilayer structures. Scripta Metall. **21**, 1587–1592 (1987)
131. M. Shinn, L. Hultman, S.A. Barnett, Growth, structure, and microhardness of epitaxial TiN/NbN superlattices. J. Mater. Res. **7**, 901–911 (1992)
132. J.E. Krzanowski, The effect of composition profile on the strength of metallic multilayer structures. Scripta Metall. **25**, 1465–1470 (1991)
133. J.B. Vella, R.C. Cammarata, T.P. Weihs, C.L. Chien, A.B. Mann, H. Kung, Nanoindentation study of amorphous metal multilayered thin films. MRS Symp. Proc. **594**, 25–29 (2000)
134. J.L. Cuy, A.B. Mann, K.J. Livi, M.F. Teaford, T.P. Weihs, Nanoindentation mapping of the mechanical properties of human molar tooth enamel. Arch. Oral Biol. **47**, 281–291 (2002)

# Chapter 10
# Computer Simulations of Nanometer-Scale Indentation and Friction

**Susan B. Sinnott, Seong-Jun Heo, Donald W. Brenner, Judith A. Harrison, and Douglas L. Irving**

**Abstract** Engines and other machines with moving parts are often limited in their design and operational lifetime by friction and wear. This limitation has motivated the study of fundamental tribological processes with the ultimate aim of controlling and minimizing their impact. The recent development of miniature apparatus, such as microelectromechanical systems (MEMS) and nanometer-scale devices, has increased interest in atomic-scale friction, which has been found to, in some cases, be due to mechanisms that are distinct from the mechanisms that dominate in macroscale friction.

Presented in this chapter is a review of computational studies of tribological processes at the atomic and nanometer scale. In particular, a review of the findings of computational studies of nanometer-scale indentation, friction and lubrication is presented, along with a review of the salient computational methods that are used in these studies, and the conditions under which they are best applied.

Engines and other machines with moving parts are often limited in their design and operational lifetime by friction and wear. This limitation has motivated the study of tribological processes with the aim of controlling and minimizing the impact of these processes. There are numerous historical examples that illustrate the importance of friction to the development of civilizations, including the ancient Egyptians who invented technologies to move the stones used to build the pyramids [1]; Coulomb, who was motivated to study friction by the need to move ships easily and without wear from land to the water [1]; and Johnson et al. [2], who developed an improved understanding of contact mechanics and surface energies through the study of automobile windshield wipers. At present, substantial research and development is aimed at microscale and nanoscale machines with moving parts that at times challenge our fundamental understanding of friction and wear. This has motivated the study of atomic-scale friction and has, consequently, led to new discoveries such as self-lubricating surfaces and wear-resistant materials. While there are similarities between friction at the macroscale and the atomic scale, in many instances the mechanisms that lead to friction at these two scales are quite different. Thus, as devices such as magnetic storage disks and microelectromechanical systems (MEMS) [3] continue to shrink in size, it is expected that new phenomena

B. Bhushan (ed.), *Nanotribology and Nanomechanics*,
DOI 10.1007/978-3-642-15283-2_10, 

associated with atomic-scale friction, adhesion and wear will dominate the functioning of these devices.

The last two decades have seen considerable scientific effort expended on the study of atomic-scale friction [4, 5, 6, 7, 8, 9, 10, 11, 12, 13, 14, 15, 16, 17]. This effort has been facilitated by the development of new advanced experimental tools to measure friction over nanometer-scale distances at low loads, rapid improvements in computer power, and the maturation of computational methodologies for the modeling of materials at the atomic scale. For example, friction-force and atomic-force microscopes (FFM and AFM) allow the frictional properties of solids to be characterized with atomic-scale resolution under single-asperity indentation and sliding conditions [18, 19, 20, 21]. In addition, the surface force apparatus (SFA) provides data about the tribological and lubrication responses of many liquid and solid systems with atomic resolution [22], and the quartz crystal microbalance (QCM) provides information about the atomic-scale origins of friction [4, 23]. These and related experimental methods allow researchers to study sliding surfaces at the atomic scale and relate the observed phenomena to macroscopically observed friction, lubrication and wear.

Analytic models and computational simulations have played an important role in characterizing and understanding friction. They can, for example, assist in the interpretation of experimental data or provide predictions that subsequent experiments can confirm or refute. Analytic models have long been used to study friction, including early studies by Tomlinson [24] and Frenkel and Kontorova [25] and more recent studies by McClelland et al. [26], Sokoloff [13, 27, 28, 29, 30, 31, 32, 33], Persson [34, 35, 36, 37] and others [38, 39, 40, 41, 42, 43, 44]. Most of these idealized models divide the complex motions that create friction into more fundamental components defined by quantities such as spring constants, the curvature and magnitude of potential wells, and bulk phonon frequencies. While these simplifications provide these approaches with some predictive capabilities, many assumptions must be made in order to be able to apply these models to study friction, which may lead to incorrect or incomplete results.

In atomic-scale molecular dynamics (MD) simulations, atom trajectories are calculated by numerically integrating coupled classical equations of motion. Interatomic forces that enter these equations are typically calculated either from total energy methods that include electronic degrees of freedom, or from simplified mathematical expressions that give the potential energy as a function of interatomic displacements. MD simulations can be considered numerical experiments that provide a link between analytic models and experiments. The main strength of MD simulations is that they can reveal unanticipated phenomena or unexpected mechanisms for well-known observations. Weaknesses include a lack of quantum effects in classical atomistic dynamics, and perhaps more importantly, the fact that meaningless results can be obtained if the simulation conditions are chosen incorrectly. The next section contains a review of MD simulations, including the approximations that are inherent in their application to the study of friction, and the conditions under which they should and should not be applied.

## 10.1 Computational Details

Molecular dynamics simulations are straightforward to describe: given a set of initial conditions and a way of mathematically modeling interatomic forces, Newton's (or equivalent) classical equation of motion is numerically integrated [45]

$$\boldsymbol{F} = m\boldsymbol{a}, \tag{10.1a}$$

$$-\nabla E = m(\partial^2 \boldsymbol{r}/\partial t^2), \tag{10.1b}$$

where $F$ is the force on each atom, $m$ is the atomic mass, $a$ is the atomic acceleration, $E$ is the potential energy felt by each atom, $r$ is the atomic position, and $t$ is time. The forces acting on any given atom are calculated, and then the atoms move a short increment $\partial t$ (called a time step) forward in time in response to these applied forces. This is accompanied by a change in atomic positions, velocities and accelerations. The process is then repeated for some specified number of time steps.

The output of these simulations includes new atomic positions, velocities, and forces that allow additional quantities such as temperature and pressure to be determined. As the size of the system increases, it is useful to render the atomic positions in animated movies that reveal the responses of the system in a qualitative manner. Quantitative data can be obtained by analyzing the numerical output directly.

The following sections review the way in which energies and forces are calculated in MD simulations and the important approximations that are used to realistically model the friction that occurs in experiments with smaller systems of only a few tens of thousands of atoms in simulations. The reader is referred to additional sources [46, 47, 48, 49, 50, 51, 52] for a more comprehensive overview of MD simulations (including computer algorithms), analysis methods, and the potentials that are used to calculate energies and forces in MD simulations.

### *10.1.1 Energies and Forces*

There are several different approaches by which interatomic energies and forces are determined in MD simulations. The most theoretically rigorous methods are those that are classified as ab initio or first principles. These techniques, which include density functional theory [53, 54] and quantum chemical ab initio [55] methods, are derived from quantum mechanical principles and are generally both the most accurate and the most computationally intensive. They are therefore limited to a small number of atoms ($<500$), which has limited their use in the study of friction. Alternatively, empirical methods are functions containing parameters that are determined by fitting to experimental data or the results of ab initio calculations [50]. These techniques can usually be relied on to correctly describe qualitative trends and are often the only choice available for modeling systems containing tens of thousands, millions, or billions of atoms. Empirical methods have therefore been widely used in studies of friction. Semi-empirical methods, including tight-binding methods, include some elements of both empirical methods and ab initio methods. For instance, they require quantum mechanical information in the form of, for example, on-site and hopping matrix elements, and include fits to experimental data [56].

Empirical methods simplify the modeling of materials by treating the atoms as spheres that interact with each other via repulsive and attractive terms that can be either pairwise additive or many-body in nature. In this approach, electrons are not treated explicitly, although it is understood that the interatomic interactions are ultimately dependent on them. As discussed in this section, some empirical methods explicitly include charge through classical electrostatic interactions, although most methods assume charge-neutral systems. The repulsive and attractive functional forms generally depend on interatomic distances and/or angles and contain adjustable parameters that are fit to ab initio results and/or experimental data.

The main strength of empirical potentials is their computational speed. Recent simulations with these approaches have modeled billions of atoms [57], something that is not possible with ab initio or semi-empirical approaches at this time. The main weakness of empirical potentials is their lack of quantitative accuracy, especially if they are poorly formulated or applied to systems that are too far removed from the fitting database used in their construction. Furthermore, because of the differences in the nature of chemical bonding in various materials, such as covalent bonding in carbon versus metallic bonding in gold, empirical methods have been historically derived for particular classes of materials. They are therefore generally nontransferable, although some methods have been shown to be theoretically equivalent [51, 58], and in recent years there has been progress towards the development of empirical methods that can model heterogeneous material systems [59, 60, 61, 62, 63, 64].

Several of the most important and common general classes of empirical methods used for calculating interatomic energies and forces in materials, the so-called potentials, are reviewed here. The first to be considered are the potentials that are used to model covalently bound materials, including the bond-order potential and the Stillinger–Weber potential.

The bond-order potential was first formulated by Abell [65] and subsequently developed and parameterized by Tersoff for silicon and germanium [66, 67], Brenner and coworkers for hydrocarbons [52, 68, 69], Dyson and Smith for carbon–silicon–hydrogen systems [70], Sinnott and coworkers for carbon–oxygen–hydrogen systems [71], and Graves and coworkers [72] and Sinnott and coworkers [73] for fluorocarbons, Schall and coworkers for pure Si [74], and Hu and coworkers on C–N–O–H [75].

The bond-order potential has the general functional form

$$E = \sum_{i} \sum_{j(>i)} \left[ V_{\mathrm{R}}(r_{ij}) - b_{ij} V_{\mathrm{A}}(r_{ij}) \right] \tag{10.2}$$

where $V_R(r)$ and $V_A(r)$ are pair-additive interactions that model the interatomic repulsion and electron–nuclear attraction, respectively. The quantity $r_{ij}$ is the distance between pairs of nearest-neighbor atoms $i$ and $j$, and $b_{ij}$ is a bond-order term that takes into account the many-body interactions between atoms $i$ and $j$, including those due to nearest neighbors and angle effects. The potential is short-ranged and only considers nearest neighbor bonds. To model long-range nonbonded interactions, the bond-order potential is combined with pair-wise potentials either directly through splines [76] or indirectly with more sophisticated functions [77].

The Stillinger–Weber potential [78] potential was formulated to model silicon, with a particular emphasis on the liquid phases of silicon. It includes many-body interactions in the form of a sum of two- and three-body interactions

$$E = \sum_{ij} V_{ij}^2(r_{ij}) + \sum_{jik} V_{jik}^3(r_{ij}, r_{ik}), \tag{10.3}$$

where $V^2$ is a pair-additive interaction and $V^3$ is a three-body term. The three-body term includes an angular interaction that minimizes the potential energy for tetrahedral angles. This term favors the formation of open structures, such as the diamond cubic crystal structure.

The second potential is the embedded atom method (EAM) approach [79, 80] and related methods [81], which were initially developed for modeling metals and alloys. The functional form in the EAM is

$$E = \sum_{i} F(\rho i) + \sum_{i>j} \Phi(r_{ij}), \tag{10.4}$$

where $F$ is called the embedding energy. This term models the energy due to embedding an atom into a uniform electron gas with a uniform compensating positive background (jellium) of density $\rho_i$ that is equal to the actual electron density of the system. The term $\Phi(r_{ij})$ is a pairwise functional form that corrects for the jellium approximation. Several parameterizations of the EAM exist (see, for example, [79, 80, 82, 83, 84]) and it has recently been extended to model nonmetallic systems. For example, the modified EAM (MEAM) approach [64, 85, 86] was developed so that EAM could be applied to metal oxides [60] and covalently bound materials [86].

The third method is the general class of Coulomb or multipole interaction potentials used to model charged ionic materials or molecules [87]. In this formalism, an energy term is given as

$$E = \sum_{i} \sum_{j(>i)} \left( \frac{q(r_i)q(r_j)}{r_{ij}} \right), \tag{10.5}$$

where $q(r_i)$ is the charge on atom $i$ and $r_{ij}$ is the distance between atoms $i$ and $j$. More complex formalisms that take into account, say, the Madelung constant in the case of ionic crystals, are used in practice. In general, the charges are held fixed, but methods that allow charge to vary in a realistic manner have been developed [61, 88].

Lastly, long-range van der Waals or related forces are typically modeled with pairwise additive potentials. A widely used approximation is the Lennard-Jones (LJ) potential [89], which has the following functional form

$$E = 4\varepsilon \sum_{i} \sum_{j(>i)} \left[ \left( \frac{\sigma}{r_{ij}} \right)^{12} - \left( \frac{\sigma}{r_{ij}} \right)^{6} \right]. \tag{10.6}$$

In this approach $\varepsilon$ and $\sigma$ are parameters and $r_{ij}$ is the distance between atoms $i$ and $j$.

All of these potentials are widely used in MD simulations of materials, including studies of friction, lubrication, and wear.

### *10.1.2 Important Approximations*

Several approximations are typically used in MD simulations of friction. The first is the use of periodic boundary conditions (PBCs) and the minimum image convention for interatomic interactions [48]. In both cases the simulation supercell is surrounded by replicas of itself so that atoms (or phonons, etc.) that exit one side of the supercell remerge into the simulation through the opposite side of the supercell. In the minimum image convention an atom interacts either with another atom in the supercell or its equivalent atom in a surrounding cell depending on which distance to the atom is shortest. This process is illustrated in Fig. 10.1. In this convention supercells must be large enough that atoms do not interact with themselves over the periodic boundaries. In computational studies of friction and wear,

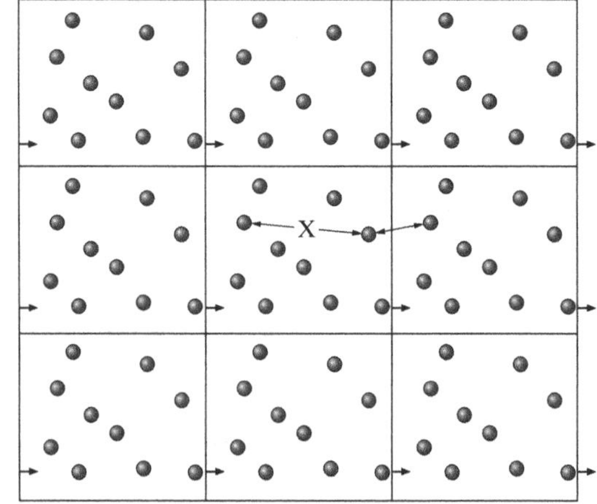


**Fig. 10.1** Illustration of periodic boundary conditions consisting of a central simulation cell surrounded by replica systems. The *solid arrows* indicate an atom leaving the central box and reentering on the opposite side. The *dotted arrows* illustrate the minimum image convention

PBCs are usually applied in the two dimensions within the plane(s) of the sliding surface(s). The strength of this approach is that it allows a finite number of atoms to model an infinite system. However, the influence of boundaries on system dynamics is not completely eliminated; for example, phonon scattering due to the periodic boundaries can influence heat transport and therefore frictional properties of sliding interfaces.

Another important tool that is often used in MD simulations of friction is thermostats to regulate system temperature. In macroscopic systems, heat that is generated from friction is dissipated rapidly from the surface to the bulk phonon modes. Because atomistic computer simulations are limited systems that are many orders of magnitude smaller than systems that are generally studied experimentally, thermostats are needed to prevent the system temperature from rising in a non-physical manner. Typically in simulations of indentation or friction, the thermostat is applied to a region of the simulation cell that is well removed from the interface where friction and indentation is taking place. In this way, local heating of the interface that occurs as work is done on the system, but excess heat is efficiently dissipated from the system as a whole. In this manner the adjustment of atomic temperatures occurs away from the processes of interest, and simplified approximations for the friction term can be used without unduly influencing the dynamics produced by the interatomic forces.

There are several different formalisms for atomistic thermostats. The simplest of these controls the temperature by intermittently rescaling the atomic velocities to values corresponding to the desired temperature [90] such that

$$\left(\frac{\boldsymbol{v}_{\text{new}}}{\boldsymbol{v}_{\text{old}}}\right)^2 = \frac{T}{T_{\text{ins}}}, \tag{10.7}$$

where $\boldsymbol{v}_{\text{new}}$ is the rescaled velocity, and $v_{\text{old}}$ is the velocity before the rescaling. This approach, which is called the velocity rescaling method, is both simple to implement and effective at maintaining a given temperature over the course of an MD simulation. It was consequently widely used in early MD simulations. The velocity rescaling approach does have some significant disadvantages, however. First, there is little theoretical basis for the adjustment of atomic velocities, and the system dynamics are not time-reversible, which is inconsistent with classical mechanics. Second, the rate and mode of heat dissipation are disconnected from system properties, which may affect system dynamics. Lastly, for typical MD simulation system sizes, the averaged quantities that are obtained, such as pressure for instance, do not correspond to values in any thermodynamic ensemble.

For these reasons, more sophisticated methods for maintaining system temperatures in MD simulations have been developed. The Langevin dynamics approach [48], which was originally developed from the theory of Brownian motion, falls into this category. In this approach, terms are added to the interatomic forces that correspond to a random force and a frictional term [46, 91, 92]. Therefore, Newton's equation of motion for atoms subjected to Langevin thermostats is given by the following equation rather than (10.1a,10.1b)

$$m\boldsymbol{a} = \boldsymbol{F} - m\xi\boldsymbol{v} + R(t), \tag{10.8}$$

where $\boldsymbol{F}$ are the forces due to the interatomic potential, $\xi$ is a friction coefficient, $m$ and $\boldsymbol{v}$ are the particle's mass and velocity, respectively, and $R(t)$ is a random force that acts as *white noise*. The friction term can be formulated in terms of a memory kernal, typically for harmonic solids [93, 94, 95], or a friction coefficient can be approximated using the Debye frequency. The random force can be given by a Gaussian distribution where the width, which is chosen to satisfy the fluctuation-dissipation theorem, is determined from the equation

$$\langle R(0)R(t)\rangle = 2mk_{\mathrm{B}}T\xi\delta(t). \tag{10.9}$$

Here, the function $R$ is the random force in (10.8), $m$ is the particle mass, $T$ is the desired temperature, $k_{\mathrm{B}}$ is Boltzmann's constant, $t$ is time, and $\xi$ is the friction coefficient. It should be noted that the random forces are uncoupled from those at previous steps, which is denoted by the delta function. Additionally, the width of the Gaussian distribution from which the random force is obtained varies with temperature. Thus, the Langevin approach does not require any feedback from the current temperature of the system as the random forces are determined solely from (10.9).

In the early 1980s, *Nosé* developed a new thermostat that corresponds directly to a canonical ensemble (system with constant temperature, volume and number of atoms) [96, 97], which is a significant advance from the methods described so far. In this approach, Nosé introduces a degree of freedom $s$ that corresponds to the heat bath and acts as a time scaling factor, and adds a parameter $Q$ that may be regarded as the heat bath *mass*. A simplified form of Nosé's method was subsequently implemented by Hoover [46] that eliminated the time scaling factor whilst introducing a thermodynamic friction coefficient $\zeta$. Hoover's formulation of Nosé's method is therefore easy to use and is commonly referred to as the Nosé–Hoover thermostat.

When this thermostat is applied to a system containing $N$ atoms, the equations of motion are written as (dots denote time derivatives)

$$\begin{aligned} \dot{\boldsymbol{r}}_i &= \frac{\boldsymbol{p}_i}{m_i}, \\ \dot{\boldsymbol{p}}_i &= \boldsymbol{F}_i - \zeta\boldsymbol{p}_i, \\ \dot{\zeta} &= \frac{1}{Q}\left(\sum_{i=1}^{N}\frac{p_i^2}{m_i} - N_{\mathrm{f}}k_{\mathrm{B}}T\right), \end{aligned} \tag{10.10}$$

where $\boldsymbol{r}_i$ is the position of atom $i$, $\boldsymbol{p}_i$ is the momentum and $\boldsymbol{F}_i$ is the force applied to each atom. The last equation in (10.10) contains the temperature control mechanism in the Nosé–Hoover thermostat. In particular, the term between the parentheses on the right-hand side of this equation is the difference between the system's instantaneous kinetic energy and the kinetic energy at the desired temperature.

If the instantaneous value is higher than the desired one, the friction force will increase to lower it and vice versa.

It should be pointed out that the choice of the heat bath *mass* $Q$ is arbitrary but crucial to the successful performance of the thermostat. For example, a small value of $Q$ leads to rapid temperature fluctuation while large $Q$ values result in inefficient sampling of phase space. Nosé recommended that $Q$ should be proportional to $N_f k_B T$ and should allow the added degree of freedom $s$ to oscillate around its averaged value at a frequency of the same order as the characteristic frequency of the physical system [96, 97]. If ergodic dynamic behavior is assumed, the Nosé–Hoover thermostat will maintain a well-defined canonical distribution in both momentum and coordinate space. However, for small systems where the dynamic is not ergodic, the Nosé–Hoover thermostat fails to generate a canonical distribution. Therefore, more sophisticated algorithms based on the Nosé–Hoover thermostat have been proposed to fix its ergodicity problem; for example, the *Nosé–Hoover chain* method of Martyna et al. [98]. However, these complex thermostats are not as easy to apply as the Nosé–Hoover thermostat due to the difficult evaluation of the coupling parameters for each different case and the significant computational cost [99]. From a practical point of view, if the molecular system is large enough that the movements of the atoms are sufficiently chaotic, ergodicity is guaranteed and the performance of the Nosé–Hoover thermostat is satisfactory [25].

In an alternative approach, Schall et al. recently introduced a hybrid continuum-atomistic thermostat [100]. In this method, an MD system is divided into grid regions, and the average kinetic energy in the atomistic simulation is used to define a temperature for each region. A continuum heat transfer equation is then solved stepwise on the grid using a finite difference approximation, and the velocities of the atoms in each grid region are scaled to match the solution of the continuum equation. To help account for a time lag in the transfer of kinetic to potential energy, Hoover constraining forces are added to those from the interatomic potential. This process is continued, leading to an ad hoc feedback between the continuum and atomistic simulations. The main advantage of this approach is that the experimental thermal diffusivity can be used in the continuum expression, leading to heat transfer behavior that matches experimental data. For example, in metals the majority of the thermal properties at room temperature arise from electronic degrees of freedom that are neglected with strictly classical potentials. This thermostat is relatively straightforward to implement, and requires only the interatomic potential and the bulk thermal diffusivity as input. It is also appropriate for nonequilibrium heat transfer, such as occurs as heat is dissipated from sliding surfaces moving at high relative velocities. Other localized phenomena, such as Joule heating and melting in current carrying applications, can also be simulated by using a recent extension to the hybrid thermostat [101]. This modification allows the ability to model degradation of interfaces under high electrical load at the atomic level. Relevant examples are hot switched radio frequency MEMS and metal/metal contacts in electromagnetic launchers.

Cushman et al. [102, 103] developed a unique alternative to the grand canonical ensemble by performing a series of grand canonical Monte Carlo simulations [48, 104] at various points along a hypothetical sliding trajectory. The results from these simulations are then used to calculate the correct particle numbers at a fixed chemical potential, which are then used as inputs to nonsliding, constant-*NVE* MD simulations at each of the chosen trajectory points. The sliding speed can be assumed to be infinitely slow because the system is fully equilibrated at each step along the sliding trajectory. This approach offers a useful alternative to continuous MD simulations that are restricted to sliding speeds that are orders of magnitude larger than most experimental studies (about 1 m/s or greater).

To summarize, this section provides a brief review and description of components that are used in atomistic, molecular dynamics simulation of many of the processes related to friction, such as indentation, sliding, and wear. The components discussed here include the potential energy expression used to calculate energies and forces in the simulations, periodic boundary conditions and thermostats. Each of these components has their own strengths and weaknesses that should be well-understood both prior to their use and in the interpretation of results. For example, general principles related to liquid lubrication in confined areas may be most easily understood and generalized from simulations that use pair potentials and may not require a thermostat. On the other hand, if one wants to study the wear or indentation of a surface of a particular metal, then EAM or other semiempirical potentials, together with a thermostat, would be expected to yield more reliable results. If one requires information on electronic effects, ab initio or semi-empirical approaches that include the evaluation of electronic degrees of freedom must be used. Thus, the best combination of components for a particular study depends on the chemical nature of the system of interest, the processes being simulated, the type of information desired, and the available computational resources.

## 10.2 Indentation

It is critical to understand the nanometer-scale properties of materials that are being considered for use as new coatings with specific friction and wear behavior. Experimental determination of these properties is most frequently done with the AFM, which provides a variety of data related to the interaction of the microscope tips with the sample surface [105, 106, 107]. In AFM experiments, the tip has a radius of about 1–100 nm and is pressed against the surface under ambient conditions (in air), ultrahigh vacuum (UHV) conditions, or in a liquid. The microscope tip can either move in the direction normal to the surface, which is the case in nanoindentation studies, or raster across the surface, which is the case in surface imaging or friction studies. Sliding rates of 1 nm/s–1 μ/s are typically used, which are many orders of magnitude slower than the rates used in MD simulations of sliding or indentation of around 1–100 m/s. As discussed in the previous section, the higher rates used in computational simulations are a consequence of modeling full atomic motion, which occurs on a femto- to picosecond timescale, and the stepwise solution of the classical equations of motion, which makes the large number of simulation steps needed to reach experimental timescales computationally impossible with current processor speeds.

As the tip moves either normal to or across the surface, the forces acting upon it as a result of its interactions with the surface are measured. When the tip is moved in the surface normal direction, it can penetrate the surface on the nanometer scale and provide information on the nanometer-scale mechanical properties of the surface [108, 109]. The indentation process also causes the force on the tip to increase, and the rate of increase is related to both the depth of indentation and the properties of the surface. The region of the force curve that reflects this high force is known as the repulsive wall region [105], or, when considered without any lateral motion of the tip, an indentation curve. When the tip is retracted after indentation, enhanced adhesion between the tip and surface relative to the initial contact can result. This phenomenon is indicated by hysteresis in the force curve.

Tip–surface adhesion can result from the formation of chemical bonds between the tip and the sample, or from the formation of liquid capillaries between the microscope tip and the surface caused by the interaction of the tip with a layer of liquid contamination on the surface. The latter case is especially prevalent in AFM studies conducted in ambient environments. In the case of clean metallic systems, the sample can wet the tip or the tip can wet the sample in the form of a connective *neck* of metal atoms between the surface and the tip that can lead to adhesion. In the case of polymeric or molecular systems, entanglement of molecules that are anchored on the tip with molecules anchored on the sample can be responsible for force curve hysteresis.

In the case of horizontal rastering of AFM tips across surfaces, the force curve data provide a map of the surface that is indicative of the surface topography [110]. If the deflection of the tip in the lateral direction is recorded while the tip is being rastered, a friction map of the surface [20] is produced.

The rest of this section discusses some of the important insights and findings that have been obtained from MD simulations of nanoindentation. These studies have not only provided insight into the physical phenomena responsible for the qualitative shapes of AFM force curves, they have also revealed a wealth of atomic-scale phenomena that occur during nanoindentation that was not previously known.

### 10.2.1 Surfaces

The nature of adhesive interactions between clean, deformable metal tips indenting metal surfaces have been identified and clarified over the course of the last decade through the use of MD simulations [107, 111, 112, 113, 114, 115]. In particular, the high surface energies associated with clean metal surfaces can lead to strongly attractive interactions between surfaces in contact. The strength of this attraction can be so large that when the tip gets close enough to the surface to interact with it, surface atoms *jump* upwards to wet the tip in a phenomena known as jump-to-contact (JC). This wetting mechanism was first discovered in MD simulations [114] and has been confirmed experimentally [108, 116, 117, 118] using the AFM, as shown in Fig. 10.2.

The MD simulations of Landman et al. [114, 119, 120, 121] using EAM potentials revealed that the JC phenomenon in metallic systems is driven by the need of the atoms at the tip–surface interface to optimize their interaction energies while maintaining their individual material cohesive binding. When the tip advances past the JC point it indents the surface, which causes the force to

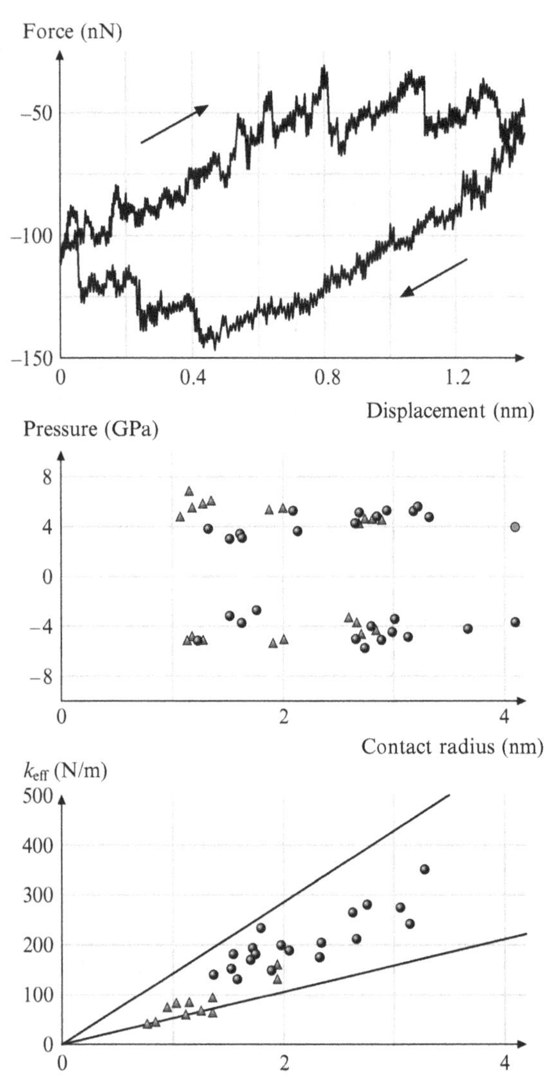


**Fig. 10.2** *Top*: The experimental values for the force between a tip and a surface that have a connective neck between them. The neck contracts and extends without breaking on the scales shown. *Bottom*: The effective spring constant $k_{eff}$ determined experimentally for the connective necks and corresponding maximum pressures, versus contact radius of the tip. The *triangles* indicate measurements taken at room temperature; the *circles* are the measurements taken at liquid He temperatures (After [118], © ACS 1996)

increase. This behavior is indicated in Fig. 10.3, points D to M. This region of the computer-generated force curve has a maximum not present in the force curve generated from experimental data (Fig. 10.3, point L). This is due to tip-induced flow of the metal atoms in the surface that causes *pile-up* of the surface atoms around the edges of the indenter. Hysteresis on the withdrawal of the tip, shown in Fig. 10.3, points M to X, is present due to adhesion between the tip and the substrate. In particular, as the tip retracts from the sample, a connective *neck* or nanowire of atoms forms between the tip and the substrate that is primarily composed of metal atoms from the surface with some atoms from the metal indenter that have diffused into the structure. A snapshot from the MD simulations that illustrates this behavior is shown in Fig. 10.4.

As the tip is withdrawn farther, the magnitude of the force increases (becomes more negative) until, at a critical force, the atoms in adjacent layers of the connective nanowire rearrange so that an additional row of atoms is created. This process causes elongation of the connective nanowire and is responsible for the fine structure

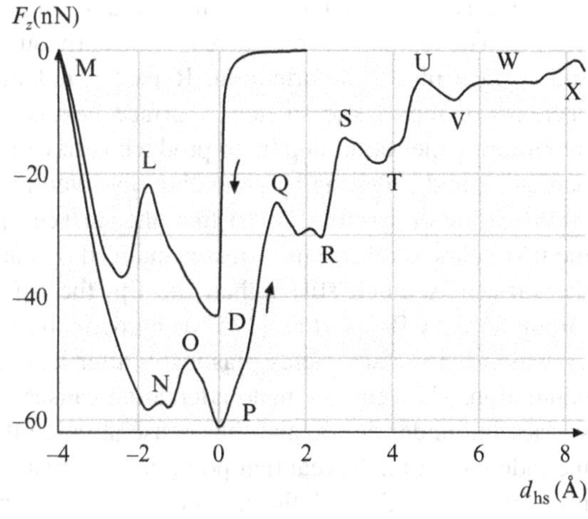


**Fig. 10.3** Computationally derived force $F_z$ versus tip-to-sample distance $d_{hs}$ curves for approach, contact, indentation, then separation using the same tip–sample system shown in Fig. 10.4. These data were calculated from an MD simulation (After [114], © AAAS 1990)

**Fig. 10.4** Illustration of atoms in the MD simulation of a Ni tip being pulled back from an Au substrate. This causes the formation of a connective neck of atoms between the tip and the surface (After [114], © AAAS 1990)

(apparent as a series of maxima) present in the retraction portion of the force curve. These elongation and rearrangement steps are repeated until the connection between the tip and the surface is broken. Similar elongation events have been observed experimentally. For example, scanning tunneling microscopy (STM) experiments demonstrate that the metal nanowires between metal tips and surfaces can elongate $\approx 2{,}500$ Å without breaking [122].

The JC process has been shown to affect the temperature at the tip–surface interface. For instance, the constant-energy MD simulations of Tomagnini et al. [123] predicted that the energy released due to the wetting of the tip by surface atoms increases the temperature of the tip by about 15 K at room temperature and is accompanied by significant structural rearrangement. At temperatures high enough to cause the first few metal surface layers to be liquid, the distance at which the JC occurs increases, as does the contact area between the tip and the surface and the amount of nanowire elongation prior to breakage.

Simulations by Komvopoulos and Yan [124] using LJ potentials showed how metallic surfaces respond to single and repeated indentation by metallic, or covalently bound, rigid tips. The simulations predicted that a single indentation event produces hysteresis in the force curve as a result of surface plastic deformation and heating. The repulsive force decreases abruptly during surface penetration by the tip and surface plastic deformation. Repeated indentation results in the continuous decrease of the elastic stiffness, surface heating, and mean contact pressures at maximum penetration depths to produce behaviors that are similar to cyclic work hardening and softening by annealing observed in metals at the macroscale.

When the tip is much stiffer than the surface, pile-up of surface atoms around the tip occurs to relieve the stresses induced by nanoindentation. In contrast, when the surface is much stiffer than the tip, the tip can be damaged or destroyed. Simulations by Belak et al. [125] using perfectly rigid tips showed the mechanism by which the surface yields plastically after its elastic threshold is exceeded. The simulations showed how nanoindentation causes surface atoms to move on to the surface but under the tip and thus cause atomic pile-up. In this study, variations in the indentation rate reveal that point defects created as a result of nanoindentation relax by moving through the surface if the rate of indentation is slow enough. If the indentation rate is too high, there is no time for the point defects to relax and move away from the indentation area and so strain builds up more rapidly. The rigid indenters considered in these MD simulations are analogous to experiments that use surface passivation to prevent JC between the tip and the surface [126, 127], the results of which agree with the predicted results of pile-up and crater formation, as shown in Fig. 10.5 [126].

In short, MD simulations are able to explain the atomic-scale mechanisms behind measured experimental force curves produced when metal tips indent homogeneous metal surfaces to nanometer-scale depths. This preliminary work has spawned much of the current interest in using the JC to produce metal nanowires [128, 129, 130].

MD simulations have also been used to examine the relationship between nanoindentation and surface structure. This is most apparent in a series of computational studies that consider the indentation of a surface with a *virtual* hard-sphere

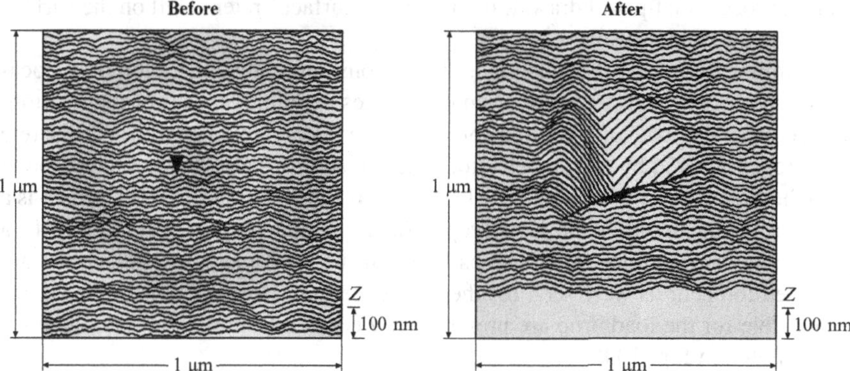


**Fig. 10.5** Images of a gold surface before and after being indented with a pyramidal shaped diamond tip in air. The indentation created a surface crater. Note the pile-up around the crater edges (After [126], © Elsevier 1993)

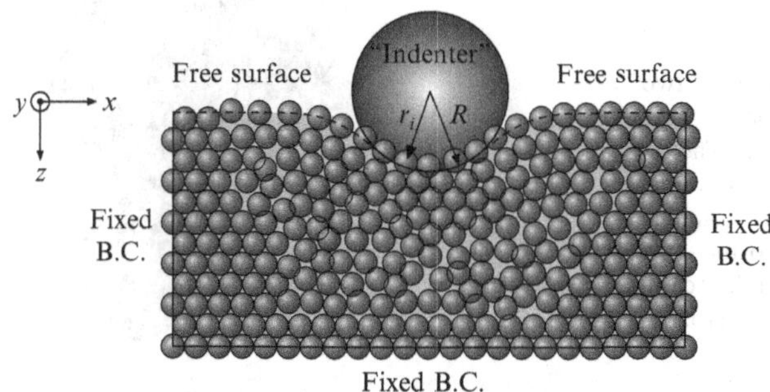


**Fig. 10.6** A schematic of a spherical, virtual tip indenting a metal surface (After [132], © Elsevier 1993) (B.C. – boundary conditions)

indenter in a manner that is independent of the rate of indentation, as shown in Fig. 10.6. The virtual indenter is modeled through the application of a repulsive force to the surface rather than through the presence of an actual atomic tip. Kelchner et al. [131], rather than use MD, pushed the indenter against the surface a short distance and then allowed the system to relax using standard energy minimization methods in combination with EAM potentials. The system is fully relaxed when the energy of the surface system is minimized. After relaxation, the tip is pushed further into the surface and the process is repeated. As the tip generates more stresses in the surface, dislocations are generated and plastic deformation occurs. If the tip is pulled back after indenting less than a specific critical value, the atoms that were plastically deformed are healed during the retraction and the surface recovers its original structure. In contrast, if the tip is indented past the critical depth, additional dislocations are created that interfere with the surface healing process on tip withdrawal. In this case, a surface crater is left on the surface following nanoindentation.

A similar study by Lilleodden et al. [132] considered the generation of dislocations in perfect crystals and near grain boundaries in gold. Analysis of the relationship between the load and the tip displacement in the perfect crystal shows discrete load drops followed by elastic behavior. These load drops are shown to correspond to the homogeneous nucleation of dislocations, as illustrated in Fig. 10.7, which is a snapshot taken just after the load drop. When nanoindentation occurs close to a grain boundary, similar relationships between the load and tip displacement are predicted to occur as were seen for the perfect crystal. However, the dislocations responsible for the load drop are preferentially emitted from the grain boundaries, as illustrated in Fig. 10.8.

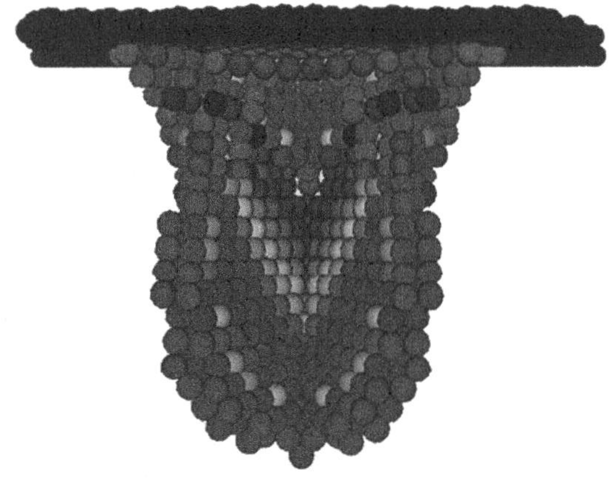

**Fig. 10.7** Snapshot of two partial dislocations separated by a stacking fault. The *dark spheres in the center* of the structure indicate atoms in perfect crystal positions after both partial dislocations have passed (After [132], © Elsevier 1993)

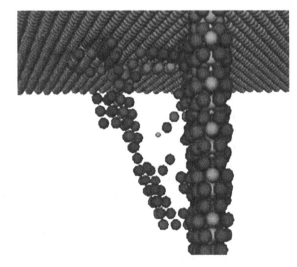

**Fig. 10.8** Snapshot of the high-energy atoms only after a load drop caused by dislocation generation during the nanoindentation of gold near a grain boundary (After [132], © Elsevier 1993)

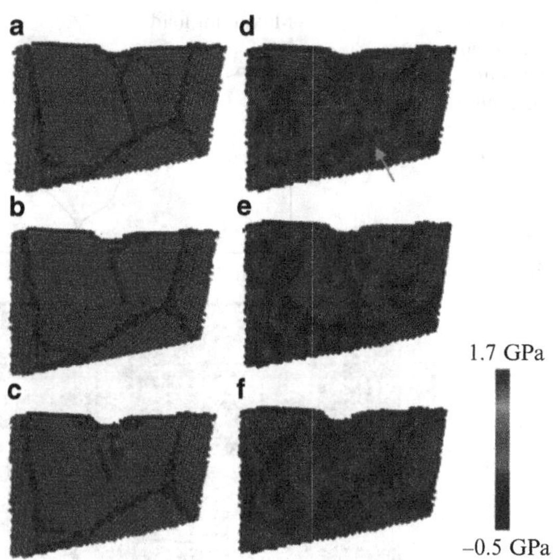


**Fig. 10.9** Snapshots showing the atomic stress distribution and atomic structures in a gold surface. Figures **(a)**–**(c)** show the atomic structure at indentation depths of 7.9, 8.6, and 9.6 Å, respectively, with a virtual spherical indenter. A dislocation is represented by the two parallel {111} planes (*two dark lines*) that show the stacking fault left behind after the leading partial dislocation has passed. Figures **(d)**–**(f)** show the atomic stress distribution of the same system at the same indentation depths. Here the *dark color* indicates compressive hydrostatic pressures of 1.7 GPa and higher while the *gray* color indicates tensile pressures of −0.5 GPa and lower. The *arrow* in **(d)** shows the region of the system where a dislocation interacts with a grain boundary (After [133], © Elsevier 2004)

Simulations can also show how atomic structure and stresses are affected by nanoindentation. For instance, MD simulations with a virtual indenter by Hasnaoui et al. [133] using semi-empirical tight-binding methods showed the interaction between the grain boundaries under the indenter and the dislocations generated by the indentation, as illustrated in Fig. 10.9. This study shows that if the size of the indenter is smaller than the grain size, the grain boundaries can emit, absorb, and reflect the dislocations in a manner that depends on atomic structure and the distribution of stresses.

Zimmerman et al. considered the indentation of a single-crystal gold substrate both near and far from a surface step [134]. The results of these simulations, which used EAM potentials, showed that the onset of plastic deformation depends to a significant degree on the distance of indentation from the step, and whether the indentation is on the plane above or below the step. In a related set of simulations, Shenderova et al. [135] examined whether ultrashallow elastic nanoindentation can nondestructively probe surface stress distributions associated with surface structures such as a trench and a dislocation intersecting a surface. The simulations carried out the nanoindentation to a constant depth. They predicted maximum loads that reflect the in-plane stresses at the point of contact between the indenter and the substrate, as illustrated in Fig. 10.10.

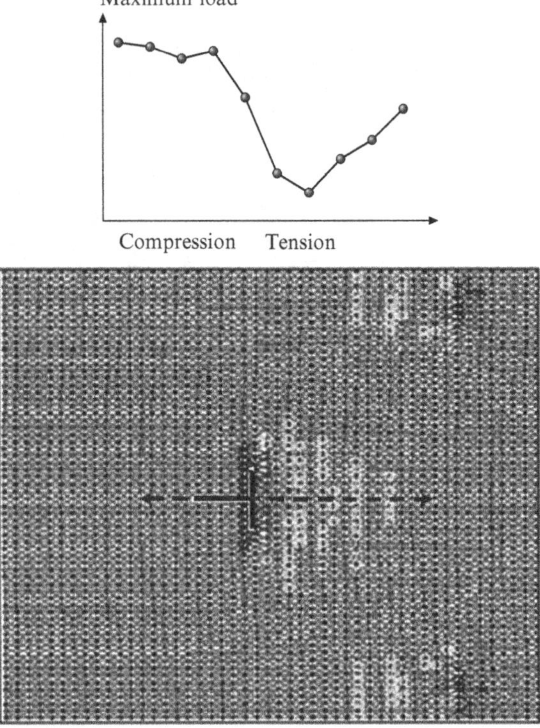


**Fig. 10.10** Data and system illustration from a simulation of a gold surface containing a dislocation. *Top*: Maximum load for simulated shallow indentation at several points along the *dotted line* in the *bottom* illustration. *Bottom*: Top view of the simulated surface. The dislocation is denoted by the *solid black lines*

Since the 1930s, studies have been performed using hardness measurement techniques [136, 137, 138, 139] and indentation methods [140] that suggest that the hardness of a material depends on applied in-plane uni- and bi-axial strain. In general, tensile strain appeared to decrease hardness while increases in hardness under compressive in-plane strain were reported. This behavior had traditionally been attributed to the contribution of stresses from the local strain from the indentation to the resolved shear stresses and the in-plane strain [137, 139]. However, in 1996, Pharr and coworkers determined that changes in elastic modulus determined from unloading curves of strained substrates using contact areas estimated via an elastic model are too large to have physical significance, a result that brought into question the interpretation of prior hardness data [140, 141]. They hypothesized that the apparent change in modulus (and hardness) with in-plane strain is mainly due to changes in contact area that are not typically taken into consideration in elastic half-space models. This hypothesis was based on experimental nanoindentation studies of a strained polycrystalline aluminum alloy and finite element calculations on an isotropic solid [140, 141]. They further suggested that in-plane compression increases pile-up around the indenter that, when not taken into account in the analysis of unloading curves, implies a nonphysical increase in modulus. Likewise, they suggested that in-plane tensile strain reduces the amount of material that is piled up around an indenter, which leads to a corresponding reduced (nonphysical) modulus when interpreting unloading curves using elastic models.

To explore in more detail the issue of pile-up and its influence on the interpretation of loading curves, Schall and Brenner used MD simulations and EAM potentials to model the plastic nanoindentation of a single-crystal gold surface under an applied in-plane strain [142]. These simulations predicted that the mean pressure, calculated from true contact areas that take into account plastic pile-up around the indenter, varies only slightly with applied pre-stress. They also predicted that the higher values occur in compression rather than in tension, and that the modulus calculated from the true contact area is essentially independent of the pre-stress level in the substrate. In contrast, if the contact area is estimated from approximate elastic formulae, the contact area is underestimated, which leads to a strong, incorrect dependence of apparent modulus on the pre-stress level. The simulations also showed larger pile-up in compression than in tension, in agreement with the Pharr model, and both regimes produced contact areas larger than those typically assumed in elastic analyses. These findings are illustrated in Fig. 10.11.

Nanometer-scale indentation of ceramic systems has also been investigated with MD simulations. Ceramics are stiffer and more brittle than metals at the macroscale and examining the nanoindentation of ceramic surfaces provides information about the nanometer-scale properties. They also reveal the manner by which defects form in covalent and ionic materials. For example, Landman et al. [113, 143] considered the interaction of a $CaF_2$ tip with a $CaF_2$ substrate in MD simulations using empirical potentials. As the tip approaches the surface, the attractive force between them steadily increases. This attractive force increases dramatically at the critical distance of 2.3 Å as the interlayer spacing of the tip increases (the tip is elongated)

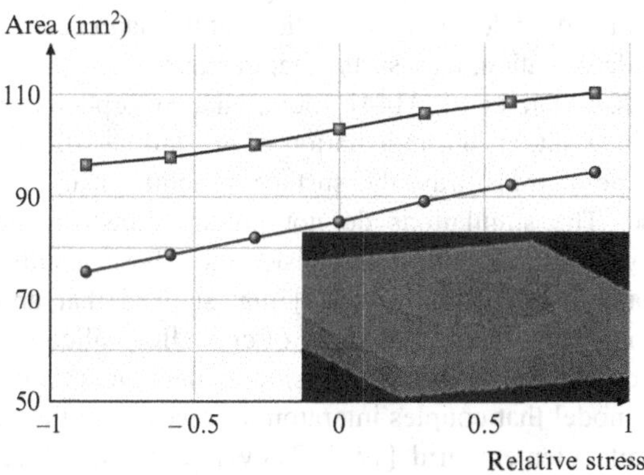


**Fig. 10.11** Contact area projected in the plane at a maximum load for simulated indention of a gold surface as a function of in-plane biaxial stress. The stress is normalized to the theoretical yield stress. The *top curve* is from an atomistic simulation; the *bottom curve* is from an elastic model. *Inset*: Illustration of the region near the indention from the simulation. The tip is not shown for clarity. Initial formation of pile-up around the edge of the indentation is apparent

in a process that is similar to the JC phenomenon observed in metals. An important difference, however, is the amount of elongation, which is 0.35 Å in the case of the ionic ceramics and several angstroms in the case of metals. As the distance between the tip and the surface decreases further, the attractive nature of their interaction increases until a maximum value is reached. Indentation beyond this point results in a repulsive tip–substrate interaction, compression of the tip, and ionic bonding between the tip and substrate. These bonds are responsible for the hysteresis predicted to occur in the force curve on retraction, which ultimately leads to plastic deformation of the tip followed by fracture.

The responses of covalently bound ceramics such as diamond and silicon to nanoindentation have been heavily studied with MD simulations. One of the first of these computational studies was carried out by Kallman et al. who used the Stillinger–Weber potential to examine the indentation of amorphous and crystalline silicon [144]. The motivation for this study came from experimental data that indicated a large change in electrical resistivity during indentation of silicon, which led to the suggestion of a load-induced phase transition below the indenter. Clarke et al., for example, reported forming an Ohmic contact under load, and using transmission electron microscopy they observed an amorphous phase at the point of contact after indentation [145]. Using micro-Raman microscopy, Kailer et al. identified a metallic $\beta$-Sn phase in silicon near the interface of a diamond indenter during hardness loading [146]. Furthermore, upon rapid unloading they detected amorphous silicon as in the Clarke et al. [145] experiments, while slow unloading resulted in a mixture of high-pressure polymorphs near the indent point. At the highest indentation rate and the lowest temperature, the simulations by Kallman et al. [144] showed that amorphous and crystalline silicon have similar yield strengths of 138 and 179 kbar, respectively. In contrast, at temperatures near the melting temperature and at the slowest indentation rate, both amorphous and crystalline silicon are predicted to have lower yield strengths of 30 kbar. The simulations thus show how the predicted yield strength of silicon at the nanometer scale depends on structure, rate of deformation, and surface temperature.

Interestingly, Kallman et al. [144] found that amorphous silicon does not crystallize upon indentation, but indentation of crystalline silicon at temperatures near the melting point transforms the surface structure near the indenter to the amorphous phase. The simulations do not predict transformation to the $\beta$-Sn structure under any of the conditions considered. These results agree with the outcomes of scratching experiments [147] that showed that amorphous silicon emerges from room-temperature scratching of crystalline silicon.

Kaxiras and coworkers revisited the silicon nanoindentation issue using a quasi-continuum model that couples interatomic forces from the Stillinger–Weber potential to a finite element grid [148]. They report good agreement between simulated loading curves and experiment provided that the curves are scaled by the indenter size. Rather than the $\beta$-Sn structure, however, atomic displacements suggest formation of a metallic structure with fivefold coordination below the indenter upon loading, and a residual simple cubic phase near the indentation site after the load is released rather than the mix of high-pressure phases characterized experimentally. Smith et al. attribute this discrepancy to shortcomings of the Stillinger–Weber potential inadequately describing the high-pressure phases of silicon. They also used a simple model for changes in electrical resistivity with loading involving contributions from both a Schottky barrier and spreading resistance. Simulated resistance-versus-loading curves agree well with experiment despite possible discrepancies between the high-pressure phases under the indenter, suggesting that the salient features of the experiment are not dependent on the details of the high-pressure phases produced.

Additional MD simulations of the indentation of silicon were carried out by Cheong and Zhang [149]. Their simulations provide further details about the phase transformations that occur in silicon as a result of nanoindentation. In particular, they find that the diamond cubic silicon is transformed into a body-centered tetragonal structure ($\beta$-Si) upon loading of the indenter, as illustrated in Fig. 10.12. Figure 10.13 shows that the coordination numbers of silicon atoms also coincide with that of the theoretical $\beta$-Si structure. The body-centered tetragonal structure is transformed into amorphous silicon during the unloading stage. A second indentation simulation again predicted that that this is a reversible process. Atomistic simulations by Sanz-Navarro et al. [150] shows the relation between the indentation

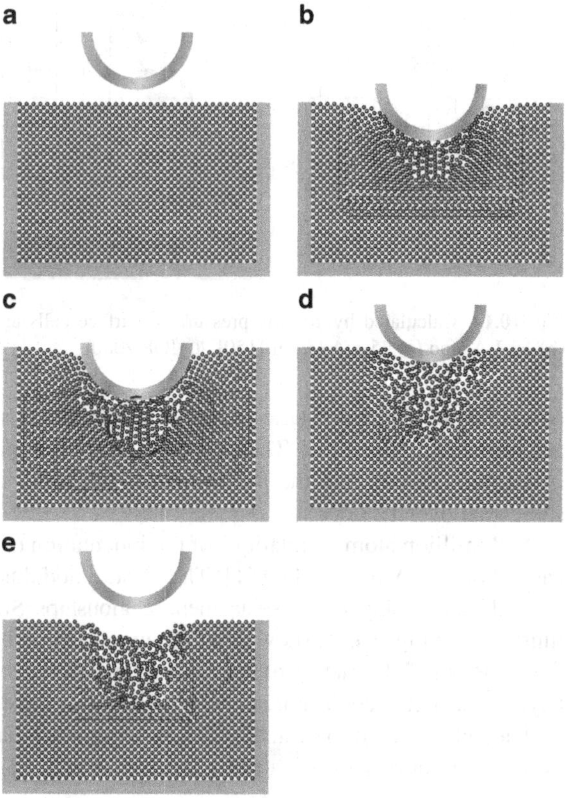

**Fig. 10.12** Snapshots of a silicon sample during indentation. The *smaller dots* are diamond atoms. **(a)** Crystalline silicon prior to indentation. **(b)** Atoms beneath the indenter are displaced as a result of indentation. **(c)** The system at maximum indentation. Some of the atoms are in a crystalline arrangement (*circled region*) that is different from the diamond structure. **(d)** The surface structure is largely amorphous as the tip is withdrawn. **(e)** The surface after indentation. Note the amorphous region at the site of the indentation process (After [133], © IOP 2000)

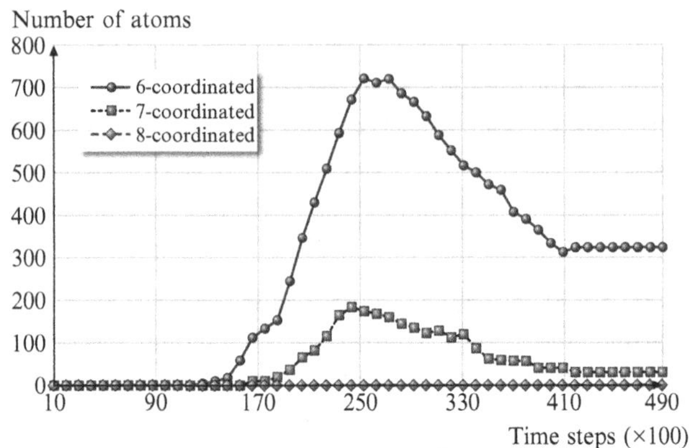


**Fig. 10.13** The coordination of the silicon atoms shown in Fig. 10.12 as a function of time during nanoindentation (After [149], © IOP 2000)

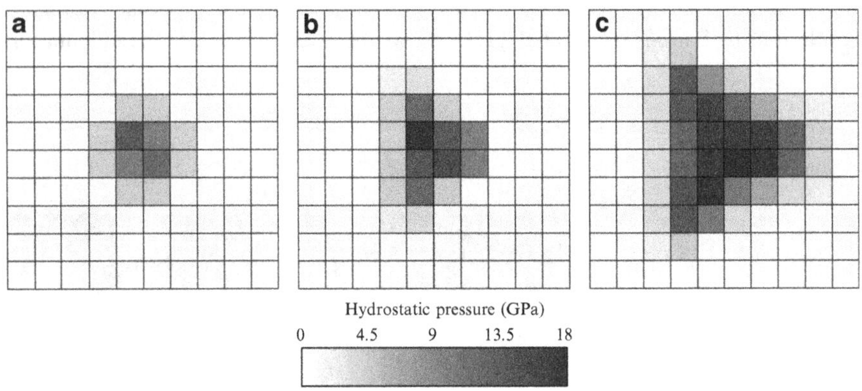


**Fig. 10.14** Calculated hydrostatic pressure of surface cells at indentation depths of (**a**) 8.9 Å, (**b**) 15.7 Å, and (**c**) 25.3 Å (After [150], © IOP 2000)

of silicon and the hydrostatic pressure on surface cells due to the nanoindentation, as illustrated in Fig. 10.14. These simulations further predict that the transformation of diamond silicon into the $\beta$-Si structure can occur if the hydrostatic pressure is somewhat over 12 GPa.

Multimillion atom simulations of the indentation of silicon nitride were recently carried out by Walsh et al. [151]. The elastic modulus and hardness of the surface was calculated using load–displacement relationships. Snapshots from the simulations, illustrated in Fig. 10.15, show that pile-up occurs on the surface along the edges of the tip. Plastic deformation of the surface is predicted to extend a significant distance beyond the actual contact area of the indenter, as illustrated in Fig. 10.15.

The indentation of bare and hydrogen-terminated diamond (111) surfaces beyond the elastic limit was investigated by Harrison et al. [152] using a hydrogen-terminated

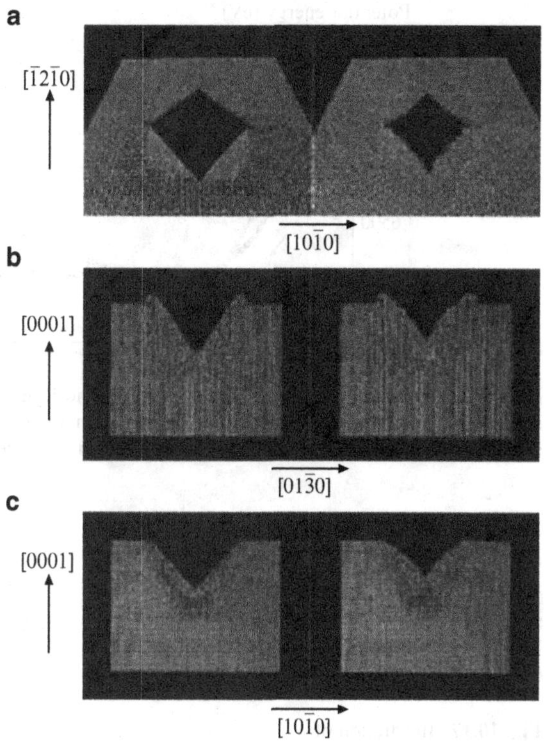


**Fig. 10.15** Snapshots of the silicon nitride (**a**) surface, (**b**) slide parallel to the edges of the indenter, and (**c**) slide across the indenter diagonal. The *left-hand side* shows the surface when it is fully loaded, while the *right-hand side* shows the surface after the tip has been withdrawn (After [151], © AIOP 2003)

$sp^3$-bonded tip in MD simulations that utilized bond-order potentials. The simulations identified the depth and applied force at which the diamond (111) substrate incurred plastic deformation due to indentation. At low indentation forces, the tip–surface interaction is purely elastic, as illustrated in Fig. 10.16. This finding agrees with the findings of Cho and Joannopoulos [153], who examined the atomic-scale mechanical hysteresis experienced by an AFM tip indenting Si(100) with density functional theory. The calculations predicted that at low rates it is possible to cycle repeatedly between two buckled configurations of the surface without adhesion.

When the nanoindentation process of diamond (111) is plastic, connective strings of atoms are formed between the tip and the surface, as illustrated in Fig. 10.17. These strings break as the distance between the tip and crystal increases and each break is accompanied by a sudden drop in the potential energy at large positive values of tip–substrate separation. The simulations further predict that the tip end twists to minimize interatomic repulsive interactions between the hydrogen atoms on the surface and the tip. This behavior is predicted to lead to new covalent bond formation between the tip and the carbon atoms below the first layer of the surface and connective strings of atoms between the tip and the surface when the tip is retracted. Not surprisingly, when the surface is bare and not terminated

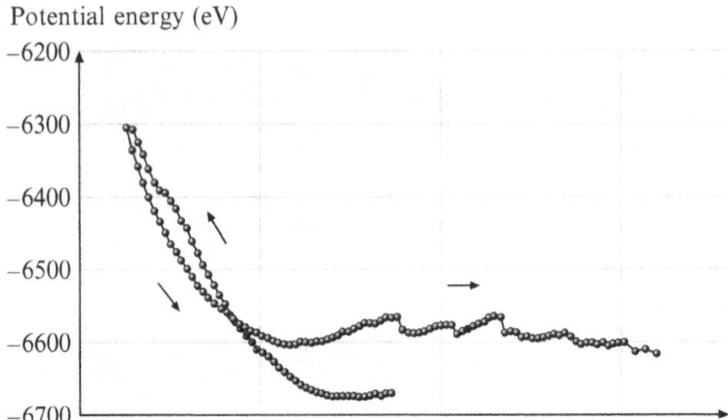


**Fig. 10.16** Potential energy as a function of rigid-layer separation generated from an MD simulation of an elastic (nonadhesive) indentation of a hydrogen-terminated diamond (111) surface using a hydrogen-terminated, $sp^3$-hybridized tip (After [152], © Elsevier 1992)

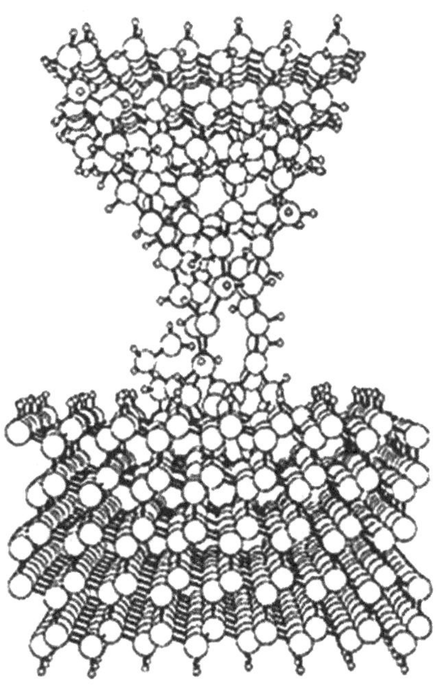

**Fig. 10.17** Illustration of atoms in the MD simulation of the indentation of a hydrogen-terminated diamond (111) substrate with a hydrogen-terminated, $sp^3$-hybridized tip at selected time intervals. The figure illustrates the tip–substrate system as the tip was being withdrawn from the sample. *Large* and *small spheres* represent carbon and hydrogen atoms, respectively (After [152], © Elsevier 1992)

with hydrogen atoms, the repulsive interactions between the tip and the surface are minimized and the tip indents the substrate without twisting [152]. Because carbon–carbon bonds are formed between the tip and the first layer of the substrate, the indentation is ordered (the surface is not disrupted as much by interacting with the tip) and the eventual fracture of the tip during retraction results in minimal damage to the substrate. The concerted fracture of all bonds in the tip gives rise to a single maximum in the potential versus distance curve at large distances.

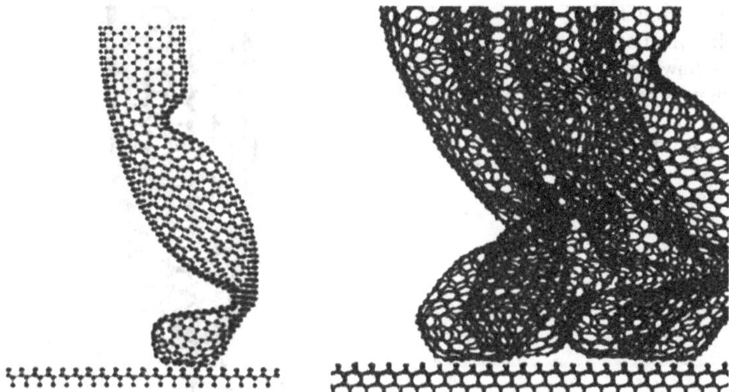

**Fig. 10.18** Snapshots of the indentation of a single-wall nanotube (*left-hand image*) and a bundle of nanotubes (*right-hand image*) on hydrogen-terminated diamond (111)

Harrison et al. [154, 155] and Garg et al. [156, 157] considered the indentation of hydrogen-terminated diamond and graphene surfaces with AFM tips of carbon nanotubes and nanotube bundles using MD simulations and bond-order potentials. Tips consisting of both single-wall nanotubes and multiwall nanotubes were considered. The simulations predicted that nanotubes do not plastically deform during tip crashes on these surfaces. Rather, they elastically deform, buckle, and slip as shown in Fig. 10.18. However, as is the case for diamond tips indenting reactive diamond surfaces discussed above, strong adhesion can occur between the nanotube and the surface that destroys the nanotube in the case of highly reactive surfaces, as illustrated in Fig. 10.19.

To summarize, MD simulations reveal the properties of ceramic tips and surfaces with covalent or ceramic bonding that are most important for nanometer-scale indentation. They predict that brittle fracture of the tip can occur that is sometimes accompanied by strong adhesion with the surface. They also reveal the conditions under which neither the tip nor the surface is affected by the nanoindentation process. The insight gained from these simulations helps in the interpretation of experimental data, and it also reveals the nanometer-scale mechanisms by which, for example, tip buckling and permanent modification of the surface occur.

### *10.2.2 Thin Films*

In many instances, surfaces are covered with thin films that can range in thickness from a few atomic layers to several μm. These films are more likely to have properties that differ from the properties of bulk materials of similar composition, and the likelihood of this increases as the film thickness decreases. Nanoindentation is one of the best approaches to determining the properties of these films. Consequently, numerous computational simulations of this process have been carried out.

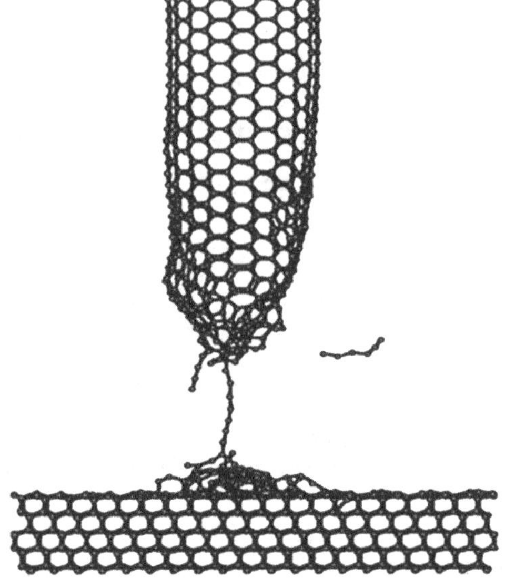

**Fig. 10.19** Snapshot of a single-wall carbon nanotube as it is withdrawn following indentation on a bare diamond (111) surface (After [156], © APS 1999)

For example, MD simulations have been used to study the indentation of metal surfaces covered with liquid *n*-hexadecane films, as illustrated in Fig. 10.20. As the metal tip touches the film, some of the molecules from the surface transfer to the tip and this causes the film to *swell*. As the tip continues to push against the surface, the hydrocarbon film wets the side of the tip. The simulations show how the hydrocarbon film passivates the surface and prevents the strong attractive interactions discussed above for clean metal surfaces and tips from occurring.

In a series of MD simulations, *Tupper* and Brenner modeled the compression of a thiol self-assembled monolayer (SAM) on a rigid gold surface using both a smooth compressing surface [158] and a compressing surface with an asperity [159]. These simulations showed that compression with the smooth surface produced a compression-induced structural change that led to a change in slope of the simulated force versus compression curve. This transition is reversible and involves a change in the ordered arrangement of the sulfur head groups on the gold surface. A similar change in slope seemed to be present in the experimental indentation curves of Houston and coworkers [160], but was not discussed by the authors. The simulations with the asperity showed that the asperity is able to penetrate the tail groups of the SAM, as illustrated in Fig. 10.21, before an appreciable load is apparent on the compressing surface. This result indicates that it is possible to image the head groups of a thiol self-assembled monolayer that are adsorbed onto the surface of a gold substrate using STM, and consequently ordered images of these systems may not be indicative of the arrangement of the tail groups.

Zhang et al. [161] used a hybrid MD simulation approach, where a dynamic element model for the AFM cantilever was merged with a MD relaxation approach

Fig. 10.20 Cutaways of the side view from molecular dynamics simulations of a Ni tip indenting a Au(001) surface covered with a hexadecane film. In (e) only the metal atoms are shown. Note how the hexadecane is forced out from between the metal surfaces (After [113], © Elsevier 1995)

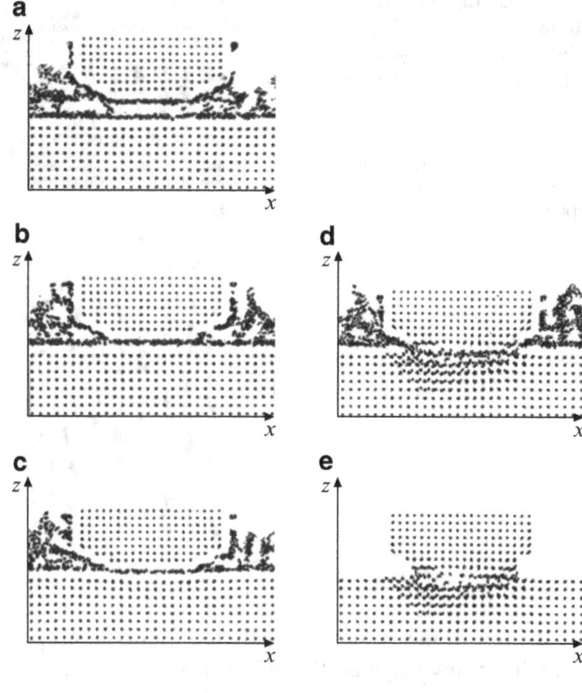


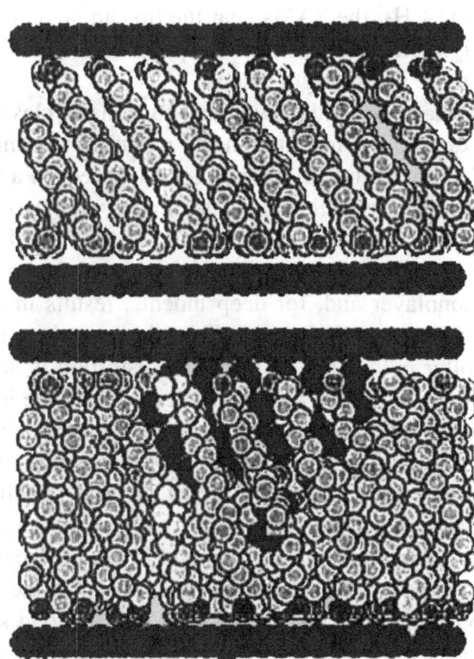

Fig. 10.21 Snapshots illustrating the compression of a self-assembled thiol film on gold for a smooth surface (*top*) and a surface containing an asperity (*bottom*). The asperity can penetrate and disorder the film tail groups before appreciable load occurs

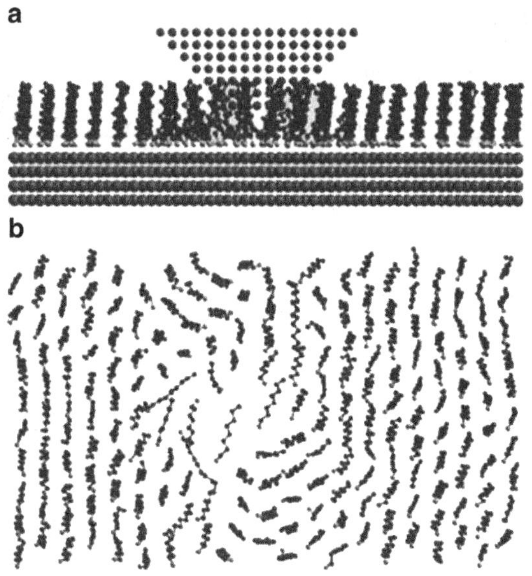


**Fig. 10.22** (**a**) Side and (**b**) top views of the final configuration of a $C_7CH_3$ self-assembled monolayer on Au(111) under a high normal load of 1.2 nN at 300 K. The tip is not shown in (**b**) for clarity (After [161], © ACS 2003)

for the rest of the system, to study the frictional properties of alkanethiol SAMs on gold. They investigated the effect of several variables like chain length, terminal group, scan direction, and scan velocity. Their results show that friction forces decrease as the chain length of the SAMs increase. In the case of shorter chains such as $C_7CH_3$, the SAMs near the tip can be deformed by indentation, as illustrated in Fig. 10.22. This behavior is predicted to be the cause of higher friction that occurs for the short-length chains.

Harrison and coworkers have used classical MD simulations [155, 162] to examine the indentation of monolayers composed of linear hydrocarbon chains that are chemically bound (or anchored) to a diamond substrate. Both flexible and rigid single-wall, capped nanotubes were used as tips. The simulations showed that indentation causes the ordering of the monolayer to be disrupted regardless of the type of tip used. Indentation results in the formation of gauche defects within the monolayer and, for deep indents, results in the pinning of selected hydrocarbon chains beneath the tube. Flexible nanotubes tilt slightly as they begin to indent the softer monolayers. This small distortion is due to the fact that nanotubes are stiff along their axial direction and more flexible in the transverse direction. In contrast, when the nanotubes encounter the hard diamond substrate, after *pushing* through the monolayer, they buckle. This process is illustrated in Fig. 10.23 and the force curves are shown in Fig. 10.24. The buckling of the nanotube was previously observed when single-wall, capped nanotubes were brought into contact with hydrogen-terminated diamond (111) surfaces [154, 156]. In the absence of the monolayer, the nanotube tips encounter the hard substrate in an almost vertical position. This interaction with the diamond substrate causes the cap of the nanotubes to be *pushed* inside the nanotube (they invert). Increasing the load on the

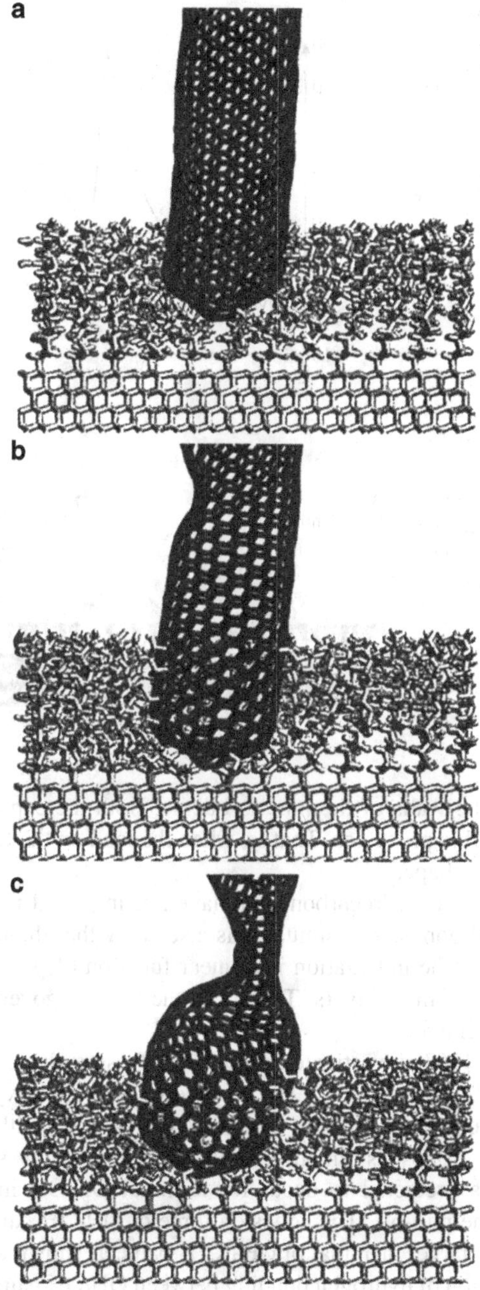


**Fig. 10.23** Snapshots from the simulation of the interaction of a flexible single-walled carbon nanotube with a monolayer of $C_{13}$ chains on diamond. The loads are **(a)** 19.8 nN, **(b)** 41.2 nN, and **(c)** 36.0 nN (After [162], © ACS 2003)

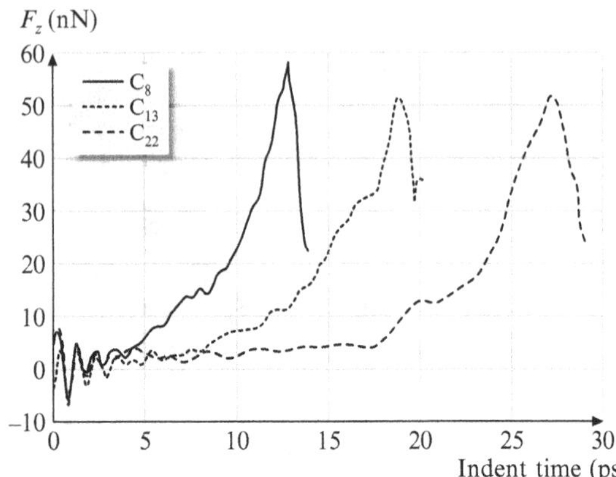


**Fig. 10.24** The load on the upper two layers of the flexible carbon nanotube indenter shown in Fig. 10.23 as a function of indentation time for the nanoindentation of the indicated hydrocarbon monolayes on diamond (After [162], © ACS 1999)

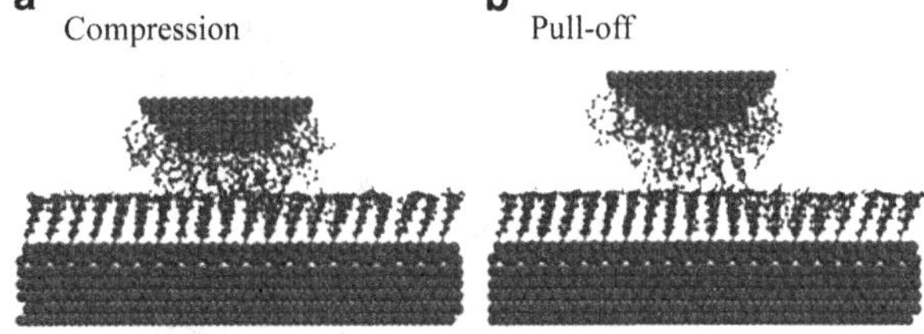


**Fig. 10.25** Snapshots from the OH/OH pair interaction during (**a**) compression and (**b**) pull-off (After [163], © ACS 2002)

nanotubes causes the walls of the tube to buckle. Both the cap inversion and the buckling are reversible processes. That is, when the load on the tube is removed, it recovers its original shape.

Deep indents of the hydrocarbon monolayers using rigid nanotubes result in rupture of chemical bonds. The simulations also show that the number of gauche defects generated by the indentation is a linear function of penetration depth and equal for $C_{13}$ and $C_{22}$ monolayers. Thus, it is the tip that governs the number of gauche defects generated.

Leng and Jiang [163] investigated the effect of using tips coated with SAMs containing hydrophobic methyl ($CH_3$) or hydrophilic hydroxyl (OH) terminal groups to nanoindent gold surfaces that also are covered with SAMs with the identical terminal groups as the tip. Figure 10.25 contains snapshots for the indentation process predicted to occur for terminal OH/OH interactions during compression and the pull-off. The adhesion force of OH/OH pairs is calculated to be about four times larger than that of $CH_3$/$CH_3$ pairs, as shown in Fig. 10.26. This is due to the formation of hydrogen bonding between OH/OH pairs. This interaction is also expected to increase the frictional force between monolayers with OH terminations.

Related MD simulations by Mate predict that the end groups on polymer lubricants have a significant influence on the lubrication properties of polymers [164].

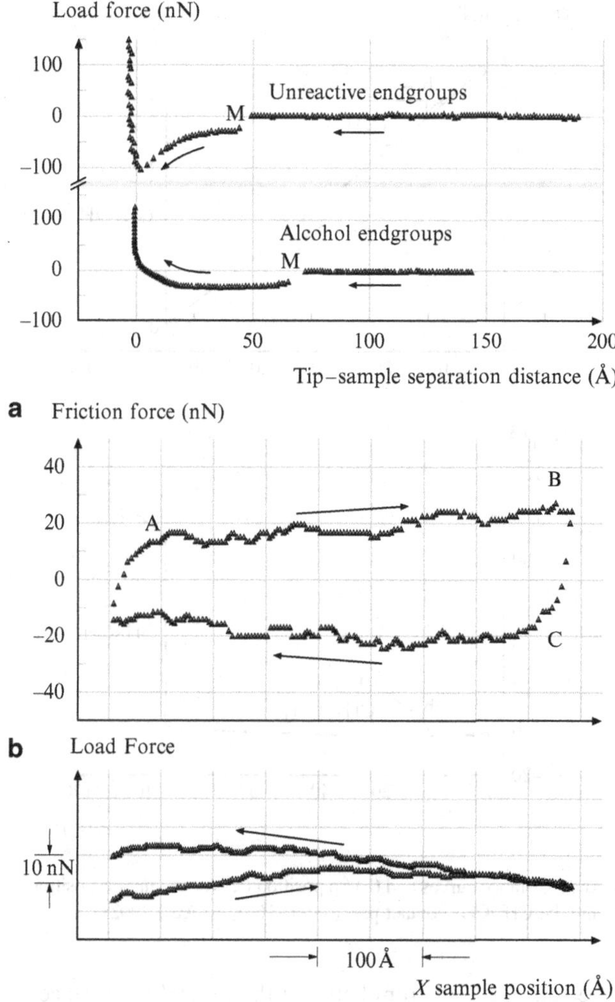


**Fig. 10.26** *Top*: The force versus distance curve (indentation part only) for unbonded perfluoropolyether on Si(100). The unreactive end groups were from a 10 Å-thick film; the reactive alcohol end groups were from a 30 Å-thick film. The negative forces represent attractive interactions between the tip and the surface. *Bottom*: Measured plots of friction and load forces of the tip as it slides over the sample with the alcohol end groups (After [164], © APS 1992)

For instance, fluorinated end groups are predicted to be less reactive than regular alcohol end groups. When fluorinated films are indented, the normal force becomes more attractive as the distance between the tip and film decreases until the hard wall limit is reached and the interactions become repulsive. In contrast, when AFM tips indent hydrogenated films, the forces become increasingly repulsive as the distance between them decreases, as shown in Fig. 10.27 and Fig. 10.28. This predicted behavior is due to the compression of the end group beneath the tip. For the lubricant

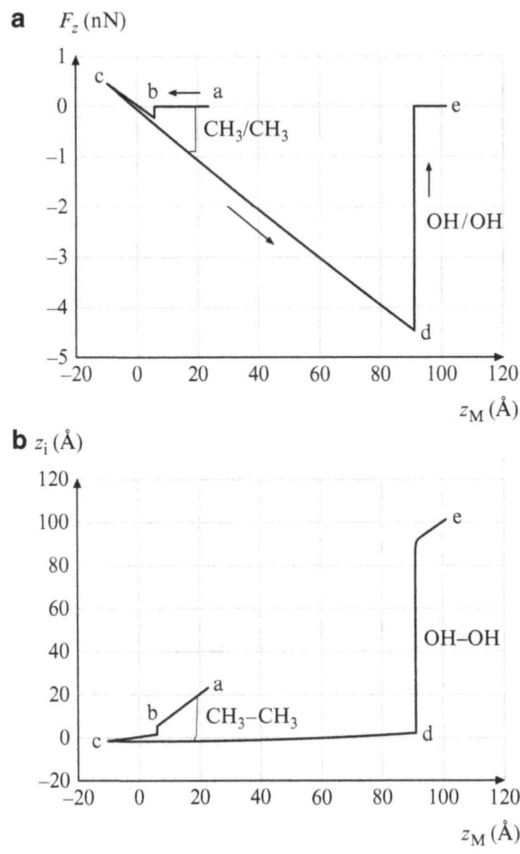


**Fig. 10.27** (**a**) Force–distance curves and (**b**) tip position ($z_i$) versus support position ($z_M$) for the OH/OH contact pair and the $CH_3/CH_3$ contact pair (After [163], © ACS 2002)

molecules to be squeezed out from between the tip and the surface, the hydrogen bonding between the two must first be broken and this increases the force needed to indent the system. As a result, a major effect of the presence of alcohol end groups is to dramatically increase the load that a liquid lubricant can support before failure (solid–solid contact) occurs.

When atomically sharp tips are used to indent solid-state thin films where there is a large mismatch in the mechanical properties of the film and the substrate, it is difficult to determine the true contact area between the tip and the surface during nanoindentation. In the case of soft films on hard substrates, pile-up can occur around the tip that effectively increases the contact area. In contrast, with hard films on soft substrates, *sink-in* is experienced around the tip that decreases the true contact area.

A class of coatings that has received much attention is diamondlike amorphous carbon (DLC) coatings. DLC coatings are almost as hard as crystalline diamond and may have very low friction coefficients (<0.01) depending upon the growth

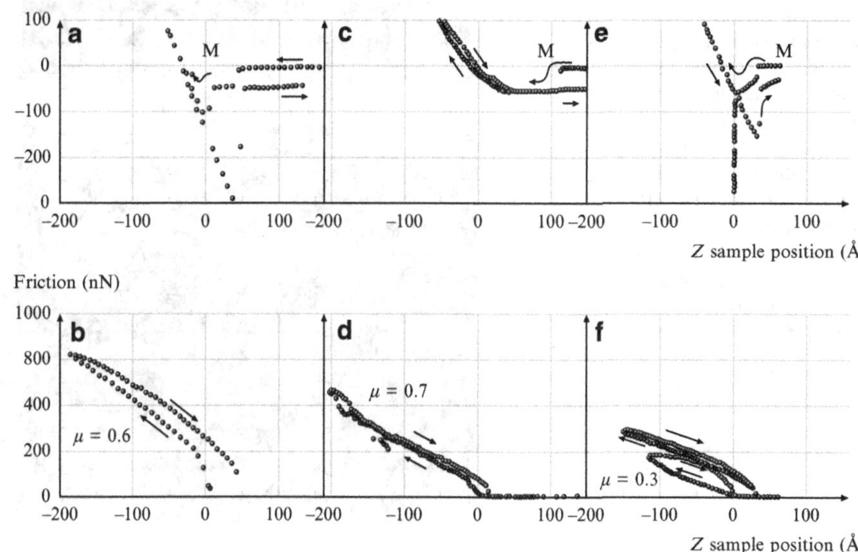


**Fig. 10.28** Measured values for friction and load as an atomic-force microscope tip is scanned across a 30 Å-thick sample of perfluoropolyether on Si(100). **(a, b)** The unbonded polymer with unreactive end groups. **(c, d)** The unbonded polymer with alcohol end groups. **(e, f)** A bonded polymer (After [164], © APS 1992)

conditions [166, 167, 168, 169]. They have therefore generated much interest in the tribological community and there have been several MD simulation studies to determine the mechanical and atomic-scale frictional properties of DLC coatings. MD simulations with bond-order potentials by Sinnott et al. [165] examined the differences in indentation behavior of a hydrogen-terminated diamond tip on hydrogen-terminated single-crystal diamond surfaces and diamond surfaces covered with DLC. In the former case, the tip goes through shear and twist deformations at low loads that change to plastic deformation and adhesion with the surface at high loads. When the surface is covered with the DLC film, the tip easily penetrates the film, as illustrated in Fig. 10.29, which *heals* easily when the tip is retracted so that no crater or other evidence of the indentation is left behind.

MD simulations by Glosli et al. [170] of the indentation of DLC films that are about 20 nm-thick give similar results. In this case a larger, rigid diamond tip was used in the indentations and was also slid across the surface. During sliding, the tip plows the surface, which causes some changes to the film not seen during indentation. However, because the tip is perfectly rigid, adhesion between the film and surface is not allowed which influences the results.

This section shows that repulsive interactions between surfaces covered with molecular films and proximal probe tips are minimized relative to interactions between bare surfaces and indentation tips. The lubrication properties of polymers and SAMs can vary with chain length, the rigidity of the tip, and the chemical

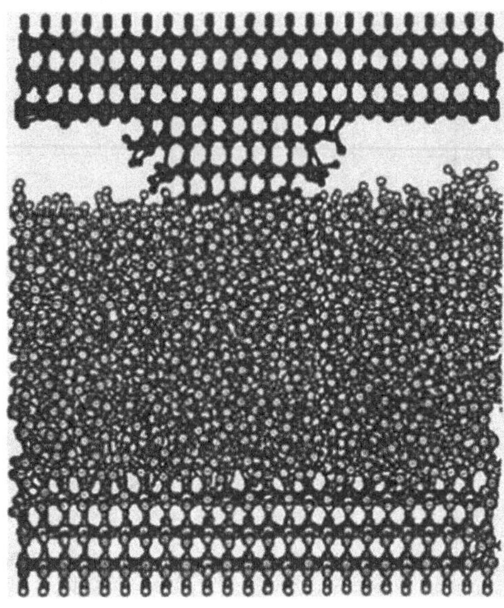

**Fig. 10.29** Snapshot from a molecular dynamics simulation where a pyramidal diamond tip indented an amorphous carbon thin film that is 20 layers thick. The simulation took place at room temperature and the carbon atoms in the film were 21% $sp^3$-hybridized and 58% $sp^2$-hybridized (the remaining atoms were on the surface and were not counted) (After [165], © AIP 1997)

properties of the end groups. In some cases, indentations can disrupt the initial ordering of polymers and SAMs, which affects their responses to nanoindentation and friction.

## 10.3 Friction and Lubrication

Work is required to slide two surfaces against one another. When the work of sliding is converted to a less ordered form, as required by the first law of thermodynamics, friction will occur. For instance, if the two surfaces are strongly adhering to one another, the work of sliding can be converted to damage that extends beyond the surfaces and into the bulk. If the adhesive force between the two surfaces is weaker, the conversion of work results in damage that is limited to the area at or near the surface and produces transfer films or wear debris [171, 172]. While the thermodynamic principles of the conversion of work to heat are well known, the mechanisms by which this takes place at sliding surfaces are much less well established despite their obvious importance for a wide variety of technological applications.

Atomic-scale simulations of friction are therefore important tools for achieving this understanding. They have consequently been applied to numerous materials in a wide variety of structures and configurations, including atomically flat and atomically rough diamond surfaces [173, 174, 175], rigid substrates covered with monolayers of alkane chains [176], perfluorocarboxylic acid and hydrocarboxylic Langmuir–Blodgett (LB) monolayers [177], between contacting copper surfaces

[178, 179], between a silicon tip and a silicon substrate [119, 143], and between contacting diamond surfaces that have organic molecules absorbed on them [180]. These and several other studies are discussed below.

### *10.3.1 Bare Surfaces*

Sliding friction that takes place between two surfaces in the absence of lubricant is termed *dry* friction even if the process occurs in an ambient environment. Simple models have been developed to model dry sliding friction that, for example, consider the motion of a single atom over a monoatomic chain [181]. Results from these models reveal how elastic deformation of the substrate from the sliding atom affects energy dissipation and how the average frictional force varies with changes in the force constant of the substrate in the direction normal to the scan direction. Much of the correct behavior involved in dry sliding friction is captured by these types of simple models. However, more detailed models and simulations, such as MD simulations, are required to provide information about more complex phenomena.

MD simulations have been used to study the sliding of metal tips across clean metal surfaces by numerous groups [179, 182, 183, 184, 185, 186]. An illustrative case is shown in Fig. 10.30 for a copper tip sliding across a copper surface [179]. Adhesion and wear occur when the attractive force between the atoms on the tip and the atoms at the surface becomes greater than the attractive forces within the tip itself. Atomic-scale stick and slip can occur through nucleation and subsequent motion of dislocations, and wear can occur if part of the tip gets left behind on the surface (Fig. 10.30). The simulations can further provide data on how the

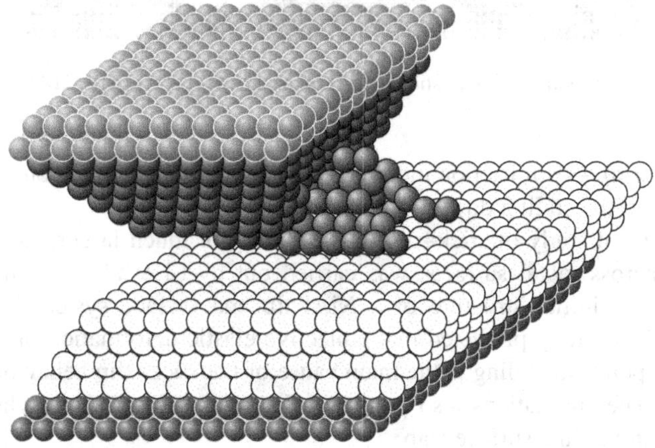

**Fig. 10.30** Snapshot from a molecular dynamics simulation of a copper tip sliding across a Cu(100) surface. A connective neck between the two is sheared during the sliding, leading to wear of the tip. The simulation was performed at a temperature of 0 K (After [179], © APS 1996)

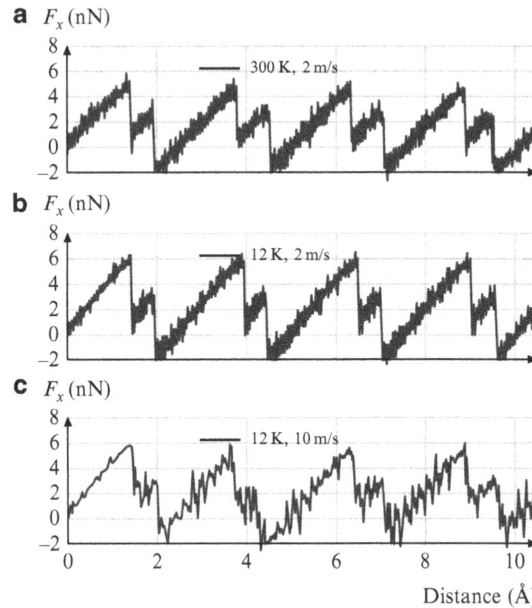


**Fig. 10.31** Plots of the lateral force versus distance from a simulation similar to that shown in Fig. 10.30. The plots illustrate the dependence of the force on temperature and sliding velocity. **(a)** Temperature of 300 K and a sliding velocity of 2 m/s, **(b)** temperature of 12 K and a sliding velocity of 2 m/s, and **(c)** temperature of 12 K and a sliding velocity of 10 m/s (After [179], © APS 1996)

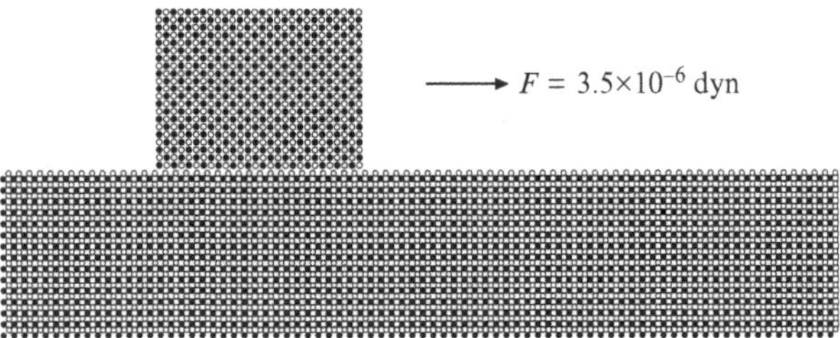


**Fig. 10.32** Starting configuration of sliding NiAl tip on a NiAl surface (After [187], © AIP 2001)

characteristic *stick–slip* friction motion can depend on the area of contact, the rate of sliding, and the sliding direction (Fig. 10.31).

An additional study of stick–slip in the sliding of much larger, square-shaped metal tips across metal surfaces was carried out by Li et al. [187] using EAM potentials. The initial structure of a NiAl tip and surface system is shown in Fig. 10.32. This study predicted that collective elastic deformation of the surface layers in response to sliding is the main cause of the stick–slip behavior shown in Fig. 10.33. The simulations also predicted that stick–slip produces phonons that propagate through the surface slab.

Large-scale simulations using pairwise Morse potentials that are similar in form to (10.6) were used to study the wear of metal surfaces caused by metal tips that plow the surface, as illustrated in Fig. 10.34. They provide insight into the wear

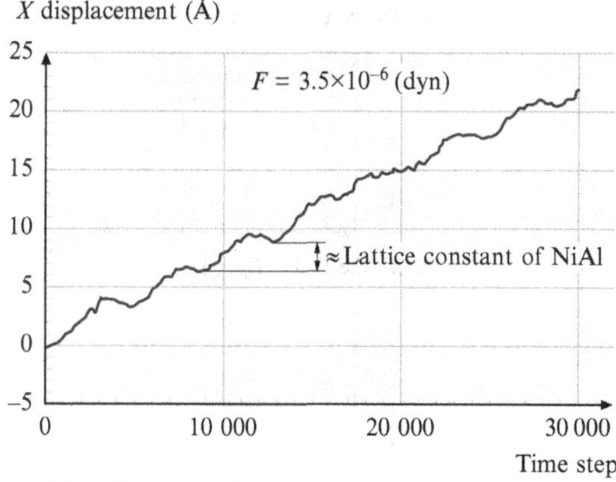


**Fig. 10.33** A structured curve of frictional dynamics of an atom in the *upper right corner* that is indicative of stick–slip (After [187], © AIP 2001)

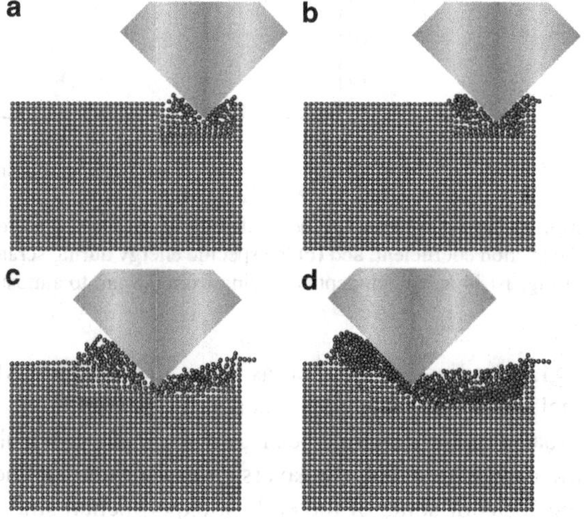


**Fig. 10.34** Snapshots of the scratching of an aluminum surface with a rigid tip at a depth of 0.8 nm (After [189], © APS 2000)

track dependence of the sliding rate [188] and how variations in the scratching force, friction coefficient, and other quantities depend on the scratch depth [189], as illustrated in Fig. 10.35.

On the whole, the results of experimental studies show good agreement with the results of the computational studies described above. This is true despite the fact that all of these MD simulations use empirical potentials that do not include electronic effects and thus effectively assume that the electronic contributions to friction on metal surfaces are negligible. However, experiments have measured a nonnegligible contribution of conduction electrons to friction [190]. Thus, future simulations of metal-tip–metal-substrate interactions using more sophisticated tight-binding or first principles methods that include electronic effects are encouraged.

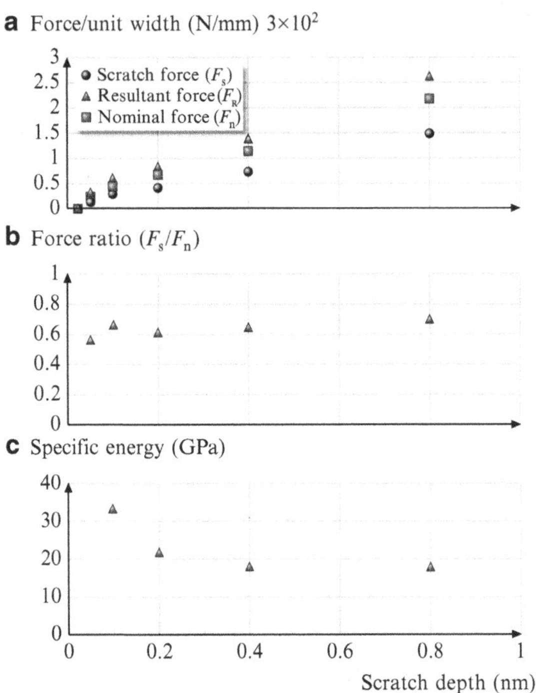


**Fig. 10.35** Variation in (**a**) the scratching force, the normal force, and the resultant force, and (**b**) the friction coefficient, and (**c**) the specific energy during scratch processes similar to those shown in Fig. 10.34 at scratch depths ranging from 0.8 nm to almost 0 nm (After [189], © APS 2000)

Layered ceramics, such as mica, graphite and $MoS_2$, that have structures that include strongly bound layers that interact with one another through weak van der Waals bonds, have long been known to have good lubricating properties because of the ease with which the layers slide over one another. They have, therefore, been the focus of some of the earliest experimental studies of nanometer-scale friction [19, 191]. The results of these early studies lead researchers to hypothesize that at high loads measured friction forces were related to *incipient sliding* [192, 193] caused by a small flake from the surface becoming attached to the end of the tip. If true, this would mean that all measured interactions were between the surface and the flake, which has a larger contact area than the clean tip. However, subsequent simulations of constant force AFM images of graphite by Tang et al. [194] showed that there is no need for the assumption of a graphite flake under the tip to reproduce the experimental images of a graphite surface.

Surprisingly strong localized fluctuations in atomic-scale friction are displayed by layered ceramics [195, 196, 197, 198]. For instance, square-well signals with sub-angstrom lateral width are obtained in FFM scans on $MoS_2(001)$ in the direction across the scan direction, while sawtooth signals are detected along the scan direction, as shown in Fig. 10.36. This finding can be explained by a stick–slip

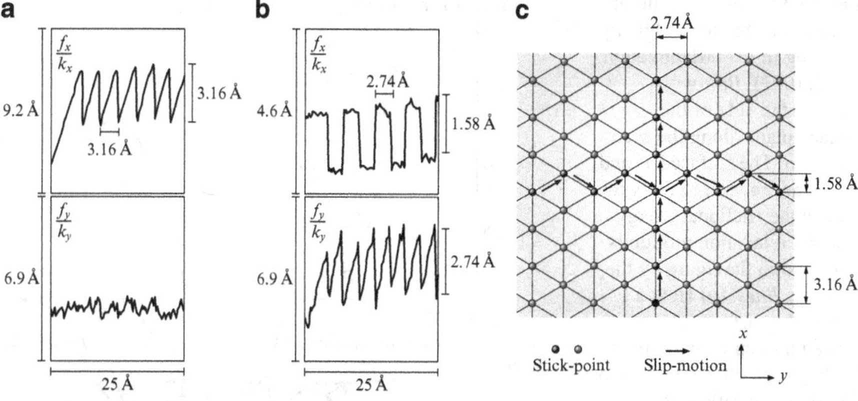


**Fig. 10.36** Displacement data from a scan across a $MoS_2$(001) surface. The data in **(a)** and **(b)** are form scans along the $x$ -and $y$-directions, respectively, on the surface shown in **(c)** (After [198], © APS 1996)

model by Mate [19] and Erlandsson [191] that assumes that the tip does a zigzag walk along the scan. Measured variations in the frictional force with the periodicity of cleavage planes [191] are consistent with the results of this simple model. However, additional experiments indicate a more complex tip–surface interaction, such as changes in the intrinsic lateral force between the substrate and the AFM tip [199] or sliding-induced chemistry between the tip and the surface [200, 201].

$MoS_2$ has proven to be a very successful solid lubricant for applications that operate in vacuum but its performance quickly deteriorates when exposed to ambient air. Recently, Liang et al. used ab initio DFT methods to examine the potential energy surfaces between sliding $MoS_2$(001) | $MoS_2$(001), $MoS_2$(001) | $MoO_3$(001), and $MoO_3$(001) | $MoO_3$(001) interface systems in an effort to understand the deterioration in lubricity due to oxidation [202]. The potential energy surfaces give information on the minimum energy path along particular sliding directions from which lateral forces needed to slide the interface can be calculated. In this work a normal force of 500 MPa was applied to all three interfacial systems before the energy surface calculations were performed. It was found that the minimum energy path for self-mated $MoS_2$(001) | $MoS_2$(001) was a zigzag path that avoided direct overlap of sulfur atoms in the topmost layers of the top and bottom surfaces. From this PE surface a lateral frictional force of 0.058 nN/atom was predicted for this interface. The lowest frictional force of 0.011 nN/atom predicted in these calculations was for the $MoS_2$(001) | $MoO_3$(001) interface along the channel direction formed by S atoms at the sliding surface. Although this doesn't explain the mechanism by which degradation occurs experimentally, it is in general agreement with experimental results that show that this interface can produce lower frictional coefficients than pure $MoS_2$. As suggested by the authors, it may be a surface defect driven process that is currently inaccessible to first principles methods employed in these calculations. The last interface of $MoO_3$(001) | $MoO_3$(001) was found to have the largest frictional force of 0.352 nN/atom.

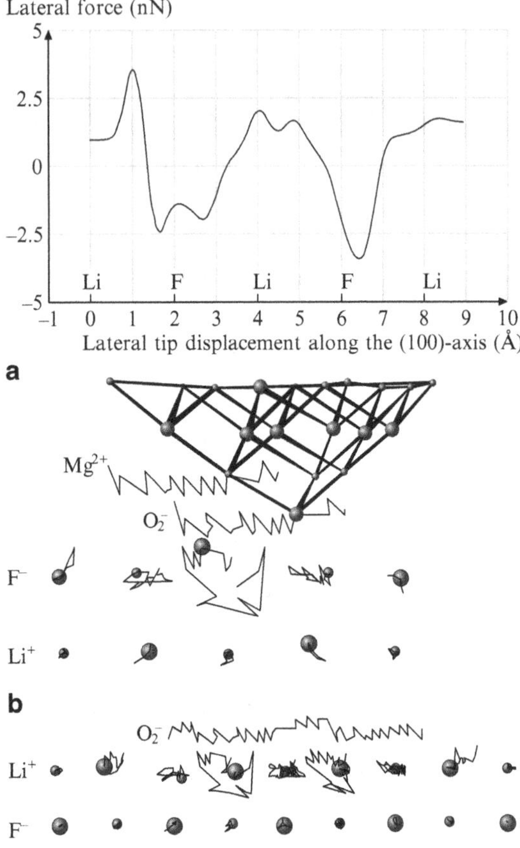


**Fig. 10.37** *Top*: The lateral force calculated for a MgO tip scanning in the 001-direction on LiF(001). *Bottom*: A view of the side of the surface plane along the scan direction. The surface $Li^+$ and $F^-$ atoms are seen to relax to relieve the frictional energy and this relaxation motion is indicated in the figure by the *category lines*. (**a**) How a $F^-$ ion on the surface can be moved into an interstitial site by the tip and then it returns to its original position. (**b**) How the relaxation of the surface atoms is reversible (After [203], © Elsevier 1995)

Crystalline ceramics differ from layered ceramics in that they are held together by relatively strong covalent or ionic bonds. In the case of ionic systems, Shluger et al. [203] used a mixture of atomistic and macroscopic modeling methods to study the interaction of a MgO tip and a LiF surface. In particular, the tip–surface interaction was treated atomistically and the cantilever deflection was treated with a macroscopic approach. The results, shown in Fig. 10.37, show that if the tip is charged and in hard contact with the surface, tip and surface distortions are possible that can lead to motion of the surface ions within the surface plane and the transfer of some of the ions onto the tip.

In the case of covalently bound ceramics, there is extensive literature related to friction of diamond [204, 205] because, while it is the hardest material known, it also exhibits relatively low friction. The *ratchet mechanism* has been proposed for energy dissipation during friction on the macroscale in diamond, where energy is released by the transfer of normal force from one surface asperity to another. The elastic mechanism is another mechanism that has been proposed, where the released energy comes from elastic strain in an asperity. Atomic-scale friction has been measured experimentally [20] for diamond tips with near atomic-scale radii sliding over hydrogen-terminated diamond surfaces. These experiments are sensitive enough to detect the 2 × 1 reconstruction on the diamond (100) surface. Furthermore, the average friction coefficient determined with an AFM on H-terminated diamond (111) surfaces is about two orders of magnitude smaller than the value measured on bare, 2 × 1 diamond (111) surfaces, indicating greater adhesion in the latter case [206]. More recently, the friction between a tungsten carbide tip and hydrogen-terminated diamond (111) was examined with AFM in UHV by Enachescu et al. [207]. The friction between these two hard surfaces was shown to obey Derjaguin–Muller–Toporov or DMT [208] contact mechanics and the shear strength of the interface was determined to be 246 MPa.

Extensive MD simulations have been carried out by Harrison and coworkers that examine the friction between hydrogen-terminated diamond (111) surfaces [173, 210] and diamond (100) surfaces [209] in sliding contact and its temperature dependence [211]. The simulations of sliding between the diamond (111) surfaces reveal that the potential energy, load, and friction are all periodic functions of the sliding distance (Fig. 10.38). Maxima in these quantities occur when the hydrogen atoms on opposing surfaces interact strongly. Recent ab initio studies by Neitola and Pakannen of the friction between hydrogen-terminated diamond (111) surfaces also show that the potential energy is periodic with sliding distance (Fig. 10.39)

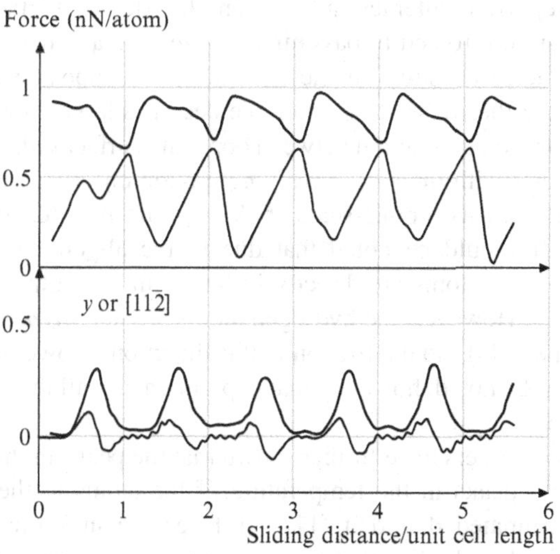


**Fig. 10.38** Calculated frictional force (*lower lines*) and normal force (*upper lines*) felt by a hydrogen-terminated (111) surface as it slides against another hydrogen-terminated diamond (111) surface in a MD simulation. The sliding direction is given in the legend. The sliding speed is 1 Å/ps and the simulation temperature is 300 K. The two plots show how the simulated stick–slip motion changes as a function of the applied load. The load is high and low in the *upper* and *lower panels*, respectively (After [209], © ACS 1995)

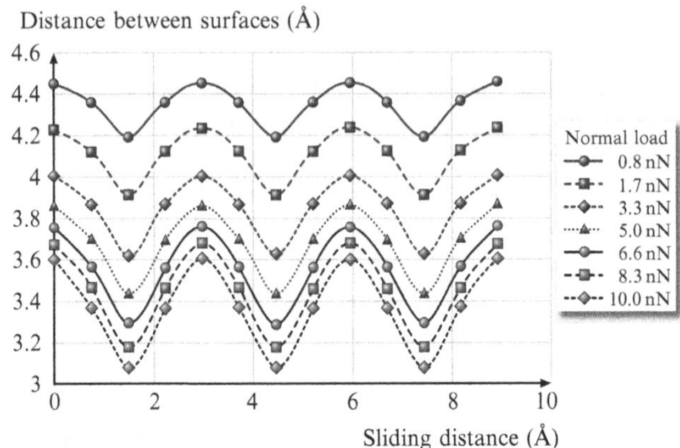


**Fig. 10.39** Distance between hydrogen-terminated (111) crystals as a function of sliding distance (After [212], © ACS 2001)

[212]. Because the results of the ab initio studies and the MD simulations are in good agreement, Neitola and Park conclude that the potential model used in the MD studies is accurate.

As mentioned previously, the maxima in the load and the friction values during sliding are caused by the interactions of hydrogen atoms on opposing surfaces. When sliding in the $[1\,1\,\overline{2}]$ direction, the H atoms *revolve* around one another, thus decreasing the repulsive interaction between the sliding surfaces because the hydrogen atoms are not forced to pass directly over one another [173]. Increasing the load causes increased stress at the interface. The opposing hydrogen atoms become *stuck*. Once the stress at the interface becomes large enough to overcome the hydrogen–hydrogen interaction between opposing surfaces, the hydrogen atoms *slip* past one another with the same *revolving* motion observed at low loads. This phenomenon is known as atomic-scale stick–slip and has the periodicity of the diamond lattice. It should be noted that due to the alignment of the opposing surfaces, the hydrogen atoms are directly in line with each other when sliding in the $[1\,1\,\overline{2}]$ direction. However, the hydrogen atoms are not *aligned* with each other when sliding in the $[1\,1\,\overline{0}]$, so the friction in this direction is lower than in the $[1\,1\,\overline{2}]$ direction. It should be noted, however, that experimentally all initial alignments are likely to be probed.

Harrison and coworkers have further shown that the peaks in the frictional force are correlated with peaks in the temperature of the atoms at the interface when two hydrogen-terminated diamond (111) surfaces are in sliding contact [210]. Figure 10.40 shows the vibrational energy (or temperature) between diamond layers as a function of sliding distance. These data clearly show that layers close to the sliding interface can be vibrationally excited during sliding. When the hydrogen atoms are *stuck* or interacting with each other strongly, the stress and friction force at the interface build up. When the hydrogen atoms *slip* past one another, the stress at the interface is relieved and the energy is transferred to the diamond in the form

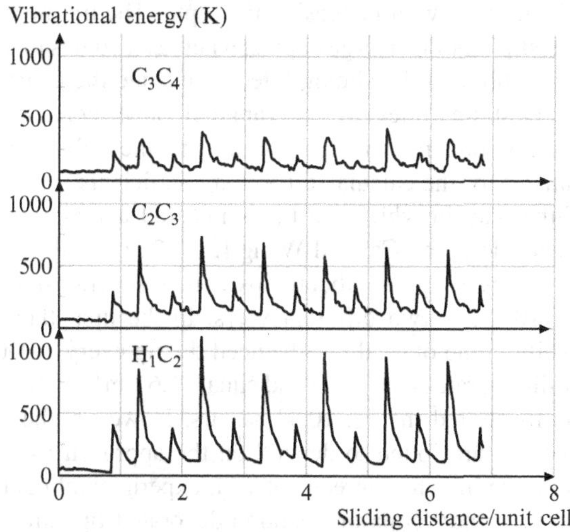


**Fig. 10.40** Average vibrational energy of oscillators between diamond layers as a function of sliding distance. These energies are derived from a molecular dynamics simulation of the sliding of a hydrogen-terminated diamond (111) surface over another hydrogen-terminated diamond (111) surface. The vibrational energy between the first and second layers of the lower diamond surface is shown in the *lower panel*, between the second and third layers in the *middle panel*, and between the third and fourth layers in the *upper panel* (After [213], © Elsevier 1995)

of vibration or heat. Thus, the peaks in the temperature occur slightly after the peaks in the frictional force.

It should be noted that atomic-scale stick slip is observed in other systems. Perry and Harrison used MD simulations to demonstrate that two hydrogen-terminated diamond (100) $(2 \times 1)$ surfaces in sliding contact also exhibit stick–slip [209]. In addition, it was shown that the shape of the friction versus sliding distance curves is influenced slightly by the speed of the sliding, with features in the curves becoming more pronounced at slower speeds. Stick–slip behavior was also observed in AFM studies of diamond (100) $(2 \times 1)$ surfaces [206]. However, in this case the stick–slip was over a much longer length scale and may be due to the fact that the surfaces were not hydrogen-terminated.

Mulliah et al. [214] used MD simulations with bond-order potentials [215] to model interactions between indenter atoms, EAM potentials [216] to model interactions between substrate atoms, and the Ziegler–Biersack–Littmack potential [217] to model interactions between indenter and substrate atoms to study the atomic-scale stick–slip phenomenon of a pyramidal diamond tip interacting with a silver surface at several sliding rates and vertical support displacements. These simulations showed that dislocations are related to the stick events emitting a dislocation in the substrate near the tip. The scratch in the substrate is discrete due to the tip jumping over the surface in the case of small vertical displacements. In contrast, large displacements of 15 Å or more result in a continuous scratch. These simulations also showed how the dynamic friction coefficient and the static friction coefficient increase with increasing tip depth. The tip moves continuously through a stick and slip motion at large depths, whereas it comes to a halt in the case of shallow indents. Although the sliding rate can change the exact points of stick and slip, the range of sliding rates over the range of values considered in this study (1.0–5.0 m/s) has no influence on the damage to the substrate, the atomistic stick–slip mechanisms, or the calculated friction coefficients.

The effect of the way in which the tip is rastered across the surface in MD simulations was considered by Cai and Wang [218, 219] using bond-order potentials. In particular, they dragged silicon tips across several silicon surfaces, as illustrated in Fig. 10.41, in two different ways. In the first, they moved the tip every MD step while in the other they advanced the tip every 1,000 steps. In both cases, the overall sliding rate is the same and equals 1.67 m/s. In both cases, wear of the tip such as is illustrated in Fig. 10.41 occurs. However, the mechanisms by which the wear occurs are found to depend on the approach used, and the latter approach is found to be in better agreement with experimental data.

In many studies, diamond tips or diamond-decorated tips are used in friction measurements. Diamond is an attractive material for an FFM tip because of its high mechanical strength and the belief that such tips are wear-resistant. However, diamond tips that were used to scratch diamond and silicon surfaces and then imaged showed significant wear that increased with the increasing hardness of the tested material [220, 221]. This wear altered the shape of the tip and hence influenced the contact area that is used to determine friction coefficients.

In summary, MD simulations provide insight into dry sliding friction and the sliding of metal tips across clean metal, crystalline ceramics, and layered ceramics surfaces. Stick–slip friction or wear can occur depending on the sliding conditions.

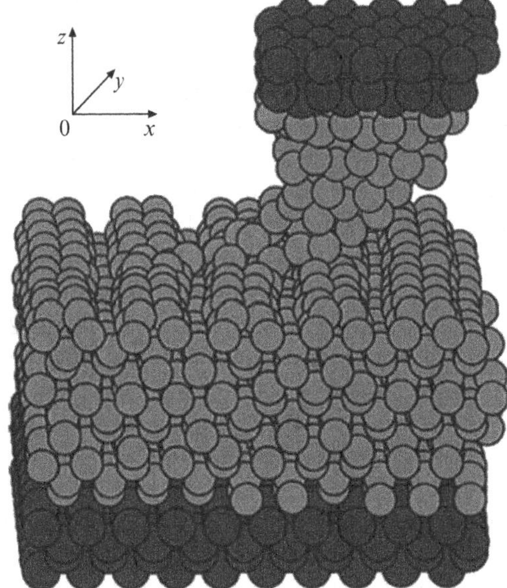


**Fig. 10.41** Snapshots of a Si(111) tip interacting with a Si(001) 2 × 1 surface. The tip is rastering along the surface in the *x*-direction and starts off at a distance of 9 Å from the surface (After [218], © CCLRC 2002)

The good lubricating properties of layered ceramics are observed in the simulations along with localized fluctuations in atomic-scale friction. Crystalline ceramics, such as diamond, exhibit relatively low friction and the simulations show how stick–slip atomic-scale motion changes with the conditions of sliding and the way in which the simulation is performed.

### 10.3.2 Decorated Surfaces

While dry sliding friction in vacuum assumes that ambient gas particles have no direct effect on the results, MD simulations show that free particles between two surfaces in sliding contact influence friction to a surprisingly large degree. These so-called third-body molecules have been studied extensively by Perry and Harrison [180, 222, 223] using MD simulations with bond-order and LJ potentials. These simulations focus on the effect of trapped small hydrocarbon molecules on the atomic-scale friction of two (111) crystal faces of diamond with hydrogen termination. These molecules might represent hydrocarbon contamination trapped between contacting surfaces prior to a sliding experiment in dry friction, or hydrocarbon debris formed during sliding.

In particular, the effects on friction of methane ($CH_4$), ethane ($C_2H_6$), and isobutane $(CH_3)_3CH$ trapped between diamond (111) surfaces in sliding contact were examined in separate studies (Fig. 10.42). The frictional force for all these

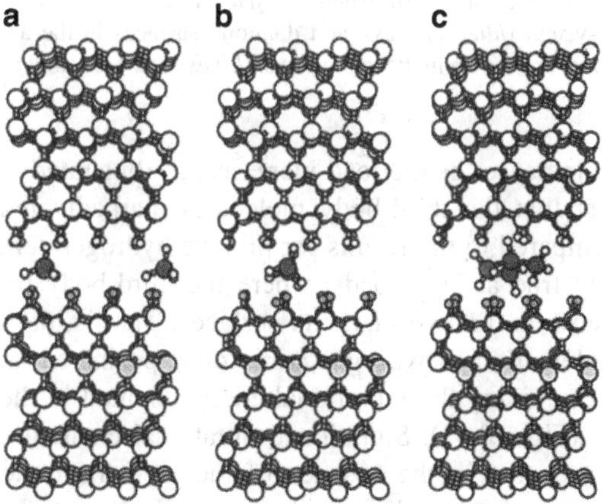


**Fig. 10.42** Initial configuration at low load for the diamond plus third-body molecule systems. These systems are composed of two diamond surfaces, viewed along the $[\bar{1}\,1\,0]$ direction, and two methane molecules in (**a**), one ethane molecule in (**b**), and one isobutane molecule in (**c**). Large *white* and *dark gray spheres* represent carbon atoms of the diamond surfaces and the third-body molecules, respectively. *Small gray spheres* represent hydrogen atoms of the lower diamond surface. Hydrogen atoms of the upper diamond surface and the third-body molecules are both represented by *small white spheres*. Sliding is achieved by moving the rigid layers of the upper surface from *left to right* in the figure (After [180], © ACS 1997)

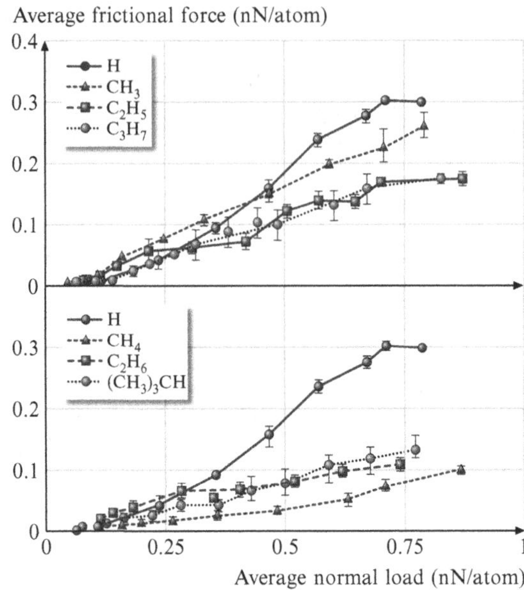


**Fig. 10.43** Average frictional force per rigid-layer atom as a function of average normal load per rigid-layer atom for sliding the upper diamond surface in the $[1\,1\,\bar{2}]$ crystallographic direction. Data for the methane ($CH_4$) system (*open triangles*), the ethane ($C_2H_6$) system (*open squares*), the isobutane $(CH_3)_3CH$ system (*filled circles*), and diamond surfaces in the absence of third-body molecules (*open circles*) are shown in the lower panel. Data for the methyl-terminated $-CH_3$ system (*open triangles*), the ethyl-terminated ($-C_2H_5$) system (*open squares*), the *n*-propyl-terminated ($-C_3H_7$) system (*filled circles*), and diamond surfaces in the absence of third-body molecules (*open circles*) are shown in the *upper panel*. *Lines* have been drawn to aid the eye (After [180], © ACS 1997)

systems generally increases as the load increases, as illustrated in Fig. 10.43. The simulations predict that the third-body molecules markedly reduce the average frictional force compared to the results for pristine hydrogen-terminated surfaces. This is particularly true at high loads, where the third-body molecules act as a boundary layer between the two diamond surfaces. That is, the third-body molecules reduce the interaction of hydrogen atoms on opposing surfaces [223]. This is demonstrated by examining the vibrational energy excited in the diamond lattice during the sliding (Fig. 10.44). Significant vibrational excitation of the diamond outer layer (C–H) occurs in the absence of the methane molecules. Thus, the friction is $\approx 3.5$ times larger when methane is not present. The application of load to the diamond surfaces causes the normal mode vibrations of the trapped methane molecules to change. Power spectra calculated from MD simulations [222, 223] show that even under low loads, the peaks in the power spectra are significantly broadened. Peaks in the low-energy region of the spectrum almost disappear with the additional application of load.

The size of the methane molecules allows them to be *pushed* in-between hydrogen atoms on the diamond surfaces while sliding [223]. However, steric considerations

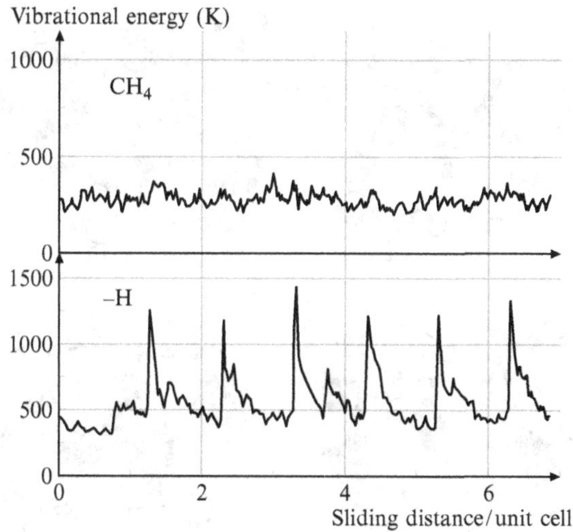


**Fig. 10.44** Average vibrational energy between the (C–H) bonds of the upper diamond (111) surface versus sliding distance for hydrogen-terminated diamond (111) surfaces, with ($CH_4$) and without methane (H), trapped between them. The average normal load is approximately the same in both simulations and is in the range 0.8–0.85 nN/atom. The average frictional force on the upper surface is about 3.5 times smaller in the presence of the methane third-body molecules (After [223], © Elsevier 1996)

cause the larger ethane and isobutane molecules to change orientation during sliding. Conformations that lead to increased interactions with the diamond surfaces increase the average frictional force. Thus, despite the fact that the two diamond surfaces are farther apart when ethane and isobutane are present compared to when methane is present, the friction is larger because these molecules do not *fit* nicely into potential energy valleys between hydrogen atoms when sliding.

When similar hydrocarbon molecules (methyl, ethyl, and *n*-propyl groups) are chemisorbed to one of the sliding diamond surfaces, instead of trapped between the surfaces, different behavior is observed by Harrison et al. [174, 175, 210, 224]. Simulations show that methyl-termination does not decrease friction significantly but results in frictional forces that are nearly the same as they are for hydrogen-terminated diamond surfaces [213]. While the methane third-body molecules decrease the frictional force to a greater extent than the chemisorbed methyl groups, friction as a function of load is comparable for the ethyl-terminated and ethane systems, with the former giving slightly higher frictional forces. Attaching the hydrocarbon groups to the diamond surfaces causes them to have less freedom to move between hydrogen atoms on opposing diamond surfaces during sliding. This generally increases their repulsive interaction with the diamond counterface.

MD simulations can also provide insight into the rich, nonequilibrium tribochemistry that occurs between surfaces in sliding contact. Harrison and Brenner examined the tribochemistry that occurs when ethane molecules are trapped between diamond surfaces in sliding contact, as illustrated in Fig. 10.45 [225]. This simulation

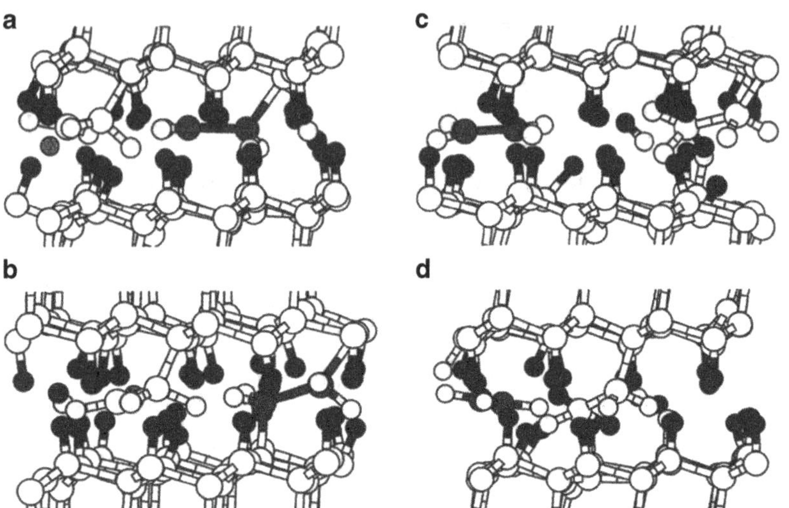


**Fig. 10.45** Snapshots from a molecular dynamics simulation of the sliding of two hydrogen-terminated diamond (111) surfaces against one another in the $[[1\,1\,\overline{2}]]$ direction. The upper surface has two ethyl fragments chemisorbed to it. The simulation shows how sliding can induce chemistry at the interface. (**a**) Initial structure of the sliding surfaces with ethyl fragments chemisorbed to the top surface, (**b**) the ethyl fragments have started to react with atoms on the bottom surface, (**c**) continued sliding modifies the reactions, and (**d**) continued reactions are occurring between the ethyl fragments and the bottom surface (After [225], © ACS 1994)

was the first to show the atomic-scale mechanisms for the degradation of lubricant molecules due to friction. The type of debris formed during the sliding simulation is similar to the types of debris molecules that were observed in macroscopic experiments that examined the friction between diamond surfaces [226].

In the case of sliding metal surfaces, impurity molecules or atoms (both adsorbed and absorbed) on thin metal films can be expected to affect the film's properties. For example, calculations have shown that resistivity changes in the metal are strongly dependent on the nature of the adsorption bond [227]. When this result is used to interpret atomic-scale friction results obtained with the QCM, the sliding of adsorbate structures on metal surfaces are shown to be a combination of electron excitation and lattice vibrations. Additionally, other interesting quantum effects can come into play when the adsorbate is very different chemically from the surface on which it is sliding. For instance, the electronic frictional forces acting on small, inert atoms and molecules, such as $C_2H_6$ and Xe, sliding on metal surfaces have been calculated by Persson and *Volokitin* [228], where the metal surface was approximated by a electron gas (jellium) model. The calculations showed that the Pauli repulsive and attractive van der Waals forces are of similar magnitudes. In addition, the calculated electronic friction contributions agree well with the values derived from surface resistivity by Grabhorn et al. [229] and QCM measurements. These studies showed that parallel friction is mainly due to electronic effects while perpendicular friction is phonic in nature in this system.

In summary, MD simulations show that the average frictional force decreases significantly in systems with third-body molecules, especially at high loads. Simulations also provide information about the details of tribochemical interactions that can occur between lubricants and sliding surfaces. Additionally, the effect of the presence of small molecules on thin metal films can influence film properties, such as resistivity.

### *10.3.3 Thin Films*

As discussed at the beginning of this section, the conversion of the work of sliding into some other less ordered form is responsible for friction at sliding solid interfaces. In the case of adhering systems, the work of sliding may be converted into damage within the bulk (plastic deformation), while in the case of weakly adhesive forces, friction can occur through the conversion of work to heat at the interface that causes no permanent damage to the surface (wearless friction). The latter case, when it is achieved through the presence of lubricating thin films, is the topic of this section.

There are several types of lubricating thin films, the simplest of which consist of small molecules that are analogous to wear debris that can *roll* between the sliding surfaces or that represent very short-chain bonded lubricants. These thin films were discussed in the previous subsection. The rest of this subsection will, therefore, focus on the effects of liquids, larger nanoparticles, self-assembled monolayers and solid thin films on lubrication and friction.

#### Liquids

Liquids are common lubricants and so they have been studied in great depth at the macroscale. At the nanoscale, the tribological response of spherical liquid molecules has been well-characterized experimentally using the SFA and computationally with MD simulations by Berman et al. [230]. The SFA experiments considered one to three liquid layers and the stick–slip motion at the interface is found to increase in a quantized fashion as the number of lubricant layers decrease. When no external forces are applied to the system, the sliding stops and the solid–lubricant interactions are strong enough to force the liquid molecules to form a close-packed structure that is ordered. The transformation of the liquid into this solidlike structure causes the two surfaces to effectively bond to each other through the lubricant. When the surfaces start to slide again, lateral shear forces are introduced that steadily increase, which causes the molecules in the liquid to undergo small lateral displacements that change the film thickness. If these shear forces become greater than a critical value, the film disorders in a manner that is analogous to melting. This allows the surfaces to slide easily past each other in a manner that is still quantized. This sequence of events is nicely illustrated in Fig. 10.46 and can be reproduced multiple times for the same system.

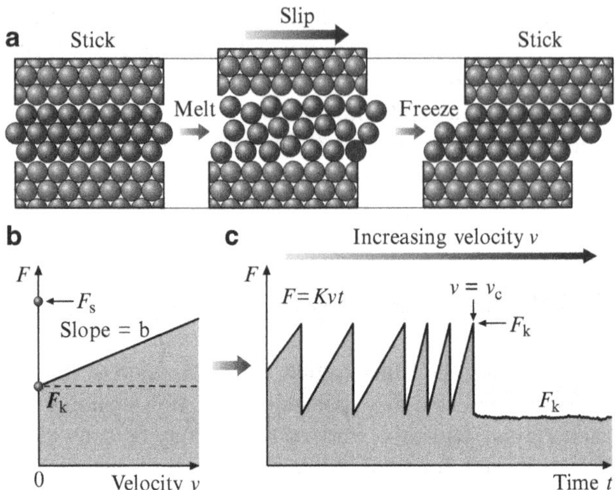


**Fig. 10.46** The stick–slip transition that occurs for thin films of liquid between two sliding solid surfaces. $F$ is the intrinsic friction and $F_s$ is the friction where the liquid is in the rigid state; $F_k$ is the friction where the liquid is in the liquidlike state. (**a**) Snapshots that indicate how the liquid melting and freezing is related to the stick–slip process. (**b**) Indicates the intrinsic friction relative to the friction values where the liquid is liquidlike and solidlike. (**c**) Illustrates how the intrinsic friction varies with time as the velocity increases (After [230], © ACS 1996)

Persson [231] used MD simulations with pairwise potentials similar to those in (10.6) to examine the mechanism by which this sharp transition occurs. They find that in the case of sliding on insulating crystal surfaces, the solid-state lubricant may be in a *superlubric* state where the friction is negligible. It is clear from the simulations, however, that any surface defects, even in low concentrations, will disrupt this state and transform the lubricant back into a fluid. In addition, when sliding occurs on metallic surfaces above cryogenic temperatures, the electronic contributions to friction are no longer zero and no superlubric state is possible.

High applied pressures can force the fluid molecules out from between the two confining surfaces [232]. The fact that liquid molecules close to a stiff surface are strongly layered in the direction perpendicular to the surface explains the experimental observation of a ($n \to n-1$) layer transition, where $n$ is number of monolayers, that is observed as the normal load increases [233]. Nucleation theory is used to calculate the critical pressure and determine the spreading dynamics of the ($n-1$) island.

The reactivity of the liquid molecules are also critically important to boundary layer friction. MD studies by Persson [234] show that inert molecules interact weakly with sliding surfaces. Consequently, as the rate of sliding increases, the molecular conversion from the solid state to the liquid state occurs in an abrupt manner. However, when the molecules interact strongly with the surfaces, they undergo a more gradual transition from the solid to the liquid state. Persson et al. [34] also considered systems where the molecules are attached to one of the surfaces, which causes the transitions to be abrupt. This is especially true if there are large separations between the chains.

While the studies discussed so far have focused on spherical liquids, most widely used liquid lubricants consist of long-chain hydrocarbons. Nonspherical liquid molecules have more difficulty aligning and solidifying. This is borne out in MD simulations by Thompson and Robbins [235] that show that spherical molecules have higher critical velocities than branched molecules. In particular, the simulations show that when the molecules are branched, the amount of time various parts of the system spend in the sticking and sliding modes changes with sliding rate. The critical velocity can also depend on the number of liquid layers in the film, the structure and relative orientation of the two sliding surfaces, the applied load and the stiffness of the surfaces.

Additional studies by Landman et al. [236] used MD simulations with bond-order and EAM potentials coupled with pair-wise potentials similar to (10.6) to study the sliding of two gold surfaces with pyramidal asperities that have straight chain $C_{16}H_{34}$ lubricant molecules trapped between them, as illustrated in Fig. 10.47. An important aspect of this study is that the sliding rate in the simulations is about 10 m/s, which is the same order of magnitude as the scanning speed in a computer disk. As the asperities approach each other, the hydrocarbon molecules begin to form layers. This is reflected in the oscillations in the frictional force shown in Fig. 10.48. When the asperities overlap in height and approach each other laterally, the pressure of the lubricant molecules increases to about 4 GPa which leads to the deformation of the gold asperities.

Glosli and McClelland [176] modeled the sliding of two ordered monolayers of alkane chains that are attached to two rigid substrates. This system is shown schematically in Fig. 10.49. The simulations predicted that energy dissipation occurs by a discontinuous plucking mechanism (sudden release of shear strain) or a viscous mechanism (continuous collisions of atoms of opposite films). The specific mechanism that occurs depends on the interfacial interaction strength. In particular, the *pluck* occurs when mechanical energy stored as strain is converted into thermal energy that leads to low friction forces at low temperatures. On the other hand, at higher temperatures some of the energy of sliding is dissipated through phonon excitations, which results in higher frictional forces. Interestingly, this trend reverses again at the highest temperatures considered when the molecules move so much that they slide easily over the surfaces, which decreases the frictional force. These results are summarized in Figs. 10.50 and 10.51.

Other studies of sliding surfaces with attached organic chains include MD simulations with LJ potentials by Müser and coworkers, [41, 237, 238] which considered friction between polymer *brushes* in sliding contact with one another. In particular, they considered the effect of sliding rate on the tilting of polymers and the effect of steady-state sliding versus nonsteady-state (*transient*) sliding. The simulations find that shear forces are lower for chains that tilt in a direction that is parallel to the shear direction. This tilting effect is significant for grafted polymers, as illustrated in Fig. 10.52, and less significant for absorbed polymers. This is due to the decrease in the differential frictional coefficient for the grafted polymers as well as the

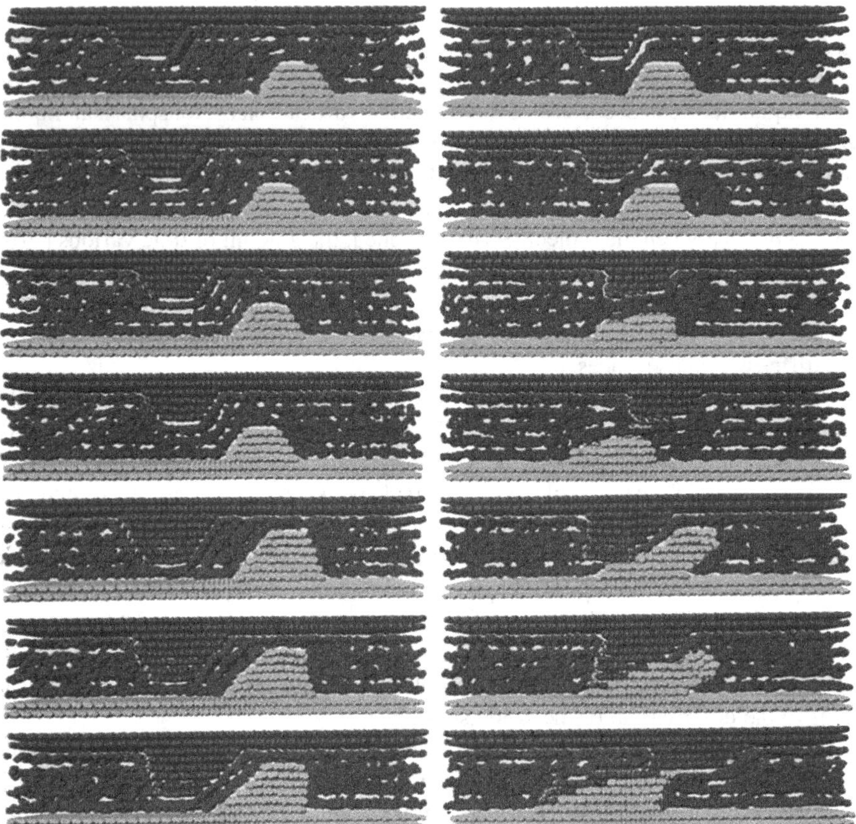

**Fig. 10.47** Stills from a molecular dynamics simulation where Au(111) surfaces with surface roughness slide over one another while separated by hexadecane molecules. The scanning velocity is 10 m/s. Layering of the lubricant and asperity deformation occurs as the sliding continues. The *top three rows* show the results when the asperity heights are separated by 4.6 Å. The *bottom three rows* show the results when the asperity heights are separated by −6.7 Å (After [236], © ACS 1996)

increase in the friction coefficient for absorbed polymers under shear. The tilting is also affected by the rate of sliding and is much larger at high sliding rates than small rates, as indicated in Fig. 10.52. The simulations further show that the inclination angle of the chains decreases much more slowly than the shear stress, and the shear stress maximum is more pronounced if there is hysteresis in the chain orientations.

Typical friction loops for tips that are functionalized and sliding against surfaces that are functionalized in the same manner as illustrated in Fig. 10.25 are shown in Fig. 10.53. The friction force between the OH/OH pairs is significantly larger than the friction force between the $CH_3/CH_3$ pairs. This is due to the formation and breaking of hydrogen bonds during the shearing for the OH/OH pairs. The mean forces versus load forces for the OH/OH and $CH_3/CH_3$ pairs given in Fig. 10.54 are reduced by the tip radius.

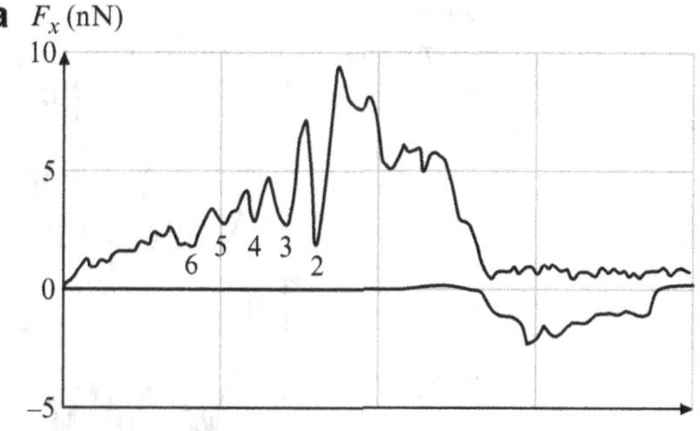


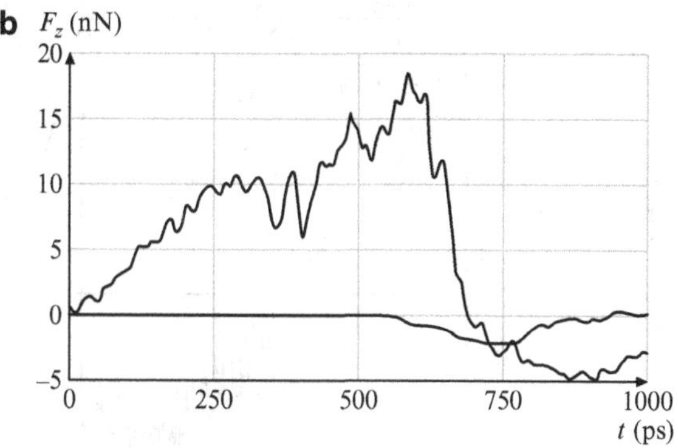


**Fig. 10.48** **(a)** The lateral force ($F_x$) and **(b)** normal force ($F_z$) from the molecular dynamics simulations shown in Fig. 10.35 as a function of time. The forces between the two metal surfaces are shown by the *dashed line*. The force oscillations correspond to the structural changes of the lubricant in Fig. 10.35 (After [236], © ACS 1996)

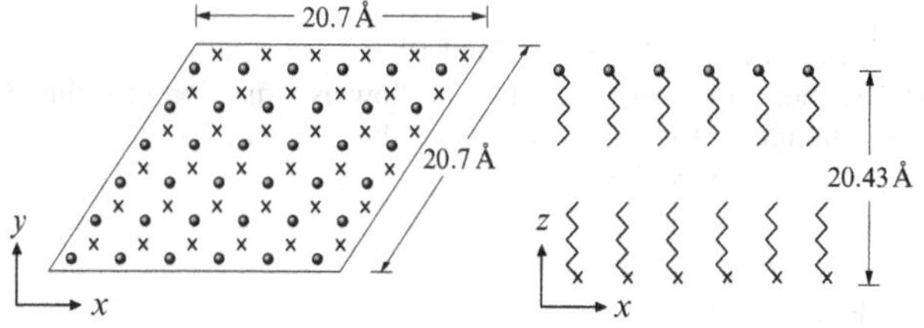


**Fig. 10.49** Top and side views of the alkane chains attached to surfaces that are sliding against each other (After [176], © APS 1993)

MD simulations by Manias et al. considered the shearing of entangled oligomer chains that are attached to sliding surfaces, as illustrated in Fig. 10.55 [239]. They find that slip takes place within the film and that this occurs through changes in the chain conformations. Increased viscosity is predicted at the film–surface interface compared to the middle of the film, which results in a range of viscosities across the film as one moves away from the points of sliding contact.

To summarize this section, experiments and MD simulations show similar stick/slip transitions that occur for thin films of liquid between two sliding solid surfaces. Frictional properties are found to depend to a significant degree on

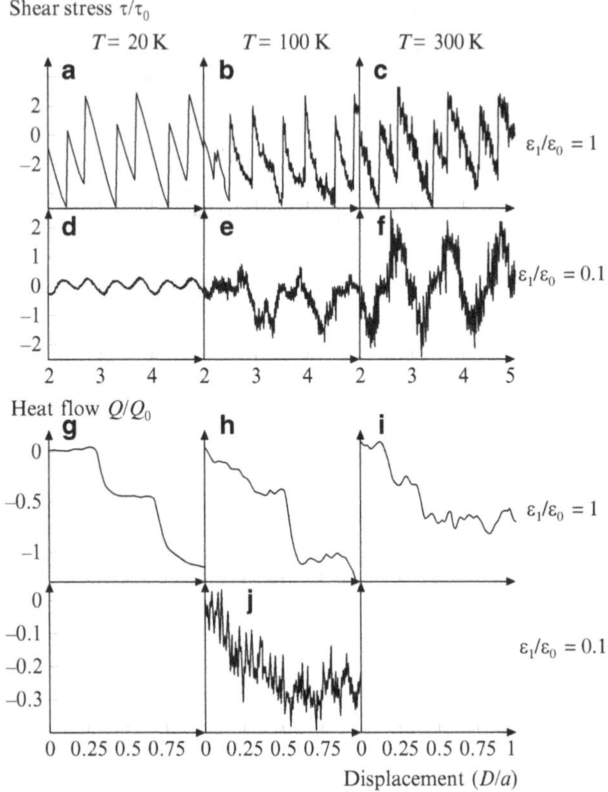


**Fig. 10.50** Data from molecular dynamics simulations of the sliding of the surfaces shown in Fig. 10.32. **(a–f)** The shear stress and **(g–j)** the heat flow as a function of sliding for normal and reduced interfacial strengths. The *plots* show how the calculated values change with system temperature (After [176], © APS 1993)

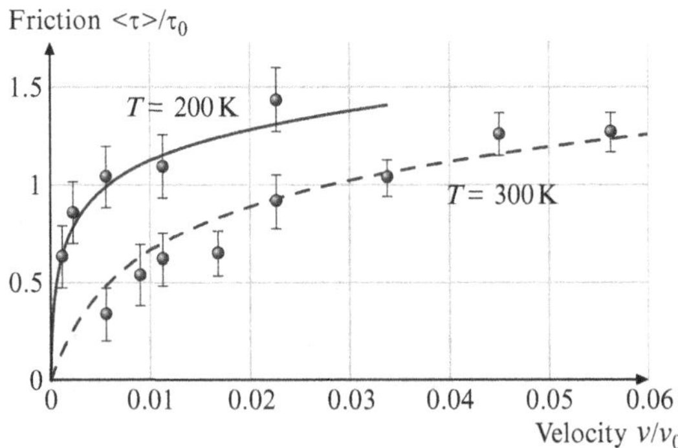


**Fig. 10.51** A plot of calculated values of the average interfacial shear stress as a function of the velocity of sliding of the two surfaces shown in Fig. 10.33 (After [176], © APS 1993)

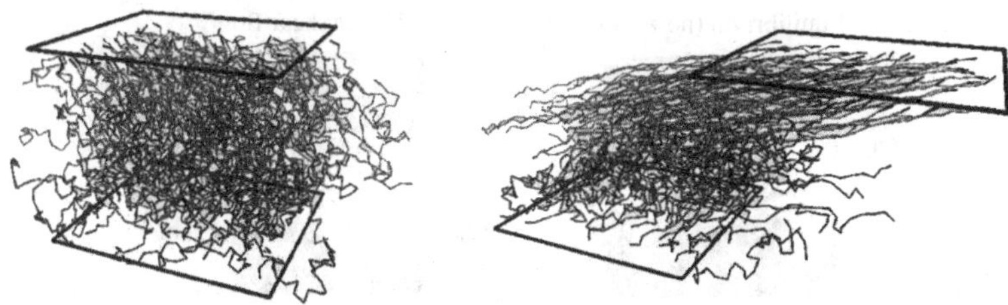

**Fig. 10.52** Snapshots of sliding walls with attached polymers in a solvent. *Right-hand figure* illustrates the sliding process at low sliding rates while the *left-hand figure* illustrates the sliding process at high sliding rates (After [41], © CCLR 2002)

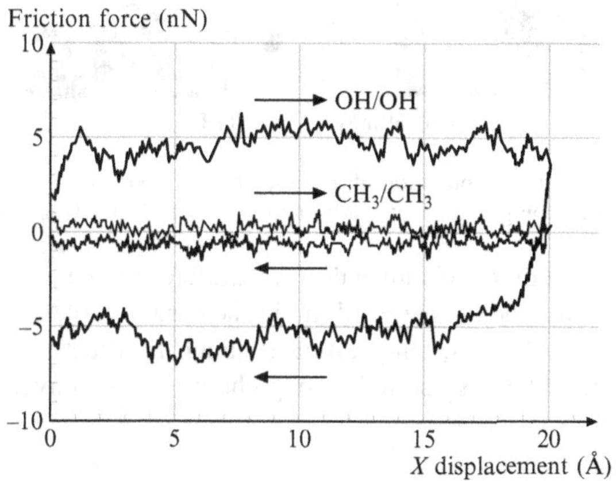


**Fig. 10.53** Typical friction loops for the systems shown in Fig. 10.25 for $CH_3/CH_3$ and OH/OH pairs under a contact load of 0.2 nN (After [163], © ACS 2002)

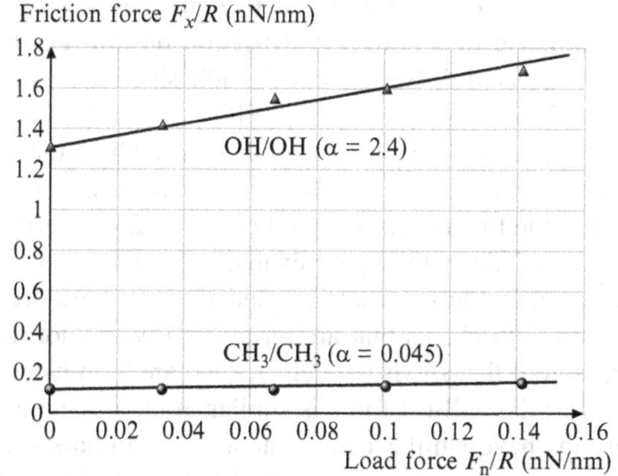


**Fig. 10.54** Friction force versus contact load from the systems shown in Fig. 10.25 for $CH_3/CH_3$ and OH/OH (After [163], © ACS 2002)

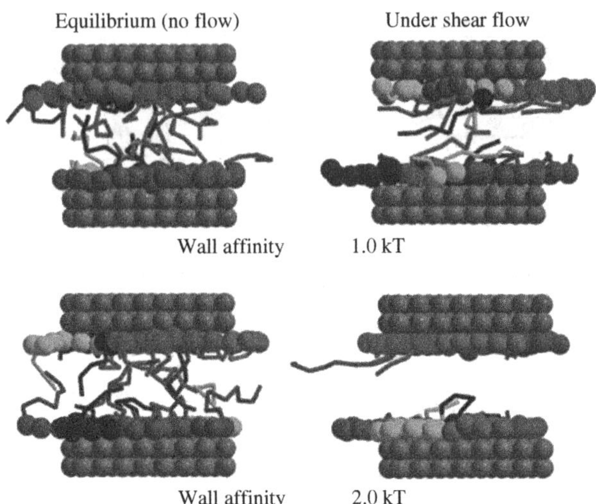


**Fig. 10.55** Changes in the conformation of adsorbed hydrocarbon chains on weakly (*top*) and strongly (*bottom*) physisorbing surfaces at equilibrium and under shear (After [239], © ACS 1996)

molecular shape, whether the molecules are grafted on the surfaces or merely absorbed on them, and the degree of tilting in the case of molecular chains. In the case of long-chain molecules, temperature is found to affect the frictional force because the mechanical energy stored in long-chains can be converted into thermal energy by friction.

## Self-Assembled and Polymer Thin-Film Structures

There have been numerous experimental studies of friction on SAMs on solid surfaces with AFM and FFM. The experimental results reveal relationships among elastic compliance, topography and friction on thin LB films [240]. For example, they have detected differences in the adhesive interactions between the microscope tips and $CH_3$ and $CF_3$ end groups [109]. Fluorocarbon domains generally exhibit higher friction than the hydrocarbon films, which the authors attribute to the lower elasticity modulus of the fluorocarbon films that results in a larger contact area between the tip and the sample [240, 240, 242]. Perry and coworkers examined the friction of alkanethiols terminated with $-CF_3$ and $-CH_3$ [243]. The lattice constants for both films are similar and the films are well-ordered. The friction of the SAMs with chains that are terminated with fluorine end groups is larger than the friction of the SAMs with chains that are terminated with hydrogen end groups. However, the pull-off force is similar in both systems, which implies that these end groups have similar contact areas. The authors speculate that the larger $-CF_3$ groups interact more strongly with adjacent chains than the $-CH_3$-terminated chains. Therefore, the fluorinated chains have more modes of energy dissipation within the plane of the monolayer and, thus, have larger friction.

Molecular disorder of the alkyl chains at the surface can also affect the frictional properties of self-assembled films if the layers are not packed too closely together [244]. Indentation can induce disorder in the chains that then compress as the tip continues to press against them. If the tip presses hard enough, the film hardens as a result of the repulsive forces between the chains. However, if the chains are tilted, they bend or deform when the tip pushes on them in a mostly elastic fashion that produces long lubrication lifetimes. At low contact loads of about $10^{-8}$ N, wear usually occurs at defect sites, such as steps. Wear can also occur if there are strong adhesive forces between the film and the surface [245].

The friction of model SAMs composed of alkane chains was examined using MD simulations with bond-order and LJ potentials by Mikulski and Harrison [246, 247]. These simulations show that periodicities observed in a number of system quantities are the result of the synchronized motion of the chains when they are in sliding contact with the diamond counterface. The tight packing of the monolayer and commensurability of the counterface are both needed to achieve synchronized motion when sliding in the direction of chain tilt. The tightly packed monolayer is composed of alkane chains attached to diamond (111) in the $(2 \times 2)$ arrangement and the loosely packed system has $\approx 30\%$ fewer chains. The average friction at low loads is similar in both the tightly and loosely packed systems at low loads. Increasing the load, however, causes the tightly packed monolayer to have significantly lower friction than the loosely packed monolayer (Fig. 10.56). While the movement of chains is somewhat restricted in both systems, the tightly packed monolayer under high loads is more constrained with respect to the movement of individual chains than the loosely packed monolayer, as illustrated in Fig. 10.57. Therefore, sliding initiates larger bond-length fluctuations in the loosely packed system, which ultimately lead to more energy dissipation via vibration and, thus, higher friction. Thus, the efficient packing of the chains is responsible for the lower friction observed for tightly packed monolayers under high loads.

Several AFM experiments have examined the friction of SAMs composed of chains of mixed lengths [248]. For example, the friction of SAMs composed of

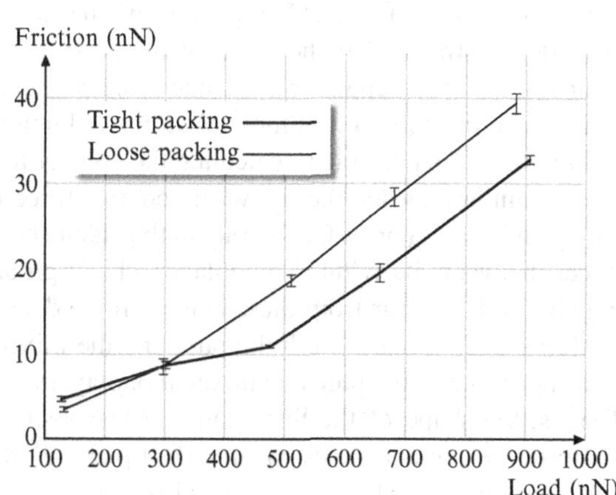


**Fig. 10.56** Friction as a function of load when a hydrogen-terminated counterface is in sliding contact with $C_{18}$ alkane monolayers (After [247], © ACS 2001)

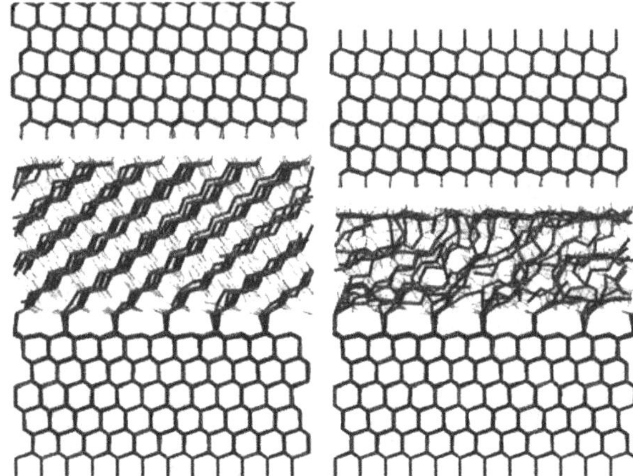

**Fig. 10.57** Snapshots of tightly packed $C_{18}$ alkane monolayers on the *left*, and loosely packed monolayers on the *right* under a load of about 500 nN. The chains in both systems are arranged in a (2 × 2) arrangement on diamond (111). The loosely packed system has 30% of the chains randomly removed. The sliding direction is from *left to right* (After [247], © ACS 2001)

spiroalkanedithiols was examined by Perry and coworkers [249]. The effects of crystalline order at the sliding interface were examined by systematically shortening some of the chains. The resulting increase in disorder at the sliding interface causes an increase in friction.

The link between friction and disorder in monolayers composed of *n*-alkane chains was recently examined using MD simulations by Harrison and coworkers [250]. The tribological behavior of monolayers of 14 carbon atom-containing alkane chains, or pure monolayers, was compared to monolayers that randomly combine equal amounts of 12 and 16 carbon-atom chains, or mixed monolayers. Pure monolayers consistently show lower friction than mixed monolayers when sliding under repulsive (positive) loads in the direction of chain tilt. These MD simulations reproduce trends observed in AFM experiments of mixed-length alkanethiols [248] and spiroalkanedithiols on Au [251]. Harrison and coworkers [252] have also examined the *odd-even* effect noted in experiment [253], where friction is found to be larger for SAMs differing by one methylene group. The MD simulations demonstrated that the effect was due to conformational differences in the chains of different length and became more pronounced at higher loads.

Because the force on individual atoms is known as a function of time in MD simulations, it is possible to calculate the contact forces between individual monolayer chain groups and the tip, where contact force is defined as the force between the tip and a $-CH_3$ or a $-CH_2$-group in the alkane chains. The distribution of contact forces between individual monolayer chain groups and the tip are shown in Fig. 10.58. It is clear from these contact force data that the magnitude, or scale of the forces, is similar in both the pure and the mixed monolayers. In addition, it is also apparent that the pure and mixed monolayers resist tip motion in the same way. That is, the shape of the histograms in the positive force intervals is similar. In contrast, the contact forces *pushing* the tip along differ in the two monolayers. The pure monolayers exhibit a high level of symmetry between resisting and pushing

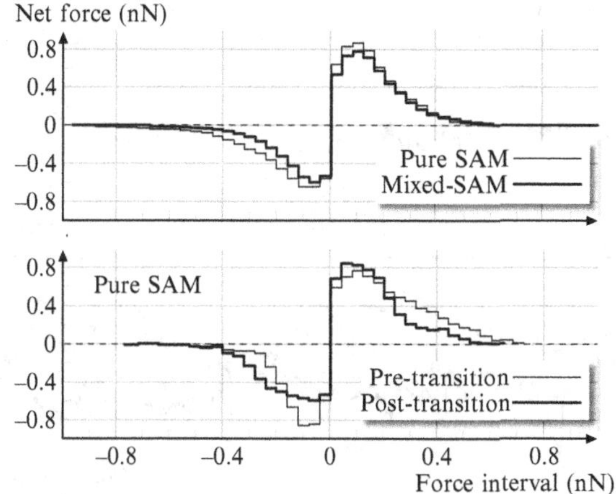


**Fig. 10.58** The distribution of contact forces along the sliding direction (friction force). In the *upper panel*, the forces for the mixed and pure system sliding in the direction of chain tilt are shown. The forces for the pure system sliding in the transverse direction to the chain tilt are shown in the *lower panel*. Positive force intervals correspond to chain groups that resist tip motion while negative intervals correspond to chain groups that *push* the tip in the sliding direction. Forces from four runs with independent starting configurations are binned for all sets of data

forces. Because the net friction is the sum of the resisting and pushing forces, the symmetry in these distributions of the pure monolayers results in a lower net friction than the mixed monolayers. Thus, the ordered, densely packed nature of the pure monolayers allows the energy stored when the monolayer is resisting tip motion (positive forces) to be regained efficiently when the monolayer *pushes* on the tip (negative forces). The distribution of negative contact forces in the mixed monolayers is different from the distribution of the positive forces. For this reason, mechanical energy is not efficiently channeled back into the mixed monolayer as the tip passes over the chains and, as a result, the friction is higher. The range of motion of the chains is monitored by computing the deviation in a chain group's position compared to its starting position, as illustrated in Fig. 10.59. It is clear from analyzing these data that the increased range of motion is linked to large contact forces. The increased range of motion of the protruding tails in the mixed system prevents the efficient recovery of energy during sliding (negative contact force distribution) and allows for the dissipation of energy.

The pure monolayers exhibit marked friction anisotropy. The contact force distribution changes dramatically as a result of the change in sliding direction, resulting in an increase in friction (Fig. 10.58). Sliding in the direction perpendicular to chain tilt can cause both types of monolayers to transition to a state where the chains are primarily tilted along the sliding direction. This transition is accompanied by a large change in the distribution of contact forces and a reduction in friction.

Recently, the response of monolayers composed of alkyne chains, which contain diacetylene moieties, to compression and shear [254] was examined using MD

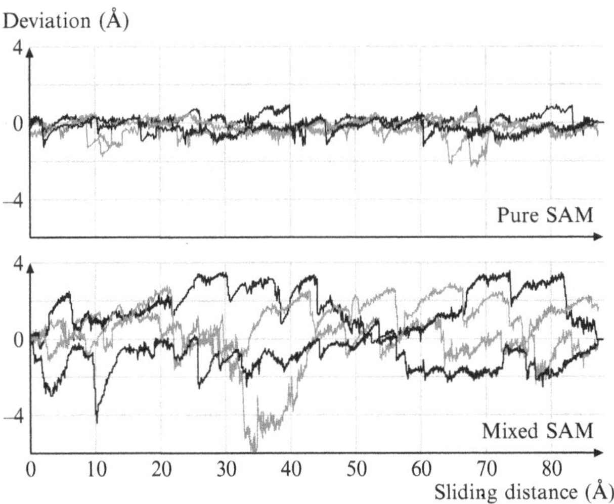


**Fig. 10.59** Trajectories of individual chain groups that generate the largest contact forces when sliding in the direction of chain tilt for both the pure and mixed monolayer systems. The deviation is defined as the change in position along the sliding direction relative to the chain group's starting position. (The positions are averaged over 2,000 simulation steps)

simulations. These are the only simulations to date that show that compression and shear can result in cross-linking, or polymerization, between chains. The vertical positioning of the diacetylene moieties within the alkyne chains (spacer length) and the sliding direction both have an influence on the pattern of cross-linking and friction. Compression and shear cause irregular polymerization patterns to be formed among the carbon backbones, as illustrated in Fig. 10.60. When diacetylene moieties are located at the ends of the chains closest to the tip, chemical reactions between the chains of the monolayer and the amorphous carbon tip occur causing the friction to increase 100 times, as indicated in Fig. 10.61. The friction between the amorphous carbon tip and all of the diacetylene-containing chains is larger than the friction between a hydrogen-terminated diamond counterface and tightly packed monolayers composed of *n*-alkane chains [247]. This is attributed to the disorder at the interface caused by the irregular counterface.

Zhang and Jiang [255] used MD simulations to study the effect of confined water between alkyl monolayers terminated with $–CH_3$ (hydrophobic) and –OH (hydrophilic) groups on Si(111), as illustrated in Fig. 10.62. For the hydrophobic molecules, the friction coefficient is almost constant independent of the number of water molecules. For the hydrophilic molecules, the friction coefficient decreases rapidly with an increase in the number of water molecules, as shown in Fig. 10.63. These results are in good agreement with surface force microscopy (SFM) experimental results. Zhang et al. [256] also studied the friction of alkanethiol SAMs on gold using hybrid molecular simulations at the same time scales as are used in AFM and FFM experiments. Various quantities were varied in the simulations, including chain length, terminal group, scan direction and scan velocity. The simulations

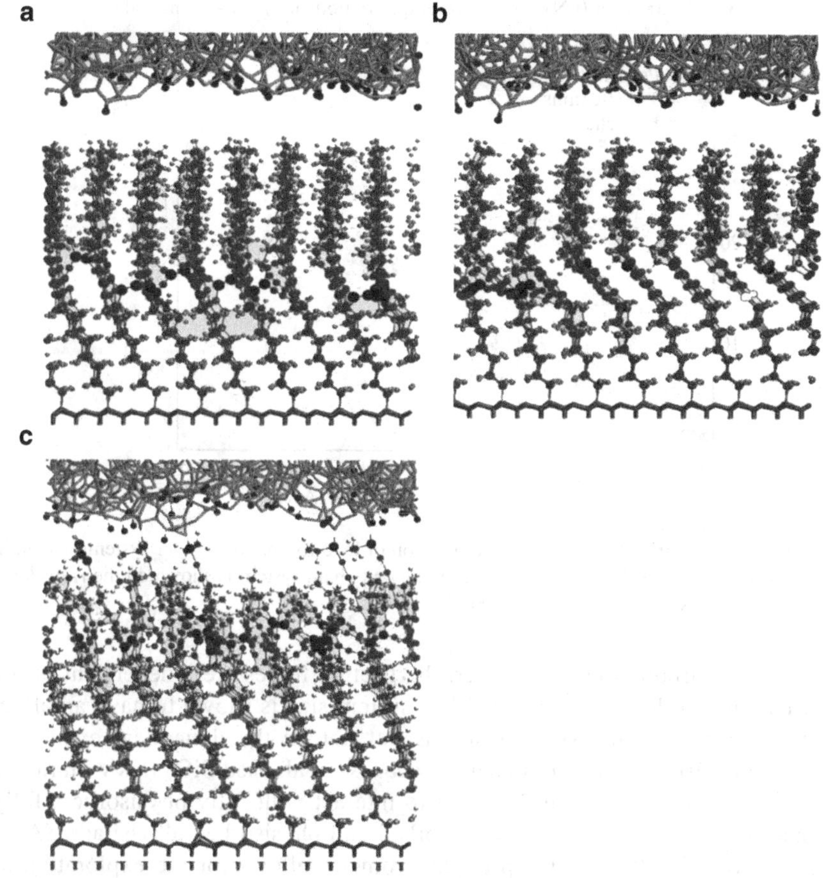


**Fig. 10.60** (**a**) Perpendicular-chain, (**b**) tilted-chain, and (**c**) end-chain monolayer systems after compression to 200 nN and pull-back of the hydrogen-terminated tip. *Large, dark spheres* in the hydrocarbon monolayers represent cross-linked atoms with $sp^2$ hybridization. *Dark, small spheres* represent hydrogen atoms that are initially on the hydrogen-terminated amorphous carbon tip (After [254], © ACS 2004)

showed that the frictional force decreases as the chain length increases and is smallest when scanned along the tilt direction. They also predicted a maximum friction coefficient for hydrophobic $-CH_3$-terminated SAMs and low friction coefficients for hydrophilic –OH-terminated SAMs as the scan velocity increases. The simulations further predicted a saturated constant value at high scan rates for both surfaces. These results are summarized in Figs. 10.64 and 10.65.

The work of Chandross et al. [257, 258] illustrates the effects of chain length on friction and stick–slip behavior between two ordered SAMs consisting of alkylsilane chains over a range of shear rates at various separation distances or pressure, as illustrated in Fig. 10.66. The adhesion forces between the two SAMs at the same separation distance decrease as the chain length increases

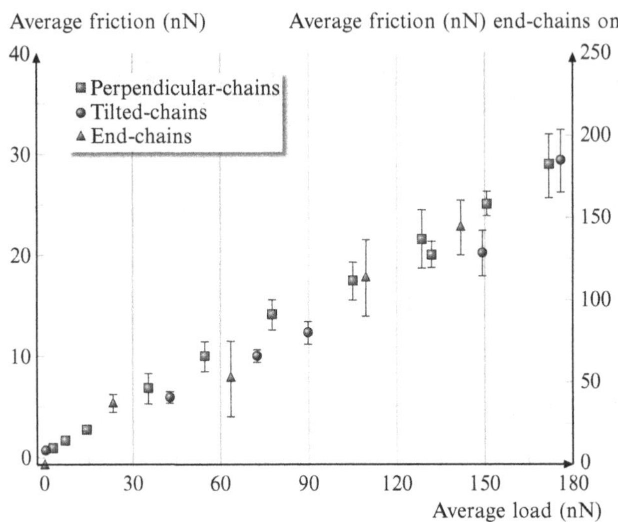


**Fig. 10.61** Average friction on the tip as a function of load for the monolayer systems shown in Fig. 10.60. The *scale* for the average friction in the end-chain system is shown on the *right-hand side* of the figure (After [254], © ACS 2004)

from 6 to 18 carbon atoms. However, the friction forces are independent of the chain length and the shear velocity. The system size is shown to have an effect on the sharpness of the slip transitions but not on the dynamical events, as shown in Fig. 10.67. In a later paper, *Chandross* et al. used $SiO_2$ tips with radius of curvatures in the range of 3–30 nm to interact with fully physisorbed, fully chemisorbed, and a mixture of chemisorbed and physisorbed alkylsilane SAMs on amorphous Si [259]. This tip-based geometry allows for the exploration of actual contact area as a function of load, which was found to be proportional to the square root of the load.

SAMs have been very successful in the lubrication of surfaces that infrequently come into sliding or normal contact. Their inability to lubricate a reciprocating contact is due, in part, to the inability to replenish the coating in situ. One proposed solution is to use a chemically bound SAM in conjunction with a physisorbed mobile molecule in a *bound + mobile* lubricant scheme, which is similar to the lubricants used to mitigate head crashes in hard drives.

Irving and Brenner used molecular dynamics to study the interfacial structure, self-diffusion, and ability of the mobile phase to incorporate into defected sites [260]. A potential bound + mobile lubricant combination of a chemically bound octadecyltrichlorosilane (ODTS) SAM together with mobile tricresyl phosphate (TCP) molecules was studied. The simulations showed that the TCP did not incorporate into the interior of the close packed defect free ODTS SAM. The TCP molecules on the surface of the SAM were also not tightly bound to a particular surface site but instead were found to readily diffuse across the surface in a random walk fashion. An estimated diffusion barrier of 0.0937 eV with an

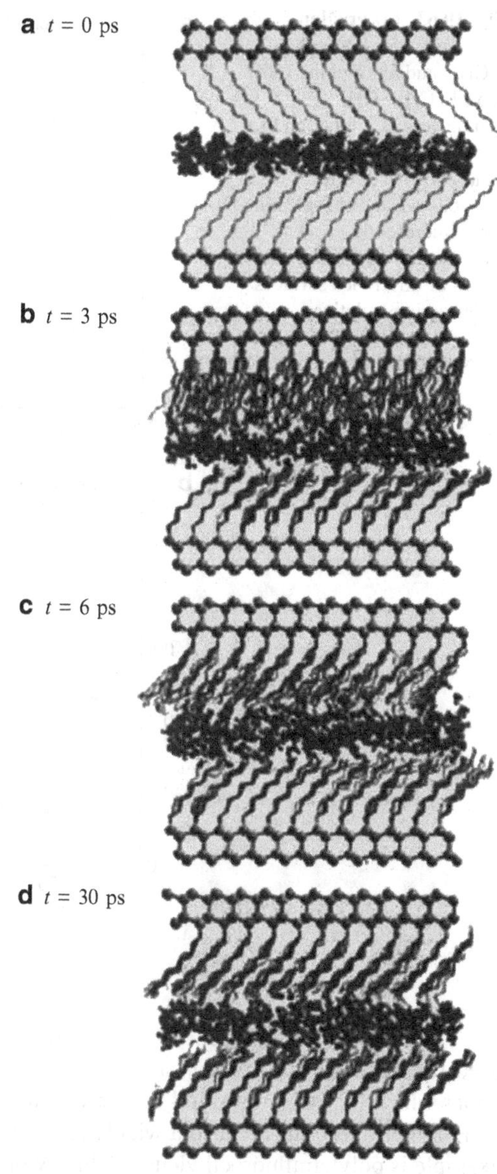


**Fig. 10.62** Snapshots of hydrophilic monolayers and confined water molecules from MD simulations at 300 K. The tilt direction of monolayers on the top plate changed after $t = 10.0$ ps. **(a–d)** illustrates how the tilting of the monolayers change as a function of time, and the way in which the water becomes increasingly less confined (After [255], © AIP 2005)

Arrhenius prefactor of $26.47 \times 10^{-4}$ $cm^2/s$ was calculated for single-molecule diffusion. It was also found that the TCP molecules would only localize in the vicinity of methylene groups ($-CH_2-$) along the backbone of the ODTS chain, which were exposed when a cylindrical defect was created in the SAM. To get to these localizing sites, however, first required the TCP molecules to overcome an anisotropic energy barrier for inclusion into the cylindrical defect. This anisotropic barrier was found to depend on the direction of ODTS chain tilt and the direction

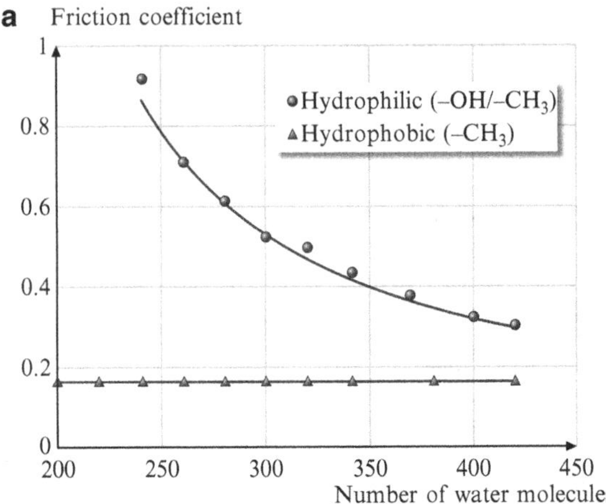


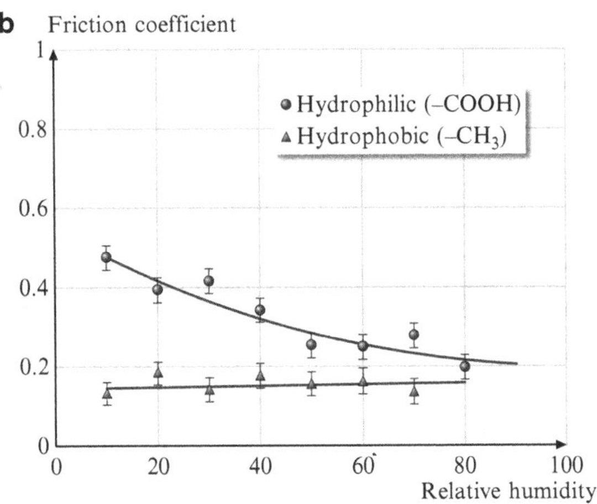


**Fig. 10.63** **(a)** Friction coefficients for hydrophobic (–$CH_3$) and hydrophilic (50% mixed –$CH_3$/–OH) monolayers as a function of water molecules from MD simulations at 300 K ($H$ =6.0 Å), and **(b)** scanning force microscopy measurements of frictional forces of difference surfaces under various relative humidities (After [255], © AIP 2005)

the TCP molecules approached the cylindrical defect in the SAM. A later study used the diffusion information and multiscale methods to analyze the conditions under which this scheme would be successful [261].

Polymer thin films are also a widely studied for their lubricating properties. An example is polytetrafluoroethylene (PTFE), which has been used in a wide range of applications from satellites to frying pans. In a joint computational and experimental work Jang et al. examined the molecular origins of friction using classical molecular dynamics as well as an AFM and microtribometer [262]. The simulations predicted an anisotropic behavior of the friction coefficient depending on whether the sliding direction was parallel or perpendicular to the PTFE chains lying on the surface. Sliding directions parallel produced lower friction coefficients and wear while sliding perpendicular to alignment produced higher coefficients and wear. The microtribometer results were in agreement with these findings. Also of interest

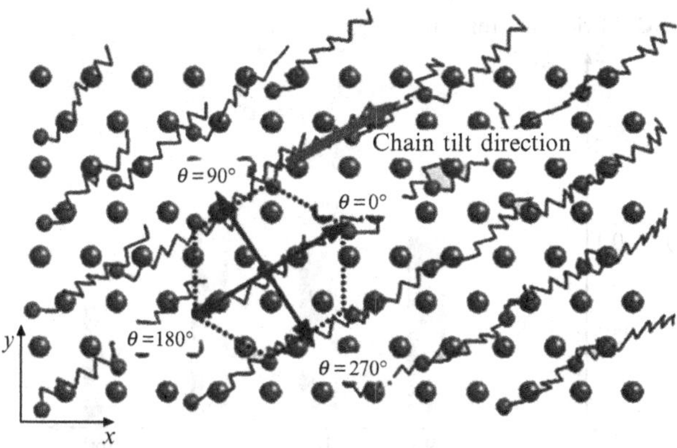


**Fig. 10.64** Schematic illustration of the chain tilt and scan directions on alkanethiol SAMs/Au(111) in hybrid molecular simulations; $\theta$ is the angle between the tip moving direction and the chain tilt direction. The *larger spheres* represent substrate Au atoms, smaller spheres sulfur atoms in molecular chains, and *zigzag lines* molecular chains (After [256], © ACS 2003)

in the AFM work was that transfer films were always produced parallel to the sliding direction.

Similar experimental findings for anisotropic tribological behavior of polyethylene (PE) in the literature motivated Heo et al. to examine this system using classical molecular dynamics to study crystalline PE interfaces [263]. Like the findings for PTFE it was found that friction and wear had an anisotropic behavior that depended on molecular orientation and sliding direction. Unlike the findings for PTFE, the PE system exhibited a stick–slip motion as the interfaces passed by one another. The reason for the differences between the two systems was attributed to increased bond scission seen in PTFE as compared to PE under sliding conditions. This scission allows collections of molecules in PTFE to move at the interface, which does not occur as readily in the PE system.

In short, atomic-scale simulations show the relationship between elastic properties, degree of molecular disorder and friction of self-assembled thin films that illuminates the origin of the properties that are measured experimentally.

## Nanoparticles

Nanoparticles are being considered for a wide variety of applications, including as fillers for nanocomposite materials, novel catalysts or catalytic supports, and components for nanometer-scale electronic devices [264]. They have also generated considerable interest as possible new lubricant materials that have the potential to function as *nanoballbearings* with exceptionally low friction coefficients. The nanoparticles of most interest for tribological applications include $C_{60}$ [265–277], carbon nanotubes [278–285], and $MoO_3$ nanoparticles [286, 287], among others [288].

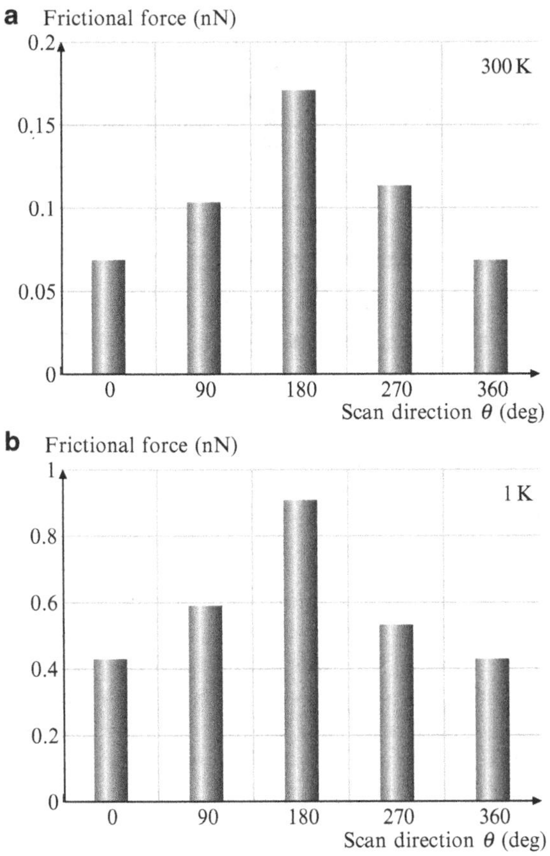


**Fig. 10.65** Frictional force as a function of scan direction from hybrid simulations for $C_{11}CH_3$ SAMs on Au(111) at **(a)** 300 K and **(b)** 1.0 K. Frictional force is the smallest when scanned along the tilt direction, the largest when scanned against the tilt direction, and between when scanned perpendicular to the tilt direction at both temperatures (After [256], © ACS 2003)

The experiments report wide variations in frictional coefficients (for instance, values of 0.06 to 0.9 have been measured for $C_{60}$) that may be caused by differences in the experimental methods used, the thickness of the nanoparticle layer or island, the atmosphere (argon versus air, levels of humidity) used, and the transfer of nanoparticles to the FFM tips. As a result, there is much that remains to be clarified about the tribological behavior of nanoparticle films.

In the case of $C_{60}$, the mechanistic response to applied shear forces has not been definitively determined. For example, some experimental studies show evidence of $C_{60}$ molecules rolling against the substrate, each other, or the sliding surfaces [265, 270, 272, 275, 277] while others hypothesize that the low friction of $C_{60}$ films is due in part to blunting of the tip by transfer of fullerene molecules to the tip apex. Fullerene films are found experimentally to have dissipation energies and shear strengths that are a full order of magnitude lower than the values that are typical for

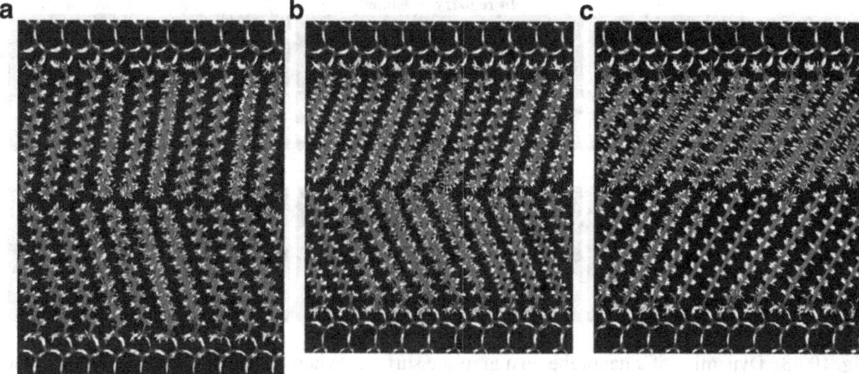

**Fig. 10.66** Wireframe images of $n = 18$ SAMs at fixed separations of (**a**) $d = -5.2$ Å (low pressure, under compression only) (**b**) $d = -10.2$ Å (high pressure, under compression only) and (**c**) $d = -10.2$ Å (high pressure, under shear) (After [258], © ACS 2005)

**Fig. 10.67** Shear stress $\sigma_s$ as a function of system size for $n = 6$ SAMs corresponding to a pressure of 200 MPa at $\boldsymbol{v} = 1.0$ m/s: (**a**) 100 chains per surface, (**b**) 400 chains per surface, (**c**) 1600 chains per surface, and (**d**) 16 point box average of system with 100 chains per surface (After [258], © ACS 2002)

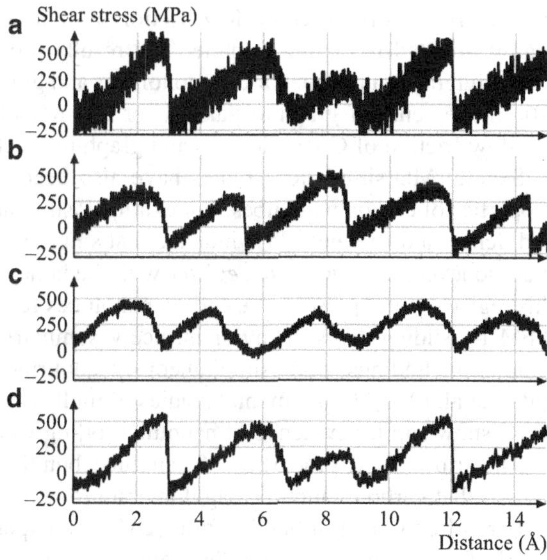


boundary lubricants [289]. Experimental testing of the frictional properties of fullerenes reveal low mechanical stability accompanied by progressive wear and transfer of fullerene materials when they are only physisorbed on a solid surface [290]. Furthermore, measurements with a FFM show that under certain conditions, adsorbed fullerene films deteriorate at pressures as low as about 0.1 GPa [291]. The challenge is therefore to obtain mechanically stable, ordered molecular films of fullerenes firmly attached to a solid substrate.

There have been several MD simulation studies to investigate the tribological properties of fullerenes. A representative study by Legoas et al. [292] investigated

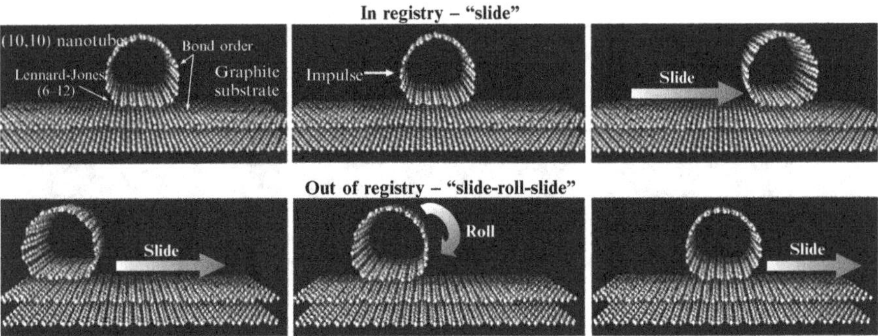


**Fig. 10.68** Dynamics of a nanotube on a graphite surface. When the nanotube and graphite plane are out of registry, the nanotube slides as it slows down from an initial impulse (*upper right panel*). When the nanotube is oriented such that it is in registry with the graphite, it slows by a combination of rolling and sliding

the experimentally observed low-friction system of $C_{60}$ molecules positioned on highly oriented pyrolytic graphite. The results show that decreasing the van der Waals interaction between a $C_{60}$ monolayer and graphite sheets, and the characteristic movements of graphite flakes over $C_{60}$ monolayers, explains the measured ultralow friction of $C_{60}$ molecules and graphite sheets.

Several MD simulation studies have also been carried out on the tribological properties of carbon nanotubes. For example, simulations by Buldum and Lu [278] and Schall and Brenner [281] indicate that single-wall carbon nanotubes roll when their honeycomb lattice is *in registry* with the honeycomb lattice of the graphite. If this registry is not present, the carbon nanotubes respond to applied forces from an AFM by sliding. This behavior is nicely summarized in Fig. 10.68. These MD simulation findings were simultaneously confirmed in experimental studies by Falvo et al. [280]. Experimental studies of multiwall carbon nanotubes on graphite [284] show similar evidence of nanotube rolling when the outer tube is pushed.

The tribological properties of nanotube bundles are important, as it is well-known that carbon nanotubes agglomerate together very readily to form bundles and are often grown in bundle form [264]. An experimental study by Miura et al. [285] of carbon nanotube bundles being pushed around on a KCl surface with an AFM tip indicates that bundles of single-wall carbon nanotubes can be induced to roll in a manner that is similar to the rolling observed for multiwall nanotubes.

MD simulations by Ni and Sinnott [282, 283] considered the responses of horizontally and vertically aligned single-wall carbon nanotubes between two hydrogen-terminated diamond surfaces, where the top surface is slid relative to the bottom surface. The movement of the carbon nanotubes in response to the shear forces was predicted to be simple sliding for both orientations. Interestingly, the simulations do not predict rolling of the horizontally arranged carbon nanotubes even when they are aligned with each other in two-layer and three-layer structures. Instead, at low compressive forces, illustrated in Fig. 10.69, the nanotube bundles

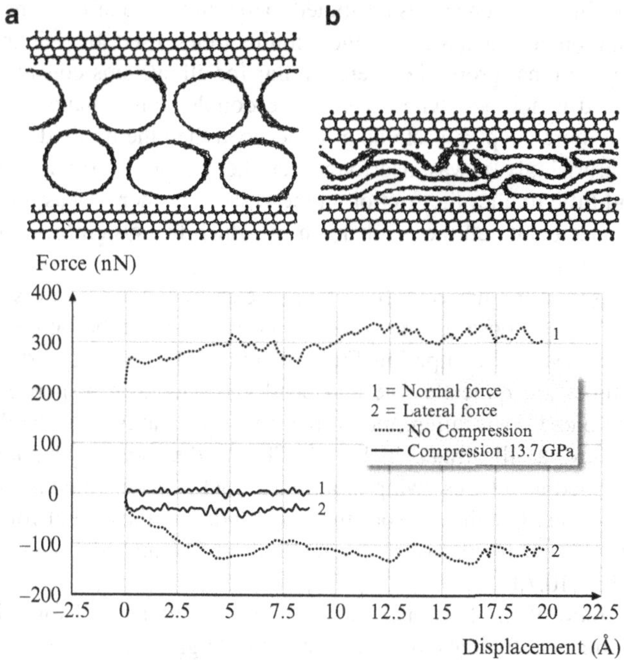


**Fig. 10.69** *Upper*: Snapshots from simulations that examine the sliding of the topmost diamond surface on horizontally arranged nanotubes at different compressions; **(a)** is at a pressure of ≈ 0 GPa; **(b)** is with a pressure of 13.7 GPa. *Lower*: Plots of the normal and lateral components of force during sliding of the top diamond's surface on horizontally arranged nanotubes as a function of the displacement of the top diamond surface with respect to the diamond surface on the bottom

slide as a single unit, and at high compressive forces, also illustrated in Fig. 10.69, the deformed carbon nanotubes closest to the topmost moving diamond surface start to slide in a motion reminiscent of the movement of a tank or bulldozer wheel belt. However, when these moving carbon nanotube atoms would have turned the first corner at the top of the ellipse, they encounter the neighboring nanotube and cannot slide past it. This causes them to deform even further, form cross-links with one another, and, in some cases, move in the reverse direction to the sliding motion of the diamond surface. This causes the large oscillations in the normal and lateral forces plotted in Fig. 10.69.

A later study by Heo and Sinnott examined the frictional properties of single wall, double wall, and filled carbon nanotubes contained between two hydrogen terminated diamondlike carbon layers [293]. It was shown that over a wide range of loads the simulations predicted a friction coefficient of 0.13 for filled as well as unfilled nanotubes. This friction coefficient was found to be constant for pressures below 5 GPa, which should include most practical applications. At pressures above 5 GPa the friction coefficient was found to increase to 0.2. Unlike the work of Ni and Sinnott it was demonstrated that the nanotubes would roll in response to the applied shear. This difference was attributed to the finite size of the nanotubes used in this simulation as compared to the infinite tubes used by Ni and Sinnott. Although the frictional properties were similar for all systems considered, it was found that filled nanotubes, whether they are double-walled nanotubes or nano-peapods, were more adept at sustaining higher load then the unfilled single-walled nanotubes. Surprisingly, it was also found that the addition of lubricating benzene molecules did little to alter the friction in the system. Rather, it was found that the addition of benzene altered the mechanism by which the system responded to the applied shear stress.

The responses of the horizontally arranged carbon nanotubes are substantially different from the responses of the vertically arranged nanotubes at high compression, as can be seen by comparing Figs. 10.69 and 10.70. The vertical, capped carbon nanotubes are quite flexible and bend and buckle in response to applied forces. As the buckle is forming, the normal force decreases then stabilizes in the buckled structure, as illustrated in Fig. 10.70. As the topmost diamond surface slides, the buckled nanotubes swing around the buckle *neck* which helps dissipate the applied stresses. For this reason, the magnitudes of the lateral forces are not significantly different for the vertical nanotubes at low and high compression, as indicated in Fig. 10.70.

When the ratio of the frictional (lateral) force to the normal force is taken to calculate friction coefficients for these systems, high, nonintuitive values were obtained. As outlined by Ni and Sinnott [282], this is because the actual contact area of the nanotubes is not proportional to the sliding force. In the case of the

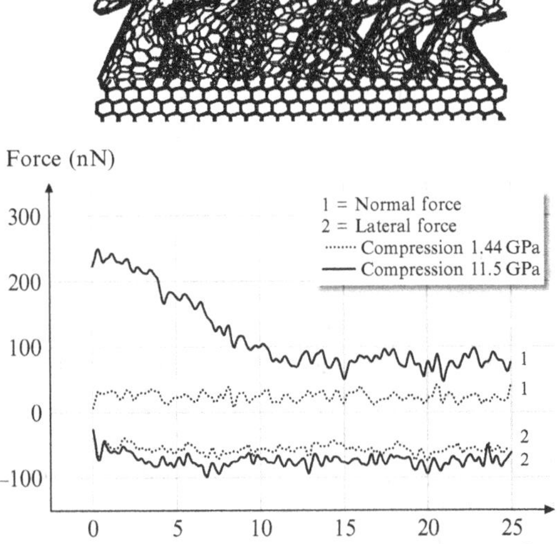


**Fig. 10.70** *Top*: Snapshots from simulations that examine the sliding of the topmost diamond surface on vertically arranged nanotubes with one set of capped ends compressed at a pressure of 11.5 GPa. *Bottom*: Plots of the normal and lateral components of force during sliding of the top diamond's surface on vertically arranged nanotubes as a function of the displacement of the top diamond surface with respect to the diamond surface on the bottom

horizontal nanotube bundles, the tubes are able to deform and significantly change their contact area with the sliding surface with minimal change in the normal force, as shown in Fig. 10.69. In the case of the vertical nanotubes, the contact area remains approximately the same regardless of the initial loading force because of the flexibility of the nanotubes. This causes the lateral forces to change only slightly with significant changes in the normal force, as shown in Fig. 10.70. This analysis indicates that care must be taken in calculating friction coefficients for nanotube systems. Recent experiments by Dickrell et al. [294] show good agreement with these predictions, as shown in Fig. 10.71.

To summarize, this section shows that nanoparticles show some promise as lubricating materials due to their exceptionally low friction coefficients in experiments and simulations. Some nanopaticles show lattice-directed sliding on substrates due

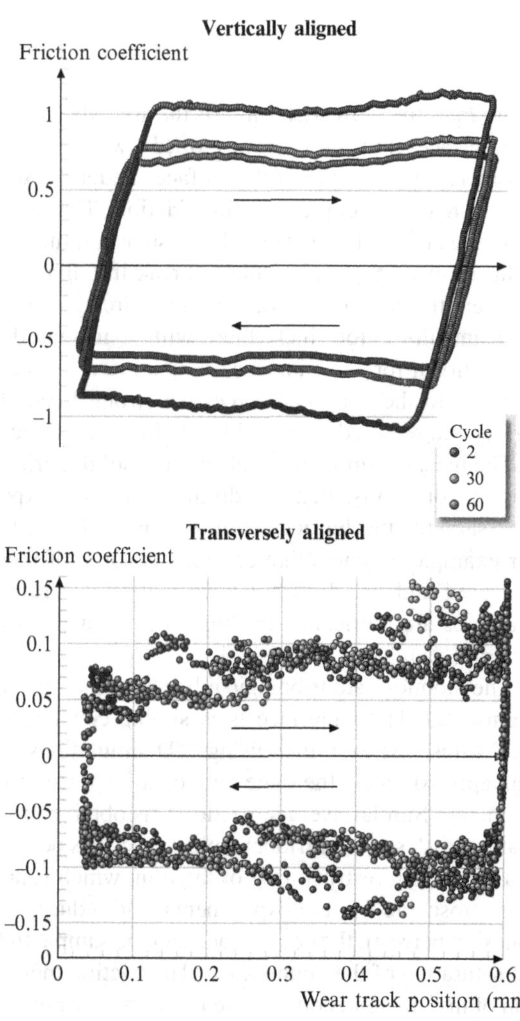


**Fig. 10.71** Coefficient of friction data versus track position collected for one full cycle of reciprocating sliding for nanotubes that are vertically and transversely aligned (After [294])

to their unique atomic structures. However, there is much that remains to be done before the nanometer-scale friction of these materials is well understood.

## Solid State

Surfaces are able to slide over each other at high loads with a minimum of resistance from friction in the presence of liquid lubricants. Some solid thin films can also fulfill these functions and, when they do, are termed solid lubricants. Solid lubricants are generally defined as having friction coefficients of 0.3 or less and low wear.

Bowden and Tabor showed how thin solid films can reduce friction as follows [295]. The total friction force $F_f$ is given as

$$F_f = A F_S + F_p, \tag{10.11}$$

where $F_p$ is the plowing term, $A$ is the area of contact and $F_S$ is the shear strength of the interface. If the surfaces are soft, $F_S$ will be reduced while the other parameters will increase. However, if the surfaces under the solid film are very stiff, $A$ and $F_p$ will decrease thereby decreasing friction. The properties specific to the film will also have an effect on friction. For instance, if the films are less than 1 μm thick, the surface asperities will be able to break through the film to eventually cause wear between the surfaces under normal circumstances. On the other hand, if the lubricant film is too thick, there will be increased plowing and wear that causes the frictional forces to increase. It is important that the lubricant not delaminate in response to the frictional forces, so strong bonds between the lubricant and the surface are required for a solid state lubricant to be effective.

The most common materials used as solid lubricants have layered structures like graphite or $MoS_2$, that, as discussed above, experience low friction. It is not necessary for the lubricant film to have a layered structure to give low friction. For example, diamondlike carbon has some of the lowest coefficients of friction measured and yet does not have a layered structure. Similarly, not all layered structures are lubricants. For instance, mica gives a relatively high coefficient of friction ($>1$).

The atomic-scale tribological behavior that occurs when a hydrogen-terminated diamond (111) counterface is in sliding contact with amorphous, hydrogen-free, DLC films was examined using MD simulations by Gao et al. [296]. Two films, with approximately the same ratio of $sp^3$–$sp^2$ carbon but different thicknesses, were examined. Similar average friction was obtained from both films in the load range examined. A series of tribochemical reactions occur above a critical load that result in a significant restructuring of the film, which is analogous to the *run-in* observed in macroscopic friction experiments, and reduces the friction. The contribution of adhesion between the counterface and the sample to friction is examined by varying the saturation of the counterface. The friction increases when the degree of saturation of the diamond counterface is reduced by randomly removing hydrogen atoms.

Lastly, two potential energy functions that differ only in their long-range forces are used to examine the contribution of long-range interactions to friction in the same system (as illustrated in Figs. 10.72 and 10.73).

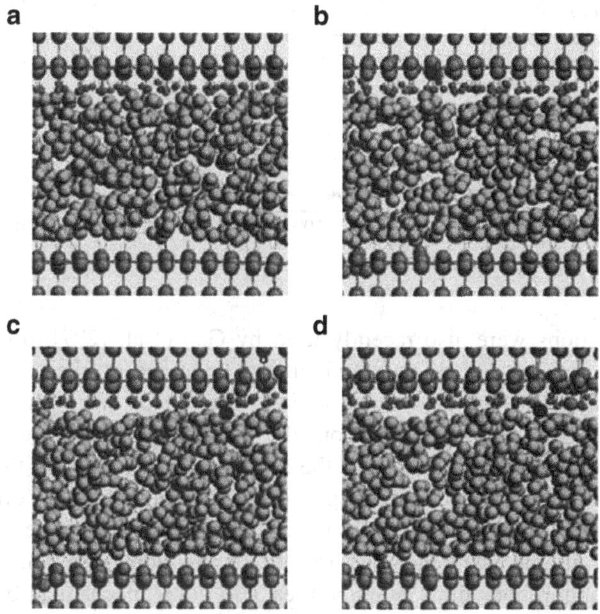


**Fig. 10.72** A series of chemical reactions induced by sliding of the counterface over the thin film under an average load of 300 nN. (**a**) The sliding causes the rupture of a carbon–hydrogen bond in the counterface. (**b**) The hydrogen atom is incorporated into the film and forms a bond to a carbon atom in the film. (**c**) A bond is formed between the unsaturated carbon atoms in the film and the carbon that suffered the bond rupture in the counterface, and continued sliding causes this carbon to be transferred into the film. (**d**) The transferred carbon forms a bond with another carbon in the counterface. The counterface has slid 0.0 (**a**), 15.9 (**b**), 26.1 (**c**), and 30.5 Å (**d**) (After [296], © ACS 2002)

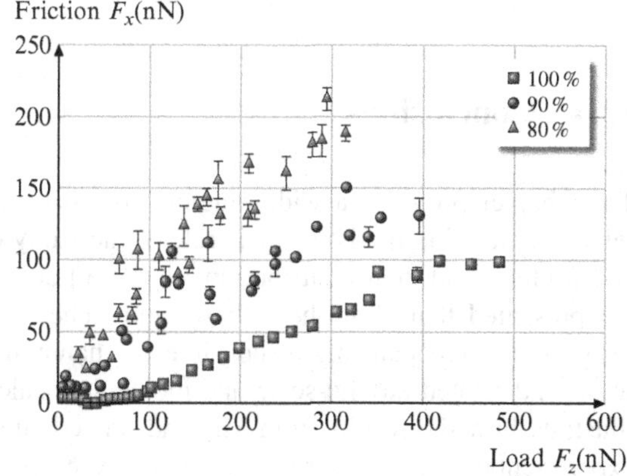


**Fig. 10.73** Friction curves for the thin film system with a counterface that is 100% hydrogen-terminated (*open squares*), 90% hydrogen-terminated (*filled squares*), and 80% hydrogen-terminated (*open circles*) (After [296], © ACS 2002)

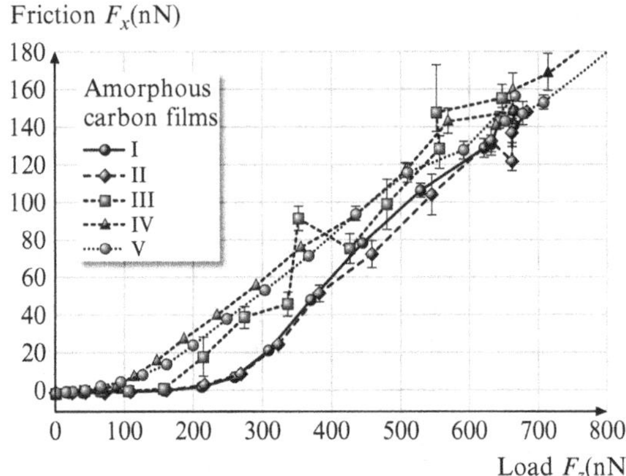


**Fig. 10.74** Average friction versus load for five amorphous carbon films. Films I–III are hydrogen-free and contain various ratios of $sp^2$-to-$sp^3$ carbon. Films IV and V are both over 90% $sp^3$ carbon and have surface hydrogenation

MD simulations were also recently used by Gao et al. [297] to examine the effects of the $sp^2$–$sp^3$ carbon ratio and surface hydrogen on the mechanical and tribological properties of amorphous carbon films. This work showed that, in addition to the $sp^2$–$sp^3$ ratio of carbon, the three-dimensional structures of the films are important when determining the mechanical properties of the films. For example, it is possible to have high $sp^2$-carbon content, which is normally associated with softer films, and large elastic constants. This occurs when $sp^2$-ringlike structures are oriented perpendicular to the compression direction. The layered nature of the amorphous films examined leads to novel mechanical behavior that influences the shape of the friction versus load data, as illustrated in Fig. 10.74. When load is applied to the films, the film layer closest to the interface is compressed. This results in the very low friction of films I and II up to $\approx$ 300 nN and the response of films IV and V up to 100 nN. Once the outer film layers have been compressed, additional application of load causes an almost linear increase in friction for films I and II as well as IV and V. Film III has an erratic friction versus load response due to the early onset of tribochemical reactions between the tip and the film.

## 10.4 Conclusions

This chapter provides a wide-ranging discussion of the background of MD and related simulation methods, their role in the study of nanometer-scale indentation and friction, and their contributions to these fields. Specific, illustrative examples are presented that show how these approaches are providing new and exciting insights into mechanisms responsible for nanoindentation, atomic-scale friction, wear, and related atomic-scale and molecular scale processes. The examples also illustrate how the results from MD and related simulations are complementary to experimental studies, serve to guide experimental work, and assist in the interpretation of experimental data. The ability of these simulations and experimental techniques such as the surface force apparatus and proximal probe microscopes to study nanometer-scale indentation and friction at approximately the same scale is revolutionizing our understanding of the origin of friction at its most fundamental atomic level.

## References


1. D. Dowson. *History of Tribology* (Longman, London, 1979).
2. K.L. Johnson, K. Kendell, A.D. Roberts: Surface energy and the contact of elastic solids, Proc. R. Soc. Lond. A **324**, 301–313 (1971).
3. M. Gad-el-Hak (ed.): *The MEMS Handbook*, Mech. Eng. Handbook (CRC, Boca Raton, 2002).
4. J. Krim: Friction at the atomic scale, Sci. Am. **275**, 74–80 (1996).
5. J. Krim: Atomic-scale origins of friction, Langmuir **12**, 4564–4566 (1996).
6. J. Krim: Progress in nanotribology: Experimental probes of atomic-scale friction, Comments Condens. Matter Phys. **17**, 263–280 (1995).
7. A.P. Sutton: Deformation mechanisms, electronic conductance and friction of metallic nanocontacts, Curr. Opin. Solid State Mater. Sci. **1**, 827–833 (1996).
8. C.M. Mate: Force microscopy studies of the molecular origins of friction and lubrication. IBM J. Res. Dev. **39**, 617–627 (1995).
9. A.M. Stoneham, M.M.D. Ramos, A.P. Sutton: How do they stick together – The statics and dynamics of interfaces. Philos. Mag. A **67**, 797–811 (1993).
10. I.L. Singer: Friction and energy dissipation at the atomic scale: A review. J. Vac. Sci. Technol. A **12**, 2605–2616 (1994).
11. B. Bhushan. J.N. Israelachvili, U. Landman: Nanotribology – Friction, wear and lubrication at the atomic scale. Nature **374**, 607–616 (1995).
12. J.A. Harrison, D.W. Brenner: Atomic-scale simulation of tribological and related phenomena, in, *Handbook of Micro/Nanotechnology*, ed. by B. Bhushan (CRC, Boca Raton, 1995) pp.397–439.
13. J.B. Sokoloff: Theory of atomic level sliding friction between ideal crystal interfaces. J. Appl. Phys. **72**, 1262–1270 (1992).
14. W. Zhong, G. Overney, D. Tomanek: Theory of atomic force microscopy on elastic surfaces, in, *The Structure of Surfaces III: Proc. 3rd Int. Conf. Struct. Surf.*, Vol.24, ed. by S.Y. Tong, M.A.V. Hove, X. Xide, K. Takayanagi (Springer, Berlin, Heidelberg, 1991) pp.243–.
15. J.N. Israelachvili: Adhesion, friction and lubrication of molecularly smooth surfaces, in, *Fundamentals of Friction: Macroscopic and Microscopic processes*, ed. by I.L. Singer, H.M. Pollock (Kluwer, Dordrecht, 1992) pp.351–385.
16. S.B. Sinnott: Theory of atomic-scale friction, in, *Handbook of Nanostructured Materials and Nanotechnology*, vol.2, ed. by H. Nalwa (Academic, San Diego, 2000) pp.571–618.
17. S.-J. Heo, S.B. Sinnott, D.W. Brenner. J.A. Harrison: Computational modeling of nanometer-scale tribology, in, *Nanotribology and Nanomechanics*, ed. by B. Bhushan (Springer, Berlin, Heidelberg, 2005).
18. G. Binnig, C.F. Quate, C. Gerber: Atomic force microscope. Phys. Rev. Lett. **56**, 930–933 (1986).
19. C.M. Mate, G.M. McClelland, R. Erlandsson, S. Chiang: Atomic-scale friction of a tungsten tip on a graphite surface. Phys. Rev. Lett. **59**, 1942–1945 (1987).
20. G.J. Germann, S.R. Cohen, G. Neubauer, G.M. McClelland, H. Seki, D. Coulman: Atomic-scale friction of a diamond tip on diamond (100) surface and (111) surface. J. Appl. Phys. **73**, 163–167 (1993).

21. R.W. Carpick, M. Salmeron: Scratching the surface: Fundamental investigations of tribology with atomic force microscopy. Chem. Rev. **97**, 1163–1194 (1997).
22. J.N. Israelachvili: *Intermolecular and surface forces: With applications to colloidal and biological systems* (Academic, London, 1992).
23. J. Krim, D.H. Solina, R. Chiarello: Nanotribology of a Kr monolayer – A quartz crystal microbalance study of atomic-scale friction. Phys. Rev. Lett. **66**, 181–184 (1991).
24. G.A. Tomlinson: A molecular theory of friction. Philos. Mag. **7**, 905–939 (1929).
25. F.C. Frenkel, T. Kontorova: On the theory of plastic deformation and twinning. Zh. Eksp. Teor. Fiz. **8**, 1340 (1938).
26. G.M. McClelland. J.N. Glosli: Friction at the atomic scale, in, *Fundamentals of friction: Macroscopic and microscopic processes*, ed. by I.L. Singer, H.M. Pollock (Kluwer, Dordrecht, 1992) pp.405–422.
27. J.B. Sokoloff: Theory of dynamical friction between idealized sliding surfaces. Surf. Sci. **144**, 267–272 (1984).
28. J.B. Sokoloff: Theory of energy dissipation in sliding crystal surfaces. Phys. Rev. B **42**, 760–765 (1990).
29. J.B. Sokoloff: Possible nearly frictionless sliding for mesoscopic solids. Phys. Rev. Lett. **71**, 3450–3453 (1993).
30. J.B. Sokoloff: Microscopic mechanisms for kinetic friction: Nearly frictionless sliding for small solids. Phys. Rev. B **52**, 7205–7214 (1995).
31. J.B. Sokoloff: Theory of electron and phonon contributions to sliding friction, in, *Physics of Sliding Friction*, ed. by B.N.J. Persson, E. Tosatti (Kluwer, Dordrecht, 1996) pp.217–229.
32. J.B. Sokoloff: Static friction between elastic solids due to random asperities. Phys. Rev. Lett. **86**, 3312–3315 (2001).
33. J.B. Sokoloff: Possible microscopic explanation of the virtually universal occurrence of static friction. Phys. Rev. B **65**, 115415 (2002).
34. B.N.J. Persson, D. Schumacher, A. Otto: Surface resistivity and vibrational damping in adsorbed layers. Chem. Phys. Lett. **178**, 204–212 (1991).
35. A.I. Volokitin, B.N.J. Persson: Resonant photon tunneling enhancement of the van der Waals friction. Phys. Rev. Lett. **91**, 106101 (2003).
36. A.I. Volokitin, B.N.J. Persson: Noncontact friction between nanostructures. Phys. Rev. B **68**, 155420 (2003).
37. A.I. Volokitin, B.N.J. Persson: Adsorbate-induced enhancement of electrostatic noncontact friction. Phys. Rev. Lett. **94**, 086104 (2005).
38. J.S. Helman, W. Baltensperger. J.A. Holyst: Simple model for dry friction. Phys. Rev. B **49**, 3831–3838 (1994).
39. T. Kawaguchi, H. Matsukawa: Dynamical frictional phenomena in an incommensurate two-chain model. Phys. Rev. B **56**, 13932–13942 (1997).
40. M.H. Müser: Nature of mechanical instabilities and their effect on kinetic friction. Phys. Rev. Lett. **89**, 224301 (2002).
41. M.H. Müser: Towards an atomistic understanding of solid friction by computer simulations. Comput. Phys. Commun. **146**, 54–62 (2002).
42. P. Reimann, M. Evstigneev: Nonmonotonic velocity dependence of atomic friction. Phys. Rev. Lett. **93**, 230802 (2004).
43. C. Ritter, M. Heyde, B. Stegemann, K. Rademann, U.D. Schwarz: Contact area dependence of frictional forces: Moving adsorbed antimony nanoparticles. Phys. Rev. B **71**, 085405 (2005).
44. C. Fusco, A. Fasolino: Velocity dependence of atomic-scale friction: A comparative study of the one- and two-dimensional Tomlinson model. Phys. Rev. B **71**, 045413 (2005).
45. C.W. Gear: *Numerical Initial Value Problems in Ordinary Differential Equations* (Prentice-Hall, Englewood Cliffs, 1971).
46. W.G. Hoover: *Molecular Dynamics* (Springer, Berlin, Heidelberg, 1986).
47. D.W. Heermann: *Computer Simulation Methods in Theoretical Physics* (Springer, Berlin, Heidelberg, 1986).

48. M.P. Allen, D.J. Tildesley: *Computer Simulation of Liquids* (Clarendon, Oxford, 1987).
49. J.M. Haile: *Molecular Dynamics Simulation: Elementary Methods* (Wiley, New York, 1992).
50. M. Finnis: *Interatomic Force's in Condensed Matter* (Oxford Univ. Press, Oxford, 2003).
51. D.W. Brenner: Relationship between the embedded-atom method and Tersoff potentials. Phys. Rev. Lett. **63**, 1022–1022 (1989).
52. D.W. Brenner: The art and science of an analytic potential. Phys. Status Solidi (b) **217**, 23–40 (2000).
53. R.G. Parr, W. Yang: *Density Functional Theory of Atoms and Molecules* (Oxford University. Press, New York, 1989).
54. R. Car, M. Parrinello: Unified approach for molecular dynamics and density functional theory. Phys. Rev. Lett. **55**, 2471–2474 (1985).
55. C. Cramer: *Essentials of Computational Chemistry, Theories and Models* (Wiley, Chichester, 2004).
56. A.P. Sutton: *Electronic Structure of Materials* (Clarendon, Oxford, 1993).
57. K. Kadau, T.C. Germann, P.S. Lomdahl: Large-scale molecular dynamics simulation of 19 billion particles. Int. J. Mod. Phys. C **15**, 193–201 (2004).
58. B.J. Thijsse: Relationship between the modified embedded-atom method and Stillinger–Weber potentials in calculating the structure of silicon. Phys. Rev. B **65**, 195207 (2002).
59. M.I. Baskes. J.S. Nelson, A.F. Wright: Semiempirical modified embedded atom potentials for silicon and germanium. Phys. Rev. B **40**, 6085–6100 (1989).
60. T. Ohira, Y. Inoue, K. Murata. J. Murayama: Magnetite scale cluster adhesion on metal oxide surfaces: Atomistic simulation study. Appl. Surf. Sci. **171**, 175–188 (2001).
61. F.H. Streitz. J.W. Mintmire: Electrostatic potentials for metal oxide surfaces and interfaces. Phys. Rev. B **50**, 11996–12003 (1994).
62. A. Yasukawa: Using an extended Tersoff interatomic potential to analyze the static fatigue strength of $SiO_2$ under atmospheric influence. JSME Int. J. **A 39**, 313–320 (1996).
63. T. Iwasaki, H. Miura: Molecular dynamics analysis of adhesion strength of interfaces between thin films. J. Mater. Res. **16**, 1789–1794 (2001).
64. B.-J. Lee, M.I. Baskes: Second nearest-neighbor modified embedded-atom method potential. Phys. Rev. B **62**, 8564–8567 (2000).
65. G.C. Abell: Empirical chemical pseudopotential theory of molecular and metallic bonding. Phys. Rev. B **31**, 6184–6196 (1985).
66. J. Tersoff: New empirical approach for the structure and energy of covalent systems. Phys. Rev. B **37**, 6991–7000 (1988).
67. J. Tersoff: Modeling solid-state chemistry: Interatomic potentials for multicomponent systems. Phys. Rev. B **39**, 5566–5569 (1989).
68. D.W. Brenner: Empirical potential for hydrocarbons for use in simulating the chemical vapor deposition of diamond films. Phys. Rev. B **42**, 9458–9471 (1990).
69. D.W. Brenner, O.A. Shenderova. J.A. Harrison, S.J. Stuart, B. Ni, S.B. Sinnott: Second generation reactive empirical bond order (REBO) potential energy expression for hydrocarbons. J. Phys. C **14**, 783–802 (2002).
70. A.J. Dyson, P.V. Smith: Extension of the Brenner empirical interactomic potential to C-Si-H. Surf. Sci. **355**, 140–150 (1996).
71. B. Ni, K.-H. Lee, S.B. Sinnott: Development of a reactive empirical bond order potential for hydrocarbon-oxygen interactions. J. Phys. C **16**, 7261–7275 (2004).
72. J. Tanaka, C.F. Abrams, D.B. Graves: New C-F interatomic potential for molecular dynamics simulation of fluorocarbon film formation. Nucl. Instrum. Methods B **18**, 938–945 (2000).
73. I. Jang, S.B. Sinnott: Molecular dynamics simulations of the chemical modification of polystyrene through $C_xF_y^+$ beam deposition. J. Phys. Chem. B **108**, 9656–9664 (2004).
74. J.D. Schall, G. Gao. J.A. Harrison: Elastic constants of silicon materials calculated as a function of temperature using a parametrization of the second-generation reactive empirical bond-order potential. Phys. Rev. B **77**(11), 115209 (2008).

75. Y. Hu: Personal communication (2008).
76. S.B. Sinnott, O.A. Shenderova, C.T. White, D.W. Brenner: Mechanical properties of nanotubule fibers and composites determined from theoretical calculations and simulations. Carbon **36**, 1–9 (1998).
77. S.J. Stuart, A.B. Tutein. J.A. Harrison: A reactive potential for hydrocarbons with intermolecular interactions. J. Chem. Phys. **112**, 6472–6486 (2000).
78. F.H. Stillinger, T.A. Weber: Computer simulation of local order in condensed phases of silicon. Phys. Rev. B **31**, 5262–5271 (1985).
79. S.M. Foiles: Application of the embedded-atom method to liquid transition metals. Phys. Rev. B **32**, 3409–3415 (1985).
80. M.S. Daw, M.I. Baskes: Semiempirical, quantum mechanical calculation of hydrogen embrittlement in metals. Phys. Rev. Lett. **50**, 1285–1288 (1983).
81. T.J. Raeker, A.E. Depristo: Theory of chemical bonding based on the atom-homogeneous electron gas system. Int. Rev. Phys. Chem. **10**, 1–54 (1991).
82. R.W. Smith, G.S. Was: Application of molecular dynamics to the study of hydrogen embrittlement in Ni–Cr–Fe alloys. Phys. Rev. B **40**, 10322–10336 (1989).
83. R. Pasianot, D. Farkas, E.J. Savino: Empirical many-body interatomic potential for bcc transition metals. Phys. Rev. B **43**, 6952–6961 (1991).
84. R. Pasianot, E.J. Savino: Embedded-atom method interatomic potentials for hcp metals. Phys. Rev. B **45**, 12704–12710 (1992).
85. M.I. Baskes. J.S. Nelson, A.F. Wright: Semiempirical modified embedded-atom potentials for silicon and germanium. Phys. Rev. B **40**, 6085–6100 (1989).
86. M.I. Baskes: Modified embedded-atom potentials for cubic materials and impurities. Phys. Rev. B **46**, 2727–2742 (1992).
87. K. Ohno, K. Esfarjani, Y. Kawazoe: *Computational Materials Science from ab initio to Monte Carlo Methods* (Springer, Berlin, Heidelberg, 1999).
88. A.K. Rappe, W.A. Goddard III: Charge equilibration for molecular dynamics simulations. J. Phys. Chem. **95**, 3358–3363 (1991).
89. D. Frenkel, B. Smit: *Understanding Molecular Simulation: From Algorithms to Applications* (Academic, San Diego 1996).
90. L.V. Woodcock: Isothermal molecular dynamics calculations for liquid salts. Chem. Phys. Lett. **10**, 257–261 (1971).
91. T. Schneider, E. Stoll: Molecular dynamics study of a three-dimensional one-component model for distortive phase transitions. Phys. Rev. B **17**, 1302–1322 (1978).
92. K. Kremer, G.S. Grest: Dynamics of entangled linear polymer melts – A molecular dynamics simulation. J. Chem. Phys. **92**, 5057–5086 (1990).
93. S.A. Adelman. J.D. Doll: Generalized Langevin equation approach for atom-solid-surface scattering – General formulation for classical scattering off harmonic solids. J. Chem. Phys. **64**, 2375–2388 (1976).
94. S.A. Adelman: Generalized Langevin equations and many-body problems in chemical dynamics. Adv. Chem. Phys. **44**, 143–253 (1980).
95. J.C. Tully: Dynamics of gas-surface interactions – 3-D generalized Langevin model applied to fcc and bcc surfaces. J. Chem. Phys. **73**, 1975–1985 (1980).
96. S. Nosé: A unified formulation of the constant temperature molecular dynamics methods. J. Chem. Phys. **81**, 511–519 (1984).
97. S. Nosé: A molecular dynamics method for simulations in the canonical ensemble. Mol. Phys. **52**, 255–268 (1984).
98. G.J. Martyna, M.L. Klein, M. Tuckerman: Nosé–Hoover chains – The canonical ensemble via continuous dynamics. J. Chem. Phys. **97**, 2635–2643 (1992).
99. M. D'Alessandro, M. D'Abramo, G. Brancato, A. Di Nola, A. Amadei: Statistical mechanics and thermodynamics of simulated ionic solutions. J. Phys. Chem. B **106**, 11843–11848 (2002).

100. J.D. Schall, C.W. Padgett, D.W. Brenner: Ad hoc continuum-atomistic thermostat for modeling heat flow in molecular dynamics simulations. Mol. Simul. **31**, 283–288 (2005).
101. C.W. Padgett, D.W. Brenner: A continuum-atomistic method for incorporating Joule heating into classical molecular dynamics simulations. Mol. Simul. **31**(11), 749–757 (2005).
102. M. Schoen, C.L. Rhykerd, D.J. Diestler. J.H. Cushman: Shear forces in molecularly thin films. Science **245**, 1223–1225 (1989).
103. J.E. Curry, F.S. Zhang. J.H. Cushman, M. Schoen, D.J. Diestler: Transient coexisting nanophases in ultrathin films confined between corrugated walls. J. Chem. Phys. **101**, 10824–10832 (1994).
104. D.J. Adams: Grand canonical ensemble Monte Carlo for a Lennard-Jones fluid. Mol. Phys. **29**, 307–311 (1975).
105. R.J. Colton, N.A. Burnham: Force microscopy. In: *Scanning Tunneling Microscopy and Spectroscopy: Theory, Techniques, and Applications*, ed. by D.A. Bonnell (VCH, New York, 1993) pp. 191–249.
106. E. Meyer: *Nanoscience: Friction and Rheology on the Nanometer Scale* (World Scientific, Hackensack, 1998).
107. G.E. Totten, H. Liang: *Mechanical Tribology: Materials Characterization and Applications* (Marcel Dekker, New York, 2004).
108. N.A. Burnham, R.J. Colton: Measuring the nanomechanical properties and surface forces of materials using an atomic force microscope. J. Vac. Sci. Technol. A **7**, 2906–2913 (1989).
109. N.A. Burnham, D.D. Dominguez, R.L. Mowery, R.J. Colton: Probing the surface forces of monolayer films with an atomic force microscope. Phys. Rev. Lett. **64**, 1931–1934 (1990).
110. E. Meyer, R. Overney, D. Brodbeck, L. Howald, R. Luthi. J. Frommer, H.J. Guntherodt: Friction and wear of Langmuir–Blodgett films observed by friction force microscopy. Phys. Rev. Lett. **69**, 1777–1780 (1992).
111. A.P. Sutton. J.B. Pethica, H. Rafii-Tabar. J.A. Nieminen: Mechanical properties of metals at the nanometer scale. In: *Electron Theory in Alloy Design*, ed. by D.G. Pettifor, A.H. Cottrell (Institute of Materials, London, 1992) pp. 191–233.
112. H. Raffi-Tabar, A.P. Sutton: Long-range Finnis–Sinclair potentials for fcc metallic alloys. Philos. Mag. Lett. **63**, 217–224 (1991).
113. U. Landman, W.D. Luedtke, E.M. Ringer: Atomistic mechanisms of adhesive contact formation and interfacial processes, Wear **153**, 3–30 (1992).
114. U. Landman, W.D. Luedtke, N.A. Burnham, R.J. Colton: Atomistic mechanisms and dynamics of adhesion, nanoindentation and fracture, Science **248**, 454–461 (1990).
115. O. Tomagnini, F. Ercolessi, E. Tosatti: Microscopic interaction between a gold tip and a Pb (110) surface, Surf. Sci. **287/288**, 1041–1045 (1991).
116. N. Ohmae: Field ion microscopy of microdeformation induced by metallic contacts, Philos. Mag. A **74**, 1319–1327 (1996).
117. N.A. Burnham, R.J. Colton, H.M. Pollock: Interpretation of force curves in force microscopy, Nanotechnology **4**, 64–80 (1993).
118. N. Agrait, G. Rubio, S. Vieira: Plastic deformation in nanometer-scale contacts, Langmuir **12**, 4505–4509 (1996).
119. U. Landman, W.D. Luedtke, A. Nitzan: Dynamics of tip-substrate interactions in atomic force microscopy, Surf. Sci. **210**, L177–L182 (1989).
120. U. Landman, W.D. Luedtke: Nanomechanics and dynamics of tip substrate interactions. J. Vac. Sci. Technol. B **9**, 414–423 (1991).
121. U. Landman, W.D. Luedtke. J. Ouyang, T.K. Xia: Nanotribology and the stability of nanostructures, Jpn. J. Appl. Phys. **32**, 1444–1462 (1993).
122. J.W.M. Frenken, H.M. Vanpinxteren, L. Kuipers: New views on surface melting obtained with STM and ion scattering, Surf. Sci. **283**, 283–289 (1993).
123. O. Tomagnini, F. Ercolessi, E. Tosatti: Microscopic interaction between a gold tip and a Pb(110) surface, Surf. Sci. **287**, 1041–1045 (1993).

124. K. Komvopoulos, W. Yan: Molecular dynamics simulation of single and repeated indentation. J. Appl. Phys. **82**, 4823–4830 (1997).
125. J. Belak, I.F. Stowers: A molecular dynamics model of the orthogonal cutting process, Proc. Am. Soc. Precis. Eng. Annu. Conf. (1990) pp.76–79.
126. T. Yokohata, K. Kato: Mechanism of nanoscale indentation, Wear **168**, 109–114 (1993).
127. M. Fournel, E. Lacaze, M. Schott: Tip-surface interactions in STM experiments on Au(111): Atomic-scale metal friction, Europhys. Lett. **34**, 489–494 (1996).
128. J.L. Costakramer, N. Garcia, P. Garciamochales, P.A. Serena: Nanowire formation in macroscopic metallic contacts – Quantum-mechanical conductance tapping a table top, Surf. Sci. **342**, L1144–L1149 (1995).
129. A.I. Yanson. J.M. van Ruitenbeek, I.K. Yanson: Shell effects in alkali metal nanowires, Low Temp. Phys. **27**, 807–820 (2001).
130. A.I. Yanson, I.K. Yanson. J.M. van Ruitenbeek: Crossover from electronic to atomic shell structure in alkali metal nanowires. Phys. Rev. Lett. **8721**, 216805 (2001).
131. C.L. Kelchner, S.J. Plimpton. J.C. Hamilton: Dislocation nucleation and defect structure during surface indentation. Phys. Rev. B **58**, 11085–11088 (1998).
132. E.T. Lilleodden. J.A. Zimmerman, S.M. Foiles, W.D. Nix: Atomistic simulations of elastic deformation and dislocation nucleation during nanoindentation. J. Mech. Phys. Solids **51**, 901–920 (2003).
133. A. Hasnaoui, P.M. Derlet, H.V. Swygenhoven: Interaction between dislocations, grain boundaries under an indenter – A molecular dynamics simulation, Acta Mater. **52**, 2251–2258 (2004).
134. O.R. de la Fuente. J.A. Zimmerman, M.A. Gonzalez. J. de la Figuera. J.C. Hamilton, W.W. Pai. J.M. Rojo: Dislocation emission around nanoindentations on a (001) fcc metal surface studied by scanning tunneling microscopy and atomistic simulations. Phys. Rev. Lett. **88**, 036101 (2002).
135. O.A. Shenderova. J.P. Mewkill, D.W. Brenner: Nanoindentation as a probe of nanoscale residual stresses. Mol. Simul. **25**, 81–92 (2000).
136. S. Kokubo: On the change in hardness of a plate caused by bending, Sci. Rep. Tohoku Imp. Univ. **21**, 256–267 (1932).
137. G. Sines, R. Calson: Hardness measurements for determination of residual stresses, ASTM Bulletin **180**, 35–37 (1952).
138. G.U. Oppel: Biaxial elasto-plastic analysis of load and residual stresses, Exp. Mech. **21**, 135–140 (1964).
139. T.R. Simes, S.G. Mellor, D.A. Hills: A note on the influence of residual stress on measured hardness. J. Strain Anal. Eng. Des. **19**, 135–137 (1984).
140. T.Y. Tsui, G.M. Pharr, W.C. Oliver, C.S. Bhatia, C.T. White, S. Anders, A. Anders, I.G. Brown: Nanoindentation and nanoscratching of hard carbon coatings for magnetic disks, Mater. Res. Soc. Symp. Proc. **383**, 447–452 (1995).
141. A. Bolshakov, W.C. Oliver, G.M. Pharr: Influences of stress on the measurement of mechanical properties using nanoindentation. 2. Finite element simulations. J. Mater. Res. **11**, 760–768 (1996).
142. J.D. Schall, D.W. Brenner: Atomistic simulation of the influence of pre-existing stress on the interpretation of nanoindentation data. J. Mater. Res. **19**, 3172–3180 (2004).
143. U. Landman, W.D. Luedtke, M.W. Ribarsky: Structural and dynamical consequences of interactions in interfacial systems. J. Vac. Sci. Technol. A **7**, 2829–2839 (1989).
144. J.S. Kallman, W.G. Hoover, C.G. Hoover, A.J. Degroot, S.M. Lee, F. Wooten: Molecular-dynamics of silicon indentation. Phys. Rev. B **47**, 7705–7709 (1993).
145. D.R. Clarke, M.C. Kroll, P.D. Kirchner, R.F. Cook, B.J. Hockey: Amorphization and conductivity of silicon and germanium induced by indentation. Phys. Rev. Lett. **60**, 2156–2159 (1988).
146. A. Kailer, K.G. Nickel, Y.G. Gogotsi: Raman microspectroscopy of nanocrystalline and amorphous phases in hardness indentations. J. Raman Spectrosc. **30**, 939–961 (1999).

147. K. Minowa, K. Sumino: Stress-induced amorphization of a silicon crystal by mechanical scratching. Phys. Rev. Lett. **69**, 320–322 (1992).
148. G.S. Smith, E.B. Tadmor, E. Kaxiras: Multiscale simulation of loading and electrical resistance in silicon nanoindentation. Phys. Rev. Lett. **84**, 1260–1263 (2000).
149. W.C.D. Cheong, L.C. Zhang: Molecular dynamics simulation of phase transformations in silicon monocrystals due to nano-indentation, Nanotechnology **11**, 173–180 (2000).
150. C.F. Sanz-Navarro, S.D. Kenny, R. Smith: Atomistic simulations of structural transformations, Nanotechnology **15**, 692–697 (2004).
151. P. Walsh, A. Omeltchenko, R.K. Kalia, A. Nakano, P. Vashishta, S. Saini: Nanoindentation of silicon nitride: A multimillion-atom molecular dynamics study, Appl. Phys. Lett. **82**, 118–120 (2003).
152. J.A. Harrison, C.T. White, R.J. Colton, D.W. Brenner: Nanoscale investigation of indentation, adhesion and fracture of diamond (111) surfaces, Surf. Sci. **271**, 57–67 (1992).
153. K. Cho. J.D. Joannopoulos: Mechanical hysteresis on an atomic-scale, Surf. Sci. **328**, 320–324 (1995).
154. J.A. Harrison, S.J. Stuart, D.H. Robertson, C.T. White: Properties of capped nanotubes when used as SPM tips. J. Phys. Chem. B **101**, 9682–9685 (1997).
155. J.A. Harrison, S.J. Stuart, A.B. Tutein: A new, reactive potential energy function to study indentation and friction of $C_{13}$ n-alkane monolayers. In: *Interfacial Properties on the Submicron Scale*, ed. by J.E. Frommer, R. Overney (ACS, Washington 2001) pp. 216–229.
156. A. Garg. J. Han, S.B. Sinnott: Interactions of carbon-nanotubule proximal probe tips with diamond and graphene. Phys. Rev. Lett. **81**, 2260–2263 (1998).
157. A. Garg, S.B. Sinnott: Molecular dynamics of carbon nanotubule proximal probe tip-surface contacts. Phys. Rev. B **60**, 13786–13791 (1999).
158. K.J. Tupper, D.W. Brenner: Compression-induced structural transition in a self-assembled monolayer, Langmuir **10**, 2335–2338 (1994).
159. K.J. Tupper, R.J. Colton, D.W. Brenner: Simulations of self-assembled monolayers under compression – Effect of surface asperities, Langmuir **10**, 2041–2043 (1994).
160. S.A. Joyce, R.C. Thomas. J.E. Houston, T.A. Michalske, R.M. Crooks: Mechanical relaxation of organic monolayer films measured by force microscopy. Phys. Rev. Lett. **68**, 2790–2793 (1992).
161. L. Zhang, Y. Leng, S. Jiang: Tip-based hybrid simulation study of frictional properties of self-assembled monolayers: Effects of chain length, terminal group, and scan direction, scan velocity, Langmuir **19**, 9742–9747 (2003).
162. A.B. Tutein, S.J. Stuart. J.A. Harrison: Indentation analysis of linear-chain hydrocarbon monolayers anchored to diamond. J. Phys. Chem. B **103**, 11357–11365 (1999).
163. Y. Leng, S. Jiang: Dynamic simulations of adhesion and friction in chemical force microscopy, J. Am. Chem. Soc. **124**, 11764–11770 (2002).
164. C.M. Mate: Atomic force microscope study of polymer lubricants on silicon surfaces. Phys. Rev. Lett. **68**, 3323–3326 (1992).
165. S.B. Sinnott, R.J. Colton, C.T. White, O.A. Shenderova, D.W. Brenner. J.A. Harrison: Atomistic simulations of the nanometer-scale indentation of amorphous carbon thin films. J. Vac. Sci. Technol. A **15**, 936–940 (1997).
166. K. Enke, H. Dimigen, H. Hubsch: Frictional properties of diamond-like carbon layers, Appl. Phys. Lett. **36**, 291–292 (1980).
167. K. Enke: Some new results on the fabrication of and the mechanical, electrical, optical properties of i-carbon layers, Thin Solid Films **80**, 227–234 (1981).
168. S. Miyake, S. Takahashi, I. Watanabe, H. Yoshihara: Friction and wear behavior of hard carbon films, ASLE Trans. **30**, 121–127 (1987).
169. A. Erdemir, C. Donnet: Tribology of diamond, diamond-like carbon, and related films. In: *Modern Tribology Handbook*, Vol. II, ed. by B. Bhushan (CRC, Boca Raton 2000) pp. 871–908.
170. J.N. Glosli, M.R. Philpott, G.M. McClelland: Molecular dynamics simulation of mechanical deformation of ultra-thin amorphous carbon films, Mater. Res. Soc. Symp. Proc. **383**, 431–435 (1995).

171. I.L. Singer: A thermochemical model for analyzing low wear-rate materials, Surf. Coat. Technol. **49**, 474–481 (1991).
172. I.L. Singer, S. Fayeulle, P.D. Ehni: Friction and wear behavior of tin in air – The chemistry of transfer films and debris formation, Wear **149**, 375–394 (1991).
173. J.A. Harrison, C.T. White, R.J. Colton, D.W. Brenner: Molecular dynamics simulations of atomic-scale friction of diamond surfaces. Phys. Rev. B **46**, 9700–9708 (1992).
174. J.A. Harrison, R.J. Colton, C.T. White, D.W. Brenner: Effect of atomic-scale surface roughness on friction – A molecular dynamics study of diamond surfaces, Wear **168**, 127–133 (1993).
175. J.A. Harrison, C.T. White, R.J. Colton, D.W. Brenner: Atomistic simulations of friction at sliding diamond interfaces, MRS Bulletin **18**, 50–53 (1993).
176. J.N. Glosli, G.M. McClelland: Molecular dynamics study of sliding friction of ordered organic monolayers. Phys. Rev. Lett. **70**, 1960–1963 (1993).
177. A. Koike, M. Yoneya: Molecular dynamics simulations of sliding friction of Langmuir–Blodgett monolayers. J. Chem. Phys. **105**, 6060–6067 (1996).
178. J.E. Hammerberg, B.L. Holian, S.J. Zhou: Studies of sliding friction in compressed copper, in *Conference of the American Physical Society Topical Group on Shock Compress*, Seattle, ed. by S.C. Schmidt, W.C. Tao (AIP, New York, 1995) p.370.
179. M.R. Sorensen, K.W. Jacobsen, P. Stoltze: Simulations of atomic-scale sliding friction. Phys. Rev. B **53**, 2101–2113 (1996).
180. M.D. Perry. J.A. Harrison: Friction between diamond surfaces in the presence of small third-body molecules. J. Phys. Chem. B **101**, 1364–1373 (1997).
181. A. Buldum, S. Ciraci: Atomic-scale study of dry sliding friction. Phys. Rev. B **55**, 2606–2611 (1997).
182. A.P. Sutton. J.B. Pithica: Inelastic flow processes in nanometre volumes of solids. J. Phys. Condens. Matter **2**, 5317–5326 (1990).
183. S. Akamine, R.C. Barrett, C.F. Quate: Improved atomic force microscope images using microcantilevers with sharp tips, Appl. Phys. Lett. **57**, 316–318 (1990).
184. J.A. Nieminen, A.P. Sutton. J.B. Pethica: Static junction growth during frictional sliding of metals, Acta Metall. Mater. **40**, 2503–2509 (1992).
185. J.A. Niemienen, A.P. Sutton. J.B. Pethica, K. Kaski: Mechanism of lubrication by a thin solid film on a metal surface, Model. Simul. Mater. Sci. Eng. **1**, 83–90 (1992).
186. V.V. Pokropivny, V.V. Skorokhod, A.V. Pokropivny: Atomistic mechanism of adhesive wear during friction of atomic sharp tungsten asperity over (114) bcc-iron surface, Mater. Lett. **31**, 49–54 (1997).
187. B. Li, P.C. Clapp. J.A. Rifkin, X.M. Zhang: Molecular dynamics simulation of stick-slip. J. Appl. Phys. **90**, 3090–3094 (2001).
188. T.-H. Fang, C.-I. Weng. J.-G. Chang: Molecular dynamics simulation of nanolithography process using atomic force microscopy, Surf. Sci. **501**, 138–147 (2002).
189. R. Komanduri, N. Chandrasekaran: Molecular dynamics simulation of atomic-scale friction. Phys. Rev. B **61**, 14007–14019 (2000).
190. A. Dayo, W. Alnasrallah. J. Krim: Superconductivity-dependent sliding friction. Phys. Rev. Lett. **80**, 1690–1693 (1998).
191. R. Erlandsson, G. Hadziioannou, C.M. Mate, G.M. McClelland, S. Chiang: Atomic scale friction between the muscovite mica cleavage plane and a tungsten tip. J. Chem. Phys. **89**, 5190–5193 (1988).
192. K.L. Johnson: *Contact Mechanics* (Cambridge Univ. Press, Cambridge 1985).
193. J.B. Pethica: Interatomic forces in scanning tunneling microscopy – Giant corrugations of the graphite surface – Comment. Phys. Rev. Lett. **57**, 3235–3235 (1986).
194. H. Tang, C. Joachim. J. Devillers: Interpretation of AFM images – The graphite surface with a diamond tip, Surf. Sci. **291**, 439–450 (1993).
195. S. Fujisawa, Y. Sugawara, S. Morita: Localized fluctuation of a two-dimensional atomic-scale friction, Jpn. J. Appl. Phys. **35**, 5909–5913 (1996).

196. S. Fujisawa, Y. Sugawara, S. Ito, S. Mishima, T. Okada, S. Morita: The two-dimensional stick-slip phenomenon with atomic resolution, Nanotechnology **4**, 138–142 (1993).
197. S. Fujisawa, Y. Sugawara, S. Morita, S. Ito, S. Mishima, T. Okada: Study on the stick-slip phenomenon on a cleaved surface of the muscovite mica using an atomic-force lateral force microscope. J. Vac. Sci. Technol. B **12**, 1635–1637 (1994).
198. S. Morita, S. Fujisawa, Y. Sugawara: Spatially quantized friction with a lattice periodicity, Surf. Sci. Rep. **23**, 1–41 (1996).
199. J.A. Ruan, B. Bhushan: Atomic-scale and microscale friction studies of graphite and diamond using friction force microscopy. J. Appl. Phys. **76**, 5022–5035 (1994).
200. R.W. Carpick, N. Agrait, D.F. Ogletree, M. Salmeron: Variation of the interfacial shear strength and adhesion of a nanometer-sized contact, Langmuir **12**, 3334–3340 (1996).
201. R.W. Carpick, N. Agrait, D.F. Ogletree, M. Salmeron: Measurement of interfacial shear (friction) with an ultrahigh vacuum atomic force microscope. J. Vac. Sci. Technol. B **14**, 1289–2772 (1996).
202. T. Liang, W.G. Sawyer, S.S. Perry, S.B. Sinnott, S.R. Phillpot: First-principles determination of static potential energy surfaces for atomic friction in $MoS_2$ and $MoO_3$. Phys. Rev. B **77** (10), 104105 (2008).
203. A.L. Shluger, R.T. Williams, A.L. Rohl: Lateral and friction forces originating during force microscope scanning of ionic surfaces, Surf. Sci. **343**, 273–287 (1995).
204. B. Samuels. J. Wilks: The friction of diamond sliding on diamond. J. Mater. Sci. **23**, 2846–2864 (1988).
205. T. Cagin. J.W. Che, M.N. Gardos, A. Fijany, W.A. Goddard: Simulation and experiments on friction, wear of diamond: A material for MEMS and NEMS application, Nanotechnology **10**, 278–284 (1999).
206. R.J.A. van den Oetelaar, C.F.J. Flipse: Atomic-scale friction on diamond(111) studied by ultra-high vacuum atomic force microscopy, Surf. Sci. **384**, L828–L835 (1997).
207. M. Enachescu, R.J.A. van den Oetelaar, R.W. Carpick, D.F. Ogletree, C.F.J. Flipse, M. Salmeron: Atomic force microscopy study of an ideally hard contact: The diamond(111) tungsten carbide interface. Phys. Rev. Lett. **81**, 1877–1880 (1998).
208. B.V. Derjaguin, V.M. Muller, Y. Toporov: Effect of contact deformations on adhesion of particles. J. Colloid Interf. Sci. **53**, 314–326 (1975).
209. M.D. Perry. J.A. Harrison: Universal aspects of the atomic-scale friction of diamond surfaces. J. Phys. Chem. B **99**, 9960–9965 (1995).
210. J.A. Harrison, C.T. White, R.J. Colton, D.W. Brenner: Investigation of the atomic-scale friction and energy dissipation in diamond using molecular dynamics, Thin Solid Films **260**, 205–211 (1995).
211. M.J. Brukman, G.G.R.J. Nemanich. J.A. Harrison: Temperature dependence of single asperity diamond-diamond friction elucidated using AFM and MD simulations. J. Phys. Chem. **112**, 9358–9369 (2008).
212. R. Neitola, T.A. Pakannen: Ab initio studies on the atomic-scale origin of friction between diamond (111) surfaces. J. Phys. Chem. B **105**, 1338–1343 (2001).
213. J.A. Harrison, R.J. Colton, C.T. White, D.W. Brenner: Atomistic simulation of the nanoindentation of diamond and graphite surfaces, Mater. Res. Soc. Symp. Proc. **239**, 573–578 (1992).
214. D. Mulliah, S.D. Kenny, R. Smith: Modeling of stick-slip phenomena using molecular dynamics. Phys. Rev. B **69**, 205407 (2004).
215. D.W. Brenner: Empirical potential for hydrocarbons for use in simulating the chemical vapor deposition of diamond films. Phys. Rev. B **42**, 9458–9471 (1990).
216. G.J. Ackland, G. Tichy, V. Vitek, M.W. Finnis: Simple n-body potentials for the noble metals and nickel, Philos. Mag. A **56**, 735–756 (1987).
217. J.P. Biersack. J. Ziegler, U. Littmack: *The Stopping and Range of Ions in Solids* (Pergamon, Oxford, 1985).

218. J. Cai. J.-S. Wang: Friction between Si tip and (001)–2 × 1 surface: A molecular dynamics simulation, Comput. Phys. Commun. **147**, 145–148 (2002).
219. J. Cai. J.S. Wang: Friction between a Ge tip and the (001)–2 × 1 surface: A molecular dynamics simulation. Phys. Rev. B **64**, 113313 (2001).
220. A.G. Khurshudov, K. Kato, H. Koide: Nano-wear of the diamond AFM probing tip under scratching of silicon, studied by AFM, Tribol. Lett. **2**, 345–354 (1996).
221. A. Khurshudov, K. Kato: Volume increase phenomena in reciprocal scratching of polycarbonate studied by atomic-force microscopy. J. Vac. Sci. Technol. B **13**, 1938–1944 (1995).
222. M.D. Perry. J.A. Harrison: Molecular dynamics studies of the frictional properties of hydrocarbon materials, Langmuir **12**, 4552–4556 (1996).
223. M.D. Perry. J.A. Harrison: Molecular dynamics investigations of the effects of debris molecules on the friction and wear of diamond, Thin Solid Films **291**, 211–215 (1996).
224. J.A. Harrison, C.T. White, R.J. Colton, D.W. Brenner: Effects of chemically-bound, flexible hydrocarbon species on the frictional properties of diamond surfaces. J. Phys. Chem. **97**, 6573–6576 (1993).
225. J.A. Harrison, D.W. Brenner: Simulated tribochemistry – An atomic-scale view of the wear of diamond, J. Am. Chem. Soc. **116**, 10399–10402 (1994).
226. Z. Feng. J.E. Field: Friction of diamond on diamond and chemical vapor deposition diamond coatings, Surf. Coat. Technol. **47**, 631–645 (1991).
227. B.N.J. Persson: Applications of surface resistivity to atomic scale friction, to the migration of hot adatoms, and to electrochemistry. J. Chem. Phys. **98**, 1659–1672 (1993).
228. B.N.J. Persson, A.I. Volokitin: Electronic friction of physisorbed molecules. J. Chem. Phys. **103**, 8679–8683 (1995).
229. H. Grabhorn, A. Otto, D. Schumacher, B.N.J. Persson: Variation of the dc-resistance of smooth and atomically rough silver films during exposure to $C_2H_6$ and $C_2H_4$, Surf. Sci. **264**, 327–340 (1992).
230. A.D. Berman, W.A. Ducker. J.N. Israelachvili: Origin and characterization of different stick-slip friction mechanisms, Langmuir **12**, 4559–4563 (1996).
231. B.N.J. Persson: Theory of friction – Dynamical phase transitions in adsorbed layers. J. Chem. Phys. **103**, 3849–3860 (1995).
232. B.N.J. Persson, E. Tosatti: Layering transition in confined molecular thin films – Nucleation and growth. Phys. Rev. B **50**, 5590–5599 (1994).
233. H. Yoshizawa. J. Israelachvili: Fundamental mechanisms of interfacial friction. 2. Stick-slip friction of spherical and chain molecules. J. Phys. Chem. **97**, 11300–11313 (1993).
234. B.N.J. Persson: Theory of friction: Friction dynamics for boundary lubricated surfaces. Phys. Rev. B **55**, 8004–8012 (1997).
235. P.A. Thompson, M.O. Robbins: Origin of stick-slip motion in boundary lubrication, Science **250**, 792–794 (1990).
236. U. Landman, W.D. Luedtke. J.P. Gao: Atomic-scale issues in tribology: Interfacial junctions and nano-elastohydrodynamics, Langmuir **12**, 4514–4528 (1996).
237. T. Kreer, M.H. Müser, K. Binder. J. Klein: Frictional drag mechanisms between polymer-bearing surfaces, Langmuir **17**, 7804–7813 (2001).
238. T. Kreer, K. Binder, M.H. Müser: Friction between polymer brushes in good solvent conditions: Steady-state sliding versus transient behavior, Langmuir **19**, 7551–7559 (2003).
239. E. Manias, G. Hadziioannou, G. ten Brinke: Inhomogeneities in sheared ultrathin lubricating films, Langmuir **12**, 4587–4593 (1996).
240. R.M. Overney, T. Bonner, E. Meyer, M. Reutschi, R. Luthi, L. Howald. J. Frommer, H.J. Guntherodt, M. Fujihara, H. Takano: Elasticity, wear, and friction properties of thin organic films observed with atomic-force microscopy. J. Vac. Sci. Technol. B **12**, 1973–1976 (1994).
241. R.M. Overney, E. Meyer. J. Frommer, D. Brodbeck, R. Luthi, L. Howald, H.J. Guntherodt, M. Fujihira, H. Takano, Y. Gotoh: Friction measurements on phase-separated thin-films with a modified atomic force microscope, Nature **359**, 133–135 (1992).

242. R.M. Overney, E. Meyer. J. Frommer, H.J. Guntherodt, M. Fujihira, H. Takano, Y. Gotoh: Force microscopy study of friction and elastic compliance of phase-separated organic thin-films, Langmuir **10**, 1281–1286 (1994).
243. H.I. Kim, T. Koini, T.R. Lee, S.S. Perry: Systematic studies of the frictional properties of fluorinated monolayers with atomic force microscopy: Comparison of $CF_3$- and $CH_3$-terminated films, Langmuir **13**, 7192–7196 (1997).
244. M. GarciaParajo, C. Longo. J. Servat, P. Gorostiza, F. Sanz: Nanotribological properties of octadecyltrichlorosilane self-assembled ultrathin films studied by atomic force microscopy: Contact and tapping modes, Langmuir **13**, 2333–2339 (1997).
245. R.M. Overney, H. Takano, M. Fujihira, E. Meyer, H.J. Guntherodt: Wear, friction and sliding speed correlations on Langmuir–Blodgett films observed by atomic force microscopy, Thin Solid Films **240**, 105–109 (1994).
246. P.T. Mikulski. J.A. Harrison: Periodicities in the properties associated with the friction of model self-assembled monolayers, Tribol. Lett. **10**, 29–35 (2001).
247. P.T. Mikulski. J.A. Harrison: Packing density effects on the friction of n-alkane monolayers, J. Am. Chem. Soc. **123**, 6873–6881 (2001).
248. E. Barrena, C. Ocal, M. Salmeron: A comparative AFM study of the structural and frictional properties of mixed and single component films of alkanethiols on Au(111), Surf. Sci. **482**, 1216–1221 (2001).
249. Y.-S. Shon, S. Lee, R. Colorado, S.S. Perry, T.R. Lee: Spiroalkanedithiol-based SAMS reveal unique insight into the wettabilities and frictional properties of organic thin films, J. Am. Chem. Soc. **122**, 7556–7563 (2000).
250. P.T. Mikulski, G. Gao, G.M. Chateauneuf. J.A. Harrison: Contact forces at the sliding interface: Mixed versus pure model alkane monolayers. J. Chem. Phys. **122**, 024701 (2005).
251. S. Lee, Y.S. Shon, R. Colorado, R.L. Guenard, T.R. Lee, S.S. Perry: The influence of packing densities, surface order on the frictional properties of alkanethiol self-assembled monolayers (SAMs) on gold: A comparison of SAMs derived from normal and spiroalkanedithiols, Langmuir **16**, 2220–2224 (2000).
252. P.T. Mikulski, L.A. Herman. J.A. Harrison: Odd and even model self-assembled monolayers: Links between friction and structure, Langmuir **21**(26), 12197–12206 (2005).
253. S.S. Wong, H. Takano, M.D. Porter: Mapping orientation differences of terminal functional groups by friction force microscopy, Anal. Chem. **70**(24), 5209–5212 (1998).
254. G.M. Chateauneuf, P.T. Mikulski, G.T. Gao. J.A. Harrison: Compression- and shear-induced polymerization in model diacetylene-containing monolayers. J. Phys. Chem. B **108**, 16626–16635 (2004).
255. L. Zhang, S. Jiang: Molecular simulation study of nanoscale friction for alkyl monolayers on Si(111). J. Chem. Phys. **117**, 1804–1811 (2002).
256. L.Z. Zhang, Y.S. Leng, S.Y. Jiang: Tip-based hybrid simulation study of frictional properties of self-assembled monolayers: Effects of chain length, terminal group, scan direction, and scan velocity, Langmuir **19**, 9742–9747 (2003).
257. M. Chandross, E.B.W. III, M.J. Stevens, G.S. Grest: Systematic study of the effect of disorder on nanotribology of self-assembled monolayers. Phys. Rev. Lett. **93**, 166103 (2004).
258. M. Chandross, G.S. Grest, M.J. Stevens: Friction between alkylsilane monolayers: Molecular simulation of ordered monolayers, Langmuir **18**, 8392–8399 (2002).
259. M. Chandross, C.D. Lorenz, M.J. Stevens, G.S. Grest: Simulations of nanotribology with realistic probe tip models, Langmuir **24**(4), 1240–1246 (2008).
260. D.L. Irving, D.W. Brenner: Diffusion on a self-assembled monolayer: Molecular modeling of a bound plus mobile lubricant. J. Phys. Chem. B **110**(31), 15426–15431 (2006).
261. D.W. Brenner, D.L. Irving, A.I. Kingon. J. Krim: Multiscale analysis of liquid lubrication trends from industrial machines to micro-electrical-mechanical systems, Langmuir **23**(18), 9253–9257 (2007).

262. I. Jang, D.L. Burris, P.L. Dickrell, P.R. Barry, C. Santos, S.S. Perry, S.R. Phillpot, S.B. Sinnott, W.G. Sawyer: Sliding orientation effects on the tribological properties of polytetrafluoroethylene. J. Appl. Phys. **102**(12), 123509 (2007).
263. S.J. Heo, I. Jang, P.R. Barry, S.R. Phillpot, S.S. Perry, W.G. Sawyer, S.B. Sinnott: Effect of the sliding orientation on the tribological properties of polyethylene in molecular dynamics simulations. J. Appl. Phys. **103**(8), 083502 (2008).
264. S.B. Sinnott, R. Andrews: Carbon nanotubes: Synthesis, properties and applications, Crit. Rev. Solid State Mater. Sci. **26**, 145–249 (2001).
265. B. Bhushan, B.K. Gupta, G.W. Van Cleef, C. Capp. J.V. Coe: Sublimed $C_{60}$ films for tribology, Appl. Phys. Lett. **62**, 3253–3255 (1993).
266. T. Thundat, R.J. Warmack, D. Ding, R.N. Compton: Atomic force microscope investigation of $C_{60}$ adsorbed on silicon and mica, Appl. Phys. Lett. **63**, 891–893 (1993).
267. C.M. Mate: Nanotribology studies of carbon surfaces by force microscopy, Wear **168**, 17–20 (1993).
268. R. Lüthi, E. Meyer, H. Haefke: Sled-type motion on the nanometer scale: Determination of dissipation and cohesive energies of $C_{60}$, Science **266**, 1979–1981 (1993).
269. R. Lüthi, H. Haefke, E. Meyer, L. Howald, H.-P. Lang, G. Gerth, H.J. Güntherodt: Frictional and atomic-scale study of $C_{60}$ thin films by scanning force microscopy, Z. Phys. B **95**, 1–3 (1994).
270. Q.-J. Xue, X.-S. Zhang, F.-Y. Yan: Study of the structural transformations of $C_{60}/C_{70}$ crystals during friction, Chin. Sci. Bull. **39**, 819–822 (1994).
271. W. Allers, U.D. Schwarz, G. Gensterblum, R. Wiesendanger: Low-load friction behavior of epitaxial $C_{60}$ monolayers, Z. Phys. B **99**, 1–2 (1995).
272. U.D. Schwarz, W. Allers, G. Gensterblum, R. Wiesendanger: Low-load friction behavior of epitaxial $C_{60}$ monolayers under Hertzian contact. Phys. Rev. B **52**, 14976–14984 (1995).
273. J. Ruan, B. Bhushan: Nanoindentation studies of sublimed fullerene films using atomic force microscopy. J. Mater. Res. **8**, 3019–3022 (1996).
274. U.D. Schwarz, O. Zworner, P. Koster, R. Wiesendanger: Quantitative analysis of the frictional properties of solid materials at low loads. I. Carbon compounds. Phys. Rev. B **56**, 6987–6996 (1997).
275. S. Okita, M. Ishikawa, K. Miura: Nanotribological behavior of $C_{60}$ films at an extremely low load, Surf. Sci. **442**, L959–L963 (1999).
276. S. Okita, K. Miura: Molecular arrangement in $C_{60}$ and $C_{70}$ films on graphite and their nanotribological behavior, Nano Lett. **1**, 101–103 (2001).
277. K. Miura, S. Kamiya, N. Sasaki: $C_{60}$ molecular bearings. Phys. Rev. Lett. **90**, 055509 (2003).
278. A. Buldum. J.P. Lu: Atomic scale sliding and rolling of carbon nanotubes. Phys. Rev. Lett. **83**, 5050–5053 (1999).
279. M.R. Falvo, R.M. Taylor, A. Helser, V. Chi, F.P. Brooks, S. Washburn, R. Superfine: Nanometer-scale rolling and sliding of carbon nanotubes, Nature **397**, 236–238 (1999).
280. M.R. Falvo. J. Steele, R.M.T. Taylor II, R. Superfine: Gearlike rolling motion mediated by commensurate contact: Carbon nanotubes on HOPG. Phys. Rev. B **62**, R10664–R10667 (2000).
281. J.D. Schall, D.W. Brenner: Molecular dynamics simulations of carbon nanotube rolling and sliding on graphite. Mol. Simul. **25**, 73–80 (2000).
282. B. Ni, S.B. Sinnott: Tribological properties of carbon nanotube bundles, Surf. Sci. **487**, 87–96 (2001).
283. B. Ni, S.B. Sinnott: Mechanical and tribological properties of carbon nanotubes investigated with atomistic simulations, Nanotubes and related materials. In: *Nanotubes and Related Materials*, MRS Proc., Vol.633, ed. by A.M. Rao (Materials Research Society, Pittsburgh 2001) pp.A17.13.11–A17.13.15.
284. K. Miura, T. Takagi, S. Kamiya, T. Sahashi, M. Yamauchi: Natural rolling of zigzag multiwalled carbon nanotubes on graphite, Nano Lett. **1**, 161–163 (2001).

285. K. Miura, M. Ishikawa, R. Kitanishi, M. Yoshimura, K. Ueda, Y. Tatsumi, N. Minami: Bundle structure and sliding of single-walled carbon nanotubes observed by friction-force microscopy, Appl. Phys. Lett. **78**, 832–834 (2001).
286. P.E. Sheehan, C.M. Lieber: Nanotribology and nanofabrication of $MoO_3$ structures by atomic force microscopy, Science **272**, 1158–1161 (1996).
287. J. Wang, K.C. Rose, C.M. Lieber: Load-independent friction: $MoO_3$ nanocrystal lubricants, J. Phys. Chem. B **103**, 8405–8408 (1999).
288. Q. Ouyang, K. Okada: Nanoballbearing effect of ultra-fine particles of cluster diamond, Appl. Surf. Sci. **78**, 309–313 (1994).
289. R. Luthi, E. Meyer, H. Haefke, L. Howald, W. Gutmannsbauer, H.J. Guntherodt: Sled-type motion on the nanometer-scale – Determination of dissipation and cohesive energies of $C_{60}$, Science **266**, 1979–1981 (1994).
290. B. Bhushan, B.K. Gupta, G.W. Vancleef, C. Capp. J.V. Coe: Fullerene ($C_{60}$) films for solid lubrication, Tribol. Trans. **36**, 573–580 (1993).
291. U.D. Schwarz, W. Allers, G. Gensterblum, R. Wiesendanger: Low-load friction behavior of epitaxial $C_{60}$ monolayers under Hertzian contact, Phys. Rev. B **52**, 14976–14984 (1995).
292. S.B. Legoas, R. Giro, D.S. Galvao: Molecular dynamics simulations of $C_{60}$ nanobearings, Chem. Phys. Lett. **386**, 425–429 (2004).
293. S. Heo, S.B. Sinnott: Effect of molecular interactions on carbon nanotube friction, J. Appl. Phys. **102**(6), 064307 (2007).
294. P.L. Dickrell, S.B. Sinnott, D.W. Hahn, N.R. Raravikar, L.S. Schadler, P.M. Ajayan, W.G. Sawyer: Frictional anisotropy of oriented carbon nanotube surfaces, Tribol. Lett. **18**, 59–62 (2005).
295. F.P. Bowden, D. Tabor: *The Friction and Lubrication of Solids, Part 2* (Clarendon, Oxford, 1964).
296. G.T. Gao, P.T. Mikulski. J.A. Harrison: Molecular-scale tribology of amorphous carbon coatings: Effects of film thickness, adhesion, and long-range interactions, J. Am. Chem. Soc. **124**, 7202–7209 (2002).
297. G.T. Gao, P.T. Mikulski, G.M. Chateauneuf. J.A. Harrison: The effects of film structure and surface hydrogen on the properties of amorphous carbon films, J. Phys. Chem. B **107**, 11082–11090 (2003).

# Chapter 11
# Mechanical Properties of Nanostructures

**Bharat Bhushan**

**Abstract** NEMS Structural integrity is of paramount importance in all devices. Load applied during the use of devices can result in component failure. Cracks can develop and propagate under tensile stresses, leading to failure. Knowledge of the mechanical properties of nanostructures is necessary for designing realistic micro-/nanoelectromechanial systems (MEMS/NEMS) and biological micro-/nanoelectromechanical systems (bioMEMS/bioNEMS) devices. Elastic and inelastic properties are needed to predict the deformation due to an applied load in the elastic and inelastic regimes, respectively. The strength property is needed to predict the allowable operating limit. Some of the properties of interest are hardness, elastic modulus, bending strength, fracture toughness, and fatigue strength. Many of the mechanical properties are scale dependent; therefore these should be measured at relevant scales. Atomic force microscopy and nanoindenters can be used satisfactorily to evaluate the mechanical properties of micro-/nanoscale structures. Commonly used materials in MEMS/NEMS are single-crystal silicon and silicon-based materials, e.g., $SiO_2$ and polysilicon films deposited by low-pressure chemical vapor deposition. Single-crystal SiC deposited on large-area silicon substrates is used for high-temperature micro-/nanosensors and actuators. Amorphous alloys can be formed on both metal and silicon substrates by sputtering and plating techniques, providing more flexibility in surface integration. Electroless-deposited Ni-P amorphous thin films have been used to construct microdevices, especially using the so-called LIGA (lithography, galvanoformung, abformung) techniques. Micro-/nanodevices need conductors to provide power, as well as electrical/magnetic signals, to make them functional. Electroplated gold films have found wide applications in electronic devices because of their ability to make thin films and be processed simply. Polymers, such as poly(methyl methacrylate) (PMMA), poly (dimethylsiloxane) (PDMS), and polystyrene are commonly used in bioMEMS/bioNEMS, such as micro-/nanofluidic devices, because of ease of manufacturing and reduced cost. Many polymers are biocompatible so they may be integrated into biomedical devices.

This chapter presents a review of mechanical property measurements on the micro-/nanoscale of various materials of interest, and stress and deformation analyses of nanostructures.

B. Bhushan (ed.), *Nanotribology and Nanomechanics*,
DOI 10.1007/978-3-642-15283-2_11, 

Microelectromechanical systems (MEMS) refer to microscopic devices that have a characteristic length of <1 mm but >100 nm (or 1 μm), and nanoelectromechanical systems (NEMS) refer to nanoscopic devices that have a characteristic length of <100 nm (or 1 μm). These are referred to as an intelligent miniaturized system comprising sensing, processing, and/or actuating functions and combining electrical and mechanical components. The acronym MEMS originated in the USA. The term commonly used in Europe is microsystem technology (MST) and in Japan the term used is micromachines. Another term generally used is micro-/nanodevices. MEMS/NEMS terms are also now used in a broad sense and include electrical, mechanical, optical, biological, and/or fluidic functions. To put the dimensions in perspective, individual atoms are typically a fraction of a nanometer in diameter, DNA molecules are ≈ 2.5 nm wide, biological cells are in the range of thousands of nm in diameter, and human hair is ≈ 75 μm in diameter. The mass of a micromachined silicon structure can be as low as 1 nN, and NEMS can be built with mass as low as $10^{-20}$ N with cross sections of ≈ 10 nm. In comparison, the mass of a drop of water is ≈ 10 μN, and the mass of an eyelash is ≈ 100 nN.

A wide variety of MEMS, including Si-based devices, chemical and biological sensors and actuators, and miniature nonsilicon structures (e.g., devices made from plastics or ceramics) have been fabricated with dimensions in the range of a few to a few thousand micrometers [1, 2, 3, 4, 5, 6, 7, 8, 9, 10, 11, 12, 13]. A variety of NEMS have also been produced [14, 15, 16, 17, 18, 19]. MEMS/NEMS technology and fabrication processes have found a variety of applications in biology and biomedicine, leading to the establishment of an entirely new field known as bioMEMS/bioNEMS [20, 21, 22, 23, 24, 25, 26, 27, 28]. The ability to use micro-/nanofabrication processes to develop precision devices that can interface with biological environments at the cellular and molecular level has led to advances in the fields of biosensor technology [28, 29, 30, 31, 32], drug delivery [33, 34, 35], and tissue engineering [36, 37, 38]. The miniaturization of fluidic systems using micro-/nanofabrication techniques has led to new and more efficient devices for medical diagnostics and biochemical analysis [39]. The largest *killer* industrial applications of MEMS include accelerometers (over a billion US dollars during 2004), pressure sensors for manifold absolute pressure sensing for engines (more than 30 million units in 2004) and tire pressure measurements, inkjet printer heads (more than 500 million units in 2004), and digital micromirror devices (about US $ 700 million revenues in 2004). BioMEMS and bioNEMS are increasingly used in commercial applications. The largest applications of bioMEMS include silicon-based disposable blood pressure sensor chips for blood pressure monitoring (more than 25 million units in 2004), and a variety of biosensors.

Structural integrity is of paramount importance in all devices. Load applied during the use of devices can result in component failure. Cracks can develop and propagate under tensile stresses leading to failure [40, 41]. Friction/stiction and wear limit the lifetimes and compromise the performance and reliability of devices involving relative motion [4, 5, 6, 42]. Most MEMS/NEMS applications demand extreme reliability. Stress and deformation analyses are carried out for an optimal design. MEMS/NEMS designers require mechanical properties on the nanoscale.

Elastic and inelastic properties are needed to predict deformation due to an applied load in the elastic and inelastic regimes, respectively. The strength property is needed to predict the allowable operating limit. Some of the properties of interest are hardness, elastic modulus, creep, bending strength (fracture stress), fracture toughness, and fatigue strength. Micro-/nanostructures have surface topography and local scratches, depending on the manufacturing process. Surface roughness and local scratches may compromise the reliability of the devices, and their effect needs to be studied.

Most mechanical properties are scale dependent [5, 6, 40, 41]. Several researchers have measured the mechanical properties of silicon and silicon-based, milli- to microscale structures including tensile tests and bending tests [43, 44, 45, 46, 47, 48, 49, 50, 51, 52], resonant structure tests for the measurement of elastic properties [53], fracture toughness tests [44, 46, 54, 55, 56, 57, 58], and fatigue tests [56, 59, 60]. Most recently, a few researchers have measured the mechanical properties of nanoscale structures using atomic force microscopy (AFM) [61, 62] and nanoindentation [63, 64, 65, 66]. For stress and deformation analyses of simple geometries and boundary conditions, analytical models can be used. For analysis of complex geometries, numerical models are needed. Conventional finite-element method (FEM) can be used down to the scale of a few tens of nanometers, although its applicability is questionable at the nanoscale. FEM has been used for simulation and prediction of residual stresses and strains induced in MEMS devices during fabrication [67], to perform fault analysis in order to study MEMS faulty behavior [68], to compute mechanical strain resulting from doping of silicon [69], and to analyze micromechanical [46, 70, 71] and nanomechanical [62] experimental data. FEM analysis of nanostructures has been performed to analyze the effect of types of surface roughness and scratches on stresses in nanostructures [72, 73].

Commonly used materials for MEMS/NEMS are single-crystal silicon and silicon-based materials, e.g., $SiO_2$ and polysilicon films deposited by low-pressure chemical vapor deposition (LPCVD) process [5, 6]. An early study showed silicon to be a mechanically resilient material in addition to its favorable electronic properties [74]. Single-crystal 3C-SiC (cubic or $\beta$-SiC) films, deposited by atmospheric-pressure chemical vapor deposition (APCVD) process on large-area silicon substrates are produced for high-temperature micro-/nanosensor and actuator applications [75, 76, 77]. Amorphous alloys can be formed on both metal and silicon substrates by sputtering and plating techniques, providing greater flexibility for surface integration. Electroless-deposited Ni-P amorphous thin films have been used to construct microdevices, especially using the so-called LIGA techniques [5, 6, 64]. Micro-/nanodevices need conductors to provide power as well as electrical/magnetic signals to make them functional. Electroplated gold films have found wide applications in electronic devices because of their ability to make thin films and process simplicity [64].

As the field of MEMS/NEMS has progressed, alternative materials, especially polymers, have established an important role in their construction. This trend has been driven by the reduced cost associated with polymer materials. Polymer microfabrication processes, including micromolding and hot embossing techniques

[78], can be orders of magnitude less expensive than traditional silicon photolithography processes. Many polymers are biocompatible, so these can be integrated into biomedical devices with minimal detrimental effects to the host or biofluids. An improvement in device functionality (relative to silicon) is also possible due to the mechanical properties of the polymer used, which have properties much closer to thoes of biological tissues. Polymer bioMEMS structures involving microbeams have been designed to measure cellular forces [65, 66]. Polymer materials most commonly used for biomedical applications include poly(methyl methacrylate) (PMMA), poly(dimethylsiloxane) (PDMS), and polystyrene [66, 78, 79, 80]. Another material of interest due to ease of fabrication is poly(propyl methacrylate) (PPMA), which has a lower glass-transition temperature ($T_g$) (35–43°C) [81] than PMMA (104–106°C) [82, 83], which permits low-temperature thermal processing [65, 66, 80].

This chapter presents a review of mechanical property measurements on the nanoscale of various materials of interest, and stress and deformation analyses of nanostructures.

## 11.1 Experimental Techniques for Measurement of Mechanical Properties of Nanostructures

### *11.1.1 Indentation and Scratch Tests Using Micro-/Nanoindenters*

A nanoindenter is commonly used to measure hardness, elastic modulus, and fracture toughness, and to perform micro-/nanoscratch studies to get a measure of scratch/wear resistance of materials [5, 6, 84].

#### Hardness and Elastic Modulus

A nanoindenter monitors and records the dynamic load and displacement of a three-sided pyramidal diamond (Berkovich) indenter during indentation with force resolution of $\approx$ 75 nN and displacement resolution of $\approx$ 0.1 nm. Hardness and elastic modulus are calculated from the load–displacement data [5, 6, 84]. The peak indentation load depends on the mechanical properties of the specimen; a harder material requires a higher load for reasonable indentation depth.

#### Fracture Toughness

The indentation technique for fracture toughness measurement of brittle samples on the microscale is based on the measurement of the lengths of median-radial cracks produced by indentation. A Vickers indenter (a four-sided diamond pyramid) is used in microhardness testers. A load on the order of 0.5 N is typically used in making Vickers indentations. The indentation impressions are examined using an optical microscope with Nomarski interference contrast to measure the length of median-radial cracks $c$. The fracture toughness ($K_{\mathrm{IC}}$) is calculated by the following relation [85]

$$K_{\mathrm{IC}} = \alpha \left(\frac{E}{H}\right)^{1/2} \left(\frac{P}{c^{3/2}}\right), \tag{11.1}$$

where $\alpha$ is an empirical constant depending on the geometry of the indenter, $H$ and $E$ are hardness and elastic moduli, and $P$ is the peak indentation load. For Vickers indenters, $\alpha$ has been found empirically based on experimental data to be 0.016 [5]. Both $E$ and $H$ values are obtained from the nanoindentation data. The crack length is measured from the center of the indent to the end of the crack using an optical microscope. For one indent, all crack lengths are measured. The crack length $c$ is obtained from the average value of several indents.

### Indentation Creep

Indentation creep tests of polymer samples are performed using a continuous stiffness measurement (CSM) technique [84]. In a study by Wei et al. [65], the indentation load was typically 30 μN and the loading rate was 3 μN/s. The tip was typically held for 600 s after the indentation load reached 30 μN. To measure the mean stress and contact stiffness, during the hold segment the indenter was oscillated at a peak-to-peak load amplitude of 1.2 μN and a frequency of 45 Hz.

### Scratch Resistance

In micro-/nanoscratch studies, in a nanoindenter, a conical diamond indenter with a tip radius of about 1 μm and an included angle of 60° is drawn over the sample surface, and the load is ramped up until substantial damage occurs [5, 6, 84]. The coefficient of friction is monitored during scratching. In order to determine scratch depths, the surface profile of the sample surface is first obtained by translating the sample at a low load of $\approx$ 0.2 mN, which is insufficient to damage a hard sample surface. The 500 μm-long scratches are made by translating the sample while ramping the load on the conical tip over different values dependent upon the material hardness. The actual depth during scratching is obtained by subtracting the initial profile from the scratch depth measured during scratching. In order to measure the scratch depth after the scratch, the scratched surface is profiled at a low load of 0.2 mN and is subtracted from the actual surface profile before scratching.

### 11.1.2 Bending Tests of Nanostructures Using an AFM

Quasistatic bending tests of fixed nanobeam arrays in the normal direction are carried out using an AFM [62, 86]. A three-sided pyramidal diamond tip (with radius of $\approx$ 200 nm) mounted on a rectangular stainless-steel cantilever is used for the bending tests. The beam stiffness is selected based on the desired load range. The stiffness of the cantilever beams for application of normal load up to 100 μN is $\approx$ 150–200 N/m.

For the bending test, the tip is brought over the nanobeam array with the help of the sample stage of the AFM and a built-in high-magnification optical microscope (Fig. 11.1) [62]. For fine positioning of the tip over a chosen beam, the array is scanned in contact mode at a contact load of about 2–4 μN, which results in negligible damage to the sample. After scanning, the tip is located at one end of a chosen beam. To position the tip at the center of the beam span, the tip is moved to the other end of the beam by giving the $x$-piezo an offset voltage. The value of this offset is determined after several such attempts have been made in order to minimize the effects of piezo drift. Half of this offset is then applied to the $x$-piezo after the tip is positioned at one end of the beam, which usually results in the tip being moved to the center of the span. Once the tip is positioned over the center of the beam span, the tip is held stationary without scanning, and the $z$-piezo is extended by a known distance, typically $\approx$ 2.5 μm, at a rate of 10 nm/s, as shown in Fig. 11.1. During this time, the vertical deflection signal ($dV_{AFM}$), which is

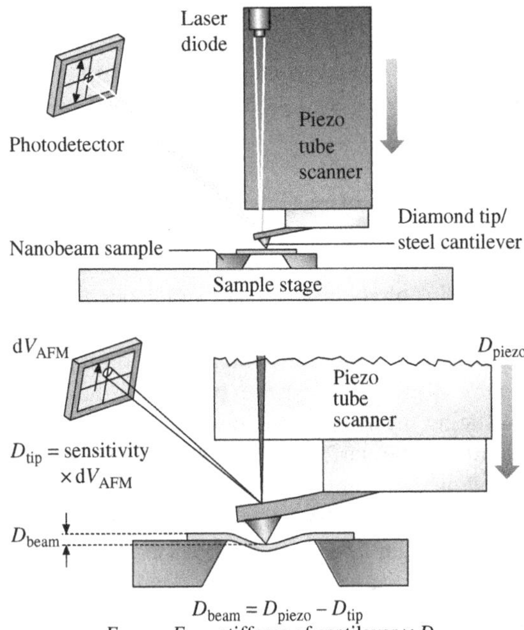


**Fig. 11.1** Schematic showing the details of a nanoscale bending test using an AFM. The AFM tip is brought to the center of the nanobeam and the piezo is extended over a known distance. By measuring the tip displacement, a load–displacement curve for the nanobeam can be obtained (After [62])

proportional to the deflection of the cantilever ($D_{tip}$), is monitored. The displacement of the piezo is equal to the sum of the displacements of the cantilever and the nanobeam. Hence the displacement of the nanobeam ($D_{beam}$) under the point of load can be determined as

$$D_{beam} = D_{piezo} - D_{tip}. \tag{11.2}$$

The load ($F_{beam}$) on the nanobeam is the same as the load on the tip/cantilever ($F_{tip}$) and is given by

$$F_{beam} = F_{tip} = D_{tip}k, \tag{11.3}$$

where $k$ is the stiffness of the tip/cantilever. In this manner, a load–displacement curve for each nanobeam can be obtained.

The photodetector sensitivity of the cantilever needs to be calibrated to obtain $D_{tip}$ in nm. For this calibration, the tip is pushed against a smooth diamond sample by moving the $z$-piezo over a known distance. For the hard diamond material, the actual deflection of the tip can be assumed to be the same as the $z$-piezo travel ($D_{piezo}$), and the photodetector sensitivity ($S$) for the cantilever setup is determined as

$$S = \frac{D_{piezo}}{dV_{AFM}} \frac{nm}{V}. \tag{11.4}$$

In the measurements, $D_{tip}$ is given as $dV_{AFM}S$.

Since a sharp tip would result in an undesirable large local indentation, Sundararajan and Bhushan [62] used a diamond tip which was worn (blunt). Indentation experiments using this tip on a silicon substrate yielded a residual depth of <8 nm at a maximum load of 120 μN, which is negligible compared with displacements of the beam (several hundred nm). Hence we can assume that negligible local indentation or damage is created during the bending process of the beams and that the displacement calculated from (11.2) is due to the beam structure.

### Elastic Modulus and Bending Strength

Elastic modulus and bending strength (fracture stress) of beams can be estimated by equations based on the assumption that the beams follow the linear elastic theory of an isotropic material. This is probably valid since the beams have high length-to-width ($\ell/w$) and length-to-thickness ($\ell/t$) ratios and also since the length direction is along the direction of principal stress during the test. For a fixed elastic beam loaded at the center of the span, the elastic modulus is expressed as

$$E = \frac{\ell^3}{192I} m, \tag{11.5}$$

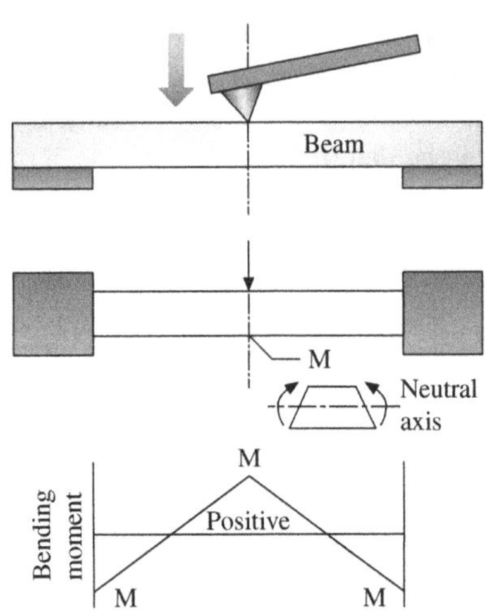


**Fig. 11.2** A schematic of the bending moments generated in the beam during a quasistatic bending experiment, with the load at the center of the span. The maximum moments occur under the load and at the fixed ends. Due to the trapezoidal cross section, the maximum tensile bending stresses occur at the top surfaces at the fixed ends

where $\ell$ is the beam length, $I$ is the area moment of inertia for the beam cross section, and $m$ is the slope of the linear region of the load–displacement curve during bending [87]. The area moment of inertia for a beam with a trapezoidal cross section is calculated from

$$I = \frac{4w_1^2 w_1 w_2 + w_2^2}{36(w_1 + w_2)} t^3, \tag{11.6}$$

where $w_1$ and $w_2$ are the upper and lower widths, respectively, and $t$ is the thickness of the beam. According to linear elastic theory, for a centrally loaded beam, the moment diagram is that shown in Fig. 11.2. The maximum moments are generated at the ends (negative moment) and under the loading point (positive moment), as shown in Fig. 11.2. The bending stresses generated in the beam are proportional to the moments and are compressive or tensile about the neutral axis (the line of zero stress). The maximum tensile stress ($\sigma_b$, which is the bending strength or fracture stress) is produced on the top surface at both the ends and is given by [87]

$$\sigma_b = \frac{F_{max} \ell e_1}{8I}, \tag{11.7}$$

where $F_{max}$ is the applied load at failure, and $e_1$ is the distance of the top surface from the neutral plane of the beam cross section, given by [87]

$$e_1 = \frac{t(w_1 + 2w_2)}{3(w_1 + w_2)}. \tag{11.8}$$

Although the moment value at the center of the beam is the same as at the ends, the tensile stresses at the center (generated on the bottom surface) are less than those generated at the ends (as per (11.7)) because the distance from the neutral axis to the bottom surface is $<e_1$. This is because of the trapezoidal cross section of the

beam, which results in the neutral axis being closer to the bottom surface than to the top (Fig. 11.2).

In the preceding analysis, the beams were assumed to have fixed ends. However, in the nanobeams used by Sundararajan and Bhushan [62], the underside of the beams was pinned over some distance on either side of the span. Hence a finite-element model of the beams was created to see if the difference in the boundary conditions affected the stresses and displacements of the beams. It was found that the difference in the stresses was $<1\%$. This indicates that the boundary conditions near the ends of the actual beams are not that different from those for fixed ends. Therefore the bending strength values can be calculated from (11.7).

## Fracture Toughness

Fracture toughness is another important parameter for brittle materials such as silicon. In the case of the nanobeam arrays, these are not best suited for fracture toughness measurements because they do not possess regions of uniform stress during bending. Sundararajan and Bhushan [62] developed a methodology for this, outlined schematically in Fig. 11.3a. First, a crack of known geometry is introduced in the region of maximum tensile bending stress, i.e., on the top surface near the ends of the beam. This is achieved by generating a scratch at high normal load across the width $w_1$ of the beam using a sharp diamond tip (radius $<100$ nm). A typical scratch thus generated is shown in Fig. 11.3b. By bending the beam as shown, a stress concentration will be formed under the scratch. This will lead to failure of the beam under the scratch once a critical load (fracture load) is attained. The fracture load and relevant dimensions of the scratch are input into the FEM model, which is used to generate the fracture stress plots. Figure 11.3c shows an

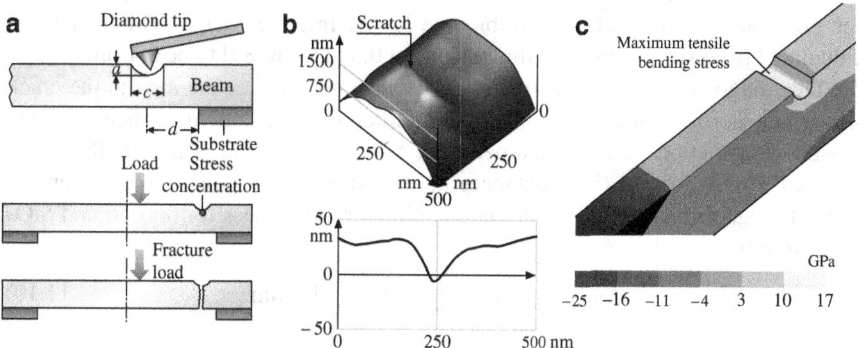


**Fig. 11.3** **(a)** Schematic of a technique to generate a defect (crack) of known dimensions in order to estimate fracture toughness. A diamond tip is used to generate a scratch across the width of the beam. When the beam is loaded as shown, a stress concentration is formed at the bottom of the scratch. The fracture load is then used to evaluate the stresses using FEM. **(b)** AFM 3-D image and two-dimensional (2-D) profile of a typical scratch. **(c)** Finite-element model results verifying that the maximum bending stress occurs at the bottom of the scratch (After [62])

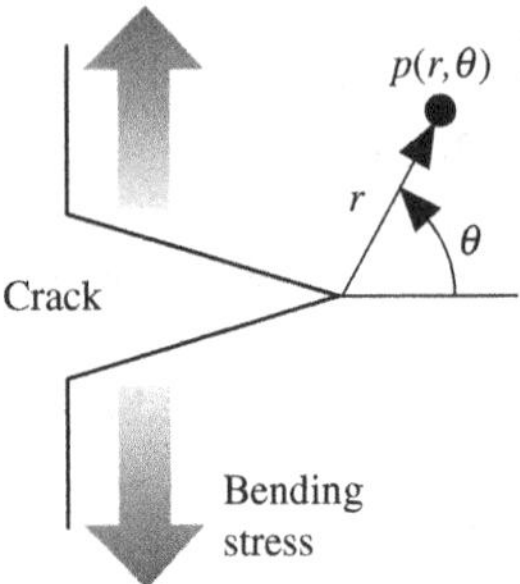


**Fig. 11.4** Schematic of crack tip and coordinate systems used in (11.9) to describe the stress field around the crack tip in terms of the stress intensity parameter $K_I$ (After [62])

FEM simulation of one such experiment, which reveals that the maximum stress does occur under the scratch.

If we assume that the scratch tip acts as a crack tip, a bending stress will tend to open the crack in mode I. In this case, the stress field around the crack tip can be described by the stress intensity parameter $K_I$ (for mode I) for linear elastic materials [88]. In particular the stresses corresponding to the bending stresses are described by

$$\sigma = \frac{K_I}{\sqrt{2\pi r}} \cos\left(\frac{\theta}{2}\right)\left[1 + \sin\left(\frac{\theta}{2}\right)\sin\left(\frac{3\theta}{2}\right)\right], \qquad (11.9)$$

for every point $p(r, \theta)$ around the crack tip as shown in Fig. 11.4. If we substitute the fracture stress ($\sigma_f$) into the left-hand side of (11.9), then $K_I$ can be substituted by its critical value, which is the fracture toughness $K_{IC}$. Now, the fracture stress can be determined for the point ($r = 0$, $\theta = 0$), i.e., immediately under the crack tip as explained above. However, we cannot substitute $r = 0$ into (11.9). The alternative is to substitute a value for $r$ which is as close to zero as possible. For silicon, a reasonable number is the distance between neighboring atoms in the (111) plane, the plane along which silicon exhibits the lowest fracture energy. This value was calculated from silicon unit cell dimensions of 0.5431 nm [89] to be 0.4 nm (half of the face diagonal). This assumes that Si displays no plastic zone around the crack tip, which is reasonable since, in tension, silicon is not known to display much plastic deformation at room temperature. Sundararajan and Bhushan [62] used values $r = 0.4$–$1.6$ nm (i.e., distances up to four times the distance between the nearest neighboring atoms) to estimate the fracture toughness for both Si and $SiO_2$ according to

$$K_{IC} = \sigma_f\sqrt{2\pi r}, \quad r = 0.4 - 1.6 \text{ nm}. \qquad (11.10)$$

## Fatigue Strength

In addition to the properties mentioned so far that can be evaluated from quasistatic bending tests, the fatigue properties of nanostructures are also of interest. This is especially true for MEMS/NEMS involving vibrating structures such as oscillators and comb drives [90] and hinges in digital micromirror devices [91]. To study the fatigue properties of the nanobeams, Sundararajan and Bhushan [62] applied monotonic cyclic stresses using an AFM (Fig. 11.5a). Similar to the bending test, the diamond tip is first positioned at the center of the beam span. In order to ensure that the tip is always in contact with the beam (rather than impacting it), the piezo is first extended by a distance $D_1$, which ensures a minimum stress on the beam. After this extension, a cyclic displacement of amplitude $D_2$ is applied continuously until failure of the beam occurs. This results in the application of a cyclic load to the beam. The maximum frequency of the cyclic load that could be attained using the

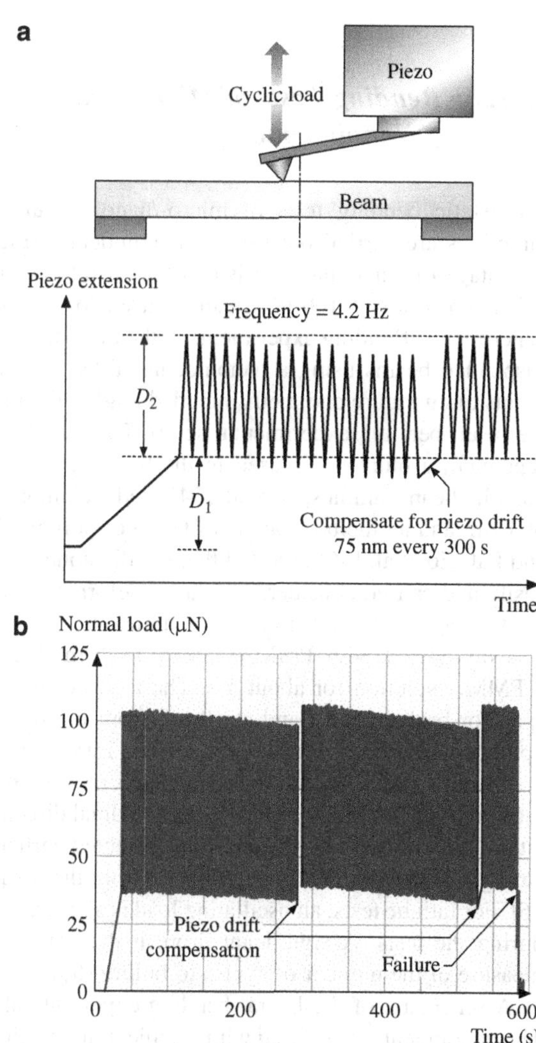


**Fig. 11.5** (**a**) Schematic showing the details of the technique to study fatigue behavior of the nanobeams. The diamond tip is located at the middle of the span and a cyclic load at 4.2 Hz is applied to the beam by forcing the piezo to move in the pattern shown. An extension is made every 300 s to compensate for the piezo drift to ensure that the load on the beam is kept fairly constant. (**b**) Data from a fatigue experiment on a nanobeam until failure. The normal load is computed from the raw vertical deflection signal. The compensations for piezo drift keep the load fairly constant (After [62])

AFM by Sundararajan and Bhushan [62] was 4.2 Hz. The vertical deflection signal of the tip is monitored throughout the experiment. The signal follows the pattern of the piezo input up to failure, which is indicated by a sudden drop in the signal. During initial runs, piezo drift was observed, which caused the piezo to gradually move away from the beam (i.e., to retract), resulting in a continuous decrease in the applied normal load. In order to compensate for this, the piezo is given a finite extension of 75 nm every 300 s as shown in Fig. 11.5a. This results in keeping the applied loads fairly constant. The normal load variation (calculated from the vertical deflection signal) from a fatigue test is shown in Fig. 11.5b. The cyclic stress amplitudes (corresponding to $D_2$) and fatigue lives are recorded for every sample tested. Values for $D_1$ are set such that minimum stress levels are about 20% of the bending strengths.

### *11.1.3 Bending Tests of Micro-/Nanostructures Using a Nanoindenter*

Quasistatic bending tests of micro-/nanostructures in both normal and lateral directions are carried out using a nanoindenter (Fig. 11.6) [63, 64, 65, 66]. The advantage of a nanoindenter is that loads up to about 400 mN, higher than that in AFM (up to about 100 μN), can be used for structures requiring high loads for experiments. Bending experiments in the normal direction have been carried out on suspended beams using a nanoindenter fitted with a conical tip that has a 1 μm radius of curvature and a 60° included angle. Figure 11.6a shows the schematic of the beam bending experimental setup. To avoid the indenter tip pushing into the specimen, a blunt tip is used in the bending and fatigue tests. For ceramic and metallic beam samples, Li et al. [64] used a diamond conical indenter with a radius of 1 μm and an included angle of 60°. For polymer beam samples, Wei et al. [65] and Palacio et al. [66] reported that the diamond tip penetrated the polymer beams easily and caused considerable plastic deformation during the bending test, which led to significant errors in the measurements. To avoid this issue, the diamond tip was dip-coated with PMMA (about 1–2 μm thick) by dipping the tip in *wt%* 2 PMMA solution for about 5 s. The load position used was at the center of the span for bridge beams and at 10 μm from the free end for cantilever beams. An optical microscope with magnification of 1500× or an in situ AFM is used to locate the loading position. Then the specimen is moved under the indenter location with a resolution of about 200 nm in the longitudinal direction and less than 100 nm in the lateral direction. Using the analysis presented earlier, elastic modulus and bending strength of the beams can be obtained from the load–displacement curves [64, 65, 66]. For fatigue tests, an oscillating load is applied and contact stiffness is measured during the tests. A significant drop in the contact stiffness during the test is a measure of the number of cycles to failure [63].

A schematic of the lateral bending experimental setup in shown in Fig. 11.6b. The nanoindenter was fitted with a conical tip which has a 1 μm radius of curvature

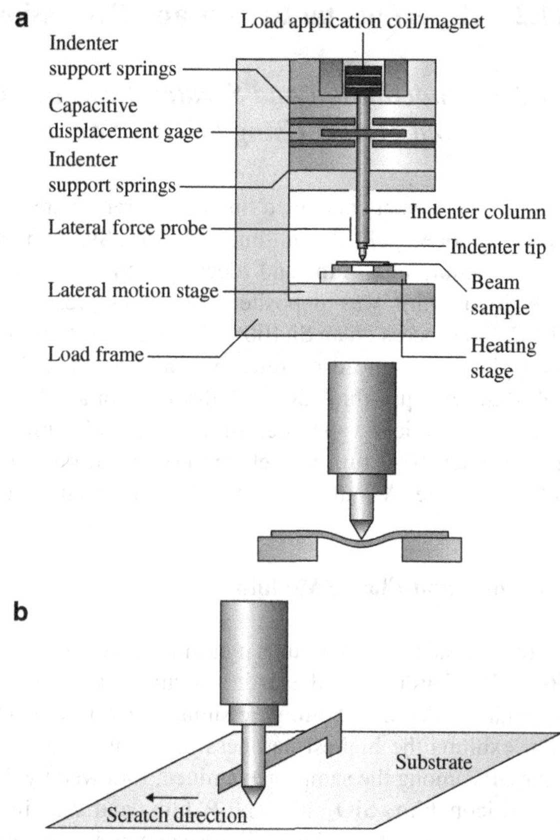


**Fig. 11.6** Schematic of micro/nanoscale bending experimental setup using a nanoindenter (**a**) in the normal direction, and (**b**) in lateral direction

and a 60° included angle. The tip was positioned at 200–300 μm away from the beam anchor and 20 μm from the edge of the beam length. The indenter was programmed to perform a scratch at a constant load of 400 μN, and the stage was moved at a rate of 10 μm/s. Since the lateral force contribution of the substrate surface is constant, any additional force recorded comes from the cantilever beam response. From the linear elastic regime in the measured lateral force as a function of lateral displacement, the elastic modulus can be evaluated. For a cantilever beam with one end clamped, it is expressed as [87]

$$E = \frac{\ell^3}{3I} m, \tag{11.11}$$

where, as described earlier, $\ell$ is the beam length, $I$ is the area moment of inertia for the beam cross section, and $m$ is the slope of the linear region of the force–displacement curve.

## 11.2 Experimental Results and Discussion

### *11.2.1 Indentation and Scratch Tests of Various Ceramic and Metals Using Micro-/Nanoindenter*

Studies have been conducted on six different materials: undoped single-crystal Si (100), undoped polysilicon film, $SiO_2$ film, SiC film, electroless-deposited Ni-*wt%* 11.5 P amorphous film, and electroplated Au film [64, 76, 77]. A 3 μm-thick polysilicon film was deposited by a low-pressure chemical vapor deposition (LPCVD) process on an Si(100) substrate. The 1 μm-thick $SiO_2$ film was deposited by a plasma-enhanced chemical vapor deposition (PECVD) process on a Si(111) substrate. A 3 μm-thick 3C-SiC film was epitaxially grown using an atmospheric-pressure chemical vapor deposition (APCVD) process on Si(100) substrate. A 12 μm-thick Ni-P film was electroless-plated on a 0.8 mm-thick Al *wt%* 4.5 Mg alloy substrate. A 3 μm-thick Au film was electroplated on an Si(100) substrate.

#### Hardness and Elastic Modulus

Hardness and elastic modulus measurements were carried out using a nanoindenter [64]. The hardness and elastic modulus values of various materials at a peak indentation depth of 50 nm are summarized in Fig. 11.7 and Table 11.1. The SiC film exhibits the highest hardness of about 25 GPa, an elastic modulus of about 395 GPa among the samples examined, followed by the undoped Si(100), undoped polysilicon film, $SiO_2$ film, Ni-P film, and Au film. The hardness and elastic modulus data of the undoped Si(100) and undoped polysilicon film are comparable. For the metal alloy films, the Ni-P film exhibits higher hardness and elastic modulus than the Au film.

#### Fracture Toughness

Optical images of Vickers indentations made using a microindenter at a normal load of 0.5 N held for 15 s on the undoped Si(100), undoped polysilicon film, and SiC film are shown in Fig. 11.8 [77]. The SiC film exhibits the smallest indentation mark, followed by the undoped polysilicon film and undoped Si(100). These Vickers indentation depths are smaller than one-third of the film thickness. Thus, the influence of the substrate on the fracture toughness of the films can be ignored. In addition to the indentation marks, radial cracks are observed, emanating from the indentation corners. The SiC film shows the longest radial crack length, followed by the undoped Si(100) and undoped polysilicon film. The radial cracks for the undoped Si(100) are straight, whereas those for the SiC and undoped polysilicon film are not straight but rather zigzag. The fracture toughness ($K_{IC}$) is calculated using (11.1).

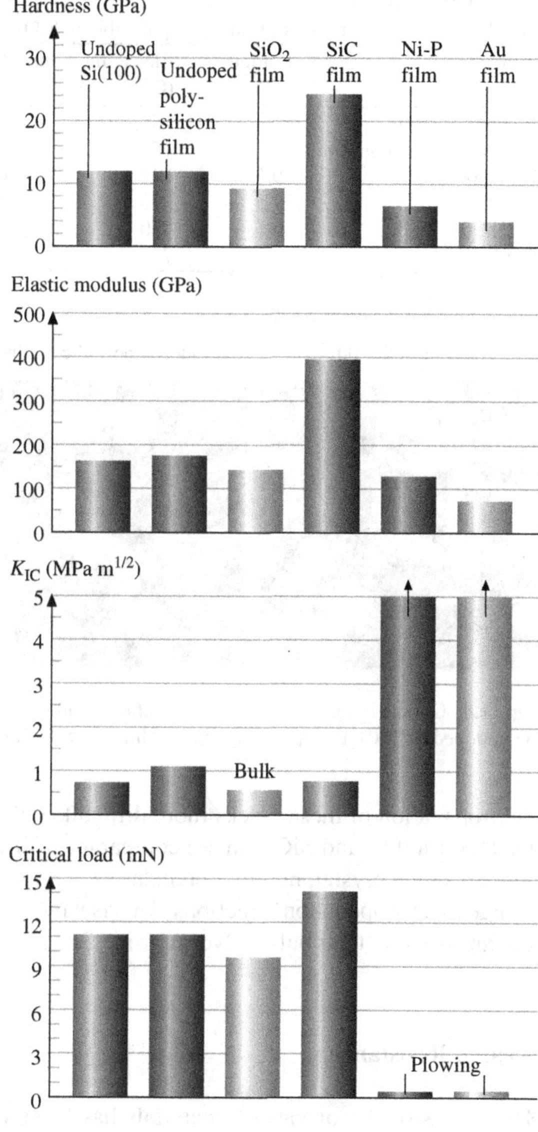


**Fig. 11.7** Bar chart summarizing the hardness, elastic modulus, fracture toughness, and critical load (from scratch tests) results of bulk undoped single-crystal Si(100) and thin films of undoped polysilicon, $SiO_2$, SiC, Ni-P, and Au (After [64])

The fracture toughness values of all samples are summarized in Fig. 11.7 and Table 11.1. The $SiO_2$ film used in this study is about 1 μm thick, which is not thick enough for fracture toughness measurement. The fracture toughness value of bulk silica is listed instead for reference. The Ni-P and Au films exhibit very high fracture toughness values that cannot be measured by indentation methods. For other samples, the undoped polysilicon film has the highest value, followed by the undoped Si(100), SiC film, and $SiO_2$ film. For the undoped polysilicon film, grain boundaries can stop radial cracks and change their propagation directions, making

**Table 11.1** Hardness, elastic modulus, fracture toughness, and critical load results of bulk single-crystal Si(100) and thin films of undoped polysilicon, $SiO_2$, SiC, Ni-P, and Au

| Samples | Hardness (GPa) | Elastic modulus (GPa) | Fracture toughness MPa $m^{1/2}$ | Critical load (mN) |
|---|---|---|---|---|
| Undoped Si(100) | 12 | 165 | 0.75 | 11 |
| Undoped polysilicon film | 12 | 167 | 1.11 | 11 |
| $SiO_2$ film | 9.5 | 144 | 0.58 (bulk) | 9.5 |
| SiC film | 24.5 | 395 | 0.78 | 14 |
| Ni-P film | 6.5 | 130 | | 0.4 (plowing) |
| Au film | 4 | 72 | | 0.4 (plowing) |

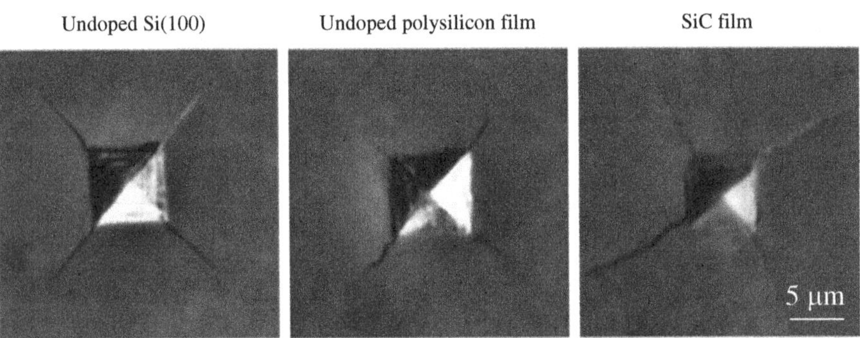


**Fig. 11.8** Optical images of Vickers indentations made at a normal load of 0.5 N held for 15 s on the undoped Si(100), undoped polysilicon film, and SiC film (After [77])

the propagation of these cracks more difficult. Values of fracture toughness for the undoped Si(100) and SiC film are comparable. Since the undoped Si(100) and SiC film are single crystal, no grain boundaries are present to stop the radial cracks and change their propagation directions. This is why the SiC film shows a lower fracture toughness value than bulk polycrystalline SiC material (3.6 MPa $m^{1/2}$) [92].

## Scratch Resistance

Scratch resistance of various materials has been studied using a nanoindenter by Li et al. [64]. Figure 11.9 compares the coefficient of friction and scratch depth profiles as a function of increasing normal load and optical images of three regions over scratches: at the beginning of the scratch (indicated by A on the friction profile), at the point of initiation of damage at which the coefficient of friction increases to a high value or increases abruptly (indicated by B on the friction profile), and towards the end of the scratch (indicated by C on the friction profile) for all samples. Note that the ramp loads for Ni-P and Au range from 0.2 to 5 mN, whereas the ramp loads for other samples range from 0.2 to 20 mN. All samples exhibit a continuous increase in the coefficient of friction with increasing normal

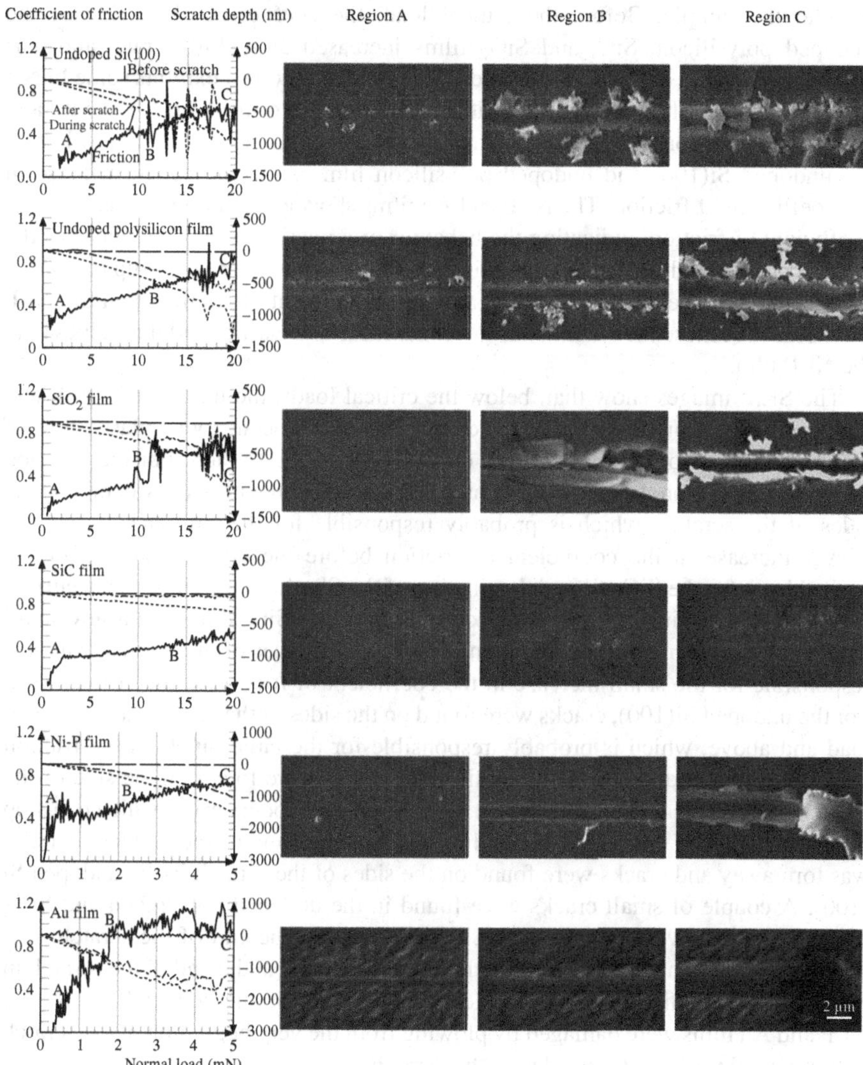


**Fig. 11.9** Coefficient of friction and scratch depth profiles as a function of increasing normal load and optical images of three regions over scratches: at the beginning of the scratch (indicated by A on the friction profile), at the point of initiation of damage at which the coefficient of friction increases to a high value or increase abruptly (indicated by B on the friction profile), and towards the end of the scratch (indicated by C on the friction profile) for all samples (After [64])

load from the very beginning of the scratch. The continuous increase in the coefficient of friction during scratching is attributed to increasing plowing of the sample by the tip with increasing normal load, as shown in the scanning electron microscopy (SEM) images in Fig. 11.9. The abrupt increase in the coefficient of friction is associated with catastrophic failure as well as significant plowing of the tip into the sample. Before the critical load, the coefficient of friction of the undoped polysilicon, SiC, and $SiO_2$ films increased at a slower rate, and was smoother than that of the other samples. The undoped Si(100) exhibits some bursts in the friction profiles before the critical load. At the critical load, the SiC and undoped polysilicon films exhibit a small increase in coefficient of friction, whereas the undoped Si(100) and undoped polysilicon film exhibit a sudden increase in the coefficient of friction. The Ni-P and Au films show a continuous increase in the coefficient of friction, indicating the behavior of a ductile metal. The bursts in the friction profile might result from plastic deformation and material pile-up in front of the scratch tip. The Au film exhibits a higher coefficient of friction than the Ni-P film. This is because the Au film has lower hardness and elastic modulus values than the Ni-P film.

The SEM images show that, below the critical loads, the undoped Si(100) and undoped polysilicon film were damaged by plowing, associated with plastic flow of the material and formation of debris on the sides of the scratch. For the SiC and $SiO_2$ films, in region A, a plowing scratch track was found without any debris on the sides of the scratch, which is probably responsible for the smoother curve and slower increase in the coefficient of friction before the critical load. After the critical load, for the $SiO_2$ film, delamination of the film from the substrate occurred, followed by cracking along the scratch track. For the SiC film, only a few small debris particles were found without any cracks on the sides of the scratch, which is responsible for the small increase in the coefficient of friction at the critical load. For the undoped Si(100), cracks were found on the sides of the scratch at the critical load and above, which is probably responsible for the large bursts in the friction profile. For the undoped polysilicon film, no cracks were found on the side of the scratch at the critical load. This might result from grain boundaries, which can stop the propagation of cracks. At the end of the scratch, some of the surface material was torn away and cracks were found on the sides of the scratch in the undoped Si (100). A couple of small cracks were found in the undoped polysilicon and $SiO_2$ films. No cracks were found in the SiC film. Even at the end of the scratch, less debris was found in the SiC film. A curly chip was found at the end of the scratch in both Ni-P and Au films. This is a typical characteristic of ductile metal alloys. The Ni-P and Au films were damaged by plowing from the very beginning of the scratch with material pile-up at the sides of the scratch.

The scratch depth profiles obtained during and after the scratch test on all samples with respect to initial profile, after the cylindrical curvature is removed, are plotted in Fig. 11.9. Reduction in scratch depth is observed after scratching as compared with that during scratching. This reduction in scratch depth is attributed to elastic recovery after removal of the normal load. The scratch depth measured after scratching indicates the final depth, which reflects the extent of permanent damage and plowing of the tip into the sample surface, and is probably more relevant for visualizing the damage that can occur in real applications. For the undoped Si(100), undoped polysilicon film, and $SiO_2$ film, there is a large scatter in the scratch depth data after the critical loads, which is associated with the generation of cracks, material removal, and debris. The scratch depth profile is smooth for the SiC film. It is noted that the SiC film exhibits the lowest scratch depth among the samples examined. The scratch depths of the undoped Si(100), undoped polysilicon film, and $SiO_2$ film are comparable. The Ni-P and Au films exhibit much larger scratch depth than the other samples. The scratch depth of the Ni-P film is smaller than that of the Au film.

The critical loads estimated from friction profiles for all samples are compared in Fig. 11.7 and Table 11.1. The SiC film exhibits the highest critical load of about 14 mN, as compared with other samples. The undoped Si(100) and undoped polysilicon film show comparable critical load of about 11 mN, whereas the $SiO_2$ film shows a low critical load of about 9.5 mN. The Ni-P and Au films were damaged by plowing from the very beginning of the scratch.

### *11.2.2 Bending Tests of Ceramic Nanobeams Using an AFM*

Bending tests have been performed on Si and $SiO_2$ nanobeam arrays [62, 86]. The single-crystal silicon bridge nanobeams were fabricated by bulk micromachining incorporating enhanced-field anodization using an AFM [61]. The Si nanobeams are oriented along the [110] direction in the (001) plane. Subsequent thermal oxidation of the beams results in the formation of $SiO_2$ beams. The cross section of the nanobeams is trapezoidal owing to the anisotropic wet etching process. SEM micrographs of Si and $SiO_2$ nanobeam arrays and a schematic of the shape of a typical nanobeam are shown in Fig. 11.10. The actual widths and thicknesses of nanobeams were measured using an AFM in tapping mode prior to tests using a standard Si tapping-mode tip (radius $<10$ nm). Surface roughness measurements of the nanobeam surfaces in tapping mode yielded a $\sigma$ of $0.7 \pm 0.2$ nm and peak-to-valley (P–V) distance of $4 \pm 1.2$ nm for Si and a $\sigma$ of $0.8 \pm 0.3$ nm and a P–V distance of $3.1 \pm 0.8$ nm for $SiO_2$. Prior to testing, the Si nanobeams were cleaned by immersing them in a *Piranha etch* solution (3:1 solution by volume of 98% sulfuric acid and 30% hydrogen peroxide) for 10 min to remove any organic contaminants.

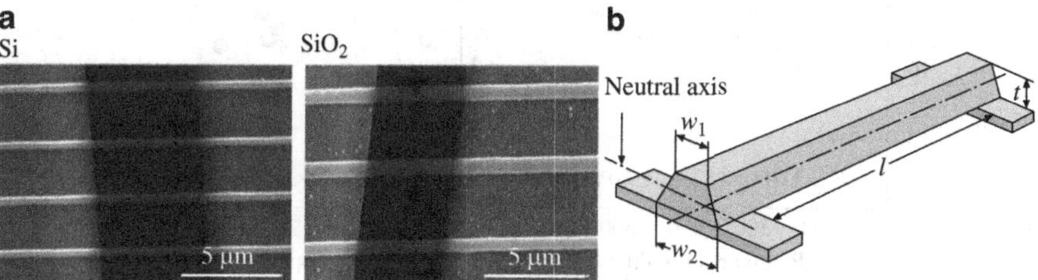


**Fig. 11.10** (**a**) SEM micrographs of nanobeam arrays, and (**b**) schematic of the shape of a typical nanobeam. The trapezoidal cross section is due to the anisotropic wet etching during the fabrication (After [86])

## Bending Strength

Figure 11.11 shows typical load–displacement curves for Si and $SiO_2$ beams that were bent to failure [62, 86]. The upper width $w_1$ of the beams is indicated in the figure. Also indicated in Fig. 11.11 are the elastic modulus values obtained from the slope of the load–displacement curve (11.5). All the beams tested showed linear elastic behavior followed by abrupt failure, which is suggestive of brittle fracture. Figure 11.12 shows the scatter in the values of elastic modulus obtained for both Si and $SiO_2$ along with the average values (± standard deviation). The scatter in the values may be due to differences in orientation of the beams with respect to the trench and the loading point being a little off-center with respect to the beam span. The average values are a little higher than the bulk values (169 GPa for Si[110] and 73 GPa for $SiO_2$ in Table 11.2). However the values of $E$ obtained from (11.5) have

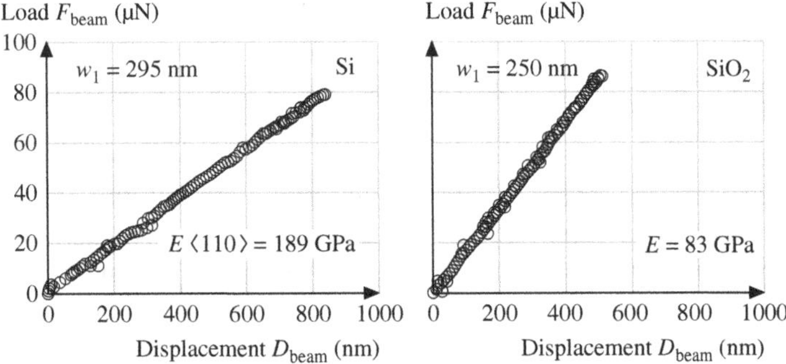


**Fig. 11.11** Typical load–displacement curves of silicon and $SiO_2$ nanobeams. The *curves* are linear until sudden failure, indicative of brittle fracture of the beams. The elastic modulus ($E$) values calculated from the *curves* are shown. The dimensions of the Si beam were $w_1 = 295$ nm, $w_2 = 484$ nm, and $t = 255$ nm, while those of the $SiO_2$ beam were $w_1 = 250$ nm, $w_2 = 560$ nm, and $t = 425$ nm (After [86])

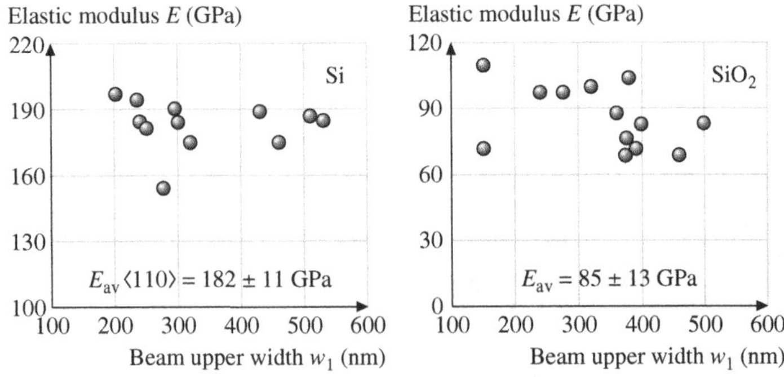


**Fig. 11.12** Elastic modulus values measured for Si and $SiO_2$. The average values are shown. These are comparable to bulk values, which shows that elastic modulus shows no specimen size dependence (After [62])

**Table 11.2** Summary of measured parameters from quasistatic bending tests

| Sample | Elastic modulus $E$(GPa) | | Bending strength $\sigma_b$(GPa) | | Fracture toughness $K_{IC}$(MPa m$^{1/2}$) | | |
|---|---|---|---|---|---|---|---|
| | Measured | Bulk value | Measured | Reported (microscale) | Estimated | Reported (microscale) | Bulk value |
| Si | 182 ± 11 | 169[a] | 18 ± 3 | <10[c] | 1.67 ± 0.4 | 0.6–1.65[e] | 0.9[f] |
| $SiO_2$ | 85 ± 13 | 73[b] | 7.6 ± 2 | <2[d] | 0.60 ± 0.2 | 0.5–0.9[4] | – |

[a]Si[110] [93]
[b][94]
[c][43, 44, 46, 46, 47, 48, 49, 95, 96]
[d][58]
[e][54, 55, 56, 57]
[f][89]

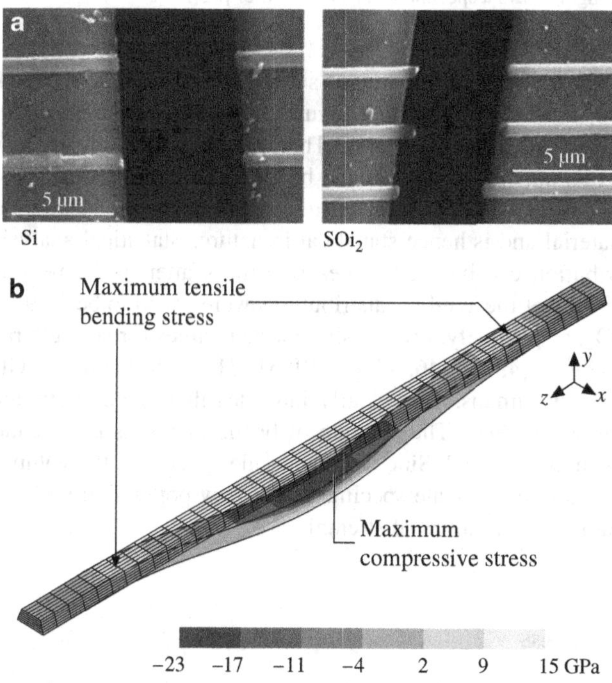


**Fig. 11.13** **(a)** SEM micrographs of nanobeams that failed during quasistatic bending experiments. The beams failed at or near the ends, which is the location of maximum tensile bending stress (After [86]), and **(b)** bending stress distribution for silicon nanobeam indicating that the maximum tensile stresses occur on the top surfaces near the fixed ends

an error of ≈ 20% due to the uncertainties in beam dimensions and spring constant of the tip/cantilever (which affects the measured load). Hence the elastic modulus values on the nanoscale can be considered to be comparable to bulk values.

Most of the beams when loaded quasistatically at the center of the span broke at the ends, as shown in Fig. 11.13a, which is consistent with the fact that the

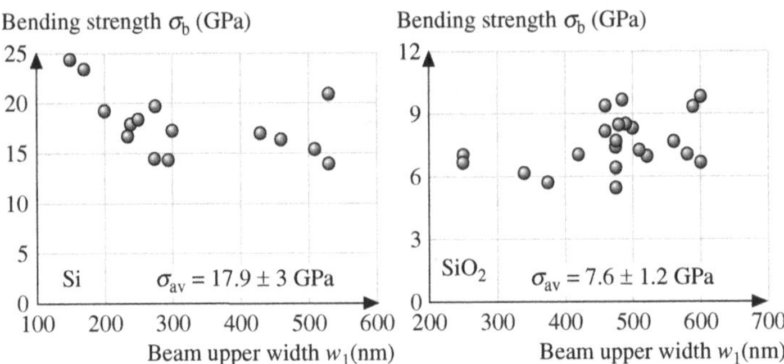


**Fig. 11.14** Bending strength values obtained from bending experiments. Average values are indicated. These values are much higher than those reported for microscale specimens, indicating that bending strength shows a specimen size effect (After [62])

maximum tensile stress occurs on the top surfaces near the ends (see the FEM stress distribution results in Fig. 11.13b). Figure 11.14 shows the values of bending strength obtained for different beams. There appears to be no trend in bending strength with the upper width $w_1$ of the beams. The large scatter is expected for the strength of brittle materials, which is dependent on the preexisting flaw population in the material and is hence statistical in nature. Statistical analysis using the Weibull distribution can be used to describe the scatter in the bending strength values. The means of the Weibull distributions were found to be 17.9 and 7.6 GPa for Si and $SiO_2$, respectively. Previously reported values for strength range from 1 to 6 GPa for silicon [43, 44, 46, 47, 48, 49, 51, 71, 95, 96] and $\approx 1$ GPa for $SiO_2$ [58] microscale specimens. This clearly indicates that bending strength shows a specimen size dependence. The strength of brittle materials is dependent on preexisting flaws in the material. Since for nanoscale specimens the volume is smaller than for micro- and macroscale specimens, the flaw population will be smaller as well, resulting in higher values of strength.

## Fracture Toughness

Estimates of fracture toughness calculated using (11.10) for Si and $SiO_2$ are shown in Fig. 11.15 [62]. The results show that the $K_{IC}$ estimate for Si is $\approx 1$–2 MPa m$^{1/2}$, whereas for $SiO_2$ the estimate is $\approx 0.5$–0.9 MPa m$^{1/2}$. These values are comparable to those reported by others on larger specimens for Si [54, 55, 56, 57] and $SiO_2$ [58]. The high values obtained for Si could be due to the fact that the scratches, despite being quite sharp, still have a finite radius of $\approx 100$ nm. The bulk value for silicon is $\approx 0.9$ MPa m$^{1/2}$ (Table 11.2). Fracture toughness is considered to be a material property and is believed to be independent of specimen size. The values obtained in this study, given its limitations, appear to show that fracture toughness is comparable, if not a little higher, on the nanoscale.

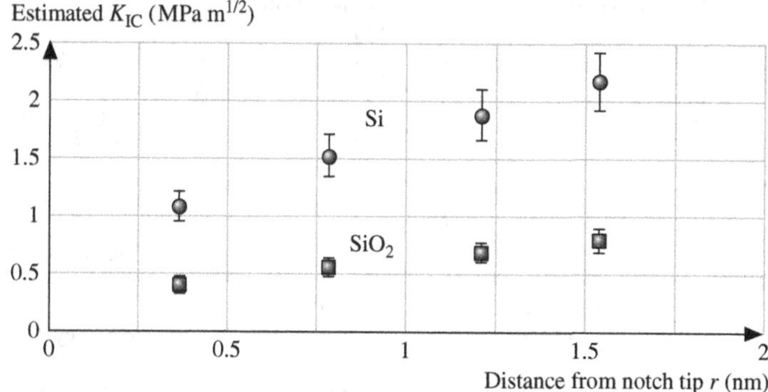


**Fig. 11.15** Fracture toughness ($K_{IC}$) values of for increasing values of $r$ corresponding to the distance between neighboring atoms in {111} planes of silicon (0.4 nm). Hence $r$ values between 0.4 and 1.6 nm are chosen. The $K_{IC}$ values thus estimated are comparable to those reported by others for both Si and $SiO_2$ (After [62])

## Fatigue Strength

Fatigue strength measurements of Si nanobeams have been carried out by Sundararajan and Bhushan [62] using an AFM, and by Li and Bhushan [63] using a nanoindenter. Various stress levels were applied to nanobeams in [62]. The minimum stress was 3.5 GPa for Si beams and 2.2 GPa for $SiO_2$ beams. The frequency of the applied load was 4.2 Hz. In general, fatigue life decreased with increasing mean stress as well as increasing stress amplitude. When the stress amplitude was <15% of the bending strength, the fatigue life was >30,000 cycles for both Si and $SiO_2$. However, the mean stress had to be <30% of the bending strength for a life of >30,000 cycles for Si whereas, even at mean stress of 43% of the bending strength, $SiO_2$ beams showed a life >30,000 cycles. During fatigue, the beams broke under the loading point or at the ends, when loaded at the center of the span. This was different from in the quasistatic bending tests, where the beams broke at the ends almost every time. This could be due to the fact that the stress levels under the load and at the ends are not that different and fatigue crack propagation could occur at either location. Figure 11.16 shows a nanoscale *S*–*N* curve of bending stress (*S*) as a function of fatigue in number of cycles (*N*) with an apparent endurance life at lower stress. This study clearly demonstrates that fatigue properties of nanoscale specimens can be studied.

## SEM Observations of Fracture Surfaces

Figure 11.17 shows SEM images of the fracture surfaces of nanobeams broken during quasistatic bending as well as fatigue [62]. In the quasistatic cases, the

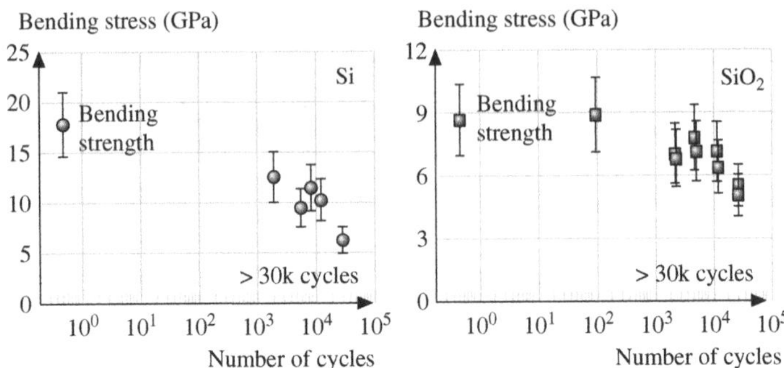


**Fig. 11.16** Fatigue test data showing applied bending stress as a function of number of cycles. A single load–unload sequence is considered as one cycle. The bending strength data points are therefore associated with $\frac{1}{2}$ cycle, since failure occurs upon loading (After [62])

maximum tensile stresses occur on the top surface, so it is reasonable to assume that fracture initiated at or near the top surface and propagated downward. The fracture surfaces of the beams suggest a cleavage type of fracture. Silicon beam surfaces show various ledges or facets, which is typical for brittle crystalline materials. Silicon usually fractures along the (111) plane, which has the lowest surface energy barrier to propagating cracks. However, failure has also been known to occur along the (110) planes in microscale specimens, despite the higher energy required as compared with the (111) planes [46]. The plane normal to the beam direction in these samples is the (110) plane, while (111) planes will be oriented at 35° from the (110) plane. The presence of facets and irregularities on the silicon surface in Fig. 11.17a suggest that it is a combination of these two types of fractures that has occurred. Since the stress levels are very high for these specimens, it is reasonable to assume that crack propagating forces will be high enough to result in (110)-type failures.

In contrast, the silicon surfaces fractured under fatigue (Fig. 11.17b) appear very smooth, without facets or irregularities. This is suggestive of low-energy fracture, i.e., (111)-type fracture. We do not see evidence of fatigue crack propagation in the form of steps or striations on the fracture surface. We believe that, for the stress levels applied in these fatigue experiments, failure in silicon occurred via cleavage associated with *static fatigue* type of failures.

$SiO_2$ shows very smooth fracture surfaces for both quasistatic bending and fatigue. This is in contrast to the irregular surface that one might expect for brittle failure of an amorphous material on the macroscale. However, in larger-scale fracture surfaces for such materials, the region near the crack initiation usually appears smooth or mirror-like. Since the fracture surface here is so small and very near the crack initiation site, it is not unreasonable to see such a smooth surface for $SiO_2$ on this scale. There appears to be no difference between the fracture surfaces obtained by quasistatic bending and fatigue for $SiO_2$.

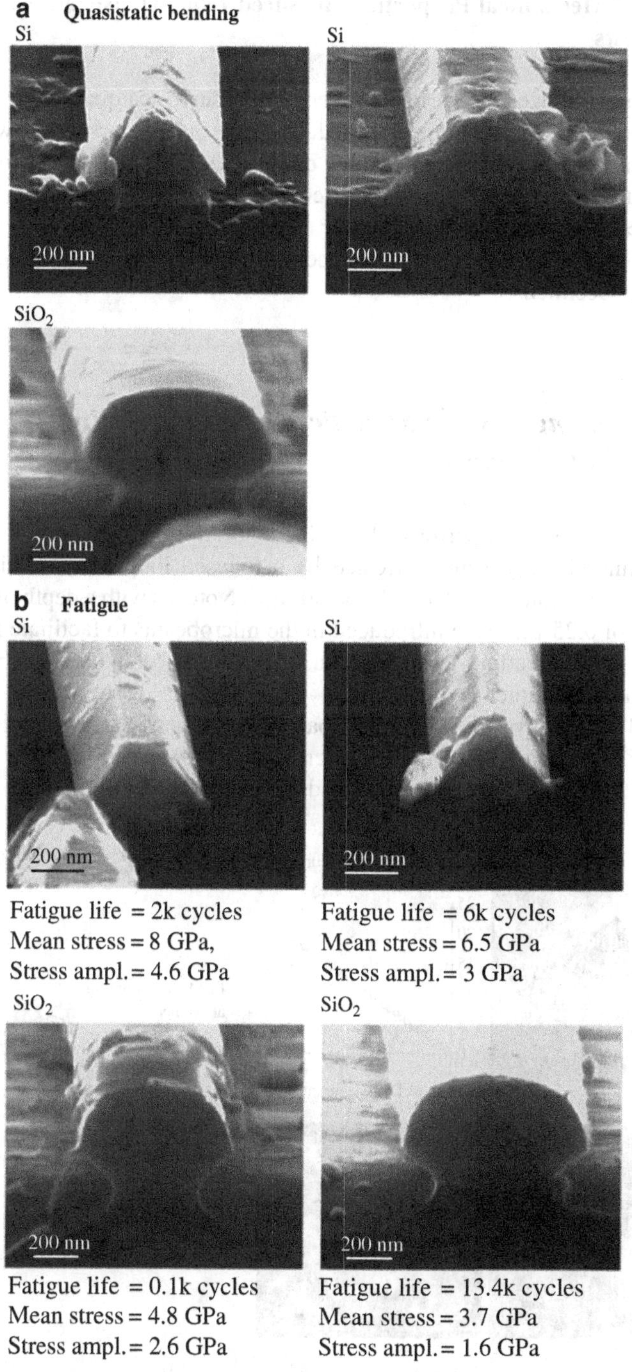


**Fig. 11.17** SEM micrographs of fracture surfaces of silicon and $SiO_2$ beams subjected to (**a**) quasistatic bending and (**b**) fatigue (After [62])

**Summary of Mechanical Properties Measured Using Quasistatic Bending Tests**

Table 11.2 summarizes the various properties measured via quasistatic bending in this study [62]. Also shown are bulk values of the parameters along with values reported on larger-scale specimens by other researchers. Elastic modulus and fracture toughness values appear to be comparable to bulk values and show no dependence on specimen size. However bending strength shows clear specimen size dependence with nanoscale values being twice as large as those reported for larger-scale specimens.

### *11.2.3 Bending Tests of Metallic Microbeams Using a Nanoindenter*

Bending tests have been performed on Ni-P and Au microbeams [64]. The Ni-P cantilever microbeams were fabricated by a focused ion-beam machining technique. The dimensions were $10 \times 12 \times 50\ \mu m^3$. Notches with a depth of 3 μm and a tip radius of 0.25 μm were introduced in the microbeams to facilitate failure at a lower load in the bending tests. The Au bridge microbeams were fabricated by electroplating technique.

Figure 11.18 shows SEM images, load–displacement curve, and FEM stress contour for a notched Ni-P cantilever microbeam that was bent to failure [64]. The distance between the loading position and the fixed end is 40 μm. The 3 μm-deep

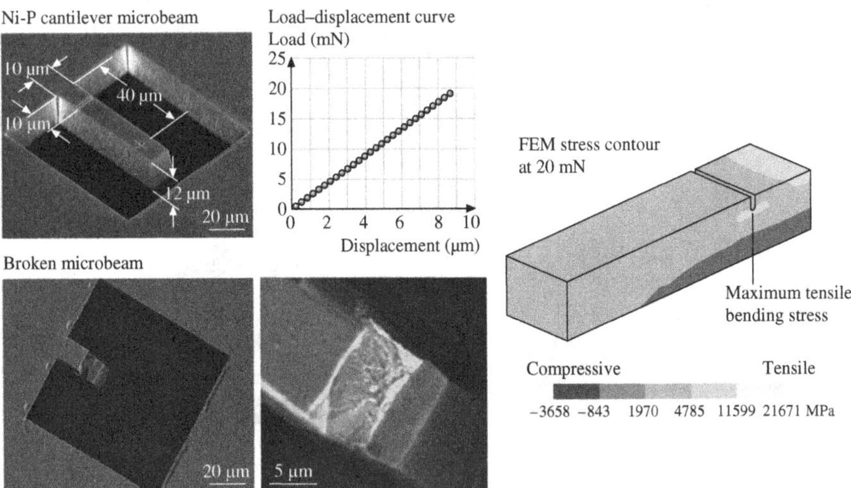


**Fig. 11.18** SEM micrographs of the new and broken beams, load–displacement curve, and FEM stress contour for the notched Ni-P cantilever microbeam (After [64])

notch is 10 μm from the fixed end. The notched beam showed linear behavior followed by abrupt failure. The FEM stress contour shows that there is a higher stress concentration at the notch tip. The maximum tensile stress $\sigma_m$ at the notch tip can be analyzed by using the Griffith fracture theory as [85]

$$\sigma_m \approx 2\sigma_o \left(\frac{c}{\rho}\right)^{1/2}, \tag{11.12}$$

where $\sigma_o$ is the average applied tensile stress on the beam, $c$ is the crack length, and $\rho$ is the crack tip radius. Therefore, elastic–plastic deformation will first occur locally at the end of the notch tip, followed by abrupt fracture failure after $\sigma_m$ reaches the ultimate tensile strength of Ni-P, even though the rest of the beam is still in the elastic regime. The SEM image of the fracture surface shows that the fracture started right from the notch tip with plastic deformation characteristics. This indicates that, although local plastic deformation occurred at the notch tip area, the whole beam failed catastrophically. The present study shows that FEM simulation can predict well the stress concentration, and helps in understanding the failure mechanism of notched beams.

Figure 11.19 shows SEM images, load–displacement curve, and FEM stress contour for a Au bridge microbeam that was deformed by the indenter [64]. The recession gap between the beam and substrate is about 7 μm, which is not large enough to break the beam at the load applied. From the load–displacement curve, we note that the beam experienced elastic–plastic deformation. The FEM stress contour shows that the maximum tensile stress is located at the fixed ends, whereas the minimum compressive stress is located around the center of the beam. The SEM image shows that the beam has been permanently deformed. No cracks were found on the beam surface. The present study shows the possibility of mechanically forming Au film into a required shape. This may help in designing/fabricating functionally complex smart micro-/nanodevices which need conductors for power supply and input/output signals.

### 11.2.4 Indentation and Scratch Tests of Polymeric Microbeams Using a Nanoindenter

Four potential structural materials for bioMEMS devices were selected for this study, namely poly(propyl methacrylate) (PPMA), poly(methyl methacrylate) (PMMA), polystyrene (PS), and a polystyrene/nanoclay composite (PS/clay) [66]. The physical properties of PPMA, PMMA, and PS are summarized in Table 11.3. PMMA and PS are widely used engineering polymers for a number of commercial applications. Historically, PMMA was the polymer of choice in ophthalmologic devices due to its high refractive index, hardness, and biocompatibility. Its surface can be functionalized with proteins, which promotes the bonding of tissues for

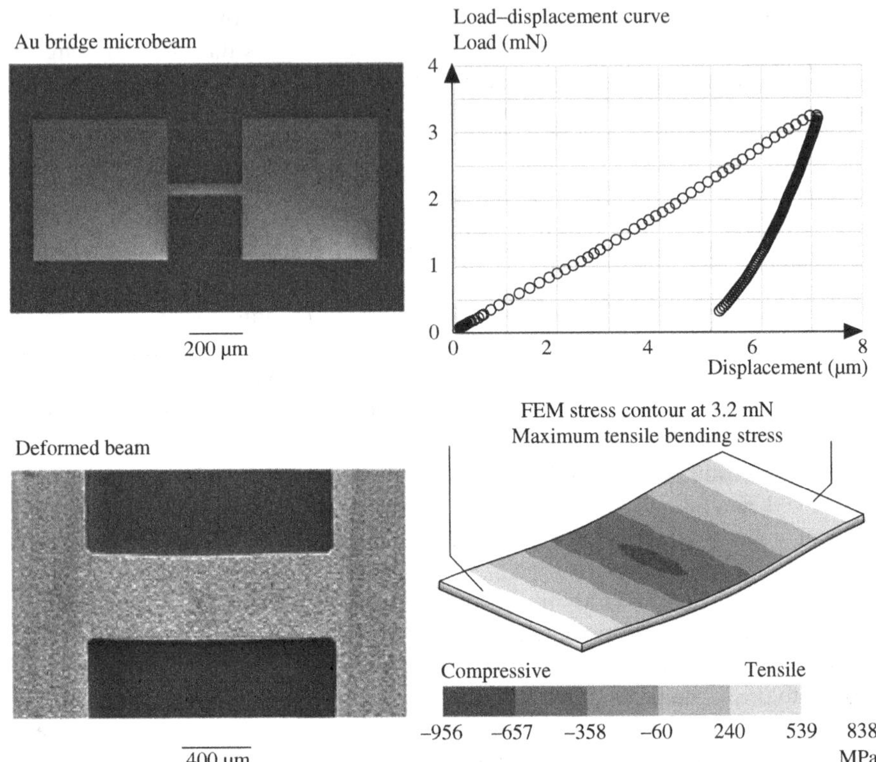


**Fig. 11.19** SEM micrographs of the new and deformed beams, load–displacement curve, and FEM stress contour for the Au bridge microbeam (After [64])

in vivo implants [97]. PMMA is also employed in chips and valve components for immunosensors and other lab-on-a-chip applications [28, 98]. PPMA has a lower glass-transition temperature ($T_g$) (35–43 °C) [81] than PMMA (104–106 °C) [82, 83], which allows for easier processing at lower temperature.

Polystyrene is particularly desirable for bioMEMS due to its ubiquitous use in tissue culture applications. Plasma-treated polystyrene is the most commonly used material for in vitro cell biology studies of adherent cells. Thus, there is a wealth of knowledge about cellular behavior on polystyrene. In addition to tissue culture polystyrene, various surface modification techniques can be employed for functionalizing the PS surface in order to promote cell attachment and proliferation [97]. Oxygen-plasma-modified polystyrene has been shown to improve cell growth, proliferation, and expression of cellular adhesion protein proportional to the surface oxygen concentration [101]. This previous knowledge of cellular interactions with polystyrene makes it a logical choice to use in bioMEMS devices for cellular interactions.

Clay nanoparticles in a polymer matrix act to improve the mechanical and thermal properties as compared with the native polymer [102, 103]. Nanoclay

**Table 11.3** Summary of the physical properties of PPMA, PMMA, and PS

| | Structure | Density ($g/cm^3$) | $T_g$ (°C) | Water absorption after 24 h (%) | Thermal expansion coefficient ($K^{-1}$) | Thermal conductivity (0–50 °C) (W/(m K)) | Heat capacity (0–100 °C) (kJ/(kg K)) |
|---|---|---|---|---|---|---|---|
| PPMA | $[-CH_2-C(CH_3)(COCH_2CH_2CH_3{=}O)-]_n$ | – | 35–43[a] | – | – | – | – |
| PMMA | $[-CH_2-C(CH_3)(COCH_3{=}O)-]_n$ | 1.188[b] | 104–106[b,c] | 0.3–0.4[d]<br>0.1–0.3[e] | 2.0–3.0 × $10^{-4}$[c] ($<T_g$)<br>6.0 × $10^{-4}$[c] ($>T_g$) | 0.193[b] | 1.255–1.720[b] |
| PS | $[-CH(C_6H_5)-CH_2-]_n$ | 1.040–1.065[c] (amorphous)<br>1.111–1.127[c] (crystalline) | 100[b,c] | 0.03–0.1[d] | 1.7–2.1 × $10^{-4}$[c] ($<T_g$)<br>5.1–6.0 × $10^{-4}$[c] ($>T_g$) | 0.105–0.116[b] | 1.185–1.838[b] |

[a][81]
[b][82]
[c][83]
[d][99]
[e][100]

composites can also be used to improve barrier resistance and improve ionic conductivity. The extent to which polymer properties are affected is determined by the clay content, polymer/clay interfacial strength, and dispersion of the clay particles. The fabrication procedure for the double-anchored and cantilever beams is presented in Sect. 11.A. SEM images of the two types of beams are shown in Fig. 11.20.

The hardness (*H*), elastic modulus (*E*), and creep of the beams were measured at the supported region of the beams [65, 66]. The indentation location, where the *H*, *E*, and creep were measured, is indicated in Fig. 11.20a.

Since the fabricated polymer beams were too narrow (width $\approx 5$ μm) to perform nanoscratch tests, the nanoscratch experiments were conducted on thin films of polymeric materials. PPMA, PMMA, PS, and PS/clay were dissolved in anisole at

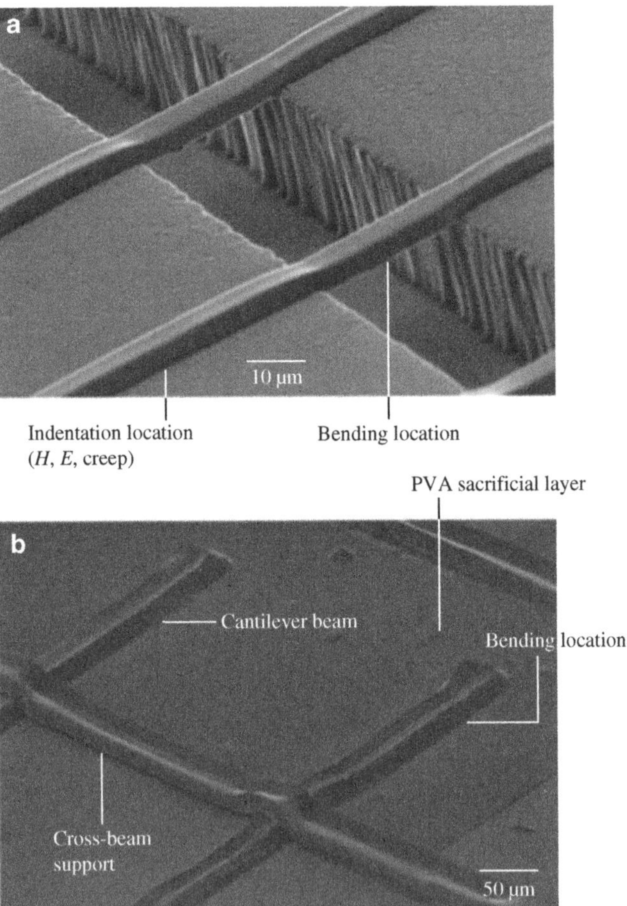


**Fig. 11.20** SEM images of polystyrene (**a**) double-anchored beams, and (**b**) cantilever beams (After [66])

various concentrations and spin-coated for 1 min at 3,000 rpm to achieve the desired film thickness of 500 nm. All film samples were then baked at 95 °C for 2 min to remove any residual solvent.

### Hardness and Elastic Modulus

Figure 11.21a shows the *H* and *E* of various beams as a function of contact depth [66]. Five beams in one wafer sample were indented once, and the average values of *H* and *E* at a contact depth of 100 nm are reported in Table 11.4. Among the

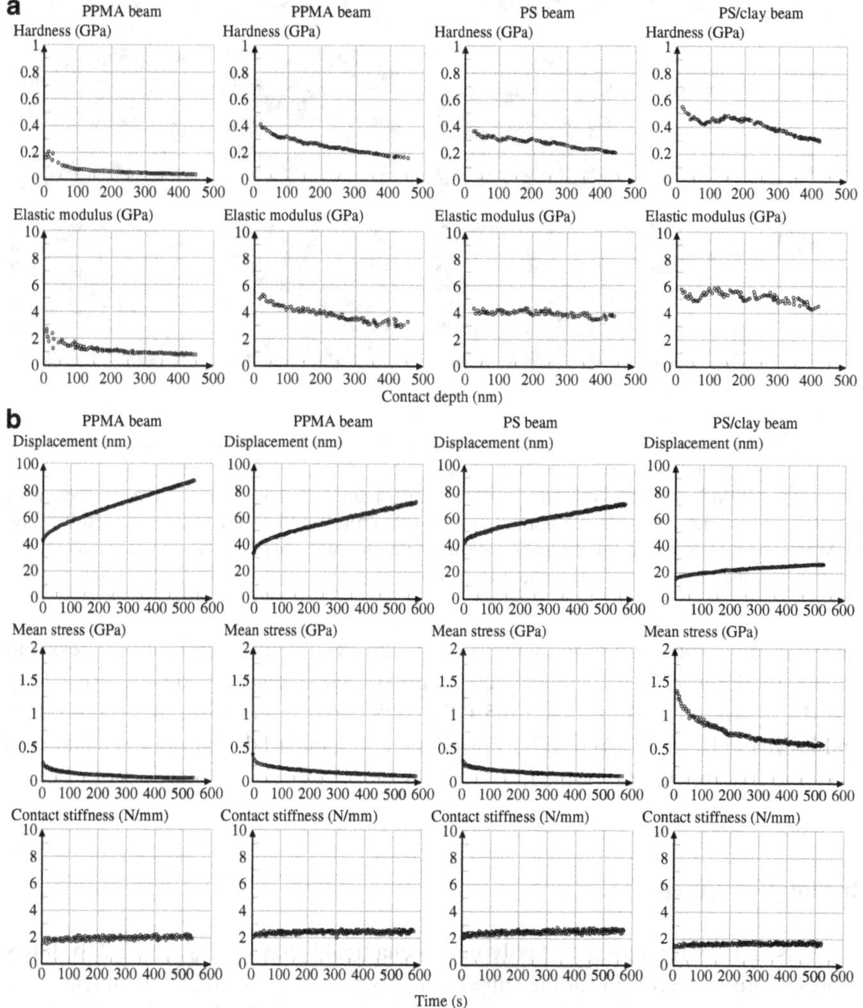


**Fig. 11.21** (continued)

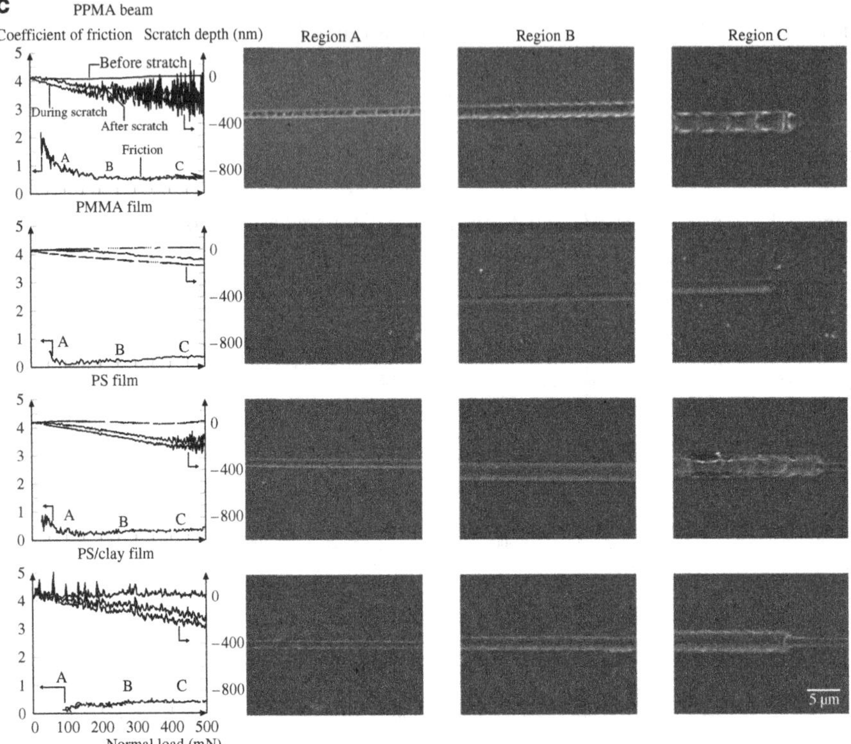


**Fig. 11.21** (**a**) Hardness and elastic modulus of PPMA, PMMA, PS, and PS/clay beams as a function of contact depth, (**b**) creep displacement, mean stress, and contact stiffness as a function of time for PPMA, PMMA, PS, and PS/clay, and (**c**) scratch depth profiles and coefficient of friction as a function of increasing normal loads for thin films of PPMA, PMMA, PS, and PS/clay. SEM images were taken at three regions: at the beginning of the scratch (A), in the middle of the scratch (B), and at the end of the scratch (C) (After [66])

materials examined, the PS/clay nanocomposite exhibited the highest hardness and elastic modulus (390 ± 50 MPa and 5.1± 0.4 GPa, respectively), followed by PMMA (340 ± 30 MPa, 4.8± 0.5 GPa), PS (290 ± 20 MPa, 3.6± 0.4 GPa), and PPMA (110 ± 30 MPa, 1.7± 0.5 GPa). The standard deviation was calculated from the five indents performed.

## Indentation Creep

Creep is an important aspect of polymer mechanical behavior. Figure 11.21b shows representative creep data for the supported polymer beams using CSM indentation, where the change in displacement, mean stress, and contact stiffness was monitored while holding the tip at the maximum imposed load [66]. Conclusions on the creep

**Table 11.4** Mechanical properties of the polymer beams under investigation. All values have $\pm\,\sigma$ of $\approx$ 10% based on measurements of five samples for nanoindentation and 3–5 samples for beam bending except for the 10–40% variation in breaking strength

| Polymer | | PPMA | PMMA | PS | PS/clay |
|---|---|---|---|---|---|
| Poisson's ratio | | – | 0.35[a] | 0.325[b] | – |
| Elastic modulus | Bulk (literature) | – | 3.1–3.3[c,d] | 3.2–3.4[c] | – |
| (*E*) (GPa) | Nanoindentation (measured) | 1.7 | 4.8 | 3.6 | 5.1 |
| | Normal beam bending (measured) | 0.7 | 1.9 | 1.9 | 4.6 |
| | Lateral beam bending (measured) | – | – | 0.2 | 0.6 |
| Hardness (*H*) | Bulk (literature) | – | 195[c] | 110[d] | – |
| (GPa) | Nanoindentation (measured) | 110 | 340 | 290 | 390 |
| Yield strength ($\sigma_{ys}$) | Bulk (literature) | – | 53.8–73.1[e] | – | – |
| (MPa) | Normal beam bending (measured) | 54 | 71 | 66 | 95 |
| | Lateral beam bending (measured) | – | – | 21 | 64 |
| Breaking strength | Tensile strength (literature) | – | 48–76[d] | 30–60[d] | – |
| ($\sigma_b$) (MPa) | Flexural strength (literature) | – | – | 95[d] | – |
| | Normal beam bending (measured)[f] | 600 | 944 | 271 | 128 |
| | Lateral beam bending (measured) | – | – | – | 110 |

[a][100]
[b][104]
[c][82]
[d][83]
[e][105]
[f]Values may be an overestimate due to limitations of the technique and the geometry of the beams

response of the materials are based mainly on the displacement change at constant loading, which is directly measured in the experiment. For each material, three beams on one sample were tested, and it was found that the variation in creep response was no more than 15%. As expected for polymers, the indentation displacement increases with time. The material with the lowest hardness and modulus (PPMA) exhibited the fastest rate of displacement change (i.e., the highest creep), while the PS/clay nanocomposite (which has the highest *H* and *E* value) showed minimal creep. This shows a direct correlation between the hardness, modulus, and creep resistance. The mean stress exerted by the tip decreases with time as the polymer deforms viscoelastically underneath the tip during the hold segment. For the PS/clay nanocomposite, it is observed that this quantity decreases more rapidly during the first few seconds of the measurement compared with in the other materials. This could be attributed to the presence of an additional phase (the clay filler). The filler–matrix interface could exhibit yield without necessarily causing further penetration into the material.

### Scratch Resistance

The scratch behavior of thin films of the four polymers studied is presented in Fig. 11.21c. Plots on the left side show the depth profile before, during, and after a scratch, along with the coefficient of friction. SEM images were taken at three areas: at the beginning of the scratch (indicated by A on the friction profile), the middle of the scratch (B), and towards the end of the scratch (C). Among the three homogeneous films, PMMA had the greatest scratch resistance as it exhibited the least penetration as well as the absence of bursts in the scratch depth profile, followed by PS, then PPMA. This order is consistent with the observed trend for the elastic modulus and hardness. The features in the scratch depth profiles can be correlated to the morphology of the scratches. In the PPMA depth profile, the tip jumped up and down during the scratch instead of continuously ramping down. This is consistent with the SEM images, which reveal deformation bands that are convex with respect to the scratch direction, indicating plastic deformation and material pileup in front of the tip. In PS, the oscillation in the depth profile occurs only towards the end of the scratch, and the deformation bands are observed only in region C. For PMMA, the smooth depth profile corresponds to a clear scratch track, indicating that minimal pileup occurred. The nanocomposite exhibits enhanced scratch resistance relative to the unfilled PS. The peaks and valleys in the depth profile come from the roughness of the surface and not from scratch-induced material displacement, as confirmed by the SEM images.

### *11.2.5 Bending Tests of Polymeric Microbeams Using a Nanoindenter*

#### Suspended Beams

For each material, five suspended beams in one sample were measured; representative load–displacement data are presented in Fig. 11.22. In this case, the displacement is due to beam bending. The loading profile can be fitted by a straight line, indicating that bending at this load range is fully elastic. From the slope of this line and (11.5), the calculated elastic moduli are $0.7 \pm 0.2$, $1.9 \pm 0.3$, $1.9 \pm 0.1$, and $4.6 \pm 0.3$ GPa for PPMA, PMMA, PS, and PS/clay, respectively. These standard deviations are based on five measurements.

These elastic moduli calculated from beam bending measurements were compared with values from nanoindentation and bulk values available from the literature. The consistency of the trend obtained in $E$ for the four materials using the three methods provides validation for the beam bending technique. In addition, the beam bending method has been successfully employed for various metal and ceramic materials, as presented earlier [64]. The slightly lower elastic moduli calculated from beam bending tests (relative to nanoindentation and bulk literature data) can be attributed to several reasons. Some deformation may have occurred in the boundary between the supported and the suspended region of the beam due to stress concentration at this junction. Also, a small amount of plastic deformation of the beam surface may occur even though the indenter tip was coated with PMMA. This could reduce the slope of the loading curve.

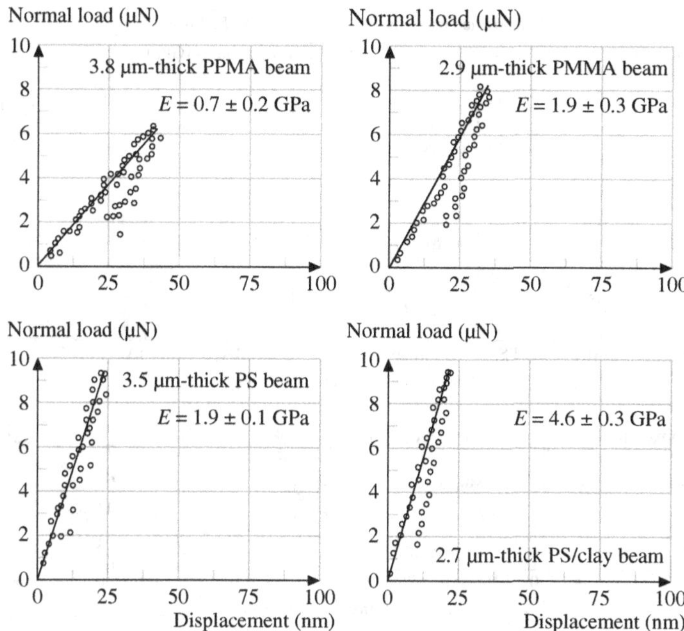


**Fig. 11.22** Normal beam bending results for PPMA, PMMA, PS, and PS/clay at a maximum load of 10 μN (After [66])

## Effect of Soaking and Temperature

The effects of aqueous medium and human body temperature (37.5 °C) were investigated on both the suspended and supported beams. Figure 11.23a shows a summary of the effects of beam soaking and elevated temperature on the hardness and elastic modulus as obtained by nanoindentation on the supported part of the beam. Among the four materials studied, soaking had the greatest effect on PPMA. On the other hand, the $H$ and $E$ of the other materials (PMMA, PS, and PS/clay) were not significantly affected. The properties of the PPMA beam were adversely affected when the indentation temperature was increased to 37.5 °C. This is because the glass-transition temperature ($T_g$) of PPMA is within 35–43 °C [81], such that the glassy-to-rubbery transition is observed here. On the other hand, no significant decrease in $H$ and $E$ was observed for the other three materials since their $T_g$ values are much higher, such that the indent was performed well within the glassy regime.

In Fig. 11.23b, beam bending data are shown for each of the four polymers investigated. The PPMA beam is affected by soaking in deionized (DI) water, as shown by the appearance of a large amount of hysteresis in the load–displacement data. Human body temperature has an adverse effect on the stiffness of the PPMA beam. These are consistent with results from indentation. From a materials selection standpoint, these are important findings as these imply that the performance of

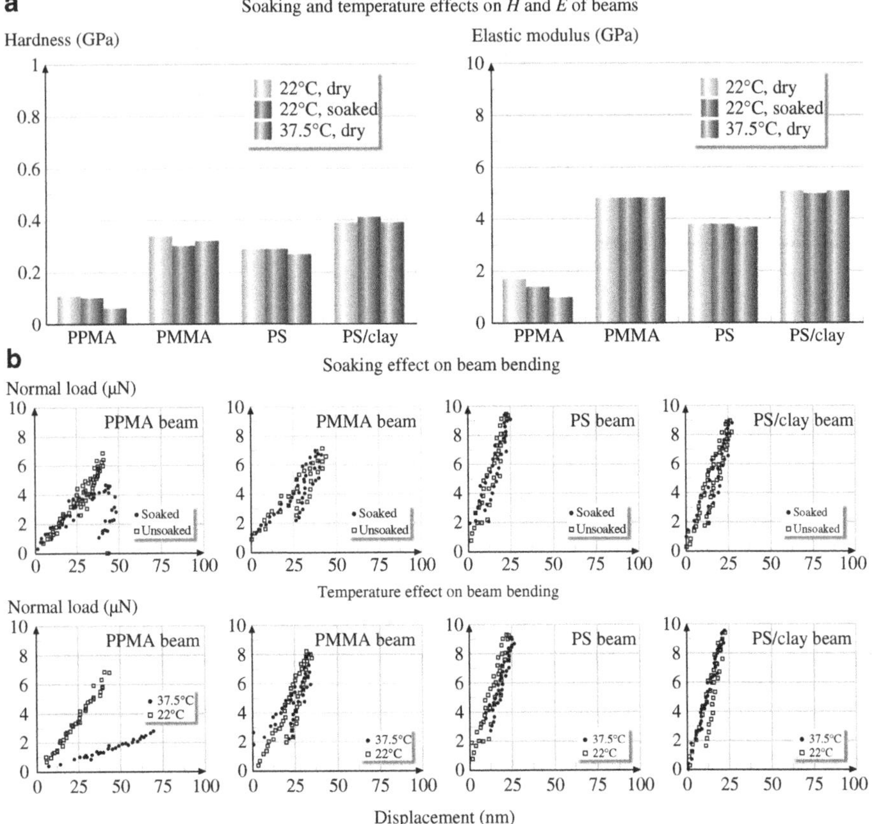


**Fig. 11.23** (**a**) Effect of soaking and elevated temperature on suspended beams and (**b**) effect of soaking and elevated temperature on the hardness and elastic modulus of supported beams (After [66])

PPMA structures will be compromised if used in a device subjected to either soaking in an aqueous medium or above ambient temperatures.

### Normal Beam Bending at Elevated Loads

Bending experiments were performed with applied loads in the mN range in order to investigate the behavior of polymer beams beyond the elastic deformation regime. The upper half of Fig. 11.24 shows normal load–displacement profiles at maximum imposed loads of 0.5, 1.5, and 3.0 mN applied to three different beams in one wafer sample. For PPMA, PMMA, and PS, the loading portion overlaps, indicating uniform beam quality. Slight variability in the loading profile is observed in the PS/clay composite, which is attributed to variable interfacial adhesion between the filler and the matrix and inhomogeneities in the dispersion of the clay particles.

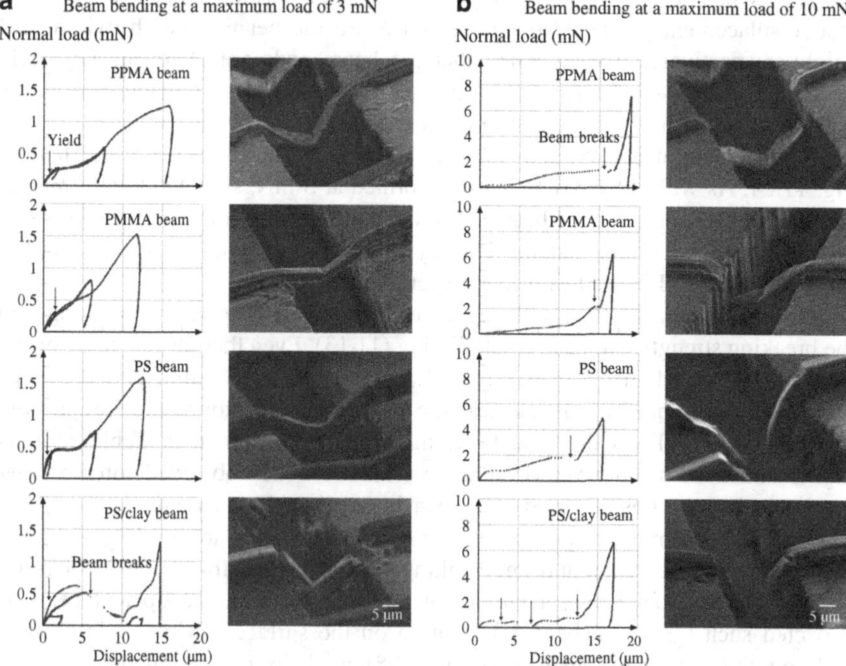


**Fig. 11.24** Normal beam bending at elevated loads. (**a**) Normal load–displacement profiles at a maximum load of 0.5, 1.5, and 3.0 mN; *arrows* indicate the onset of yield and breaking of the beams. For plots with multiple profiles (**a**), the SEM image corresponds to the beam subjected to the highest load (3 mN). (**b**) Normal load–displacement profiles and SEM images for beams subjected to 10 mN maximum load (After [66])

The first inflection in the load–displacement profiles implies the onset of yield, and its location is indicated by arrows. The three unfilled polymers (PPMA, PMMA, and PS) exhibit ductility, as seen in the SEM images for the beams tested with applied load of 3 mN. The beams stretch at the ends and underneath the center of the beam. These correspond to regions subjected to maximum tensile stress [64]. It is worth noting that the beams deformed symmetrically, indicating that the load was applied equidistant from the clamping points. The PS/clay nanocomposite beam exhibited slight yielding, followed by breaking, as indicated by an arrow in the load–displacement profile. After breaking, the load continued to increase because a predefined load is imposed by the nanoindenter. The corresponding SEM image for PS/clay shows a flat fracture surface, indicating that the crack propagated perpendicular to the direction of the applied stress.

From the load–displacement curves, the yield strength was evaluated by using the following equation [87]

$$\sigma = \frac{3}{4}\frac{F\ell}{bh^2}, \tag{11.13}$$

where $F$ is the load corresponding to yield of the beam (as indicated by arrows in the load–displacement profiles), and $\ell$, $b$, and $h$ are the beam's length, width, and height, respectively. It is assumed that the beam ends are clamped. The yield strength values listed in Table 11.4 are of the same order of magnitude as the bulk yield strength data available in the literature.

Bending data at a maximum load of 10 mN are shown in the lower half of Fig. 11.24. As with the bending tests performed at 3 mN, symmetric deformation and failure is observed in the beams. The three unfilled polymers stretched considerably prior to fracture, and the PS/clay nanocomposite beam exhibited brittle failure as observed in the lower load experiment.

For brittle materials, the elastic regime is immediately followed by breaking, and the breaking strength can be evaluated using (11.13). Even though plastic deformation was observed in all materials, (11.13) was assumed valid, and the force corresponding to breaking is taken as the load observed prior to the abrupt jump in displacement. The calculated breaking strength (as shown in Table 11.4) is higher and does not appear to correlate with the trends observed for the other mechanical properties reported. This can be attributed to a number of reasons. The load–displacement response may not be purely from beam bending. A higher load can be imposed without beam breaking due to plastic deformation on the beam surface. At elevated loads, a larger contact area between the tip and beam is expected such that permanent deformation on the surface can be significant. In addition, the compliant trench material (SU-8) can help to accommodate a higher stress prior to breaking of the beam.

### Lateral Bending of the Cantilever Beams

Figure 11.25 presents the results of lateral bending experiments on PS and PS/clay cantilever beams. The left column shows plots of lateral force recorded as a function of lateral displacement. From the linear elastic regime (as indicated by the straight line drawn over the data points), the modulus can be evaluated using (11.11). The resulting elastic modulus of PS/clay is three times that of PS, as shown in Table 11.4. This was obtained from measurements on two beams for each material. However, the elastic modulus values obtained from this experiment are lower than those obtained from nanoindentation and beam bending in the normal direction. This is attributed to the fact that the pressure applied due to the addition of the crossbeam reinforcement induces deformation of the anchor. In addition, a small curvature is observed on the beam. This diminishes the beam width, which is the dominant term in the equation for evaluating the modulus in (11.5) and (11.11).

From the load–displacement curves, the yield strength was evaluated by using [87]

$$\sigma = \frac{6F\ell}{bh^2}, \qquad (11.14)$$

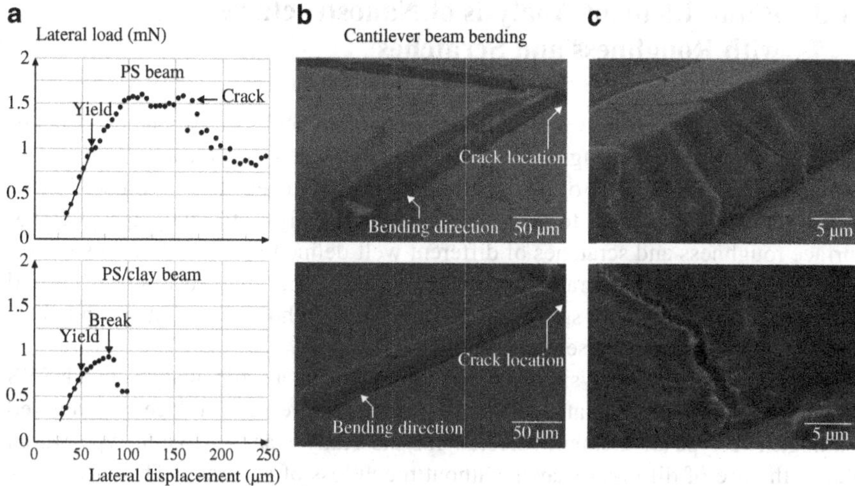


**Fig. 11.25** Lateral bending of PS and PS/clay cantilever beams. **(a)** Lateral force plotted as a function of lateral displacement; *arrows* indicate the onset of yield for both PS and PS/clay, and cracking for PS and breaking for PS/clay. **(b)** SEM images of the cantilever beams indicate the bending direction and the location of the cracks. The images in the *right-hand column* **(c)** show the cracks at higher magnification (After [66])

where $F$, $\ell$, $b$, and $h$ are defined similarly as in (11.13). The force used for calculating the yield strength was taken as the point where a change in slope was observed (as indicated by the arrows on the left in the figure). The results are listed in Table 11.4. They are lower than the values obtained from normal bending, which could be attributed to the same causes mentioned above.

For PS, a steady decrease in lateral load was observed at 160 μm, while for the PS/clay beam, the drop at 80 μm was abrupt (as indicated by the arrows on the right in the figure). SEM images reveal that, for PS, this corresponds to crack formation at the base of the beam that did not run through the entire thickness of the beam. However, for PS/clay, the beam broke. This implies that the addition of the nanoclay filler induces embrittlement of the beam.

Equation (11.14) is assumed to be valid for evaluating the breaking strength of this beam. The force used in the calculation was the load prior to the abrupt decrease. The breaking strength for PS/clay obtained from this technique (listed in Table 11.4) is lower than the value from the normal bending experiment. This is possibly due to the beam geometry and processing conditions, as described earlier. However, the large difference in the calculated yield strength between the unfilled and filled PS demonstrates that the technique can adequately differentiate between material compositions.

## 11.3 Finite-Element Analysis of Nanostructures with Roughness and Scratches

Micro-/nanostructures have some surface topography and local scratches, depending upon the manufacturing process. Surface roughness and local scratches may compromise the reliability of devices, and their effect needs to be studied. Finite-element modeling is used to perform parametric analysis to study the effect of surface roughness and scratches of different well-defined forms on tensile stresses, which are responsible for crack propagation [72, 73]. The analysis has been carried out on trapezoidal beams supported at the bottom whose data (on Si and $SiO_2$ nanobeams) have been presented earlier.

The finite-element analysis was carried out by using a static analysis in ANSYS 5.7 software which calculated the deflections and stresses produced by applied loading. The type of element selected for the study was SOLID95, type which allows the use of different shapes without much loss of accuracy. This element is three-dimensional (3-D) with 20 nodes, each node having three degrees of freedom: translation in the $x$-, $y$-, and $z$-directions. The nanobeam cross section was divided into 6 elements along both its width and thickness, and 40 elements along its length. SOLID95 has plasticity, creep, stress stiffening, large deflection, and large-strain capabilities. The large-displacement analysis is used for large loads. The mesh is kept finer near the asperities and the scratches in order to take into account variation in the bending stresses. The beam materials studied were single-crystal silicon(110) and $SiO_2$ films whose data have been presented earlier. Based on the bending experiments presented earlier, the beam materials can be assumed to be linearly elastic, isotropic materials. The Young's modulus of elasticity ($E$) and Poisson's ratio ($\nu$) for Si and $SiO_2$ are 169 GPa [93] and 0.28 [89], and 73 GPa and 0.17 [94], respectively. A sample nanobeam of silicon was chosen for performing most of the analysis, as silicon is the most widely used MEMS/NEMS material. The cross section of the fabricated beams used in the experiment was trapezoidal and supported at the bottom so nanobeams with trapezoidal cross section were modeled (Fig. 11.10). The following dimensions are used $w_1 = 200$ nm, $w_2 = 370$ nm, $t = 255$ nm, and $P = 6$ μm. In the boundary conditions, displacements were constrained in all directions on the bottom surface for 1 μm from each end. A point load applied at the center of the beam was simulated, with the load being applied at three closely located central nodes on the beam. It has been observed from the experimental results that the Si nanobeam breaks at around 80 μN. Therefore, in this analysis, a nominal load of 70 μN was selected. At this load, deformations are large, so the large-displacement option was used.

To study the effect of surface roughness and scratches on the maximum bending stresses the following cases were studied. First semicircular and grooved asperities in the longitudinal direction with defined geometrical parameters were analyzed (Fig. 11.26a). Next semicircular asperities and scratches placed along the transverse direction at a distance $c$ from the end and separated by pitch $p$ from each other were analyzed (Fig. 11.26b). Lastly the beam material was assumed to be either purely elastic, elastic–plastic or elastic–perfectly plastic. In the following, we begin with the stress distribution in smooth nanobeams, followed by the effect of surface roughness in the longitudinal and transverse directions and scratches in the transverse direction.

### 11.3.1 Stress Distribution in a Smooth Nanobeam

Figure 11.27 shows the stress and vertical displacement contours for a nanobeam supported at the bottom and loaded at the center [72, 73]. As expected, the maximum tensile stress occurs at the ends, whereas the maximum compressive stress occurs

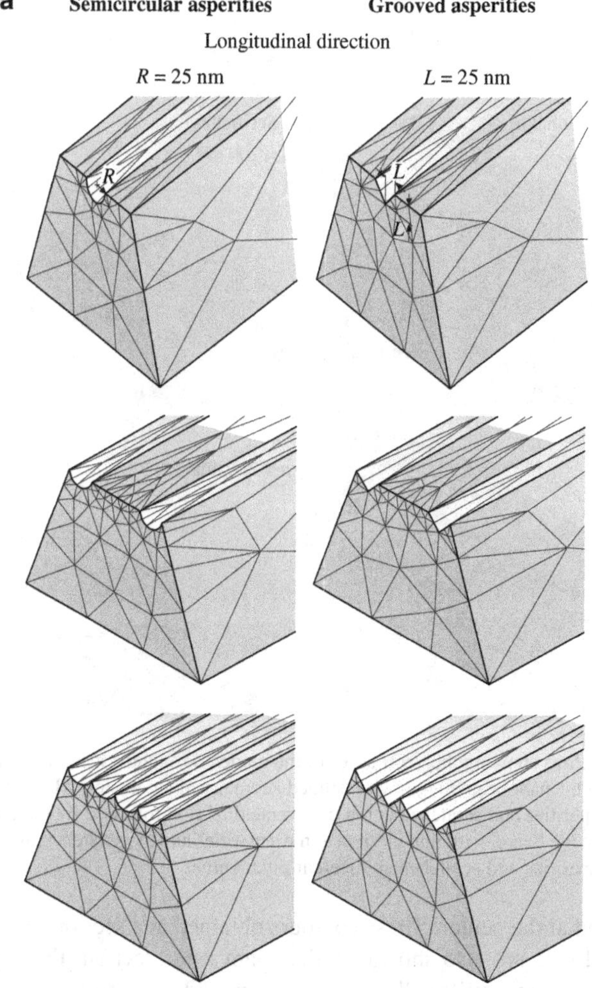


**Fig. 11.26** (continued)

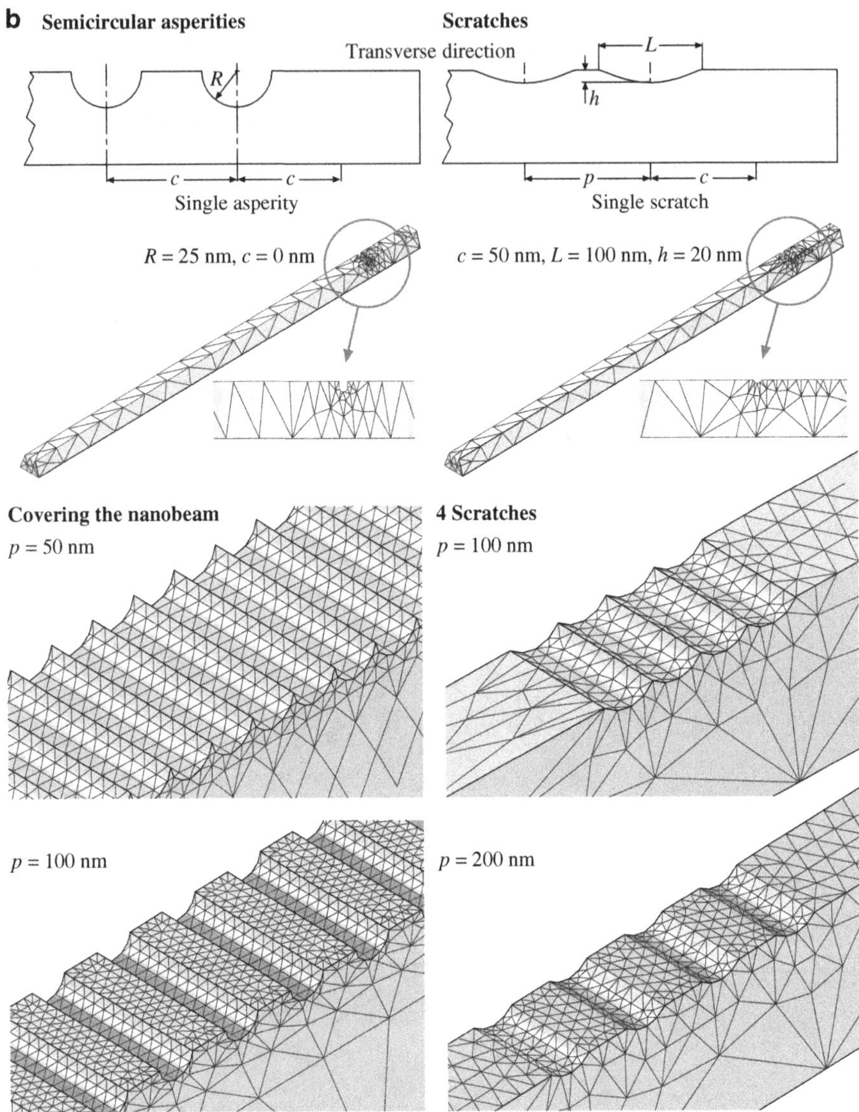


**Fig. 11.26** (**a**) Plots showing the geometries of the modeled roughness: semicircular and grooved asperities along the nanobeam length with defined geometrical parameters. (**b**) Schematic showing semicircular asperities and scratches in the transverse direction followed by the illustration of the mesh created on the beam, with fine mesh near the asperities and the scratches. Also shown are the semicircular asperities and scratches at different pitch values

under the load at the center. Stress contours obtained at a section of the beam from the front and side are also shown. In the beam cross section, the stresses remain constant at a given vertical distance from one side to another and change with vertical location. This can be explained due to the fact that the bending moment is

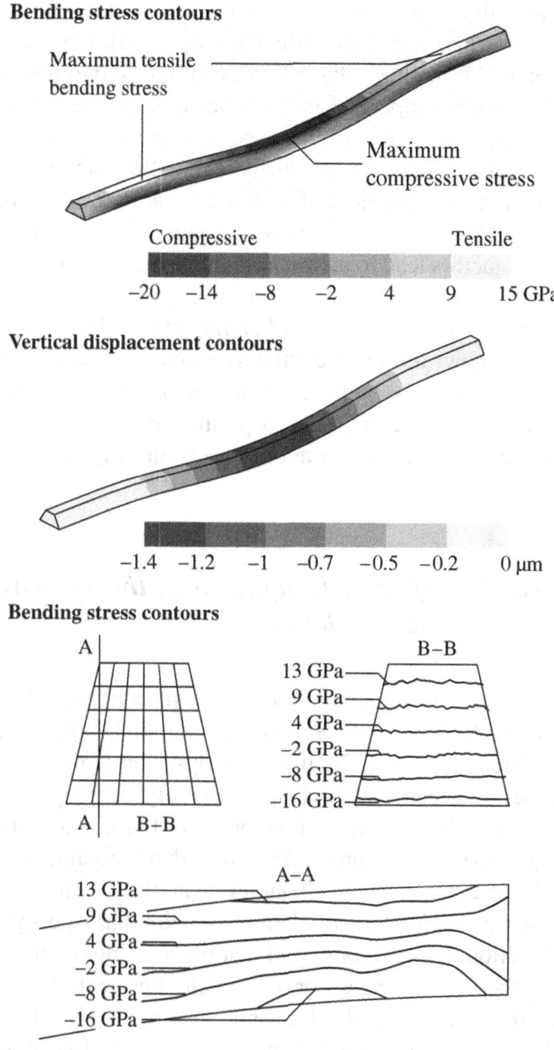


**Fig. 11.27** Bending stress contours, vertical displacement contours, and bending stress contours after loading a trapezoidal Si nanobeam ($w_1 = 200$ nm, $w_2 = 370$ nm, $t = 255$ nm, $\ell = 6$ μm, $E = 169$ GPa, $\nu = 0.28$) at 70 μN load (After [73])

constant at a particular cross section so the stress is only dependent on the distance from the neutral axis. However, along the cross section A–A the high tensile and compressive stresses are localized near the end of the beam, at the top and bottom, respectively, whereas the lower values are spread out away from the ends. High values of tensile stress occur near the ends because of the high bending moment.

### 11.3.2 Effect of Roughness in the Longitudinal Direction

The effect of roughness in the form of semicircular and grooved asperities in the longitudinal direction on the maximum bending stresses was analyzed [73].

The radius $R$ and depth $L$ were kept fixed at 25 nm, while the number of asperities was varied, and their effect on the maximum bending stresses was observed. Figure 11.28 shows the variation of the maximum bending stresses as a function of asperity shape and the number of asperities. The maximum bending stresses increase as the number of asperities increases for both semicircular and grooved asperities. This can be attributed to the fact that, as the number of asperities increases, the moment of inertia decreases for that cross section. Also the distance from the neutral axis increases because the neutral axis shifts downwards. Both of these factors lead to an increase in the maximum bending stresses, and this effect is more pronounced in the case of the semicircular asperities, as this shape exhibits a higher value of maximum bending stress than for the grooved asperities. Figure 11.28 shows the stress contours obtained at a section of the beam from the front and from the side for both cases when we have a single semicircular asperity and when four adjacent semicircular asperities are present. The trends are similar to those observed earlier for a smooth nanobeam (Fig. 11.27).

### 11.3.3 Effect of Roughness in the Transverse Direction and Scratches

We analyzed semicircular asperities placed along the transverse direction followed by the effect of scratches on the maximum bending stresses for various numbers and different pitches [73]. In the analysis of semicircular transverse asperities three cases were considered: a single asperity, and asperities over the nanobeam surface separated by a pitch of either 50 or 100 nm. In all three cases, the $c$-value was kept equal to 0 nm. Figure 11.29 shows that the value of the maximum tensile stress is 42 GPa, which is much larger than the maximum tensile stress value with no asperities (16 GPa) or when a semicircular asperity is present in the longitudinal direction. It is also observed that the maximum tensile stress does not vary with the number of asperities or the pitch, but that the maximum compressive stress increases dramatically for asperities present over the beam surface compared with the value when a single asperity is present. The maximum tensile stress occurs at the ends. An increase in $p$ does not add any asperities at the ends, whereas asperities are added in the central region where compressive stresses are maximum. The semicircular asperities present at the center cause local perturbation in the stress distribution at the center of the asperity where load is being applied, leading to a high value of maximum compressive stress [106]. Figure 11.29 also shows the stress contours obtained for a section of the beam from the front and from the side for both cases, i.e., for a single semicircular asperity and when asperities are present over the beam surface at a pitch of 50 nm. The trends are similar to those observed earlier for a smooth nanobeam (Fig. 11.27).

In the study pertaining to scratches, the number of scratches was varied as well as their pitch. Furthermore, the load was applied at the center of the beam as well as

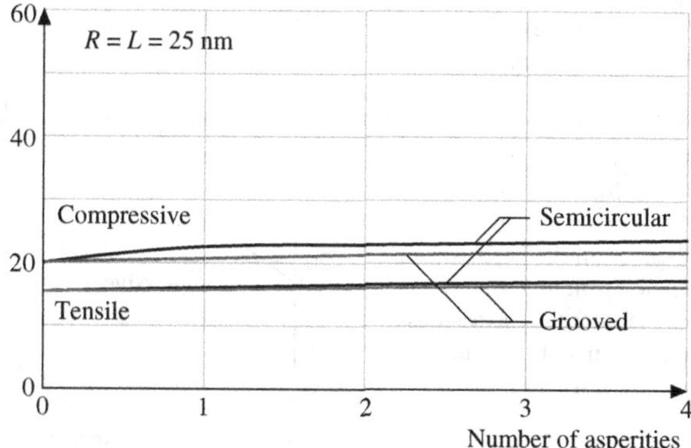


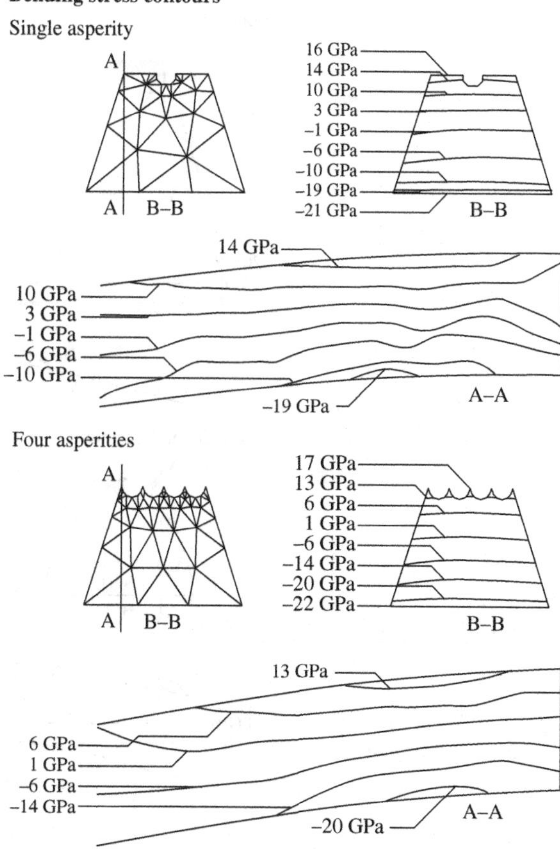


**Fig. 11.28** Effect of different numbers of longitudinal semicircular and grooved asperities on the maximum bending stresses after loading trapezoidal Si nanobeams ($w_1 = 200$ nm, $w_2 = 370$ nm, $t = 255$ nm, $\ell = 6$ μm, $E = 169$ GPa, $\nu = 0.28$, load = 70 μN). Bending stress contours obtained in the beam with a single semicircular asperity and four adjacent asperities of $R = 25$ nm (After [73])

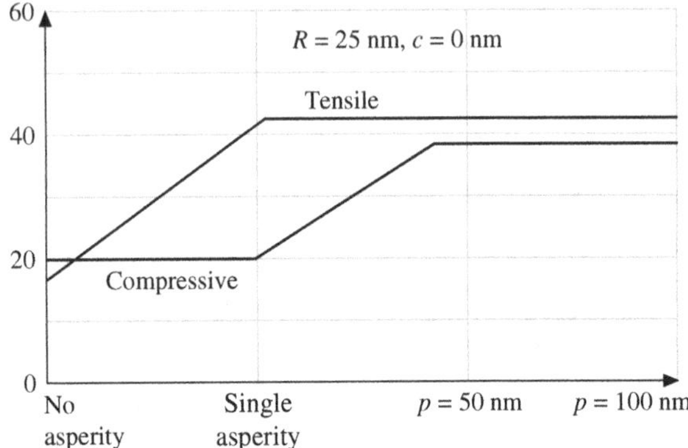


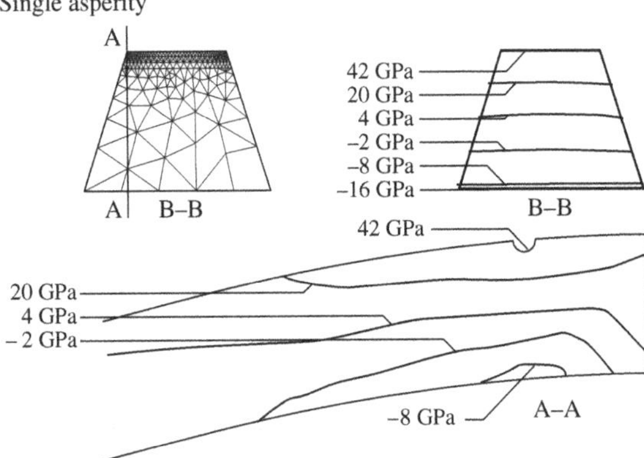


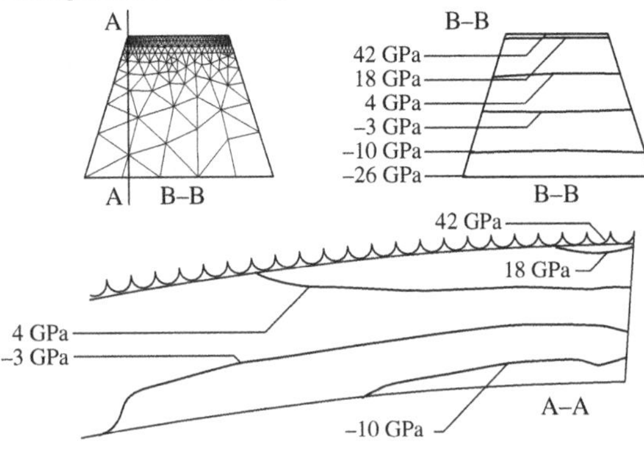


**Fig. 11.29** Effect of transverse semicircular asperities at different pitch values on the maximum bending stresses after loading trapezoidal Si nanobeams ($w_1 = 200$ nm, $w_2 = 370$ nm, $t = 255$ nm, $\ell = 6$ μm, $E = 169$ GPa, $\nu = 0.28$, load = 70 μN). Bending stress contours obtained in the beam with a single semicircular asperity and semicircular asperities over the nanobeam surface at $p = 50$ nm (After [73])

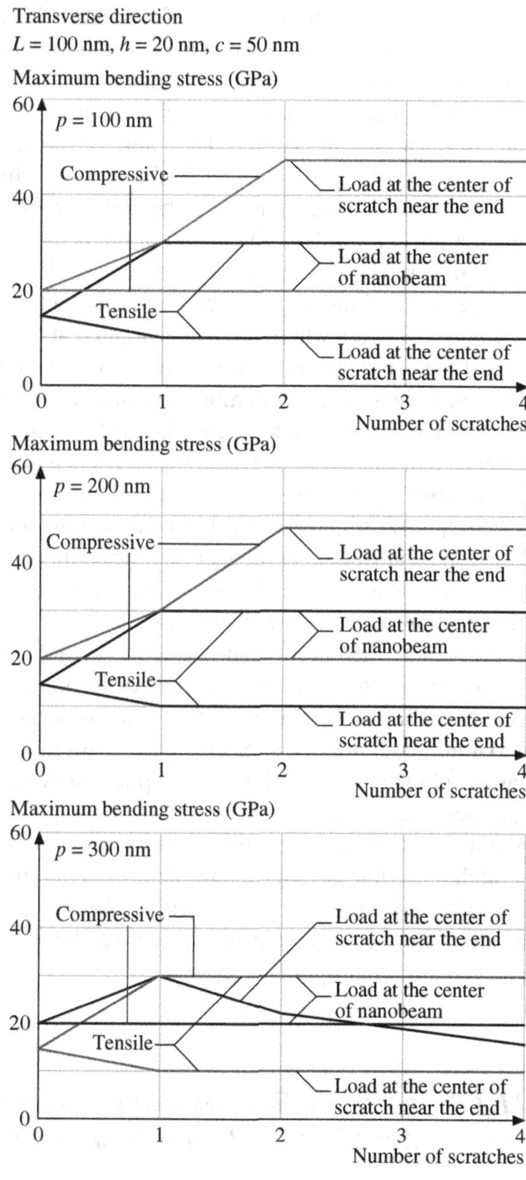


**Fig. 11.30** Effect of number of scratches and variation in pitch on the maximum bending stresses after loading trapezoidal Si nanobeams ($w_1 = 200$ nm, $w_2 = 370$ nm, $t = 255$ nm, $\ell = 6$ μm, $E = 169$ GPa, $\nu = 0.28$, load = 70 μN). Also shown is the effect of load when applied at the center of the beam and at the center of the scratch near the end (After [73])

at the center of the scratch near the end for all cases. In all of these cases, $c$ was kept equal to 50 nm, $L$ was set to 100 nm, and $h$ was 20 nm. Figure 11.30 shows that the value of the maximum tensile stress remained almost the same irrespective of the number of scratches for both types of loading, i.e., when the load was applied at the center of the beam or at the center of the scratch near the end. This is because the maximum tensile stress occurs at the beam ends, no matter where the load is applied. However, the presence of a scratch does increase the maximum tensile stress as compared with its value for a smooth nanobeam, although the number of scratches no longer matters as the maximum tensile stress occurring at the nanobeam end is unaffected by the presence of more scratches beyond the first one in the direction towards the center. The value of the tensile stress is much lower when the load is applied at the center of the scratch, which can be explained as follows. The negative bending moment at the end near the applied load decreases with load offset after two-thirds of the length of the beam [107]. Since this negative bending moment is responsible for tensile stresses, their behavior with the load offset is the same as that of the negative bending moment. Also, the value of the maximum compressive stress when the load is applied at the center of the nanobeam remains almost the same, as the center geometry is unchanged by the number of scratches, and hence the maximum compressive stress occurring below the load at the center is same. On the other hand, when the load is applied at the center of the scratch, we observe that the maximum compressive stress increases dramatically due to local perturbation in the stress distribution at the center of the scratch where load is applied [106]. It increases further with the number of scratches and then levels off. This can be attributed to the fact that, when there is another scratch present close to the scratch near the end, the stress concentration is greater, as the effect of the local perturbation in the stress distribution is more significant. However, this effect is insignificant when more than two scratches are present.

Now we address the effect of pitch on the maximum compressive stress when the load is applied at the center of the scratch near the end. Up to a pitch of 200 nm the maximum compressive stress increases with the number of scratches, as discussed earlier. On the other hand, when the pitch value is higher than 225 nm this effect is reversed. This is because the presence of another scratch no longer affects the local perturbation in the stress distribution at the scratch near the end. Instead, more scratches at a fair distance distribute the maximum compressive stress at the scratch near the end, and the stress starts to decrease. Such observations of maximum bending stresses can help in identifying the number of asperities and scratches allowed and their optimum separation.

### *11.3.4 Effect on Stresses and Displacements for Materials Which Are Elastic, Elastic–Plastic or Elastic–Perfectly Plastic*

This section deals with the beam modeled as elastic, elastic–plastic, and elastic–perfectly plastic to observe the change in the stresses and displacements compared with the elastic model used so far [73]. Figure 11.31 shows typical stress–strain curves for the three types of deformation regimes and their corresponding load–displacement curves obtained from the model of an Si nanobeam, which was found to exhibit the same trends.

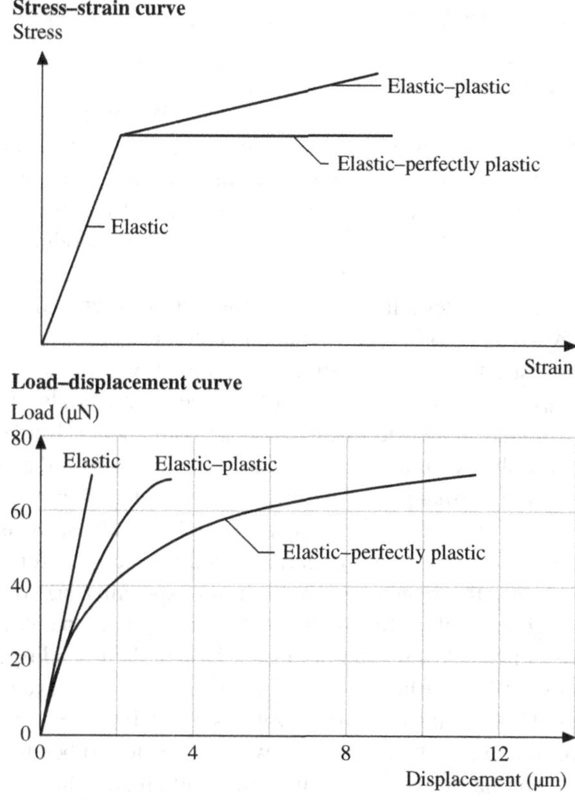


**Fig. 11.31** Schematic representation of stress–strain curves and load–displacement curves for material that is elastic, elastic–plastic or elastic–perfectly plastic for a Si nanobeam ($w_1 = 200$ nm, $w_2 = 370$ nm, $t = 255$ nm, $\ell = 6$ μm, $E = 169$ GPa, tangent modulus in plastic range = 0.5 E, $\nu = 0.28$) (After [73])

**Table 11.5** Stresses and displacements for materials which are elastic, elastic–plastic or elastic–perfectly plastic (load = 70 μN, $w_1 = 200$ nm, $w_2 = 370$ nm, $t = 255$ nm, $\ell = 6$ μm, $R = 25$ nm, $E = 169$ GPa, tangent modulus in plastic range = 0.5 E, $\nu = 0.28$)

| | Elastic | | Elastic–plastic | | Elastic–perfectly plastic | |
|---|---|---|---|---|---|---|
| | Smooth nanobeam | Single semicircular longitudinal asperity | Smooth nanobeam | Single semicircular longitudinal asperity | Smooth nanobeam | Single semicircular longitudinal asperity |
| Maximum von Mises stress (GPa) | 18.2 | 19.3 | 13.5 | 15.2 | 7.8 | 9.1 |
| Maximum displacement (μm) | 1.34 | 1.40 | 3.35 | 3.65 | 11.5 | 12.3 |

Table 11.5 shows a comparison of the maximum von Mises stress and maximum displacements for both a smooth nanobeam and a nanobeam with defined roughness corresponding to a single semicircular longitudinal asperity with $R = 25$ nm for the three different models. It is observed that the maximum value of stress is obtained at a given load for elastic material, whereas the displacement is maximum for elastic–perfectly plastic material. Also the pattern that the maximum bending stress value increases for a rough nanobeam still holds true in the other models as well.

## 11.4 Summary

Mechanical properties of nanostructures are necessary for designing realistic MEMS/NEMS and bioMEMS/bioNEMS devices. Most mechanical properties are scale dependent. Micro-/nanomechanical properties, hardness, elastic modulus, and scratch resistance of bulk undoped single-crystal silicon (Si) and thin films of undoped polysilicon, $SiO_2$, SiC, Ni-P, and Au are presented. It is found that the SiC film exhibits higher hardness, elastic modulus, and scratch resistance as compared with other materials.

Bending tests have been performed on Si and $SiO_2$ nanobeams – AfM, PPMA, PMMA, PS and PS/clay – nanoindenter, respectively. The bending tests were used to evaluate the elastic modulus, bending strength (fracture stress), fracture toughness ($K_{IC}$), and fatigue strength of the beam materials. The Si and $SiO_2$ nanobeams exhibited an elastic linear response with sudden brittle fracture. The notched Ni-P beam showed linear deformation behavior, followed by abrupt failure. The Au beam showed elastic–plastic deformation behavior. Measured elastic modulus values for Si(110) and $SiO_2$ were comparable to bulk values. Measured bending strength for Si and $SiO_2$ were twice as large as values reported for larger-scale specimens. This indicates that bending strength shows a specimen size dependence. Measured fracture toughness values for Si and $SiO_2$ were also comparable to those obtained for larger specimens. At stress amplitudes $<15\%$ of their bending strength and at mean stresses of $<30\%$ of the bending strength, Si and $SiO_2$ displayed an apparent endurance life of $>30{,}000$ cycles. SEM observations of the fracture surfaces revealed a cleavage type of fracture for both materials when subjected to bending as well as fatigue.

The hardness, elastic modulus, and creep behavior of PPMA, PMMA, PS, and PS/clay nanocomposite microbeams were evaluated using nanoindentation in continuous stiffness mode. The $H$, $E$, and creep resistance decreased in the following order: PS/clay, PMMA, PS, PPMA. The scratch resistance was observed to decrease in the order: PMMA, PS/clay, PS, PPMA.

The elastic modulus of double-anchored beams obtained from normal beam bending is lower than that measured from nanoindentation due to stress concentration at the beam ends. The test environment affects the mechanical properties as well. After 36 h of soaking in DI water, the hardness and elastic modulus of PPMA was affected, while the properties of the other three tested polymers were unaffected due to their low water absorption properties. Similarly, in tests conducted at human body temperature (37.5 °C), only PPMA exhibited a decrease in hardness and modulus whereas the other materials did not exhibit any change; this is because the temperature applied is within the glass-transition temperature range of PMMA but well below the $T_g$ of the other tested polymers. By performing bending experiments at higher imposed loads, permanent deformation can be attained such that the yield and breaking strength can be evaluated. Adding nanoclay filler improves bending strength. It also modifies the fracture behavior of the polymer microbeams. The unfilled polymers deformed in a ductile manner whereas the PS/clay nanocomposite exhibited minimal yielding followed by brittle failure.

Lateral bending has been demonstrated to provide a measure for the elastic, plastic, and fracture properties of cantilever beams.

The AFM and nanoindenters used in this study can be satisfactorily used to evaluate the mechanical properties of micro-/nanoscale structures for use in MEMS/NEMS.

FEM simulations are used to predict stress and deformation in nanostructures. FEM has been used to analyze the effect of the type of surface roughness and scratches on stresses and deformation of nanostructures. We find that roughness affects the maximum bending stresses. The maximum bending stresses increase as the number of asperities increases for both semicircular and grooved asperities in the longitudinal direction. When a semicircular asperity is present in the transverse direction, the value of the maximum tensile stress is much larger than with no asperity or when a semicircular asperity is present in the longitudinal direction. This observation suggests that the asperity in the transverse direction is more detrimental. The presence of scratches increases the maximum tensile stress. The maximum tensile stress remains almost the same with the number of scratches for the two types of loading tested, i.e., load applied at the center of the beam or at the center of the scratch near the end, although the value of the tensile stress is much lower when the load is applied at the center of the scratch. This means that load applied at the ends is less damaging. This analysis shows that FEM simulations can be useful to designers to develop the most suitable geometry for nanostructures.

## 11.5 Fabrication Procedure for the Double-Anchored and Cantilever Beams

The starting materials for the polymer microbeams are PPMA ($M_w = 25{,}0000$, Scientific Polymer Products), PMMA ($M_w = 75{,}000$, Sigma-Aldrich), PS (melt flow index 4.0, Sigma-Aldrich), and PS/clay solutions in anisole (Acros Organics). The clay additive in PS/clay is Cloisite 20A surface-modified natural montmorillonite (Southern Clay Products, Inc.) with a thickness of $\approx$ 1 nm and lateral dimensions of 70–150 nm. The surface modification of the clay additive improves the dispersion of the nanoparticles in the polymer matrix, thus improving the properties of the composite. To prepare the nanocomposite, the clay was first dispersed in the PS matrix by melt compounding at a concentration of $wt\%10$ (clay/PS). The composite was then dissolved in anisole and sonicated for at least 8 h to dissolve the polymer and redisperse the particles.

Double-anchored polymer microbeams were fabricated using a soft-lithography-based micromolding process along with standard photolithography [66]. The process involves selectively filling a poly(dimethylsiloxane) (PDMS) mold with the polymer of interest, followed by transfer of the resulting structures to a prefabricated substrate. This substrate is a silicon wafer with a layer of SU-8 25 negative tone photoresist (MicroChem Corp.) patterned by photolithography to create

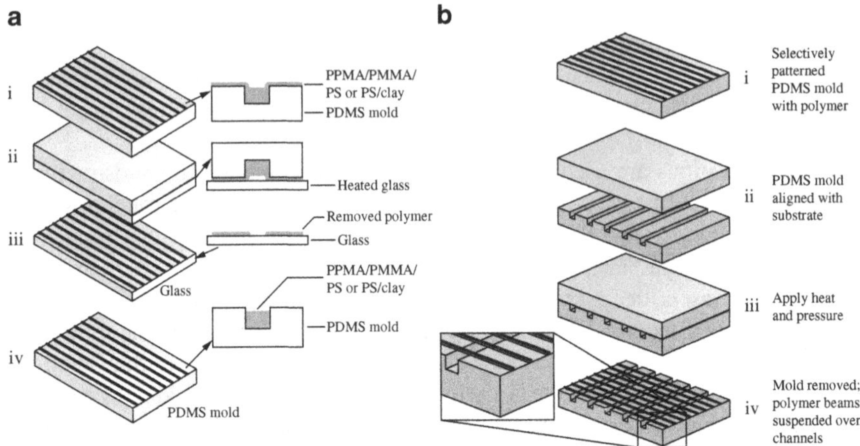


**Fig. 11.32** Schematic of the micromolding process used in polymer double-anchored beam fabrication. (**a**) Selective patterning of PDMS mold: (i) mold is spin-coated with a polymer layer, (ii) mold is inverted and brought into contact with a heated glass plate, (iii) mold is removed from the plate, transferring the surface polymer onto the glass, (iv) mold is left with polymer only in the recessed features. (**b**) Stamping of the polymer beams: (i) selectively patterned mold, (ii) mold is inverted and aligned with the photolithographically patterned substrate, (iii) mold and substrate are brought into contact under heat and pressure, (iv) polymer is transferred onto the photoresist and mold is removed (After [66])

25 μm-wide channels, which is the resulting length of the suspended beams. The second number 25 in the SU-8 designation relates to a viscosity appropriate for a given thickness.

A patterned PDMS mold with the desired polymer beam geometry was fabricated from a photoresist master. Briefly, a layer of SU-8 5 photoresist was spin-coated on silicon, and photolithography was used to define 5 μm-wide photoresist features separated by 45 μm gaps. A 10:1 ratio of T-2 PDMS translucent base and curing agent (Dow Corning) was mixed thoroughly and poured over the photoresist master to transfer the pattern to the PDMS. The mold was then placed in a vacuum desiccator to remove bubbles. The sample was removed from the vacuum periodically, and a razor blade was used to remove surface bubbles. After the bubbles were completely removed, the PDMS mold was allowed to cure at room temperature for 48 h before removing it from the wafer.

Next, the PDMS mold was selectively coated with the polymer solution to form the microbeams and then transferred to the substrate. Figure 11.32 shows a schematic of the fabrication process used for making the polymer microbeams. As shown in Fig. 11.32a, the polymer solutions were spin-coated onto the PDMS mold for 1 min. A 10% solution of polymer was spin-coated on a mold with 5.3 μm-deep features at 3,000 rpm for fabrication of the beams in the bending experiments. After spin-coating, the mold was brought into contact with a heated glass plate to promote adhesion of the contacting polymer materials. This process removed the polymer material from the raised surfaces of the mold, resulting in the polymer remaining only in the recessed portions of the PDMS mold. The glass plate was heated to 175 °C for all four materials. As shown in Fig. 11.32b, the selectively coated mold was then aligned with the photolithographically patterned silicon substrate so that the beams of interest (PPMA, PMMA, PS, PS/clay) ran perpendicular to the channels defined in the photoresist. The substrate was then heated and pressure was applied to the top of the mold to transfer the material onto the substrate. The transfer temperature for PPMA, PS, PS/clay and PMMA were 95, 125, 125, and 175 °C, respectively, and the transfer pressure for all materials was ≈ 0.21 MPa.

After removal of the mold, two types of polymer beams were transferred onto the wafer sample. The first type was supported beams, which were used to determine the hardness, elastic modulus, and creep response. The second type was suspended beams, on which the bending experiments were performed. The double-anchored beam samples studied by Palacio et al. [66] were 3–5 μm thick, 5 μm wide, and nominally 25 μm long.

The process for fabrication of the polymer cantilever for lateral bending tests is shown in Fig. 11.33 [66]. A PDMS mold was first cast from a photolithographically patterned SU-8/silicon master. The resulting molds consisted of 50 μm-wide channels that were ≈ 27 μm deep. The PDMS mold was then coated with the polymer of interest (PS or PS/clay). The polymers were spin-coated on the mold at 3,000 rpm for 1 min at concentrations of 15% and 10% for PS and PS/clay, respectively. The polymer on the raised surface of the mold was removed by contacting the surface with a glass slide heated to 180 °C. The mold was then inverted and manually aligned with a silicon substrate coated with patches of polyvinyl alcohol (PVA),

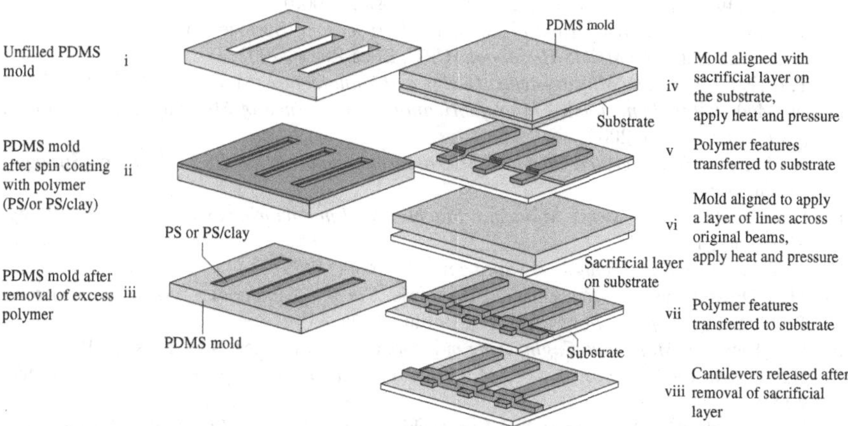


**Fig. 11.33** Schematic of the polymer cantilever beam fabrication process: (i) PDMS mold, (ii) mold spin-coated with the polymer of interest, (iii) selectively filled mold after removal of surface polymer, (iv) mold aligned with sacrificial layer, (v) polymer features transfer to the substrate using heat and pressure, (vi) mold aligned for a second time to add another layer of lines across original beams for reinforcement, (vii) polymer features transfer to the substrate using heat and pressure, (viii) cantilever structures after removal of the sacrificial layer (After [66])

which acts as a sacrificial layer. The PVA was patterned using photolithography and reactive-ion etching with an oxygen plasma. The substrate was heated and pressure was applied to the top side of the mold to transfer the polymer from the recessed features of the mold onto the substrate such that a 350 μm portion of it is attached to the PVA and the remainder is attached to the bare silicon surface. Transfer temperatures for PS and PS/clay were 150 and 175 °C, respectively. The process was repeated to apply another layer of beams across the length of the original layer. This was performed to provide reinforcement as it was observed that the single-layer design was not robust. The cantilever beam samples studied by Palacio et al. [66] were 12–27 μm thick, 60–80 μm wide, and 350 μm long.

## References

1. R.S. Muller, R.T. Howe, S.D. Senturia, R.L. Smith, R.M. White (eds.): *Microsensors* (IEEE Press, New York, 1990)
2. I. Fujimasa: *Micromachines: A New Era in Mechanical Engineering* (Oxford Univ. Press, Oxford, 1996)
3. W.S. Trimmer (ed.): *Micromachines and MEMS, Classic and Seminal Papers to 1990* (IEEE Press, New York, 1997)
4. B. Bhushan: *Tribology Issues and Opportunities in MEMS* (Kluwer, Dordrecht, 1998)
5. B. Bhushan: *Handbook of Micro-/Nanotribology*, 2nd edn. (CRC, Boca Raton, 1999)
6. B. Bhushan: *Nanotribology and Nanomechanics*, 2nd edn. (Springer, Berlin, Heidelberg, 2008)
7. G.T.A. Kovacs: *Micromachined Transducers Sourcebook* (WCB McGraw-Hill, Boston, 1998)
8. S.D. Senturia: *Microsystem Design* (Kluwer, Boston, 2000)
9. M. Elwenspoek, R. Wiegerink: *Mechanical Microsensors* (Springer, Berlin, 2001)
10. M. Gad-el-Hak: *The MEMS Handbook* (CRC, Boca Raton, 2002)
11. T.R. Hsu: *MEMS and Microsystems: Design and Manufacture* (McGraw-Hill, Boston, 2002)
12. M. Madou: *Fundamentals of Microfabrication: The Science of Miniaturization*, 2nd edn. (CRC, Boca Raton, 2002)
13. A. Hierlemann: *Integrated Chemical Microsensor Systems in CMOS Technology* (Springer, Berlin, 2005)
14. K.E. Drexler: *Nanosystems: Molecular Machinery, Manufacturing and Computation* (Wiley, New York, 1992)
15. G. Timp (ed.): *Nanotechnology* (Springer, New York, 1999)
16. M.S. Dresselhaus, G. Dresselhaus, P. Avouris: *Carbon Nanotubes – Synthesis, Structure, Properties and Applications* (Springer, Berlin, 2001)
17. E.A. Rietman: *Molecular Engineering of Nanosystems* (Springer, New York, 2001)
18. H.S. Nalwa (ed.): *Nanostructures Materials and Nanotechnology* (Academic, San Diego, 2002)
19. W.A. Goddard, D.W. Brenner, S.E. Lyshevski, G.J. Iafrate (ed.): *Handbook of Nanoscience, Engineering, and Technology* (CRC, Boca Raton, 2002)
20. A. Manz, H. Becker (eds.): *Microsystem Technology in Chemistry and Life Sciences*, Top. Curr. Chem., vol. 194 (Springer, Heidelberg, 1998)
21. J. Cheng, L.J. Kricka (eds.): *Biochip Technology* (Harwood Academic, Philadelphia, 2001)
22. M.J. Heller, A. Guttman (eds.): *Integrated Microfabricated Biodevices* (Marcel Dekker, New York, 2001)

23. C. Lai Poh San, E.P.H. Yap (eds.): *Frontiers in Human Genetics* (World Scientific, Singapore, 2001)
24. C.H. Mastrangelo, H. Becker (eds.): Microfluidics and BioMEMS, Proc. SPIE, vol. 4560 (SPIE, Bellingham, 2001)
25. H. Becker, L.E. Lacascio: Polymer microfluidic devices. Talanta **56**, 267–287 (2002)
26. D.J. Beebe, G.A. Mensing, G.M. Walker: Physics and applications of microfluidics in biology. Annu. Rev. Biomed. Eng. **4**, 261–286 (2002)
27. C.P. Poole, F.J. Owens: *Introduction to Nanotechnology* (Wiley, Hoboken, 2003)
28. A. van den Berg (ed.): *Lab-on-a-Chip: Chemistry in Miniaturized Synthesis and Analysis Systems* (Elsevier, Amsterdam, 2003)
29. J.V. Zoval, M.J. Madou: Centrifuge-based fluidic platforms. Proc. IEEE **92**, 140–153 (2000)
30. R. Raiteri, M. Grattarola, H. Butt, P. Skladal: Micromechanical cantilever-based biosensors. Sens. Actuators B **79**, 115–126 (2001)
31. W.C. Tang, A.P. Lee: Defense applications of MEMS. MRS Bulletin **26**, 318–319 (2001)
32. M.R. Taylor, P. Nguyen, J. Ching, K.E. Peterson: Simulation of microfluidic pumping in a genomic DNA blood-processing cassette. J. Micromech. Microeng. **13**, 201–208 (2003)
33. K. Park (ed.): *Controlled Drug Delivery: Challenges and Strategies* (American Chemical Society, Washington, 1997)
34. R.S. Shawgo, A.C.R. Grayson, Y. Li, M.J. Cima: BioMEMS for drug delivery. Curr. Opin. Solid State Mater. Sci. **6**, 329–334 (2002)
35. P.Å. Öberg, T. Togawa, F.A. Spelman: *Sensors in Medicine and Health Care* (Wiley, New York, 2004)
36. S.N. Bhatia, C.S. Chen: Tissue engineering at the micro-scale, Biomed. Microdevices **2**, 131–144 (1999)
37. R.P. Lanza, R. Langer, J. Vacanti (eds.): *Principles of Tissue Engineering* (Academic, San Diego, 2000)
38. E. Leclerc, K.S. Furukawa, F. Miyata, T. Sakai, T. Ushida, T. Fujii: Fabrication of microstructures in photosensitive biodegradable polymers for tissue engineering applications, Biomaterials **25**, 4683–4690 (2004)
39. T.H. Schulte, R.L. Bardell, B.H. Weigl: Microfluidic technologies in clinical diagnostics. Clin. Chim. Acta **321**, 1–10 (2002)
40. B. Bhushan: *Principles and Applications of Tribology* (Wiley, New York, 1999)
41. B. Bhushan: *Introduction to Tribology* (Wiley, New York, 2002)
42. B. Bhushan: Macro- and microtribology of MEMS materials, in *Modern Tribology Handbook*, ed. by B. Bhushan (CRC, Boca Raton, 2001) pp. 1515–1548
43. S. Johansson, J.A. Schweitz, L. Tenerz, J. Tiren: Fracture testing of silicon microelements in-situ in a scanning electron microscope. J. Appl. Phys. **63**, 4799–4803 (1988)
44. F. Ericson, J.A. Schweitz: Micromechanical fracture strength of silicon. J. Appl. Phys. **68**, 5840–5844 (1990)
45. E. Obermeier: Mechanical and thermophysical properties of thin film materials for MEMS: Techniques and devices. Micromech. Struct. Mater. Res. Symp. Proc., vol. 444 (Materials Research Society, Pittsburgh, 1996) pp. 39–57
46. C.J. Wilson, A. Ormeggi, M. Narbutovskih: Fracture testing of silicon microcantilever beams. J. Appl. Phys. **79**, 2386–2393 (1996)
47. W.N. Sharpe Jr., B. Yuan, R.L. Edwards: A new technique for measuring the mechanical properties of thin films. J. Microelectromech. Syst. **6**, 193–199 (1997)
48. K. Sato, T. Yoshioka, T. Anso, M. Shikida, T. Kawabata: Tensile testing of silicon film having different crystallographic orientations carried out on a silicon chip. Sens. Actuators A **70**, 148–152 (1998)
49. S. Greek, F. Ericson, S.S. Johansson, M. Furtsch, A. Rump: Mechanical characterization of thick polysilicon films: Young's modulus and fracture strength evaluated with microstructures. J. Micromech. Microeng. **9**, 245–251 (1999)

50. D.A. LaVan, T.E. Buchheit: Strength of polysilicon for MEMS devices. Proc. SPIE **3880**, 40–44 (1999)
51. E. Mazza, J. Dual: Mechanical behavior of a μ m-sized single crystal silicon structure with sharp notches. J. Mech. Phys. Solids **47**, 1795–1821 (1999)
52. T. Yi, C.J. Kim: Measurement of mechanical properties for MEMS materials. Meas. Sci. Technol. **10**, 706–716 (1999)
53. H. Kahn, M.A. Huff, A.H. Heuer: Heating effects on the Young's modulus of films sputtered onto micromachined resonators. Microelectromech. Struct. Mater. Res. Symp. Proc., vol. 518 (Materials Research Society, Pittsburgh, 1998) pp. 33–38
54. S. Johansson, F. Ericson, J.A. Schweitz: Influence of surface-coatings on elasticity, residual-stresses, and fracture properties of silicon microelements. J. Appl. Phys. **65**, 122–128 (1989)
55. R. Ballarini, R.L. Mullen, Y. Yin, H. Kahn, S. Stemmer, A.H. Heuer: The fracture toughness of polysilicon microdevices: A first report. J. Mater. Res. **12**, 915–922 (1997)
56. H. Kahn, R. Ballarini, R.L. Mullen, A.H. Heuer: Electrostatically actuated failure of microfabricated polysilicon fracture mechanics specimens. Proc. R. Soc. Lond. Ser. A **455**, 3807–3823 (1999)
57. A.M. Fitzgerald, R.H. Dauskardt, T.W. Kenny: Fracture toughness and crack growth phenomena of plasma-etched single crystal silicon. Sens. Actuators A **83**, 194–199 (2000)
58. T. Tsuchiya, A. Inoue, J. Sakata: Tensile testing of insulating thin films: Humidity effect on tensile strength of $SiO_2$ films. Sens. Actuators A **82**, 286–290 (2000)
59. J.A. Connally, S.B. Brown: Micromechanical fatigue testing. Exp. Mech. **33**, 81–90 (1993)
60. K. Komai, K. Minoshima, S. Inoue: Fracture and fatigue behavior of single-crystal silicon microelements and nanoscopic AFM damage evaluation. Microsyst. Technol. **5**, 30–37 (1998)
61. T. Namazu, Y. Isono, T. Tanaka: Evaluation of size effect on mechanical properties of single-crystal silicon by nanoscale bending test using AFM. J. Microelectromech. Syst. **9**, 450–459 (2000)
62. S. Sundararajan, B. Bhushan: Development of AFM-based techniques to measure mechanical properties of nanoscale structures. Sens. Actuators A **101**, 338–351 (2002)
63. X. Li, B. Bhushan: Fatigue studies of nanoscale structures for MEMS/NEMS applications using nanoindentation techniques. Surf. Coat. Technol. **163/164**, 521–526 (2003)
64. X. Li, B. Bhushan, K. Takashima, C.W. Baek, Y.K. Kim: Mechanical characterization of micro-/nanoscale structures for MEMS/NEMS applications using nanoindentation techniques. Ultramicroscopy **97**, 481–494 (2003)
65. G. Wei, B. Bhushan, N. Ferrell, D. Hansford: Microfabrication and nanomechanical characterization of polymer MEMS for biological applications. J. Vac. Sci. Technol. A **23**, 811–819 (2005)
66. M. Palacio, B. Bhushan, N. Ferrell, D. Hansford: Nanomechanical characterization of polymer beam structures for bioMEMS applications. Sens. Actuators A **135**, 637–650 (2007)
67. T. Hsu, N. Sun: Residual stresses/strains analysis of MEMS, in Proceedings of the International Conference on Modeling and Simulation of Microsystems, Semiconductors, Sensors and Actuators, Santa Clara, ed. by M. Laudon, B. Romanowicz (Computational Publications, Cambridge, 1998) pp. 82–87
68. A. Kolpekwar, C. Kellen, R.D. Blanton: Fault model generation for MEMS, Proc. Int. Conf. Model. Simul. Microsyst. Semicond. Sens. Actuators, ed. by M. Laudon, B. Romanowicz (Computational Publications, Cambridge, 1998) pp. 111–116
69. H.A. Rueda, M.E. Law: Modeling of strain in boron-doped silicon cantilevers, in Proceedings of the International Conference on Modeling and Simulation of Microsystems, Semiconductors, Sensors and Actuators, Santa Clara, ed. by M. Laudon, B. Romanowicz (Computational Publications, Cambridge, 1998) pp. 94–99
70. M. Heinzelmann, M. Petzold: FEM analysis of microbeam bending experiments using ultramicro indentation. Comput. Mater. Sci. **3**, 169–176 (1994)

71. C.J. Wilson, P.A. Beck: Fracture testing of bulk silicon microcantilever beams subjected to a side load. J. Microelectromech. Syst. **5**, 142–150 (1996)
72. B. Bhushan, G.B. Agrawal: Stress analysis of nanostructures using a finite element method. Nanotechnology **13**, 515–523 (2002)
73. B. Bhushan, G.B. Agrawal: Finite element analysis of nanostructures with roughness and scratches. Ultramicroscopy **97**, 495–507 (2003)
74. K.E. Petersen: Silicon as a mechanical material. Proc. IEEE **70**, 420–457 (1982)
75. B. Bhushan, S. Sundararajan, X. Li, C.A. Zorman, M. Mehregany: Micro-/nanotribological studies of single-crystal silicon and polysilicon and SiC films for use in MEMS devices, in *Tribology Issues and Opportunities in MEMS*, ed. by B. Bhushan (Kluwer, Dordrecht, 1998) pp. 407–430
76. S. Sundararajan, B. Bhushan: Micro-/nanotribological studies of polysilicon and SiC films for MEMS applications. Wear **217**, 251–261 (1998)
77. X. Li, B. Bhushan: Micro-/nanomechanical characterization of ceramic films for microdevices. Thin Solid Films **340**, 210–217 (1999)
78. H. Becker, C. Gärtner: Polymer microfabrication methods for microfluidic analytical applications. Electrophoresis **21**, 12–26 (2000)
79. J.C. McDonald, D.C. Duffy, J.R. Anderson, D.T. Chiu, H. Wu, O.J.A. Schueller, G.M. Whitesides: Fabrication of microfluidic systems in poly(dimethylsiloxane). Electrophoresis **21**, 27–40 (2000)
80. M. Palacio, B. Bhushan, N. Ferrell, D. Hansford: Adhesion properties of polymer/silicon interfaces for biological micro-/nanoelectromechanical applications. J. Vac. Sci. Technol. A **25**, 1275–1284 (2007)
81. B. Ellis: *Polymers: A Property Database* (CRC, Boca Raton, 2000), available on compact disk, also see http://www.polymersdatabase.com/
82. J. Brandrup, E.H. Immergut, E.A. Grulke: *Polymer Handbook* (Wiley, New York, 1999)
83. J.E. Mark: *Polymers Data Handbook* (Oxford Univ. Press, New York, 1999)
84. B. Bhushan, X. Li: Nanomechanical characterization of solid surfaces and thin films. Int. Mater. Rev. **48**, 125–164 (2003)
85. B.R. Lawn, A.G. Evans, D.B. Marshall: Elastic/plastic indentation damage in ceramics: the median/radial system. J. Am. Ceram. Soc. **63**, 574 (1980)
86. S. Sundararajan, B. Bhushan, T. Namazu, Y. Isono: Mechanical property measurements of nanoscale structures using an atomic force microscope. Ultramicroscopy **91**, 111–118 (2002)
87. W.C. Young, R.G. Budynas: *Roark's Formulas for Stress and Strain* (McGraw-Hill, New York, 2002)
88. R.W. Hertzberg: *Deformation and Fracture Mechanics of Engineering Materials*, 3rd edn. (Wiley, New York, 1989) pp. 277–278
89. Anonymous: *Properties of Silicon*, EMIS Datarev. Ser., vol. 4 (INSPEC Institution of Electrical Engineers, London, 1988)
90. C.T.-C. Nguyen, R.T. Howe: An integrated CMOS micromechanical resonator high-Q oscillator. IEEE J. Solid-State Circuits **34**, 440–455 (1999)
91. L.J. Hornbeck: A digital light processing update – status and future applications. Proc. Soc. Photo-Opt. Eng., Projection Displ. V, vol. 3634 (1999) pp. 158–170
92. M. Tanaka: Fracture toughness and crack morphology in indentation fracture of brittle materials. J. Mater. Sci. **31**, 749 (1996)
93. B. Bhushan, S. Venkatesan: Mechanical and tribological properties of silicon for micromechanical applications: A review. Adv. Inf. Storage Syst. **5**, 211–239 (1993)
94. B. Bhushan, B.K. Gupta: *Handbook of Tribology: Materials, Coatings, and Surface Treatments* (McGraw-Hill, New York, 1991), reprint with corrections (Krieger, Malabar, 1997)
95. T. Tsuchiya, O. Tabata, J. Sakata, Y. Taga: Specimen size effect on tensile strength of surface-micromachined polycrystalline silicon thin films. J. Microelectromech. Syst. **7**, 106–113 (1998)

96. T. Yi, L. Li, C.J. Kim: Microscale material testing of single crystalline silicon: Process effects on surface morphology and tensile strength. Sens. Actuators A **83**, 172–178 (2000)
97. I.H. Loh, M.S. Sheu, A.B. Fischer: Biocompatible polymer surfaces, in *Functional Polymers: Syntheses and Applications*, ed. by R. Arshady (American Chemical Society, Washington, 1997)
98. D.B. Holt, P.R. Gauger, A.W. Kusterbech, F.S. Ligler: Fabrication of a capillary immunosensor in polymethyl methacrylate. Biosens. Bioelectron. **17**, 95–103 (2002)
99. F.W.J. Billmeyer: *Textbook of Polymer Science* (Wiley, New York, 1984)
100. Anonymous: Rohm and Haas General Information on PMMA (Philadelphia)
101. T.G. van Kooten, H.T. Spijker, H.H. Busscher: Plasma-treated polystyrene surfaces: Model surface for studying cell-biomaterial interactions. Biomaterials **25**, 1735–1747 (2004)
102. M. Alexandre, P. Dubois: Polymer-layered silicate nanocomposites: preparation, properties and uses of a new class of materials. Mater. Sci. Eng. **28**, 1–63 (2000)
103. S.S. Ray, M. Okamoto: Polymer/layered silicate nanocomposites: A review from preparation to processing. Prog. Polym. Sci. **28**, 1539–1641 (2003)
104. R.H. Boundy, R.F. Boyer (eds.): *Styrene, Its Polymers, Copolymers and Derivatives* (Reinhold, New York, 1952)
105. Anonymous: *Modern Plastics Encyclopedia* (McGraw-Hill, New York, 1996)
106. S.P. Timoshenko, J.N. Goodier: *Theory of Elasticity*, 3rd edn. (McGraw-Hill, New York, 1970)
107. J.E. Shigley, L.D. Mitchell: *Mechanical Engineering Design*, 4th edn. (McGraw-Hill, New York, 1993)

## Editor's Vita

Dr. Bharat Bhushan received an M.S. in mechanical engineering from the Massachusetts Institute of Technology in 1971, an M.S. in mechanics and a Ph.D. in mechanical engineering from the University of Colorado at Boulder in 1973 and 1976, respectively, an MBA from Rensselaer Polytechnic Institute at Troy, NY in 1980, Doctor Technicae from the University of Trondheim at Trondheim, Norway in 1990, a Doctor of Technical Sciences from the Warsaw University of Technology at Warsaw, Poland in 1996, and Doctor Honouris Causa from the National Academy of Sciences at Gomel, Belarus in 2000. He is a registered professional engineer. He is presently an Ohio Eminent Scholar and The Howard D. Winbigler Professor in the College of Engineering, and the Director of the Nanoprobe Laboratory for Bio- & Nanotechnology and Biomimetics (NLB$^2$) at the Ohio State University, Columbus, OH. His research interests include fundamental studies with a focus on scanning probe techniques in the interdisciplinary areas of bio/nanotribology, bio/nanomechanics and bio/nanomaterials characterization, and applications to bio/nanotechnology and biomimetics. He is an internationally recognized expert of bio/nanotribology and bio/nanomechanics using scanning probe microscopy, and is one of the most prolific authors. He is considered by some a pioneer of the tribology and mechanics of magnetic storage devices. He has authored seven scientific books, more than 90 handbook chapters, more than 700 scientific papers (h-index – 45+; ISI Highly Cited in Materials Science, since 2007), and more than 60 technical reports, edited more than 45 books, and holds 17 US and foreign patents. He is co-editor of Springer NanoScience and Technology Series and co-editor of Microsystem Technologies. He has given more than 400 invited presentations on six continents and more than 140 keynote/plenary addresses at major international conferences.

Dr. Bhushan is an accomplished organizer. He organized the first symposium on Tribology and Mechanics of Magnetic Storage Systems in 1984 and the first international symposium on Advances in Information Storage Systems in 1990, both of which are now held annually. He is the founder of an ASME Information Storage and Processing Systems Division founded in 1993 and served as the founding chair during 1993–1998. His biography has been listed in over two dozen Who's Who books including Who's Who in the World and has received more than two dozen awards for his contributions to science and technology from professional societies, industry, and US government agencies. He is also the recipient of various international fellowships including the Alexander von Humboldt Research Prize for Senior Scientists, Max Planck Foundation Research Award for Outstanding Foreign Scientists, and the Fulbright Senior Scholar Award. He is a foreign member of the International Academy of Engineering (Russia), Byelorussian Academy of Engineering and Technology and the Academy of Triboengineering of

B. Bhushan (ed.), *Nanotribology and Nanomechanics*,
DOI 10.1007/978-3-642-15283-2, © Springer-Verlag Berlin Heidelberg 2011

Ukraine, an honorary member of the Society of Tribologists of Belarus, a fellow of ASME, IEEE, STLE, and the New York Academy of Sciences, and a member of ASEE, Sigma Xi and Tau Beta Pi.

Dr. Bhushan has previously worked for the R&D Division of Mechanical Technology Inc., Latham, NY; the Technology Services Division of SKF Industries Inc., King of Prussia, PA; the General Products Division Laboratory of IBM Corporation, Tucson, AZ; and the Almaden Research Center of IBM Corporation, San Jose, CA. He has held visiting professor appointments at University of California at Berkeley, University of Cambridge, UK, Technical University Vienna, Austria, University of Paris, Orsay, ETH Zurich and EPFL Lausanne.

# Index

**Note: Page references in Roman denote Vol. I and Italic page references denote Vol. II.**

**P**

## Q

**Z**